Theory of Multicodimensional $(n+1)$-Webs

Vladislav V. Goldberg

Department of Mathematics,
New Jersey Institute of Technology, U.S.A.

Theory of Multicodimensional $(n+1)$-Webs

KLUWER ACADEMIC PUBLISHERS
DORDRECHT / BOSTON / LONDON

Library of Congress Cataloging in Publication Data

CIP

Goldberg, V. V. (Vladislav Viktorovich)
Theory of multicodimensional (n+1)-webs / Vladislav V. Goldberg.
p. cm. -- (Mathematics and its applications)
Bibliography: p.
Includes index.
ISBN 90-277-2756-2
1. Webs (Differential geometry) I. Title. II. Series:
Mathematics and its applications (D. Reidel Publishing Company)
QA648.5.G65 1988
516.3'6--dc19

88-15895
CIP

ISBN 90-277-2756-2

Published by Kluwer Academic Publishers,
P.O. Box 17, 3300 AA Dordrecht, The Netherlands.

Kluwer Academic Publishers incorporates
the publishing programmes of
D. Reidel, Martinus Nijhoff, Dr W. Junk and MTP Press.

Sold and distributed in the U.S.A. and Canada
by Kluwer Academic Publishers,
101 Philip Drive, Norwell, MA 02061, U.S.A.

In all other countries, sold and distributed
by Kluwer Academic Publishers Group,
P.O. Box 322, 3300 AH Dordrecht, The Netherlands.

Printed in The Netherlands

SERIES EDITOR'S PREFACE

Approach your problems from the right end and begin with the answers. Then one day, perhaps you will find the final question.

'The Hermit Clad in Crane Feathers' in R. van Gulik's *The Chinese Maze Murders.*

It isn't that they can't see the solution. It is that they can't see the problem.

G.K. Chesterton. *The Scandal of Father Brown* 'The point of a Pin'.

Growing specialization and diversification have brought a host of monographs and textbooks on increasingly specialized topics. However, the "tree" of knowledge of mathematics and related fields does not grow only by putting forth new branches. It also happens, quite often in fact, that branches which were thought to be completely disparate are suddenly seen to be related.

Further, the kind and level of sophistication of mathematics applied in various sciences has changed drastically in recent years: measure theory is used (non-trivially) in regional and theoretical economics; algebraic geometry interacts with physics; the Minkowsky lemma, coding theory and the structure of water meet one another in packing and covering theory; quantum fields, crystal defects and mathematical programming profit from homotopy theory; Lie algebras are relevant to filtering; and prediction and electrical engineering can use Stein spaces. And in addition to this there are such new emerging subdisciplines as "experimental mathematics", "CFD", "completely integrable systems", "chaos, synergetics and large-scale order", which are almost impossible to fit into the existing classification schemes. They draw upon widely different sections of mathematics. This programme, Mathematics and Its Applications, is devoted to new emerging (sub)disciplines and to such (new) interrelations as exempla gratia:

- a central concept which plays an important role in several different mathematical and/or scientific specialized areas;
- new applications of the results and ideas from one area of scientific endeavour into another;
- influences which the results, problems and concepts of one field of enquiry have and have had on the development of another.

The Mathematics and Its Applications programme tries to make available a careful selection of books which fit the philosophy outlined above. With such books, which are stimulating rather than definitive, intriguing rather than encyclopaedic, we hope to contribute something towards better communication among the practitioners in diversified fields.

A web is a collection of d foliations in general position of the same codimension. For example one has the case of 3-webs of curves in the plane (the first interesting case, and the subject of numerous early studies). The French and German names for the concept are respectively "tissu" and "Gewebe", which mean tissue, fabric, texture (in its original meaning of woven fabric), web; and these words with these meanings convey a good initial intuitive picture of the kind of geometric structure involved. An easy example of a 3-web of curves in the plane is given by the three systems of lines x = const., y = const., $x + y$ = const.. By definition a codimension foliation is locally like a set of parallel hyperplanes. However, whether several foliations, i.e. a web, can be (locally) 'straightened out' simultaneously is a much tougher question. For example the trivial 3-web of lines just mentioned has the following *closure property*. Take a point 0 and draw the three leaves of the three foliations through that point. Take a neighboring point A on one of these leafs and walk around along the leafs and the foliations as indicated in the figure below

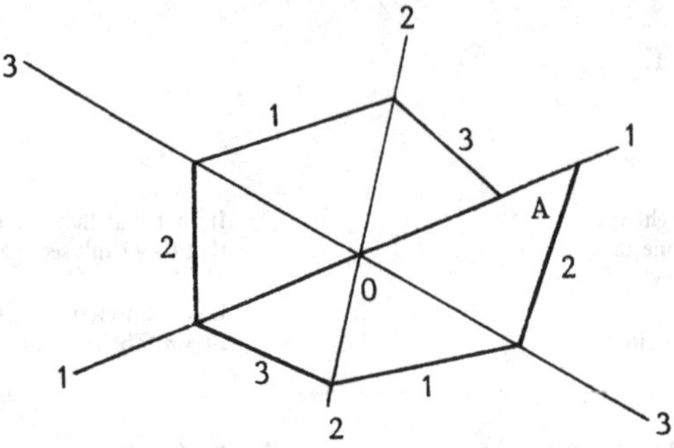

In the case of the given trivial foliation one finishes back in the point A after traversing a hexagon. This turns out to be a necessary and sufficient condition for a 3-web of curves in the plane to look locally like our trivial example, i.e. to be *linearizable*.

There are, of course, a good many places in mathematics where multiple foliations come up naturally, and where such results would thus be important. For example in control theory, although, as far as I know, this particular potential application of web theory remains unexplored.

It is an old theorem (1924) that a linearizable web in the plane consists of the tangents to a curve of degree 3 in the projective plane, and thus the question arises whether every web arises in some such algebraic manner.

This also provided a rather unexpected link with algebraic geometry, one of the first of many interrelations of the theory of webs with various parts of the theories of symmetric spaces, differential equations, algebraic geometry, integral geometry, singularities, and holomorphic mappings. The author sums up some 11 such interrelations in his own preface.

After the initial flowering of the subject which resulted in the classics by Bol-Blaschke and Blaschke there was a dormant period. More recently starting around 1970 there was a substantial resurge of interest, resurge of interest, which still continues, led and/or sparked by Akivis, the author himself, Chern and Griffiths.

This is a modern, up-to-date, comprehensive book on webs a topic rich in interrelations and (potential) application, unique in its coverage of higher codimensional webs, and, as such, a volume I am more than glad to welcome in this series.

The unreasonable effectiveness of mathematics in science ...

Eugene Wigner

Well, if you know of a better 'ole, go to it.

Bruce Bairnsfather

What is now proved was once only imagined.

William Blake

As long as algebra and geometry proceeded along separate paths, their advance was slow and their applications limited.

But when these sciences joined company they drew from each other fresh vitality and thenceforward marched on at a rapid pace towards perfection.

Joseph Louis Lagrange.

Bussum, May 1988

Michiel Hazewinkel

Table of Contents

CHAPTER 2

Almost Grassmann Structures Associated with Webs $W(n+1,n,r)$

CHAPTER 3

Local Differentiable n-Quasigroups Associated with a Web $W(n+1,n,r)$

CHAPTER 4

Special Classes of Multicodimensional $(n+1)$-Webs

CHAPTER 5

Realisations of Multicodimensional $(n+1)$-Webs

CHAPTER 6

Applications of the Theory of $(n+1)$-Webs

CHAPTER 7

The Theory of Four-Webs $W(4,2,r)$

CHAPTER 8

Rank Problems for Webs $W(d,2,r)$

Preface

The systematic study of webs was initiated by Blaschke in the 1930's. The subject started when in 1926-1927 Blaschke and Thomsen realised that the configuration of three foliations of curves in the plane has local invariants. Then Blaschke, his students and co-workers in a short span of time (1927-1938) published 66 articles under the general heading "Topologische Fragen der Differentialgeometrie". These and other results were summarised in the well-known monograph "Geometrie der Gewebe" by Blaschke and Bol [BB 38].

A d-web $W(d,n,r)$ of codimension r is given in an open domain D of a differentiable manifold X^{nr} of dimension nr by a set of d foliations of codimension r which are in general position.

The articles and the book were mostly devoted to webs of curves in the plane and webs of surfaces in the three-dimensional space; both cases are of codimension one. Webs of multicodimensional surfaces were studied only in two papers: Bol in [B 35] considered three-webs $W(3,2,2)$ in a four-dimensional space and Chern in [C 36b] investigated three-webs $W(3,2,r)$ of codimension r in a $(2r)$-dimensional space.

During this period the relationship between web geometry and quasigroups was discovered (see [Rei 28], [Kn 32], [Mo 35], and [B 37]) and closure conditions for webs and algebraic properties of quasigroups were obtained (see [BB 38], [Ac 65], and [Be 67]).

Note that at that time Chern [C 36b] recognised that while in the case of webs of codimension one ($r = 1$) all the analytical conditions defining three-webs with different closure conditions (hexagonal, Reidemeister, Thomsen) coincide, they are different for webs of codimension $r > 1$.

From 1938 to 1969 there were only a few publications on webs of codimension one and no publications on multicodimensional webs. Since 1969 Akivis [Ak 69a, 69b] and his school began a systematic study of three-webs of codimension r on a $(2r)$-dimensional manifold. In 1973 the author [G 73, 74a, 74b, 75a, 75b, 75c] began the study of $(n+1)$-webs $W(n+1,n,r)$ of codimension r on a (nr)-dimensional manifold. This study was very fruitful because the close relationship of web geometry with many branches of contemporary mathematics was revealed. In particular, a local differential quasigroup or loop (n-quasigroup or n-loop) was associated with any three-web (respectively $(n+1)$-web). Akivis and the author have developed techniques for webs using the intrinsic geometric structure imposed on the ambient space by a web, have defined the affine connections generated by a web and their torsion and curvature tensors and have investigated webs placing restrictions on the associated tensors.

During this study Chern's idea of distinguishing the cases $r = 1$ and $r > 1$ was substantially developed: corresponding to various closure conditions for three-webs and $(n+1)$-webs, $n \geq 2$, different geometric realisations were given by Akivis and the author.

In the course of this study Akivis also generalised the notion of Lie algebra [Ak 76]: in the tangent space to the identity of an arbitrary real analytic loop he introduced the so-called *triple system* or *local W-algebra* that is associated with such a loop exactly as a Lie algebra is associated with a Lie group. Later on this algebra was named the *Akivis algebra* by Hofmann and Strambach [HS 86a, 86b], who also added some original work to this theory. In particular, they gave the analogue to Lie's Third Fundamental Theorem for real analytic loops (see [HS 86a] or [CPS 88], Chapter IX). The author (see [CPS 88], Problem X.3.9 or [Sc 84], p. 16) has raised the problem of finding an algebraic construction in the tangent bundle of the coordinate n-loop of a web $W(n+1, n, r)$ similar to the construction of the Akivis algebra for three-webs or binary loops. Smith [Sm 88] proposed such a construction in general, and in detail for the case $n = 3$ of ternary loops. He also proved Lie's Third Fundamental Theorem for n-ary analytic loops.

Another important relationship is between web geometry and almost Grassmann structures. Almost Grassmann structures were studied by Hangan [Ha 66] and Mikhailov [Mi 78] and in connection with webs by the author [G 75c] and Akivis [Ak 80, 82]. Namely this connection allows Akivis and the author to solve the Grassmannisation and algebraisation problems for webs $W(d, n, r)$, $d \geq n+1$, $n \geq 2$, $r \geq 2$.

Thus in the 1970-1980's the main emphasis has been on the study of multicodimensional webs.

During the last few years web geometry has been growing rapidly and finding more and more applications.

We have already mentioned the connection between web geometry and quasigroups and web geometry and almost Grassmann structures. In addition to this, the connections between web geometry and

i) Abel's differential equations (see [Gr 76, 77] and [CG 78a]),
ii) Almost complex structures [Ak 75b],
iii) Symmetric spaces [Ak 78],
iv) The theory of point correspondences among $n+1$ $(n \geq 2)$ spaces endowed with the same structure [Bo 82,83,84,85],

v) Characteristic classes [D 83],
vi) Surfaces of translation [Lit 83],
vii) The theory of holomorphic mappings [Bau 82],
viii) The geometric theory of differential equations [V 83], [Ki 84],
ix) The theory of singularities [Car 83],
x) Double fibrations and integral geometry [GS 83],
xi) Four-dimensional pseudo-conformal structures [Ak 83a]

were established during this period.

Moreover, Chern published a brilliantly written expository paper [C 82]; Chern and Griffiths wrote a few papers on rank problems where they found the boundary for the r-rank of a web $W(d,n,r)$ (see [CG 78b]), described so-called normal webs $W(d,n,1)$ of maximum 1-rank and solved the Grassmannisation and algebraisation problems for such webs (see [CG 77, 78a, 81]); the author solved 1- and r-rank problems for almost Grassmannisable webs $W(d,2,r)$ [G 83,84,85b] and constructed examples of non-algebraisable webs $W(4,2,2)$ of maximum 2-rank [G 85a, 86, 87]; Akivis and the author solved the Grassmannisation and algebraisation problems for webs $W(d,n,r)$, $d \geq n+1$, $n \geq 2$, $r \geq 2$ (see [Ak 74, 80, 82, 83b] and [G 74b, 75c, 82b]).

Note also that web geometry was instrumental in the solution of the algebraisation problem for d submanifolds in a real projective space. This problem was solved for four hypersurfaces by the author [G 82c], for d $(d \geq 4)$ hypersurfaces by Wood [Wo 82,84] and for d surfaces of any codimension by Akivis [Ak 83b].

The connection between web geometry and algebraic geometry was noted in the 1930's. In the 1970 – 1980's this connection was exhibited with a new strength.

Projective algebraic varieties allow the construction of some examples of multi-codimensional webs. Moreover, multicodimensional webs $W(d,n,r)$ can be considered as a generalisation of projective algebraic varieties [C 82].

Recognising the growing importance of web geometry, the Mathematisches Forschungsinstitut in West Germany organised in 1984 a session in web geometry in Oberwolfach [Sc 84]. Following this, the second conference in web geometry was held in Szeged, Hungary, in 1987.

Unfortunately, no English text exists in the theory of *multicodimensional* webs. The above mentioned classical monograph [BB 38] as well as the Blaschke book [Bl 55] deal with *one-codimensional* webs. In 1981 Akivis and Shelekhov wrote a short 81-page textbook [AS 81] on the theory of multicodimensional webs. Only 500 xerox copies of this book were produced. In addition, this text deals with multicodimensional *three-webs* only and it is available only in Russian.

Thus, the necessity of a text introducing d-webs $W(d,n,r)$, $d \geq n+1$, $n \geq 2$, of codimension r on an (nr)-dimensional manifold is obvious. As to the theory of three-webs $W(3,2,r)$, I hope that the extended version of the book [AS 81] will be written some day and it will be available in English.

In this book we develop the theory of d-webs $W(d,n,r)$, $d \geq n+1$, $n \geq 2$.

We use exterior differential and tensor calculus and E. Cartan's method of moving frames. The reader is assumed to be familiar with the general theory of differential geometry as can be found in [KN 63]. However, in the text and notes we give most of the required background material and/or make necessary references. In some places to prove existence theorems for some classes of webs we use E. Cartan's test for Pfaffian differential systems to be involutive (see [Ca 45] or the recently published book [GJ 87]). However, these parts of the text are not essential for the understanding of subsequent material, and thus the reader who is unfamiliar with the above theory will not be unduly penalised.

Sections are numbered within each chapter and subsections are numbered within each section. Thus, Section 5.2 is the second section of Chapter 5. and Section 1.2.1 is the first subsection of Section 2 in Chapter 1. Definitions, theorems, lemmas, propositions, corollaries, examples and remarks are numbered consecutively within each section, with chapter and section numbers given to faciliate cross reference. The same numbering is incorporated for formulae; tables and figures are numbered consequtively within each chapter. Thus Theorem 8.3.6 refers to the sixth theorem of Section 8.3 and equation (3.1.5) refers to the fifth numbered formula in Section 3.1. The letters and numbers in square brackets indicate the first letter (letters) of the last name (names) of the author (authors) and the year of publication and refer to the bibliography at the end of the book.

In Chapter 1 we construct the differential geometry of webs $W(n+1,n,r)$. In Sections 1.1–1.3 we give the main definitions, illustrate them by examples, derive the structure equations of a web $W(n+1,n,r)$, $n \geq 2$, introduce its torsion and curvature tensors and give the conditions defining a web $W(n+1,n,r)$ up to an analytical transformation. After this we study the $n+1$ affine connections induced by a web $W(n+1,n,r)$ and in the case of a three-web $W(3,2,r)$ construct the so-called canonical affine connection introduced by Chern [C 36b] and used by Akivis [Ak 69a] and his students. This allows us to present the information on three-webs $W(3,2,r)$ that we will use in the sequel. In Section 1.3.4 we study three affine connections associated with 3-subwebs of the web $W(n+1,n,r)$. In Sections 1.4–1.11 we study special classes of webs $W(n+1,n,r)$ which are most important from the differential geometry point of view: parallelisable webs $W(n+1,n,r)$; webs $W(n+1,n,r)$ with paratactical 3-subwebs, with integrable diagonal distributions and those of its 4-subwebs; hexagonal, transversally geodesic, in particular, $(2n+2)$-hedral, and isoclinic webs $W(n+1,n,r)$.

In Chapter 2 we introduce almost Grassmann structures, show that such a structure $AG(n-1,r+n-1)$ is associated with any web $W(n+1,n,r)$, that almost Grassmann structures associated with transversally geodesic and isoclinic webs $W(n+1,n,r)$ are semiintegrable and prove that a web $W(n+1,n,r)$ is Grassmannisable, i.e., equivalent to a Grassmann web, if and only if it is transversally geodesic and isoclinic. This solves the Grassmannisation problem for webs $W(n+1,n,r)$. In Section 2.5 we consider double webs: they occur when the same almost Grassmann structure $AG(n-1, r+n-1)$ is associated with a web $W(n+1,n,r)$ and a web $W(r+1,r,n)$. In the final Section 2.6 we present the solution of the Grassmannisation and algebraisation problems for webs $W(d,n,r)$, $d \geq n+1$, $n \geq 2$, $r \geq 2$.

In Chapter 3 we give the basics of the theory of local differentiable n-quasigroups as well as their link with webs $W(n+1,n,r)$, derive the canonical expansions of the equations of a local analytical n-quasigroup and use them to find necessary and sufficient conditions for the existence of one-parameter subquasigroups or subloops of a local analytical n-quasigroup for any direction. In addition, in this chapter we present recent results of Smith who found in the tangent bundle of a real analytic n-loop, $n > 2$, an algebraic construction (that of Akivis and comtrans algebras) similar to the construction of the Akivis algebras for binary loops and proved Lie's Third Fundamental Theorem for analytic n-loops, $n \geq 3$ using this algebraic construction.

The link between webs $W(n+1,n,r)$ and n-quasigroups established in Chapter 3 allows us to present in Chapter 4 new special classes of webs (and corresponding local n-quasigroups): reducible, multiple and completely reducible, group, $(2n+2)$-hedral and Bol webs $W(n+1,n,r)$. In addition, this link allows to give a new characterisation of some of the webs which were introduced in Chapter 1 (for example, the $(2n+2)$-hedral webs $W(n+1,n,r)$).

In Chapter 5 we construct projective realisations of different classes of webs $W(n+1,n,r)$. Grassmann webs $GW(n+1,n,r)$ and their special classes (reducible, hexagonal, group, parallelisable, algebraic, Bol, Moufang etc.) in a projective space P^{r+n-1} of dimension $r+n-1$ are studied in Sections 5.1–5.5, and a diagonal four-web $W(4,3,r)$ formed by four pencils of $(2r)$-planes with $(2r-1)$-dimensional vertices in a projective space P^{3r} of dimension $3r$ is studied in Sections 5.6–5.7. In Section 5.2 we give the proof of the Grassmannisation theorem for webs $W(n+1,n,r)$ which is different from the proof given in Section 2.6.

In Chapter 6 we give applications of the theory of $(n+1)$-webs to the theory of point correspondences of $n+1$ projective lines and $n+1$ projective spaces (Sections 6.1–6.2) and the theory of holomorphic mappings betwen polyhedral domains (Section 6.3).

The subject of Chapter 7 is webs $W(d,n,r)$, $d > n+1$. As the first step in constructing the theory of such webs we consider webs $W(4,2,r)$. We show that a pair of orthogonal binary quasigroups is associated with a $W(4,2,r)$, introduce the basis affinor of a $W(4,2,r)$, find canonical expansions of equations of $W(4,2,r)$, study special classes of webs $W(4,2,r)$, in particular, webs satisfying the Desargues or triangle closure conditions, Grassmann and algebraic webs and their different subclasses and give a classification of group webs $W(4,2,3)$. While considering algebraic webs $AW(n+1,n,r)$, we give the proof of algebraisation theorem for webs $W(4,2,r)$ which is different from the proof given in Section 2.6.

The final Chapter 8 deals with the 1- and r-rank problems for certain webs $W(d,2,r)$. We introduce and study here almost Grassmannisable and almost algebraisable webs which play an important role in the study of rank problems, find the boundary for the 1- and r-rank of an almost Grassmannisable web $AGW(d,2,r)$, $r \geq 2$, and describe maximum 1- and r-rank webs $AGW(d,2,r)$. Note that since a web $W(4,2,2)$ of maximum 2-rank and a web $W(d,n,r)$, $d > r(n-1)+2$, of maximum r-rank are almost Grassmannisable [Lit 86], in these cases our description is concerned with general webs $W(4,2,2)$ and $W(d,n,r)$. In Section 8.4 we construct examples of non-algebraic webs $W(4,2,2)$ of maximum 2-rank. The results of Sections 8.3 and 8.4 show that the Griffiths' conjecture that maximum r-rank webs $W(d,n,r)$ are algebraisable is true for webs $W(d,2,2)$, $d > 4$, and is not true for webs $W(4,2,2)$ which are not necessarily algebraisable. In the final Section 8.5 we study geometry of the webs $W(4,2,2)$ of maximum 2-rank: we define and study functions, vector fields, and exterior forms which can be given on some foliation of a web and using the idea of double fibrations can be carried to other foliations.

The results of all chapters except Chapter 6 are mostly due to the author. The results of Chapter 6 are due to Bolodurin (Sections 6.1-6.2) and Baumann (Section 6.3) and the results of Section 5.5 are due to Smith. In addition, we included in Sections 1.3, 1.9, 3.3, 4.5, 5.4.2 the recent results of Gerasimenko, in Sections 1.11, 2.6 the results of Akivis, in Section 2.6 the result of Wood, and in Section 7.2 the result of Tolstikhina. For details, see Notes at the end of each chapter. These notes contain references to source material and also some explanatory material marked in the text by numbers enclosed in parentheses.

In concluding the preface, the author would like to thank M.A. Akivis who involved him in the field of web geometry, made him excited about the field and with whom cooperation has always been a pleasure. The author also likes to thank J. Baumann, J.D.H. Smith, and J.A. Wood for materials which they provided during the work on the book, K.W. Johnson for his numerous suggestions, and D. Reidel Publishing Company for its patience and kind cooperation. Special thanks are due to the author's wife Ludmila for her encouragement and cooperation during the tough period of writing the book. Finally the author thanks the administration of New Jersey Institute of Technology for its continuous support.

Livingston, New Jersey, U.S.A. *Vladislav V. Goldberg*

Acknowledgements

The publisher and author would like to thank the following for giving permission to reproduce previously copyrighted material:

American Mathematical Society for extracts from [G87]*
Birkhauser Verlag for extracts from [Sm88]
Duke University Press for extracts from [Wo84]
Gauthier-Villars for extracts from [G83b] and [G85a]
D. Reidel Publishing Company for extracts from [G82c] and [G84]
Springer-Verlag New York for extracts from [G86]
Tensor Society for extracts from [G82a], [G82b] and [G82d]

* The references given here can be found in the list of references appearing on pages 440–451.

Chapter 1

Differential Geometry of Multicodimensional $(n+1)$-Webs

1.1 Fibrations, Foliations, and d-Webs $W(d,n,r)$ of Codimension r on a Differentiable Manifold X^{nr}

1.1.1 Definitions and Examples

We will give here some concepts which are necessary to introduce the notion of a web.

Let S and X be differentiable manifolds of respective dimensions M and N where $M > N$. In what follows we shall suppose that all manifolds and submanifolds are of class C^k, $k \geq 3$. We will be interested only in the local structure of such manifolds. Because of this, in many cases it is convenient to think about such manifolds as connected domains of a Euclidean space of the same dimension.

A *locally trivial fibration of class* C^k (see [Hu 66] or [KN 63]) is a triple $\lambda = (S, X, \pi)$ where $\pi : S \to X$ is a projection of S onto X satisfying the following conditions:

(i) For each point $x \in X$ the set $\pi^{-1}(x) \subset S$ is a submanifold of dimension $M - N$ which is diffeomorphic to a manifold F;

(ii) For each point $x \in X$ there exists a neighbourhood $U_x \subset X$ such that $\pi^{-1}(U_x)$ is diffeomorphic to the product $U_x \times F$, and the diffeomorphism between $\pi^{-1}(U_x)$ and $U_x \times F$ is compatible with π and the projection $\mathrm{pr}_{U_x} : U_x \times F \to U_x$.

Since we will consider only local trivial fibrations, we will simply call them *fibrations.* The manifold S is the *space* of the fibration λ, the manifold X is its *base*, and $\pi^{-1}(x) = F$ is a *fibre* of λ. If $p \in S$ and $\pi(p) = x$, then there exists a unique fibre F_x passing through p (Figure 1.1). We will also use the notation $\lambda(S)$ for λ.

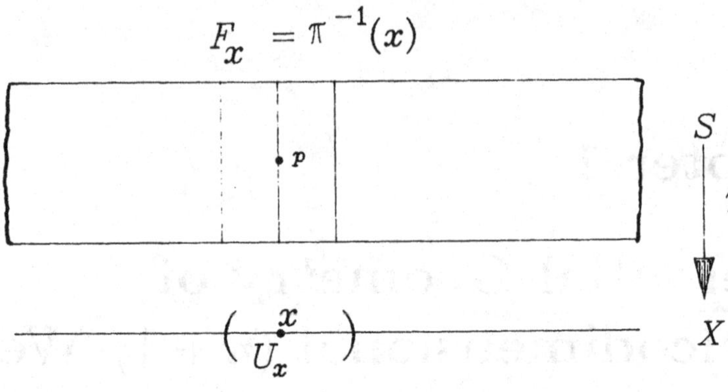

Figure 1.1

An example of a fibration is a *standard trivial fibration* $(X \times F, X, \pi)$ where π is the natural projection $X \times F \to X$. All its fibres are diffeomorphic to F.

Let $T_p(S)$ be the tangent space of S at a point $p \in S$. A fibre F passing through p determines in $T_p(S)$ a subspace $T_p(F_x)$, of codimension r say, tangent to F_x at p.

We associate the space $T_p(S)$ with a local vectorial moving frame $\{e_i, e_\sigma;$ $i = 1, \ldots, r; \sigma = r+1, \ldots, M\}$ where $e_\sigma \in T_p(F_x)$ (Figure 1.2). We have now $\dim F = M - r$ and $\dim X = r$.

Let $\{\omega^i, \omega^\sigma\}$ be a co-frame dual to the frame $\{e_i, e_\sigma\}$. Then $\omega^i(e_\sigma) = 0$, and the fibres F_x of the fibration λ are integral manifolds of the system of equations

$$\omega^i = 0. \tag{1.1.1}$$

Since there exists a unique fibre F through a point $p \in S$, the system (1.1.1) is completely integrable. According to the Frobenius theorem, its integrability conditions are

$$d\omega^i = \omega^j \wedge \phi^i_j \tag{1.1.2}$$

where ϕ^i_j are differential forms containing the differentials of parameters which determine the location of the frame $\{e_i, e_\sigma\}$. Note that in (1.1.2) and in the sequel we shall adhere to the summation convention proposed by A.Einstein. In (1.1.2) it is understood that summation from 1 to r is carried out over the index j which appears twice: once above and once below. Hence for (1.1.2) it gives

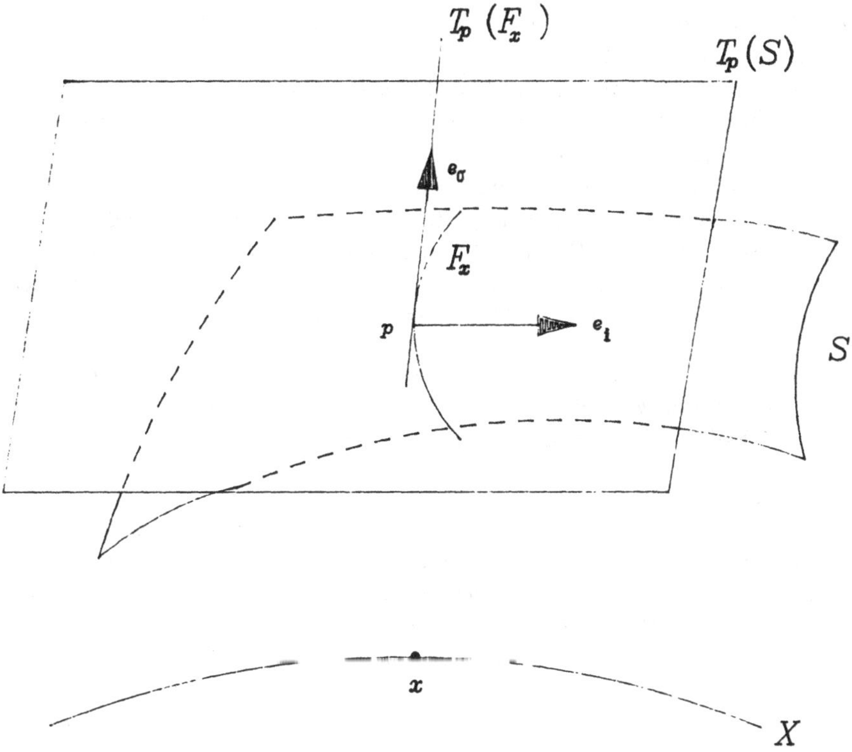

Figure 1.2

$$\omega^j \wedge \phi^i_j = \omega^1 \wedge \phi^i_1 + \omega^2 \wedge \phi^i_2 + \ldots + \omega^r \wedge \phi^i_r.$$

A *k-dimensional distribution* θ on an M-dimensional manifold S, $0 \leq k \leq M$, is a smooth field of k-dimensional tangential directions, i.e., a function which assigns a linear k-dimensional subspace of the tangent space $T_p(S)$ to each point $p \in S$. A distribution is said to be *integrable* if through each point $p \in S$ there passes a unique k-dimensional integral surface of θ which is tangent to the distribution at any of its points. A distribution θ is integrable if and only if a system of differential equations defining θ is completely integrable (see [Ch 73] or [L 53]).

We say that we are given a *k-dimensional foliation* on S if the manifold S is fibrated into k-dimensional surfaces. This means that through each point $p \in S$ there passes one and only one smooth k-dimensional surface which smoothly depends on p. These surfaces are called the *leaves* of the foliation, the numbers k and $M - k$ are correspondingly the *dimension* and *codimension* of the foliation.

In other words, we should be able to introduce locally in a neighbourhood of each point $p \in S$ coordinates $x^1, \ldots, x^k, y^1, \ldots, y^{M-k}$ such that level surfaces $y^1 = a^1, \ldots, y^{M-k} = a^{M-k}$, where $a^1, \ldots, a^{M-k}$ are constants, determine the leaves of the foliation and $x^1, \ldots, x^k$ are local coordinates of the leaves.

The foliations are given by integrable distributions.

A smooth fibration $\lambda = (S, X, \pi)$ is a trivial example of a foliation: the decomposition of the manifold S into fibres F_x of λ is an $(M-r)$-dimensional foliation. Further details and examples concerning foliations see in [DFN 85] or in [La 74].

As follows from the above definitions, locally in a sufficiently small domain, a foliation becomes a fibration. Because of this, we can speak about the base of the foliation and locally apply the apparatus known for fibrations. In particular, locally a foliation can be given by equations (1.1.1) and (1.1.2).

Definition 1.1.1 Let $S = X^{nr}$ be a differentiable manifold of dimension $M = nr$. We shall say that a *d-web* $W(d,n,r)$ *of codimension* r is given in an open domain $D \subset X^{nr}$ by a set of d foliations of codimension r which are in general position.

Let us make a few remarks:

1) All functions and differential forms which we will introduce on X^{nr} will be real of class C^k, $k \geq 3$, or holomorphic in the complex case. Since in the complex analytic case we will not use the conjugate variables or Cauchy-Riemann equations, it will not be necessary to distinguish the real and complex cases.

2) In the notation $W(d,n,r)$ the number d is the number of foliations forming a web, r is the web codimension and n is the ratio of the dimension nr of the manifold X^{nr} and the web codimension. Of course, we may define a d-web of codimension r without having r as a divisor of the dimension of the ambient manifold. However, when $M = nr$, we are getting a wide generalisation of projective algebraic varieties (see Example 1.1.5 below). This is a justification for considering webs $W(d,n,r)$.

3) Since the codimension of a web $W(d,n,r)$ and each of its foliations is the codimension of a leaf of a foliation, the dimension of a leaf is $nr - r = (n-1)r$.

4) Note also that d foliations are *in general position* if the tangent $(n-1)r$-planes to the leaves of different foliations through a point $p \in D$ are in general position ([1]).

5) To study webs globally, one cannot use the apparatus developed for fibrations. There are some results of global nature for three-webs obtained by Nishimori [N 80,81].

6) A brief history of study of webs $W(d,n,r)$ see in ([2]).

Definition 1.1.2 Two webs $W(d,n,r)$ and $\tilde{W}(d,n,r)$ with domains $D \subset X^{nr}$ and $\tilde{D} \subset \tilde{X}^{nr}$ are *locally equivalent* if there exists a local diffeomorphism $\phi : D \to \tilde{D}$ of their domains such that $\phi(\lambda_\xi) = \tilde{\lambda}_\xi$, $\xi = 1, \ldots, d$.

In this definition a *local diffeomorphism* ϕ is a differentiable mapping $\phi : D \to \tilde{D}$ which induces at each point $p \in D$ a diffeomorphism of some open neighbourhood of p onto an open neighbourhood of $\tilde{p} = \phi(p)$.

The study of webs is concerned with local invariants (under the group of diffeomorphisms) of a set of d foliations in general position.

Let us consider three examples of webs.

Example 1.1.3 Let λ_ξ, $\xi = 1, \ldots, d$, be d foliations of parallel $(n-1)r$-planes in an affine space A^{nr} of dimension nr. Suppose that the $(n-1)r$-planes of different foliations are in general position. Such a d-web is called *parallel.* A web $W(d,n,r)$ which is equivalent to a parallel $W(d,n,r)$ is called *parallelisable.* We will denote parallelisable webs by $PW(d,n,r)$.

Example 1.1.4 Let C be an algebraic curve of class d in a complex projective plane $\mathbf{CP}^2$, i.e., a set of straight lines in the dual space $\mathbf{CP}^{2^*}$ of lines whose tangential (line) coordinates satisfy a homogeneous equation of the dth order. One can always find an open domain $D \subset \mathbf{CP}^{2^*}$ such that there exist d different straight lines of C through each point of D. The domain D admits a covering by neighbourhoods and the indicated straight lines form d foliations in general position in each of these neighbourhoods. Therefore the curve C defines a web $W(d,2,1)$. It is worthy to note that this $W(d,2,1)$ cannot be represented as a set of d fibrations (or even foliations) in the whole domain D.

Note that a curve C of class d defines a web $W(d,2,1)$ in a real projective plane $\mathbf{RP}^2$ if and only if there exists at least one point in $\mathbf{RP}^2$ through which there pass three real tangents to C.

In the sequel, when we speak of d foliations forming a web $W(d,n,r)$, we shall mean d foliations in each of the covering neighbourhoods.

Let V be a complex vector space of dimension t. The *Grassmannian* $G(k,V)$ is the set of k-dimensional linear subspaces of V; we write $G(k,t)$ for $G(k,\mathbf{C}^t)$.

Example 1.1.5 ([C 82]). Let V_d^r be an algebraic variety of dimension r and degree d in a projective space P^{r+n-1} of dimension $r+n-1$. A linear subspace P^{n-1} of dimension $n-1$ meets V_d^r in d points. There are $\infty^{r(n-1)}$ subspaces P^{n-1} passing through each of these d points. This gives, in the Grassmannian $G(n-1,r+n-1)$ of all subspaces P^{n-1} in P^{r+n-1}, d foliations of dimension $(n-1)r$. Since $\dim\, G(n-1,r+n-1) = nr$ (see [HP 52]), the variety V_d^r gives rise to d foliations of codimension r in $G(n-1,r+n-1)$. In other words, V_d^r defines a web $W(d,n,r)$.

Such a web is called *algebraic.* Example 1.1.5 shows why it is worth considering webs $W(d,n,r)$ of codimension r in X^{nr} although as we noted above, it is possible to consider d-webs of codimension r in X^M where M is not necessarily a multiple of r. An algebraic web is a particular case of a Grassmann web which we will define and study in Chapter 2. In what follows we will see another reason for the importance of considering webs $W(d,n,r)$: the possibility of relating the theory of webs $W(d,n,r)$ in X^{nr} to the theory of n-quasigroups.

1.1.2 Closed Form Equations of a Web $W(n+1,n,r)$ and Further Examples

Now assume $d > n$. Let λ_ξ, $\xi = 1,\dots,d$, be the foliations of a web $W(d,n,r)$. They are in general position in $D \subset X^{nr}$.

Let p be any point of D and U_p be a neighbourhood of p. Let x_ξ, $\xi = 1,\dots,d$, be the points of the bases X_ξ of the foliations λ_ξ determining the leaves $F_{x_\xi} \subset \lambda_\xi$ passing through the point p.

On X_ξ in neighbourhoods of x_ξ we introduce local coordinates x^i_ξ, $\xi = 1,\dots,d$; $i = 1,\dots,r$. As coordinates in U_p one can take any n systems $x^{j_1}_{\xi_1},\dots,x^{j_n}_{\xi_n}$, $\xi_1,\dots,\xi_n \in \{1,\dots,d\}$; $j_n \in \{1,\dots,r\}$. Then in U_p each λ_ξ can be defined by equations $x^i_\xi = c^i_\xi, i = 1,\dots,r$, where c^i_ξ are constants, variables x^i_α for $\alpha = 1,\dots,n$, are appropriately chosen coordinates in U_p and the variables x^i_τ for $\tau = n+1,\dots,d$, can be expressed in terms of x^i_α by

$$x^i_\tau = f^i_\tau(x^{j_1}_1,\dots,x^{j_n}_n). \tag{1.1.3}$$

The functions f^i_τ in (1.1.3) are of class $C^k, k \geq 3$, and the following inequalities hold:

$$\det\left(\frac{\partial(f^{i_1}_{\tau_1},\dots,f^{i_s}_{\tau_s})}{\partial(x^{j_1}_{\alpha_1},\dots,x^{j_s}_{\alpha_s})}\right) \neq 0, \tag{1.1.4}$$

where

$$\tau_1,\dots,\tau_s = n+1,\dots,d; \quad \alpha_1,\dots,\alpha_s = 1,\dots,n; \quad s \leq \min(n, d-n).$$

The inequalities (1.1.4) make possible to solve equations (1.1.3) with respect to any $x^{j_1}_{\alpha_1}; x^{j_1}_{\alpha_1}, x^{j_2}_{\alpha_2}; \dots; x^{j_1}_{\alpha_1},\dots,x^{j_n}_{\alpha_n}$. Geometrically this corresponds to the fact that any n leaves of different foliations $\lambda_{\xi_1},\dots,\lambda_{\xi_n}$, $\xi_1,\dots,\xi_n = 1,\dots,d$, of the web $W(d,n,r)$ have one and only one common point ([3]) .

Let us consider now a few more examples of webs $W(d,n,r)$ defined by (1.1.3) and (1.1.4). To make geometrical interpretations of some of them, it is convenient to assume that X^{nr} is a connected domain of $\mathbf{R}^{nr}$.

Example 1.1.6 For a web $W(n+1,n,1)$ defined in $\mathbf{R}^n$ by

$$x^{n+1} = \sqrt{\sum_\alpha (x^\alpha)^2} \tag{1.1.5}$$

conditions (1.1.4) become

$$x^\alpha \neq 0, \quad \alpha = 1,\dots,n. \tag{1.1.6}$$

As a domain D one can take for example the domain $D = \{(x^1,\dots,x^n) : x^\alpha > 0\}$. The leaves of λ_α, $\alpha = 1,\dots,n$, are the coordinate hyperplanes. The leaves of λ_{n+1} of the web $W(n+1,n,1)$ are parts of concentric hyperspheres with the centre at the origin lying in D.

Example 1.1.7 Let us consider a $W(n+1,n,1)$ defined in $\mathbf{R}^n$ by

$$x^{n+1} = \frac{\sum_\alpha x^\alpha - 2x^1}{\sum_\alpha \left((x^\alpha)^2 - x^\alpha\right)}. \tag{1.1.7}$$

If we introduce the notation

$$\sum_\alpha x^\alpha - 2x^1 = B, \ \sum_\alpha \left((x^\alpha)^2 - x^\alpha\right) = A, \tag{1.1.8}$$

then first of all we see that $A \neq 0$. Conditions (1.1.4) give

$$A + (2x^1 - 1)B \neq 0, \ A - (2x^{\hat{\alpha}} - 1)B \neq 0, \ \hat{\alpha} \neq 1. \tag{1.1.9}$$

All the conditions (1.1.9) will be satisfied inside the n-dimensional cube $D = \{(x^1, \ldots, x^n) : 0 < x^1 < 1/2, 1/2 < x^{\hat{\alpha}} < 1, \hat{\alpha} \neq 1\}$ since in this cube $A < 0$, $B > 0$, $2x^1 - 1 < 0$, and $2x^{\hat{\alpha}} - 1 > 0$, $\hat{\alpha} \neq 1$.

The foliation λ_{n+1} of the $W(n+1,n,1)$ consists of parts of hyperspheres lying inside D. The radius of the hyperspheres is

$$R = \frac{1}{2}\sqrt{(C-1)^2 + (n-1)(C+1)}$$

and their centres are located at the points $((1-C)/2, (1+C)/2, \ldots, (1+C)/2)$ filling a part of the straight line $x^2 = x^3 = \ldots = x^n = 1 - x^1$. Those points of this line should be taken as centres for which an intersection of the corresponding hypersphere and the cube D is not empty.

Example 1.1.8 Suppose that a $W(3,2,2)$ is given in $\mathbf{R}^4$ by

$$x_3^1 = x_1^1 + x_2^1, \ x_3^2 = (x_2^1 - x_1^1)(x_1^2 + x_2^2) \tag{1.1.10}$$

(see [B 35]). Conditions (1.1.4) become

$$x_2^1 - x_1^1 \neq 0. \tag{1.1.11}$$

As a domain D one can take, for example, the domain where $x_2^1 - x_1^1 > 0$. The leaves of λ_3 are two-dimensional quadrics.

Example 1.1.9 Suppose that a web $W(4,3,r)$ is defined by ([4])

$$x_4^i = \frac{x_1^i + x_2^i x_3^i}{\sum_k x_1^k}, \quad i,k = 1,\ldots,r. \tag{1.1.12}$$

Let us introduce the notations:

$$z = \sum_k x_1^k, \ a^i = \sum_{\alpha=1}^{3} x_\alpha^i, \quad i,k = 1,\ldots,r. \tag{1.1.13}$$

It is easy to see that

$$\Delta_{\hat{\alpha}} = \det\left(\frac{\partial x_4^i}{\partial x_{\hat{\alpha}}^j}\right) = \frac{1}{z^r}, \quad \hat{\alpha} = 2, 3;$$

$$\Delta_1 = \det\left(\frac{\partial x_4^i}{\partial x_1^j}\right) = \frac{1}{z^{r+1}}(z - \sum_i a^i).$$

Therefore conditions (1.1.4) are

$$z \neq \sum_i a^i. \tag{1.1.14}$$

As a domain D one can take, for example, the domain given by $z > 0$ and $z > \sum_i a^i$.

The leaves of λ_4 are defined by $a^i = C^i z$, C^i are constants, and therefore they are $(2r)$-planes. They form a pencil with $(2r-1)$-dimensional vertex since each of the $(2r)$-planes passes through a $(2r-1)$-plane defined by equations $z = 0$ and $a^i = 0$.

Example 1.1.10 For a web $W(4,2,2)$ defined by (see [G 87])

$$\begin{aligned} x_3^1 &= x_1^1 + x_2^1 + \tfrac{1}{2}(x_1^1)^2 x_2^2, & x_4^1 &= -x_1^1 + x_2^1 + \tfrac{1}{2}(x_1^1)^2 x_2^2, \\ x_3^2 &= x_1^2 + x_2^2 - \tfrac{1}{2}x_1^1(x_2^2)^2, & x_4^2 &= x_1^1 - x_2^1 - \tfrac{1}{2}x_1^1(x_2^2)^2 \end{aligned} \tag{1.1.15}$$

the determinants from (1.1.4) are

$$\det\left(\frac{\partial x_3^i}{\partial x_1^j}\right) = -\det\left(\frac{\partial x_4^i}{\partial x_2^j}\right) = 1 + x_1^1 x_2^2,$$

$$\det\left(\frac{\partial x_3^i}{\partial x_2^j}\right) = -\det\left(\frac{\partial x_4^i}{\partial x_1^j}\right) = 1 - x_1^1 x_2^2,$$

$$\det\left(\frac{\partial(x_3^i, x_4^j)}{\partial(x_1^k, x_2^l)}\right) = 2.$$

Therefore conditions (1.1.4) become

$$x_1^1 x_2^2 \neq \pm 1,$$

and a domain D can be defined by $x_1^1 x_2^2 > 1$ or $-1 < x_1^1 x_2^2 < 1$ or $x_1^1 x_2^2 < -1$.

1.2 The Structure Equations and Fundamental Tensors of a Web $W(n+1,n,r)$

1.2.1 Moving Frames Associated with a Web $W(n+1,n,r)$

We will consider now the case $d = n+1$, i.e., an $(n+1)$-web $W(n+1,n,r)$ given in an open domain D of a differentiable manifold of dimension nr. In a neighbourhood of a point $p \in D$ a web $W(n+1,n,r)$ is defined by $n+1$ foliations λ_ξ, $\xi = 1,...,n+1$. According to (1.1.1) each foliation λ_ξ can be defined by a completely integrable system of Pfaffian equations

$$\underset{\xi}{\bar{\omega}}^i = 0, \quad \xi = 1,\dots,n+1; \quad i = 1,\dots,r. \tag{1.2.1}$$

Conditions (1.1.2) of complete integrability of each of the systems (1.2.1) have the form

$$d\underset{\xi}{\bar{\omega}}^i = \underset{\xi}{\bar{\omega}}^j \wedge \underset{\xi}{\bar{\omega}}^i_j. \tag{1.2.2}$$

There are $(n+1)r$ forms $\underset{\xi}{\bar{\omega}}^i$. On X^{nr} only nr of them are linearly independent. Hence the forms $\underset{\xi}{\bar{\omega}}^i$ are connected by r relations:

$$\sum_{\xi} \underset{\xi}{x}^i_j \underset{\xi}{\bar{\omega}}^j = 0, \tag{1.2.3}$$

where $\det(\underset{\xi}{x}^i_j) \neq 0$ for any $\xi = 1,\dots,n+1$, since each group of the forms $\underset{\xi}{\bar{\omega}}^i$, where ξ is fixed, should be linearly independent and equations (1.2.3) should be solvable with respect to each of $n+1$ groups of forms $\underset{\xi}{\bar{\omega}}^i$.

For each foliation λ_ξ we make the following change of the basis forms $\underset{\xi}{\bar{\omega}}^i$:

$$\underset{\xi}{\omega}^i = \underset{\xi}{x}^i_j \underset{\xi}{\bar{\omega}}^j. \tag{1.2.4}$$

Using (1.2.4), we can write the relations (1.2.3) in the form

$$\sum_{\xi} \underset{\xi}{\omega}^i = 0. \tag{1.2.5}$$

Equations (1.2.2) and (1.2.4) imply the following structure equations for the forms $\underset{\xi}{\omega}^i$:

$$d\underset{\xi}{\omega}^i = \underset{\xi}{\omega}^j \wedge \underset{\xi}{\omega}^i_j, \tag{1.2.6}$$

where

$$\underset{\xi}{\omega}^i_j = \underset{\xi}{\tilde{x}}^l_j(-d\underset{\xi}{x}^i_l + \underset{\xi}{x}^i_k \underset{\xi}{\bar{\omega}}^k_l)$$

and $(\tilde{x}^l_{\xi j})$ is the inverse matrix of $(\tilde{x}^i_{\xi l})$, i.e., $\underset{\xi}{x}^i_l \underset{\xi}{\tilde{x}}^l_j = \underset{\xi}{x}^l_j \underset{\xi}{\tilde{x}}^i_l = \delta^i_j$.

The forms $\underset{\xi}{\omega}^i$ satisfying (1.2.5) are defined up to the following concordant transformations which preserve (1.2.5):

$$'\underset{\xi}{\omega}^i = A^i_j \underset{\xi}{\omega}^j, \quad \det(A^i_j) \neq 0. \tag{1.2.7}$$

The transformations (1.2.7) single out a subfamily of moving frames from all the moving frames associated with X^{nr}. To describe this subfamily, first note that since any n foliations $\lambda_{\xi_1}, \ldots, \lambda_{\xi_n}$ are in general position, any nr forms $\underset{\xi_1}{\omega}^{i_1}, \ldots, \underset{\xi_n}{\omega}^{i_n}$ are linearly independent and can be taken as a co-frame $\{\underset{\xi_1}{\omega}^{i_1}, \ldots, \underset{\xi_n}{\omega}^{i_n}\}$ dual to a vectorial frame $\{e^{\xi_1}_{i_1}, \ldots, e^{\xi_n}_{i_n}\}$ in the tangent space $T_p(X^{nr})$. Let us take the co-frame $\{\underset{\alpha}{\omega}^i\}, \alpha = 1, ..., n; i = 1, ..., r$, dual to a frame $\{e^\alpha_i\}$. When the forms $\underset{\alpha}{\omega}^i$ are transformed to $'\underset{\alpha}{\omega}^i$ according to (1.2.7), the corresponding transformations of the vectors e^α_i are:

$$'e^\alpha_i = \tilde{A}^j_i e^\alpha_j \tag{1.2.8}$$

where $(\tilde{A}^j_i)$ is the inverse matrix of (A^j_i). If ξ is a tangent vector to X^{nr} at p, then

$$\xi = \sum_{\alpha=1}^{n} \underset{\alpha}{\omega}^i e^\alpha_i. \tag{1.2.9}$$

It follows from (1.2.9) that

$$\underset{\alpha}{\omega}^i(e^\beta_j) = \delta^\beta_\alpha \delta^i_j. \tag{1.2.10}$$

Since the foliation λ_α, α is fixed, is defined by $\underset{\alpha}{\omega}^i = 0$, equation (1.2.9) shows that the vectors $e^{\hat{\alpha}}_i, \hat{\alpha} \neq \alpha$, are tangent at p to the leaf F_α of λ_α passing through p and form a basis in $T_p(F_\alpha)$. The foliation λ_{n+1} is defined by $\underset{n+1}{\omega}{}^i = -\sum_\alpha \underset{\alpha}{\omega}^i = 0$. This and (1.2.9) imply that the vectors $e^{\hat{\alpha}}_i - e^\alpha_i, \hat{\alpha} \neq \alpha, \alpha$ fixed, form a basis in $T_p(F_{n+1})$.

The vectors e^α_i, α fixed, are tangent to the r-surface $U_\alpha = \bigcap_{\hat{\alpha} \neq \alpha} F_{\hat{\alpha}}$. In a neighbourhood of p the leaves of λ_{n+1} intersect each U_α at the point and produce a point correspondence among the U_α in which corresponding lines of U_α are tangent to vectors with equal coordinates. These corresponding lines are defined by the equations

$$\underset{\alpha}{\omega}^i = \xi^i dt, \; \underset{\hat{\alpha}}{\omega}^i = 0, \; \hat{\alpha} \neq \alpha, \; \alpha \text{ fixed}, \tag{1.2.11}$$

and the vectors

$$\xi^\alpha = \xi^i e^\alpha_i \tag{1.2.12}$$

are tangent to these lines. These n vectors ξ^α define a transversal n-vector $[\xi^1 \wedge \ldots \wedge \xi^n](\xi^i)$.

Definition 1.2.1 An n-vector $\xi^1 \wedge \ldots \wedge \xi^n$ is called a *transversal n-vector* of the web $W(n+1,n,r)$.

Since r parameters ξ^i are homogeneous coordinates of a transversal n-vector $[\xi^1 \wedge \ldots \wedge \xi^n](\xi^i)$, the set of transversal n-vectors at the point p depends on $r-1$ parameters.

Proposition 1.2.2 *The set of n-vectors $\xi^1 \wedge \ldots \wedge \xi^n$ is invariant under the admissible transformations* (1.2.8).

Proof: Using (1.2.12) and (1.2.8), we can write the following relations:

$$\begin{aligned}[]['\eta^1 \wedge \ldots \wedge' \eta^n](\eta^i) &= \eta^{i_1} \ldots \eta^{i_n}\ {}'e^1_{i_1} \wedge \ldots \wedge' e^n_{i_n} \\ &= \eta^{i_1} \ldots \eta^{i_n} \tilde{A}^{j_1}_{i_1} \ldots \tilde{A}^{j_n}_{i_n} e^1_{j_1} \wedge \ldots \wedge e^n_{j_n} \\ &= \xi^{j_1} \ldots \xi^{j_n} e^1_{j_1} \wedge \ldots \wedge e^n_{j_n} = [\xi^1 \wedge \ldots \wedge \xi^n](\xi^i)\end{aligned}$$

where $\xi^i = \tilde{A}^i_j \eta^j$. They prove Proposition 1.2.2. ■

Proposition 1.2.2 show that a subfamily of moving frames singled out by the transformations (1.2.7) (or (1.2.8)) preserving equation (1.2.5) can be characterized as the set of moving frames which keep invariant the set of n-vectors $\xi^1 \wedge \ldots \wedge \xi^n$.

Equations (1.2.8) also show that the group of admissible transformations of a moving frame which preserve equation (1.2.5) is the general linear group $\mathbf{GL}(r)$ of r parameters. It defines on X^{nr} carrying a $W(n+1,n,r)$ a G-structure (see [St 83]) with the structure group $G = \mathbf{GL}(\mathbf{r})$ which is a subgroup of the general linear group $\mathbf{GL}(nr)$ of nr parameters acting in an (nr)-dimensional tangent space $T_p(X^{nr})$.

1.2.2 The Structure Equations and Fundamental Tensors of a Web $W(n+1,n,r)$

A web $W(n+1,n,r)$ is defined by $n+1$ completely integrable systems of 1-forms $\{\underset{\xi}{\omega}^i\}$, $\xi = 1, ..., n+1$, satisfying (1.2.5) and (1.2.6). In (1.2.6) the forms $\underset{\xi}{\omega}^i$ contain the differentials of parameters defining a set of moving frames in $D \subset X^{nr}$.

The forms $\underset{\xi}{\omega}^i_j$ are not independent. Exterior differentiation of (1.2.5) by means of (1.2.6) leads to

$$\sum_\alpha \underset{\alpha}{\omega}^j \wedge (\underset{\alpha}{\omega}^i_j - \omega^i_j) = 0 \tag{1.2.13}$$

where $\omega^i_j = \underset{n+1}{\omega}{}^i_j$. Equation (1.2.13) show that the forms $\underset{\alpha}{\omega}^i_j - \omega^i_j$ (and therefore their differences $\underset{\alpha}{\omega}^i_j - \underset{\beta}{\omega}^i_j, \alpha \neq \beta$) are linear combinations of the basis forms $\underset{\alpha}{\omega}^j$. We will write their decompositions in the form

$$\underset{\alpha}{\omega}^i_j - \omega^i_j = \sum_{\beta=1}^n \underset{\alpha\beta}{a}{}^i_{jk} \underset{\beta}{\omega}^k. \tag{1.2.14}$$

Since the forms $\underset{\alpha}{\omega}^i$ are linearly independent, the exterior products $\underset{\alpha}{\omega}^j \wedge \underset{\beta}{\omega}^k$, $\alpha \leq \beta$, are also independent. Substituting (1.2.14) into (1.2.13) and using this independence, we obtain

$$\underset{\alpha\alpha}{a}{}^i_{jk} = \underset{\alpha\alpha}{a}{}^i_{kj}, \quad \underset{\alpha\beta}{a}{}^i_{jk} = \underset{\beta\alpha}{a}{}^i_{kj}, \quad \beta \neq \alpha. \tag{1.2.15}$$

Equations (1.2.6), (1.2.14), and (1.2.15) imply

$$d\underset{\alpha}{\omega}^i = \underset{\alpha}{\omega}^j \wedge \omega^i_j + \sum_{\beta\neq\alpha} \underset{\alpha\beta}{a}{}^i_{jk} \underset{\alpha}{\omega}^j \wedge \underset{\beta}{\omega}^k \tag{1.2.16}$$

where

$$\underset{\alpha\beta}{a}{}^i_{jk} = \underset{\beta\alpha}{a}{}^i_{kj}. \tag{1.2.17}$$

It is worthy noting that equations (1.2.5) and (1.2.17) imply

$$d\underset{n+1}{\omega}{}^i = \underset{n+1}{\omega}{}^j \wedge \omega^i_j. \tag{1.2.18}$$

Exterior differentiation of (1.2.16) gives

$$-\underset{\alpha}{\omega}^j \wedge \Omega^i_j + \sum_{\beta\neq\alpha} \Delta \underset{\alpha\beta}{a}{}^i_{jk} \wedge \underset{\alpha}{\omega}^j \wedge \underset{\beta}{\omega}^k = 0 \tag{1.2.19}$$

where

$$\begin{aligned} \Omega^i_j &= d\omega^i_j - \omega^k_j \wedge \omega^i_k, \\ \Delta \underset{\alpha\beta}{a}{}^i_{jk} &= \nabla \underset{\alpha\beta}{a}{}^i_{jk} - \sum_{\gamma=1}^{n} (\underset{\alpha\beta}{a}{}^i_{mk} \underset{\gamma\alpha}{a}{}^m_{lj} + \underset{\alpha\beta}{a}{}^i_{jm} \underset{\beta\gamma}{a}{}^m_{kl}) \underset{\gamma}{\omega}^l, \\ \nabla \underset{\alpha\beta}{a}{}^i_{jk} &= d\underset{\alpha\beta}{a}{}^i_{jk} - \underset{\alpha\beta}{a}{}^i_{lk}\omega^l_j - \underset{\alpha\beta}{a}{}^i_{jl}\omega^l_k + \underset{\alpha\beta}{a}{}^l_{jk}\omega^i_l \end{aligned}$$

and $\underset{\alpha\beta}{a}{}^i_{jk} = 0$. To solve the cubic exterior equation (1.2.19) with respect to Ω^i_j and $\nabla \underset{\alpha\beta}{a}{}^i_{jk}$, we suppose that

$$\begin{cases} \Omega^i_j = \sum_{\alpha,\beta} \underset{\alpha\beta}{b}{}^i_{jkl} \underset{\alpha}{\omega}^k \wedge \underset{\beta}{\omega}^l + \sum_{\alpha} \underset{\alpha}{\theta}{}^i_{jk} \wedge \underset{\alpha}{\omega}^k + \Theta^i_j, \\ \Delta \underset{\alpha\beta}{a}{}^i_{jk} = \sum_{\gamma} \underset{\alpha\beta\gamma}{a}{}^i_{jkl} \underset{\gamma}{\omega}^l + \underset{\alpha\beta}{\sigma}{}^i_{jk} \end{cases} \tag{1.2.20}$$

where the 1-forms $\underset{\alpha}{\Theta}{}^i_{jk}$ and $\underset{\alpha\beta}{\sigma}{}^i_{jk}$ are linear combinations of some 1-forms θ^u complementing the co-basis $\{\underset{\alpha}{\omega}^i\}$ and a decomposition of the 2-forms Θ^i_j contains only exterior products of the form $\theta^u \wedge \theta^v$.

Substituting Ω^i_j and Δa^i_{jk} from (1.2.20) into (1.2.19), we find

$$\underset{\alpha\beta}{\sigma}{}^i_{jk} = \sigma^i_{jk} = \sigma^i_{kj}, \quad \underset{\alpha}{\theta}{}^i_{jk} = -\sigma^i_{jk}, \quad \Theta^i_j = 0. \tag{1.2.21}$$

Let us now consider the quantities

$$a^{i}_{jk} = \sum_{\alpha,\beta} \underset{\alpha\beta}{a}{}^{i}_{jk} = 2 \sum_{(\alpha,\beta)} \underset{\alpha\beta}{a}{}^{i}_{(jk)} \qquad (1.2.22)$$

Equations (1.2.17) show that $a^{i}_{jk} = a^{i}_{kj}$. In (1.2.22) $\sum_{(\alpha,\beta)}$ means that the summation is carried out over all combinations of $1, \ldots, n$ taken two at a time.

Lemma 1.2.3 *Each of the quantities a^{i}_{jk} and the 1-forms σ^{i}_{jk} can be reduced to* 0.

Proof. In order to show this, let us fix a point $p \in D \subset X^{nr}_{*}$. Then $\underset{\alpha}{\omega}{}^{i} = 0$. We will use the notations δ, π^{i}_{j}, ∇_{δ}, $\sigma^{i}_{jk}(\delta)$ for d, ω^{i}_{j}, ∇ and σ^{i}_{jk} when $\underset{\alpha}{\omega}{}^{i} = 0$. It follows from (1.2.20) and (1.2.21) that

$$\nabla_{\delta} a^{i}_{jk} = n(n-1)\sigma^{i}_{jk}(\delta). \qquad (1.2.23)$$

Taking all $\pi^{i}_{j} = 0$, we reduce (1.2.23) to

$$\delta a^{i}_{jk} = n(n-1)\sigma^{i}_{jk}(\delta) \ .$$

For any i, j, k fixed, only one parameter is varying. It can be always taken in such a way that

$$\sigma^{i}_{jk}(\delta) = \frac{1}{n(n-1)} \delta t^{i}_{jk}.$$

Then

$$\delta a^{i}_{jk} = \delta t^{i}_{jk} \text{ and } a^{i}_{jk} = t^{i}_{jk} + \text{const}.$$

Taking $t^{i}_{jk} = -$ const, we reduce a^{i}_{jk} to 0. After these equations (1.2.20) and (1.2.22) imply

$$\sum_{\alpha,\beta,\gamma} (\underset{\alpha\beta\gamma}{a}{}^{i}_{jkl} + \underset{\alpha\beta}{a}{}^{i}_{mk} \underset{\gamma\alpha}{a}{}^{m}_{lj} + \underset{\alpha\beta}{a}{}^{i}_{jm} \underset{\beta\gamma}{a}{}^{m}_{kl}) \underset{\alpha}{\omega}{}^{l} + n(n-1)\sigma^{i}_{jk} = 0.$$

Since the σ^{i}_{jk} are linear combinations of the θ^{u} (not the $\underset{\alpha}{\omega}{}^{i}$), it follows from the last equation that $\sigma^{i}_{jk} = 0$. ■

In the sequel we will assume that such a reduction has been done, i.e., we have

$$\sum_{(\alpha,\beta)} \underset{\alpha\beta}{a}{}^{i}_{(jk)} = 0 \qquad (1.2.24)$$

and

$$\sigma^{i}_{jk} = 0. \qquad (1.2.25)$$

The equations (1.2.19) and (1.2.21) also give

$$\underset{\alpha\beta}{b}{}^{i}_{jkl} = \frac{1}{2}\big(\underset{\gamma\alpha\beta}{a}{}^{i}_{jkl} - \underset{\beta\gamma\alpha}{a}{}^{i}_{ljk}\big), \quad \gamma \neq \alpha, \beta, \tag{1.2.26}$$

$$\underset{\alpha\beta}{b}{}^{i}_{[jkl]} = 0, \tag{1.2.27}$$

$$\underset{\alpha\alpha}{b}{}^{i}_{ljk} + 2\,\underset{\alpha\beta}{b}{}^{i}_{[jk]l} = 0. \tag{1.2.28}$$

The equations (1.2.21) and (1.2.25) allow us to write the equations (1.2.20) in the form

$$d\omega^i_j - \omega^k_j \wedge \omega^i_k = \sum_{\alpha,\beta=1}^{n} \underset{\alpha\beta}{b}{}^{i}_{jkl}\,\underset{\alpha}{\omega}{}^{k} \wedge \underset{\beta}{\omega}{}^{l}, \tag{1.2.29}$$

$$\nabla \underset{\alpha\beta}{a}{}^{i}_{jk} = \sum_{\gamma=1}^{n}\big(\underset{\alpha\beta\gamma}{a}{}^{i}_{jkl} + \underset{\alpha\beta}{a}{}^{i}_{mk}\,\underset{\gamma\alpha}{a}{}^{m}_{lj} + \underset{\alpha\beta}{a}{}^{i}_{jm}\,\underset{\beta\gamma}{a}{}^{m}_{kl}\big)\underset{\gamma}{\omega}{}^{l}, \quad \alpha \neq \beta \tag{1.2.30}$$

where $\underset{\alpha\alpha}{a}{}^{i}_{jk} = 0$.

Equations (1.2.30) show that the quantities $\{\underset{\alpha\beta}{a}{}^{i}_{jk}\}$ form a tensor (see [L 53]) ([5]). It follows from this that the quantities $\{\underset{\alpha\beta\gamma}{a}{}^{i}_{jkl}\}$ also form a tensor. Equations (1.2.26) then show that the quantities $\{\underset{\alpha\beta}{b}{}^{i}_{jkl}\}$ form a tensor.

Definition 1.2.4 The tensors $\{\underset{\alpha\beta}{a}{}^{i}_{jk}\}$ and $\{\underset{\alpha\beta}{b}{}^{i}_{jkl}\}$ are called respectively the *torsion* and *curvature tensors* of a web $W(n+1,n,r)$.

Equations (1.2.26) show that if $n > 2$, the curvature tensor of a $W(n+1,n,r)$ is expressed in terms of the Pfaffian derivatives of its torsion tensor and the torsion tensor itself. This is not the case for $n = 2$ because in formulae (1.2.26) $\gamma \neq \alpha, \beta$.

Let us summarise the results obtained so far.

Theorem 1.2.5 *The structure equations of a web* $W(n+1,n,r)$ *can be reduced to the form* (1.2.16) *and* (1.2.29). *The coefficients* $\underset{\alpha\beta}{a}{}^{i}_{jk}$ *and* $\underset{\alpha\beta}{b}{}^{i}_{jkl}$ *of these equations are respectively the components of the torsion and curvature tensors of the web* $W(n+1,n,r)$. *They satisfy equations* (1.2.17), (1.2.24), (1.2.30), (1.2.26), (1.2.27), *and* (1.2.28). ■

The following proposition give some additional relations between the components of the torsion tensor, their Pfaffian derivatives, and the components of the curvature tensor of the $W(n+1,n,r)$. In the sequel we will use some of these relations.

Proposition 1.2.6 *The quantities* $\underset{\alpha\beta}{a}{}^i_{jk}$, $\underset{\alpha\beta\gamma}{a}{}^i_{jkl}$, $\gamma \neq \alpha, \beta$, *and* $\underset{\alpha\beta\gamma}{b}{}^i_{jkl}$ *are connected by the following relations:*

$$\underset{\alpha\beta\gamma}{a}{}^i_{jkl} = \underset{\beta\alpha\gamma}{a}{}^i_{kjl}, \quad \alpha \neq \beta, \tag{1.2.31}$$

$$\sum_{\substack{\alpha,\beta=1\\ \alpha\neq\beta}} \big(\underset{\alpha\beta\gamma}{a}{}^i_{jkl} + \underset{\alpha\beta}{a}{}^i_{mk}\underset{\gamma\alpha}{a}{}^i_{lj} + \underset{\alpha\beta}{a}{}^i_{jm}\underset{\beta\gamma}{a}{}^m_{kl}\big) = 0, \tag{1.2.32}$$

$$\underset{\alpha\beta}{b}{}^i_{jkl} = \underset{\beta\alpha}{b}{}^i_{jlk}, \tag{1.2.33}$$

$$\underset{\alpha\alpha}{b}{}^i_{j[kl]} = \underset{\gamma\alpha\alpha}{a}{}^i_{j[kl]}, \quad \gamma \neq \alpha, \tag{1.2.34}$$

$$\underset{\alpha\beta\beta}{a}{}^i_{[jkl]} = 0, \quad \alpha \neq \beta, \tag{1.2.35}$$

$$2\,\underset{\alpha\beta}{b}{}^i_{[jk]l} + \underset{\alpha\beta\alpha}{a}{}^i_{[j|k|l]} = 0, \quad \alpha \neq \beta, \tag{1.2.36}$$

$$\underset{\alpha\alpha}{b}{}^i_{jkl} + \underset{\alpha\alpha}{b}{}^i_{klj} + \underset{\alpha\alpha}{b}{}^i_{ljk} = 0, \tag{1.2.37}$$

$$\underset{\alpha\alpha}{b}{}^i_{[jk]l} = \underset{\alpha\beta}{b}{}^i_{[jk]l}, \tag{1.2.38}$$

$$\underset{\alpha\beta}{b}{}^i_{jkl} = \underset{\alpha\gamma}{b}{}^i_{lkj} + \underset{\gamma\beta}{b}{}^i_{kjl}, \quad \gamma \neq \alpha, \beta, \quad \alpha \neq \beta, \tag{1.2.39}$$

$$\underset{\alpha\beta}{b}{}^i_{jkl} = \underset{\alpha\gamma}{b}{}^i_{jkl} + \underset{\gamma\beta}{b}{}^i_{jkl} - \underset{\gamma\gamma}{b}{}^i_{jkl}. \tag{1.2.40}$$

Proof. (1.2.31) follows from (1.2.17) and (1.2.30); (1.2.32) can be obtained by exterior differentiation of (1.2.24); (1.2.26) implies (1.2.33) and (1.2.34); (1.2.35) is a consequence of (1.2.27) and (1.2.34); (1.2.37) follows from (1.2.33) and (1.2.27); (1.2.28), (1.2.33), and (1.2.37) imply (1.2.38); to obtain (1.2.36), one should use (1.2.38), (1.2.26), and (1.2.35); (1.2.39) follows from (1.2.26) and (1.2.31); and finally (1.2.40) is obtained from (1.2.39), (1.2.33), (1.2.28), and (1.2.37). ■

We give now the fundamental theorem on the determination of a web $W(n+1,n,r)$ on an analytic manifold X^{nr}.

Theorem 1.2.7 *Let there be given in an open domain D of an analytic manifold X^{nr}, the 1-forms $\underset{\alpha}{\omega}{}^i$, ω^i_j and tensors $\{\underset{\alpha\beta}{a}{}^i_{jk}\}$ and $\{\underset{\alpha\beta}{b}{}^i_{jkl}\}$ satisfying the relations* (1.2.17) *and* (1.2.24), *the structure equations* (1.2.16) *and* (1.2.29) *and the integrability conditions* (1.2.30), (1.2.26), (1.2.27), *and* (1.2.28). *Then in the domain D the systems of Pfaffian equations $\underset{\alpha}{\omega}{}^i = 0$, $\alpha = 1, \ldots, n$, α is fixed, $i = 1, \ldots, r$, and $\sum_\alpha \underset{\alpha}{\omega}{}^i = 0$ define up to an analytic transformation a unique web $W(n+1,n,r)$ for which the indicated tensors are respectively the torsion and curvature tensors .*

Proof. The theorem is a consequence of E. Cartan's theorem on equivalence of two systems of 1-forms [Ca 08]. ■

1.2.3 The Structure Equations and Fundamental Tensors of a Web $W(3,2,r)$

Suppose now that $n = 2$. Let us denote $\underset{12}{a}{}^i_{jk}$ by a^i_{jk} :

$$\underset{12}{a}{}^i_{jk} = a^i_{jk}. \tag{1.2.41}$$

By (1.2.17),

$$\underset{21}{a}{}^i_{jk} = a^i_{kj}. \tag{1.2.42}$$

By (1.2.24), $\underset{12}{a}{}^i_{jk} + \underset{21}{a}{}^i_{jk} = 0$; therefore

$$a^i_{jk} + a^i_{kj} = 0. \tag{1.2.43}$$

Using (1.2.41), (1.2.42), and (1.2.43), we can write the structure equations (1.2.16) in the form

$$\begin{cases} d\underset{1}{\omega}{}^i = \underset{1}{\omega}{}^j \wedge \omega^i_j + a^i_{jk}\underset{1}{\omega}{}^j \wedge \underset{2}{\omega}{}^k, \\ d\underset{2}{\omega}{}^i = \underset{2}{\omega}{}^j \wedge \omega^i_j - a^i_{jk}\underset{1}{\omega}{}^j \wedge \underset{2}{\omega}{}^k. \end{cases} \tag{1.2.44}$$

Let us introduce the forms

$$\overline{\omega}^i_j = \omega^i_j + a^i_{jk}\underset{2}{\omega}{}^k - a^i_{jk}\underset{1}{\omega}{}^k. \tag{1.2.45}$$

Substituting (1.2.45) into (1.2.4) reduces (1.2.44) to

$$\begin{cases} d\underset{1}{\omega}{}^i = \underset{1}{\omega}{}^j \wedge \overline{\omega}^i_j + a^i_{jk}\underset{1}{\omega}{}^j \wedge \underset{1}{\omega}{}^k, \\ d\underset{2}{\omega}{}^i = \underset{2}{\omega}{}^j \wedge \overline{\omega}^i_j - a^i_{jk}\underset{2}{\omega}{}^j \wedge \underset{2}{\omega}{}^k. \end{cases} \tag{1.2.46}$$

where according to (1.2.43) the a^i_{jk} are skew-symmetric in the lower indices.

For a web $W(3,2,r)$ the structure equations in the form of (1.2.46) were found by Chern [C 36b] and systematically used by Akivis [Ak 69a].

Equations (1.2.44) and (1.2.46) show that for a $W(3,2,r)$ the torsion tensor introduced by Chern is the same as our torsion tensor.

Let us find the expression of the curvature tensor b^i_{jkl} of a web $W(3,2,r)$ used in [C 36b] and [Ak 69a] in terms of $\underset{\alpha\beta}{b}{}^i_{jkl}$, $\underset{\alpha\beta}{a}{}^i_{jk}$, and $\underset{\alpha\beta\beta}{a}{}^i_{jkl}$. The tensor b^i_{jkl} is defined by

$$\overline{\Omega}^i_j = d\overline{\omega}^i_j - \overline{\omega}^k_j \wedge \overline{\omega}^i_k = \overline{b}^i_{jkl}\underset{1}{\omega}{}^k \wedge \underset{2}{\omega}{}^l. \tag{1.2.47}$$

Using (1.2.45), (1.2.29), (1.2.30), and (1.2.46), we find that

$$\overline{\Omega}^i_j = (2\underset{12}{b}{}^i_{jkl} + \underset{121}{a}{}^i_{jlk} - \underset{122}{a}{}^i_{kjl} + a^m_{jk}a^i_{ml} - a^m_{jl}a^i_{mk})\underset{1}{\omega}{}^k \wedge \underset{2}{\omega}{}^l. \tag{1.2.48}$$

Comparison of (1.2.47) and (1.2.48) gives

$$\overline{b}^i_{jkl} = 2\underset{12}{b}{}^i_{jkl} + \underset{121}{a}{}^i_{jlk} - \underset{122}{a}{}^i_{kjl} + a^m_{jk}a^i_{ml} - a^m_{jl}a^i_{mk}. \tag{1.2.49}$$

Since now $\alpha, \beta, \gamma = 1, 2$, the components $\underset{12}{b}{}^i_{jkl}$, and therefore $\overline{b}^i_{jkl}$, can no longer be expressed by (1.2.26) in terms of $\underset{\alpha\beta\gamma}{a}{}^i_{jkl}$.

If we alternate (1.2.49) with respect to lower indices and use (1.2.27) and (1.2.35), we obtain

$$\overline{b}^i_{jkl} = 2a^m_{[jk}a^i_{|m|l]}. \tag{1.2.50}$$

The relation (1.2.50) is well-known in the theory of multidimensional three-webs (see [C 36b]and[Ak 69a]).

Substituting expressions of ω^i_j from (1.2.45) into (1.2.30), we obtain

$$\overline{\nabla} a^i_{jk} = \underset{121}{\overline{a}}{}^i_{jkl}\underset{1}{\omega}{}^l + \underset{122}{\overline{a}}{}^i_{jkl}\underset{2}{\omega}{}^l, \tag{1.2.51}$$

where

$$\overline{\nabla} a^i_{jk} = da^i_{jk} - a^i_{mk}\overline{\omega}^m_j - a^i_{jm}\overline{\omega}^m_k + a^m_{jk}\overline{\omega}^i_m, \tag{1.2.52}$$

$$\underset{121}{\overline{a}}{}^i_{jkl} = \underset{121}{a}{}^i_{[jk]l} + a^m_{jk}a^i_{lm} - a^m_{l[j}a^i_{|m|k]}, \tag{1.2.53}$$

$$\underset{122}{\overline{a}}{}^i_{jkl} = \underset{122}{a}{}^i_{[jk]l} + a^m_{jk}a^i_{ml} + a^m_{[j|l|}a^i_{k]m}. \tag{1.2.54}$$

It follows from (1.2.49), (1.2.53), (1.2.54), (1.2.36), (1.2.33), and (1.2.31) that

$$\underset{121}{\overline{a}}{}^i_{jkl} = \overline{b}^i_{[j|l|k]}, \qquad \underset{122}{\overline{a}}{}^i_{jkl} = \overline{b}^i_{[jk]l}. \tag{1.2.55}$$

Using (1.2.55), we can rewrite equations (1.2.52) in the form

$$\overline{\nabla} a^i_{jk} = \overline{b}^i_{[j|l|k]}\underset{1}{\omega}{}^l + \overline{b}^i_{[jk]l}\underset{2}{\omega}{}^l. \tag{1.2.56}$$

Equations (1.2.56) have the form in which they appeared in [C 36b] and [Ak 69a].

1.2.4 Special Classes of 3-Webs $W(3,2,r)$

Definition 1.2.8 We shall say that on a web $W(3,2,r)$ the *closure condition* (T) (respectively $(R), (B_l), (B_r), (B_m)$, and (H)) holds if on $W(3,2,r)$ all sufficiently small figures (T) (respectively $(R), (B_l), (B_r), (B_m)$, and (H)) are closed (see Figures 1.3–1.8).

It is worthwhile to note that since our study of webs is of local nature, we have to restrict ourselves to *sufficiently small closure figures that are completely located in* D.

In the notation above $(T), (R), (B)$, and (H) are the first letters of the words "Thomsen", "Reidemeister", "Bol", and hexagonal and in $(B_l), (B_r)$, and (B_m) the subscripts l, r, and m refer to the words "left", "right", and "middle".

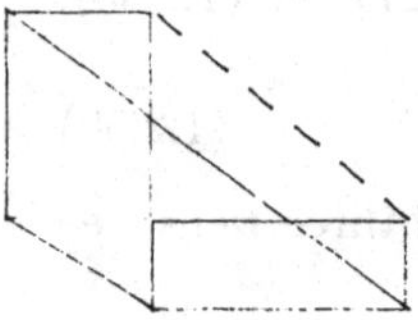
(T)

Figure 1.3

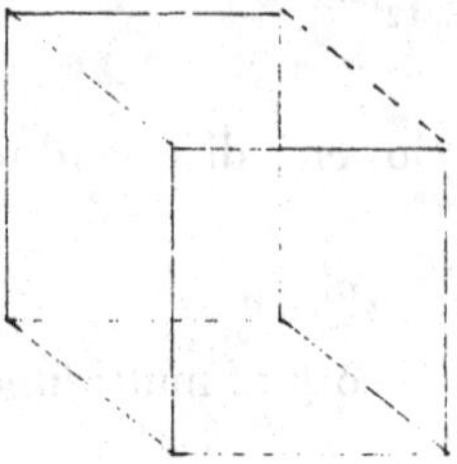
(R)

Figure 1.4

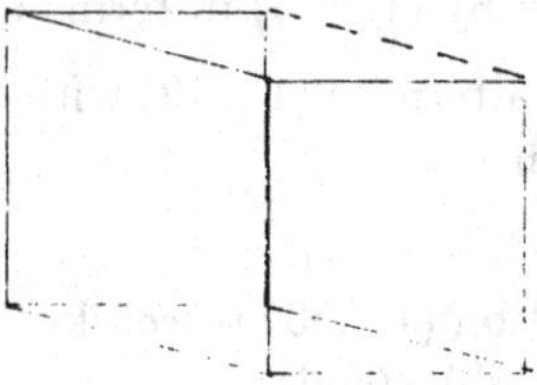
(B_l)

Figure 1.5

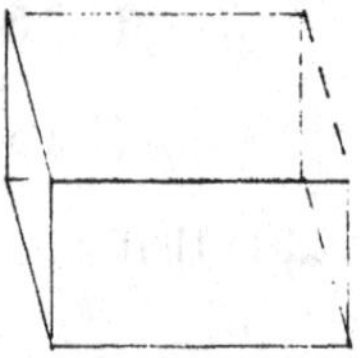
(B_r)

Figure 1.6

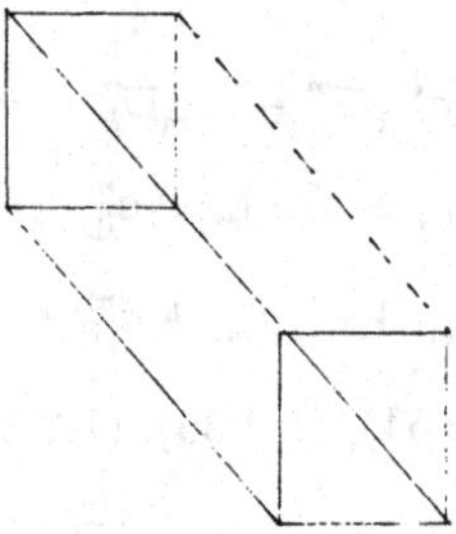
(B_m)

Figure 1.7

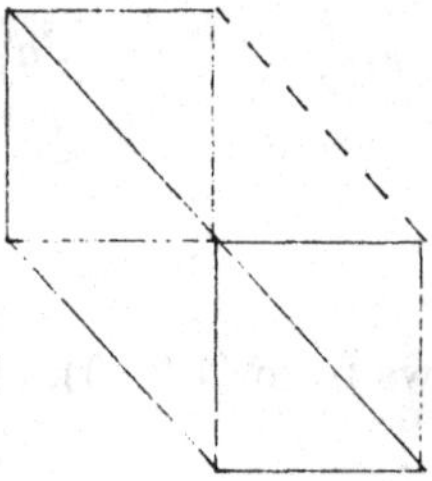
(H)

Figure 1.8

Definition 1.2.9 A web $W(3,2,r)$ is *group* (respectively *Bol* or *hexagonal*) web if on it the closure condition (R) (respectively any of $(B_l), (B_r), (B_m)$, or (H)) holds. If on $W(3,2,r)$ all the Bol figures $(B_l), (B_r)$, and (B_m) are closed, then it is called a *Moufang* web and denoted by (M).

Let us clarify Definitions 1.2.8 and 1.2.9. We consider, for example, the hexagonal condition (H) and hexagonal webs. For the condition (H) Definition 1.2.8 means the following. Let M be a point of $D \subset X^{2r}$ (see Figure 1.9) and F_1, F_2, and F_3 be the three leaves of the foliations λ_1, λ_2 and λ_3 passing through M. Let M be any point of F_1 and F_2' be a leaf of λ_2 through M. Then $F_2' \cap F_3 = M_2$. In the same way we subsequently obtain $F_1' \ni M, F_1' \cap F_2 = M_3, F_3' \ni M_3, F_3' \cap F_1 = M_4, F_2'' \ni M_4, F_2'' \cap F_3 = M_5, F_1'' \ni M_5, F_1'' \cap F_2 = M_6, F_3'' \ni M_6$, and $F_3'' \cap F_1 = M_7$. If $M_7 = M_1$, the hexagonal figure is closed.

According to Definition 1.2.9, a web $W(3,2,r)$ is hexagonal if such hexagonal figures are closed for any choice of M and M_1 sufficiently close to each other.

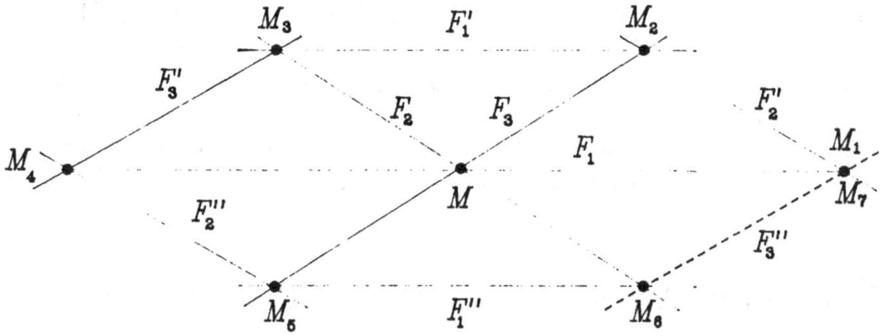

Figure 1.9

Closure Condition	Analytic Characterisation	Web $W(3,2,r)$
(T)	$a^i_{jk}=0,\ \ b^i_{jkl}=0$	Parallelisable
(R)	$b^i_{jkl}=0$	Group
(M)	$b^i_{jkl}=b^i_{[jkl]}$	Moufang
(B_l)	$b^i_{(jk)l}=0$	(Left) Bol
(B_r)	$b^i_{(j\|k\|l)}=0$	(Right) Bol
(B_m)	$b^i_{j(kl)}=0$	Bol
(H)	$b^i_{(jkl)}=0$	Hexagonal

Table 1.1: Classes of 3-Webs

Table 1.1 summarises the results of [C 36b], [Ak 69a], [Fe 78], and [AS 71a] (see also [AS 81]).

Note that the analytical conditions presented in the second column of Table 1.1 are both necessary and sufficient for a web $W(3,2,r)$ to be of the type indicated in the third column. Note also that for webs $W(3,2,1)$ all these conditions are equivalent and if one of the closure conditions holds, then so do the six others (see [BB 38]).

1.3 Invariant Affine Connections Associated with a Web $W(n+1,n,r)$

1.3.1 The Geometrical Meaning of the Forms $\omega_j^i(\delta)$

Suppose that a point $p \in D \subset X^{nr}$ is fixed, i.e., $\underset{\alpha}{\omega}^i = 0$. Let us denote by δ the symbol of differentiation with respect to secondary parameters, i.e., parameters of the stationary subgroup of admissible transformation of moving frames in $T_p(X^{nr})$, and by π_i^j the values of the forms ω_j^i when $\underset{\alpha}{\omega}^i = 0$:

$$\pi_j^i = \omega_j^i(\delta) = \omega_j^i\Big|_{\underset{\alpha}{\omega}^i = 0}.$$

The following proposition gives a geometrical meaning of π_j^i .

Proposition 1.3.1 *The forms π_j^i define concordant transformations of the moving frames $\{e_i^\alpha\}, \alpha$ is fixed, in the tangent spaces $T_p(F_\alpha)$ to the leaves F_α passing through p.*

Proof. If a point p is fixed, i.e.,$\underset{\alpha}{\omega}^i = 0$, it follows from (1.2.16) that

$$\delta \underset{\alpha}{\omega}^i = \underset{\alpha}{\omega}^j \pi_j^i. \tag{1.3.1}$$

Since an arbitrary vector $\xi \in T_p(X^{nr})$ is expressed by (1.2.9), we obtain from (1.3.1) that

$$\delta e_i^\alpha = \pi_i^j e_j^\alpha. \tag{1.3.2}$$

The equations (1.3.2) prove Proposition 1.3.1. ■

Note that the forms π_i^j can be expressed in terms of A_i^j (see (1.2.7)) and δA_i^j.

1.3.2 Affine Connections Associated with an $(n+1)$-Web

We already mentioned that the quantities $\{a^i_{\alpha\beta\,jk}\}$ and $\{b^i_{\alpha\beta\,jkl}\}$ are tensors. Equations (1.2.16) and (1.2.29) show that the forms $\omega^I = \{\underset{\alpha}{\omega}^i\}$ and

$$\omega^I_J = \begin{pmatrix} \omega^i_j & 0 & \dots & 0 \\ 0 & \omega^i_j & \dots & 0 \\ \dots & \dots & \dots & \dots \\ \dots & \dots & \dots & \dots \\ 0 & 0 & \dots & \omega^i_j \end{pmatrix}, \quad I, J = 1, \dots, nr, \tag{1.3.3}$$

define an affine connection (see [Ca 23] or [KN 63]). This connection is induced on X^{nr} by a web $W(n+1,n,r)$. When we derived equations (1.2.16) and (1.2.29), we took the forms $\underset{\alpha}{\omega}^i$ as a co-frame (see Section 1.2). This distinguishes the foliation λ_{n+1}. However, we could distinguish any of other foliations λ_α, $\alpha = 1, \dots, n$. Thus there are $n+1$ affine connections naturally associated with a web $W(n+1,n,r)$. All of them are G-connections (see [St 83]) consistent with the G-structure induced on X^{nr} by a web $W(n+1,n,r)$.

Let us denote these affine connections by γ_{n+1} and γ_α, $\alpha = 1, \dots, n$. The subscript shows which of the $n+1$ foliations is distinguished while constructing a connection.

Let us continue to study the connection γ_{n+1} defined by (1.2.16) and (1.2.29). The torsion tensor of γ_{n+1} has the form

$$R^I_{JK} = [A_1, \dots, A_n] \tag{1.3.4}$$

where

$$A_1 = \begin{pmatrix} 0 & \frac{1}{2} a^i_{12\,jk} & \dots & \frac{1}{2} a^i_{1n\,jk} \\ -\frac{1}{2} a^i_{21\,jk} & 0 & \dots & 0 \\ \dots & \dots & \dots & \dots \\ \dots & \dots & \dots & \dots \\ -\frac{1}{2} a^i_{n1\,jk} & 0 & \dots & 0 \end{pmatrix}, \dots, A_n = \begin{pmatrix} 0 & 0 & \dots & -\frac{1}{2} a^i_{1n\,jk} \\ 0 & 0 & \dots & -\frac{1}{2} a^i_{2n\,jk} \\ \dots & \dots & \dots & \dots \\ \dots & \dots & \dots & \dots \\ \frac{1}{2} a^i_{n1\,jk} & \frac{1}{2} a^i_{n2\,jk} & \dots & 0 \end{pmatrix}$$

and the curvature tensor of γ_{n+1} has the form

$$R^I_{JKL} = \begin{pmatrix} A & 0 & \dots & 0 \\ 0 & A & \dots & 0 \\ \dots & \dots & \dots & \dots \\ \dots & \dots & \dots & \dots \\ 0 & 0 & \dots & A \end{pmatrix}, \tag{1.3.5}$$

where

$$A = \begin{pmatrix} \underset{11}{b}{}^i_{jkl} & \underset{12}{b}{}^i_{jkl} & \cdots & \underset{1n}{b}{}^i_{jkl} \\ -\underset{12}{b}{}^i_{jlk} & \underset{22}{b}{}^i_{jkl} & \cdots & \underset{2n}{b}{}^i_{jkl} \\ \cdots & \cdots & \cdots & \cdots \\ \cdots & \cdots & \cdots & \cdots \\ -\underset{1n}{b}{}^i_{jlk} & -\underset{2n}{b}{}^i_{jlk} & \cdots & \underset{nn}{b}{}^i_{jkl} \end{pmatrix}.$$

Let us now find the equations of the geodesic lines of the manifold X^{nr} carrying a web $W(n+1, n, r)$ in the connection γ_{n+1}. In general, the equations of the geodesic lines of a space X^{nr} with affine connection have the form

$$d\omega^I + \omega^J \omega^I_J = \varphi\omega^I, \quad I, J = 1, \ldots, nr, \tag{1.3.6}$$

where ω^I form a cobasis in $T_p(X^{nr})$ and ω^I_J are the connection forms of the space X^{nr}

In our case $\omega^I = \{\underset{\alpha}{\omega}^i\}$ and the forms ω^I_J have the structure (1.3.3). Because of this, equations (1.3.6) fall into n groups:

$$d\underset{\alpha}{\omega}^i + \underset{\alpha}{\omega}^j \omega^i_j = \varphi\underset{\alpha}{\omega}^i, \quad \alpha = 1, \ldots, n; \quad i = 1, \ldots, r, \tag{1.3.7}$$

where φ is a 1-form. Note that in (1.3.6) and (1.3.7) d is the symbol of ordinary (not exterior) differentiation.

Let us recall the following definition (cf. [KN 63]).

Definition 1.3.2 . A submanifold V of a manifold X^{nr} with an affine connection is said to be *totally geodesic* at a point $p \in X^{nr}$ if for every vector $X \in T_p(V)$ the geodesic line determined by (p, X) lies locally in V. If V is totally geodesic at each of these points, it is called a *totally geodesic submanifold* of X^{nr}.

The leaves $F_\alpha \subset \lambda_\alpha$ are defined by the equations $\underset{\alpha}{\omega}^i = 0$ and the leaves $F_{n+1} \subset \lambda_{n+1}$ are defined by the equations $\sum_\alpha \underset{\alpha}{\omega}^i = 0$. Equations (1.3.7) show that all these leaves are totally geodesic in the γ_{n+1}.

Of course, similar statements can be proved for the connections γ_α.

The following theorem summarises the results on the affine connections which we have obtained so far.

Theorem 1.3.3 *A web $W(n+1, n, r)$ induces on the manifold X^{nr} the $n+1$ affine connections $\gamma_\xi, \xi = 1, \ldots, n, n+1$. The leaves F_ξ are totally geodesic in each of these connections. The connection γ_{n+1} has the connection forms* (1.3.3), *the torsion tensor* (1.3.4), *and the curvature tensor* (1.3.5). ■

1.3.3 The Affine Connections Induced by the Connection γ_{n+1} on Leaves

Theorem 1.3.4 *The affine connection γ_η induces the affine connections on leaves F_ξ of each foliation λ_ξ. The torsion and curvature tensors of the induced connections are subtensors of the torsion and curvature tensors of γ_η. The connections induced by γ_η on the leaves F_η are connections with absolute parallelism (curvature-free connections).*

Proof. For definiteness let us take the connection γ_{n+1} : the proof for the connections γ_α is similar. Let us consider a leaf $F_\alpha \subset \lambda_\alpha$. It is defined by the equations $\underset{\alpha}{\omega^i} = 0$. Let us recall that codim $F_\alpha = r$. At a point p of each F_α the web $W(n+1,n,r)$ defines naturally n different normals of the first kind (r-planes having only the point p in common with $T_p(F_\alpha)$). These normals are r-planes defined by vectors a) e_i^α, α fixed, or b) $f_i^\beta = e_i^{\hat\beta} - e_i^\beta$, $\beta \neq \alpha$, $\hat\beta \neq \beta$, β fixed. Each such normal defines an affine connection on F_α . If we choose the normalisation a), we can deduce from (1.2.16) and (1.2.29) that

$$\left\{\begin{array}{l} d\underset{\hat a}{\omega^i} = \underset{\hat a}{\omega^j} \wedge \omega_j^i + \sum\limits_{\hat\beta\neq\hat\alpha} \underset{\hat\alpha\hat\beta}{a}{}^i_{jk} \underset{\hat\alpha}{\omega^j} \wedge \underset{\hat\beta}{\omega^k}, \\ \Omega_j^i = \sum\limits_{\hat\alpha} \underset{\hat\alpha\hat\alpha}{b}{}^i_{jkl} \underset{\hat\alpha}{\omega^k} \wedge \underset{\hat\alpha}{\omega^l} + 2 \sum\limits_{(\alpha,\hat\beta)} \underset{\hat\alpha\hat\beta}{b}{}^i_{jkl} \underset{\hat\alpha}{\omega^k} \wedge \underset{\hat\beta}{\omega^l}, \quad \hat\alpha, \hat\beta \neq \alpha. \end{array}\right. \tag{1.3.8}$$

Similarly, the web $W(n+1,n,r)$ defines n different normals of the first kind to all leaves $F_{n+1} \subset \lambda_{n+1}$ defined by equations $\underset{n+1}{\omega}{}^i = 0$. They are r-planes defined by vectors e_i^α, where α is fixed. Let us fix an integer $\gamma, 1 \leq \gamma \leq n$, and assume in this proof that $\alpha \neq \gamma$ and $\beta \neq \gamma$ in all summations.

Using (1.2.5), we obtain the following expression for $\underset{\gamma}{\omega^i}$:

$$\underset{\gamma}{\omega^i} = -\underset{n+1}{\omega}{}^i - \sum_{\beta=1}^{n} \underset{\beta}{\omega^i}, \tag{1.3.9}$$

and substituting (1.3.9) into (1.2.29) gives

$$\begin{aligned} \Omega_j^i - \underset{\gamma\gamma}{b}{}^i_{jkl} \underset{n+1}{\omega}{}^k \wedge \underset{n+1}{\omega}{}^l + 2 \sum_{\alpha=1}^{n} (\underset{\gamma\gamma}{b}{}^i_{jkl} - \underset{\gamma\alpha}{b}{}^i_{jkl}) \underset{n+1}{\omega}{}^k \wedge \underset{\alpha}{\omega^l} \\ + \sum_{\alpha,\beta=1}^{n} (\underset{\gamma\gamma}{b}{}^i_{jkl} - \underset{\alpha\gamma}{b}{}^i_{jkl} - \underset{\gamma\beta}{b}{}^i_{jkl} + \underset{\alpha\beta}{b}{}^i_{jkl}) \underset{\alpha}{\omega^k} \wedge \underset{\beta}{\omega^l}. \end{aligned} \tag{1.3.10}$$

The relation (1.2.40) allow us to rewrite (1.3.10) in the form

$$\Omega_j^i = \underset{\gamma\gamma}{b}{}^i_{jkl} \underset{n+1}{\omega}{}^k \wedge \underset{n+1}{\omega}{}^l + 2 \sum_{\alpha=1}^{n} (\underset{\gamma\gamma}{b}{}^i_{jkl} - \underset{\gamma\alpha}{b}{}^i_{jkl}) \underset{n+1}{\omega}{}^k \wedge \underset{\alpha}{\omega^l}. \tag{1.3.11}$$

Since on F_{n+1} we have $\underset{n+1}{\omega}{}^i = 0$, (1.3.11) implies

$$\Omega^i_j = 0. \tag{1.3.12}$$

In addition, on F_{n+1} the equations (1.2.16) can be written in the form

$$d\underset{\alpha}{\omega}{}^i = \underset{\alpha}{\omega}{}^j \wedge \omega^i_j + \sum_{\beta=1}^{n} (\underset{\alpha\beta}{a}{}^i{}_{jk} - \underset{\alpha\gamma}{a}{}^i{}_{jk}) \underset{\alpha}{\omega}{}^j \wedge \underset{\beta}{\omega}{}^k. \tag{1.3.13}$$

Equations (1.3.12) mean that each of the affine connections induced by γ_{n+1} on F_{n+1} is a connection with absolute parallelism (curvature-free connection). Equations (1.3.8) and (1.3.13) prove that the torsion and curvature tensors of the connections induced by γ_{n+1} on F_ξ are subtensors of the corresponding tensors of γ_{n+1}. ■

1.3.4 Affine Connections Associated with 3-Subwebs of an $(n+1)$-Web

Let us fix two integers α and β, $1 \leq \alpha$, $\beta \leq n$, and consider $(2r)$-dimensional intersection $M_{\alpha\beta}$ of leaves λ_γ, $\gamma \neq \alpha$, β, defined by equations

$$\underset{\gamma}{\omega}{}^i = 0, \quad \gamma = 1, \ldots, n, \quad \gamma \neq \alpha, \beta. \tag{1.3.14}$$

On $M_{\alpha\beta}$ the foliations $\lambda_{n+1}, \lambda_\alpha$ and λ_β cut a 3-*subweb* $[n+1, \alpha, \beta]$. We will study three affine connections $\gamma_{\alpha\beta}, \bar{\gamma}_{\alpha\beta}$ and $\tilde{\gamma}_{\alpha\beta}$ associated with the 3-subweb $[n+1, \alpha, \beta]$. The connection $\gamma_{\alpha\beta}$ is induced by γ_{n+1} and the two other connections are defined by the 3-subweb $[n+1, \alpha, \beta]$ itself.

Let us find the torsion and curvature tensors of $\gamma_{\alpha\beta}$. Substituting (1.3.14) into (1.2.18), (1.2.16), (1.2.29), and (1.2.30), we obtain

$$\underset{n+1}{\omega}{}^i + \underset{\alpha}{\omega}{}^i + \underset{\beta}{\omega}{}^i = 0, \tag{1.3.15}$$

$$d\underset{n+1}{\omega}{}^i = \underset{n+1}{\omega}{}^j \wedge \underset{\alpha\beta}{\omega}{}^i_j, \tag{1.3.16}$$

$$\left\{ \begin{array}{l} d\underset{\alpha}{\omega}{}^i = \underset{\alpha}{\omega}{}^j \wedge \underset{\alpha\beta}{\omega}{}^i_j + \underset{\alpha\beta}{a}{}^i{}_{jk} \underset{\alpha}{\omega}{}^j \wedge \underset{\beta}{\omega}{}^k, \\ d\underset{\beta}{\omega}{}^i = \underset{\beta}{\omega}{}^j \wedge \underset{\alpha\beta}{\omega}{}^i_j + \underset{\beta\alpha}{a}{}^i{}_{jk} \underset{\beta}{\omega}{}^j \wedge \underset{\alpha}{\omega}{}^k, \end{array} \right. \tag{1.3.17}$$

$$d\underset{\alpha\beta}{\omega}{}^i_j = \underset{\alpha\beta}{\omega}{}^k_j \wedge \underset{\alpha\beta}{\omega}{}^i_k + \underset{\alpha\alpha}{b}{}^i{}_{jkl} \underset{\alpha}{\omega}{}^k \wedge \underset{\alpha}{\omega}{}^l + 2\, \underset{\alpha\beta}{b}{}^i{}_{jkl} \underset{\alpha}{\omega}{}^k \wedge \underset{\beta}{\omega}{}^l + \underset{\beta\beta}{b}{}^i{}_{jkl} \underset{\beta}{\omega}{}^k \wedge \underset{\beta}{\omega}{}^l, \tag{1.3.18}$$

$$\underset{\alpha\beta}{\nabla} \underset{\alpha\beta}{a}{}^i{}_{jk} = (\underset{\alpha\beta\alpha}{a}{}^i{}_{jkl} + \underset{\alpha\beta}{a}{}^i{}_{jm} \underset{\beta\alpha}{a}{}^m{}_{kl}) \underset{\alpha}{\omega}{}^l + (\underset{\alpha\beta\beta}{a}{}^i{}_{jkl} + \underset{\alpha\beta}{a}{}^i{}_{mk} \underset{\alpha\beta}{a}{}^m{}_{jl}) \underset{\beta}{\omega}{}^l, \tag{1.3.19}$$

where

$$\underset{\alpha\beta}{\omega}{}^i_j = \omega^i_j \Big|_{\underset{\gamma}{\omega}{}^k = 0}, \quad \underset{\alpha\beta}{\nabla} = \nabla \Big|_{\underset{\gamma}{\omega}{}^k = 0}, \quad \gamma = 1, \ldots, n, \quad \gamma \neq \alpha, \beta.$$

Therefore, the torsion and curvature tensors of $\gamma_{\alpha\beta}$ are subtensors of the corresponding tensors of γ_{n+1}.

Next let us consider on the 3-subweb $[n+1,\alpha,\beta]$ the connection described by equations (1.2.46) and (1.2.47). This connection is called the *canonical* connection of a 3-web. We will denote this connection by $\bar\gamma_{\alpha\beta}$. Comparing (1.2.46) and (1.3.17). we find the following expressions for the connection forms $\underset{\alpha\beta}{\bar\omega}{}^i_j$ of $\bar\gamma_{\alpha\beta}$:

$$\underset{\alpha\beta}{\bar\omega}{}^i_j = \underset{\alpha\beta}{\omega}{}^i_j + \underset{\beta\alpha}{a}{}^i_{jk}\underset{\alpha}{\omega}{}^k + \underset{\alpha\beta}{a}{}^i_{jk}\underset{\beta}{\omega}{}^k. \tag{1.3.20}$$

To find the expressions for the torsion and curvature tensors of $\bar\gamma_{\alpha\beta}$, we reduce the structure equations of the 3-subweb $[n+1,\alpha,\beta]$ to the form (1.2.46) and (1.2.47). Substituting $\underset{\alpha\beta}{\omega}{}^i_j$ from (1.3.20) into (1.3.16) and (1.3.17), we rewrite equations (1.3.16) and (1.3.17) in the form:

$$d\underset{n+1}{\omega}{}^i = \underset{n+1}{\omega}{}^j \wedge \underset{\alpha\beta}{\bar\omega}{}^i_j + \underset{\alpha\beta}{\bar a}{}^i_{jk}\underset{n+1}{\omega}{}^j \wedge (\underset{\alpha}{\omega}{}^k - \underset{\beta}{\omega}{}^k), \tag{1.3.21}$$

$$\begin{cases} d\underset{\alpha}{\omega}{}^i = \underset{\alpha}{\omega}{}^j \wedge \underset{\alpha\beta}{\bar\omega}{}^i_j + \underset{\alpha\beta}{\bar a}{}^i_{jk}\underset{\alpha}{\omega}{}^j \wedge \underset{\alpha}{\omega}{}^k, \\ d\underset{\beta}{\omega}{}^i = \underset{\beta}{\omega}{}^j \wedge \underset{\alpha\beta}{\bar\omega}{}^i_j - \underset{\alpha\beta}{\bar a}{}^i_{jk}\underset{\beta}{\omega}{}^j \wedge \underset{\beta}{\omega}{}^k, \end{cases} \tag{1.3.22}$$

where

$$\underset{\alpha\beta}{\bar a}{}^i_{jk} = \underset{\alpha\beta}{a}{}^i_{[jk]}, \tag{1.3.23}$$

is the torsion tensor of $[n+1,\alpha,\beta]$ in the connection $\bar\gamma_{\alpha\beta}$.

Exterior differentiation of (1.3.20) leads to the equation

$$d\underset{\alpha\beta}{\bar\omega}{}^i_j = \underset{\alpha\beta}{\bar\omega}{}^k_j \wedge \underset{\alpha\beta}{\bar\omega}{}^i_k + \underset{\alpha\beta}{\bar b}{}^i_{jkl}\underset{\alpha}{\omega}{}^k \wedge \underset{\beta}{\omega}{}^l, \tag{1.3.24}$$

where

$$\underset{\alpha\beta}{\bar b}{}^i_{jkl} = 2\underset{\alpha\beta}{b}{}^i_{jkl} - \underset{\beta\alpha\beta}{a}{}^i_{jkl} + \underset{\alpha\beta\alpha}{a}{}^i_{jlk} - \underset{\beta\alpha}{a}{}^m_{jk}\underset{\alpha\beta}{a}{}^i_{ml} + \underset{\alpha\beta}{a}{}^m_{jl}\underset{\beta\alpha}{a}{}^i_{mk} \tag{1.3.25}$$

is the curvature tensor of $[n+1,\alpha,\beta]$ in the connection $\bar\gamma_{\alpha\beta}$.

Let us denote by $\bar\nabla_{\alpha\beta}$ the symbol of covariant differentiation in the connection $\bar\gamma_{\alpha\beta}$. i.e., for example,

$$\underset{\alpha\beta}{\bar\nabla}\underset{\alpha\beta}{\bar a}{}^i_{jk} = d\underset{\alpha\beta}{\bar a}{}^i_{jk} - \underset{\alpha\beta}{\bar a}{}^i_{mk}\underset{\alpha\beta}{\bar\omega}{}^m_j - \underset{\alpha\beta}{\bar a}{}^i_{jm}\underset{\alpha\beta}{\bar\omega}{}^m_k + \underset{\alpha\beta}{\bar a}{}^m_{jk}\underset{\alpha\beta}{\bar\omega}{}^i_m. \tag{1.3.26}$$

Then, using (1.3.20) and (1.3.23), we obtain from (1.3.18) that

$$\underset{\alpha\beta}{\bar\nabla}\underset{\alpha\beta}{\bar a}{}^i_{jk} = \underset{\alpha\beta\alpha}{\bar a}{}^i_{jkl}\underset{\alpha}{\omega}{}^l + \underset{\alpha\beta\beta}{\bar a}{}^i_{jkl}\underset{\beta}{\omega}{}^l, \tag{1.3.27}$$

where

$$\begin{aligned} \underset{\alpha\beta\alpha}{\bar a}{}^i_{jkl} &= \underset{\alpha\beta\alpha}{a}{}^i_{[jk]l} + \underset{\alpha\beta}{a}{}^m_{[jk]}\underset{\beta\alpha}{a}{}^i_{ml} - \underset{\beta\alpha}{a}{}^m_{[j|l|}\underset{\alpha\beta}{a}{}^i_{m|k]}, \\ \underset{\alpha\beta\beta}{\bar a}{}^i_{jkl} &= \underset{\alpha\beta\beta}{a}{}^i_{[jk]l} + \underset{\alpha\beta}{a}{}^m_{[jk]}\underset{\alpha\beta}{a}{}^i_{ml} + \underset{\alpha\beta}{a}{}^m_{[j|l|}\underset{\beta\alpha}{a}{}^i_{m|k]}. \end{aligned} \tag{1.3.28}$$

Now considerations similar to ones at the end of Section 1.2 show that

$$\underset{\alpha\beta}{\bar b}{}^{i}_{[jkl]} = 2\,\underset{\alpha\beta}{\bar a}{}^{m}_{[jk}\,\underset{\alpha\beta}{\bar a}{}^{i}_{|m|l]}, \tag{1.3.29}$$

$$\underset{\alpha\beta\alpha}{\bar a}{}^{i}_{jkl} = \underset{\alpha\beta}{\bar b}{}^{i}_{[j|l|k]}, \quad \underset{\alpha\beta\beta}{\bar a}{}^{i}_{jkl} = \underset{\alpha\beta}{\bar b}{}^{i}_{[jk]l}. \tag{1.3.30}$$

Finally, we will consider one more affine connection $\tilde\gamma_{\alpha\beta}$ associated with the 3-subweb $[n+1,\alpha,\beta]$. The connection $\tilde\gamma_{\alpha\beta}$ is given by the following connection forms:

$$\underset{\alpha\beta}{\tilde\omega}{}^{i}_{j} = \underset{\alpha\beta}{\bar\omega}{}^{i}_{j} + \underset{\alpha\beta}{\bar a}{}^{i}_{jk}\underset{\alpha}{\omega}{}^{k} - \underset{\alpha\beta}{\bar a}{}^{i}_{jk}\underset{\beta}{\omega}{}^{k}. \tag{1.3.31}$$

For $\tilde\gamma_{\alpha\beta}$ the structure equations (1.3.21), (1.3.22), (1.3.24) and (1.3.27) of $[n+1,\alpha,\beta]$ are reduced to the form:

$$d\underset{n+1}{\omega}{}^{i} = \underset{n+1}{\omega}{}^{j}\wedge\underset{\alpha\beta}{\tilde\omega}{}^{i}_{j}, \tag{1.3.32}$$

$$\begin{cases} d\underset{\alpha}{\omega}{}^{i} = \underset{\alpha}{\omega}{}^{j}\wedge\underset{\alpha\beta}{\tilde\omega}{}^{i}_{j} + \underset{\alpha\beta}{\tilde a}{}^{i}_{jk}\underset{\alpha}{\omega}{}^{j}\wedge\underset{\beta}{\omega}{}^{k}, \\ d\underset{\beta}{\omega}{}^{i} = \underset{\beta}{\omega}{}^{j}\wedge\underset{\alpha\beta}{\tilde\omega}{}^{i}_{j} + \underset{\beta\alpha}{\tilde a}{}^{i}_{jk}\underset{\beta}{\omega}{}^{j}\wedge\underset{\alpha}{\omega}{}^{k}, \end{cases} \tag{1.3.33}$$

$$d\underset{\alpha\beta}{\tilde\omega}{}^{i}_{j} = \underset{\alpha\beta}{\tilde\omega}{}^{k}_{j}\wedge\underset{\alpha\beta}{\tilde\omega}{}^{i}_{k} + \underset{\alpha\alpha}{\tilde b}{}^{i}_{jkl}\underset{\alpha}{\omega}{}^{k}\wedge\underset{\alpha}{\omega}{}^{l} + \underset{\alpha\beta}{\tilde b}{}^{i}_{jkl}\underset{\alpha}{\omega}{}^{k}\wedge\underset{\beta}{\omega}{}^{l} + \underset{\beta\alpha}{\tilde b}{}^{i}_{jkl}\underset{\beta}{\omega}{}^{k}\wedge\underset{\alpha}{\omega}{}^{l} + \underset{\beta\beta}{\tilde b}{}^{i}_{jkl}\underset{\beta}{\omega}{}^{k}\wedge\underset{\beta}{\omega}{}^{l}, \tag{1.3.34}$$

$$\underset{\alpha\beta}{\tilde\nabla}\underset{\alpha\beta}{\tilde a}{}^{i}_{jk} = (\underset{\alpha\beta\alpha}{\tilde a}{}^{i}_{jkl} + \underset{\alpha\beta}{\tilde a}{}^{i}_{jm}\underset{\beta\alpha}{\tilde a}{}^{m}_{kl})\underset{\alpha}{\omega}{}^{l} + (\underset{\alpha\beta\beta}{\tilde a}{}^{i}_{jkl} + \underset{\alpha\beta}{\tilde a}{}^{i}_{mk}\underset{\alpha\beta}{\tilde a}{}^{m}_{jl})\underset{\beta}{\omega}{}^{l}, \tag{1.3.35}$$

where $\tilde\nabla$ is the symbol of covariant differentiation in $\tilde\gamma_{\alpha\beta}$ and the following identities hold:

$$\underset{\alpha\beta}{\tilde a}{}^{i}_{jk} = \underset{\beta\alpha}{\tilde a}{}^{i}_{kj}, \quad \underset{\alpha\beta}{\tilde b}{}^{i}_{jkl} = -\underset{\beta\alpha}{\tilde b}{}^{i}_{jlk}, \quad \underset{\alpha\alpha}{\tilde b}{}^{i}_{j(kl)} = 0, \quad \underset{\beta\beta}{\tilde b}{}^{i}_{j(kl)} = 0, \tag{1.3.36}$$

$$\underset{\alpha\beta\alpha}{\tilde a}{}^{i}_{jkl} = \underset{\beta\alpha\alpha}{\tilde a}{}^{i}_{kjl}, \quad \underset{\alpha\beta\beta}{\tilde a}{}^{i}_{jkl} = \underset{\beta\alpha\beta}{\tilde a}{}^{i}_{kjl}, \tag{1.3.37}$$

$$\underset{\alpha\beta}{\tilde a}{}^{i}_{jk} = \underset{\alpha\beta}{\bar a}{}^{i}_{jk}, \tag{1.3.38}$$

$$\underset{\alpha\alpha}{\tilde b}{}^{i}_{jkl} = -\frac{1}{2}\underset{\alpha\beta}{\bar b}{}^{i}_{[l|j|k]} + \underset{\alpha\beta}{\bar a}{}^{m}_{j[k}\underset{\alpha\beta}{\bar a}{}^{i}_{m|l]}, \tag{1.3.39}$$

$$\underset{\alpha\beta}{\tilde b}{}^{i}_{jkl} = \frac{1}{4}\underset{\alpha\beta}{\bar b}{}^{i}_{lkj} + \frac{1}{4}\underset{\alpha\beta}{\bar b}{}^{i}_{kjl} + \underset{\alpha\beta}{\bar a}{}^{m}_{j[k}\underset{\alpha\beta}{\bar a}{}^{i}_{|m|l]}, \tag{1.3.40}$$

$$\underset{\beta\beta}{\tilde b}{}^{i}_{jkl} = -\frac{1}{2}\underset{\alpha\beta}{\bar b}{}^{i}_{[kl]j} + \underset{\alpha\beta}{\bar a}{}^{m}_{j[k}\underset{\alpha\beta}{\bar a}{}^{i}_{|m|l]}, \tag{1.3.41}$$

$$\underset{\alpha\beta\alpha}{\tilde a}{}^{i}_{jkl} = \underset{\alpha\beta\alpha}{\bar a}{}^{i}_{jkl} + 2\,\underset{\alpha\beta}{\bar a}{}^{m}_{j[k}\underset{\alpha\beta}{\bar a}{}^{i}_{|m|l]}, \tag{1.3.42}$$

$$\underset{\alpha\beta\beta}{\tilde a}{}^{i}_{jkl} = \underset{\alpha\beta\beta}{\bar a}{}^{i}_{jkl} + 2\,\underset{\alpha\beta}{\bar a}{}^{m}_{[l|k}\underset{\alpha\beta}{\bar a}{}^{i}_{|m|j]}. \tag{1.3.43}$$

Equations (1.3.29), (1.3.30), and (1.3.36) – (1.3.43) allow us to check that the torsion tensor $\{\underset{\alpha\beta}{\tilde a}{}^{i}_{jk}\}$ and the curvature tensor $\{\underset{\alpha\alpha}{\tilde b}{}^{i}_{jkl}, \underset{\alpha\beta}{\tilde b}{}^{i}_{jkl}\}$ of $[n+1,\alpha,\beta]$ satisfy relations of the form (1.2.26)–(1.2.28) and (1.2.31)–(1.2.40).

Note that the connection γ_{n+1} for $W(n+1,n,r)$ is a generalisation of the similar connection for $W(3,2,r)$. They coincide if $n=2$. However, if $n>2$, the restriction of γ_{n+1} to the 3-subweb $[n+1,\alpha,\beta]$, i.e., the connection $\gamma_{\alpha\beta}$, is different from $\tilde{\gamma}_{\alpha\beta}$.

Let us find how the curvature tensor of the connection $\tilde{\gamma}_{\alpha\beta}$ is expressed in terms of the fundamental tensors of the web $W(n+1,n,r)$.

If we substitute (1.3.20) into (1.3.31) and use (1.3.23), we obtain

$$\underset{\alpha\beta}{\tilde{\omega}}{}^i_j = \underset{\alpha\beta}{\omega}{}^i_j + \underset{\alpha\beta}{s}{}^i_{jk}(\underset{\alpha}{\omega}{}^k + \underset{\beta}{\omega}{}^k), \tag{1.3.44}$$

where

$$\underset{\alpha\beta}{s}{}^i_{jk} = \underset{\alpha\beta}{a}{}^i_{(jk)}. \tag{1.3.45}$$

Equation (1.2.17) implies that

$$\underset{\alpha\beta}{s}{}^i_{jk} = \underset{\beta\alpha}{s}{}^i_{jk}. \tag{1.3.46}$$

Using (1.3.19), we find

$$\underset{\alpha\beta}{\nabla}\underset{\alpha\beta}{s}{}^i_{jk} = \underset{\alpha\beta\alpha}{s}{}^i_{jkl}\underset{\alpha}{\omega}{}^l + \underset{\alpha\beta\beta}{s}{}^i_{jkl}\underset{\beta}{\omega}{}^l, \tag{1.3.47}$$

where

$$\underset{\alpha\beta\alpha}{s}{}^i_{jkl} = \underset{\alpha\beta\alpha}{a}{}^i_{(jk)l} + \underset{\alpha\beta}{a}{}^i_{(j|m|}\underset{\beta\alpha}{a}{}^m_{k)l}, \qquad \underset{\alpha\beta\beta}{s}{}^i_{jkl} = \underset{\alpha\beta\beta}{a}{}^i_{(jk)l} + \underset{\alpha\beta}{a}{}^i_{m(k}\underset{\alpha\beta}{a}{}^m_{j)l}. \tag{1.3.48}$$

Equations (1.3.46) and (1.3.48) give

$$\underset{\alpha\beta\alpha}{s}{}^i_{jkl} = \underset{\beta\alpha\alpha}{s}{}^i_{jkl}, \qquad \underset{\alpha\beta\beta}{s}{}^i_{jkl} = \underset{\beta\alpha\beta}{s}{}^i_{jkl}. \tag{1.3.49}$$

Exterior differentiation of (1.3.44) by means of (1.3.18) and (1.3.19) gives

$$\left\{\begin{array}{l} \underset{\alpha\alpha}{\tilde{b}}{}^i_{jkl} = \underset{\alpha\alpha}{b}{}^i_{jkl} - \underset{\beta\alpha\alpha}{s}{}^i_{j[kl]} - \underset{\alpha\beta}{s}{}^m_{j[k}\underset{\alpha\beta}{s}{}^i_{|m|l]} \\ \underset{\beta\beta}{\tilde{b}}{}^i_{jkl} = \underset{\beta\beta}{b}{}^i_{jkl} - \underset{\alpha\beta\beta}{s}{}^i_{j[kl]} - \underset{\alpha\beta}{s}{}^m_{j[k}\underset{\alpha\beta}{s}{}^i_{|m|l]}, \\ \underset{\alpha\beta}{\tilde{b}}{}^i_{jkl} = \underset{\alpha\beta}{b}{}^i_{jkl} - \dfrac{1}{2}\underset{\alpha\beta\beta}{s}{}^i_{jkl} + \dfrac{1}{2}\underset{\beta\alpha\alpha}{s}{}^i_{jlk} - \underset{\alpha\beta}{s}{}^m_{j[k}\underset{\alpha\beta}{s}{}^i_{|m|l]}. \end{array}\right. \tag{1.3.50}$$

It follows from (1.3.39) and (1.3.40) that

$$\underset{\alpha\alpha}{\tilde{b}}{}^i_{jkl} - \underset{\alpha\beta}{\tilde{b}}{}^i_{jkl} = -\frac{1}{2}\underset{\alpha\beta}{\bar{b}}{}^i_{l(jk)}. \tag{1.3.51}$$

Equations (1.3.49) lead to

$$\underset{\alpha\alpha}{\tilde{b}}{}^i_{jkl} - \underset{\alpha\beta}{\tilde{b}}{}^i_{jkl} = \underset{\alpha\alpha}{b}{}^i_{jkl} - \underset{\alpha\beta}{b}{}^i_{jkl} - \frac{1}{2}\underset{\beta\alpha\alpha}{s}{}^i_{jkl} + \frac{1}{2}\underset{\alpha\beta\beta}{s}{}^i_{jkl}. \tag{1.3.52}$$

Comparison of (1.3.51) and (1.3.52) shows that

$$\underset{\alpha\alpha}{b}{}^i_{jkl} - \underset{\alpha\beta}{b}{}^i_{jkl} - \frac{1}{2}\underset{\beta\alpha\alpha}{s}{}^i_{jkl} + \frac{1}{2}\underset{\alpha\beta\beta}{s}{}^i_{jkl} + \frac{1}{2}\underset{\alpha\beta}{\bar{b}}{}^i_{l(jk)} = 0. \tag{1.3.53}$$

Akivis and Shelekhov [AS 81] introduced a bundle of affine connections on the manifold X^{2r} carrying a 3-web. In our notation affine connections of this bundle on the manifold $M_{\alpha\beta}$ carrying a 3-subweb $[n+1,\alpha,\beta]$ are given by the 1-forms $\{\underset{\alpha}{\omega}^i, \underset{\beta}{\omega}^j\}$ and

$$\begin{pmatrix} \underset{\alpha\beta}{\Theta}{}^i_j & 0 \\ 0 & \underset{\alpha\beta}{\Theta}{}^i_j \end{pmatrix}$$

where

$$\underset{\alpha\beta}{\Theta}{}^i_j = \underset{\alpha\beta}{\bar{\omega}}{}^i_j + \underset{\alpha\beta}{\bar{a}}{}^i_{jk}(p\underset{\alpha}{\omega}^k + q\underset{\beta}{\omega}^k). \tag{1.3.54}$$

Let us denote the bundle by $\gamma_{\alpha\beta}(\Theta)$ and let us denote the affine connection of it corresponding to parameters p and q by $\gamma_{\alpha\beta}(p,q)$. The connections $\bar{\gamma}_{\alpha\beta}$ and $\tilde{\gamma}_{\alpha\beta}$ belong to the bundle since equations (1.3.54) and (1.3.31) show that $\bar{\gamma}_{\alpha\beta} = \gamma_{\alpha\beta}(0,0)$ and $\tilde{\gamma}_{\alpha\beta} = \gamma_{\alpha\beta}(1,-1)$.

Theorem 1.3.5 *The affine connection $\gamma_{\alpha\beta}$ belongs to the bundle $\gamma_{\alpha\beta}(\Theta)$ if and only if*

$$\underset{\alpha\beta}{s}{}^i_{jk} = \underset{\alpha\beta}{a}{}^i_{(jk)} = 0. \tag{1.3.55}$$

Proof. Substituting (1.3.20) and (1.3.23) into (1.3.54) and using (1.2.17), we obtain

$$\underset{\alpha\beta}{\Theta}{}^i_j = \underset{\alpha\beta}{\omega}{}^i_j + (\underset{\beta\alpha}{a}{}^i_{jk} - p\underset{\beta\alpha}{a}{}^i_{[jk]})\underset{\alpha}{\omega}^k + (\underset{\alpha\beta}{a}{}^i_{jk} + q\underset{\alpha\beta}{a}{}^i_{[jk]})\underset{\beta}{\omega}^k. \tag{1.3.56}$$

Since the forms $\underset{\alpha}{\omega}^i$ are linearly independent, equations (1.3.56) imply that the connection $\gamma_{\alpha\beta}$ belongs to the bundle $\gamma_{\alpha\beta}(\Theta)$ if and only if

$$\underset{\beta\alpha}{a}{}^i_{jk} - p\underset{\beta\alpha}{a}{}^i_{[jk]} = 0, \quad \underset{\alpha\beta}{a}{}^i_{jk} + q\underset{\alpha\beta}{a}{}^i_{[jk]} = 0. \tag{1.3.57}$$

It follows from (1.3.57) that in general the connection $\gamma_{\alpha\beta}$ does not belong to the bundle $\gamma_{\alpha\beta}(\Theta)$ since equations (1.3.57) do not hold on a general $W(n+1,n,r)$ for any values of p and q.

Let us denote the left members of (1.3.57) by $\underset{\beta\alpha}{c}{}^i_{jk}$ and $\underset{\alpha\beta}{c}{}^i_{jk}$:

$$\underset{\beta\alpha}{c}{}^i_{jk} = \underset{\beta\alpha}{a}{}^i_{jk} - p\underset{\beta\alpha}{a}{}^i_{[jk]}, \quad \underset{\alpha\beta}{c}{}^i_{jk} = \underset{\alpha\beta}{a}{}^i_{jk} + q\underset{\alpha\beta}{a}{}^i_{[jk]}. \tag{1.3.58}$$

Calculating the symmetric and skew-symmetric parts of the tensors $\underset{\beta\alpha}{c}{}^i_{jk}$ and $\underset{\alpha\beta}{c}{}^i_{jk}$, we get

$$\underset{\beta\alpha}{c}{}^i_{[jk]} = (1-p)\underset{\beta\alpha}{a}{}^i_{[jk]}, \quad \underset{\alpha\beta}{c}{}^i_{[jk]} = (1+q)\underset{\alpha\beta}{a}{}^i_{[jk]}, \tag{1.3.59}$$

$$\underset{\beta\alpha}{c}{}^i_{(jk)} = \underset{\alpha\beta}{c}{}^i_{(jk)} = \underset{\alpha\beta}{a}{}^i_{(jk)}. \tag{1.3.60}$$

Equations (1.3.59) show that the skew-symmetric parts of the tensors $\underset{\beta\alpha}{c}{}^i_{jk}$ and $\underset{\alpha\beta}{c}{}^i_{jk}$ vanish if we take $p=1$ and $q=-1$. Equations (1.3.60) show that the symmetric parts of these tensors vanish if and only if condition (1.3.55) holds. ∎

The following theorem gives analytical conditions for coincidence of the connections $\gamma_{\alpha\beta}, \bar{\gamma}_{\alpha\beta}$, and $\tilde{\gamma}_{\alpha\beta}$ defined on the 3-subweb $[n+1,\alpha,\beta]$.

Theorem 1.3.6 *The necessary and sufficient conditions for coincidence of two or three of connections $\gamma_{\alpha\beta}$, $\bar{\gamma}_{\alpha\beta}$, and $\tilde{\gamma}_{\alpha\beta}$ are:*

(i) *For $\gamma_{\alpha\beta}$ and $\tilde{\gamma}_{\alpha\beta}$ the condition* (1.3.55).

(ii) *For $\bar{\gamma}_{\alpha\beta}$ and $\tilde{\gamma}_{\alpha\beta}$ the condition*

$$\underset{\alpha\beta}{a}{}^i_{[jk]} = 0. \tag{1.3.61}$$

(iii) *For $\gamma_{\alpha\beta}$ and $\bar{\gamma}_{\alpha\beta}$ as well as for all three connections $\gamma_{\alpha\beta}$, $\bar{\gamma}_{\alpha\beta}$ and $\tilde{\gamma}_{\alpha\beta}$ the condition*

$$\underset{\alpha\beta}{a}{}^i_{jk} = 0. \tag{1.3.62}$$

Proof. Part (i) follows from equations (1.3.44) and (1.3.45). Equations (1.3.31) and (1.3.23) imply part (ii). Equations (1.3.20) show that condition (1.3.62) is necessary and sufficient for $\gamma_{\alpha\beta}$ and $\bar{\gamma}_{\alpha\beta}$ to coincide. Since (1.3.62) implies (1.3.55), condition (1.3.62) is necessary and sufficient for all three connections to coincide. ∎

Corollary 1.3.7 *The connection $\gamma_{\alpha\beta}$ belongs to the bundle $\gamma_{\alpha\beta}(\Theta)$ if and only if it coincides with the connection $\underset{\alpha\beta}{\tilde{\gamma}}$.*

Proof. It follows from theorems 1.3.3 and 1.3.4. ∎

In conclusion of this section let us consider the 3-subweb $[\alpha,\beta,\gamma]$ and the canonical connection on it. The 3-subweb $[\alpha,\beta,\gamma]$ is defined by the leaves of the foliations $\lambda_\alpha, \lambda_\beta, \lambda_\gamma$ on the $(2r)$-dimensional intersection of leaves $F_\xi \subset \lambda_\xi$, $\xi = 1,\ldots,n,n+1$, $\xi \neq \alpha,\beta,\gamma$. Its equations are

$$\underset{\xi}{\omega}{}^i = 0, \;\; \xi = 1,\ldots,n,n+1, \;\; \xi \neq \alpha,\beta,\gamma. \tag{1.3.63}$$

Substituting (1.3.63) into (1.2.5), (1.2.16), (1.2.18), and (1.2.29), we obtain

$$\underset{\alpha}{\omega}{}^i + \underset{\beta}{\omega}{}^i + \underset{\gamma}{\omega}{}^i = 0, \tag{1.3.64}$$

$$\begin{cases} d\underset{\alpha}{\omega}{}^i = \underset{\alpha}{\omega}{}^j \wedge \underset{\alpha\beta\gamma}{\omega}{}^i_j + \underset{\alpha\beta}{a}{}^i_{jk}\underset{\alpha}{\omega}{}^j \wedge \underset{\beta}{\omega}{}^k + \underset{\alpha\gamma}{a}{}^i_{jk}\underset{\alpha}{\omega}{}^j \wedge \underset{\gamma}{\omega}{}^k, \\ d\underset{\beta}{\omega}{}^i = \underset{\beta}{\omega}{}^j \wedge \underset{\alpha\beta\gamma}{\omega}{}^i_j + \underset{\beta\alpha}{a}{}^i_{jk}\underset{\beta}{\omega}{}^j \wedge \underset{\alpha}{\omega}{}^k + \underset{\beta\gamma}{a}{}^i_{jk}\underset{\beta}{\omega}{}^j \wedge \underset{\gamma}{\omega}{}^k, \\ d\underset{\gamma}{\omega}{}^i = \underset{\gamma}{\omega}{}^j \wedge \underset{\alpha\beta\gamma}{\omega}{}^i_j + \underset{\gamma\alpha}{a}{}^i_{jk}\underset{\gamma}{\omega}{}^j \wedge \underset{\alpha}{\omega}{}^k + \underset{\gamma\beta}{a}{}^i_{jk}\underset{\gamma}{\omega}{}^j \wedge \underset{\beta}{\omega}{}^k, \end{cases} \tag{1.3.65}$$

$$d\underset{\alpha\beta\gamma}{\omega}{}^i_j = \underset{\alpha\beta\gamma}{\omega}{}^k_j \wedge \underset{\alpha\beta\gamma}{\omega}{}^i_k, \tag{1.3.66}$$

where

$$\underset{\alpha\beta\gamma}{\omega}{}^i_j = \omega^i_j \Big|_{\underset{\xi}{\omega}^i = 0}, \quad \xi = 1, \ldots, n, n+1, \ \ \xi \neq \alpha, \beta, \gamma.$$

To find the torsion and curvature tensors of the canonical connection of $[\alpha, \beta, \gamma]$ in terms of the corresponding tensors of λ_{n+1}, we will take forms $\underset{\alpha}{\omega}^i$, $\underset{\beta}{\omega}^i$ as a co-basis and denote by $\underset{\alpha\beta\gamma}{\bar{\omega}}{}^i_j$, $\underset{\alpha\beta\gamma}{\bar{a}}{}^i_{jk}$, $\underset{\alpha\beta\gamma}{\bar{b}}{}^i_{jkl}$ the connection form and the torsion and curvature tensors of this connection. Then equations (1.3.65) can be reduced to the form (1.2.16):

$$\begin{cases} d\underset{\gamma}{\omega}^i = \underset{\gamma}{\omega}^j \wedge \underset{\alpha\beta\gamma}{\bar{\omega}}{}^i_j + \underset{\alpha\beta\gamma}{\bar{a}}{}^i_{jk}\underset{\gamma}{\omega}^j \wedge (\underset{\alpha}{\omega}^k - \underset{\beta}{\omega}^k), \\ d\underset{\alpha}{\omega}^i = \underset{\alpha}{\omega}^j \wedge \underset{\alpha\beta\gamma}{\bar{\omega}}{}^i_j + \underset{\alpha\beta\gamma}{\bar{a}}{}^i_{jk}\underset{\alpha}{\omega}^j \wedge \underset{\alpha}{\omega}^k, \\ d\underset{\beta}{\omega}^i = \underset{\beta}{\omega}^j \wedge \underset{\alpha\beta\gamma}{\bar{\omega}}{}^i_j - \underset{\alpha\beta\gamma}{\bar{a}}{}^i_{jk}\underset{\beta}{\omega}^j \wedge \underset{\beta}{\omega}^k, \end{cases} \tag{1.3.67}$$

where

$$\underset{\alpha\beta\gamma}{\bar{\omega}}{}^i_j = \underset{\alpha\beta\gamma}{\omega}{}^i_j + (\underset{\beta\alpha}{a}{}^i_{jk} - \underset{\beta\gamma}{a}{}^i_{jk})\underset{\alpha}{\omega}^k + (\underset{\alpha\beta}{a}{}^i_{jk} - \underset{\alpha\gamma}{a}{}^i_{jk})\underset{\beta}{\omega}^k, \tag{1.3.68}$$

$$\underset{\alpha\beta\gamma}{\bar{a}}{}^i_{jk} = \underset{\alpha\beta}{a}{}^i_{[jk]} + \underset{\beta\gamma}{a}{}^i_{[jk]} + \underset{\gamma\alpha}{a}{}^i_{[jk]} \tag{1.3.69}$$

are the expressions of the connection forms and the torsion tensor of the considered connection. It is obvious from (1.3.69) that the torsion tensor is skew-symmetric.

Exterior differentiation of (1.3.67) gives

$$d\underset{\alpha\beta\gamma}{\bar{\omega}}{}^i_j = \underset{\alpha\beta\gamma}{\bar{\omega}}{}^k_j \wedge \underset{\alpha\beta\gamma}{\bar{\omega}}{}^i_k + \underset{\alpha\beta\gamma}{\bar{b}}{}^i_{jkl}\underset{\alpha}{\omega}^k \wedge \underset{\beta}{\omega}^l, \tag{1.3.70}$$

where

$$\begin{aligned} \underset{\alpha\beta\gamma}{\bar{b}}{}^i_{jkl} = & - \underset{\beta\alpha\beta}{a}{}^i_{jkl} + \underset{\alpha\beta\alpha}{a}{}^i_{jlk} - \underset{\beta\gamma\gamma}{a}{}^i_{jkl} + \underset{\beta\gamma\beta}{a}{}^i_{jkl} - \underset{\alpha\gamma\alpha}{a}{}^i_{jlk} + \underset{\alpha\gamma\gamma}{a}{}^i_{jlk} - \underset{\alpha\beta\gamma}{a}{}^i_{jlk} + \underset{\beta\alpha\gamma}{a}{}^i_{jkl} \\ & + (\underset{\beta\alpha}{a}{}^m_{lj} + \underset{\gamma\beta}{a}{}^m_{lj} - \underset{\gamma\alpha}{a}{}^m_{lj})(\underset{\beta\alpha}{a}{}^i_{mk} - \underset{\beta\gamma}{a}{}^i_{mk}) + (\underset{\beta\alpha}{a}{}^m_{jk} - \underset{\beta\gamma}{a}{}^m_{jk} + \underset{\alpha\gamma}{a}{}^m_{jk})(\underset{\alpha\gamma}{a}{}^i_{ml} - \underset{\alpha\beta}{a}{}^i_{ml}) \\ & + (-\underset{\alpha\beta}{a}{}^m_{kl} + \underset{\gamma\beta}{a}{}^m_{kl} + \underset{\alpha\gamma}{a}{}^m_{kl})(\underset{\alpha\beta}{a}{}^i_{mj} - \underset{\gamma\alpha}{a}{}^i_{mj}) \end{aligned} \tag{1.3.71}$$

is the curvature tensor of our connection.

Equations (1.3.71) and (1.3.69) allow us to verify the identity (1.2.51) for tensors $\underset{\alpha\beta\gamma}{\bar{a}}{}^i_{jk}$ and $\underset{\alpha\beta\gamma}{\bar{b}}{}^i_{jkl}$.

Theorem 1.3.8 *The expressions for the torsion and curvature tensors of the 3-subweb $[\alpha, \beta, \gamma]$, i.e. of the canonical connection of this 3-web, in terms of the corresponding tensors of the 3-subwebs $[n+1, \alpha, \beta]$, $[n+1, \beta, \gamma]$, and $[n+1, \gamma, \alpha]$ are:*

$$\underset{\alpha\beta\gamma}{\bar{a}}{}^i_{jk} = \underset{\alpha\beta}{\bar{a}}{}^i_{jk} + \underset{\beta\gamma}{\bar{a}}{}^i_{jk} + \underset{\gamma\alpha}{\bar{a}}{}^i_{jk}, \tag{1.3.72}$$

$$\underset{\alpha\beta\gamma}{\bar{b}}{}^{i}_{jkl} = \underset{\alpha\beta}{\bar{b}}{}^{i}_{jkl} + \underset{\beta\gamma}{\bar{b}}{}^{i}_{ljk} + \underset{\gamma\alpha}{\bar{b}}{}^{i}_{klj} - 2\underset{\gamma\gamma}{b}{}^{i}_{jkl} - \underset{\alpha\beta\gamma}{a}{}^{i}_{jlk} + \underset{\beta\alpha\gamma}{a}{}^{i}_{jkl}$$
$$+ (\underset{\beta\alpha}{a}{}^{m}_{jk} - \underset{\beta\gamma}{a}{}^{m}_{jk})\underset{\alpha\gamma}{a}{}^{i}_{ml} + (\underset{\beta\gamma}{a}{}^{m}_{jk} - \underset{\alpha\gamma}{a}{}^{m}_{jk})\underset{\alpha\beta}{a}{}^{i}_{ml} + (\underset{\beta\alpha}{a}{}^{m}_{lk} - \underset{\beta\gamma}{a}{}^{m}_{lk})\underset{\alpha\gamma}{a}{}^{i}_{jm}. \qquad (1.3.73)$$

Proof. Formula (1.3.72) follows from (1.3.69) and (1.3.23) and formula (1.3.73) is obtained from (1.3.71) by means of (1.3.25), (1.2.33), (1.2.28), and (1.2.40). ■

1.4 Webs $W(n+1,n,r)$ with Vanishing Curvature

One of important classes of affinely connected spaces are *projectively Euclidean spaces.* They can be defined as affinely connected spaces for which, in a neighbourhood of each of its points, there exists such a coordinate system x^I where the geodesic lines are given by linear parametric equations

$$x^I = a^I t + b^I, \quad a^I, b^I = \text{ const.}$$

The condition for a space X^{nr}, $nr > 2$, with an affine connection to be projectively Euclidean is the following structure of its curvature tensor (see [SS 38] or [Ra 59]):

$$\begin{aligned} R^I_{JKL} &= \frac{1}{nr+1}\delta^I_J(R_{LK} - R_{KL}) + \frac{1}{n^2r^2-1}\delta^I_K(nrR_{LJ} + R_{JL}) \\ &-\frac{1}{n^2r^2-1}\delta^I_L(nrR_{KJ} + R_{JK}). \end{aligned} \qquad (1.4.1)$$

where

$$R_{JK} = R^I_{JKI} \qquad (1.4.2)$$

is the Ricci tensor of the affine connection. We will call an affine connection of such a space *projectively Euclidean.* In the case $nr = 2$, the condition has the form:

$$\nabla_L P_{KI} = \nabla_K P_{LI}, \qquad (1.4.3)$$

where ∇_L is the operator of the covariant derivative and

$$P_{KI} = -\frac{1}{3}(2R_{KI} + R_{IK}). \qquad (1.4.4)$$

We will prove the following theorem:

Theorem 1.4.1 *For the affine connection γ_ξ induced by a web $W(n+1,n,r)$, $n \geq 2$, $r \geq 1$, to be projectively Euclidean, it is necessary and sufficient that γ_ξ be curvature-free.*

Proof. We will prove the statement for the connection γ_{n+1}. Since the vanishing of the curvature tensor of a web is of an invariant nature, if one of the affine connections induced by a web is curvature-free, then so are the other n affine connections induced by the web.

According to the conditions we discussed for a connection to be projectively Euclidean, we should consider separately the cases $nr > 2$ and $nr = 2$.

Suppose first that $nr > 2$. Let us write down explicit expressions for the curvature tensor R^I_{JKL}, $I, \ J, \ K, \ L = 1, \ \ldots, nr$. It follows from (1.3.5) that the non-vanishing components of this tensor can be conveniently written in the form:

$$\underset{\alpha\beta\beta}{\overset{\alpha}{R}}{}_{jkl} = \underset{\beta\beta}{b}{}^{i}_{jkl}, \quad \underset{\alpha\beta\gamma}{\overset{\alpha}{R}}{}^{i}_{jkl} = \underset{\beta\gamma}{b}{}^{i}_{jkl} (\beta < \gamma), \quad \underset{\alpha\beta\gamma}{\overset{\alpha}{R}}{}^{i}_{jkl} = -\underset{\gamma\beta}{b}{}^{i}_{jlk} \ (\beta > \gamma). \tag{1.4.5}$$

We can find from equations (1.4.2) and (1.4.5) the Ricci tensor of our connection γ_{n+1}. Only the following its components can be differrent from zero:

$$\underset{\alpha\alpha}{R}{}_{jk} = \underset{\alpha\alpha\alpha}{\overset{\alpha}{R}}{}^{p}_{jkp} = \underset{\alpha\alpha}{b}{}^{p}_{jkp}, \quad \underset{\alpha\beta}{R}{}_{jk} = \underset{\alpha\beta\alpha}{\overset{\alpha}{R}}{}^{p}_{jkp} = \underset{\beta\alpha}{b}{}^{p}_{jkp} \ (\beta < \alpha), \quad \underset{\alpha\beta}{R}{}_{jk} = -\underset{\alpha\beta}{b}{}^{p}_{jpk} \ (\beta > \alpha). \tag{1.4.6}$$

Using (1.4.5) and (1.4.6), we can deduce conditions (1.4.1). For $\alpha > \gamma$ they have the form:

$$\underset{\alpha\gamma}{b}{}^{i}_{jkl} = \frac{1}{nr+1}\delta^i_j(\underset{\alpha\gamma}{b}{}^{p}_{lkp} + \underset{\alpha\gamma}{b}{}^{p}_{kpl}) + \frac{1}{n^2r^2-1}\delta^i_k(nr\,\underset{\alpha\gamma}{b}{}^{p}_{ljp} - \underset{\alpha\gamma}{b}{}^{p}_{jpl}), \tag{1.4.7}$$

and for $\alpha < \gamma$ they have the form:

$$\underset{\alpha\gamma}{b}{}^{i}_{jkl} = \frac{1}{nr+1}\delta^i_j(\underset{\alpha\gamma}{b}{}^{p}_{kpl} + \underset{\alpha\gamma}{b}{}^{p}_{lkp}) + \frac{1}{n^2r^2-1}\delta^i_k(nr\,\underset{\alpha\gamma}{b}{}^{p}_{kpj} - \underset{\alpha\gamma}{b}{}^{p}_{jkp}). \tag{1.4.8}$$

If $r > 1$, it follows from (1.4.7) and (1.4.8) that

$$\underset{\alpha\gamma}{b}{}^{p}_{jpl} = nr\,\underset{\alpha\gamma}{b}{}^{p}_{ljp}, \quad \underset{\alpha\gamma}{b}{}^{p}_{ljp} = nr\,\underset{\alpha\gamma}{b}{}^{p}_{jpl},$$

which implies

$$\underset{\alpha\gamma}{b}{}^{p}_{jpl}(1 - n^2r^2) = 0. \tag{1.4.9}$$

Since $nr > 2$, we obtain from (1.4.9) that

$$\underset{\alpha\gamma}{b}{}^{p}_{jpl} = \underset{\alpha\gamma}{b}{}^{p}_{jlp} = 0. \tag{1.4.10}$$

Equations (1.4.9), (1.4.7), (1.4.8), and (1.2.28) show that

$$\underset{\alpha\gamma}{b}{}^{i}_{jkl} = 0, \quad \gamma \neq \alpha, \quad \underset{\alpha\alpha}{b}{}^{i}_{jkl} = 0. \tag{1.4.11}$$

If $r = 1$, then (1.4.7) and (1.4.8) yield the same equation

$$\underset{\alpha\gamma}{b}{}^{1}_{111} = \frac{3}{n+1}\underset{\alpha\gamma}{b}{}^{1}_{111}. \tag{1.4.12}$$

Since $nr > 2$, we have $n > 2$, and (1.4.12) again yields (1.4.11).

It remains to consider the case $nr = 2$, i.e. $n = 2$ and $r = 1$. In this case equations (1.2.29) takes the form

$$d\omega = 2b \underset{1}{\omega} \wedge \underset{2}{\omega}, \tag{1.4.13}$$

where we set $\omega_1^1 = \omega$, $\underset{1}{\omega}^1 = \underset{1}{\omega}$, $\underset{2}{\omega}^1 = \underset{2}{\omega}$, $\underset{12}{b}{}^1_{111} = b$. It follows from (1.4.13) that the curvature tensor of the web $W(n+1,n,1)$ has a unique non-zero component: $2\underset{12}{b}{}^1_{111} = 2b$.

The curvature tensor R^I_{JKL} of our affine connection γ_{n+1} in this case can have the following components non-zero:

$$\underset{112}{\overset{1}{R}}{}^1_{111} = \underset{212}{\overset{2}{R}}{}^1_{111} = b, \quad \underset{121}{\overset{1}{R}}{}^1_{111} = \underset{221}{\overset{2}{R}}{}^1_{111} = -b. \tag{1.4.14}$$

Equation (1.4.14) shows that the Ricci tensor of this connection can have the following components non-zero:

$$\underset{12}{R}{}_{11} = b, \quad \underset{21}{R}{}_{11} = -b. \tag{1.4.15}$$

From (1.4.15) and (1.4.4) we obtain

$$P_{11} = \underset{11}{R}{}_{11} = 0, \quad P_{22} = \underset{22}{R}{}_{22} = 0, \quad P_{12} = \underset{12}{R}{}_{11} = -\frac{1}{3}b, \quad P_{21} = \underset{21}{R}{}_{11} = \frac{1}{3}b. \tag{1.4.16}$$

Since

$$\nabla P_{ki} = dP_{ki} - P_{li}\omega_k^l - P_{kl}\omega_i^l = P_{kil}\omega^l,$$

then $\nabla_l P_{ki} = P_{kil}$, and (1.4.3) takes the form

$$P_{kil} = P_{lik}. \tag{1.4.17}$$

It follows from (1.4.16) that $P_{112} = P_{221} = 0$, and so by virtue of (1.4.17), we have $P_{212} = P_{122} = 0$, $P_{121} = P_{211} = 0$,. Consequently, $\nabla P_{12} = \nabla P_{21} = 0$,whence, by virtue of (1.4.16), we obtain

$$db - 2b\omega = 0. \tag{1.4.18}$$

Equation (1.4.18) implies that $b = 0$, since for $b = 0$ exterior differentiation of (1.4.18) yields $d\omega = 0$, which contradicts (1.4.13). This finishes the proof of the necessity of our assertion; the sufficiency is obvious.

Thus, in all cases (for $n \geq 2$, $r \geq 1$) the affine connection γ_{n+1} induced by the web $W(n+1,n,r)$ is projectively Euclidean if and only if the web is curvature-free.

As we mentioned in the beginning of the proof, the theorem is valid for the connections γ_α, $\alpha = 1, \ldots, n$, since it is valid for γ_{n+1}. ■

Corollary 1.4.2 *An affine connection induced by a 3-web is projectively Euclidean if and only if the 3-web is a group web.*

Proof. This follows from the fact that for a 3-web the vanishing of the curvature tensor is equivalent to the fact that the 3-web is a group web (see Table 1.1, p. 18). ■

In the case of a 3-web, in addition to the three mentioned affine connections, three more, which are defined by the following connection forms, are treated (see [AS 71b, 81]):

$$\begin{aligned}\underset{12}{\omega}{}^i_j &= \omega^i_j + \underset{12}{a}{}^i_{jk}\underset{2}{\omega}{}^k + \underset{12}{a}{}^i_{kj}\underset{1}{\omega}{}^k,\\ \underset{23}{\omega}{}^i_j &= \omega^i_j + \underset{12}{a}{}^i_{kj}\underset{1}{\omega}{}^k - \underset{12}{a}{}^i_{jk}\underset{2}{\omega}{}^k,\\ \underset{31}{\omega}{}^i_j &= \omega^i_j + \underset{12}{a}{}^i_{jk}\underset{2}{\omega}{}^k + \underset{12}{a}{}^i_{jk}\underset{1}{\omega}{}^k.\end{aligned}$$

The following corollary follows from our proof.

Corollary 1.4.3 *If one of these six affine connections induced by a 3-web is projectively Euclidean, then so are the other five.* ■

Remark 1.4.4 For $n > 2$ the vanishing of the curvature tensor of the web $W(n+1, n, r)$ does not imply that the components of the torsion tensor are constant, i.e., does not lead to the group $(n+1)$-web (see its definition in Section 4.3).

1.5 Parallelisable $(n+1)$-Webs

Definition 1.5.1 A web $W(n+1, n, r)$ formed by $n+1$ *foliations* of parallel $(n-1)r$-dimensional planes of (nr)-dimensional affine space A^{nr} is called *parallel.* A web $W(n+1, n, r)$ is said to be *parallelisable* if it is equivalent to a parallel web (see Example 1.1.3).

We will denote a parallelisable web by $PW(n+1, n, r)$.

The following theorem gives an analytical necessary and sufficient condition for a web $W(n+1, n, r), n > 2$, to be parallelisable.

Theorem 1.5.2 *A web $W(n+1, n, r), n > 2$, is parallelisable if and only if its torsion tensor $\underset{\alpha\beta}{a}{}^i_{jk}$ vanishes.*

Proof. To prove the necessity of the condition, let us suppose that a web $W(n+1, n, r)$ is parallelisable, i.e., it can be mapped onto $n+1$ foliations of $(n-1)r$-planes of an (nr)-dimensional affine space A^{nr}. With any point $p \in A^{nr}$ we associate a moving frame $\{e^\alpha_i,\ i = 1, \dots, r; \alpha = 1, \dots, n\}$ in such a way that its vectors $e^{\hat{\alpha}}_i, \hat{\alpha} \neq \alpha$,

α, $\hat{\alpha} = 1, \ldots, n$, are parallel to the plane F_α of the foliation λ_α passing through the point p, and vectors $f_i^{\hat{\alpha}} = e_i^{\hat{\alpha}} - e_i^\alpha$, $\hat{\alpha} \neq \alpha$, $\alpha = 1, \ldots, n$, α is fixed, are parallel to the plane F_{n+1} of the foliation λ_{n+1} passing through p. In this case equations (1.2.5) are satisfied, and the foliations λ_ξ, $\xi = 1, \ldots, n+1$, are defined by equations

$$\underset{\alpha}{\omega}^i = 0, \ \ \alpha = 1, \ldots, n; \ \ \sum_\alpha \underset{\alpha}{\omega}^i = 0.$$

Since the $(n-1)r$-planes of the foliation λ_α are parallel to each other,

$$de_i^\alpha = \underset{\alpha}{\omega}_i^j e_j^\alpha \qquad \text{(no summation in } \alpha). \tag{1.5.1}$$

It follows from (1.5.1) that

$$d(e_i^{\hat{\alpha}} - e_i^\alpha) = \frac{1}{2}(\underset{\hat{\alpha}}{\omega}_i^j + \underset{\alpha}{\omega}_i^j)(e_j^{\hat{\alpha}} - e_j^\alpha) + \frac{1}{2}(\underset{\hat{\alpha}}{\omega}_i^j - \underset{\alpha}{\omega}_i^j)(e_j^{\hat{\alpha}} + e_j^\alpha) \tag{1.5.2}$$

(no summation in α and $\hat{\alpha}$).The planes of the foliation λ_{n+1} are also parallel to each other and the vectors $e_j^{\hat{\alpha}} - e_j^\alpha$ and $e_j^{\hat{\alpha}} + e_j^\alpha$ are linearly independent.Because of this, equations (1.5.2) imply

$$\underset{\hat{\alpha}}{\omega}_i^j = \underset{\alpha}{\omega}_i^j \stackrel{\text{def}}{=} \theta_i^j. \tag{1.5.3}$$

By virtue of (1.5.3), we can rewrite equations (1.5.1) in the form

$$d\bar{e}_i^\alpha = \theta_i^j \bar{e}_j^\alpha. \tag{1.5.4}$$

Exterior differentiation of equations (1.2.5) and (1.5.4) leads to the following structure equations of a parallelisable web $W(n+1, n, r)$:

$$d\underset{\alpha}{\omega}^i = \underset{\alpha}{\omega}^j \wedge \theta_j^i, \qquad d\theta_i^j = \theta_i^k \wedge \theta_k^j. \tag{1.5.5}$$

Thus, the structure equations (1.2.16) and (1.2.29) of a general web $W(n+1, n, r)$ can be reduced to the form (1.5.5) when the web $W(n+1, n, r)$ is parallelisable.

Equations (1.2.6) and (1.2.16) imply

$$\underset{\alpha}{\omega}_i^j = \sum_{\beta \neq \alpha} (\underset{\alpha\beta}{a}{}^i_{jk} \underset{\beta}{\omega}^k + \underset{\alpha\beta}{a}{}^i_{(jk)} \underset{\alpha}{\omega}^k) + \omega_i^j. \tag{1.5.6}$$

According to (1.5.3), the forms $\underset{\alpha}{\omega}_j^i$ are equal to each other. This and equation (1.5.7) give

$$\underset{\alpha\beta}{a}{}^i_{jk} = \underset{\beta}{a}{}^i_{jk}, \ \ \underset{\alpha\beta}{a}{}^i_{[jk]} = 0. \tag{1.5.7}$$

It follows from (1.2.17) and (1.5.7) that

$$\underset{\alpha\beta}{a}{}^i_{(jk)} = a^i_{jk}. \tag{1.5.8}$$

Now equations (1.5.8) and (1.2.24) imply

$$\underset{\alpha\beta}{a}{}^{i}_{(jk)} = 0. \tag{1.5.9}$$

Therefore from (1.5.7) and (1.5.9) we get

$$\underset{\alpha\beta}{a}{}^{i}_{jk} = 0. \tag{1.5.10}$$

Moreover, since in the case $n > 2$ the curvature tensor $\underset{\alpha\beta}{b}{}^{i}_{jkl}$ is expressed in terms of the torsion tensor $\underset{\alpha\beta}{a}{}^{i}_{jk}$ and its Pfaffian derivatives, equations (1.2.16) and (1.2.29) are reduced to the form (1.5.5).

Conversely, if equations (1.5.10) hold, then the structure equations of a web have the form (1.5.5). However, we proved that a parallel web has such the structure equations. Applying Theorem 1.2.7, we conclude that our (n+1)-web is parallelisable. ∎

We will give another geometrical criteria for a web $W(n+1, n, r)$ to be parallelisable. For this we will need the following definition of a $(k+1)$-subweb of of $W(n+1, n, r)$ which generalises definition of a 3-subweb (see Section 1.3).

Definition 1.5.3 Consider the intersection of $n-k$ leaves F_s, $s = k+1, \ldots, n$, $1 < k < n$, of the web $W(n+1, n, r)$. It is defined by the system of equations

$$\underset{\xi_{k+1}}{\omega}{}^{i} = 0, \ldots, \underset{\xi_n}{\omega}{}^{i} = 0 \tag{1.5.11}$$

and has dimension kr. The leaves F_t, $t = 1, \ldots, k, n+1$, of the other $k+1$ foliations cut on this intersection a $(k+1)$-web $W(k+1, k, r)$ of codimension r which is called a $(k+1)$-*subweb* of $W(n+1, n, r)$.

We will denote such a subweb by $[\xi_{n+1}, \xi_1, \ldots, \xi_k]$. A web $W(n+1, n, r)$ has $\binom{n+1}{k+1}$ $(k+1)$-subwebs.

Corollary 1.5.4 *If a web $W(n+1, n, r)$ is parallelisable, then so all its $(k+1)$-subwebs.*

Proof. In fact, according to (1.5.11), a $(k+1)$-subweb $[n+1, \alpha_{n-k+1}, \ldots, \alpha_n]$ is defined by the system

$$\underset{\alpha_1}{\omega}{}^{i} = 0, \ldots, \underset{\alpha_{n-k}}{\omega}{}^{i} = 0, \tag{1.5.12}$$

and a $(k+1)$-subweb $[\alpha_{n-k}, \ldots, \alpha_n]$ by the system

$$\underset{n+1}{\omega}{}^{i} = 0, \ \underset{\alpha_1}{\omega}{}^{i} = 0, \ \ldots \ \underset{\alpha_{n-k-1}}{\omega}{}^{i} = 0. \tag{1.5.13}$$

In the case of (1.5.12) we can see from the structure equations (1.2.16) that the torsion tensor of $[n+1, \alpha_{n-k+1}, \ldots, \alpha_n]$ is a subtensor of the torsion tensor of the web $W(n+1, n, r)$.

In the case of (1.5.13) we can write the structure equations of $[\alpha_{n-k}, \ldots, \alpha_n]$ in the form

$$d\underset{\hat{\alpha}}{\omega^i} = \underset{\hat{\alpha}}{\omega^j} \wedge \underset{\hat{\beta}}{\omega^i_j} + \sum_{\gamma \neq \hat{\beta}, \hat{\alpha}} (\underset{\hat{\alpha}\gamma}{a^{\,i}_{jk}} - \underset{\hat{\beta}\gamma}{a^{\,i}_{jk}} - \underset{\hat{\alpha}\hat{\beta}}{a^{\,i}_{jk}}) \underset{\hat{\alpha}}{\omega^j} \wedge \underset{\gamma}{\omega^k}, \tag{1.5.14}$$

where

$$\underset{\hat{\beta}}{\omega^i_j} = \omega^i_j + \sum_{\hat{\gamma} \neq \hat{\beta}} \underset{\hat{\beta}\hat{\gamma}}{a^{\,i}_{jk}} \underset{\hat{\gamma}}{\omega^k}$$

and $\hat{\alpha}, \hat{\beta}, \hat{\gamma}$ take those values from $1 \ldots, n$ which are different from $\alpha_1, \ldots, \alpha_{n-k-1}$; $\hat{\beta}$ is fixed and $\hat{\alpha} \neq \hat{\beta}$. Equations (1.5.14) and (1.5.10) prove our Corollary 1.5.4 for a subweb $[\alpha_{n-k}, \ldots, \alpha_n]$. ■

Now we are going to prove the theorem which is converse to Corollary 1.5.4. However, if we assume that all $(k+1)$-subwebs of $W(n+1, n, r)$ are parallelisable, the parallelisability of $W(n+1, n, r)$ is almost obvious. So, the problem is to find a minimal number of $(k+1)$-subwebs whose parallelisability implies the parallelisability of the whole $W(n+1, n, r)$.

Theorem 1.5.5 *The $n-k+2$ leaves of the different foliations, $2 < k \leq n-1$, give a rise to $\binom{n-k+2}{2}$ $(k+1)$-subwebs. If all these $(k+1)$-subwebs are parallelisable, then so is the whole web $W(n+1, n, r)$.*

Proof. The indicated $n-k+2$ leaves of the web are defined by systems

$$\underset{\alpha_1}{\omega} = 0, \ldots, \underset{\alpha_{n-k+2}}{\omega} = 0 \tag{1.5.15}$$

or by systems

$$\underset{n+1}{\omega^{\,i}} = 0, \quad \underset{\alpha_1}{\omega^i} = 0, \quad \underset{\alpha_{n-k+1}}{\omega} = 0. \tag{1.5.16}$$

In the case of (1.5.15) the parallelisability of the indicated subwebs implies $\underset{\hat{\alpha}\hat{\beta}}{a^{\,i}_{jk}} = 0$ where $\hat{\alpha}, \hat{\beta}$ take values $\alpha_{n-k+1}, \ldots, \alpha_n$ and any two of the values $\alpha_1, \ldots, \alpha_{n-k+2}$. So, in the case (1.5.15) all $\underset{\alpha\beta}{a^{\,i}_{jk}} = 0$.

In the case of (1.5.16), first we will take those $(k+1)$-subwebs which are cut on the intersection of any $n-k$ leaves from $n-k+1$ leaves of the foliations $\underset{\alpha_1}{\omega^i} = 0, \ldots, \underset{\alpha_{n-k+1}}{\omega^{\,i}} = 0$. Then from the parallelisability of such $(k+1)$-subwebs and (1.2.16) we obtain that $\underset{\hat{\alpha}\hat{\beta}}{a^{\,i}_{jk}} = 0$ where at least one of the numbers of $\hat{\alpha}$ and $\hat{\beta}$ is different from $\alpha_1, \ldots, \alpha_{n-k+1}$. Next we will take those $(k+1)$-subwebs which are cut on the intersection of a leaf of $\underset{n+1}{\omega^{\,i}} = 0$ and $n-k-1$ leaves from $n-k$ leaves of $\underset{\alpha_1}{\omega^i} = 0, \ldots, \underset{\alpha_{n-k+1}}{\omega^{\,i}} = 0$. Then we will write equations (1.2.14) in the form (1.5.16).

Using the parallelisability of these subwebs and the fact that $\underset{\hat{\alpha}\hat{\beta}}{a}{}^{i}_{jk} = 0$, we obtain that the others $\underset{\alpha\beta}{a}{}^{i}_{jk}$ also vanish. ■

Corollary 1.5.6 *If n-subwebs which are cut on three fixed leaves of different foliations of a web $W(n+1,n,r)$ by the leaves of the other foliations are parallelisable, then the whole web $W(n+1,n,r)$ is parallelisable.*

Proof. To prove this, we take $k = n-1$ and apply Theorem 1.5.5. ■

We will conclude this section by the following corollary which follows directly from Theorem 1.3.6 (iii) and Theorem 1.5.2.

Corollary 1.5.7 *A web $W(n+1,n,r)$, $n > 2$, is parallelisable if and only if the connection γ_{n+1} induces the canonical connections $\bar{\gamma}_{\alpha\beta}$ on all 3-subwebs $[n+1,\alpha,\beta]$.* ■

1.6 $(n+1)$-Webs with Paratactical 3-Subwebs

Let us consider now 3-subwebs $[\xi,\eta,\zeta]$, $\xi,\ \eta,\ \zeta = 1,\dots,n+1$, of an $(n+1)$-web $W(n+1,n,r)$. There are $\binom{n+1}{3}$ such 3-subwebs. As we saw in Section 1.3, for $\binom{n}{2}$ 3-subwebs $[n+1,\alpha,\beta]$ defined by equations (1.3.14) the structure equations, the connection forms and the torsion tensor in the canonical connection $\bar{\gamma}_{\alpha\beta}$ have the form (1.3.22), (1.3.20), and (1.3.23). For other $\binom{n}{3}$ 3-subwebs $[\alpha,\beta,\gamma]$ defined by (1.3.63) similar equations in the canonical connection have the form (1.3.67), (1.3.68), and (1.3.69).

Definition 1.6.1 3-webs with vanishing torsion tensor are called *paratactical* (see [Ak 69a]).

Formulae (1.3.23) and (1.3.69) imply the following theorem.

Theorem 1.6.2 *The 3-subwebs $[n+1,\alpha,\beta]$ and $[\alpha,\beta,\gamma]$ of a web $W(n+1,n,r)$ are paratactical if and only if the following conditions are satisfied:*

$$\underset{\alpha\beta}{a}{}^{i}_{[jk]} = 0, \quad \underset{\alpha\beta}{a}{}^{i}_{[jk]} + \underset{\beta\gamma}{a}{}^{i}_{[jk]} + \underset{\gamma\alpha}{a}{}^{i}_{[jk]} = 0, \tag{1.6.1}$$

where the first set of conditions refers to $[n+1,\alpha,\beta]$ and the second set refers to $[\alpha,\beta,\gamma]$. ■

Corollary 1.6.3 *For a web $W(n+1,n,1)$ any of the $\binom{n+1}{3}$ 3-subwebs are paratactical.*

Proof. In this case $r = 1$, and all equations (1.6.1) are satisfied identically. ■

Corollary 1.6.4 *The 3-subweb $[n+1,\alpha,\beta]$ is paratactical if and only if the affine connections $\bar{\gamma}_{\alpha\beta}$ and $\tilde{\gamma}_{\alpha\beta}$ coincide.*

Proof. This follows from Theorem 1.3.6 (ii) and Theorem 1.6.2. ■

Let us consider now a 4-subweb $[n+1,\alpha,\beta,\gamma]$ of $W(n+1,n,r)$ defined by the equations

$$\underset{\delta}{\omega}^i = 0, \quad \delta \neq \alpha,\beta,\gamma. \tag{1.6.2}$$

Equations (1.3.23) and (1.3.69) lead us to the following theorem.

Theorem 1.6.5 *The paratacticity of any three out of four 3-subwebs of a 4-web $[\xi_1,\xi_2,\xi_3,\xi_4]$, ξ_1, ξ_2, ξ_3, $\xi_4 = 1,\ldots,n+1$, implies the paratacticity of the 4th 3-subweb.*

Proof. In fact, the indicated equations prove Theorem 1.6.5 for the 4-web $[n+1,\alpha,\beta,\gamma]$. Similar calculations prove that Theorem 1.6.5 is valid also for the 4-webs $[\alpha,\beta,\gamma,\delta]$ defined by equations

$$\underset{n+1}{\omega}^i = 0, \quad \underset{\varepsilon}{\omega}^i = 0, \quad \varepsilon \neq \alpha,\beta,\gamma,\delta. \ ■ \tag{1.6.3}$$

1.7 $(n+1)$-Webs with Integrable Diagonal Distributions of 4-Subwebs

Let us consider a 4-subweb $[n+1,\alpha,\beta,\gamma]$ of an $(n+1)$-web $W(n+1,n,r)$. It has three $(2r)$-dimensional distributions defined by the systems

$$\underset{n+1}{\omega}^i + \underset{\varepsilon}{\omega}^i = 0, \quad \varepsilon = \alpha,\beta,\gamma. \tag{1.7.1}$$

Definition 1.7.1 The $(2r)$-dimensional distributions defined by the systems (1.7.1) are called the *diagonal distributions* of a 4-subweb $[n+1,\alpha,\beta,\gamma]$.

We will find conditions of integrability of each of these distributions. First note that $4r$ forms $\underset{n+1}{\omega}^i$, $\underset{\alpha}{\omega}^i$, $\underset{\beta}{\omega}^i$, $\underset{\gamma}{\omega}^i$ are connected on the 4-web $[n+1,\alpha,\beta,\gamma]$ by the r relations

$$\underset{n+1}{\omega}^i + \underset{\alpha}{\omega}^i + \underset{\beta}{\omega}^i + \underset{\gamma}{\omega}^i = 0. \tag{1.7.2}$$

So there are only $3r$ independent among them. It is convenient to introduce $3r$ forms $\tau_\alpha^i, \tau_\beta^i, \tau_\gamma^i$ defined by the following equations:

$$\begin{cases} \underset{n+1}{\omega}^i = \tau_\alpha^i + \tau_\beta^i + \tau_\gamma^i, & \underset{\beta}{\omega}^i = -\tau_\alpha^i + \tau_\beta^i - \tau_\gamma^i, \\ \underset{\alpha}{\omega}^i = \tau_\alpha^i - \tau_\beta^i - \tau_\gamma^i, & \underset{\gamma}{\omega}^i = -\tau_\alpha^i - \tau_\beta^i + \tau_\gamma^i. \end{cases} \tag{1.7.3}$$

It follows from (1.7.3) that

$$\left\{\begin{array}{l} 2\tau^i_\alpha = \underset{n+1}{\omega}{}^i + \underset{\alpha}{\omega}{}^i = -\underset{\beta}{\omega}{}^i - \underset{\gamma}{\omega}{}^i, \\ 2\tau^i_\beta = \underset{n+1}{\omega}{}^i + \underset{\beta}{\omega}{}^i = -\underset{\gamma}{\omega}{}^i - \underset{\alpha}{\omega}{}^i, \\ 2\tau^i_\gamma = \underset{n+1}{\omega}{}^i + \underset{\gamma}{\omega}{}^i = -\underset{\alpha}{\omega}{}^i - \underset{\beta}{\omega}{}^i. \end{array}\right. \tag{1.7.4}$$

The forms $\tau^i_\alpha, \tau^i_\beta$ and τ^i_γ are linearly independent on the 4-web $[n+1, \alpha, \beta, \gamma]$.

The following theorem establishes an analytical condition of integrability of the diagonal distribution defined by equations

$$2\tau^i_\alpha = \underset{n+1}{\omega}{}^i + \underset{\alpha}{\omega}{}^i = 0. \tag{1.7.5}$$

Theorem 1.7.2 *The diagonal distribution* (1.7.5) *is integrable if and only if the torsion tensor of a web* $W(n+1, n, r)$ *satisfies the following conditions:*

$$\underset{\alpha\beta}{a}{}^i_{jk} = \underset{\alpha\gamma}{a}{}^i_{jk}. \tag{1.7.6}$$

Proof. Exterior differentiation of (1.7.5) gives

$$\begin{aligned} d\tau^i_\alpha \;=\; & \tau^j_\alpha \wedge \tau^i_{\alpha j} + \frac{1}{2}(\underset{\alpha\beta}{a}{}^i_{[jk]} + \underset{\gamma\alpha}{a}{}^i_{[jk]})(\tau^j_\gamma \wedge \tau^k_\gamma - \tau^j_\beta \wedge \tau^k_\beta) \\ & + (\underset{\alpha\beta}{a}{}^i_{(jk)} - \underset{\gamma\alpha}{a}{}^i_{(jk)})\tau^j_\beta \wedge \tau^k_\gamma, \end{aligned} \tag{1.7.7}$$

where

$$\tau^i_{\alpha j} = \omega^i_j + \frac{1}{2}(-\underset{\alpha\beta}{a}{}^i_{[jk]} + \underset{\gamma\alpha}{a}{}^i_{[jk]})\tau^k_\alpha + (\underset{\alpha\beta}{a}{}^i_{[jk]} - \underset{\gamma\alpha}{a}{}^i_{(jk)})\tau^k_\beta - (\underset{\alpha\beta}{a}{}^i_{(kj)} + \underset{\gamma\alpha}{a}{}^i_{[jk]})\tau^k_\gamma,$$

It follows from (1.7.7) that the distribution $\tau^i_\alpha = 0$ is integrable if and only if

$$\underset{\alpha\beta}{a}{}^i_{(jk)} = \underset{\gamma\alpha}{a}{}^i_{(jk)}, \tag{1.7.8}$$

$$\underset{\alpha\beta}{a}{}^i_{[jk]} + \underset{\gamma\alpha}{a}{}^i_{[jk]} = 0. \tag{1.7.9}$$

By virtue of (1.2.17) equation (1.7.9) implies

$$\underset{\alpha\beta}{a}{}^i_{[jk]} = \underset{\alpha\gamma}{a}{}^i_{[jk]}. \tag{1.7.10}$$

Therefore equations (1.7.8) and (1.7.9) are equaivalent to equation (1.7.6). ■

The following two theorems give necessary and sufficient conditions of parallelisability of $W(n+1, n, r)$ in terms of the integrability of diagonal distributions of its 4-subwebs.

Theorem 1.7.3 *If on each of 4-subwebs $[n+1, \alpha, \beta, \gamma]$ of a web $W(n+1, n, r)$ all three diagonal distributions are integrable, then the web $W(n+1, n, r)$ is parallelisable.*

Proof. First note that conditions of integrability of the two other diagonal distributions $\tau^i_\beta = 0$ and $\tau^i_\gamma = 0$ are similar to (1.7.6) and have the form:

$$\underset{\beta\gamma}{a}{}^i_{jk} = \underset{\beta\alpha}{a}{}^i_{jk}. \tag{1.7.11}$$

$$\underset{\gamma\alpha}{a}{}^i_{jk} = \underset{\gamma\beta}{a}{}^i_{jk}. \tag{1.7.12}$$

Under assumption of the theorem, all three equations (1.7.6), (1.7.11), and (1.7.12) are valid for any α, β, and γ. These equations together with (1.2.17) imply

$$\underset{\alpha\beta}{a}{}^i_{[jk]} = \underset{\beta\gamma}{a}{}^i_{[jk]} = \underset{\gamma\alpha}{a}{}^i_{[jk]} = 0, \tag{1.7.13}$$

and together with (1.2.24) they imply

$$\underset{\alpha\beta}{a}{}^i_{(jk)} = \underset{\beta\gamma}{a}{}^i_{(jk)} = \underset{\gamma\alpha}{a}{}^i_{(jk)} = 0. \tag{1.7.14}$$

It follows from (1.7.13) and (1.7.14) that $\underset{\alpha\beta}{a}{}^i_{jk} = 0$, i.e. the web $W(n+1, n, r)$ is parallelisable. The converse is trivial. ■

Theorem 1.7.4 *If on each of the 4-subwebs $[n+1, \alpha, \beta, \gamma]$ of a web $W(n+1, n, r)$ two diagonal distributions are integrable and one of 3-subwebs of $[n+1, \alpha, \beta, \gamma]$ is paratactical, then the web $W(n+1, n, r)$ is parallelisable.*

Proof. Suppose that on a 4-subweb $[n+1, \alpha, \beta, \gamma]$ only two diagonal distributions, let say, $\tau^i_\alpha = 0$ and $\tau^i_\beta = 0$, are integrable. Then we have equations (1.7.6) and (1.7.11). It follows from (1.7.6), (1.7.11), and (1.2.24) that

$$\underset{\alpha\beta}{a}{}^i_{(jk)} = \underset{\beta\gamma}{a}{}^i_{(jk)} = \underset{\gamma\alpha}{a}{}^i_{(jk)} = 0, \tag{1.7.15}$$

and

$$\underset{\gamma\alpha}{a}{}^i_{[jk]} = \underset{\beta\gamma}{a}{}^i_{[jk]} = -\underset{\alpha\beta}{a}{}^i_{[jk]}. \tag{1.7.16}$$

If, in addition, one of the 3-subwebs of $[n+1, \alpha, \beta, \gamma]$ is paratactical, then equations (1.7.16) and (1.6.1) imply

$$\underset{\alpha\beta}{a}{}^i_{jk} = \underset{\beta\gamma}{a}{}^i_{jk} = \underset{\gamma\alpha}{a}{}^i_{jk} = 0, \tag{1.7.17}$$

i.e. the 4-web $[n+1, \alpha, \beta, \gamma]$ is parallelisable. Since equations (1.7.17) hold for any α, β and γ, by virtue of Theorem 1.5.2 our web $W(n+1, n, r)$ is parallelisable. The converse is trivial. ■

Corollary 1.7.5 *If two diagonal distributions of a 4-web $[n+1,\alpha,\beta,\gamma]$ are integrable and one of its 3-subwebs is paratactical, then the third diagonal distribution of the 4-web is integrable and the 4-web is parallelisable.*

Proof. This follows from the proof of Theorem 1.7.4 for fixed α,β, and γ. ■

Remark 1.7.6 If $r=1$, the torsion tensor of $W(n+1,n,1)$ is always zero, and Theorem 1.7.4 and Corollary 1.7.5 are trivial.

We will consider in conclusion a 4-subweb $[\delta,\alpha,\beta,\gamma]$ and its diagonal distributions defined by the equations

$$\underset{\lambda}{\omega}^i + \underset{\mu}{\omega}^i = 0, \quad \lambda,\mu = \delta,\alpha,\beta,\gamma, \quad \lambda \neq \mu. \tag{1.7.18}$$

If we introduce forms $\tau_\alpha^i, \tau_\beta^i$ and τ_γ^i using (1.7.3) where the index $n+1$ is replaced by δ , we obtain formulas similar to (1.7.7):

$$d\tau_\alpha^i = \tau_\alpha^j \wedge \sigma_{\alpha j}^i + \frac{1}{2}B^i_{[jk]}(\tau_\gamma^j \wedge \tau_\gamma^k - \tau_\beta^j \wedge \tau_\beta^k) + \frac{1}{2}B^i_{(jk)}\tau_\beta^j \wedge \tau_\gamma^k, \tag{1.7.19}$$

where $B^i_{jk} = \underset{\alpha\beta}{a}{}^i_{jk} - \underset{\alpha\gamma}{a}{}^i_{jk} + \underset{\delta\gamma}{a}{}^i_{jk} - \underset{\delta\beta}{a}{}^i_{jk}$ and $\sigma^i_{\alpha j}$ are 1-forms.

It follows from (1.7.19) that the diagonal distribution $\underset{\delta}{\omega}^i + \underset{\alpha}{\omega}^i = 0$ is integrable if and only if the torsion tensor of $W(n+1,n,r)$ satisfies the relations

$$\underset{\alpha\beta}{a}{}^i_{jk} - \underset{\alpha\gamma}{a}{}^i_{jk} + \underset{\delta\gamma}{a}{}^i_{jk} - \underset{\delta\beta}{a}{}^i_{jk} = 0.$$

If we write the conditions of integrability of the two other diagonal distributions of the 4-web $[\delta,\alpha,\beta,\gamma]$, then we can easily prove the statements similar to Theorems 1.7.3 and 1.7.4.

1.8 $(n+1)$-Webs with Integrable Diagonal Distributions

In this section we will generalise our considerations in Section 1.7. The equations

$$\underset{n+1}{\omega}^i + \sum_{s=1}^{h} \underset{\alpha_s}{\omega} = 0, \quad h < n, \tag{1.8.1}$$

or the equations

$$\sum_{t=h+1}^{n} \underset{\alpha_t}{\omega} = 0, \tag{1.8.2}$$

equivalent to (1.8.1) via (1.2.5), define on a whole web $W(n+1,n,r)$ an $(n-1)r$-dimensional distribution.

Definition 1.8.1 A distribution defined by equations (1.8.1) (or equations (1.8.2)) is called *diagonal.*

Equations (1.2.9) and (1.8.2) show that each of $(n-1)r$-planes of such a distribution is determined by the vectors $e_i^{\alpha_s}$ and $e_i^{\alpha_t} - e_i^{\delta_t}$, $s \neq t$, $\gamma_t \neq \delta_t$, δ_t is fixed. We will denote such a diagonal distribution by $\{\alpha_1, \ldots, \alpha_h\}$.

Theorem 1.8.2 *The distribution $\{\alpha_1, \ldots, \alpha_h\}$ is integrable if and only if the torsion tensor of $W(n+1, n, r)$ satisfies the conditions:*

$$\underset{\alpha_s \gamma_t}{a}{}^i_{jk} = \underset{\alpha_s \delta_t}{a}{}^i_{jk}, \quad s = 1, \ldots, h; \quad t = h+1, \ldots, n. \tag{1.8.3}$$

Proof. It follows from equations (1.2.16) and (1.8.2). ■

Theorem 1.8.3 *Integrability of the diagonal distribution $\{\alpha\}$ is necessary and sufficient for integrability of one of the diagonal distribution of each of $\binom{n-1}{2}$ the 4-subwebs $\{n+1, \alpha, \beta, \gamma\}$.*

Proof. To prove this result, one should compare equations (1.8.2) corresponding to the distribution $\{\alpha\}$ with equations (1.7.6). ■

The following theorem indicates the consequences implied by integrability of one or more diagonal distributions of type $\{\alpha\}$.

Theorem 1.8.4 (i) *If diagonal distributions $\{\alpha\}$ and $\{\beta\}$ are integrable and the three-web $[n+1, \alpha, \beta]$ is paratactical, then the diagonal distribution $\{\alpha_1, \ldots, \alpha_{n-2}\}$, $\alpha_s \neq \alpha, \beta$; $s = 1, \ldots, n-2$, is integrable and each of the 4-subwebs $[n+1, \alpha, \beta, \gamma]$ and $[\delta, \alpha, \beta, \gamma]$ is parallelisable.*

(ii) *If the diagonal distributions $\{\alpha\}$, $\{\beta\}$, and $\{\gamma\}$ are integrable, then the diagonal distribution $\{\alpha_1, \ldots, \alpha_{n-3}\}$, $\alpha_s \neq \alpha, \beta, \gamma$; $s = 1, \ldots, n-3$, is integrable and the 4-subweb $[n+1, \alpha, \beta, \gamma]$ is parallelisable.*

(iii) *If the diagonal distributions $\{\alpha_1, \ldots, \alpha_h\}$, $h < n$, are integrable, then so is the diagonal distribution $\{\alpha_1, \ldots, \alpha_{n-h}\}$, $\alpha_s \neq \alpha_1, \ldots, \alpha_h$; $s = 1, \ldots, n-h$.*

Proof. (i) We have

$$\underset{\alpha\beta}{a}{}^i_{jk} = \underset{\alpha\gamma}{a}{}^i_{jk}, \quad \underset{\beta\alpha}{a}{}^i_{jk} = \underset{\beta\gamma}{a}{}^i_{jk}, \quad \gamma \neq \alpha, \beta; \quad \alpha, \beta \text{ fixed.}$$

it follows from this that

$$\underset{\gamma\alpha}{a}{}^i_{jk} = \underset{\beta\alpha}{a}{}^i_{jk} = \underset{\beta\alpha}{a}{}^i_{kj} = \underset{\beta\gamma}{a}{}^i_{kj} = \underset{\gamma\beta}{a}{}^i_{jk},$$

We used here the fact that the 3-subweb $[n+1, \alpha, \beta]$ is paratactical which implies that $\underset{\beta\alpha}{a}{}^i_{jk} = \underset{\beta\alpha}{a}{}^i_{kj}$. Therefore, $\underset{\gamma\alpha}{a}{}^i_{jk} = \underset{\gamma\beta}{a}{}^i_{jk}$. From (1.8.3) this is equivalent to the integrability of the distribution $\{\alpha_1, \ldots, \alpha_{n-2}\}$; $\alpha_s \neq \alpha, \beta$; $s = 1, \ldots, n-2$. Each 4-subweb $[n+1, \alpha, \beta, \gamma]$ has three integrable diagonal distributions: $\underset{n+1}{\omega}{}^i + \underset{\alpha}{\omega}{}^i = 0$, $\underset{n+1}{\omega}{}^i + \underset{\beta}{\omega}{}^i = 0$, and $\underset{n+1}{\omega}{}^i + \underset{\gamma}{\omega}{}^i = 0$ (this follows from (1.7.6)), and each 4-subweb $[\delta, \alpha, \beta, \gamma]$ has two integrable diagonal distributions: $\underset{\delta}{\omega}{}^i + \underset{\alpha}{\omega}{}^i = 0$, and $\underset{\delta}{\omega}{}^i + \underset{\beta}{\omega}{}^i = 0$ (this follows from (1.7.20)). In addition, by Theorem 1.6.5, the 3-subweb $[\alpha, \beta, \gamma]$ is paratactical since it is easy to see from the identities obtained above that the 3-subwebs $[n+1, \alpha, \beta], [n+1, \beta, \gamma]$, and $[n+1, \gamma, \alpha]$ are paratactical ($\underset{\alpha\beta}{a}{}^i_{[jk]} = 0$, $\underset{\beta\gamma}{a}{}^i_{[jk]} = 0$, $\underset{\gamma\alpha}{a}{}^i_{[jk]} = 0$). To complete the proof, one should apply Theorems 1.7.3 and 1.7.4.

(ii) In this case we have

$$\underset{\alpha\beta}{a}{}^i_{jk} = \underset{\alpha\gamma}{a}{}^i_{jk} = \underset{\alpha\delta}{a}{}^i_{jk}, \quad \underset{\beta\alpha}{a}{}^i_{jk} = \underset{\beta\gamma}{a}{}^i_{jk} = \underset{\beta\delta}{a}{}^i_{jk}, \quad \underset{\gamma\alpha}{a}{}^i_{jk} = \underset{\gamma\beta}{a}{}^i_{jk} = \underset{\gamma\delta}{a}{}^i_{jk},$$

α, β and γ fixed. It follows from this that

$$\underset{\delta\alpha}{a}{}^i_{jk} = \underset{\beta\alpha}{a}{}^i_{jk} = \underset{\beta\delta}{a}{}^i_{jk}, \quad \underset{\delta\alpha}{a}{}^i_{jk} = \underset{\gamma\alpha}{a}{}^i_{jk} = \underset{\gamma\delta}{a}{}^i_{jk}.$$

Therefore, $\underset{\beta\delta}{a}{}^i_{jk} = \underset{\gamma\delta}{a}{}^i_{jk}$ or $\underset{\delta\beta}{a}{}^i_{jk} = \underset{\delta\gamma}{a}{}^i_{jk}$. The equations $\underset{\delta\alpha}{a}{}^i_{jk} = \underset{\gamma\alpha}{a}{}^i_{jk} = \underset{\gamma\delta}{a}{}^i_{jk} = 0$ by (1.8.3) lead to the integrability of the distribution $\{\alpha_1, \ldots, \alpha_{n-3}\}$; $\alpha_s \neq \alpha, \beta, \gamma$; $s = 1, \ldots, n-3$. According to Theorem 1.8.2, the 4-web $[n+1, \alpha, \beta, \gamma]$ has three integrable diagonal distributions, and by Theorem 1.7.3, it is parallelisable.

(iii) The proof is similar to the first part of the proof of (ii). ■

Theorem 1.8.5 *A web $W(n+1, n, r)$ is parallelisable if and only if its $n-1$ diagonal distributions $\{\hat{\alpha}\}, \hat{\alpha} \neq \alpha, \alpha$ is fixed, are integrable.*

Proof. The theorem follows from equivalence of conditions (1.5.11) and equations (1.8.3) written for all $\hat{\alpha} \neq \alpha$. ■

In conclusion of this section we generalise all the preceding considerations. The system

$$\underset{n+1}{\omega}{}^i + \sum_{\alpha_1=1}^{k_1} \underset{\alpha_1}{\omega}{}^i = 0, \quad \sum_{\alpha_2=k_1+1}^{k_1+k_2} \underset{\alpha_2}{\omega}{}^i = 0, \quad \ldots, \quad \sum_{\alpha_s=k_1+\ldots+k_{s-1}+1}^{k_1+\ldots+k_s} \underset{\alpha_s}{\omega}{}^i = 0, \tag{1.8.4}$$

where $k_1 + \ldots + k_s < n$, defines an $(n-s)r$-dimensional *diagonal distribution.* Using (1.2.16), we can easily prove the following theorem.

Theorem 1.8.6 *The distribution* (1.8.4) *is integrable if and only if the torsion tensor of the web* $W(n+1,n,r)$ *satisfies the following conditions:*

$$\underset{\alpha_1\beta_h}{a}{}^{i}_{jk} = \underset{\alpha_1\gamma_h}{a}{}^{i}_{jk}, \quad \underset{\beta_t\delta_{t+1}}{a}{}^{i}_{jk} - \underset{\beta_t\varepsilon_{t+1}}{a}{}^{i}_{jk} = \underset{\gamma_t\delta_{t+1}}{a}{}^{i}_{jk} - \underset{\delta_t\varepsilon_{t+1}}{a}{}^{i}_{jk};$$
$$h = 2,\ldots,s; \;\; t = 2,\ldots,s-1; \;\; \beta_t \neq \gamma_t, \;\; \beta_h \neq \gamma_h, \;\; \delta_{t+1} \neq \varepsilon_{t+1}.$$
$$\alpha_t, \beta_t, \ldots = k_1 + \ldots + k_{t-1} + 1, \ldots, k_1 + \ldots, +k_t, \;\; k_0 = 0. \quad \blacksquare$$

We can also consider the $(n-s)r$-dimensional diagonal distribution defined by the equations:

$$\underset{n+1}{\omega}{}^{i} + \sum_{\alpha_1=1}^{k_1} \underset{\alpha_1}{\omega}{}^{i} = 0, \quad \underset{n+1}{\omega}{}^{i} + \sum_{\alpha_2=k_1+1}^{k_1+k_2} \underset{\alpha_2}{\omega}{}^{i} = 0, \quad \ldots, \underset{n+1}{\omega}{}^{i} + \sum_{\alpha_s=k_1+\ldots+k_{s-1}+1}^{k_1+\ldots+k_s} \underset{\alpha_s}{\omega}{}^{i} = 0,$$

where $k_1 + \ldots + k_s < n$, and find under what conditions it is integrable.

As an example, let us consider a 4-web $[n+1, \alpha, \beta, \gamma]$ and its r-dimensional diagonal distribution defined by equations

$$\underset{n+1}{\omega}{}^{i} + \underset{\beta}{\omega}{}^{i} = 0, \quad \underset{n+1}{\omega}{}^{i} + \underset{\gamma}{\omega}{}^{i} = 0. \tag{1.8.5}$$

The system (1.8.5) is equivalent to the system

$$\underset{\beta}{\omega}{}^{i} = \underset{\gamma}{\omega}{}^{i} = -\underset{\alpha}{\omega}{}^{i}. \tag{1.8.6}$$

The conditions of integrability of the diagonal distribution (1.8.5) are

$$\underset{\alpha\gamma}{a}{}^{i}_{[jk]} = \underset{\beta\gamma}{a}{}^{i}_{[jk]} = \underset{\beta\alpha}{a}{}^{i}_{[jk]}. \tag{1.8.7}$$

If the $(2r)$-dimensional diagonal distributions $2\tau^i_\beta = \underset{n+1}{\omega}{}^{i} + \underset{\beta}{\omega}{}^{i} = 0$ and $2\tau^i_\gamma = \underset{n+1}{\omega}{}^{i} + \underset{\gamma}{\omega}{}^{i} = 0$ are integrable, then their intersection, an r-dimensional diagonal distribution, is also integrable: (1.7.11) and (1.7.12) imply (1.8.7).

Remark 1.8.7 Equations (1.3.7) show that all $(n-s)r$-dimensional diagonal distributions considered in Sections 1.7 and 1.8 are geodesically parallel in the affine connection γ_{n+1} and as we have noted before in any other affine connection γ_α induced by the web $W(n+1,n,r)$.

1.9 Transversally Geodesic $(n+1)$-Webs

As we already noted in Section 1.2, the leaves of the foliation λ_{n+1} intersect each of the surfaces $U_\alpha = \bigcap_{\hat{\alpha}\neq\alpha} F_{\hat{\alpha}}$ at a point and establish a point correspondence among U_α. The corresponding lines of U_α are defined by the equations (1.2.11):

$$\underset{\alpha}{\omega}{}^i = \xi^i dt, \;\; \underset{\hat{\alpha}}{\omega}{}^i = 0, \;\; \hat{\alpha} \neq \alpha, \;\; \alpha \text{ is fixed} \tag{1.9.1}$$

and the vectors defined by (1.2.12):

$$\xi^\alpha = \xi^i \underset{\alpha}{e}{}_i, \tag{1.9.2}$$

are tangent to these lines. Here $\{\underset{\alpha}{e}{}_i\}$ is the moving frame dual to the co-frame $\{\underset{\alpha}{\omega}{}^i\}$.

The equations (1.3.7) give conditions of geodesicity of each of the lines (1.9.1) in the affine connection γ_{n+1}:

$$\nabla\xi^i \stackrel{\text{def}}{=} d\xi^i + \xi^j \omega_j^i = \varphi\xi^i. \tag{1.9.3}$$

Proposition 1.9.1 *If one of the lines* (1.9.1) *is geodesic in* γ_{n+1}, *then so are the other* $n-1$ *lines.*

Proof. It follows from equations (1.3.7) and (1.9.3). ■

Let us consider an n-dimensional surface V^n passing through n corresponding lines (1.9.1), i.e. an envelope of the field of n-vectors $[\xi^1, \ldots, \xi^n]$. The equations of V^n are

$$\underset{\alpha}{\omega}{}^i = \xi^i \theta_\alpha, \tag{1.9.4}$$

and its tangent n-plane is defined by vectors ξ^α.

Thus, at each point $p \in D \subset X^{nr}$ we have a transversal n-vector attached and this n-vector is defined by the quantities ξ^i defined up to a factor. We can consider ξ^i as homogeneous coordinates of the n-vector $[\xi^1, \ldots, \xi^n]$. This gives us another foliation μ in X^{nr}. A typical leaf of μ is a projective space of dimension $r-1$. The forms $\underset{\alpha}{\omega}{}^i$ and $\nabla\xi^i$ are the structure forms of the foliation μ.

Exterior differentiation of equations (1.9.4) of V^n leads to

$$(\nabla\xi^i + \sum_{\beta\neq\alpha} \underset{\alpha\beta}{a}{}^i \theta_\beta) \wedge \theta_\alpha = -\xi^i d\theta_\alpha, \tag{1.9.5}$$

where

$$\underset{\alpha\beta}{a} = \underset{\alpha\beta}{a}{}^i_{jk} \xi^j \xi^k, \tag{1.9.6}$$

and

$$\underset{\alpha\beta}{a}{}^i = \underset{\beta\alpha}{a}{}^i, \;\; \sum_{\alpha,\beta} \underset{\alpha\beta}{a}{}^i = 0, \tag{1.9.7}$$

If we add up equations (1.9.5) written for all $\alpha = 1, \ldots, n$, and use conditions (1.9.7), we obtain

$$\nabla \xi^i \wedge (\sum_\alpha \theta_\alpha) = -\xi^i d(\sum_\alpha \theta_\alpha). \tag{1.9.8}$$

Equation (1.9.8) shows that

$$d(\sum_\alpha \theta_\alpha) = (\sum_\alpha \theta_\alpha) \wedge \theta. \tag{1.9.9}$$

Equations (1.9.8) and (1.9.9) imply

$$(\nabla \xi^i - \xi^i \theta) \wedge (\sum_\alpha \theta_\alpha) = 0. \tag{1.9.10}$$

By Cartan's lemma we obtain from (1.9.10) that

$$\nabla \xi^i = \xi^i \theta + \lambda^i \sum_\alpha \theta_\alpha. \tag{1.9.11}$$

Let us now write the equations (1.2.16) and (1.2.24) for the web $W(n+1, n, 1)$ defined on V^n :

$$d\theta_\alpha = \theta_\alpha \wedge \omega + \sum_{\beta \neq \alpha} \underset{\alpha\beta}{a}\, \theta_\alpha \wedge \theta_\beta, \tag{1.9.12}$$

$$\sum_{(\alpha,\beta)} \underset{\alpha\beta}{a} = 0. \tag{1.9.13}$$

From (1.9.12), (1.9.11), and (1.9.5) we obtain that $\omega = \theta$ and

$$\lambda^i + \underset{\alpha\beta}{a} = \xi^i \underset{\alpha\beta}{a}. \tag{1.9.14}$$

Summing up all the equations (1.9.14) in α and β and using (1.9.13) and (1.9.7), we get that

$$\lambda^i = 0. \tag{1.9.15}$$

By (1.9.15) and (1.9.6), equations (1.9.11) and (1.9.14) take the form:

$$\nabla \xi^i = \xi^i \theta, \tag{1.9.16}$$

$$\underset{\alpha\beta}{a}{}^i_{jk} \xi^j \xi^k = \underset{\alpha\beta}{a}\, \xi^i. \tag{1.9.17}$$

Theorem 1.9.2 *The n-dimensional surfaces V^n, intersecting r-surfaces U_α along lines corresponding to each other in the correspondence established among them by leaves of the foliation λ_{n+1}, are totally geodesic surfaces in γ_{n+1}. The leaves F_ξ cut out on V^n an $(n+1)$-web $W(n+1, n, 1)$ of codimension one, and $(n-1)$-dimensional leaves of the latter one are also totally geodesic in γ_{n+1}.*

Proof. Total geodesicity of V^n follows from the fact that any line on U^n can be given by the equations $\underset{\alpha}{\omega}{}^i = \underset{\alpha}{a}\xi^i\varphi$ and, by (1.9.16), is geodesic on X^{nr} in γ_{n+1}. Analogously, any line of the intersection of V^n and $F_{\hat\alpha}$ is given by the equations $\underset{\alpha}{\omega}{}^i = 0$, $\underset{\hat\alpha}{\omega}{}^i = \underset{\hat\alpha}{a}\xi^i\varphi$, $\hat\alpha \neq \alpha$, and again, by (1.9.16), is geodesic on X^{nr} in γ_{n+1}. ■

Definition 1.9.3 The surfaces V^n satisfying Theorem 1.9.2, are called *transversally geodesic surfaces* or Tg-surfaces of the web $W(n+1,n,r)$.

Definition 1.9.4 A vector ξ^i is said to be an *eigenvector* for a tensor $a^i_{j_1,\ldots,j_h}$ if

$$a^i_{j_1,\ldots,j_h}\xi^{j_1}\ldots\xi^{j_h} = a\xi^i, \quad i, j_1, \ldots, j_h = 1, \ldots, r. \tag{1.9.18}$$

Theorem 1.9.5 *The vectors ξ^α tangent to a Tg-surface V^n are eigenvectors for the torsion and curvature tensors of the web $W(n+1,n,r)$ and for their consequent Pfaffian derivatives.*

Proof. For the torsion tensor $\underset{\alpha\beta}{a}{}^i_{jk}$ this follows from comparing (1.9.17) and (1.9.18). To prove the theorem for the curvature tensor, let us write the prolongation of equations (1.9.12):

$$d\theta = \sum_{(\alpha,\beta)} \underset{\alpha\beta}{b}\,\theta_\alpha \wedge \theta_\beta, \quad \underset{\alpha\beta}{b} = \underset{\gamma\alpha\beta}{a} - \underset{\beta\gamma\alpha}{a}, \tag{1.9.19}$$

$$\begin{aligned} d\underset{\alpha\beta}{a} - \underset{\alpha\beta}{a}\,\theta &= (\underset{\alpha\beta\alpha}{a} + \underset{\alpha\beta}{a}{}^2)\theta_\alpha + (\underset{\alpha\beta\beta}{a} + \underset{\alpha\beta}{a}{}^2)\theta_\beta + \\ &\quad \sum_{\gamma\neq\alpha,\beta} [\underset{\alpha\beta\gamma}{a} + \underset{\alpha\beta}{a}(\underset{\beta\gamma}{a} + \underset{\gamma\alpha}{a})\theta_\gamma. \end{aligned} \tag{1.9.20}$$

Differentiating (1.9.17) by means of (1.9.16),(1.2.30), and (1.9.20) and using linear independence of 1-forms θ_γ, we obtain

$$\underset{\alpha\beta\gamma}{a}{}^i_{jkl}\xi^i\xi^k\xi^l = \underset{\alpha\beta\gamma}{a}\xi^i, \quad \alpha \neq \beta. \tag{1.9.21}$$

Equations (1.2.26) and (1.9.21) imply

$$\underset{\alpha\beta}{b}{}^i_{jkl}\xi^i\xi^k\xi^l = \underset{\alpha\beta}{b}\xi^i, \quad \alpha \neq \beta, \tag{1.9.22}$$

where $2\underset{\alpha\beta}{b} = \underset{\gamma\alpha\beta}{a} - \underset{\beta\gamma\alpha}{a}$. Comparison of (1.9.21) and (1.9.22) with (1.9.18) proves the theorem for $\underset{\alpha\beta\gamma}{a}{}^i_{jkl}$ and $\underset{\alpha\beta}{b}{}^i_{jkl}$. The proof for consecutive Pfaffian derivatives of $\underset{\alpha\beta}{b}{}^i_{jkl}$ is similar. ■

Definition 1.9.6 A web $W(n+1,n,r)$ is said to be *transversally geodesic* if any of its transversal n-vectors defines a Tg-surface of the web.

Theorem 1.9.7 *For a web* $W(n+1,n,r)$, $n>2$, *to be transversally geodesic, it is necessary and sufficient that the symmetric part of its torsion tensor has the following form:*

$$\underset{\alpha\beta}{a}{}^{i}_{(jk)} = \delta^{i}_{(j}\underset{\alpha\beta}{a}{}_{k)}, \tag{1.9.23}$$

where

$$\underset{\alpha\beta}{a}{}_{k} = \frac{2}{r+1}\underset{\alpha\beta}{a}{}^{i}_{(ik)}. \tag{1.9.24}$$

Proof. For such a web equation (1.9.17) has to be satisfied identically. Then, according to Theorem 1.9.5, the same will be true automatically for equation (1.9.22) etc. In equation (1.9.17) both members should be second degree polynomials in ξ^i. Therefore, $\underset{\alpha\beta}{a}$ is a linear form in ξ^i. Differentiation of (1.9.17) with respect to ξ^i gives

$$2\underset{\alpha\beta}{a}{}^{i}_{(jk)}\xi^k = \underset{\alpha\beta}{a}\delta^i_j + \frac{\partial \underset{\alpha\beta}{a}}{\partial \xi^j \xi^i}. \tag{1.9.25}$$

Applying contraction of the indices i and j in (1.9.25), we obtain

$$2\underset{\alpha\beta}{a}{}^{i}_{(ik)}\xi^k = r\underset{\alpha\beta}{a} + \underset{\alpha\beta}{a}.$$

It follows from this that

$$\underset{\alpha\beta}{a} = \underset{\alpha\beta}{a}{}_{k}\xi^k, \tag{1.9.26}$$

where $\underset{\alpha\beta}{a}{}_{k} = \frac{2}{r+1}\underset{\alpha\beta}{a}{}^{i}_{(ik)}$. Substituting (1.9.26) into (1.9.17) gives

$$(\underset{\alpha\beta}{a}{}^{i}_{jk} - \delta^i_j\underset{\alpha\beta}{a}{}_{k})\xi^j\xi^k = 0. \tag{1.9.27}$$

Since equations (1.9.27) hold for any ξ^i, they are satisfied if and only if condition (1.9.23) holds. ■

Corollary 1.9.8 *A web* $W(n+1,n,r)$ *is transversally geodesic if and only if the system*

$$\underset{\alpha}{\omega}{}^{i} = \xi^i\theta_\alpha, \quad \nabla\xi^i = \xi^i\theta, \tag{1.9.28}$$

is completely integrable.

Proof. The *necessity* follows from the fact that, as we saw early, for a transversally geodesic web $W(n+1,n,r)$ the system (1.9.28) for each ξ^i defines an n-dimensional Tg-surface V^n, and therefore, (1.9.28) is completely integrable.

To prove the *sufficiency*, first note that complete integrability of (1.9.28) implies the existence of the foliation μ. Each leaf of μ consists of n-dimensional planes whose points of tangency with the manifold X^{nr} form an n-dimensional surface V^n satisfying, according to (1.9.28), the system (1.9.4). The conditions $\nabla\xi^i = \xi^i\theta$ fix a transversal n-plane tangent to X^{nr} at a fixed point p. Thus, the indicated leaf consists of n-dimensional planes taken by one for each point of the n-surface V^n. Equations (1.9.4) show that a vector tangent to V^n has the form

$$\xi = \sum_{\alpha=1}^{n} \underset{\alpha}{\omega}^i e_i^\alpha = \theta_\alpha \xi^i e_i^\alpha = \theta_\alpha \xi^\alpha,$$

i.e., it belongs to the considered transversal n-plane. Therefore, the n-surface V^n is an envelope of the transversal n-planes considered. This means that the web $W(n+1,n,1)$ is transversally geodesic. ■

Since the equations (1.9.28) define an n-dimensional horizontal distribution Δ_n on the foliation μ, we can give another formulation of Corollary 1.9.8:

Corollary 1.9.9 *A web $W(n+1,n,r)$, $n > 2$, is transversally geodesic if and only if the distribution Δ_n defined on μ by the web is integrable. Moreover, the Tg-surfaces of the web $W(n+1,n,r)$ are projections of integral manifolds of the distribution Δ_n on the manifold μ.*

The forms θ_α are linearly independent on the distribution Δ_n defined by (1.9.28). The distribution Δ_n can also be defined on μ by vector fields η_α such that $\theta_\alpha(\eta_\beta) = \delta_{\alpha\beta}$. The projection of the vector η_α, taken at a point $(\underset{\alpha}{x}^i, \xi^i)$ of the manifold μ, onto the manifold μ is the vector ξ^α defined by (1.9.2). Here $\{e_i^\alpha\}$ is a frame of the tangent space of μ at the point with coordinates $\{\underset{\alpha}{x}^i\}$.

The following condition of transversal geodesicity of the 3-subweb $[n+1,\alpha,\beta]$ follows from Corollary 1.9.9.

Corollary 1.9.10 *The 3-subweb $[n+1,\alpha,\beta]$ is transversally geodesic if and only if the system*

$$\underset{\gamma}{\omega}^i = 0, \;\; \gamma = 1,\ldots,n, \;\; \gamma \neq \alpha,\beta, \;\; \underset{\alpha}{\omega}^i = \xi^i\theta_\alpha, \;\; \underset{\beta}{\omega}^i = \xi^i\theta_\beta, \;\; \nabla\xi^i = \xi^i\theta, \qquad (1.9.29)$$

where α and β are fixed, is completely integrable. ■

The system (1.9.29) defines on μ a two-dimensional distribution $\Delta_2^{\alpha\beta}$. This is generated by the vector fields η_α and η_β induced by the distribution Δ_n. We can formulate Corollary 1.9.10 in terms of the distribution Δ_n.

Corollary 1.9.11 *The 3-subweb $[n+1,\alpha,\beta]$ is transversally geodesic if and only if the distribution $\Delta_2^{\alpha\beta}$ is integrable.* ■

Suppose now that a k-dimensional distribution Δ_k, $2 \leq k < nr$, is given on X^{nr} by vector fields $a_1,\ldots,a_k$. Let us denote by Δ_2^{st} a 2-dimensional distribution generated by vector fields a_s and a_t, $s,t = 1,\ldots,k$.

Lemma 1.9.12 *If all distributions Δ_2^{st} are integrable, then so is the distribution Δ_k.*

Proof. According to Frobenius' theorem, the distribution Δ_2^{st} generated by vector fields a_s and a_t is integrable if and only if $[a_s, a_t] \in \Delta_2^{st}$, where $[a_s, a_t]$ is the Poisson bracket of the fields a_s and a_t. Since the vector fields $a_s, s = 1, \ldots, k$, generate the distribution Δ_k, we have $\Delta_2^{st} \subset \Delta_k$ where $s, t = 1, \ldots, k$. Therefore, $[a_s, a_t] \in \Delta_k$, $s, t = 1, \ldots, k$, and using Frobenius' theorem, we conclude that the distribution Δ_k is integrable. ■

Theorem 1.9.13 *A web $W(n+1, n, r)$, $n > 2$, is transversally geodesic if and only if all its 3-subwebs $[n+1, \alpha, \beta]$ are transversally geodesic.*

Proof. *Sufficiency.* Let all 3-subwebs of $W(n+1, n, r)$ be transversally geodesic. Then, by Corollary 1.9.11, on the manifold μ all distributions $\Delta_2^{\alpha\beta}$ generated by the vector fields η_α and η_β are integrable. Since the distribution Δ_n is generated by the vector fields η_α, $\alpha = 1, \ldots, n$, then, by Lemma 1.9.12, it is also integrable. Therefore the web $W(n+1, n, r)$ is transversally geodesic.

Necessity. Suppose now that a web $W(n+1, n, r)$ is transversally geodesic. Then we have (1.9.23) which, by (1.2.30), implies that

$$\underset{\alpha\beta\gamma}{a}{}^{i}_{(jkl)} = \delta^i_{(j} \underset{\alpha\beta\gamma}{a}{}_{kl)}. \tag{1.9.30}$$

The curvature tensor $\underset{\alpha\beta}{\bar{b}}{}^{i}_{jkl}$ of the 3-subweb $[n+1, \alpha, \beta]$ is expressed in terms of the torsion and curvature tensors of the web $W(n+1, n, r)$ by formulae (1.3.25). Since the web $W(n+1, n, r)$ is transversally geodesic, by (1.9.23), (1.9.30), and (1.2.26), we obtain from (1.3.25) that

$$\underset{n+1\alpha\beta}{b}{}^{i}_{(jkl)} = \delta^i_{(j} \underset{n+1,\alpha\beta}{b}{}_{kl)}, \tag{1.9.31}$$

where $\underset{n+1\alpha\beta}{b}{}_{kl} = \underset{\gamma\alpha\beta}{a}{}_{kl} - \underset{\gamma\beta\alpha}{a}{}_{kl} + \underset{\alpha\beta\alpha}{a}{}_{kl} - \underset{\beta\alpha\beta}{a}{}_{kl}$. The equations (1.9.31) are the conditions of transversal geodesicity of the 3-web $[n+1, \alpha, \beta]$ (see [Ak 69a]). ■

Note that using the same method, one can prove that 3-subwebs $[n+1, \alpha, \beta]$ are transversally geodesic provided that the web $W(n+1, n, r)$ is transversally geodesic.

Let us find some new characteristics of transversally geodesic webs in terms of affine connections which we considered in Section 1.3.

A 3-subweb $[n+1, \alpha, \beta]$ is cut out by leaves of the foliations λ_{n+1}, λ_α, and λ_β on the manifold $M_{\alpha\beta}$:

$$M_{\alpha\beta} = \bigcap_{\substack{\gamma=1 \\ (\gamma\neq\alpha,\beta)}}^{n} F_\gamma, \tag{1.9.32}$$

where $F_\gamma \subset \lambda_\gamma$. It is obvious that the manifold

$$M_\alpha = \bigcap_{\substack{\gamma=1 \\ (\gamma\neq\alpha)}}^{n} F_\gamma = M_{\alpha\beta} \bigcap F_\beta \tag{1.9.33}$$

is a foliation of the 3-subweb $[n+1, \alpha, \beta]$.

Proposition 1.9.14 ([AS 81]). *All affine connections of the bundle $\gamma_{\alpha\beta}(\Theta)$ have the same geodesic lines on leaves of a 3-web $W(3,2,r)$.*

Proof. For $W(3,2,r)$ we have equations (1.2.46) and (1.2.43). It follows from (1.3.54) that the connection forms of the affine connections of our bundle have the form:

$$\underset{\alpha\beta}{\Theta}{}^i_j = \underset{\alpha\beta}{\bar{\omega}}{}^i_j + \underset{\alpha\beta}{\bar{a}}{}^i_{jk}(p\underset{\alpha}{\omega}{}^k + q\underset{\beta}{\omega}{}^k). \tag{1.9.34}$$

The equations of the geodesic lines of affine connections of the bundle have the form:

$$d\omega^I + \omega^J \Theta^I_J = \Theta\omega^I, \quad I, J = 1, \ldots, nr. \tag{1.9.35}$$

and will be separated into two groups:

$$d\underset{1}{\omega}{}^i + \underset{1}{\omega}{}^j \Theta^i_j = \Theta\underset{1}{\omega}{}^i, \quad d\underset{2}{\omega}{}^i + \underset{2}{\omega}{}^j \Theta^i_j = \Theta\underset{2}{\omega}{}^i. \tag{1.9.36}$$

Equations (1.9.36) and (1.2.5) imply

$$d\underset{3}{\omega}{}^i + \underset{3}{\omega}{}^j \Theta^i_j = \Theta\underset{3}{\omega}{}^i. \tag{1.9.37}$$

Substituting (1.9.34) into equations (1.9.36) and (1.9.37) and using (1.2.43), we obtain

$$\begin{cases} d\underset{1}{\omega}{}^i + \underset{1}{\omega}{}^j\omega^i_j + qa^i_{jk}\underset{1}{\omega}{}^j\underset{2}{\omega}{}^k = \Theta\underset{1}{\omega}{}^i, \\ d\underset{2}{\omega}{}^i + \underset{2}{\omega}{}^j\omega^i_j + pa^i_{jk}\underset{2}{\omega}{}^j\underset{1}{\omega}{}^k = \Theta\underset{2}{\omega}{}^i, \\ d\underset{3}{\omega}{}^i + \underset{3}{\omega}{}^j\omega^i_j + (q-p)a^i_{jk}\underset{1}{\omega}{}^j\underset{2}{\omega}{}^k = \Theta\underset{3}{\omega}{}^i. \end{cases} \tag{1.9.38}$$

To find, for example, the equations of the geodesic lines on leaves of the first foliation of $W(3,2,r)$, we substitute $\underset{1}{\omega}{}^i = 0$ and $\underset{2}{\omega}{}^i = \xi^i dt$ into (1.9.38) and use (1.2.43). This gives

$$d\xi^i + \xi^i\omega^i_j = \Theta\xi^i. \tag{1.9.39}$$

The relations (1.9.39) do not depend on p and q. Therefore all connections of the bundle $\gamma_{\alpha\beta}(\Theta)$ have the same geodesic lines. ∎

The equations of geodesic lines of the manifold M_α in the connection $\bar{\gamma}_{\alpha\beta}$ belonging to the bundle $\gamma_{\alpha\beta}(\Theta)$ have the form:

$$\begin{cases} \underset{\gamma}{\omega}{}^i = 0, \quad \gamma = 1, \ldots, n, \quad \gamma \neq \alpha, \\ \underset{\alpha}{\omega}{}^i = \xi^i dt, \\ d\xi^i + \xi^j \underset{\alpha\beta}{\bar{\omega}}{}^i_j = \xi^i \underset{\alpha\beta}{\bar{\omega}}. \end{cases} \tag{1.9.40}$$

Substituting $\underset{\alpha\beta}{\omega}{}^i_j$ from (1.3.20) into (1.9.40) and using the fact that $\underset{\alpha\beta}{\omega}{}^i_j\Big|_{\underset{\gamma}{\omega}{}^i = 0} = \omega^i_j$, we get

$$\begin{cases} \underset{\gamma}{\omega}{}^i = 0, \quad \gamma = 1, \dots, k, \quad \gamma \neq \alpha, \\ \underset{\alpha}{\omega}{}^i = \xi^i dt, \\ d\xi^i + \xi^j \omega^i_j + \underset{\beta\alpha}{a}{}^i_{jk} \xi^j \xi^k dt = \xi^i \underset{\alpha\beta}{\bar{\omega}}. \end{cases} \tag{1.9.41}$$

Since the connection $\gamma_{\alpha\beta}$ is induced on $M_{\alpha\beta}$ by the connection γ_{n+1}, on $M_{\alpha\beta}$, and therefore also on M_α, the connections $\gamma_{\alpha\beta}$ and γ_{n+1} have the same geodesic lines. Therefore the equations of the geodesic lines of M_α in γ_{n+1} have the form:

$$\begin{cases} \underset{\gamma}{\omega}{}^i = 0, \quad \gamma = 1, \dots, n, \quad \gamma \neq \alpha, \\ \underset{\alpha}{\omega}{}^i = \xi^i dt, \\ d\xi^i + \xi^j \omega^i_j = \xi^i \omega. \end{cases} \tag{1.9.42}$$

Comparison of systems (1.9.41) and (1.9.42) shows that, in general, they are different. It means that in the general case the connection γ_{n+1} and connections of the bundle $\gamma_{\alpha\beta}(\Theta)$ have on the submanifolds M_α different geodesic lines.

The following theorem gives a criterion for these connections to have the same geodesic lines on M_α.

Theorem 1.9.15 *A web $W(n+1, n, r)$ is transversally geodesic if and only if the connection γ_{n+1} and all the connections of all the bundles $\gamma_{\alpha\beta}(\Theta)$, $\alpha, \beta = 1, \dots, n$, determined by the web have the same geodesic lines on the submanifolds M_α, $\alpha = 1, \dots, n$.*

Proof. The equations (1.9.41) and (1.9.42) coincide if and only if we have

$$\underset{\beta\alpha}{a}{}^i_{jk} \xi^j \xi^k = \xi^i \underset{\beta\alpha}{a}. \tag{1.9.43}$$

Equations (1.9.43) should be homogeneous with respect to ξ^i. Therefore

$$\underset{\beta\alpha}{a} = \underset{\beta\alpha}{a}{}_k \xi^k. \tag{1.9.44}$$

Substituting (1.9.44) into (1.9.43), we obtain

$$\underset{\beta\alpha}{a}{}^i_{jk} \xi^j \xi^k = \xi^i \underset{\beta\alpha}{a}{}_k \xi^k. \tag{1.9.45}$$

As we saw early, since (1.9.45) should be identities in ξ^i, it follows from this that (1.9.45) holds if and only if we have (1.9.23), i.e., the web $W(n+1, n, r)$ is transversally geodesic. ■

Theorem 1.9.16 *If a web $W(n+1, n, r)$ is transversally geodesic, then the geodesic parameters of the geodesic lines of a submanifold M_α in the connection γ_{n+1} and the connections of the bundles $\gamma_{\alpha\beta}(\Theta)$, $\alpha, \beta = 1, \dots, n$, coincide.*

Proof. An n-dimensional Tg-surface V^n determined by the vector $\xi^\alpha = \xi^i e_i^\alpha$ is given by the following equations:

$$\begin{cases} \underset{\beta}{\omega^i} = \xi^i \theta_\beta, \quad \beta = 1, \ldots, n, \\ d\xi^i + \xi^j \omega_j^i = \xi^i \omega. \end{cases} \tag{1.9.46}$$

The leaves of a transversally geodesic web $W(n+1, n, r)$ cut out on V^n a web $W(n+1, n, 1)$ with the structure equations (1.9.12) where

$$\underset{\gamma\delta}{a} = \underset{\gamma\delta}{a_i} \xi^i, \quad \gamma, \delta = 1, \ldots, n. \tag{1.9.47}$$

The equations of a geodesic line g_{n+1} of M_α in γ_{n+1} have the form:

$$\begin{cases} \underset{\gamma}{\omega^i} = 0, \quad \gamma = 1, \ldots, n, \quad \gamma \neq \alpha, \\ \underset{\alpha}{\omega^i} = \xi^i \theta_\alpha, \\ d\xi^i + \xi^j \omega_j^i = \xi^i \omega. \end{cases} \tag{1.9.48}$$

The equations of the two-dimensional Tg-surface V^2 determined by the same vector ξ^α have the form

$$\begin{cases} \underset{\gamma}{\omega^i} = 0, \quad \gamma = 1, \ldots, n, \quad \gamma \neq \alpha, \beta, \\ \underset{\alpha}{\omega^i} = \xi^i \theta_\alpha, \quad \underset{\beta}{\omega^i} = \xi^i \theta_\beta, \\ d\xi^i + \xi^j \underset{\alpha\beta}{\bar{\omega}_j^i} = \xi^i \underset{\alpha\beta}{\bar{\omega}}. \end{cases} \tag{1.9.49}$$

The leaves of a 3-subweb $[n+1, \alpha, \beta]$ cut out on V^2 a 3-web $W(3, 2, 1)$ with the following structure equations:

$$d\theta_\alpha = \theta_\alpha \wedge \underset{\alpha\beta}{\bar{\omega}}, \quad d\theta_\beta = \theta_\beta \wedge \underset{\alpha\beta}{\bar{\omega}} \tag{1.9.50}$$

(if $r = 1$, the torsion tensor vanishes).

Since, by Theorem 1.9.15, all connections of the bundle $\gamma_{\alpha\beta}(\theta)$ have the same geodesic lines, we can consider one of them. Let us take, for example, the connection $\bar{\gamma}_{\alpha\beta}$. The equations of a geodesic line $\bar{g}$ of M_α in $\bar{\gamma}_{\alpha\beta}$ have the form:

$$\begin{cases} \underset{\gamma}{\omega^i} = 0, \quad \gamma = 1, \ldots, n, \quad \gamma \neq \alpha, \\ \underset{\alpha}{\omega^i} = \xi^i \theta_\alpha, \\ d\xi^i + \xi^j \underset{\alpha\beta}{\bar{\omega}_j^i} = \xi^i \underset{\alpha\beta}{\bar{\omega}}. \end{cases} \tag{1.9.51}$$

Let us find relation between $\underset{\alpha\beta}{\omega}$ and $\underset{\alpha\beta}{\bar{\omega}}$ where

$$\underset{\alpha\beta}{\omega} = \omega\Big|_{\theta_\gamma = 0}, \quad \gamma = 1, \ldots, n, \quad \gamma \neq \alpha, \beta.$$

Equations (1.9.12) on V^2 have the form:

$$
\begin{aligned}
d\theta_\alpha &= \theta_\alpha \wedge \underset{\alpha\beta}{\omega} + \underset{\alpha\beta}{a}\,\theta_\alpha \wedge \theta_\beta,\\
d\theta_\beta &= \theta_\beta \wedge \underset{\alpha\beta}{\omega} + \underset{\beta\alpha}{a}\,\theta_\beta \wedge \theta_\alpha.
\end{aligned}
\tag{1.9.52}
$$

Comparison of (1.9.50) and (1.9.52) leads to

$$
\underset{\alpha\beta}{\bar{\omega}} = \underset{\alpha\beta}{\omega} + \underset{\beta\alpha}{a}\,\theta_\alpha + \underset{\alpha\beta}{a}\,\theta_\beta.
\tag{1.9.53}
$$

Substituting (1.9.53) and (1.3.20) into (1.9.51) and using the fact that $\underset{\alpha\beta}{\omega}{}^i_j\Big|_{\underset{\gamma}{\omega}{}^i = 0} = \omega^i_j$, $\underset{\alpha\beta}{\omega}\Big|_{\underset{\gamma}{\omega} = 0} = \omega$, we obtain

$$
\left\{
\begin{array}{l}
\underset{\gamma}{\omega}{}^i = 0, \quad \gamma = 1, \ldots, n, \quad \gamma \neq \alpha,\\
\underset{\alpha}{\omega}{}^i = \xi^i \theta_\alpha,\\
d\xi^i + \xi^j \omega^i_j + \underset{\beta\alpha}{a}{}^i_{jk} \xi^j \xi^k \theta_\alpha = \xi^i \omega + \xi^i \underset{\beta\alpha}{a}\,\theta_\alpha.
\end{array}
\right.
\tag{1.9.54}
$$

Since the web $W(n+1, n, r)$ is transversally geodesic, the last equation of (1.9.54) can be written in the form:

$$
d\xi^i + \xi^j \omega^i_j + \xi^i \underset{\beta\alpha}{a}{}_k \xi^k \theta_\alpha = \xi^i \omega + \xi^i \underset{\beta\alpha}{a}\,\theta_\alpha.
\tag{1.9.55}
$$

Using (1.9.47) and (1.9.55), we can write the equations (1.9.51) of a geodesic line of M_α in $\bar{\gamma}_{\alpha\beta}$ in the form:

$$
\left\{
\begin{array}{l}
\underset{\gamma}{\omega}{}^i = 0, \quad \gamma = 1, \ldots, n, \quad \gamma \neq \alpha,\\
\underset{\alpha}{\omega}{}^i = \xi^i \theta_\alpha,\\
d\xi^i + \xi^j \omega^i_j = \xi^i \omega.
\end{array}
\right.
\tag{1.9.56}
$$

Comparing equations (1.9.48) and (1.9.56), we see that they coincide. This means that geodesic parameters of the connection γ_{n+1} and a connection of the bundle $\gamma_{\alpha\beta}(\Theta)$ coincide on the geodesic lines of a submanifold M_α. ■

The following definition introduces a class of transversally geodesic $(n+1)$-webs, so-called $(2n+2)$-hedral webs.

Definition 1.9.17 A web $W(n+1, n, r)$, $n > 2$, with the vanishing symmetric part of its torsion tensor:

$$
\underset{\alpha\beta}{a}{}^i_{(jk)} = 0
\tag{1.9.57}
$$

is called $(2n+2)$-*hedral.*

We will present now a geometrical criterion for a web $W(n+1,n,r)$ to be $(2n+2)$-hedral.

Theorem 1.9.18 *A web $W(n+1,n,r)$, $n > 2$, is $(2n+2)$-hedral if and only if it is transversally geodesic and all webs $W(n+1,n,1)$ of codimension one cut out on n-dimensional Tg-surfaces V^n by leaves F_α are parallelisable.*

Proof. In fact, equations (1.9.12) show that all the webs $W(n+1,n,1)$ indicated in theorem are parallelisable if and only if

$$\underset{\alpha\beta}{a} = 0. \tag{1.9.58}$$

It follows from (1.9.58) and (1.9.26) that $\underset{\alpha\beta}{a}{}_k = 0$ and then (1.9.23) implies (1.9.57). Conversely, (1.9.57) shows that a web $W(n+1,n,1)$ is transversally geodesic and $\underset{\alpha\beta}{a}{}_k = 0$ i.e. $\underset{\alpha\beta}{a} = 0$ and all webs $W(n+1,n,1)$ indicated in theorem are parallelisable. ■

Corollary 1.9.19 *A web $W(n+1,n,r)$, $n > 2$, is parallelisable if and only if it is transversally geodesic, all webs $W(n+1,n,1)$ indicated in Theorem 1.9.18 are parallelisable, and $\binom{n}{2}$ 3-subwebs $[n+1,\alpha,\beta]$ are paratactical.*

Proof. This follows from Theorems 1.6.2 and 1.9.18. ■

Theorem 1.3.4 and Definition 1.9.17 give the following proposition.

Proposition 1.9.20 *A web $W(n+1,n,r)$, $n > 2$, is $(2n+2)$-hedral if and only if the affine connection γ_{n+1} induces connections $\tilde{\gamma}_{\alpha\beta}$ on all 3-subwebs $[n+1,\alpha,\beta], \alpha,\beta = 1,\dots,n$.* ■

1.10 Hexagonal $(n+1)$-Webs

Definition 1.10.1 A web $W(n+1,n,r)$ is said to be *hexagonal* if all $\binom{n+1}{3}$ its 3-subwebs are hexagonal.

We know (see the end of Section 1.2) that a necessary and sufficient condition for hexagonality of a multicodimensional 3-web is the vanishing of the symmetric part of its curvature tensor.

Theorem 1.10.2 *If a web $W(n+1,n,r)$ is hexagonal, then it is transversally geodesic.*

Proof. For $n = 2$ this follows from the fact that the analytical condition (1.9.31) of transversal geodesicity of a web $W(3,2,r)$ follows from the analytical condition of hexagonality of the web $W(3,2,r)$ (see Table 1.1 in Section 1.2).

Suppose that $n > 2$. According to Definition 1.10.1, all 3-subwebs $[n+1,\alpha,\beta]$ of a hexagonal web $W(n+1,n,r)$ are hexagonal. Therefore, as we mentioned above, they are transversally geodesic. Using Theorem 1.9.12 , we conclude that the whole web $W(n+1,n,r)$ is transversally geodesic. ■

Lemma 1.10.3 *If a web $W(n+1,n,r)$ is transversally geodesic, the symmetric parts of the curvature tensors of the 3-subwebs $[\alpha,\beta,\gamma]$, $[n+1,\alpha,\beta]$, $[n+1,\beta,\gamma]$, and $[n+1,\gamma,\alpha]$ are connected by the following relation:*

$$\underset{\alpha\beta\gamma}{\bar{b}}{}^{i}_{(jkl)} = \underset{\alpha\beta}{\bar{b}}{}^{i}_{(jkl)} + \underset{\beta\gamma}{\bar{b}}{}^{i}_{(jkl)} + \underset{\gamma\alpha}{\bar{b}}{}^{i}_{(jkl)}. \tag{1.10.1}$$

Proof. If $r = 1$, then $\underset{\alpha\beta}{a}{}^{1}_{11} = 0$ and $\underset{\alpha\beta\gamma}{a}{}^{1}_{111} = 0$. In addition, by virtue of (1.2.28), $\underset{\alpha\alpha}{\bar{b}}{}^{1}_{111} = 0$. Therefore, equation (1.3.73) becomes

$$\underset{\alpha\beta\gamma}{\bar{b}} = \underset{\alpha\beta}{\bar{b}} + \underset{\beta\gamma}{\bar{b}} + \underset{\gamma\alpha}{\bar{b}}$$

where

$$\underset{\alpha\beta}{\bar{b}} = \underset{\alpha\beta}{\bar{b}}{}^{1}_{111}$$

and

$$\underset{\alpha\beta\gamma}{\bar{b}} = \underset{\alpha\beta\gamma}{\bar{b}}{}^{1}_{111}$$

Suppose now that $r > 1$. Since the web $W(n+1,n,r)$ is transversally geodesic, condition (1.9.23) holds for its torsion tensor. Equation (1.3.73) by virtue of (1.9.23) leads to (1.10.1). ■

Theorem 1.10.4 (*Generalised Dubourdieu Theorem*) *If the 3-subwebs $[\xi,\eta,\zeta]$, $\xi,\eta,\zeta = 1,\ldots,n+1$, ξ is fixed, of a web $W(n+1,n,r)$ are hexagonal, then the web $W(n+1,n,r)$ is hexagonal.*

Proof. Suppose, for example, that $\xi = n+1$ is fixed. According to the conditions of the theorem, all 3-subwebs $[n+1,\alpha,\beta]$ are hexagonal. Therefore, according to Table 1.1 (see the end of Section 1.2), we have

$$\underset{\alpha\beta}{\bar{b}}{}^{i}_{(jkl)} = 0, \quad \alpha,\beta = 1,\ldots,n.$$

Using Lemma 1.10.3, we obtain that $\underset{\alpha\beta\gamma}{\bar{b}}{}^{i}_{(jkl)} = 0$ This means that the other $\binom{n}{3}$ 3-subwebs $[\alpha,\beta,\gamma]$ and the whole web $W(n+1,n,r)$ are hexagonal. ■

Definition 1.10.5 The tensor

$$\underset{\alpha\beta}{h}{}^{i}_{jkl} = 2\,\underset{\alpha\beta}{b}{}^{i}_{(jkl)} + \underset{\alpha\beta\alpha}{a}{}^{i}_{(jkl)} - \underset{\beta\alpha\beta}{a}{}^{i}_{(jkl)} - \underset{\alpha\beta}{a}{}^{m}_{(jk}\underset{\alpha\beta}{a}{}^{i}_{|m|l)} + \underset{\alpha\beta}{a}{}^{m}_{(jk}\underset{\beta\alpha}{a}{}^{i}_{|m|l)}, \tag{1.10.2}$$

$\alpha, \beta = 1, \ldots, n,\ \ \alpha \neq \beta$, is called the *hexagonality tensor* of a web $W(n+1, n, r)$.

Corollary 1.10.6 *A web $W(n+1, n, r)$ is hexagonal if and only if its hexagonality tensor $\underset{\alpha\beta}{h}{}^{i}_{jkl}$ vanishes:*

$$\underset{\alpha\beta}{h}{}^{i}_{jkl} = 0, \ \ \alpha, \beta = 1, \ldots, n, \ \ \alpha \neq \beta. \tag{1.10.3}$$

Proof. In fact, by virtue of (1.3.25), equation (1.10.3) is equivalent to

$$\underset{\alpha\beta}{\bar{b}}{}^{i}_{(jkl)} = 0, \ \ \alpha, \beta = 1, \ldots, n, \ \ \alpha \neq \beta. \tag{1.10.4}$$

This means that all 3-subwebs $[n+1, \alpha, \beta]$, $\alpha, \beta = 1, \ldots, n$, are hexagonal. Theorem 1.10.4 shows that in this case the web $W(n+1, n, r)$ is hexagonal. ■

1.11 Isoclinic $(n+1)$-Webs

Let us consider an r-dimensional distribution given by the system

$$\underset{\alpha}{\omega}{}^{i} = \xi_\alpha \theta^i, \ \ \alpha = 1, \ldots, n; \ \ i = 1, \ldots, r, \tag{1.11.1}$$

where the θ^i are r linearly independent 1-forms and ξ_α are functions defined in the domain $D \subset X^{nr}$. Since the parameters ξ_α are determined up to a multiplicative factor, we may choose them so that

$$\sum_\alpha \xi_\alpha = -1. \tag{1.11.2}$$

From the equations (1.2.5), (1.11.1), and (1.11.2) it follows that

$$\theta^i = \underset{n+1}{\omega}{}^{i}. \tag{1.11.3}$$

By virtue of (1.11.3) the equations (1.11.1) become

$$\underset{\alpha}{\omega}{}^{i} = \xi_\alpha \underset{n+1}{\omega}{}^{i}. \tag{1.11.4}$$

The equations (1.2.9) which hold for every point $p \in D \subset X^{nr}$ for points of our distribution become

$$\xi = \xi_\alpha \underset{n+1}{\omega}{}^{i} e_i^\alpha. \tag{1.11.5}$$

by virtue of (1.11.4). Thus, the vectors

$$\eta_i = \xi_\alpha e_i^\alpha \tag{1.11.6}$$

belong to the r-plane of the distribution passing through p.

Therefore, attached to each point $p \in D \subset X^{nr}$ we have an r-vector $[\eta_1, \ldots, \eta_r]$. We can consider ξ_α as coordinates of the r-vector $[\eta_1, \ldots, \eta_r]$. Equations (1.11.2) show that in general $n-1$ out of n quantities ξ_α are independent. This gives us one more foliation ν in X^{nr}. A typical leaf of ν is of dimension $n-1$.

Definition 1.11.1 An r-vector $[\eta_1, \ldots, \eta_r]$ is called an *isoclinic r-vector* of a web $W(n+1, n, r)$.

If the distribution given by equations (1.11.4) and (1.11.2) is integrable, then at each point$p \in D$ it determines an r-dimensional integral surface V^r. The vectors η_i are tangent to V^r at the point p.

Definition 1.11.2 An integral surface V^r defined by an integrable distribution (1.11.2) and (1.11.4) at each point p is said to be an *isoclinic surface* of a web $W(n+1, n, r)$.

Example 1.11.3 The surfaces of dimension r defined by equations

$$\underset{\alpha}{\omega}{}^i = -\underset{n+1}{\omega}{}^i, \quad \underset{\beta}{\omega}{}^i = 0, \quad \beta \neq \alpha, \quad \alpha \text{ is fixed},$$

are isoclinic. For them $\xi_\alpha = -1, \quad \xi_\beta = 0, \quad \beta \neq \alpha$. These surfaces envelop r-planes determined by the vectors e_i^α, α is fixed, and are the intersections of leaves $F_\beta, \beta \neq \alpha$.

Another example of isoclinic surfaces are the r-dimensional surfaces determined by the equations

$$\underset{\alpha}{\omega}{}^i = \xi_\alpha \theta^i, \quad \underset{\beta}{\omega}{}^i = \xi_\beta \theta^i, \quad \xi_\alpha + \xi_\beta = 0, \quad \underset{\gamma}{\omega}{}^i = 0, \quad \gamma \neq \alpha, \beta, \quad \underset{n+1}{\omega}{}^i = 0,$$

where α and β are fixed. These surfaces envelop the r-planes determined by the vectors $e_i^\alpha - e_i^\beta$ and are intersections of the leaves F_{n+1} and F_γ, $\gamma \neq \alpha, \beta$; α, β are fixed.

The following theorem gives conditions for the torsion tensor $\underset{\alpha\beta}{\omega}{}^i_{jk}$ of the web $W(n+1, n, r)$ to admit an $(n-1)$-parameter family of isoclinic surfaces V^r through each point $p \in D \subset X^{nr}$.

Theorem 1.11.4 *If an $(n-1)$-parameter family of isoclinic surfaces V^r passes through each point $p \in D \subset X^{nr}$, then the skew-symmetric part of the torsion tensor of the web $W(n+1, n, r)$ is of the form:*

$$\underset{\alpha\beta}{a}{}^i_{[jk]} = \underset{\alpha\beta}{b}{}_{[j}\delta^i_{k]} \tag{1.11.7}$$

where

$$\underset{\alpha\beta}{b}{}_j = \frac{2}{r-1}\underset{\alpha\beta}{a}{}^l_{[jl]}. \tag{1.11.8}$$

Proof. By differentiating equations (1.11.4) and taking into account (1.2.9) and (1.11.4), we obtain

$$\xi_\alpha \sum_{\beta\neq\alpha} \underset{\alpha\beta}{a}{}^i_{jk}\xi_\beta \underset{n+1}{\omega}{}^j \wedge \underset{n+1}{\omega}{}^k = d\xi_\alpha \wedge \underset{n+1}{\omega}{}^i. \tag{1.11.9}$$

On the isoclinic surface V^r we have

$$d\xi_\alpha = \xi_\alpha \underset{\alpha}{a}{}_j \underset{n+1}{\omega}{}^j, \tag{1.11.10}$$

and because of (1.11.2)

$$\sum_\alpha \xi_\alpha \underset{\alpha}{a}{}_j = 0. \tag{1.11.11}$$

From (1.11.9) and (1.11.10) we obtain

$$\sum_{\beta\neq\alpha} \underset{\alpha\beta}{a}{}^i_{[jk]}\xi_\beta = \underset{\alpha}{a}{}_{[j}\delta^i_{k]}. \tag{1.11.12}$$

It follows from the relations (1.11.12) that the quantities $\underset{\alpha}{a}{}_j$ must be 1-forms with respect to ξ_β, i.e. they have the form:

$$\underset{\alpha}{a}{}_j = \sum_\gamma \underset{\alpha\gamma}{b}{}_j \xi_\gamma. \tag{1.11.13}$$

If an $(n-1)$-parameter family of isoclinic surfaces passes through each point $p \in D \subset X^{nr}$, then among the quantities ξ_α there must be exactly $n-1$ which are independent, i.e. no other relation can hold apart from (1.11.2).

The quantities ξ_α being independent, by substituting (1.11.12) into (1.11.13) we get (1.11.7) and

$$\underset{\alpha\alpha}{b}{}_j = 0. \tag{1.11.14}$$

Note that (1.11.7) implies (1.11.8), by (1.11.14) equation (1.11.13) becomes

$$\underset{\alpha}{a}{}_j = \sum_{\gamma\neq\alpha} \underset{\alpha\gamma}{b}{}_j \xi_\gamma, \tag{1.11.15}$$

and (1.11.7) and (1.2.17) lead to

$$\underset{\alpha\beta}{b}{}_j = -\underset{\beta\alpha}{b}{}_j. \quad \blacksquare \tag{1.11.16}$$

Definition 1.11.5 A web $W(n+1,n,r)$ is called *isoclinic* if an $(n-1)$-parameter family of isoclinic surfaces passes through each point $p \in D \subset X^{nr}$.

Theorem 1.11.6 *For a web $W(n+1, n, r)$ to be isoclinic, it is necessary and, if $r \neq 2$, sufficient that the skew-symmetric part of its torsion tensor has the structure* (1.11.7).

Proof. Since the isoclinity of a web $W(n+1, n, r)$ is equivalent to the complete integrability of equations (1.11.4) and (1.11.10), we will find necessary and sufficient conditions for the latter.

First of all note that by Theorem 1.11.4 these conditions must include the relations (1.11.7). Substituting $\underset{\alpha}{a}_j$ from (1.11.15) into (1.11.10), we obtain

$$d\xi_\alpha = \xi_\alpha \sum_{\gamma \neq \alpha} \underset{\alpha\gamma}{b}_j \xi_\gamma \underset{n+1}{\omega}^j. \tag{1.11.17}$$

Exterior differentiation of (1.11.17) taking into account (1.11.17), (1.2.5), and (1.2.16) gives

$$\sum_{\gamma \neq \alpha} \xi_\gamma \Big(\nabla \underset{\alpha\gamma}{b}_j + \underset{\alpha\gamma}{b}_j \sum_{\beta \neq \gamma} \underset{\gamma\beta}{b}_k \xi_\beta \underset{n+1}{\omega}^k\Big) \wedge \underset{n+1}{\omega}^j = 0, \tag{1.11.18}$$

where

$$\nabla \underset{\alpha\gamma}{b}_j = d \underset{\alpha\gamma}{b}_j - \underset{\alpha\gamma}{b}_k \omega_j^k.$$

It follows from (1.11.7) that the quantities $\underset{\alpha\gamma}{b}_j$ define a tensor. Hence on the whole web we have

$$\nabla \underset{\alpha\gamma}{b}_j = \underset{\alpha\gamma\alpha}{c}_{jk} \underset{\alpha}{\omega}^k + \underset{\alpha\gamma\gamma}{c}_{jk} \underset{\gamma}{\omega}^k + \sum_{\beta \neq \alpha,\gamma} \underset{\alpha\gamma\beta}{c}_{jk} \underset{\beta}{\omega}^k. \tag{1.11.19}$$

In view of (1.11.4), on the isoclinic surface V^r equations (1.11.19) become

$$\nabla \underset{\alpha\gamma}{b}_j = \Big(\underset{\alpha\gamma\alpha}{c}_{jk} \xi_\alpha + \underset{\alpha\gamma\gamma}{c}_{jk} \xi_\gamma + \sum_{\beta \neq \alpha,\gamma} \underset{\alpha\gamma\beta}{c}_{jk} \xi_\beta\Big) \underset{n+1}{\omega}^k. \tag{1.11.20}$$

Substituting (1.11.20) into (1.11.18) and taking into account the independence of the forms $\underset{n+1}{\omega}^l$, we get

$$\sum_{\gamma \neq \alpha} \Big[\underset{\alpha\gamma\alpha}{c}_{[jk]} \xi_\gamma \xi_\alpha + \underset{\alpha\gamma\gamma}{c}_{[jk]} \xi_\gamma \xi_\gamma + \sum_{\beta \neq \alpha,\gamma} \big(\underset{\alpha\gamma\beta}{c}_{[jk]} + \underset{\alpha\gamma}{b}_{[j} \underset{\gamma\beta}{b}_{k]}\big) \xi_\gamma \xi_\beta\Big] = 0. \tag{1.11.21}$$

Since the functions ξ_γ are connected only by (1.11.2), it follows from (1.11.21) that

$$\underset{\alpha\gamma\alpha}{c}_{[jk]} = 0, \quad \underset{\alpha\gamma\gamma}{c}_{[jk]} = 0, \tag{1.11.22}$$

$$\underset{\alpha\gamma\beta}{c}_{[jk]} + \underset{\alpha\beta\gamma}{c}_{[jk]} + \underset{\alpha\gamma}{b}_{[j} \underset{\gamma\beta}{b}_{k]} + \underset{\alpha\beta}{b}_{[j} \underset{\beta\gamma}{b}_{k]} = 0. \tag{1.11.23}$$

Conditions (1.11.7), (1.11.22), and (1.11.23) are necessary and sufficient for the complete integrability of equations (1.11.4) and (1.11.10). Now let us show that, if

$r \neq 2$, conditions (1.11.7) are not only necessary (their necessity follows from Theorem 1.11.4) but also sufficient for a web $W(n+1, n, r)$ to be isoclinic. In other words, let us show that, if $r \neq 2$, conditions (1.11.7) and equations (1.2.16) imply conditions (1.11.22) and (1.11.23).

To prove this, we substitute conditions (1.11.7) into (1.2.16). This implies that

$$d\underset{\alpha}{\omega}^i = \underset{\alpha}{\omega}^j \wedge \omega^i_j + \sum_{\beta \neq \alpha} \underset{\alpha\beta}{a}{}^i_{(jk)} \underset{\alpha}{\omega}^j \wedge \underset{\beta}{\omega}^k + \frac{1}{2} \sum_{\beta \neq \alpha} \Big(\underset{\alpha\beta}{b}{}_j \underset{\alpha}{\omega}^j \wedge \underset{\beta}{\omega}^i - \underset{\alpha\beta}{b}{}_j \underset{\alpha}{\omega}^i \wedge \underset{\beta}{\omega}^j \Big). \qquad (1.11.24)$$

Since (1.11.24) is a special type of the system (1.2.16), we may assume that the prolongation of (1.11.24) has the form (1.2.29), (1.2.30), and (1.11.19).

Having exterior differentiated equations (1.11.24) let us substitute (1.2.29), (1.2.30), and (1.11.19) into the cubic equations thus obtained and equate the coefficients of $\underset{\alpha}{\omega}^j \wedge \underset{\alpha}{\omega}^k \wedge \underset{\beta}{\omega}^l$ and $\underset{\alpha}{\omega}^j \wedge \underset{\beta}{\omega}^k \wedge \underset{\beta}{\omega}^l$ to zero:

$$\begin{aligned}
- \quad & \underset{\alpha\beta\alpha}{c}{}_{[kj]}\delta^i_l + \underset{\alpha\beta\alpha}{c}{}_{l[j}\delta^i_{k]} - \frac{1}{2} \underset{\alpha\beta}{b}{}_l \underset{\alpha\beta}{b}{}_{[k}\delta^i_{j]} + \underset{\alpha\beta}{A}{}^i_{jkl} = 0, \\
& \underset{\alpha\beta\gamma}{c}{}_{jl}\delta^i_k - \underset{\alpha\gamma\beta}{c}{}_{jk}\delta^i_l - \underset{\alpha\beta\gamma}{c}{}_{kl}\delta^i_j + \underset{\alpha\gamma\beta}{c}{}_{lk}\delta^i_j \\
+ \quad & \tfrac{1}{2}\big(- \underset{\alpha\beta}{b}{}_j \underset{\beta\gamma}{b}{}_k \delta^i_l + \underset{\alpha\beta}{b}{}_i \underset{\beta\gamma}{b}{}_l \delta^i_k + \underset{\alpha\beta}{b}{}_l \underset{\beta\gamma}{b}{}_k \delta^i_j - \underset{\alpha\beta}{b}{}_k \underset{\beta\gamma}{b}{}_l \delta^i_j \\
+ \quad & \underset{\alpha\gamma}{b}{}_j \underset{\gamma\beta}{b}{}_l \delta^i_k - \underset{\alpha\gamma}{b}{}_j \underset{\gamma\beta}{b}{}_k \delta^i_l - \underset{\alpha\gamma}{b}{}_k \underset{\gamma\beta}{b}{}_l \delta^i_j + \underset{\alpha\gamma}{b}{}_l \underset{\gamma\beta}{b}{}_k \delta^i_j\big) + \underset{\alpha\beta\gamma}{B}{}^i_{jkl} = 0,
\end{aligned} \qquad (1.11.25)$$

where $\underset{\alpha\beta}{A}{}^i_{jkl}$ and $\underset{\alpha\beta\gamma}{B}{}^i_{jkl}$ are some quadratic forms of $\underset{\varepsilon\delta}{a}{}^i_{jk}$, $\underset{\varepsilon\delta\lambda}{a}{}^i_{jkl}$, $\underset{\varepsilon\delta}{b}{}_j$ and

$$\underset{\alpha\beta}{A}{}^i_{[jkl]} = 0, \quad \underset{\alpha\beta\gamma}{B}{}^i_{[jkl]} = 0. \qquad (1.11.26)$$

Having alternated relations (1.11.25) with respect to j, k, l by means of (1.11.26), let us sum on i and j equations so obtained. As a result we get

$$(r-2)\underset{\alpha\beta\alpha}{c}{}_{[kl]} = 0, \quad (r-2)\big(\underset{\alpha\beta\gamma}{c}{}_{[kl]} + \underset{\alpha\gamma\beta}{c}{}_{[kl]} + \underset{\alpha\beta}{b}{}_{[k} \underset{\beta\gamma}{b}{}_{l]} + \underset{\alpha\gamma}{b}{}_{[k} \underset{\gamma\beta}{b}{}_{l]}\big) = 0. \qquad (1.11.27)$$

If $r \neq 2$, then from (1.11.27) we obtain the first relation in (1.11.22) and relations (1.11.23). As for the second relation in (1.11.22), they follow from the first, since by (1.11.16) and (1.11.19) we have $\underset{\beta\alpha\alpha}{c}{}^i_{jk} = -\underset{\alpha\beta\alpha}{c}{}_{jk}$. ■

Note that for $r = 2$ conditions (1.11.7) always hold since any skew-symmetric tensor has the structure (1.11.7) in a binary domain. However, whether these conditions are sufficient for a web $W(n+1, n, 2)$ to be isoclinic remains an open question.

Corollary 1.11.7 *For a web $W(n+1, n, r)$ to be isoclinic it is necessary and, if $r \neq 2$ sufficient, that its $\binom{n}{2}$ 3-subwebs $[n+1, \alpha, \beta]$ be isoclinic.*

Proof. In fact, the torsion tensor of a 3-subweb $[n+1,\alpha,\beta]$ is the tensor $\underset{\alpha\beta}{a}{}^i_{[jk]}$. Akivis [Ak 74] (see also [AS 81]) showed that for this 3-subweb to be isoclinic it is necessary, and if $r \neq 2$ sufficient, that the skew-symmetric part of this tensor be of the structure (1.11.7). ■

Example 1.11.8 A web $W(n+1,n,r)$ for which all its 3-subwebs $[n+1,\alpha,\beta]$ are paratactical (see Section 1.6) is an example of an isoclinic $(n+1)$-web. Such webs are characterized by conditions

$$\underset{\alpha\beta}{a}{}^i_{[jk]} = 0. \tag{1.11.28}$$

From this, using (1.11.7), (1.11.13), and (1.11.10), we successively obtain

$$\underset{\alpha\beta}{b}{}_j = 0, \quad \underset{\alpha}{a}{}_j = 0, \quad d\xi_\alpha = 0,$$

i.e. the ξ_α are constant.

The equations (1.3.7) of the geodesic lines of X^{nr} in the connection γ_{n+1} show that the isoclinic surfaces V^r defined by (1.11.4) are totally geodesic in this connection. This property justifies the following definition.

Definition 1.11.9 A web $W(n+1,n,r)$ with vanishing skew-symmetric part of its torsion tensor is called *isoclinicly geodesic.*

In the case $n=2$, i.e. for 3-webs, an isoclinicly geodesic 3-web is just another name for a paratactical 3-web.

For isoclinicly geodesic webs $W(n+1,n,r)$ and only for them all 3-subwebs $[n+1,\alpha,\beta]$ are isoclinicly geodesic. It is obvious that for $r=1$ any web $W(n+1,n,1)$ is isoclinicly geodesic.

The next two theorems give criteria for a web $W(n+1,n,r)$ to be isoclinicly geodesic.

Theorem 1.11.10 *For a web $W(n+1.n.r)$ to be isoclinicly geodesic, it is necessary and sufficient that its $\binom{n}{2}$ 3-subwebs $[n+1,\alpha,\beta]$ be isoclinicly geodesic.*

Proof. Theorem follows from Theorem 1.6.2 where, according to Definition 1.11.9, we call paratactical three-webs isoclinicly geodesic. ■

Theorem 1.11.11 *For a web $W(n+1,n,r)$ to be isoclinicly geodesic, it is necessary and sufficient that it be isoclinic and its isoclinic surfaces V^r be totally geodesic.*

Proof. *Necessity* of this was proved before Definition 1.11.9. To prove its *sufficiency,* we note that since our web is isoclinic, we have (1.11.7). Substituting (1.11.4) into (1.3.7) and claiming that the equations so obtained are satisfied identically, we will get $d\xi_\alpha = 0$ which implies $\underset{\alpha\beta}{b}{}_j = 0$ by virtue of (1.11.17). Now (1.11.7) leads to (1.11.28). ■

The following theorem gives the structure of the torsion tensor of a web $W(n+1,n,r)$ which is both isoclinic and transversally geodesic.

Theorem 1.11.12 *A web $W(n+1,n,r)$ is both isoclinic and transversally geodesic if and only if its torsion tensor has the following structure:*

$$\underset{\alpha\beta}{a}{}^{i}_{jk} = \delta^i_k \underset{\alpha\beta}{\lambda}{}_j + \delta^i_j \underset{\beta\alpha}{\lambda}{}_k, \tag{1.11.29}$$

where

$$\underset{\alpha\beta}{\lambda}{}_i = \frac{1}{r^2-1}(r\,\underset{\alpha\beta}{a}{}^{l}_{jl} - \underset{\alpha\beta}{a}{}^{l}_{lj}). \tag{1.11.30}$$

Proof. In fact, for such a web we obtain from (1.11.7) and (1.9.23) that the torsion tensor has the structure (1.11.29) where

$$\underset{\alpha\beta}{\lambda}{}_j = \frac{1}{2}(\underset{\alpha\beta}{a}{}_j + \underset{\alpha\beta}{b}{}_j). \tag{1.11.31}$$

Substituting $\underset{\alpha\beta}{a}{}_j$ and $\underset{\alpha\beta}{b}{}_j$ from (1.9.24) and (1.11.8) into (1.11.31), we obtain (1.11.30).

The converse is evident: from (1.11.29) it is easy to deduce the relations (1.11.7) and (1.9.23) which mean that the web $W(n+1,n,r)$ is both isoclinic and transversally geodesic. ■

In conclusion of this section we give a sufficient condition for a web $W(n+1,n,r)$, $r>2$, to be isoclinic.

Let us consider an $(n-1)r$-dimensional (or r-codimensional) distribution Δ associated with a web $W(n+1,n,r)$, $r>2$, that is defined by 1-forms ω^i (vanishing on Δ) where

$$\omega^i = \sum_{\alpha=1}^{n} \rho_\alpha \underset{\alpha}{\omega}{}^i, \tag{1.11.32}$$

and that is in general position with respect to the foliations λ_ξ Geometrically for the distribution Δ to be in general position with respect to λ_ξ means that the subspace Δ_p of Δ passing through the point p does not contain non-empty intersections of tangent subspaces $T_p(F_\xi)$, $\xi = 1,\dots,n+1$, passing through the point p. Analytically this implies the following inequalities for the coefficients ρ_α in (1.11.32):

$$\rho_\alpha \neq 0, \;\; \rho_\alpha \neq \rho_\beta, \;\; \text{for } \alpha \neq \beta, \;\; \alpha,\beta = 1,\dots,n. \tag{1.11.33}$$

Proposition 1.11.13 *If an $(n-1)r$-dimensional distribution Δ associated with a web $W(n+1,n,r)$, $r>2$, and defined by the 1-forms of type* (1.11.32), (1.11.33) *is integrable, then the web $W(n+1,n,r)$ is isoclinic.*

Proof. By Frobenius theorem, the conditions of integrability of the distribution Δ have the form

$$d\omega^i = \omega^j \wedge \phi^i_j. \tag{1.11.34}$$

Applying exterior differentiation to (1.11.32) and using (1.2.16), we obtain:

$$d\omega^i = \omega^j \wedge \phi^i_j + \sum_\alpha d\rho_\alpha \wedge \underset{\alpha}{\omega}^i + \sum_{\alpha,\beta} \rho_\alpha \underset{\alpha\beta}{a}{}^i_{jk} \underset{\alpha}{\omega}^j \wedge \underset{\beta}{\omega}^k. \tag{1.11.35}$$

Since equations (1.11.35) must have the form (1.11.34),

$$d\rho_\alpha = \rho_\alpha \phi + \sum_\beta \underset{\alpha\beta}{\rho}{}_j \underset{\beta}{\omega}^j. \tag{1.11.36}$$

Substituting expansions (1.11.36) into (1.11.35), we get

$$d\omega^i = \omega^j \wedge (\omega^i_j - \delta^i_j \phi) + \sum_{\alpha,\beta} (\underset{\beta\alpha}{\rho}{}_j \delta^i_k + \rho_\alpha \underset{\alpha\beta}{a}{}^i_{jk}) \underset{\alpha}{\omega}^j \wedge \underset{\beta}{\omega}^k. \tag{1.11.37}$$

By virtue of (1.11.34), the second term in the right member of (1.11.37) must also be of the form $\omega^j \wedge \sigma^i_j$ where $\sigma^i_j = \sum_\beta \underset{\beta}{u}{}^i_{jk} \underset{\beta}{\omega}^k$. This and (1.11.32) gives the following expression for this second term:

$$\sum_{\alpha,\beta} (\underset{\beta\alpha}{\rho}{}_j \delta^i_k + \rho_\alpha \underset{\alpha\beta}{a}{}^i_{jk}) \underset{\alpha}{\omega}^j \wedge \underset{\beta}{\omega}^k = \sum_{\alpha,\beta} \rho_\alpha \underset{\beta}{u}{}^i_{jk} \underset{\alpha}{\omega}^j \wedge \underset{\beta}{\omega}^k. \tag{1.11.38}$$

Equating alternated coefficients in (1.11.38) and using (1.2.17) and (1.2.24), we obtain

$$(\rho_\alpha - \rho_\beta) \underset{\alpha\beta}{a}{}^i_{jk} = \underset{\alpha\beta}{\rho}{}_k \delta^i_j - \underset{\beta\alpha}{\rho}{}_j \delta^i_k + \rho_\alpha \underset{\beta}{u}{}^i_{jk} - \rho_\beta \underset{\alpha}{u}{}^i_{kj}. \tag{1.11.39}$$

For $\alpha = \beta$ it follows from (1.11.39) that

$$\underset{\alpha}{u}{}^i_{[jk]} = \frac{1}{\rho_\alpha} \underset{\alpha\alpha}{\rho}{}_{[j} \delta^i_{k]}. \tag{1.11.40}$$

Alternating (1.11.39) in j and k and using (1.11.40), we get

$$\underset{\alpha\beta}{a}{}^i_{[jk]} = \underset{\alpha\beta}{a}{}_{[j} \delta^i_{k]} \tag{1.11.41}$$

where

$$\underset{\alpha\beta}{a}{}_j = \frac{1}{\rho_\alpha - \rho_\beta} \Big(\frac{\rho_\beta}{\rho_\alpha} \underset{\alpha\alpha}{\rho}{}_j + \frac{\rho_\alpha}{\rho_\beta} \underset{\beta\beta}{\rho}{}_j - 2 \underset{(\alpha\beta)}{\rho}{}_j \Big). \tag{1.11.42}$$

Equations (1.11.42), (1.11.43) and Theorem 1.11.6 imply that the web $W(n+1, n, r)$, $r > 2$, is isoclinic. ■

NOTES

1.1. (1) The foliations λ_ξ, $\xi = 1, \dots, d$, are in *general position* if their tangent spaces at any point $p \in D \subset X^{nr}$ are in general position. The latter notion should be clarified. We will give here a clarification due to Chern and Griffiths [CG 78b] for the notion of general position of straight lines.

Let us agree to identify a point $u \in P^N = P(\mathbf{R}^{N+1})$ and its homogeneous coordinate vector $u \in \mathbf{R}^{N+1} - \{0\}$. A set of points $u_1, \ldots, u_d \in P^N$ is in general position if any $h \leq N+1$ of these points span a P^{h-1}, i.e., $u_{\xi_1} \wedge \ldots \wedge u_{\xi_h} \neq 0$ for $1 \leq \xi_1 < \ldots < \xi_h \leq d, \ h \leq N+1$.

Next let us clarify what does it mean "to be in general position" for straight lines. Let $G(1, 2n)$ be the Grassmannian of lines in $P^{2n-1} = P(\mathbf{R}^{2n})$. We will identify $G(1, 2n-1)$ with its image under the Plucker embedding

$$G(1, 2n-1) \longrightarrow P^{\binom{2n}{2}} - 1 = P(\wedge^2 \mathbf{R}^{2n})$$

given by sending the line spanned by points $u, v \in P^{2n-1}$ into $u \wedge v \in P(\wedge^2 R^{2n})$. For a set of lines $\Omega_1, \ldots, \Omega_h \in G(1, 2n-1)$ to be in general position first of all is necessary that any $h \leq n$ of them span P^{2n-1}, i.e.

$$\Omega_{\xi_1} \wedge \ldots \Omega_{\xi_h} \neq 0, \ 1 \leq \xi_1 < \ldots < \xi_{i_h} \leq d. \tag{i}$$

However, this condition may not be sufficient.

For example, suppose that we are given a set of four lines $\Omega_1, \ \Omega_2, \ \Omega_3, \ \Omega_4$ in P^3. The condition (i) is equivalent to those lines being pairwise skew.There is a unique non-singular quadric Q containing $\Omega_1, \ \Omega_2, \ \Omega_3$ as A-lines (the first family of generators of Q). For the line Ω_4 there are three possibilities: a) Ω_4 meets Q in distinct points u_1 and u_2; b) Ω_4 is tangent to Q at u; and c) Ω_4 is an A-line lying in Q. In the case a) each of the B-lines (the second family of generators of Q) through u_1 and u_2 meets all four $\Omega_1, \ \Omega_2, \ \Omega_3, \ \Omega_4$ once. The case b) is the limiting case of a) when $u_1 = u_2$. However, in the case c) there are infinitely many lines meeting the four pairwise skew lines Ω_ξ. Now we are able to give a definition of general position of straight lines. For this we assume first the condition (i). Given any $n-1$ points of the Ω_ξ, $\xi = 1, \ldots, d$, say $\Omega_1, \ldots, \Omega_{n-1}$ spanning a P^{2n-3} we consider any P^{2n-5} contained in this P^{2n-3} and the linear projection $\pi : P^{2n-1} - P^{2n-5} \rightarrow P^3$.The second requirement is:

$$\textit{The lines } \pi(\Omega_n), \ldots, \pi(\Omega_d) \textit{ do not have a common point.} \tag{ii}$$

A set of lines satisfying (i) and (ii) is said to be in *general position*. It is clear that it is not the case that lines in general position have as Plucker images points in general position in $P^{\binom{2n}{2}} - 1$.

(^{2}) The webs $W(3,2,1)$, $W(4,2,1)$, and $W(4,3,1)$ – all of them of codimension one – were studied by Blaschke, Bol, and their students and co-workers in the 1930s. Their results were summed up in the well-known monograph of Blaschke and Bol [BB 38].

The webs $W(n+1, n, 1)$ and $W(d, n, 1)$ – again of codimension one – were considered by Aue [Au 38], Bartoshevich [Bar 59], Bartsch [Ba 51a, 51b, 53], Chern [C 36a], Jeger [J 50], and Shulikovsky [Sh 66].

The study of multicodimensional webs $W(3,2,r)$, $r > 1$, was initiated by Bol [B 35] and Chern [C 36b] and was substantially developed by Akivis and his school starting in 1969 (see [Ak 69a, 69b, 73, 74, 83c], [AS 71a, 71b, 81], [AGe 86]). In [Ak 83c] one can find a review of the results and further references in this direction. The author (see [G 73, 74a, 74b, 75a–75f, 76] and [AG 74]) initiated the study of multicodimensional webs $W(n+1, n, r)$, $n > 2$, $r > 1$, following Akivis' lines. An essential contribution in the last direction has been recently made by Gerasimenko (see [Ge 84a, 84b, 85a, 85b] and [AGe 86]). Some papers were devoted to webs $W(d,n,r)$, $d > n+1$, $r > 1$ (see [Ak 80, 82, 83b, 83c], [C 82], [CG 78b], [G 77, 80, 82c, 82d, 83, 84, 85a, 85b, 86, 87], [W 82, 84]).

(^{3}) If a web $W(d,n,r)$ is given implicitly:

$$F^i_\tau(x_1^{j_1}, \dots, x_d^{j_d}) = 0, \quad \tau = n+1, \dots, d, \quad i, j_1, \dots, j_d = 1, \dots, r,$$

then the functions F^i_τ must satisfy the inequalities:

$$\det \left(\frac{\partial(F_{n+1}^{j_{n+1}}, \dots, F_d^{j_d})}{\partial(x_{\xi_1}^{j_1}, \dots, x_{\xi_{d-n}}^{j_{d-n}})} \right) \neq 0, \quad j_1, \dots, j_d = 1, \dots, r; \quad \xi_1, \dots, \xi_{d-n} = 1, \dots, d.$$

These inequalities are reduced to (1.1.4) in the case when a web $W(d,n,r)$ is given explicitly by the equations (1.1.3).

(^{4}) Example 1.1.9 generalises an example in [C 36b], p. 357.

1.2. This section is mostly based on the author's papers [G 73, 74a]. In [G 73, 74a] the notation of the curvature tensor of $W(n+1,n,r)$ is slightly different from the notation in the text. If we denote by $'\underset{\alpha\beta}{b}{}^i_{jkl}$ the curvature tensor used in [G 73, 74a], then

$$'\underset{\alpha\alpha}{b}{}^i_{jkl} = \underset{\alpha\alpha}{b}{}^i_{jkl}, \quad '\underset{\alpha\beta}{b}{}^i_{jkl} = 2\,\underset{\alpha\beta}{b}{}^i_{jkl}.$$

The relations (1.2.37)–(1.2.40) were found by Gerasimenko [Ge 84a].

(^{5}) A criterion for the quantities $\underset{\alpha\beta}{a}{}^i_{jk}$ to form a tensor field which we used in the text, we will often use in the sequel. We will describe here this criterion for the quantities $\{t^i_{jk}\}$ (or for some other quantities) to form a tensor field on a differentiable manifold X^M of dimension M (see [L 53]). First consider a vector field $a = a(p)$ given on a manifold X^M. The tangent space $T_p(X^M)$ at p is an M-dimensional vector space. Let $\{e_u, \ u = 1, \dots, M\}$ be a basis of $T_p(X^M)$. Then its infinitesimal displacements are described by equations $\delta e_u = \pi_u^v e_v, \ u, v = 1, \dots, M$, where $\delta = d\big|_{\omega^u = 0}$ and $\pi_u^v = \omega_u^v(\delta)$. The vector field $a = a^u e_u$ is invariant if and only if $\delta a = 0$, or since $\delta e_u = \pi_u^v e_v$, if $(\delta a^u + a^v \pi_v^u) e_u = 0$. Since vectors e_u are linearly independent, the conditions of invariance of a are

$$\delta a^u + a^v \pi_v^u = 0. \tag{iii}$$

The system (iii) is completely integrable. In fact, we obtain the identity if we take exterior derivatives of (iii) and substitute from (iii). Integrating (iii), we find the law of coordinate transformation of the vector $a : a^{u'} = \gamma_u^{u'} a^u$.

Using the same way, we can show that the conditions of invariance of a field of linear form $\alpha(p) = \alpha_u a^u$ (a co-vector field) have the form

$$\delta \alpha_u - \alpha_v \pi_u^v = 0. \tag{iv}$$

Consider further a tensor field $t = t(p)$ of the type $\binom{1}{2}$. Suppose that t^i_{jk} are components of the tensor t with respect to some basis in $T_p(X^M)$. Then its value on vectors a, b and a co-vector α is $t(a,b,\alpha) = t^w_{uv}a^u b^v \alpha_w$. Suppose that the point p is fixed. Since the form $t(a,b,\alpha)$ defined by the tensor t is invariant, we have $\delta t = 0$. Using (iii) and (iv), we obtain the latter condition in the form

$$(\delta t^w_{uv} - t^w_{xv}\pi^x_u - t^w_{ux}\pi^x_v + t^x_{uv}\pi^w_x)a^u b^v \alpha_w = 0.$$

It follows from this that

$$\nabla_\delta t^w_{uv} = \delta t^w_{uv} - t^w_{xv}\pi^x_u - t^w_{ux}\pi^x_v + t^x_{uv}\pi^w_x = 0. \tag{v}$$

The equations (v) are the differential equations of transformations of the components t^w_{uv} of the tensor t at the point p. The equations (v) are completely integrable and their integration leads to formulas for transformations of the components of the tensor t:

$$t^{w'}_{u'v'} = \gamma^{w'}_w \gamma^u_{u'} \gamma^v_{v'} t^w_{uv},$$

where $(\gamma^{w'}_w)$ is the transpose of the inverse matrix of $(\gamma^w_{w'})$.

If a point p is not fixed, then equations (v) have the form

$$\nabla t^w_{uv} = \delta t^w_{uv} - t^w_{xv}\omega^x_u - t^w_{ux}\omega^x_v + t^x_{uv}\omega^w_x = t^w_{uvx}\omega^x. \tag{vi}$$

The operator ∇ just designates the left member of equations (vi), and in general the quantities t^w_{uvx} are not components of a tensor field.

For a vector field $a = a^u e_u$ equations (vi) have the form

$$da^u - a^v\omega^u_v = a^u_v\omega^v,$$

and for a co-vector field $\alpha = \alpha_u a^u$ they have the form

$$d\alpha_u + \alpha_v\omega^v_u = \alpha_{uv}\omega^v.$$

1.3. Theorem 1.3.4 is due to the author [G 74a]. The last statement of this theorem has been recently proved by Gerasimenko [Ge 84a]. The affine connection $\tilde{\gamma}_{\alpha\beta}$ for 3-subwebs of $W(n+1,n,r)$ was studied by the author [G 74a]. The results on all three affine connections $\gamma_{\alpha\beta}, \bar{\gamma}_{\alpha\beta}, \tilde{\gamma}_{\alpha\beta}$ as well as on the canonical connection of the 3-subwebs $[\alpha,\beta,\gamma]$ are due to Gerasimenko [Ge 85b].

In formulas (1.2.45), (1.3.20), and (1.3.25) we incorporate Gerasimenko's corrections [Ge 85b].

1.4. The results of this section are from [G 75d].

1.5–1.8. All the results of these sections are due to the author [G 73, 74a].

1.9. The results of Proposition 1.9.1 and Theorems 1.9.2, 1.9.5, 1.9.7, 1.9.18 are due to the author [G 73, 74a]. Gerasimenko [Ge 85a, 85b] recently complemented these results. His results are given in Corollaries 1.9.8–1.9.11 and in Theorems 1.9.13, 1.9.15, 1.9.16.

1.10–1.11. All the results of these sections except Proposition 1.11.13 are due to the author [G 73, 74a, 74b]. A sufficient condition of isoclinity of $W(n+1,n,r)$ in Proposition 1.11.13 has been found by Akivis [Ak 81].

Chapter 2

Almost Grassmann Structures Associated with Webs $W(n+1,n,r)$

2.1 Almost Grassmann Structures on a Differentiable Manifold

2.1.1 The Segre Variety and the Segre Cone

Let us consider some notion which we need when constructing an almost Grassmann structure associated with a web $W(n+1,n,r)$. Let P^{r-1} and $P^{(n-1)*}$ be projective spaces of dimension $r-1$ and $n-1$ and let x and y denote elements of these respective spaces, so that (x,y) is an element of the direct product $P^{r-1} \times P^{(n-1)*}$. In the spaces P^{r-1} and $P^{(n-1)*}$ we introduce homogeneous coordinates in such a way that $x = \{x^i\}, y = \{y_\alpha\}$, where $i = 1,\ldots,r$ and $\alpha = 1,\ldots,n$. The space $P^{r-1} \times P^{(n-1)*}$ can be embedded into a space P^{rn-1} (i.e. projective space of dimension $rn-1$) as an algebraic variety. The precise embedding is $(x,y) \rightarrow Z = (z^i_\alpha), i = 1,\ldots,r$; $\alpha = 1,\ldots,n$ where

$$z^i_\alpha = x^i y_\alpha \qquad (2.1.1)$$

and the z^i_α are homogeneous coordinates of a point z in P^{rn-1}.

Definition 2.1.1 The variety determined by equations (2.1.1) is said to be the *Segre variety* and is denoted by $S(r-1,n-1)$ (see [HP 52]).

The dimension of the Segre variety is $r+n-2$. It carries two families of flat generators of dimension $r-1$ and $n-1$, and these two families depend correspondingly on $n-1$ and $r-1$ parameters. One of the generators of each family passes through each point of $S(r-1,n-1)$. The Segre variety remains invariant under the projective transformations of the space P^{rn-1} that are determined by the equations

$$'z^i_\alpha = a^i_j b^\beta_\alpha z^j_\beta, \qquad (2.1.2)$$

These transformations form the group that is locally isomorphic to the group $\mathbf{SL}(r) \times \mathbf{SL}(n)$. Since we need not distinguish the cases $\mathbf{SL}(r, \mathbf{R})$ or $\mathbf{SL}(r, \mathbf{C})$, we use $\mathbf{SL}(r)$ to denote either of these groups, both here and in the sequel (see Section 1.1).

In $P^{(n-1)*}$ we consider a hyperplane $P^{(n-2)*}$ defined by the equation $\sum_{\alpha=1}^{n} \rho_\alpha y_\alpha = 0$. On the Segre variety $S(r-1, n-1)$ this hyperplane generates a subvariety $S(r-1, n-2)$ which constitutes a representation of the direct product $P^{r-1} \times P^{(n-2)*}$. This subvariety is the intersection of the variety $S(r-1, n-1)$ and a subspace of codimension r determined in P^{rn-1} by the equations

$$\sum_{\alpha=1}^{n} \rho_\alpha z^i_\alpha = 0. \tag{2.1.3}$$

Such subspaces are in a special position with respect to the Segre variety $S(r-1, n-1)$. Any subspace of codimension r is defined in P^{rn-1} by a system of equations

$$\sum_{\alpha=1}^{n} a^i_{j\alpha} z^j_\alpha = 0. \tag{2.1.4}$$

It is easy to see that for the subspace (2.1.4) to be in the special position indicated above with respect to the variety $S(r-1, n-1)$ defined by equations (2.1.1), it is necessary and sufficient that $a^i_{j\alpha} = \rho_\alpha a^i_j$. By means of a transformation of coordinates in the space P^{r-1}, the coefficients in (2.1.4) can be reduced to the form $a^i_{j\alpha} = \rho_\alpha \delta^i_j$ and the equations (2.1.4) themselves to the form (2.1.3).

Now let T be a linear subspace of dimension rn. Its projectivisation PT is a projective space P of dimension $rn-1$.

Definition 2.1.2 The cone $C(r, n)$ in P whose projectivisation coincides with the Segre variety $S(r-1, n-1) : PC(r, n) = S(r-1, n-1)$, is called the *Segre cone*.

This cone carries two families of flat generators whose projectivisations coincide with the the flat generators of the Segre variety. The generators of the first family are of dimension r and depend on $n-1$ parameters. The generators of the second family are of dimension n and depend on $r-1$ parameters. All these generators pass through the vertex O of the cone $C(r, n)$. We will denote directing r- and n-vectors of these generators by ξ^r_0 and η^n_0, respectively. It is easy to see that two generators of the same family of $C(r, n)$ meet only in the point O, two generators of different families have a common straight line passing through O, and one and only one generator of each family passes through each of the straight lines belonging to $C(r, n)$.

In the space T we introduce coordinates z^i_α which under projectivisation become the homogeneous coordinates of the space P introduced above. In these coordinates the cone $C(r, n)$ is defined by the same equation (1.2.1) as that defining the Segre variety $S(r-1, n-1)$ in the space P. This cone is invariant under the transformations of T defined by equations (1.2.2). But since here the coordinates z^i_α are no longer homogeneous, these transformations form a group $G(r,n) = \mathbf{GL}(r) \times \mathbf{SL}(n)$.

Definition 2.1.3 The subspaces of codimension r in P^{rn-1} which are in special position with respect to the Segre variety $S(r-1, n-1)$ generate in the linear space T subspaces of codimension r intersecting the Segre cone $C(r,n)$ along the cone $C(r, n-1)$. Such subspaces are called *conconical* to the cone $C(r,n)$.

Lemma 2.1.4 *Let T be a linear space of dimension rn and let T_ξ, $\xi = 1, \ldots, n+1$, be subspaces of T of codimension r in general position, i.e., $T_{\xi_1} \cap \ldots \cap T_{\xi_n} = 0$, for all $\xi_1, \ldots, \xi_n$ where $\xi_i \neq \xi_j$ for any $i, j; i \neq j$. Then there exists a unique Segre cone $C(r,n)$ to which the subspaces T_ξ are conconical.*

Proof. Since the subspaces T_ξ are in general position, it is possible to choose coordinates u^i_α, $i = 1, \ldots, r$; $\alpha = 1, \ldots, n$, in T, in such a way that the equations of the subspace T_α can be written in the form

$$u^i_\alpha = 0, \quad \alpha \text{ fixed}, \tag{2.1.5}$$

and the equation of the subspace T_{n+1} in the form

$$\sum_{\alpha=1}^{n} \underset{\alpha}{a^i_j} u^j_\alpha = 0, \tag{2.1.6}$$

where $(\underset{\alpha}{a^i_j})$ are nonsingular matrices. Therefore in T one can introduce new coordinates

$$z^i_\alpha = \underset{\alpha}{a^i_j} u^j_\alpha, \tag{2.1.7}$$

and equations (2.1.5) and (2.1.6) of T_α and T_{n+1} in new coordinates (2.1.7) can be written, respectively, as follows:

$$z^i_\alpha = 0, \quad \sum_{\alpha=1}^{n} z^i_\alpha = 0. \tag{2.1.8}$$

After such a canonisation in the space T only concordant transformations of coordinates

$$'z^i_\alpha = a^i_j z^j_\alpha. \tag{2.1.9}$$

are admissible, and the equations (2.1.8) are invariant under transformations (1.2.9). These transformations form the group $G = \mathbf{GL}(r)$.

Now let us consider the Segre cone $C(r,n)$ that is defined in coordinates z^i_α by equations (2.1.1). Since each of the equations (2.1.8) is of form (2.1.3), the subspaces T_ξ defined by them are conconical to this cone.

The uniqueness of the Segre cone thus constructed follows from the fact that the system of coordinates z^i_α introduced in T by (2.1.7) is defined up to the transformations (2.1.9) which transfer the Segre cone (2.1.1) into itself. ■

Note that for $n = 2$ Lemma 2.1.4 may be proved purely geometrically. Indeed, let PT and PT_ξ, $\xi = 1, 2, 3$, be projectivisations of the corresponding linear spaces and $z \in PT_3$. The r-planes $[z, PT_1]$ and $[z, PT_2]$ meet along a straight line which, while moving the point z in PT_3, will generate the Segre variety $S(r-1, 1)$. In addition, the subspaces PT_ξ belong to the family of $(r-1)$-dimensional flat generators of this variety. The variety $S(r-1, 1)$ determines in T the Segre cone $C(r, 2)$ whose projectivisation is $S(r-1, 1)$ itself.

Note also that if $C(r, n)$ is the Segre cone defined by the subspaces T_ξ, $\xi = 1, \ldots, n+1$, then all the $(n-1)$-fold intersections $T_{\xi_1} \cap \ldots \cap T_{\xi_{n-1}}$ of these subspaces are r-dimensional flat generators of the cone $C(r, n)$.

2.1.2 Grassmann and Almost Grassmann Structures

Let us study now the structure of the Grassmannian $G(m-r, m)$ of planes of dimension $m - r$ in a projective space P^m. Its dimension is equal to $r(m-r+1)$: $\dim G(m-r, m) = r(m-r+1)$. It is an algebraic variety in a projective space P^N of dimension $N = \binom{m+1}{m-r+1}$ (see [HP 52]). From now on we suppose that $m = n + r - 1$. Then $\dim G(n-1, n+r-1) = nr$ and $N = \binom{n+r}{n}$.

Let P and Q be two $(n-1)$-planes in P^{r+n-1} meeting in the $(n-2)$-plane P^{n-2}. They generate a linear pencil of $(n-1)$-planes $\lambda P + \mu Q$. A rectilinear generator of the Grassmannian $G(n-1, n+r-1)$ corresponds to this pencil. All the $(n-1)$-planes of this pencil belong to an n-plane P^n, and the pair of planes $P^{n-2} \subset P^n$ completely determines this pencil and consequently a straight line on $G(n-1, n+r-1)$.

We consider the r-bundle of $(n-1)$-planes P passing through a fixed $(n-2)$-plane P^{n-2}. To this bundle on the Grassmannian $G(n-1, n+r-1)$ there corresponds an r-dimensional flat generator ξ^n. Since the projective space P^{r+n-1} contains an $(r+1)(n-1)$-family of planes P^{n-2}, the Grassmannian $G(n-1, n+r-1)$ carries a family of r-dimensional generators ξ^r which depends on $(r+1)(n-1)$ parameters.

Let P^n be a fixed n-plane in P^{r+n-1}. We consider all $(n-1)$-planes P belonging to P^n. They form a field of planes of dimension n. On the manifold $G(n-1, n+r-1)$ there is an n-dimensional flat generator ξ^n corresponding to this field. Since $\dim G(n, n+r-1) = (r-1)(n+1)$, the manifold $G(n-1, n-r+1)$ carries a family of flat generators ξ^n which depends on $(r-1)(n+1)$ parameters.

If the planes P^{n-2} and P^n of the space P^{r+n-1} are incident (i.e., they satisfy the condition $P^{n-2} \subset P^n$), the corresponding flat generators ξ^r and ξ^n of $G(n-1, n+r-1)$ meet along a straight line. If they are not incident, then ξ^r and ξ^n have no common points.

Let us consider a fixed plane $P^{n-1} = P$ in P^{r+n-1}. It contains an $(n-1)$-

-parameter family of planes P^{n-2}. Therefore the family of generators ξ^r depending on $n-1$ parameters passes through the point $p \in G(n-1, n+r-1)$ corresponding to it. In addition, an $(r-1)$-parameter family of planes P^n passes through the same plane P in P^{r+n-1}. Consequently, an $(r-1)$-parameter family of generators ξ^n passes through the point p of $G(n-1, n+r-1)$. Furthermore, any two generators ξ^r and ξ^n passing through p meet along a straight line. It follows from this that all the flat generators ξ^r and ξ^n passing through $p \in G(n-1, n+r-1)$ are flat generators of the Segre cone $C_p(r,n)$ with vertex in p, and this cone lies on $G(n-1, n+r-1)$. This Segre cone is the intersection of $G(n-1, n+r-1)$ and its tangent space $T_p(G)$, whose dimension is the same as that of $G(n-1, n+r-1)$, namely nr.

In the space P^{r+n-1} the set of all $(n-1)$-dimensional planes intersecting a fixed plane $P = P^{n-1}$ in $(n-2)$-planes corresponds to the cone $C_p(r,n)$. Each of these $(n-1)$-planes together with the plane P belong to an n-plane. The dimension of the cone $C_p(r,n)$ is equal to $n+r-1$.

Now we will define an almost Grassmann structure on a differentiable manifold X^{nr}. In the tangent space $T_p(X^{nr})$ of any point $p \in X^{nr}$ we consider the Segre cone $C_p(r,n)$ with the vertex at p. We will assume that the field of the Segre cones on X^{nr} is differentiable.

Definition 2.1.5 The differential geometric structure on X^{nr} defined by the field of Segre cones is called an *almost Grassmann structure* and denoted by $AG(n-1, n+r-1)$. The manifold X^{nr} endowed with an almost Grassmann structure is said to be an *almost Grassmann manifold.*

Its structural group is a subgroup of the general linear group $\mathbf{GL}(rn)$ of transformations of the space $T_p(X^{nr})$, and the cone $C_p(r,n)$ is invariant under transformations of this subgroup. As shown above, this subgroup is isomorphic to the product $\mathbf{GL}(r) \times \mathbf{SL}(n)$. Following Hangan [Ha 66], we will denote it by $GL(r,n)$. Hangan's definition of an almost Grassmann structure is based on the structural group $\mathbf{GL}(r,n)$. In contrast , following Akivis [Ak 80, 82], we based our Definition 2.1.5 on the field of Segre cones $C_p(r,n)$ given on the manifold X^{nr}.

The Segre cone $C_p(r,n)$ attached to a point $p \in X^{nr}$ determines in $T_p(X^{nr})$ an $(n-1)$-parameter family of r-vectors ξ_p^r and an $(r-1)$-parameter family of n-vectors ξ_p^n, directed along the flat generators of the cone. On X^{nr} the r-vectors ξ_p^r form an $(rn+n-1)$-parameter family , and the n-vectors ξ_p^n form an $(rn+r-1)$-parameter family.

Definition 2.1.6 An almost Grassmann structure $AG(n-1, r+n-1)$ is called *r-semiintegrable* if on X^{nr} there is an $(r+1)(n-1)$-parameter family of r-dimensional subvarieties V^r such that $T_p(V^r) = \xi_p^r$ and each r-vector ξ_p^r is tangent to one and only one subvariety V^r. A structure $AG(n-1, r+n-1)$ is called *n-semiintegrable*, if on X^{nr} there is an $(r-1)(n+1)$-parameter family of n-dimensional subvarieties U^n such that $T_p(U^n) = \xi_p^n$ and each n-vector ξ_p^n is tangent to one and only one subvariety U^n.

Definition 2.1.7 An almost Grassmann structure which is both r- and n-semi integrable is called *integrable.*

Theorem 2.1.8 *An integrable almost Grassmann structure is locally Grassmann, i.e., the neighborhood of every point of X^{nr} admits a differentiable mapping into the Grassmannian $G(n-1, r+n-1)$, and the flat generators ξ^r and ξ^n of the Grassmannian $G(n-1, n+r-1)$ will correspond respectively to subvarieties V^r and U^n of the manifold X^{nr}.*

Proof. See [Mi 78]. ■

2.2 Structure Equations and Torsion Tensor of an Almost Grassmann Manifold

Suppose that a differentiable manifold X^{nr} is an almost Grassmann manifold $AG(n-1, r+n-1)$. According to Definition 2.1.5, this means that a G-structure is defined on X^{nr} whose group G is the direct product of $\mathbf{GL}(r)$ and $\mathbf{SL}(n)$: $G = \mathbf{GL}(r) \times \mathbf{SL}(n)$. Initially we will suppose that $G = \mathbf{GL}(r) \times \mathbf{GL}(n)$. Later on we will discuss how to reduce $\mathbf{GL}(n)$ to $\mathbf{SL}(n)$ and introduce a geometrical interpretation of this reduction.

Let $T_p(X^{nr})$ be the tangent space of X^{nr} at a point p and $\{p, e_i^\alpha\}$, $i = 1, \ldots, r;\ \ \alpha = 1, \ldots, n$, be a moving frame in $T_p(X^{nr})$.

It follows from the definition of an almost Grassmann manifold that transformations of G transform the basis vectors e_i^α as follows:

$$'e_i^\alpha = A_\beta^\alpha A_i^j e_j^\beta, \ \ i, j = 1, \ldots, r; \ \ \alpha, \beta = 1, \ldots, n, \tag{2.2.1}$$

where (A_β^α) and (A_j^i) are nonsingular square matrices of orders n and r, respectively. For each such a transformation the matrices (A_β^α) and (A_i^j) are not defined uniquely since they admit a multiplication by reciprocal scalars. However they can be made unique by restricting to unimodular matrices (A_β^α) and (A_i^j): det $(A_\beta^\alpha) = 1$ or det $(A_i^j) = 1$. Another possibility is to connect matrices (A_β^α) and (A_i^j) by a symmetric condition: det $(A_i^j) = |$ det (A_β^α) $|$. All these possibilities imply that $G = \mathbf{GL}(r) \times \mathbf{SL}(n)$. We will discuss the last possibility and its necessity at the appropriate moment.

Solving (2.2.1), we obtain

$$e_i^\alpha = \tilde{A}_\beta^\alpha \tilde{A}_i^j \, 'e_j^\beta, \tag{2.2.2}$$

where $(\tilde{A}_\beta^\alpha)$ and $(\tilde{A}_j^i)$ are the matrices inverse to (A_β^α) and (A_j^i),i.e.,

$$A_\gamma^\alpha \tilde{A}_\beta^\gamma = A_\beta^\gamma \tilde{A}_\gamma^\alpha = \delta_\beta^\alpha, \ \ A_k^i \tilde{A}_j^k = A_j^k \tilde{A}_k^i = \delta_j^i. \tag{2.2.3}$$

It follows from (2.2.3) that

$$A^{\gamma}_{\beta}\delta\tilde{A}^{\alpha}_{\gamma} = -\tilde{A}^{\alpha}_{\gamma}\delta A^{\gamma}_{\beta}, \quad A^{i}_{k}\delta\tilde{A}^{k}_{j} = -\tilde{A}^{k}_{j}\delta A^{i}_{k}, \tag{2.2.4}$$

where δ denotes differentiation for a fixed $p \in X^{nr}$ (see Section 1.2).

We shall assume that $\{p, {}'e^{\alpha}_{i}\}$ is a fixed frame, i.e., $d({}'e^{\alpha}_{i}) = 0$. Then, differentiating (2.2.2) for a fixed $p \in X^{nr}$ and using (2.2.1) and (2.2.4), we obtain

$$\delta e^{\alpha}_{i} = (\delta^{\alpha}_{\beta}\lambda^{j}_{i} - \delta^{j}_{i}\lambda^{\alpha}_{\beta})e^{\beta}_{j}, \tag{2.2.5}$$

where

$$\lambda^{j}_{i} = A^{j}_{l}\delta\tilde{A}^{l}_{i}, \quad \lambda^{\alpha}_{\beta} = \tilde{A}^{\alpha}_{\gamma}\delta A^{\gamma}_{\beta}. \tag{2.2.6}$$

Since, in general, on X^{nr} we have

$$\delta e^{\alpha}_{i} = \lambda^{j\alpha}_{\beta i} e^{\beta}_{j}, \tag{2.2.7}$$

it follows from (2.2.5) and (2.2.7) that

$$\lambda^{j\alpha}_{\beta i} = \delta^{\alpha}_{\beta}\lambda^{j}_{i} - \delta^{j}_{i}\lambda^{\alpha}_{\beta}. \tag{2.2.8}$$

If $p \in X^{nr}$ is assumed to be variable, then equations (2.2.8) have the form

$$\omega^{j\alpha}_{\beta i} = \delta^{\alpha}_{\beta}\omega^{j}_{i} - \delta^{j}_{i}\omega^{\alpha}_{\beta} + \bar{u}^{j\alpha\gamma}_{\beta ik}\omega^{k}_{\gamma}, \tag{2.2.9}$$

where $\bar{u}^{j\alpha\gamma}_{\beta ik}$ are certain functions defined on X^{nr} and ω^{k}_{γ} are basis forms of this manifold.

In the general case the first group of structure equations of X^{nr} has the form:

$$d\omega^{i}_{\alpha} = \omega^{j}_{\beta} \wedge \omega^{i\beta}_{\alpha j}. \tag{2.2.10}$$

Substituting the values of $\omega^{j\alpha}_{\beta i}$ from (2.2.9) into (2.2.10), we obtain

$$d\omega^{i}_{\alpha} = \omega^{j}_{\alpha} \wedge \omega^{i}_{j} + \omega^{\beta}_{\alpha} \wedge \omega^{i}_{\beta} + u^{i\beta\gamma}_{\alpha jk}\omega^{j}_{\beta} \wedge \omega^{k}_{\gamma}, \tag{2.2.11}$$

where $u^{i\beta\gamma}_{\alpha jk}$ denotes the result of alternating the quantities $\bar{u}^{i\beta\gamma}_{\alpha jk}$ with respect to pairs of indices $\binom{\beta}{j}$ and $\binom{\gamma}{k}$, i.e.,

$$u^{i\beta\gamma}_{\alpha jk} = -u^{i\gamma\beta}_{\alpha kj}. \tag{2.2.12}$$

We obtain the remaining structure equations of an almost Grassmann manifold X^{nr} by exterior differentiation of (2.2.11). This gives

$$\Omega^{\beta}_{\alpha} \wedge \omega^{i}_{\beta} - \Omega^{i}_{j} \wedge \omega^{j}_{\alpha} + \nabla u^{i\beta\gamma}_{\alpha jk} \wedge \omega^{j}_{\beta} \wedge \omega^{k}_{\gamma} + 2u^{i\epsilon\gamma}_{\alpha mk}u^{m\delta\beta}_{\epsilon lj}\omega^{l}_{\delta} \wedge \omega^{j}_{\beta} \wedge \omega^{k}_{\gamma} = 0, \tag{2.2.13}$$

where

$$
\begin{aligned}
\Omega_\alpha^\beta &= d\omega_\alpha^\beta - \omega_\alpha^\gamma \wedge \omega_\gamma^\beta, \\
\Omega_j^i &= d\omega_j^i - \omega_j^k \wedge \omega_k^i, \\
\nabla u_{\alpha jk}^{i\beta\gamma} &= du_{\alpha jk}^{i\beta\gamma} - u_{\delta jk}^{i\beta\gamma}\omega_\alpha^\delta - u_{\alpha lk}^{i\beta\gamma}\omega_j^l - u_{\alpha jl}^{i\beta\gamma}\omega_k^l \\
&\quad + u_{\alpha jk}^{l\beta\gamma}\omega_l^i + u_{\alpha jk}^{i\delta\gamma}\omega_\delta^\beta + u_{\alpha jk}^{i\beta\delta}\omega_\delta^\gamma.
\end{aligned}
$$

Solving (2.2.13), we obtain

$$
\left\{
\begin{array}{l}
\Omega_\alpha^\beta = b_{\alpha lm}^{\beta\delta\varepsilon}\omega_\delta^l \wedge \omega_\varepsilon^m + \omega_\gamma^k \wedge \omega_{\alpha k}^{\beta\gamma} + \delta_\alpha^\beta \theta, \\
\Omega_j^i = b_{ilm}^{j\delta\varepsilon}\omega_\delta^l \wedge \omega_\varepsilon^m + \omega_{ik}^{j\gamma} \wedge \omega_\gamma^k + \delta_i^j \theta, \\
\nabla u_{\alpha jk}^{i\beta\gamma} = b_{\alpha jkl}^{i\beta\gamma\delta}\omega_\delta^l + \phi_{\alpha jk}^{i\beta\gamma},
\end{array}
\right. \tag{2.2.14}
$$

where $\theta, \omega_{\alpha k}^{\beta\gamma}, \omega_{ik}^{j\gamma}, \phi_{\alpha jk}^{i\beta\gamma}$ are certain 2- and 1-forms not expressed in terms of principal forms ω_α^i, and

$$
\delta_{[j}^i \omega_{|\alpha|k]}^{[\beta\gamma]} + \delta_\alpha^{[\beta}\omega_{[jk]}^{|i|\gamma]} + \phi_{\alpha jk}^{i\beta\gamma} = 0, \tag{2.2.15}
$$

$$
\delta_{[j}^i b_{|\alpha|lm]}^{[\beta\delta\varepsilon]} - \delta_\alpha^{[\beta} b_{[jlm]}^{|i|\delta\varepsilon]} + b_{\alpha[mjl]}^{i[\varepsilon\beta\delta]} + 2u_{\alpha k[j}^{i\gamma[\beta} u_{|\gamma|lm]}^{|k|\delta\varepsilon]} = 0. \tag{2.2.16}
$$

Note that the alternation in (2.2.12) and (2.2.13) is carried out with respect to pairs of indices $\binom{\beta}{j}, \binom{\gamma}{k}, \binom{\delta}{l}, \binom{\varepsilon}{m}$.

Equations (2.2.14) show that the quantities $u_{\alpha jk}^{i\beta\gamma}$ do not form a tensor (see [L 53] or the footnote $(^5)$ in Notes to Chapter 1). Using the $u_{\alpha jk}^{i\beta\gamma}$, let us construct a tensor $\overset{*}{u}{}_{\alpha jk}^{i\beta\gamma}$ satisfying the following conditions:

$$
\overset{*}{u}{}_{\delta jk}^{i\delta\gamma} = 0, \quad \overset{*}{u}{}_{\alpha lk}^{l\beta\gamma} = 0. \tag{2.2.17}
$$

In (2.2.17) and in what follows we agree to denote by γ and l Greek and Latin indices with respect to which a summation must be carried out.

This very tensor has been obtained by Mikhailov [Mi 78]. He achieved this by using a specialised frame. Keeping in mind subsequent problems, we refrain from using the specialised frames and will obtain an expression of this tensor in the general moving frame.

Let us rewrite equation (2.2.11) in the form

$$
\begin{aligned}
d\omega_\alpha^i &= \omega_\alpha^j \wedge (\omega_j^i + h_{jk}^{i\gamma}\omega_\gamma^k) + (\omega_\alpha^\beta + h_{\alpha k}^{\beta\gamma}\omega_\gamma^k) \wedge \omega_\beta^i \\
&\quad + (u_{\alpha jk}^{i\beta\gamma} - \delta_\alpha^{[\beta} h_{[jk]}^{|i|\gamma]} + \delta_{[j}^i h_{|\alpha|k]}^{[\beta\gamma]})\omega_\beta^j \wedge \omega_\gamma^k,
\end{aligned} \tag{2.2.18}
$$

where $h_{jk}^{i\gamma}$ and $h_{\alpha k}^{\beta\gamma}$ are certain arbitrary functions on X^{nr}. We introduce the notation:

$$
\overset{*}{u}{}_{\alpha jk}^{i\beta\gamma} = u_{\alpha jk}^{i\beta\gamma} - \delta_\alpha^{[\beta} h_{[jk]}^{|i|\gamma]} + \delta_{[j}^i h_{|\alpha|k]}^{[\beta\gamma]} \tag{2.2.19}
$$

and show that we can choose $h^{i\gamma}_{jk}$ and $h^{\beta\gamma}_{\alpha k}$ so that conditions (2.2.17) be satisfied. In this case we must have

$$\begin{cases} 2u^{i\delta\gamma}_{\delta jk} = nh^{i\gamma}_{jk} - h^{i\gamma}_{kj} - \delta^i_j h^{\delta\gamma}_{\delta k} + \delta^i_k h^{\gamma\delta}_{\delta j}, \\ 2u^{l\beta\gamma}_{\alpha lk} = \delta^\beta_\alpha h^{i\gamma}_{lk} - \delta^\gamma_\alpha h^{l\beta}_{kl} - rh^{\beta\gamma}_{\alpha k} + h^{\gamma\beta}_{\alpha k}. \end{cases} \tag{2.2.20}$$

As was mentioned above, when an almost Grassmann manifold is studied, usually the group $\mathbf{GL}(n)$ is reduced to $\mathbf{SL}(n)$. This can be achieved, for example, by imposing an additional condition

$$\det\,(A^j_i) = \pm\,\det\,(A^\beta_\alpha), \tag{2.2.21}$$

which after being exteriorly differentiated implies

$$\omega^l_l + \omega^\delta_\delta = 0. \tag{2.2.22}$$

Usually after this it is required that the transformed forms

$$\overset{*}{\omega}{}^i_j = \omega^i_j + h^{i\gamma}_{jk}\omega^k_\gamma, \quad \overset{*}{\omega}{}^\beta_\alpha = \omega^\beta_\alpha + h^{\beta\gamma}_{\alpha k}\omega^k_\gamma \tag{2.2.23}$$

satisfy a condition similar to (2.2.22):

$$\overset{*}{\omega}{}^l_l + \overset{*}{\omega}{}^\delta_\delta = 0. \tag{2.2.24}$$

Comparison (2.2.23) and (2.2.24) gives the following relation for $h^{i\gamma}_{jk}$ and $h^{\beta\gamma}_{jk}$:

$$h^{\delta\gamma}_{\delta k} = h^{l\gamma}_{lk} = h^\gamma_k. \tag{2.2.25}$$

To find a geometric meaning of condition (2.1.20), let us consider the tangent space $T_p(X^{nr})$ at the point p. Since its dimension is the product rn of two natural numbers r and n, we can always represent it as a tensor product $L_r \otimes L_n$ of two linear spaces of dimension r and n respectively. To do this, with any point $p \in X^{nr}$ we will associate linear spaces L_r and L_n where correspondingly frames $\{p, e_i\}$ and $\{p, e^\alpha\}$ are fixed. In $T_p(X^{nr})$ itself we will consider the frame $\{p, e^\alpha_i\}$ that is smoothly spreaded along X^{nr}. We define the correspondence for the basis vectors of these three frames by

$$e_i \otimes e^\alpha \to e^\alpha_i$$

and extend it to other vectors of L_r, L_n and $T_p(X^{nr})$ by defining vectors with equal coordinates with respect to their bases as corresponding vectors. Since this correspondence is an isomorphism, we can put $e^\alpha_i = e_i \otimes e^\alpha$. Then any other basis $\{'e^\alpha_i\}$ which is obtained from $\{e^\alpha_i\}$ by a transformation of the group G, has the form $'e^\alpha_i = A^j_i A^\alpha_\beta (e_j \otimes e^\beta) = {}'e_i \otimes {}'e^\alpha$ where $'e_i = A^j_i e_j$, $'e^\alpha = A^\alpha_\beta e^\beta$. In other words, a set of frames $\{p, {}'e^\alpha_i\}$ is a set of frames of the form $\{p, {}'e_i \otimes {}'e^\alpha\}$ where $'e_i$ and $'e^\alpha$ are vectors of bases of L_r and L_n transferred by transformations from $\mathbf{GL}(r)$ and

$\mathbf{GL}(n)$ that are concordant among each other by condition (2.2.21). This restriction does not contract the set of bases $\{'e_i \otimes {}'e^\alpha\}$ but only makes this set unique. This is not essential for us now and we can ignore condition (2.2.21).

Keeping in mind this reason and subsequent problems, we will not impose condition (2.2.21). However, to be able solve the system (2.2.20), we will assume that $h^{i\gamma}_{jk}$ and $h^{\beta\gamma}_{\alpha k}$ satisfy (2.2.29).

We substitute $h^{\delta\gamma}_{\delta k}$ and $h^{l\gamma}_{lk}$ from (2.2.25) into (2.2.20) and contract the first equation so obtained with respect to i and j and the second one with respect to α and β As result, we obtain the following system with respect to $h^{l\gamma}_{kl}$ and $h^{\gamma\delta}_{\delta k}$:

$$\begin{cases} -h^{l\gamma}_{kl} + h^{\gamma\delta}_{\delta k} = 2u^{l\delta\gamma}_{\delta lk} + (n+r)h^\gamma_k, \\ nh^{l\gamma}_{kl} + rh^{\gamma\delta}_{\delta k} = -2u^{l\gamma\delta}_{\delta lk}. \end{cases} \tag{2.2.26}$$

The determinant of the system (2.2.26) is equal to $-(r+n)$ and different from 0, and the solution has the form:

$$\begin{cases} h^{l\gamma}_{kl} = -\dfrac{2}{n+r}(ru^{l\delta\gamma}_{\delta lk} + u^{l\gamma\delta}_{\delta lk}) - rh^\gamma_k, \\ h^{\gamma\delta}_{\delta k} = \dfrac{2}{n+r}(u^{l\delta\gamma}_{\delta lk} - u^{l\gamma\delta}_{\delta lk}) + nh^\gamma_k. \end{cases} \tag{2.2.27}$$

Now, by virtue of (2.2.25) and (2.2.27), we can write the first equations of the system (2.2.20) in the form:

$$nh^{i\gamma}_{jk} - h^{i\gamma}_{kj} = 2u^{i\delta\gamma}_{\delta jk} - \delta^i_j h^\gamma_k - \frac{2}{n+r}\delta^i_k(nu^{l\delta\gamma}_{\delta lj} - u^{l\gamma\delta}_{\delta lj}) - n\delta^i_k h^\gamma_j. \tag{2.2.28}$$

The system (2.2.28) for $i \neq j, k$ gives

$$nh^{i\gamma}_{jk} - h^{i\gamma}_{kj} = 2u^{i\delta\gamma}_{\delta jk}. \tag{2.2.29}$$

Interchanging indices j and k in (2.2.29), we get

$$-h^{i\gamma}_{jk} + nh^{i\gamma}_{kj} = 2u^{i\delta\gamma}_{\delta kj}. \tag{2.2.30}$$

From (2.2.29) and (2.2.30) we find

$$h^{i\gamma}_{jk} = \frac{2}{n^2-1}(nu^{i\delta\gamma}_{\delta jk} + u^{i\delta\gamma}_{\delta kj}), \quad i \neq j, k. \tag{2.2.31}$$

Let us further take in equations (2.2.28) firstly $i = j \neq k$ and secondly $i = k \neq j$. Then we obtain a system with respect to $h^{i\gamma}_{ik}$ and $h^{i\gamma}_{ki}$:

$$\begin{cases} nh^{i\gamma}_{ik} - h^{i\gamma}_{ki} = 2u^{i\delta\gamma}_{\delta ik} + h^\gamma_k, \\ -h^{i\gamma}_{ik} + nh^{i\gamma}_{ki} = 2u^{i\delta\gamma}_{\delta ki} - \dfrac{2}{n+r}(nu^{l\delta\gamma}_{\delta lk} - u^{l\gamma\delta}_{\delta lk}) - nh^\gamma_k \end{cases} \tag{2.2.32}$$

(no summation over i). The solution of the system (2.2.32) is

$$\begin{cases} h^{i\gamma}_{ik} = \dfrac{2}{n^2-1}(nu^{i\delta\gamma}_{\delta ik} + u^{i\delta\gamma}_{\delta ki}) - \dfrac{2}{(n+r)(n^2-1)}(nu^{l\delta\gamma}_{\delta lk} - u^{l\gamma\delta}_{\delta lk}), \\ h^{i\gamma}_{ki} = \dfrac{2}{n^2-1}(u^{i\delta\gamma}_{\delta ik} + nu^{i\delta\gamma}_{\delta ki}) - \dfrac{2}{(n+r)(n^2-1)}(nu^{l\delta\gamma}_{\delta lk} - u^{l\gamma\delta}_{\delta lk}) - h^{\gamma}_{k}. \end{cases} \tag{2.2.33}$$

Finally, let us take in (2.2.28) $i = j = k$. Then we obtain

$$h^{k\gamma}_{kk} = \frac{2}{n-1}u^{k\delta\gamma}_{\delta kk} - \frac{2}{(n+r)(n-1)}(nu^{l\delta\gamma}_{\delta lk} - u^{l\gamma\delta}_{\delta lk}) - h^{\gamma}_{k}. \tag{2.2.34}$$

Formulas (2.2.31), (2.2.33), and (2.2.34) can be combined as follows:

$$\begin{aligned} h^{i\gamma}_{jk} &= -\delta^i_k h^\gamma_j + \frac{2}{(n^2-1)}(nu^{i\delta\gamma}_{\delta jk} + u^{i\delta\gamma}_{\delta kj}) \\ &\quad + \frac{2}{(n+r)(n^2-1)}\Big[\delta^i_j(u^{l\gamma\delta}_{\delta lk} - nu^{l\delta\gamma}_{\delta lk}) + n\delta^i_k(u^{l\gamma\delta}_{\delta lj} - nu^{l\delta\gamma}_{\delta lj})\Big]. \end{aligned} \tag{2.2.35}$$

Next, again by virtue of (2.2.25) and (2.2.27) the second equation of the system (2.2.20) can be written in the form

$$rh^{\beta\gamma}_{\alpha k} - h^{\gamma\beta}_{\alpha k} = 2u^{l\beta\gamma}_{\alpha lk} - \delta^\beta_\alpha h^\gamma_k + r\delta^\gamma_\alpha h^\beta_k + \frac{2}{n+r}\delta^\gamma_\alpha(ru^{l\delta\beta}_{\delta lk} + u^{l\beta\delta}_{\delta lk}). \tag{2.2.36}$$

Using a similar reduction to that we used to solve the system (2.2.28), we obtain the solution of (2.2.36) in the form:

$$\begin{aligned} h^{\beta\gamma}_{\alpha k} &= \delta^\gamma_\alpha h^\beta_k - \frac{2}{r^2-1}(ru^{l\beta\gamma}_{\alpha lk} + u^{l\gamma\beta}_{\alpha lk}) \\ &\quad + \frac{2}{(n+r)(n^2-1)}\Big[\delta^\beta_\alpha(ru^{l\delta\gamma}_{\delta lk} + u^{l\gamma\delta}_{\delta lk}) + \delta^\gamma_\alpha(ru^{l\delta\beta}_{\delta lk} + u^{l\beta\delta}_{\delta lk})\Big]. \end{aligned} \tag{2.2.37}$$

Let us substitute expressions of $h^{i\gamma}_{jk}$ and $h^{\beta\gamma}_{\alpha k}$ from (2.2.35) and (2.2.37) into (2.2.19). This gives

$$\begin{aligned} \overset{*}{u}{}^{i\beta\gamma}_{\alpha jk} &= u^{i\beta\gamma}_{\alpha jk} - \frac{2}{r^2-1}\delta^i_{[j}(ru^{l[\beta\gamma]}_{|\alpha l|k]} - u^{l[\beta\gamma]}_{|\alpha|k]l}) \\ &\quad - \frac{2}{n^2-1}\delta^{[\beta}_\alpha(nu^{|i\delta|\gamma]}_{\delta[jk]} + u^{|i\delta|\gamma]}_{\delta[kj]}) \\ &\quad + \frac{2}{(r^2-1)(n^2-1)}\Big[(nr-1)(\delta^i_{[j}\delta^{[\beta}_{|\alpha|}u^{|l\delta|\gamma]}_{|\delta l|k]} - \delta^i_{[k}\delta^{[\beta}_{|\alpha|}u^{|l\gamma]\delta}_{|\delta l|j]}) \\ &\quad + (r-n)(\delta^i_{[j}\delta^{[\beta}_{|\alpha|}u^{|l\delta|\gamma]}_{|\delta|k]l} - \delta^i_{[k}\delta^{[\beta}_{|\alpha|}u^{|l|\gamma]\delta}_{|\delta|j]l}\Big], \end{aligned} \tag{2.2.38}$$

where, as earlier, the alternation is carried out with respect to pairs of indices $\binom{\beta}{j}, \binom{\gamma}{k}$ or $\binom{\beta}{k}$ $\binom{\gamma}{j}$.

With the help of (2.2.14), it can be readily verified that

$$\begin{aligned}\nabla_\delta \overset{*}{u}{}^{i\beta\gamma}_{\alpha jk} &\equiv \delta \overset{*}{u}{}^{i\beta\gamma}_{\alpha jk} - \overset{*}{u}{}^{i\beta\gamma}_{\alpha jk} h^\delta_\alpha - \overset{*}{u}{}^{i\beta\gamma}_{\alpha lk} h^l_j \\ &- \overset{*}{u}{}^{i\beta\gamma}_{\alpha jl} h^l_k + \overset{*}{u}{}^{l\beta\gamma}_{\alpha jk} h^i_l + \overset{*}{u}{}^{i\delta\gamma}_{\alpha jk} h^\beta_\delta + \overset{*}{u}{}^{i\beta\delta}_{\alpha jk} h^\gamma_\delta \\ &= 0. \end{aligned} \tag{2.2.39}$$

Equations (2.2.39) prove that the quantities $\overset{*}{u}{}^{i\beta\gamma}_{\alpha jk}$ form a tensor (see [L 53] or the footnote $(^5)$ in Notes to Chapter 1). As follows from its construction, this tensor satisfies condions (2.2.17). Moreover, equations (2.2.12) and (2.2.38) show that the tensor $\overset{*}{u}{}^{i\beta\gamma}_{\alpha jk}$ satisfies conditions:

$$\overset{*}{u}{}^{i\beta\gamma}_{\alpha jk} = - \overset{*}{u}{}^{i\gamma\beta}_{\alpha kj} . \tag{2.2.40}$$

This very tensor Mikhailov [Mi 78] used in place of $u^{i\beta\gamma}_{\alpha jk}$ in equations (2.2.11).

Definition 2.2.1 The tensor $\overset{*}{u}{}^{i\beta\gamma}_{\alpha jk}$ defined by (2.2.42) is said to be the *torsion tensor* of an almost Grassmann manifold $AG(n-1, r+n-1)$.

The torsion tensor of $AG(n-1, r+n-1)$ satisfying conditions (2.2.17) and (2.2.39) can be represented in the form:

$$\overset{*}{u}{}^{i\beta\gamma}_{\alpha jk} = \overset{*}{u}{}^{i\beta\gamma}_{\alpha(jk)} + \overset{*}{u}{}^{i\beta\gamma}_{\alpha[jk]}, \tag{2.2.41}$$

where

$$\overset{*}{u}{}^{i\beta\gamma}_{\alpha(jk)} = \overset{*}{u}{}^{i[\beta\gamma]}_{\alpha jk}, \quad \overset{*}{u}{}^{i\beta\gamma}_{\alpha[jk]} = \overset{*}{u}{}^{i(\beta\gamma)}_{\alpha jk} . \tag{2.2.42}$$

The following theorem combines the results obtained in this section.

Theorem 2.2.2 *The structure equations of an almost Grassmann manifold $AG(n-1, r+n-1)$ can be reduced to the form* (2.2.11), (2.2.14),(2.2.15), *and* (2.2.16). *Its torsion tensor is expressed by* (2.2.38) *and satisfy the conditions* (2.2.17) *and* (2.2.40). ■

Mikhailov [Mi 78] proved the theorem giving an analytical condition for an almost Grassmann structure $AG(n-1, r+n-1)$ defined on X^{nr} to be r- or n-semiintegrable or integrable. We will state this theorem. The conditions will be expressed in terms of the tensor $\overset{*}{u}{}^{i\beta\gamma}_{\alpha jk}$ and its symmetric and skew-symmetric parts $\overset{*}{u}{}^{i\beta\gamma}_{\alpha(jk)}$ and $\overset{*}{u}{}^{i\beta\gamma}_{\alpha[jk]}$ determined by (2.2.41).

Theorem 2.2.3 *For an almost Grassmann structure $AG(n-1, r+n-1)$ to be n-semiintegrable (r-semiintegrable or integrable) it is necessary and sufficient that the tensor $\overset{*}{u}{}^{i\beta\gamma}_{\alpha(jk)}$ (respectively $\overset{*}{u}{}^{i\beta\gamma}_{\alpha[jk]}$ or $\overset{*}{u}{}^{i\beta\gamma}_{\alpha jk}$) vanishes.*

Proof See [Mi 78]. ■

2.3 An Almost Grassmann Structure Associated with a Web $W(n+1, n, r)$

Suppose now that a web $W(n+1, n, r)$, $n \geq 2$, $r \geq 2$, is given in an open domain D of a differentiable manifold X^{nr}.

Theorem 2.3.1 *A web $W(n+1, n, r)$, $n \geq 2$, $r \geq 2$, defines on X^{nr} an almost Grassmann structure $AG(n-1, r+n-1)$.*

Proof. Let p be a point in $D \subset X^{nr}$ and let $T_p(X^{nr})$ be its tangent space at p. Passing through p are the $n+1$ leaves F_ξ, $\xi = 1, \ldots, n+1$, belonging to different foliations λ_ξ generating the web $W(n+1, n, r)$. The tangent spaces $T_p(F_\xi)$ of the leaves F_ξ are in general position in $T_p(X^{nr})$ and, by virtue of Lemma 2.1.4, they determine in $T_p(X^{nr})$ the Segre cone $C_p(r, n)$, with the vertex at the point p. Therefore on X^{nr} there arises a differentiable field of cones $C_p(r, n)$ which defines an almost Grassmann structure $AG(n-1, r+n-1)$ on it. ■

Note that a subspace Δ_p of codimension r passing through the point p of each of the $(n-1)r$-dimensional distributions Δ introduced at the end of Section 1.11, Chapter 1, and any n tangent spaces $T_p(F_{\hat{\xi}})$, $\hat{\xi} \neq \xi$, $\hat{\xi}, \xi = 1, \ldots, n+1$, ξ is fixed, determines the same Segre cone $C_p(r, n)$. According to Definition 2.1.3, each of these subspaces is conconical to the cone $C_p(r, n)$. On X^{nr} a distribution Δ is defined by 1-forms ω^i expressed by (1.11.32), (1.11.33). In $T_p(X^{nr})$, in local coordinates z^i_α (see Section 2.1) a subspace Δ_p is defined by equations (2.1.3).

Definition 2.3.2 A distribution Δ of codimension r given on a differentiable manifold X^{nr} endowed with an almost Grassmann structure is said to be *compatible* with the almost Grassmann structure if the subspaces Δ_p composing Δ are conconical to the Segre cones $C_p(r, n)$ defining the structure.

Thus, each of the distributions Δ defined by the 1-forms (1.11.32) is compatible with the almost Grassmann structure associated with the web $W(n+1, n, r)$.

In Section 1.2, we noted that a G-structure is also associated with the the web $W(n+1, n, r)$ and that the structural group of this G-structure is a subgroup of the group $\mathbf{GL}(rn)$ acting in $T_p(X^{nr})$ and leaving invariant the subspaces $T_p(F_\xi)$ tangent to the web leaves passing through p. In view of (2.1.9), the structural group G of this G-structure coincides with $\mathbf{GL}(r)$. The G-structure is a substructure of an almost Grassmann structure $AG(n-1, r+n-1)$ whose structure group, as was indicated above, is the group $\mathbf{GL}(r, n) = \mathbf{GL}(r) \times \mathbf{SL}(n)$.

Let us find now relations between the coefficients of the structure equations (2.2.10) and (2.2.14) of an almost Grassmann structure $AG(n-1, r+n-1)$ connected with the web $W(n+1, n, r)$ and the torsion and curvature tensors $\underset{\alpha\beta}{a}{}^i_{jk}$ and $\underset{\alpha\beta}{b}{}^i_{jkl}$ of the web.

To find them, first we will slightly change the notation we have used in Chapter 1: namely we will write θ^i_j instead of ω^i_j and ω^i_α instead of $\underset{\alpha}{\omega^i}$ in equations (1.2.16). So the foliations λ_α and λ_{n+1} are defined now by

$$\omega^i_\alpha = 0, \quad \alpha = 1, \ldots, n; \quad \omega^i_{n+1} \equiv -\sum_\alpha \omega^i_\alpha = 0,$$

and equations (1.2.16) are

$$d\omega^i_\alpha = \omega^j_\alpha \wedge \theta^i_j + \sum_{\beta \neq \alpha} \underset{\alpha\beta}{a}{}^i_{jk} \omega^j_\alpha \wedge \omega^k_\beta. \tag{2.3.1}$$

Comparing (2.3.1) with (2.2.11), first we find that

$$\theta^i_j = \omega^i_j - \delta^i_j \omega, \tag{2.3.2}$$

where

$$\omega = \omega^\alpha_\alpha \quad \text{(no summation over } \alpha) \tag{2.3.3}$$

and

$$\sum_{\beta \neq \alpha} \omega^\beta_\alpha \wedge \omega^i_\beta = 0. \tag{2.3.4}$$

In addition, this comparison gives

$$u^{i\beta\gamma}_{\alpha jk} = \delta^\beta_\alpha \underset{\alpha\gamma}{a}{}^i_{jk}, \quad \gamma \neq \alpha. \tag{2.3.5}$$

Note that it follows from (2.3.5) and (2.2.42) that on the web $\overset{*}{u}{}^{i\beta\gamma}_{\alpha jk} = 0$ for $\beta, \gamma \neq \alpha$ and the quantities $u^{i\alpha\beta}_{\alpha jk}$ form a tensor with respect to indices i, j, and k.

Relations (1.2.17), (1.2.24), (2.2.11), (2.3.5), and (2.3.42) imply

$$u^{i\alpha\beta}_{\alpha jk} = -u^{i\alpha\beta}_{\beta jk} \quad \text{(no summation over } \alpha \text{ and } \beta), \tag{2.3.6}$$

$$\sum_{\alpha,\beta} u^{i\alpha\beta}_{\alpha jk} = 0. \tag{2.3.7}$$

Since $r > 1$, then it follows from the quadratic exterior equations (2.3.4) that

$$\omega^\beta_\alpha = 0, \quad \beta \neq \alpha. \tag{2.3.8}$$

It was shown in Section 1.2, that the prolongation of equations (2.3.1) has the form (1.2.29), (1.2.30), (1.2.26), (1.2.27), and (1.2.28). From (2.2.14), (2.3.5), (2.3.6), (2.3.8) and (1.2.29), (1.2.30), (1.2.26), (1.2.27), (1.2.28) we deduce that

$$\varphi^{i\beta\gamma}_{\alpha jk} = 0, \quad \beta, \gamma \neq \alpha, \quad \varphi^{i\alpha\gamma}_{\alpha jk} = 0 \tag{2.3.9}$$

and

$$b^{i\beta\gamma\delta}_{\alpha jkl} = 0, \quad \beta, \gamma \neq \alpha, \tag{2.3.10}$$

$$\begin{cases} b^{i\alpha\beta\gamma}_{\alpha jkl} = \underset{\alpha\beta\gamma}{a}{}^{i}_{jkl} + \underset{\alpha\beta}{a}{}^{i}_{mk}\underset{\gamma\alpha}{a}{}^{m}_{lj} + \underset{\alpha\beta}{a}{}^{i}_{jm}\underset{\beta\gamma}{a}{}^{m}_{kl}, & \gamma \neq \alpha, \beta, \\ b^{i\alpha\beta\alpha}_{\alpha jkl} = \underset{\alpha\beta\alpha}{a}{}^{i}_{jkl} + \underset{\alpha\beta}{a}{}^{i}_{jm}\underset{\beta\alpha}{a}{}^{m}_{kl}, & \beta \neq \alpha, \\ b^{i\alpha\beta\beta}_{\alpha jkl} = \underset{\alpha\beta\beta}{a}{}^{i}_{jkl} + \underset{\alpha\beta}{a}{}^{i}_{mk}\underset{\beta\alpha}{a}{}^{m}_{lj}, & \beta \neq \alpha. \end{cases} \tag{2.3.11}$$

It follows from (2.3.9) and (2.2.15) that

$$\omega^{\beta\gamma}_{\alpha k} = 0, \quad \beta, \gamma \neq \alpha; \quad \omega^{i\gamma}_{jk} = 0, \quad i \neq j, k, \tag{2.3.12}$$

$$\omega^{\beta\alpha}_{\alpha k} = \omega^{i\beta}_{ki}, \quad \omega^{\alpha\beta}_{\alpha k} = -\omega^{i\beta}_{ik}, \quad \alpha \neq \beta, \ \ k \neq i; \ \ \omega^{i\beta}_{ii} = -\omega^{\alpha\beta}_{\alpha i}, \quad \alpha \neq \beta. \tag{2.3.13}$$

Note that equations (2.3.12) assume that $n > 2$ and $r > 2$.

From (2.3.2), (2.3.3), (1.2.29), (2.2.14), and (2.2.15) for $i \neq j$ we obtain

$$\omega^{i\beta}_{ki} = 0, \quad k \neq i, \tag{2.3.14}$$

$$\underset{\alpha\alpha}{b}{}^{i}_{jkl} = b^{i\alpha\alpha}_{j[kl]}, \quad \underset{[\alpha\beta]}{b}{}^{i}_{j[kl]} = b^{i[\alpha\beta]}_{j[kl]}, \tag{2.3.15}$$

where the alternation is carried out with respect to pairs $\binom{\alpha}{k}, \binom{\beta}{l}$. For $i = j$ we have

$$\begin{cases} \underset{\alpha\alpha}{b}{}^{i}_{ikl} = b^{i\alpha\alpha}_{i[kl]} - b^{\varphi\alpha\alpha}_{\varphi[kl]}, \\ \underset{\alpha\beta}{b}{}^{i}_{i[kl]} = b^{i[\alpha\beta]}_{i[kl]} - b^{\varphi[\alpha\beta]}_{\varphi[kl]}, \end{cases} \tag{2.3.16}$$

where there is no summation over i and φ, φ takes any value from $1, 2, \ldots, n$, and the alternation in the second formula is carried out with respect to pairs $\binom{\alpha}{k}, \binom{\beta}{l}$.

From (2.3.12) and (2.3.13) we deduce that

$$\omega^{\beta\alpha}_{\alpha k} = 0, \quad -\omega^{k\alpha}_{kk} = -\omega^{i\alpha}_{ik} = \omega^{\alpha\gamma}_{\alpha k} = \omega^{\gamma}_{k}. \tag{2.3.17}$$

Finally, since it follows from (2.3.3) that $\omega^{\alpha}_{\alpha} = \omega^{\beta}_{\beta}$ (no summation over α and β), using (2.2.14) and (2.3.8), we obtain that

$$b^{\beta\gamma\varphi}_{\alpha kl} = 0, \quad \beta \neq \alpha; \quad b^{\alpha\gamma\varphi}_{\alpha kl} = b^{\beta\gamma\varphi}_{\beta kl} \quad \text{(no summation over } \alpha \text{ and } \beta). \tag{2.3.18}$$

We proved the following theorem.

Theorem 2.3.3 *An almost Grassmann structure $AG(n-1, r+n-1)$ is associated with a web $W(n+1,n,r)$. It is distinguished from a general almost Grassmann structure by the relations* (2.3.5) – (2.3.8), (2.3.10), (2.3.12), (2.3.14), (2.3.17), *and* (2.3.18). *The coefficients of the structure equations* (2.2.10) *and* (2.2.14) *of this almost Grassmann structure are connected with the torsion and curvature tensors $\underset{\alpha\beta}{a}{}^{i}_{jk}$ and $\underset{\alpha\beta}{b}{}^{i}_{jkl}$ of the web $W(n+1,n,r)$ by* (2.3.5), (2.3.11), (2.3.15),*and* (2.3.16). ■

2.4 Semiintegrable Almost Grassmann Structures and Transversally Geodesic and Isoclinic $(n+1)$-Webs

Let us clarify the webs $W(n+1,n,r)$ for which the corresponding almost Grassmann structure $AG(n-1,r+n-1)$ is n- or r-semiintegrable. To do so, we first find an expression for components of the tensors $\overset{*}{u}{}^{i\beta\gamma}_{\alpha(jk)}$ and $\overset{*}{u}{}^{i\beta\gamma}_{\alpha[jk]}$ into which the torsion tensor $\overset{*}{u}{}^{i\beta\gamma}_{\alpha jk}$ of $AG(n-1,r+n-1)$ splits (see formula (2.2.41)). From (2.2.42), (2.3.5), and (2.3.6) we obtain

$$\begin{aligned}\overset{*}{u}{}^{i\alpha\gamma}_{\alpha(jk)} &= 2\underset{\alpha\gamma}{a}{}^{i}_{(jk)} - \frac{2}{r+1}\delta^i_{(j}\big(\underset{\alpha\gamma}{a}{}^{l}_{k)l} + \underset{\alpha\gamma}{a}{}^{l}_{|l|k)}\big)\\ &-\frac{2}{n-1}\sum_{\delta\neq\gamma}\underset{\delta\gamma}{a}{}^{i}_{(jk)} + \frac{2}{(r+1)(n-1)}\delta^i_{(j}\sum_{\delta\neq\gamma}\big(\underset{\delta\gamma}{a}{}^{l}_{k)l} + \underset{\delta\gamma}{a}{}^{l}_{|l|k)}\big), \end{aligned} \tag{2.4.1}$$

$$\begin{aligned}\overset{*}{u}{}^{i\alpha\gamma}_{\alpha[jk]} &= 2\underset{\alpha\gamma}{a}{}^{i}_{[jk]} - \frac{2}{r-1}\delta^i_{[j}\big(\underset{\alpha\gamma}{a}{}^{l}_{|l|k]} - \underset{\alpha\gamma}{a}{}^{l}_{k]l}\big)\\ &-\frac{2}{n+1}\sum_{\delta\neq\gamma}\underset{\alpha\gamma}{a}{}^{i}_{[jk]} + \frac{2}{(r+1)(n+1)}\delta^i_{[j}\sum_{\delta\neq\gamma}\big(\underset{\delta\gamma}{a}{}^{l}_{|l|k]} - \underset{\delta\gamma}{a}{}^{l}_{k]l}\big). \end{aligned} \tag{2.4.2}$$

Theorem 2.4.1 *For a web $W(n+1,n,r)$ to be transversally geodesic, it is necessary and sufficient that the corresponding almost Grassmann structure be n-semiintegrable.*

Proof. Suppose that an almost Grassmann structure associated with the web $W(n+1,n,r)$ is n-semiintegrable. According to Theorem 2.2.3, this means that

$$\overset{*}{u}{}^{i\alpha\gamma}_{\alpha(jk)} = 0. \tag{2.4.3}$$

We introduce the notation

$$\underset{\alpha\gamma}{A}{}^{i}_{jk} = \underset{\alpha\gamma}{a}{}^{i}_{(jk)} - \frac{1}{r+1}\delta^i_{(j}\big(\underset{\alpha\gamma}{a}{}^{l}_{k)l} + \underset{\alpha\gamma}{a}{}^{l}_{|l|k)}\big), \quad \gamma\neq\alpha. \tag{2.4.4}$$

By virtue of (2.4.1) and (2.4.4), equation (2.4.3) yields

$$(n-2)\underset{\alpha\beta}{A}{}^{i}_{jk} - \sum_{\delta\neq\alpha,\beta}\underset{\delta\beta}{A}{}^{i}_{jk} = 0. \tag{2.4.5}$$

It follows from (2.4.4), (1.2.17), and (1.2.24) that

$$\sum_{\alpha,\beta}\underset{\alpha\beta}{A}{}^{i}_{jk} = 0. \tag{2.4.6}$$

Let us consider the system (2.4.5) and (2.4.6) for fixed i, j, and k. It contains $n(n-1)/2$ unknowns $\underset{\alpha\beta}{A}{}^i_{jk}$. Let us show that the rank of the matrix of coefficients of this system is equal to $n(n-1)/2$, which implies that the solution is trivial.

An algorithm for calculation of this rank is:

i) Put the coefficients of equation (2.4.6) into the first row of the matrix of coefficients.

ii) Since for any fixed β the sum of all equations of the subsystem (2.4.5) corresponding to the fixed β vanishes, delete from the matrix of coefficients the row corresponding to the last equation of each such a subsystem.

iii) Add all columns to the first one. The new first column will consist of numbers $n(n-1)/2, 0, \ldots, 0$.

iv) Using this new first column, reduce all elements of the first row to 0.

v) Using the second diagonal entry of the matrix obtained, reduce to 0 all elements but the diagonal of the second column and the second row.

While applying such an algorithm, the 0-row can occur if at some moment a row used for creating zeros and a row where zeros are created are proportional. However, this is possible only if in both rows the element $n-2$ is missing. It occurs once in the third subsystem, twice in the fourth, three times in the fifth etc. Thus, the number of independent equations is equal to $1 + (n-2) + (n-3) + \ldots + 0 = n(n-1)/2$.

As result we obtain $\underset{\alpha\beta}{A}{}^i_{jk} = 0$ or by virtue of (2.4.4)

$$\underset{\alpha\beta}{a}{}^i_{(jk)} = \underset{\alpha\beta}{a}{}_{(j}\delta^i_{k)}, \tag{2.4.7}$$

where

$$\underset{\alpha\beta}{a}{}_j = \frac{2}{r+1}\underset{\alpha\beta}{a}{}^l_{(jl)}. \tag{2.4.8}$$

Equalities (2.4.7) and (2.4.8) coincide with (1.9.23) and (1.9.24). By Theorem 1.9.7, the web $W(n+1, n, r)$ is transversally geodesic. We have proved the sufficiency of our assertion. The necessity can be verified immediately. ■

Theorem 2.4.2 *For a web $W(n+1, n, r)$ to be isoclinic, it is necessary and sufficient that the corresponding almost Grassmann structure be r-semiintegrable.*

Proof. Suppose that an almost Grassmann structure associated with the web $W(n+1, n, r)$ is r-semiintegrable. According to Theorem 2.2.3, it means that

$$\overset{*}{u}{}^{i\beta\gamma}_{\alpha[jk]} = 0. \tag{2.4.9}$$

We introduce the notation

$$\underset{\alpha\gamma}{B}{}^i_{jk} = \underset{\alpha\gamma}{a}{}^i_{[jk]} - \frac{1}{r-1}\delta^i_{[j}\big(\underset{\alpha\gamma}{a}{}^l_{|l|k]} - \underset{\alpha\gamma}{a}{}^l_{k]l}\big). \tag{2.4.10}$$

By virtue of (2.4.10) and (2.4.2), equality (2.4.9) assumes the form

$$n \underset{\alpha\gamma}{B}{}^i_{jk} - \sum_{\delta \neq \alpha,\gamma} \underset{\alpha\gamma}{B}{}^i_{jk} = 0. \tag{2.4.11}$$

We consider a subsystem of (2.4.11) corresponding to fixed values of γ, i, j, and k. It contains $n-1$ equations in $n-1$ unknowns $\underset{\alpha\gamma}{B}{}^i_{jk}, \gamma \neq \alpha$. Its determinant is of order $n-1$ and has the form:

$$B_\gamma = \begin{vmatrix} n & -1 & \dots & -1 \\ -1 & n & \dots & -1 \\ \dots & \dots & \dots & \dots \\ \dots & \dots & \dots & \dots \\ -1 & -1 & \dots & n \end{vmatrix}.$$

For calculation of this determinant we will add all its rows starting from the second one to the first row. Then the first row becomes: $2, 2, \dots, 2$. We take out the factor 2 from the first row and add the first row to each other row. Then we obtain:

$$B_\gamma = 2 \begin{vmatrix} 1 & 1 & 1 & \dots & 1 \\ 0 & n+1 & 0 & \dots & 0 \\ 0 & 0 & n+1 & \dots & 0 \\ \dots & \dots & \dots & \dots & \dots \\ \dots & \dots & \dots & \dots & \dots \\ 0 & 0 & 0 & \dots & n+1 \end{vmatrix} = 2(n+1)^{n-2} \neq 0.$$

Therefore, this subsystem has only the trivial solution. Since γ was fixed arbitrarily, we have $\underset{\alpha\gamma}{B}{}^i_{jk} = 0$ or, by virtue of (2.4.10),

$$\underset{\alpha\gamma}{a}{}^i_{[jk]} = \underset{\alpha\gamma}{b}{}_{[j} \delta^i_{k]}, \tag{2.4.12}$$

where

$$\underset{\alpha\gamma}{b}{}_j = \frac{2}{r-1} \underset{\alpha\gamma}{a}{}^l_{[jl]}. \tag{2.4.13}$$

Equations (2.4.12) and (2.4.13) coincide with (1.11.7) and (1.11.8). By Theorem 1.11.6, the web $W(n+1, n, r)$, $r > 2$, is isoclinic. We have proved the sufficiency of our assertion. Its necessity can be verified immediately. ■

Corollary 2.4.3 *If an $(n-1)r$-dimensional distribution Δ associated with a web $W(n+1, n, r), r > 2$, and defined by 1-forms (1.11.32) is integrable, then the almost Grassmann structure associated with the web is r-semiintegrable.*

Proof. If the distribution Δ is integrable, then, by Proposition 1.11.13, the web $W(n+1, n, r)$ is isoclinic. Therefore, by Theorem 1.2.2, the almost Grassmann structure associated with the web is r-semiintegrable. ■

Theorem 2.4.4 *For a web $W(n+1,n,r)$, $n>2$, $r>2$, to be both transversally geodesic and isoclinic, it is necessary and sufficient that the corresponding almost Grassmann structure be Grassmann.*

Proof. Suppose that an almost Grassmann structure associated with the web $W(n+1,n,r)$ is integrable, i.e., it is a Grassmann structure. According to Theorem 3.2.3, in this case we have

$$\overset{*}{u}{}^{i\beta\gamma}_{\alpha jk}=0 \tag{2.4.14}$$

and therefore conditions (2.4.3) and (2.4.9) hold simultaneously. By Theorems 2.4.1 and 2.4.2, the web $W(n+1,n,r)$ is both transversally geodesic and isoclinic. Such a web has been considered at the end of Section 1.11. Its torsion tensor has the structure given by (1.11.29) and (1.11.30). Conversely, if a web $W(n+1,n,r)$ is both transversally geodesic and isoclinic, its torsion tensor has this structure, and for the corresponding almost Grassmann structure we have the condition (2.4.14). Therefore, an almost Grassmann structure connected with such a web is Grassmann, and the manifold X^{nr} itself is the Grassmannian $G(n-1,r+n-1)$. ■

2.5 Double Webs

Definition 2.5.1 We say that a *double web* is defined on a differentiable manifold X^{nr} if the same almost Grassmann structure is associated with a web $W(n+1,n,r)$ and with a web $W(r+1,r,n)$.

Suppose that the $r+1$ foliations of the $W(r+1,r,n)$ are given by the following systems of Pfaffian equations:

$$\omega^i_\alpha=0,\quad i=1,\ldots,r;\quad \omega^{r+1}_\alpha\equiv-\sum_i\omega^i_\alpha=0. \tag{2.5.1}$$

Requiring that each of the systems (2.5.1) be completely integrable, we can reduce the structure equations of the web $W(r+1,r,n)$ to a form analogous to (1.2.16), (1.2.17), and (1.2.24):

$$d\omega^i_\alpha=\omega^i_\beta\wedge\theta^\beta_\alpha+\sum_{j\neq i}\underset{ij}{a}{}^{\beta\gamma}_\alpha\omega^i_\beta\wedge\omega^j_\gamma, \tag{2.5.2}$$

where

$$\underset{ij}{a}{}^{\beta\gamma}_\alpha=\underset{ji}{a}{}^{\gamma\beta}_\alpha,\qquad \sum_{i,j}\underset{ij}{a}{}^{\beta\gamma}_\alpha=0. \tag{2.5.3}$$

Comparing (2.5.2) and (2.5.3) with (2.2.11), we obtain

$$\theta^\beta_\alpha=-\omega^\beta_\alpha+\delta^\beta_\alpha\tilde{\omega}, \tag{2.5.4}$$

$$\tilde{\omega}=\omega^i_i\ \text{(no summation over } i), \tag{2.5.5}$$

$$u^{i\beta\gamma}_{\alpha jk} = 0, \quad j,k \neq i, \tag{2.5.6}$$
$$u^{i\beta\gamma}_{\alpha jk} = \underset{ik}{a}{}^{\beta\gamma}_{\alpha} \quad \text{(no summation over } i\text{)}, \tag{2.5.7}$$
$$u^{i\beta\gamma}_{\alpha ij} = -u^{j\beta\gamma}_{\alpha ij} \quad \text{(no summation over } i \text{ and } j\text{)}, \tag{2.5.8}$$
$$\omega^i_j = 0, \quad j \neq i. \tag{2.5.9}$$

If the webs $W(n+1,n,r)$ and $W(r+1,r,n)$ are associated with the same almost Grassmann structure, then it follows from (2.3.5), (2.3.6), (2.5.6), and (2.5.7) that only the following components of the torsion tensors of these webs are nonzero:

$$\underset{\alpha\gamma}{a}{}^{i}_{ik} = \underset{ik}{a}{}^{\alpha\gamma}_{\alpha} = u^{i\alpha\gamma}_{\alpha ik} \quad \text{(no summation over } i \text{ and } \alpha\text{)}, \tag{2.5.10}$$

where, by virtue of (2.3.7) and (2.5.8), we have

$$u^{i\alpha\gamma}_{\alpha ij} = -u^{j\alpha\gamma}_{\alpha ij} = -u^{i\alpha\gamma}_{\gamma ij} \quad \text{(no summation over } i, j, \text{ and } \gamma\text{)}. \tag{2.5.11}$$

We rewrite equations (2.2.11), using (2.3.2), (2.3.5), (2.3.8), (2.5.4), (2.5.6), (2.5.9), and (2.5.10). We then obtain, on the one hand,

$$d\omega^i_\alpha = \omega^i_\alpha \wedge \omega + \sum_{k\neq i,\gamma\neq\alpha} (\underset{\alpha\gamma}{a}{}^{i}_{ik}\omega^i_\alpha \wedge \omega^k_\gamma + \underset{\alpha\gamma}{a}{}^{i}_{ki}\omega^k_\alpha \wedge \omega^i_\gamma), \tag{2.5.12}$$

and, on the other hand,

$$d\omega^i_\alpha = \tilde{\omega} \wedge \omega^i_\alpha + \sum_{k\neq i,\gamma\neq\alpha} (\underset{ik}{a}{}^{\alpha\gamma}_{\alpha}\omega^i_\alpha \wedge \omega^k_\gamma + \underset{ik}{a}{}^{\gamma\alpha}_{\alpha}\omega^i_\gamma \wedge \omega^k_\alpha). \tag{2.5.13}$$

Comparing (2.5.11) and (2.5.12) yields $\omega = -\tilde{\omega}$ or

$$\omega^\alpha_\alpha + \omega^i_i = 0, \quad \text{(no summation over } \alpha \text{ and } i\text{)}. \tag{2.5.14}$$

Theorem 2.5.2 *One of the two webs, $W(n+1,n,r)$ and $W(r+1,r,n)$, that compose a double web, is isoclinic (respectively transversally geodesic) if and only if the second web is transversally geodesic (respectively isoclinic).*

Proof. The conditions for the web $W(r+1,r,n)$ to be isoclinic or transversally geodesic have the following forms, respectively:

$$\underset{ik}{a}{}^{[\beta\gamma]}_{\alpha} = \frac{1}{n-1}\delta^{[\beta}_\alpha(\underset{ik}{a}{}^{|\delta|\gamma]}_{\delta} - \underset{ik}{a}{}^{\gamma]\delta}_{\delta}), \tag{2.5.15}$$

$$\underset{ik}{a}{}^{(\beta\gamma)}_{\alpha} = \frac{1}{n+1}\delta^{(\beta}_\alpha(\underset{ik}{a}{}^{|\delta|\gamma)}_{\delta} + \underset{ik}{a}{}^{\gamma)\delta}_{\delta}). \tag{2.5.16}$$

In the case of a double web, for $\beta = \alpha$ the conditions $\overset{*}{u}{}^{i\beta\gamma}_{\alpha(jk)} = 0$, from which the conditions (2.4.7) – (2.4.8) for the web $W(n+1,n,r)$ to be transversally geodesic were obtained, assume the form:

$$\overset{*}{u}{}^{i\alpha\gamma}_{\alpha(ik)} - \frac{1}{r+1}u^{l\alpha\gamma}_{\alpha(kl)} - \frac{1}{n-1}u^{i\delta\gamma}_{\delta(ik)} + \frac{1}{(r+1)(n-1)}u^{l\delta\gamma}_{\delta(kl)} = 0 \tag{2.5.17}$$

(no summation over i, l, α and δ). Note that (2.5.17) brings us to (2.4.7) – (2.4.8), i.e., for a double web it follows from (2.5.17) that

$$u^{i\alpha\gamma}_{\alpha(ik)} = \frac{1}{r-1} \sum_{l} u^{l\alpha\gamma}_{\alpha(lk)} \qquad (2.5.18)$$

(no summation over i and α). Let us show that for a double web there is another consequence of (2.5.17). We introduce the notation

$$\underset{\alpha\gamma}{A}_{ik} = u^{i\alpha\gamma}_{\alpha ik} - \frac{1}{n-1} u^{i\delta\gamma}_{\delta ik} \quad \text{(no summation over } i \text{ and } \alpha). \qquad (2.5.19)$$

By virtue of (2.5.19), we can write (2.5.17) in the form

$$r \underset{\alpha\gamma}{A}_{ik} - \sum_{l \neq k, i} \underset{\alpha\gamma}{A}_{kl} = 0. \qquad (2.5.20)$$

By analogy with Section 2.4, it can be shown that from (2.5.20) it follows that

$$\underset{\alpha\gamma}{A}_{ik} = 0. \qquad (2.5.21)$$

From (2.5.19) and (2.5.21) we obtain

$$u^{i\alpha\gamma}_{\alpha(ik)} = \frac{1}{n-1} \sum_{\delta \neq \gamma} u^{i\delta\gamma}_{\delta(ik)}. \qquad (2.5.22)$$

Since $u^{i\alpha\gamma}_{\alpha(ik)} = u^{i[\alpha\gamma]}_{\alpha ik}$, the relations (2.5.15) corresponding to $\beta = \alpha$ follow from (2.5.22) by means of (2.5.10), i.e., in this case the web $W(r+1, r, n)$ is isoclinic.

In the case of a double web, for $\overset{*}{u}{}^{i\beta\gamma}_{\alpha(jk)} = 0$ we have written the quantities $u^{i\alpha\gamma}_{\alpha(ik)} = u^{i[\alpha\gamma]}_{\alpha ik}$ in two ways: in the form of (2.5.18) and (2.5.22). Both (2.4.7) for $j = i$ and (2.5.15) for $\beta = \alpha$ follow from this.

Note that (2.5.18) and (2.5.22) yield

$$\frac{1}{n-1} \sum_{\delta \neq \gamma} u^{i[\delta\gamma]}_{\delta ik} = \frac{1}{r+1} \sum_{l} u^{l\alpha\gamma}_{\alpha(lk)}. \qquad (2.5.23)$$

In exactly the same way it can also be shown that for a double web two expressions for $u^{i\alpha\gamma}_{\alpha[ik]} = u^{i\alpha\gamma}_{\alpha ik}$ can be deduced from the relation $\overset{*}{u}{}^{i\beta\gamma}_{\alpha[jk]} = 0$:

$$u^{i\alpha\gamma}_{\alpha[ik]} = \frac{1}{r-1} \sum_{l} u^{l\alpha\gamma}_{\alpha[lk]}, \qquad (2.5.24)$$

$$u^{i(\alpha\gamma)}_{\alpha ik} = \frac{1}{n+1} \sum_{\delta} u^{i(\delta\gamma)}_{\delta ik}. \qquad (2.5.25)$$

Equations (2.5.24) and (2.5.25) are equivalent to (2.4.12) – (2.4.13) and (2.5.16). ■

Theorem 2.5.2 yields the following corollaries:

Corollary 2.5.3 *If one of the two webs that compose a double web is isoclinic and transversally geodesic simultaneously, then the second web is of the same type.* ■

Corollary 2.5.4 *If both webs that compose a double web are isoclinic, then both are also transversally geodesic and conversely.* ■

2.6 Problems of Grassmannisation and Algebraisation and Their Solution for Webs $W(d,n,r)$, $d \geq n+1$

2.6.1 The Grassmannisation Problem for a Web $W(n+1,n,r)$

Definition 2.6.1 Let P^{r+n-1} be a projective space of dimension $r+n-1$ and P^{n-1} be a linear subspace of P^{r+n-1}. It is well-known that all P^{n-1}'s can be represented as points of the Grassmannian $G(n-1, r+n-1)$ in a projective space P^N of dimension $N = \binom{r+n}{n} - 1$ and $\dim G(n-1, r+n-1) = nr$ (see [HP 52]). If we take all P^{n-1}'s through a point $p \in P^{r+n-1}$, then their images are points of the Schubert variety $\sum(p)$ and $\dim \sum(p) = (n-1)r$, i.e., $\operatorname{codim} \sum(p) = r$ in $G(n-1, r+n-1)$. If a point p varies on a smooth surface $U \subset P^{r+n-1}$ of dimension r, we obtain a foliation of Schubert varieties of codimension r in a domain $D \subset G(n-1, r+n-1)$. If we take in P^{r+n-1} d smooth surfaces $U_1, \ldots, U_d$, $d \geq n+1$, of dimension r in general position, we obtain d foliations of Schubert varieties on $G(n-1, r+n-1)$ or a d-web $W(d,n,r)$. Such a web is said to be a *Grassmann* web and will be denoted by $GW(d,n,r)$.

Definition 2.6.2 A web $W(d,n,r)$ is called *algebraic* (cf. Example 1.1.5) if it is a Grassmann web and if the varieties $U_1, \ldots, U_d$ generating it belong to a single algebraic variety V_d^r of dimension r and degree d. Such a web will be denoted by $AW(d,n,r)$.

Definition 2.6.3 A web $W(d,n,r)$ equivalent to a Grassmann web $GW(d,n,r)$ (an algebraic web $AW(d,n,r)$) is said to be *Grassmannisable* (respectively *algebraisable*).

As is pointed by Chern and Griffiths (see[CG 78a]), two natural problems which arise from these definitions are:

(1) The Grassmannisation problem. *Under what conditions is a given web $W(d,n,r)$ Grassmannisable ?*

(2) The algebraisation problem. *Under what conditions is a given web $W(d,n,r)$ algebraisable ?*

The following theorem gives a solution of the Grassmannisation problem for a web $W(n+1,n,r)$.

Theorem 2.6.4 *A web $W(n+1, n, r)$, $n \geq 2$, $r \geq 2$, is Grassmannisable if and only if it is both transversally geodesic and isoclinic.*

Proof. *Sufficiency.* For $n = 2$, i.e. for three-webs $W(3, 2, r)$, this theorem was proved by Akivis [Ak 74,80,82]. Suppose that $n > 2$. If a web $W(n+1, n, r)$ is both transversally geodesic and isoclinic, then, according to Theorem 2.4.4, its associated almost Grassmann structure $AG(n-1, r+n-1)$ is integrable. Therefore, by Theorem 1.6.8, it is a locally Grassmann structure. If we map a manifold X^{nr} carrying this structure on the Grassmannian $G(n-1, r+n-1)$, the web $W(n+1, n, r)$ will be mapped onto a Grassmann web $GW(n+1, n, r)$.

The *necessity* of the condition of the theorem is obvious: by Theorem 2.4.4, for a Grassmann web $W(n+1, n, r)$ its associated almost Grassmann structure is integrable, and therefore, by Theorems 2.4.1 and 2.4.2, the web $GW(n+1, n, r)$ is both transversally geodesic and isoclinic. ■

We will give another, analytical proof of this theorem in Chapter 5 where we will study in detail Grassmann and algebraic $(n+1)$-webs.

2.6.2 The Grassmannisation Problem for a Web $W(d, n, r)$, $d > n+1$

Now let us turn to the study of the Grassmannisability conditions of a web $W(d, n, r)$ for $d > n+1$. Each $(n+1)$-subweb of the web $W(d, n, r)$, formed by $n+1$ out of a total number of d foliations constituting $W(d, n, r)$, defines an almost Grassmann structure on X^{nr}. It is evident that, if $W(d, n, r)$ is Grassmannisable, then all these almost Grassmann structures coincide with a locally Grassmann structure determined on X^{nr} by this web.

Therefore, the theorem containing the condition of Grassmannizability of $W(d, n, r)$ may be easily formulated for $d > n+1$.

Theorem 2.6.5 *A web $W(d, n, r)$, $d > n+1$, $n \geq 2$, $r \geq 2$, is Grassmannisable if and only if the following two conditions are met:*

(i) *All almost Grassmann structures determined on X^{nr} by the $(n+1)$-subwebs of the web $W(d, n, r)$ coincide;*

(ii) *At least one of the $(n+1)$-subwebs of $W(d, n, r)$ is both transversally geodesic and isoclinic.*

Proof. The validity of this theorem immediately follows from Theorem 2.6.4. ■

In the case $r > 2$ the condition (ii) of Theorem 2.6.5 can be weakened to the following:

(ii′) *At least one of the $(n+1)$-subwebs of $W(n+1, n, r)$ is transversally geodesic.*

To prove it, let us consider the $(n+1)$-subweb $W(n+1, n, r)$ of the web $W(d, n, r)$ formed by the foliations $\lambda_1, \ldots, \lambda_{n+1}$. Then any other foliation λ_s, $s = n+2, \ldots, d$, of $W(d,n,r)$ defines on the manifold X^{nr} an integrable distribution Δ of codimension r compatible with the almost Grassmann structure induced on X^{nr} by the subweb $W(n+1,n,r)$. By Proposition 1.11.13, the subweb $W(n+1,n,r)$ is isoclinic and we obtained the condition (ii).

2.6.3 The Algebraisation Problem for a Web $W(3,2,r)$

Let us recall some theorems on rectilinear three-webs $W(3,2,1)$ in a projective plane P^2 which are well-known (see [Sa 25] or [BB 38] or [Bl 55]). First of all the Graf–Sauer theorem is valid for such webs:

Theorem 2.6.6 *(Graf and Sauer). A rectilinear web $W(3,2,1)$ in a plane is hexagonal if and only if all the straight lines generating it belong to an algebraic curve of the third class.*

Lemma 2.6.7 *A Grassmann web $GW(3,2,1)$ is hexagonal if and only if it is algebraic.*

Proof. Let us consider a rectilinear three-web in P^2. A correlative transformation transfers three foliations of this web onto three curves U_1, U_2, and U_3 of the plane P^{2^*}, and transfers the whole web onto a Grassmann web $GW(3,2,1)$ on the Grassmannian $G(1,2)$. In addition, each hexagonal figure of the rectilinear three-web $W(3,2,1)$ (Figure 2.1) is transferred into the figure represented on Figure 2.2, and its closure condition becomes the condition of coincidence of points z_3' and z_3'' on Figure 2.2. To the hexagonal rectilinear three-web under the mapping in consideration there corresponds a web $GW(3,2,1)$ on which all the figures shown on Figure 2.2 are closed. According to Definition 1.2.8, such webs are hexagonal. Now Lemma 2.6.7 follows from Theorem 2.6.6. ■

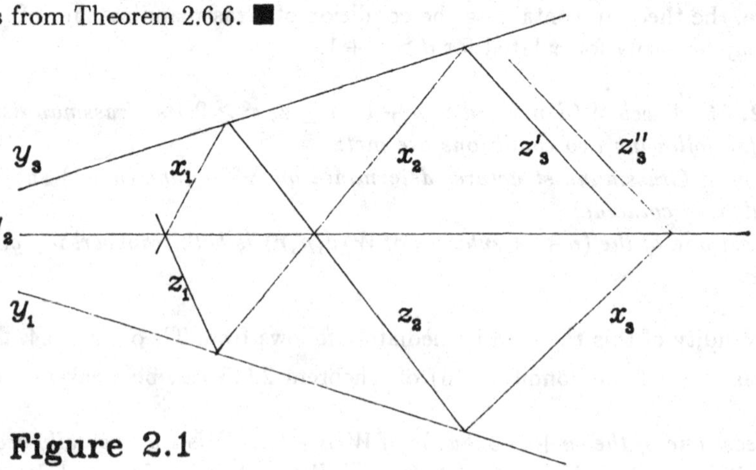

Figure 2.1

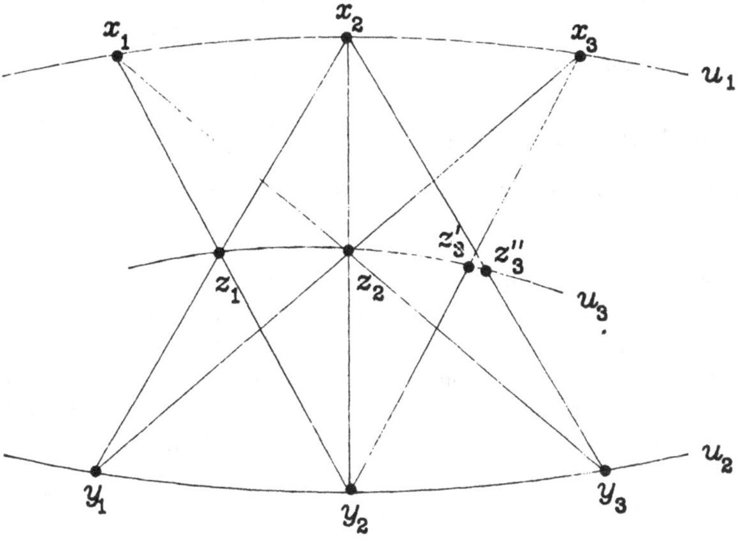

Figure 2.2

Theorem 2.6.8 *A web $W(3,2,r)$ is algebraisable if and only if it is both isoclinic and hexagonal.*

Proof. First, we will prove that for a Grassmann hexagonal web $GW(3,2,r)$ three hypersurfaces generating it belong to a cubic hypersurface. Let us consider a Grassmann three-web $GW(3,2,r)$ on the Grassmannian $G(1,r+1)$. It is generated by three hypersurfaces U_ξ, $\xi = 1,2,3$, of a space P^{r+1}. By Theorem 2.6.4, such a web is transversally geodesic and isoclinic. Its transversal geodesic surfaces coincide with two-dimensional flat generators of the Grassmannian $G(1,r+1)$ carrying this web. In the projective space P^{r+1} to these generators there correspond planar fields of straight lines lying in two-dimensional planes of P^{r+1}. Each plane P^2 intersects hypersurfaces U_ξ at a triple of curves defining a web $GW(3,2,1)$ in P^2. If a web $GW(3,2,r)$ is hexagonal, as follows from the above, every web $GW(3,2,1)$ cut out on it is also hexagonal. Then, by virtue of Lemma 2.6.7, the curves $U_\xi \cap P^2$ belong to a curve of the third order. Consequently, every plane P^2 of P^{r+1} intersects the hypersurfaces U_ξ along curves belonging to one curve of the third order. Because of this, all the hypersurfaces U_ξ belong to one hypersurface V_3^r of the third order in P^{r+1}.

If a web $W(3,2,r)$ is isoclinic and hexagonal, then, by Theorem 1.10.2, it is isoclinic and transversally geodesic. Therefore, by Theorem 2.6.4, it is Grassmannisable. But the Grassmann hexagonal web, as it has been proved, is generated by a cubic hypersurface $V_3^r \subset P^{r+1}$, i.e., it is algebraisable. Thus, the sufficiency of the condition of the theorem has been proved. Its necessity is trivial, since each algebraic three-web is isoclinic and hexagonal. ■

2.6.4 The Algebraisation Problem for a Web $W(n+1,n,r)$

Theorem 2.6.9 *A web $W(n+1,n,r)$ is algebraisable if and only if it is isoclinic and hexagonal.*

Proof. First let us recall some facts which we will use in our proof.

Let $W(n+1,n,r)$ be a web of codimension r given in an open domain $D \subset X^{nr}$ by foliations λ_ξ, $\xi = 1,\ldots,n+1$, with leaves $F_\xi \subset \lambda_\xi$. We consider the $(2r)$-dimensional submanifold $F_{\xi_1} \cap \ldots \cap F_{\xi_{n-1}} \subset D$ constituting an $(n-2)$-fold intersection of the web leaves. On this manifold the foliations $\lambda_{\xi_{n-1}}, \lambda_{\xi_n}, \lambda_{\xi_{n+1}}$ cut out a three-web $W(3,2,r)$ which is a 3-subweb of $W(n+1,n,r)$. By Definition 1.10.1, the web $W(n+1,n,r)$ is hexagonal if all its 3-subwebs $W(3,2,r)$ are hexagonal.

Next, let us indicate a way to construct a 3-subweb $GW(3,2,r)$ of a Grassmann web $GW(n+1,n,r)$.

Suppose we have a Grassmann web $GW(n+1,n,r)$ generated by the surfaces U_ξ, $\xi = 1,\ldots,n+1$, of codimension r in P^{r+n-1}. Denote by p_ξ a point of U_ξ. The leaves of the foliations λ_ξ are Schubert varieties $\sum(p_\xi)$ of subspaces $P^{n-1} \subset P^{n+r-1}$ passing through the point p_ξ. A 3-subweb of the web $GW(n+1,n,r)$ is constructed as follows. Take $n-2$ leaves $\sum(p_{\xi_1}),\ldots,\sum(p_{\xi_{n-2}})$. Their intersection is a bundle $\sum(p_{\xi_1},\ldots,p_{\xi_{n-2}})$ of $(n-1)$-planes in P^{r+n-1} the centre of which is the $(n-3)$-plane $[p_{\xi_1},\ldots,p_{\xi_{n-2}}] = P^{n-3}$. Let P^{r+1} be a subspace of P^{r+n-1} that is complimentary to P^{n-3}. Consider a projection $\pi : P^{r+n-1} - P^{n-3} \to P^{r+1}$ from the centre P^{n-3}. Under this projection, to an $(n-1)$-plane of the bundle $\sum(p_{\xi_1},\ldots,p_{\xi_{n-2}})$ there corresponds a straight line in P^{r+1}, and to the whole bundle there corresponds the Grassmannian $G(1,r+1)$. The submanifolds $U_{\xi_{n-1}}, U_{\xi_n}$, and $U_{\xi_{n+1}}$ under projection π will become three hypersurfaces $\pi(U_{\xi_{n-1}}), \pi(U_{\xi_n})$, and $\pi(U_{\xi_{n+1}})$ of the subspace P^{r+1}. Because of this, the 3-subweb cut out by the foliations $\lambda_{\xi_{n-1}}, \lambda_{\xi_n}, \lambda_{\xi_{n+1}}$ on the bundle $\sum(p_{\xi_1},\ldots,p_{\xi_{n-2}})$ is equivalent to a Grassmann web $GW(3,2,r)$.

Now it is not so difficult to prove the theorem.

Necessity. Suppose that a web $W(n+1,n,r)$ is algebraisable. Then it is Grassmannisable and therefore, by Theorem 2.6.4, transversally geodesic and isoclinic. In addition, up to equivalence, such a web is defined by an algebraic surface V_{n+1}^r of dimension r and degree $n+1$ in P^{r+n-1}. Its 3-subwebs are determined by the projections of this surface from an $(n-3)$-dimensional centre generated by $n-2$ of its points. These projections constitute cones of the third order. This implies that the three-webs defined by these projections are hexagonal, and therefore, by Definition 1.10.1, the web $W(n+1,n,r)$ is hexagonal.

Sufficiency. Suppose that a web $W(n+1,n,r)$ is isoclinic and hexagonal. By Theorem 1.10.2, it is transversally geodesic, and consequently, by Theorem 1.6.4, it is Grassmannisable. Besides, the hexagonality of $W(n+1,n,r)$ implies that the projection of any three surfaces $U_{\xi_{n-1}}, U_{\xi_n}$,and $U_{\xi_{n+1}}$ from a centre defined by points $p_{\xi_1},\ldots,p_{\xi_{n-2}}$, lying on the remaining $n-2$ surfaces $U_{\xi_1},\ldots,U_{\xi_{n-2}}$, constitutes a third order cone. It follows from the above that all surfaces U_ξ, $\xi=1,\ldots,n+1$, belong to an r-dimensional algebraic surface V^r_{n+1} of degree $n+1$, i.e. the web $W(n+1,n,r)$ is algebraisable. ■

In Chapter 5 we will give an analytical proof of this theorem.

2.6.5 The Algebraisation Problem for Webs $W(d,n,r)$, $d>n+1$

The algebraisation problem for a web $W(d,n,r)$ is equivalent to the algebraisation problem for $d\geq n+1$ smooth submanifolds of dimension r (codimension $n-1$) of a real projective space P^{r+n-1}, $n\geq 2$.

Definition 2.6.10 Let U_ξ, $\xi=1,\ldots,d$, be a system of smooth submanifolds of dimension r in P^{r+n-1}. This system is called *algebraisable* if in P^{r+n-1} there exists a projective variety V^r_d of degree d and dimension r to which all submanifolds U_ξ belong.

Theorems 2.6.8 and 2.6.9 give the algebraisation conditions correspondingly for 3 and $n+1$ smooth submanifolds of P^{r+1} and P^{r+n-1}, $n>2$, in terms of a Grassmann web generated by surfaces U_ξ.

First we will find algebraisation conditions for d smooth hypersurfaces in a projective space P^{r+1}.

Let U_ξ, $\xi=1,\ldots,d$, be smooth hypersurfaces in P^{r+1} and suppose there exists a straight line l^0 that intersect each of them at a single point $p^0_\xi=l^0\cap U_\xi$. Let D be a neighbourhood of the line l^0 on the Grassmannian $G(1,r+1)$ of straight lines of the space P^{r+1}, such that for any straight line $l\in D$ there exists a unique point $p_\xi=l\cap U_\xi$, and let $\tilde U_\xi$ be a neighbourhood of the point p^0_ξ on U_ξ such that at least one straight line from D passes through any of its points. In order not to complicate the notation, we will suppose that $\tilde U_\xi=U_\xi$.

Let us fix affine coordinates $(x_0,\ldots,x_r)$ in P^{r+1} and fix line coordinates $(m_1,\ldots,m_r,b_1,\ldots,b_r)$ (i.e., local coordinates on $G(1,r+1)$ where a line l is given by $x_i=m_ix_0+b_0$, $i=1,\ldots,r$. We will assume that l^0 has line coordinates $m_i=0$, $b_i=0$ for all i. For convenience, we will write $m=(m_1,\ldots,m_r)$, $b=(b_1,\ldots,b_r)$.

A line $l=l(m,b)\in D$ intersects each U_ξ in a point $p_\xi=p_\xi(m,b)$.
Let $X_\xi=X_\xi(m,b)$ be the 0th coordinate of p_ξ.

Theorem 2.6.11 *There exists an algebraic hypersurface V^r_d of degree d containing each $U_\xi,\xi=1,\ldots,d$, if and only if*

$$\sum_\xi \frac{\partial^2 X_\xi}{\partial b_k\partial b_l}=0 \tag{2.6.1}$$

for all $k, l = 1, \ldots, r$.

Proof. *Necessity.* Suppose there exists an algebraic hypersurface V_d^r of degree d which contains each U_ξ, and which satisfies the degree d polynomial equation

$$p(x_0, x_1, \ldots, x_r) = 0. \tag{2.6.2}$$

The points $p_\xi = p_\xi(m, b)$ are the intersection points of $l = l(m, b)$ with V_d^r. Since p_ξ is on l, $p_\xi = (X_\xi, m_1 X_\xi + b_1, \ldots, m_r X_\xi + b_r)$ where $X_\xi = X_\xi(m, b)$ is the $0th$ coordinate of p_ξ. p_ξ also lies on V_d^r, so that

$$p(p_\xi) = p(X_\xi,\ m_1 X_\xi + b_1, \ldots, m_r X_\xi + b_r) = 0. \tag{2.6.3}$$

Equation (2.6.3) holds for all (m, b) near $(0, 0)$.

If we set

$$p_0(t) = p(t, m_1 t + b_1, \ldots, m_r t + b_r), \tag{2.6.4}$$

then p_0 is a polynomial of degree d in t, with coefficients depending on (m, b), such that the X_ξ, $\xi = 1, \ldots, d$, are the d roots of the equation $p_0(t) = 0$.

To be more concrete, let

$$p_0(t) = a_0 t^d - a_1 t^{d-1} + \ldots \pm a_d. \tag{2.6.5}$$

Then $p_0(t) = a_0(t - X_1)(t - X_2) \ldots (t - X_d)$, the X_ξ, $\xi = 1, \ldots, d$, being the roots of the degree d polynomial equation $p_0(t) = 0$. We see that $\sum_{\xi=1}^d X_\xi = a_1/a_0$.

We claim that $\sum_{\xi=1}^d X_\xi$ is (at most) linear in the b's. To see this, we examine the polynomials p and p_0 more carefully. Let $\alpha = (\alpha_0, \alpha_1, \ldots, \alpha_r)$ be an $(r+1)$-tuple of non-negative integers, and let $|\alpha| = \alpha_0 + \alpha_1 + \ldots + \alpha_r$. Define x^α to mean $x_0^{\alpha_0} x_1^{\alpha_1} \ldots x_r^{\alpha_r}$. Then p has the form

$$p(x) = \sum_{|\alpha| \leq d} c_\alpha x^\alpha, \tag{2.6.6}$$

for some constants c_α. Now recall that $p_0(t)$ is expressed by (2.6.4), so that

$$p_0(t) = \sum_{|\alpha| \leq d} c_\alpha t^{\alpha_0} (m_1 t + b_1)^{\alpha_1} \ldots (m_r t + b_r)^{\alpha_r}. \tag{2.6.7}$$

Using the binomial theorem to expand, we see that the coefficient of t^d in $p_0(t)$ is

$$a_0 = \sum_{|\alpha| = d} c_\alpha m_1^{\alpha_1} m_2^{\alpha_2} \ldots m_r^{\alpha_r}, \tag{2.6.8}$$

while the coefficient of t^{d-1} in $p_0(t)$ is

$$\begin{aligned} -a_1 \;=\; & \sum_{|\alpha| = d-1} c_\alpha m_1^{\alpha_1} m_2^{\alpha_2} \ldots m_r^{\alpha_r} \\ & + \sum_{|\alpha| = d} c_\alpha (\alpha_1 m_1^{\alpha_1 - 1} b_1 m_2^{\alpha_2} \ldots m_n^{\alpha_n} + \alpha_2 m_1^{\alpha_1} m_2^{\alpha_2 - 1} b_2 m_3^{\alpha_3} \ldots m_r^{\alpha_r} + \ldots \\ & + \alpha_n m_1^{\alpha_1} \ldots m_{r-1}^{\alpha_{r-1}} m_r^{\alpha_r - 1} b_r). \end{aligned} \tag{2.6.9}$$

a_0 is independent of b, while a_1 is at most linear in b. Thus $\sum_{\xi=1}^{d} X_\xi$ is at most linear in b. Hence the second b-partials of $\sum_{\xi=1}^{d} X_\xi$ vanish.

Sufficiency. Suppose that U_ξ is defined by the equation

$$\phi_\xi(x_0, \ldots, x_r) = 0. \tag{2.6.10}$$

Then $X_\xi = X_\xi(m, b)$ satisfies the equation

$$\phi_\xi(X_\xi, m_1 X_\xi + b_1 \ldots, m_r X_\xi + b_r) = 0, \tag{2.6.11}$$

for all (m, b) near $(0,0)$. Differentiating (2.6.11) with respect to m_j and b_j , we correspondingly get

$$\frac{\partial \phi_\xi}{\partial x_0}\frac{\partial X_i}{\partial m_j} + \sum_{k=1}^{r}\left[\frac{\partial \phi_\xi}{\partial x_k}\left(m_k \frac{\partial X_i}{\partial m_j} + \delta_{kj} X_i\right)\right] = 0 \tag{2.6.12}$$

and

$$\frac{\partial \phi_\xi}{\partial x_0}\frac{\partial X_i}{\partial b_j} + \sum_{k=1}^{r}\left[\frac{\partial \phi_\xi}{\partial x_k}\left(m_k \frac{\partial X_i}{\partial b_j} + \delta_{kj}\right)\right] = 0. \tag{2.6.13}$$

Multiply equation (2.6.13) by X_i and subtract equation (2.6.12) to obtain

$$X_i \frac{\partial X_i}{\partial b_j}\left[\frac{\partial \phi_\xi}{\partial x_0} + \sum_{k=1}^{r} m_k \frac{\partial \phi_\xi}{\partial x_k}\right] = \frac{\partial X_i}{\partial m_j}\left[\frac{\partial \phi_\xi}{\partial x_0} + \sum_{k=1}^{r} m_k \frac{\partial \phi_\xi}{\partial x_k}\right]. \tag{2.6.14}$$

Our assumption that the line l^0 intersects each U_ξ transversely means that the factor $\frac{\partial \phi_\xi}{\partial x_0} + \sum_{k=1}^{r} m_k \frac{\partial \phi_\xi}{\partial x_k} \neq 0$ for m near 0. We conclude that

$$\frac{\partial X_i}{\partial m_j} = X_i \frac{\partial X_i}{\partial b_j}. \tag{2.6.15}$$

To prove sufficiency, we assume that (2.6.1) holds for all k, l. This implies that $\sum_{\xi=1}^{d} X_\xi$ is (at most) linear in b.

Lemma 2.6.12 *For $q \geq 1$, $\sum_{\xi=1}^{d} X_\xi^q$ is of degree (at most) q in b.*

Proof. We will prove by induction. The case $q = 1$ is covered in the preceding paragraph.

We next remark that if $f(m, b)$ is of degree q in b with coefficients which are functions of m, then $\frac{\partial f}{\partial m}$ is also of degree q in b.

For the induction step, we note, using (2.6.15), that

$$\frac{1}{q}\frac{\partial(\sum_{\xi=1}^{d} X_\xi^q)}{\partial b_j} = \sum_{a=1}^{d} X_\xi^{q-1}\frac{\partial X_\xi}{\partial b_j} = \sum_{q=1}^{d} X_\xi^{q-2}\frac{\partial X_\xi}{\partial m_j} = \frac{1}{q-1}\frac{\partial(\sum_{\xi=1}^{d} X_\xi^{q-1})}{\partial m_j}. \tag{2.6.16}$$

By the induction hypothesis, $\sum_{\xi=1}^{d} X_\xi^{q-1}$ is of degree $q-1$ in b. The remark above implies that $\partial(\sum_{\xi=1}^{d} X_\xi^{q-1})/\partial m_j$ is of degree $q-1$ in b, so that $\sum_{\xi=1}^{d} X_\xi^q$ is of degree q in b.

We next define the function $A(\zeta, m, b)$ by

$$A(\zeta, m, b) = \prod_q (\zeta - X_\xi(m, b)). \tag{2.6.17}$$

In (2.6.17) A is a polynomial of degree d in ζ, with coefficients which are functions of m and b:

$$A(\zeta, m, b) = \zeta^d - A_1(m, b)\zeta^{d-1} + \ldots \pm A_d(m, b) \tag{2.6.18}$$

In (2.6.18) $(-1)^q A_q(m, b)$ is just the qth elementary symmetric function in the X_ξ's.

Lemma 2.6.13 *$A_q(m, b)$ is of degree q in b, for $q \geq 1$.*

Proof. $A_1 = -\sum_{\xi=1}^{d} X_\xi$, which is of degree 1 in b, as above.

For the general induction step, we recall that the A_q's are related to the $(\sum_{\xi=1}^{d} X_\xi^q)$'s by the Newton identities. To be more precise, let $T_q = \sum_{\xi=1}^{d} X_\xi^q$. Then the Newton identities (see [Her 75]) state

(i) $T_q + A_1 T_{q-1} + A_2 T_{q-2} + \ldots + A_{q-1} T_1 + q A_q = 0, \quad q = 1, 2, \ldots, d,$

(ii) $T_q + A_1 T_{q-1} + \ldots + A_d T_{q-d} = 0, \quad q > d.$

Only statement (i) is relevant for us. A_q is expressed in terms of T_q or expressions of the form $A_k T_{q-k}$. By Lemma 2.6.12, T_q has degree s in b, and, by induction, we have that A_k is of degree k in b. A product $A_k T_{q-k}$ has degree $k + (q-k) = q$ in b. The result follows. ■

We introduce now new variables $\xi = (\xi_1, \ldots, \xi_r)$ and let $\xi_i = m_i\zeta + b_i$. We substitute the last expression into (2.6.12) and define $B = B(\zeta, \xi, m)$ by

$$B(\zeta, \xi, m) = A(\zeta, m, \xi - m\zeta). \tag{2.6.19}$$

Thus $B(\zeta, \xi, m) = \zeta^d + A_1(m, \xi - m\zeta)\zeta^{d-1} + \ldots + A_d(m, \xi - m\zeta)$. Since $A_q(m, b)$ is of degree q in b, $A_q(m, \xi - m\zeta)\zeta^{d-q}$ is of degree d in ξ and ζ. Consequently, B is a polynomial of degree d in ξ and ζ, with coefficients that are functions of m.

We will prove that, if $(\zeta, \xi) \in U_\eta$, then $B(\zeta, \xi, m) = 0$, for all m. We have

$$B(\zeta, \xi, m) = A(\zeta, m, \xi - m\zeta) = \prod_\eta (\zeta - X_\eta(m, \xi - m\zeta)). \tag{2.6.20}$$

In order to understand what $X_\eta(m, \xi - m\zeta)$ is, we recall that $X_\eta(m, b)$ is the 0th coordinate of the intersection point of U_η with the line $l(m, b)$. The line $l(m, b)$ satisfies $x_i = m_i x_0 + b$. Thus $X_\eta(m, \xi - m\zeta)$ is the 0th coordinate of the intersection point of U_η with the line $l(m, \xi - m\zeta)$, which satisfies $x_i = m_i x_0 + \xi_i - m_i\zeta$. That is, l satisfies $x_i - \xi_i = m_i(x_0 - \zeta)$. This says that l is any of the lines which pass through (ζ,ξ). Since $(\zeta,\xi)\in U_\eta,(\zeta,\xi)$ itself is the intersection of U_η and the line l. Thus $X_\eta(m,\xi-m\zeta)=\zeta$. This in turn forces $B(\zeta,\xi,m)=0$, because its ath factor vanishes.

If we define $V(m)=\{(\zeta,\xi):B(\zeta,\xi,m)=0\}$, then $V(m)$ is an algebraic hypersurface of degree d. In addition, it has just been shown that $V(m)$ contains each U_η. Because d is both the degree of $V(m)$ and the number of hypersurfaces U_η that $V(m)$ contains, $V(m)$ cannot actually depend on m. So, we have produced the desired algebraic hypersurface V_d^r. ■

Let $\sigma_\xi=T_{p_\xi}(U_\xi)$ be the tangent hyperplane to U_ξ at p_ξ, and let $\phi(p_\xi)=<\sigma_\xi,\;\; d^2p_\xi>$ be the second fundamental quadratic form of U_ξ at p_ξ. If the forms $\phi(p_\xi)$ are normalised in an appropriate way and calculated for directions on the hypersurfaces U_ξ that correspond to each other in a correspondence given by two-dimensional planes passing through the straight line l, then condition (2.6.1) is equivalent to

$$\sum_{\xi=1}^{d}\phi(p_\xi)=0, \tag{2.6.21}$$

and we obtain the following result.

Corollary 2.6.14 *The hypersurfaces U_ξ belong to the same algebraic hypersurface V_d^r of degree d if and only the relation* (2.6.21) *holds.* ■

We will give another, analytical proof of Corollary 2.6.14 for the case $d=4$ and $n=2, r\geq 2$ in Section 7.8.

Theorem 2.6.11 and the projection method used in our proof of Theorem 2.6.9 allow us to find a condition of algebraisability of a system of submanifolds U_ξ for $r\geq 2,\;\; n\geq 2$, and $d\geq n+1$.

Let $U_\xi,\;\; \xi=1,\dots,d,\;\; r\geq 2,\;\; n\geq 2,\;\; d\geq n+1$, be a system of smooth submanifolds (of class C^{d_1+1}, where $d_1=d-n+2$) of dimension r of the space P^{r+n-1}. Suppose there exists a subspace $L^0\subset P^{r+n-1}$ of dimension $n-1$ that intersects each of the U_ξ at the single point $p_\xi^0=L^0\cap U_\xi$, and the points p_ξ^0 are in general position in L^0. Also, let D be a neighbourhood of L on the Grassmannian $G(n-1,r+n-1)$ that is chosen in the same way as was indicated above for the case $n=2$. In addition, let $L\subset D$ and $p_\xi=L\cap U_\xi$. We consider a subsystem consisting of $n-2$ points $p_{\xi_s},\;\; s=1,\dots,n-2$, their linear span $Z=[p_{\xi_1},\dots,p_{\xi_{n-2}}]$ and submanifolds $U_{\xi_u},\;\; u=n-1,\dots,d$, not containing these points. Let T_{ξ_u} be the tangent subspace to U_{ξ_u} at the point $p_{\xi_u}=L\cap U_{\xi_u}$, let $\sigma_{\xi_u}=[Z,T_{\xi_u}]$ be the tangent hyperplane passing through Z and tangent to U_{ξ_u} at the point p_{ξ_u}, and finally let $\phi(p_{\xi_u})=<\sigma_{\xi_u},\; d^2p_{\xi_u}>$ be the second fundamental quadratic form of the submanifold U_{ξ_u} at p_{ξ_u} with respect to the hyperplane σ_{ξ_u}. Then we have the following theorem:

Theorem 2.6.15 *A system of submanifolds U_ξ is algebraisable if and only if the condition*

$$\sum_{u=n-2}^{d-1} \phi(p_{\xi_u}) = 0. \qquad (2.6.22)$$

is satisfied for any subspace $L \subset D$ and any distinct values ξ_u.

Proof. Note that dim $Z = n-3$. We consider a subspace $P^{r+1} \subset P^{r+n-1}$ complimentary to Z and the projection $\pi : P^{r+n-1} - Z \to P^{r+1}$ from the centre Z. Under this projection, subspaces $L \subset D$ passing through Z have corresponding lines $l \subset P^{r+1}, l = \pi(L)$. The submanifolds U_{ξ_u} under this projection become hypersurfaces $\pi(U_{\xi_u}) \subset P^{r+1}$, and the quadratic forms $\phi(p_{\xi_u})$ are the second fundamental forms of these hypersurfaces at the points $\pi(p_\xi) = l \cap \pi(U_{\xi_u})$. By Theorem 2.6.11, it follows from the above that the hypersurfaces $\pi(U_{\xi_u})$ belong to an algebraic hypersurface $V^r_{d_1} \subset P^{r+1}$, $d_1 = d - n + 2 > 2$. Since the manifolds U_{ξ_u} were taken arbitrarily from U_ξ and the points p_{ξ_u} were taken arbitrarily on the rest of U_ξ, it follows from this that all of the manifolds U_ξ belong to a projective algebraic variety V^r_d of dimension r and degree d. The sufficiency of the condition of Theorem 2.6.15 is proved. Its necessity follows easily from the necessity of condition (2.6.21) for hypersurfaces. ■

We add the following to this. Let L be a fixed subspace of dimension $n-1$ passing through Z, and let P^n be an arbitrary subspace containing L. The subspaces P^n establish a bijective correspondence among one-dimensional subspaces of the tangent spaces T_{ξ_u} to the manifolds U_{ξ_u} at the points $p_{\xi_u} = L \cap U_{\xi_u}$. Because of this, all of the quadratic forms $\phi(p_{\xi_u})$ can be represented in the form

$$\phi(p_{\xi_u}) = K_{\xi_u ij}\omega^i\omega^j, \qquad (2.6.23)$$

where $i, j = 1, \ldots, r$, and ω^i are common coordinates of corresponding tangent vectors in T_{ξ_u}. Thus, condition (2.6.22) can be rewritten in the form

$$\sum_{u=n-2}^{d} K_{\xi_u ij} = 0, \qquad (2.6.24)$$

and condition (2.6.21) in the form

$$\sum_{\xi=1}^{d} K_{\xi ij} = 0. \qquad (2.6.25)$$

A system of submanifolds $U_\xi \subset P^{r+n-1}$, $\xi = 1, \ldots, d$, $d \geq n+1$, determines a Grassmann web $GW(d, n, r)$ in a domain D of the Grassmannian $G(n-1, r+n-1)$. Therefore, Theorems 2.6.11 and 2.6.15 represent a condition for algebraisability of a web $GW(d, n, r)$ expressed in terms of submanifolds U_ξ generating the web.

For some values of d, n, and r not all of the relations (2.6.22) are independent. For instance, for $d = n + 1$ it follows from Theorem 1.10.4 that only $\binom{n}{2}$ out of the total number of $\binom{n}{3}$ relations (2.6.22) are independent.

NOTES

2.1. In our geometric introduction of almost Grassmann structures on a differentiable manifold which is based on a field of Segre cones given on the manifold we have followed the exposition in Akivis' papers [Ak 80, 82]. Lemma 1.2.4 is also due to Akivis [Ak 80, 82].

2.2. The presented way of finding the structure equations of an almost Grassmann manifold and its torsion tensor is due to Goldberg [G 75c]. In [G 75c] the expression (2.2.42) of the torsion tensor $\overset{*}{u}{}^{i\beta\gamma}_{\alpha jk}$ in a general (not specialized) frame has been constructed at the first time. Using another method, Hangan [Ha 80] deduced this expression again.

2.3. The results of this section are due to Goldberg [G 75c]. The proof of Theorem 2.3.1 follows [Ak 82].

2.4-2.5. All the results of these sections except Corollary 2.4.3 are due to Goldberg [G 75c]. Corollary 2.4.3 is due to Akivis [Ak 81].

2.6. The solution of the Grassmannisation problem for webs $W(n+1, n, r)$, $n > 2, r > 2$ (Theorem 2.6.4), is due to Goldberg [G 75c, 82b]. For $n = 2$, i.e. for three-webs $W(3, 2, r)$, this theorem was proved by Akivis [Ak 74]. The proof of Theorem 2.6.4 follows [Ak 82]. The Grassmannisation problem for webs $W(d, n, r)$, $d > n + 1$ (Theorem 2.6.5), was solved by Akivis [Ak 80, 81, 82].

For complex projective space $P^{r+n-1}(\mathbf{C})$ a condition of algebraisability is contained in Abel's well-known theorem [Gr 76]. In the case of real projective space the methods used in [Gr 76] in the proof of Abel's theorem are not applicable.

In real projective space a local condition of algebraisability of webs $W(3, 2, r)$ (Theorem 2.6.8) has been found by Akivis [Ak 73, 82]. By means of the same methods this condition was extended by Akivis and Goldberg [AG 74] for webs $W(4, 3, r)$ and by Goldberg [G 75e] (see also [G 82b]) for webs $W(n+1, n, r)$ (Theorem 2.6.9) and for webs $W(4, 2, r)$ [G 82d] (see also Section 7.9). The proof of Theorem 2.6.9 follows [Ak 82]. Wood in his dissertation [W 82] established a similar criterion of algebraisability for webs $W(d, 2, r)$, $d > 4$ (Theorem 2.6.11). Wood's proof in [Wo 82] is a generalisation of the proof in [G 82d]. He gave another much shorter and simplier proof in [Wo 84]. Finally Akivis [Ak 83b] solved the algebraisation problem for webs $W(d, n, r)$, $d > n + 1$ (Theorem 2.6.15). In the text we reproduced his proof. Its essential part is Theorem 2.6.11. In the proof of this theorem we used an expanded version of [Wo 84] kindly provided by the author.

Chapter 3

Local Differentiable n-Quasigroups Associated with a Web $W(n+1,n,r)$

3.1 Local Coordinate n-Quasigroups of a Web $W(n+1,n,r)$

Definition 3.1.1 Let $X_\xi, \xi = 1,\ldots,n+1$, be differentiable manifolds of the same dimension r. Let

$$f : X_1 \times X_2 \times \ldots \times X_n \to X_{n+1} \tag{3.1.1}$$

be a mapping satisfying the following conditions: if $a_{n+1} = f(a_1,\ldots,a_n)$, then

i) for any neighbourhood U_{n+1} of a_{n+1} there exist neighbourhoods U_α of a_α, $\alpha = 1,\ldots,n$, such that for any $x_\alpha \in U_\alpha$ the value of the function $f(x_1,\ldots,x_n)$ is defined and $f(x_1,\ldots,x_n) = x_{n+1} \in U_{n+1}$;

ii) for any neighbourhood U_α of a_α, α fixed, there exist neighbourhoods $U_{\hat\alpha}$ of $a_{\hat\alpha}$, $\hat\alpha \neq \alpha$, and a neighbourhood U_{n+1} of a_{n+1} such that for any $x_{\hat\alpha} \in U_{\hat\alpha}$ and $x_{n+1} \in U_{n+1}$ the equation $f(x_1,\ldots,x_n) = x_{n+1}$ is solvable for x_α, and $x_\alpha \in U_\alpha$.

If the manifolds $X_\xi, \xi = 1,\ldots,n+1$, and the function f are of class C^p, we say that there is given an $(n+1)$*-base local differentiable n-quasigroup* Q_r (abbreviation l.d. n-quasigroup).

Note that in the sequel the term "l.d. n-quasigroup" always means "$(n+1)$-base l.d. n-quasigroup".

Definition 3.1.2 Suppose that $X_1 = \ldots = X_n = X_{n+1} = X$ and there exists at least one element $e \in X$ such that

$$f(x,e,\ldots,e) = f(e,x,e,\ldots,e) = \ldots = f(e,e,\ldots,e,x) = x,$$

Then the l.d. n-quasigroup Q_r is called a *local differentiable n-loop* (abbreviation l.d. n-loop).

Definition 3.1.3 If $X_1 = \ldots = X_n = X_{n+1} = X$, and for any $x_1, \ldots, x_n$ we have

$$\begin{aligned} f[f(x_1,\ldots,x_n),x_{n+1},\ldots,x_{2n-1}] &= f[x_1,f(x_2,\ldots,x_{n+1}),x_{n+2},\ldots,x_{2n-1}] = \\ \ldots &= f[x_1,\ldots,x_{n-1},f(x_n,\ldots,x_{2n-1})], \end{aligned}$$

then the l.d. n-quasigroup Q_r is said to be an *l.d. n-group.*

Let us consider a few examples of l.d. n-quasigroups.

Example 3.1.4 An equation $x_{n+1} = f(x_1, \ldots, x_n)$ defines an l.d.n-quasigroup in some domain $D \subset \mathbf{R}^n$ if $f_{x_\alpha} \neq 0$ in D (f_{x_α} is the partial derivative of the function f with respect to x_α). Here X_α is the projection of D onto the axis Ox_α and X_{n+1} is the range of the function f.

Example 3.1.5 The equations

$$x^i_{n+1} = f^i(x_1^{j_1}, \ldots, x_n^{j_n}), \quad i, j_1, \ldots, j_n = 1, \ldots, r, \tag{3.1.2}$$

in some domain $D \subset \mathbf{R}^{nr}$ define an l.d. n-quasigroup if in D

$$\det\left(\frac{\partial f^i}{\partial x_\alpha^{j_\alpha}}\right) \neq 0, \qquad \alpha = 1, 2, \ldots, n. \tag{3.1.3}$$

In this case X_α is the projection of D onto the subspace defined by the axes $Ox^1_\alpha, \ldots, Ox^r_\alpha$ and X_{n+1} is the range of the functions f^i.

Example 3.1.6 Let $X_\xi, \xi = 1, \ldots, n+1$, be complex planes and let the mapping (3.1.1) is given. Then $x_{n+1} = f(x_1, \ldots, x_n)$. If $x_\xi = a_\xi + ib_\xi$, then $a_{n+1} = \varphi(a_\alpha, b_\beta)$ and $b_{n+1} = \psi(a_\alpha, b_\beta), \alpha, \beta = 1, \ldots, n$. In order to define a local n-quasigroup, the functions φ and ψ must satisfy Cauchy — Riemann equations with respect to each pair of the variables a_α and b_β.

Example 3.1.7 Suppose that P^{r+n-1} is a projective space of dimension $r+n-1$ and $X_\xi, \xi = 1, \ldots, n+1$, are surfaces of dimension r (codimension $n-1$) in P^{r+n-1}. These r-surfaces define an l.d. n-quasigroup if we give a correspondence among them as follows. We will take n arbitrary points $p_1, \ldots, p_n$ on surfaces $X_1, \ldots, X_n$, one point on each of the surfaces, provided that these points are in general position (do not belong to a space of $n-2$ dimension). These points define an $(n-1)$-plane $P^{n-1} = [p_1, \ldots, p_n]$. The plane P^{n-1} and the surface X_{n+1} have the only common point p_{n+1}. So, we defined a mapping (3.1.1). It is easy to see that all requirements of Definition 3.1.1 are satisfied.

The web $W(n+1,n,r)$ defined in Example 3.1.7 is a Grassmann web (see Definition 2.6.1, p.89). We will study Grassmann webs in more detail in Chapter 5.

Definition 3.1.8 Let $Q(f)$ be an n-quasigroup given on a set Q. On the set $S = Q \times \ldots \times Q$ it defines a web $W(n+1,Q)$ formed by $n+1$ foliations λ_ξ whose leaves are determined by equations

$$\begin{aligned} F_\alpha &= \{(x_1,\ldots,x_n): \ x_\alpha = \text{const}, \ \alpha = 1,\ldots,n\}, \\ F_{n+1} &= \{(x_1,\ldots,x_n): \ f(x_1,\ldots,x_n) = \text{const}\}. \end{aligned}$$

Such a web is called *abstract.*

Conversely, let an abstract $(n+1)$-web W be formed on a set S by $n+1$ foliations λ_ξ with bases X_ξ, $\xi = 1,\ldots,n+1$, such that $S = X_{\xi_1} \times \ldots \times X_{\xi_n}, \xi_1,\ldots,\xi_n \in \{1,\ldots,n+1\}$. Then we have a mapping

$$q_{\xi_1,\ldots,\xi_n} : X_{\xi_1} \times \ldots \times X_{\xi_n} \to X_{\xi_{n+1}} \tag{3.1.4}$$

for which the corresponding leaves pass through the same point $p \in S$. This mapping gives an $(n+1)$-base n-quasigroup.

In particular, if each foliation λ_ξ is an r-dimensional differentiable manifold, then $S = X^{nr}$, and we obtain an l.d. n-quasigroup Q_r defined by a web $W(n+1,n,r)$. Such a web defines a differentiable mapping of the direct product of any n of foliations X_ξ onto the $(n+1)$th one: leaves $V_{\xi_1},\ldots,V_{\xi_n}$ of distinct foliations $\lambda_{\xi_1},\ldots,\lambda_{\xi_n}$ have a unique common point $p = \cap_{\alpha=1}^{n} V_{\xi_\alpha}$ and to these leaves there corresponds a leaf V_{n+1} of the foliation λ_{n+1} passing through the point p. This correspondence can be written as in (3.1.4) and therefore $V_{\xi_{n+1}} = q_{\xi_1\ldots\xi_n}(V_{\xi_1},\ldots,V_{\xi_n})$. If the leaves $V^0_{\xi_1},\ldots,V^0_{\xi_n}$ have a common point p^0 and $V^0_{n+1} = q_{\xi_1\ldots\xi_n}(V^0_{\xi_1},\ldots,V^0_{\xi_n})$, then the mapping $q_{\xi_1\ldots\xi_n}$ is defined in some neighbourhood of the point p^0 and in this neighbourhood it admits the inverse mappings $q_{\xi_{n+1}\xi_2\ldots\xi_n}, \ q_{\xi_1\xi_{n+1}\ldots\xi_n}, \ \ldots, \ q_{\xi_1\xi_2\ldots\xi_{n-1}\xi_{n+1}}$.

Definition 3.1.9 An l.d. n-quasigroup $q_{\xi_1\ldots\xi_n}$ determined by this mapping is called a *coordinate* n-quasigroup of a web $W(n+1,n,r)$.

Our consideration shows that with a web $W(n+1,n,r)$ there are associated $(n+1)!$ coordinate n-quasigroups.

Definition 3.1.10 Any two coordinate n-quasigroups associated with a web $W(n+1,n,r)$ are said to be *parastrophic* to each other and the transition from one of them to another is called a *parastrophy.*

Proposition 3.1.11 *The set of* $(n+1)!$ *coordinate n-quasigroups of an* $(n+1)$*-web is closed under parastrophies.*

Proof. Let σ and τ be permutations of $\xi_1\ldots\xi_n$, and let h_α, $\alpha = 1,\ldots,n$, be the parastrophies:

$$h_\alpha q_{\xi_1\ldots\xi_{\alpha-1}\xi_\alpha\xi_{\alpha+1}\ldots\xi_n} = q_{\xi_1\ldots\xi_{\alpha-1}\xi_{n+1}\xi_{\alpha+1}\ldots\xi_n}. \tag{3.1.5}$$

Then Proposition 3.1.11 follows from the following identities that can be easily checked:

$$h_\alpha(h_\alpha q_{\xi_1\ldots\xi_n}) = q_{\xi_1\ldots\xi_n}, \qquad h_\beta(h_\alpha q_{\xi_1\ldots\xi_n}) = \bar\sigma(h_\beta q_{\xi_1\ldots\xi_n}),$$
$$\tau(\sigma q_{\xi_1\ldots\xi_n}) = (\tau\sigma) q_{\xi_1\ldots\xi_n}, \quad h_\alpha(\sigma q_{\xi_1\ldots\xi_n}) = \sigma(h_\alpha q_{\xi_1\ldots\xi_n}),$$

where $\bar\sigma$ is a fixed permutation of $\xi_1\ldots\xi_n$. ■

Definition 3.1.12 Suppose that there are given $n+1$ differentiable and locally invertible mappings (diffeomorphisms) $g_\xi, \xi = 1,\ldots,n+1$, of the manifolds X_ξ by means of which there is defined an l.d.n-quasigroup $Q_r : g_\xi : X_\xi \to \bar X_\xi$. If we put

$$\bar x_{n+1} = g_{n+1}(x_{n+1}) = g_{n+1}[f(x_1,\ldots,x_n)] = g_{n+1}\{f[g_1^{-1}(\bar x_1),\ldots,g_n^{-1}(\bar x_n)\},$$

then we obtain a mapping

$$\bar f : \bar X_1 \times \bar X_2 \times \ldots \times \bar X_n \to \bar X_{n+1}.$$

The mapping f is an l.d. n-quasigroup $Q_r(f)$ which is called *isotopic* to the l.d. n-quasigroup $Q_r(f)$.

Proposition 3.1.13 *Two abstract $(n+1)$-webs are equivalent if and only if their corresponding coordinate n-quasigroups are isotopic.*

Proof. This follows from the definitions of equivalent webs and isotopic n-quasigroups. ■

Thus, there exists a family of isotopic l.d. n-quasigroups corresponding to a given abstract $(n+1)$-web. Web geometry studies properties of webs corresponding to those properties of n-quasigroups which are invariant under isotopy.

Let us find out what kind of coordinate n-quasigroups a parallelisable web $PW(n+1,n,r)$ has.

Theorem 3.1.14 *A web $W(n+1,n,r)$ is parallelisable if and only if its corresponding coordinate n-quasigroups are abelian n-groups.*

Proof. According to Example 1.1.3, a parallelisable web $PW(n+1,n,r)$ is equivalent to a web formed by $n+1$ foliations of parallel $(n-1)r$-planes of an affine space A^{nr}.

If in A^{nr} we place vectors $e_i^{\hat\alpha}$, $\hat\alpha \neq \alpha$, of a moving frame $\{e_i^\alpha;\ i = 1,\ldots,r;\ \alpha = 1,\ldots,n\}$ in the planes of λ_α, then for any point $p \in A^{nr}$ we have $p = x_\alpha^i e_i^{\hat\alpha}$. In this case the planes of λ_α are defined by equations $x_\alpha^i = a_\alpha^i$, and the planes of λ_{n+1} by equations $\sum_\alpha p_{\alpha_j}^i x_\alpha^j = c^i$, where det $(p_{\alpha_j}^i) \neq 0$ for any $\alpha = 1,\ldots,n$ and a_α^i and c^i are constants. If we take an isotopic n-quasigroup setting $\bar x_{n+1}^i = x_{n+1}^i$ and $\bar x_0^j = p_{\alpha_j}^i x_\alpha^j$, then the equations of the $(n+1)$-web become $\bar x_{n+1}^i = \sum_\alpha \bar x_\alpha^i$. Therefore, a coordinate n-quasigroup of a web $PW(n+1,n,r)$ is isotopic to an abelian n-group.

Conversely, if a coordinate n-quasigroup of a web $W(n+1,n,r)$ is isotopic to abelian n-group, then $W(n+1,n,r)$ is parallelisable since the equations $x_{n+1}^i = \sum_\alpha x_\alpha^i$ define a parallel $(n+1)$-web.

Suppose that there is given an n-quasigroup (3.1.1) and $f(a_1,\dots,a_n) = a_{n+1}$ where $a_\alpha \in X_\alpha$ are arbitrary fixed elements in X_α. We map the sets X_ξ into X_{n+1} as follows:

$$u_\alpha = f(a_1,\dots,a_{\alpha-1},x_\alpha,a_{\alpha+1},\dots,a_n), \quad u_{n+1} = x_{n+1}. \tag{3.1.6}$$

As a result of this isotopy we obtain an n-loop $L(a_1,\dots,a_n)$ with $e = a_{n+1}$. To prove this, let us denote the n inverse operations of the n-quasigroup f by f_α. Then it follows from (3.1.6) that

$$x_\alpha = f_\alpha(a_1,\dots,a_{\alpha-1},u_\alpha,a_{\alpha+1},\dots,a_n). \tag{3.1.7}$$

Now we can introduce on the set X_{n+1} in a neighbourhood of a_{n+1} an operation F in the following way:

$$\begin{aligned} u_{n+1} = F(u_1,\dots,u_n) \quad &= \quad f(x_1,\dots,x_n) \\ = f[f_1(u_1,a_2,\dots,a_n), \quad \dots, \quad &f_n(a_1,a_2,\dots,a_{n-1},u_n)]. \end{aligned} \tag{3.1.8}$$

For the operation F:

$$F(a_{n+1},\dots,a_{n+1},u_\alpha,a_{n+1},\dots,a_{n+1}) = u_\alpha. \tag{3.1.9}$$

In fact we have

$$\begin{aligned} &F(a_{n+1},\dots,a_{n+1},u_\alpha,a_{n+1},\dots,a_{n+1}) \\ &= f[f_1(a_{n+1},a_2,\dots,a_n),\dots,f_{\alpha-1}(a_1,\dots,a_{\alpha-2},a_{n+1},a_\alpha,\dots,a_n), \\ &f_\alpha(a_1,\dots,a_{\alpha-1},u_\alpha,a_{\alpha+1},\dots,a_n),f_{\alpha+1}(a_1,\dots,a_\alpha,a_{n+1},a_{\alpha+2},\dots,a_n), \\ &\dots,f_n(a_1,\dots,a_{n-1},a_{n+1})] = f(a_1,\dots,a_{\alpha-1},x_\alpha,a_{\alpha+1},\dots,a_n) = u_\alpha. \end{aligned}$$

The above isotopy maps the n-quasigroup Q_r into an n-loop $L(a_1,\dots,a_n)$.

Definition 3.1.15 An n-loop $L(a_1,\dots,a_n)$ which can be defined for any set $(a_1,\dots,a_n)$, $a_\alpha \in X_\alpha$, is said to be a *principal isotope* of Q_r.

Example 3.1.16 Let us construct a principal isotope for the n-quasigroup defined by

$$x_{n+1} = x_1^2 x_2 \dots x_n, \quad x_\alpha > 0, \quad x_{n+1} > 0.$$

For this we fix points $a_1,\dots,a_n$ and set $a_{n+1} = a_1^2 a_2 \dots a_n$. Then

$$u_1 = x_1^2 a_2 \dots a_n, \quad u_2 = a_1^2 x_2 a_3 \dots a_n, \dots, u_n = a_1^2 a_2 \dots a_{n-1} x_n.$$

Further we find

$$f_1: \quad x_1 = \sqrt{\frac{u_1}{a_2 \dots a_n}} = a_1 \sqrt{\frac{u_1}{a_{n+1}}},$$

$$f_{\hat\alpha}: \quad x_{\hat\alpha} = \frac{u_{\hat\alpha}}{a_1^2 a_2 \dots a_{\hat\alpha-1} a_{\hat\alpha+1} \dots a_n} = \frac{a_{\hat\alpha} u_{\hat\alpha}}{a_{n+1}}, \quad \hat\alpha \neq 1.$$

Therefore we have

$$F(u_1,\dots,u_n) = f(x_1,\dots,x_n) = x_1^2x_2\dots x_n = \frac{u_1\dots u_n}{a_{n+1}^n}a_1^2a_2\dots a_n = \frac{u_1\dots u_n}{a_{n+1}^{n-1}}.$$

It follows from this that

$$\frac{F(u_1,u_2,\dots,u_n)}{a_{n+1}} = \frac{u_1}{a_{n+1}}\cdot\frac{u_2}{a_{n+1}}\cdots\frac{u_n}{a_{n+1}}.$$

We obtained an n-loop which is a principal isotope of the given n-quasigroup. The n-loop obtained is isotopic to the multiplicative group of real numbers.

3.2 Structure of a Web $W(n+1,n,r)$ and Its Coordinate n-Quasigroups in a Neighbourhood of a Point

Let us consider now the coordinate n-quasigroup $q_{12\dots n} = f$ of an $(n+1)$-web $W(n+1,n,r)$. We choose in X_α, $\alpha = 1,\dots,n$, a point a_α and set $a_{n+1} = f(a_1,\dots,a_n)$. If in the neighbourhoods U_ξ, $\xi = 1,\dots,n+1$, of the points a_ξ we introduce differentiable coordinates, then the mapping (3.1.1) can be written in the form

$$x_{n+1}^i = f^i(x_1^{j_1},\dots,x_n^{j_n}). \tag{3.2.1}$$

As we know from Section 1.1, the functions f^i are subject to conditions (1.1.4) which for $d = n+1$ have the form:

$$\det\left(\frac{\partial f^i}{\partial x_\alpha^j}\right) \neq 0, \quad i,j = 1,\dots,r; \quad \alpha = 1,\dots,n. \tag{3.2.2}$$

We now pass from the n-quasigroup $Q = q_{12\dots n}$ to the n-loop $L(a_1,\dots,a_n)$. Then we have (3.1.6) and (3.1.8) or in a coordinate form

$$u_\alpha^i = f^i(a_1^{j_1},\dots,a_{\alpha-1}^{j_{\alpha-1}},x_\alpha^{j_\alpha},a_{\alpha+1}^{j_{\alpha+1}},\dots,a_n^{j_n}), \quad u_{n+1}^i = x_{n+1}^i, \tag{3.2.3}$$

$$u_{n+1}^i = F^i(u_1^{j_1},\dots,u_n^{j_n}). \tag{3.2.4}$$

We shall assume that the points a_ξ correspond to zero coordinates. Then, by means of (3.1.9),

$$F^i(0,\dots,0,u_\alpha^{j_\alpha},0,\dots,0) = u_\alpha^i. \tag{3.2.5}$$

We now write the Taylor expansions of the functions (3.2.4). Taking into account (3.2.5) and restricting ourselves to third order terms, we shall have

$$u_{n+1}^i = \sum_\alpha u_\alpha^i + \frac{1}{2}\sum_{\alpha,\beta}\lambda_{\alpha\beta}{}^i{}_{jk}u_\alpha^j u_\beta^k + \frac{1}{6}\sum_{\alpha,\beta,\gamma}\lambda_{\alpha\beta\gamma}{}^i{}_{jkl}u_\alpha^j u_\beta^k u_\gamma^l + o(\rho^3) \tag{3.2.6}$$

where $\rho = \max \mid u^i_\alpha \mid$, $\frac{o(t)}{t} \to 0$ if $t \to 0$ and

$$\begin{cases} \underset{\alpha\alpha}{\lambda}{}^i_{jk} = 0, \quad \underset{\alpha\alpha\alpha}{\lambda}{}^i_{jkl} = 0, \\ \underset{\alpha\beta}{\lambda}{}^i_{jk} = \underset{\beta\alpha}{\lambda}{}^i_{kj}, \quad \underset{\alpha\beta\gamma}{\lambda}{}^i_{jkl} = \underset{\sigma(\alpha\beta\gamma)}{\lambda}{}^i_{\sigma(jkl)}, \quad \sigma \text{ a permutation.} \end{cases} \tag{3.2.7}$$

The expansions (3.2.6) are changed if we change local coordinates. The form of (3.2.6) is preserved under the transformations

$$u^i_\xi = \gamma^i(\tilde{u}^j_\xi) \tag{3.2.8}$$

where the γ^i are functions of class C^p, $p \geq 3$, and

$$\gamma^i(0) = 0, \quad \det\left(\frac{\partial \gamma^i}{\partial \tilde{u}^j_{n+1}}\right)\Bigg|_{\tilde{u}^j_{n+1}=0} \neq 0.$$

We will consider only transformations of this form and call them *admissible*.

The coefficients of (3.2.6) are equal to partial derivatives of the functions F^i at $u^k_\alpha = 0$, for example,

$$\underset{\alpha\beta}{\lambda}{}^i_{jk} = \frac{\partial^2 F^i}{\partial u^j_\alpha \partial u^k_\beta}\Bigg|_{u^l_\gamma=0}. \tag{3.2.9}$$

These coefficients are not tensors under (3.2.8), but they generate some tensors as will be shown below.

For elements $u_1, \ldots, u_n$ we introduce the concept of their alternator.

Definition 3.2.1 The *alternator* of elements $u_1, \ldots, u_n$ of the n-loop $L(a_1, \ldots, a_n)$ is an expression of the form

$$\{u_1, \ldots, u_n\} = n!\; u_{[1} \ldots u_{n]} \tag{3.2.10}$$

where $u_1 \ldots u_n = F(u_1, \ldots, u_n)$ and the signs + and − in $u_{[1} \ldots u_{n]}$ denote that the corresponding operations are performed on coordinates.

Theorem 3.2.2 *For $n > 3$, the alternator* (3.2.10) *vanishes up to second order terms. For $n = 2$ and $n = 3$ it is equal respectively to*

$$\{u_1, u_2\}^i = 2\underset{12}{\lambda}{}^i_{[jk]} u^j_1 u^k_2 + o(\rho^2) \tag{3.2.11}$$

and to

$$\{u_1, u_2, u_3\}^i = 2\alpha^i_{jk}(u^j_1 u^k_2 + u^j_2 u^k_3 + u^j_3 u^k_1) + o(\rho^2), \tag{3.2.12}$$

where

$$\alpha^i_{jk} = \underset{12}{\lambda}{}^i_{[jk]} + \underset{23}{\lambda}{}^i_{[jk]} + \underset{32}{\lambda}{}^i_{[jk]}. \tag{3.2.13}$$

It satisfies the relations (up to second order terms): for $n=2$

$$\{u_1,u_3\} = \{u_3,u_2\} + \{u_1,u_2\}, \tag{3.2.14}$$

where $u_3 = u_1u_2$ and for $n=3$

$$\{u_1,u_2,u_3\} = \{u_1,u_2,u_4\} + \{u_1,u_4,u_3\} + \{u_4,u_2,u_3\} \tag{3.2.15}$$

where $u_4 = u_1u_2u_3$. If for $n=3$, the ternary loop $L(a_1,a_2,a_3)$ is reducible in some way, for example if

$$F(u_1,u_2,u_3) = (u_1\cdot u_2)\circ u_3, \tag{3.2.16}$$

then the coefficients of its alternator are equal to those of the commutator of the first of the binary quasigroups occuring in the definition (3.2.16) of 3-reducibility.

Proof. First , if $n>3$ we find from (3.2.65) and (3.2.10) that

$$\{u_1,u_2,\ldots,u_n\}^i = o(\rho^2),$$

If $n=2$, we have $[u_1,u_2] = u_1u_2 - u_2u_1$ which implies (3.2.11). In addition, for $n=2$ if we take $u_3 = u_1u_2$, we find by means of (3.2.10) that

$$\{u_1,u_3\}^i = \{u_3,u_2\}^i = 2\underset{12}{\lambda}{}^i_{[jk]}u_1^ju_2^k + o(\rho^2).$$

This and (3.2.11) implies (3.2.14).

Next, if $n=3$, equations (3.2.6) and (3.2.10) lead to (3.2.12). If $u_4 = u_1u_2u_3$, then a simple calculation by means of (3.2.10) shows that

$$\begin{cases} \{u_1,u_2,u_4\}^i = 2\alpha^i_{jk}(-u_1^ju_2^k + u_2^ju_3^k + u_3^ju_1^k) + o(\rho^2), \\ \{u_1,u_4,u_3\}^i = 2\alpha^i_{jk}(u_1^ju_2^k - u_2^ju_3^k + u_3^ju_1^k) + o(\rho^2), \\ \{u_4,u_2,u_3\}^i = 2\alpha^i_{jk}(u_1^ju_2^k + u_2^ju_3^k - u_3^ju_1^k) + o(\rho^2), \end{cases} \tag{3.2.17}$$

Equations (3.2.12) and (3.2.17) imply (3.2.15).

Finally, suppose that the ternary loop $L(a_1,a_2,a_3)$ is 3-reducible, i.e., $u_1u_2u_3 = (u_1\cdot u_2)\circ u_3$ where $\cdot$ and $\circ$ denote operations in two binary quasigroups. If in these quasigroups

$$\begin{aligned} (u_1\cdot u_2)^i &= u_1^i + u_2^i + \mu^i_{jk}u_1^ju_2^k + o(\rho^2), \\ (v_1\circ u_3)^i &= v_1^i + u_3^i + \nu^i_{jk}v_1^ju_3^k + o(\rho^2), \end{aligned}$$

then it is easy to see that

$$\{u_1,u_2,u_3\}^i = \{(u_1\cdot u_2)\circ u_3\}^i = 2\mu^i_{[jk]}(u_1^ju_2^k + u_2^ju_3^k + u_3^ju_1^k) + o(\rho^2),$$

i.e., $\alpha^i_{jk} = \mu^i_{[jk]}$ which proves the last statement of the theorem. Note that in the cases of 1- and 2-reducibility when we have correspondingly

$$u_1u_2u_3 = u_1\circ(u_2\cdot u_3), \qquad u_1u_2u_3 = u_2\circ(u_1\cdot u_3)$$

we obtain the same result, i.e.,

$$\{(u_1\cdot u_2)\circ u_3\}^i = \{u_1\circ(u_2\cdot u_3)\}^i = \{u_2\circ(u_1\cdot u_3)\}^i. \quad \blacksquare$$

3.3 Computation of the Components of the Torsion and Curvature Tensors of a Web $W(n+1,n,r)$ in Terms of Its Closed Form Equations

In Section 1.2, we showed that a web $W(n+1,n,r)$ can be given by $n+1$ completely integrable systems of Pfaffian equations:

$$\underset{\alpha}{\omega}^i = 0, \quad \alpha = 1,\ldots,n; \quad \underset{n+1}{\omega}{}^i \equiv -\sum_\alpha \underset{\alpha}{\omega}^i = 0, \tag{3.3.1}$$

where the 1-forms $\underset{\alpha}{\omega}^i$ satisfied equations (1.2.16), (1.2.17), and (1.2.24). On the other hand, in Section 2.2 (see also Section 1.1. for $d = n+1$) we showed that a web $W(n+1,n,r)$ can be defined by the closed form equations (3.2.1) subject to conditions (3.2.2).

Note that integrals of the system (3.1.1) have the form (3.2.1).

In the principal isotope $L(a_1,\ldots,a_n)$ equations (3.2.1) and conditions (3.2.2) will have the form

$$u^i_{n+1} = F^i(u_1^{j_1},\ldots,u_n^{j_n}), \quad i, j_1,\ldots,j_n = 1,\ldots,r, \tag{3.3.2}$$

$$\det\left(\frac{\partial F^i}{\partial u_\alpha^{j_\alpha}}\right) \neq 0, \quad \alpha = 1,\ldots,n. \tag{3.3.3}$$

We will find expressions of the torsion and curvature tensors and the alternator of a web $W(n+1,n,r)$ in terms of the coefficients of expansions (3.2.6) of the functions F^i.

Theorem 3.3.1 *The torsion and curvature tensors of a web $W(n+1,n,r)$ in the unit of the loop $L(a_1,\ldots,a_n)$, i.e., for $u^i_\alpha = 0$, have the following expressions:*

$$\underset{\alpha\beta}{a}{}^i_{jk} = -\underset{\alpha\beta}{\lambda}{}^i_{jk} + \frac{1}{n(n-1)}\sum_{\varepsilon\mu}\underset{\varepsilon\mu}{\lambda}{}^i_{jk}, \tag{3.3.4}$$

$$\underset{\alpha\beta}{b}{}^i_{jkl} = \frac{1}{2n(n-1)}\sum_{\substack{\gamma,\delta=1\\(\gamma\neq\delta)}}^{n}(\underset{\gamma\delta\beta}{\mu}{}^i_{jkl} - \underset{\gamma\delta\alpha}{\mu}{}^i_{jlk}) - \mu^i_{[j|k|}\mu^i_{|m|l]}, \tag{3.3.5}$$

where

$$\left.\begin{aligned} \mu^i_{jk} &= \frac{1}{n(n-1)}\sum_{\alpha,\beta=1}^{n}\underset{\alpha\beta}{\lambda}{}^i_{jk}, \\ \underset{\gamma\delta\alpha}{\mu}{}^i_{jkl} &= \underset{\gamma\delta\alpha}{\lambda}{}^i_{jkl} - \underset{\gamma\delta}{\lambda}{}^i_{mk}\underset{\gamma\alpha}{\lambda}{}^m_{jl} - \underset{\gamma\delta}{\lambda}{}^i_{jm}\underset{\delta\alpha}{\lambda}{}^m_{kl}. \end{aligned}\right\} \tag{3.3.6}$$

The coefficients α^i_{jk} of the alternator of the coordinate ternary loop $L(a_1,a_2,a_3)$ of $W(4,3,r)$ are

$$\alpha^i_{jk} = -(\underset{12}{a}{}^i_{[jk]} + \underset{23}{a}{}^i_{[jk]} + \underset{31}{a}{}^i_{[jk]}). \tag{3.3.7}$$

Proof. Differentiation of (3.3.4) gives

$$du^i_{n+1} = \sum_\alpha \frac{\partial F^i}{\partial u^j_\alpha} du^j_\alpha. \tag{3.3.8}$$

We introduce the notation

$$\frac{\partial F^i}{\partial u^j_\alpha} = \underset{\alpha}{F}{}^i_j \tag{3.3.9}$$

and set

$$\underset{\alpha}{\omega}{}^i = \underset{\alpha}{F}{}^i_j du^j_\alpha, \quad \underset{n+1}{\omega}{}^i = -du^i_{n+1}. \tag{3.3.10}$$

Then (3.3.8) gives $\underset{n+1}{\omega}{}^i + \sum_\alpha \underset{\alpha}{\omega}{}^i = 0$ which coincides with (1.2.5). By virtue of (3.3.3), the matrices $(\underset{\alpha}{F}{}^i_j)$ defined by (3.3.9) are invertible. Denoting by $(\underset{\alpha}{G}{}^i_j)$ the matrices inverse to them, we obtain

$$du^i_\alpha = \underset{\alpha}{G}{}^i_j \underset{\alpha}{\omega}{}^j. \tag{3.3.11}$$

Applying exterior differentiation to (3.3.10) and using (3.3.11), we get

$$d\underset{\alpha}{\omega}{}^i = \sum_{\beta\neq\alpha} \underset{\alpha\beta}{\Gamma}{}^i_{jk} \underset{\alpha}{\omega}{}^j \wedge \underset{\beta}{\omega}{}^k, \tag{3.3.12}$$

where

$$\underset{\alpha\beta}{\Gamma}{}^i_{jk} = \frac{\partial^2 F^i}{\partial u^l_\alpha \partial u^m_\beta} \underset{\alpha}{G}{}^l_j \underset{\beta}{G}{}^m_k. \tag{3.3.13}$$

It follows from (3.3.13) that

$$\underset{\alpha\beta}{\Gamma}{}^i_{jk} = \underset{\beta\alpha}{\Gamma}{}^i_{kj}. \tag{3.3.14}$$

Equating the right sides of equations (1.2.16) and (3.3.12) and applying Cartan's lemma, we obtain

$$\omega^i_j + \sum_{\beta\neq\alpha} (\underset{\alpha\beta}{a}{}^i_{jk} - \underset{\alpha\beta}{\Gamma}{}^i_{jk}) \underset{\beta}{\omega}{}^k = \underset{\alpha}{\lambda}{}^i_{jk} \underset{\alpha}{\omega}{}^k \tag{3.3.15}$$

where $\underset{\alpha}{\lambda}{}^i_{jk} = \underset{\alpha}{\lambda}{}^i_{kj}$. On the other hand, from (3.3.10) it follows that $d\underset{n+1}{\omega}{}^i = 0$. This and (1.2.18) imply that $\underset{n+1}{\omega}{}^j \wedge \omega^i_j = 0$ and, by Cartan's lemma, we find from this that

$$\omega^i_j = -\lambda^i_{jk} \underset{n+1}{\omega}{}^k, \quad \lambda^i_{jk} = \lambda^i_{kj}. \tag{3.3.16}$$

Substituting (3.3.16) into (3.3.15), we obtain, by virtue of the linear independence of forms $\underset{\alpha}{\omega}{}^k$, that

$$\underset{\alpha}{\lambda}{}^i_{jk} = \lambda^i_{jk}, \tag{3.3.17}$$

$$\underset{\alpha\beta}{a}{}^i_{jk} - \underset{\alpha\beta}{\Gamma}{}^i_{jk} = -\lambda^i_{jk}. \tag{3.3.18}$$

It follows from (3.3.18) that

$$\underset{\alpha\beta}{a}{}^{i}_{[jk]} = \underset{\alpha\beta}{\Gamma}{}^{i}_{[jk]}. \tag{3.3.19}$$

Symmetrizing equations (3.3.18) in j and k and adding them, we find, by (1.2.24), that

$$\lambda^{i}_{jk} = \frac{1}{n(n-1)} \sum_{\varepsilon,\mu} \underset{\varepsilon\mu}{\Gamma}{}^{i}_{jk}. \tag{3.3.20}$$

It follows from (3.3.18) and (3.3.20) that

$$\underset{\alpha\beta}{a}{}^{i}_{jk} = \underset{\alpha\beta}{\Gamma}{}^{i}_{jk} - \frac{1}{n(n-1)} \sum_{\varepsilon,\mu} \underset{\varepsilon\mu}{\Gamma}{}^{i}_{jk}. \tag{3.3.21}$$

When $u^{i}_{\alpha} = 0$, we find from (3.2.6) that

$$\begin{aligned} \underset{\alpha}{F}{}^{i}_{j}\Big|_{u^{k}_{\gamma}=0} = \underset{\alpha}{G}{}^{i}_{j}\Big|_{u^{k}_{\gamma}=0} &= \delta^{i}_{j}, \\ \frac{\partial^2 F^i}{\partial u^{l}_{\alpha} \partial u^{m}_{\beta}}\Big|_{u^{k}_{\gamma}=0} &= \underset{\alpha\beta}{\lambda}{}^{i}_{lm} \quad \alpha \neq \beta. \end{aligned} \tag{3.3.22}$$

In this case we obtain from (3.3.13) when $u^{i}_{\alpha} = 0$ that

$$\underset{\alpha\beta}{\Gamma}{}^{i}_{jk} = -\underset{\alpha\beta}{\lambda}{}^{i}_{lm} \delta^{l}_{j} \delta^{m}_{k} = -\underset{\alpha\beta}{\lambda}{}^{i}_{jk}. \tag{3.3.23}$$

Now equation (3.3.4) follows from (3.3.21) and (3.3.23).

For the coordinate ternary loop $L(a_1, a_2, a_3)$ of a web $W(4,3,r)$ we have the alternator whose coefficients α^{i}_{jk} are expressed by (3.2.13). By virtue of (3.3.19), (3.3.23), and (3.3.4), the equation (3.2.13) becomes (3.3.7).

It follows from (3.3.7) that the coefficients of the alternator, up to the sign, coincide with the components of the torsion tensor of the 3-subweb [1,2,3] (see formula (1.3.69)).

Note that the alternator does not have an invariant meaning in the n-loop $L(a_1, a_2, a_3)$. However, equations (3.3.7) show that its principal part is invariant.

Now we will find an expression of the curvature tensor $\underset{\alpha\beta}{b}{}^{i}_{jkl}$ of a web $W(n+1, n, r)$ at the point $u^{i}_{\alpha} = 0$. Substituting $\underset{n+1}{\omega}{}^{i}$ from (3.3.1) into (3.3.16), we obtain

$$\omega^{i}_{j} = \lambda^{i}_{jk} \sum_{\alpha=1}^{n} \underset{\alpha}{\omega}{}^{k}. \tag{3.3.24}$$

After exterior differentiation of (3.3.24) and (3.3.20), and using the fact that $d\underset{n+1}{\omega}{}^{i} = -\sum_{\alpha} d\underset{\alpha}{\omega}{}^{i} = 0$, we arrive at

$$d\omega^{i}_{j} = d\lambda^{i}_{jk} \wedge \sum_{\alpha=1}^{n} \underset{\alpha}{\omega}{}^{k}, \tag{3.3.25}$$

$$d\lambda^i_{jk} = \frac{1}{n(n-1)} \sum_{\alpha,\beta=1}^{n} d\underset{\alpha\beta}{\Gamma}{}^i_{jk}. \qquad (3.3.26)$$

It follows from (3.3.9) that

$$d\underset{\alpha}{F}{}^i_j = \sum_{\beta=1}^{n} \frac{\partial^2 F^i}{\partial u^j_\alpha \partial u^k_\beta} \underset{\beta}{du}{}^k. \qquad (3.3.27)$$

Using (3.3.11) and (3.3.13), we rewrite (3.3.27) in the form

$$d\underset{\alpha}{F}{}^i_j = -\sum_{\beta=1}^{n} \underset{\alpha\beta}{\Gamma}{}^i_{lm} \underset{\alpha}{F}{}^l_j \underset{\beta}{\omega}{}^m. \qquad (3.3.28)$$

Next, since $(\underset{\alpha}{G}{}^i_j)$ is inverse to $(\underset{\alpha}{F}{}^i_j)$ we have

$$\underset{\alpha}{G}{}^i_k \underset{\alpha}{F}{}^k_j = \delta^i_j. \qquad (3.3.29)$$

Differentiating (3.3.29) and using (3.3.28), we get

$$d\underset{\alpha}{G}{}^k_j = \sum_{\beta=1}^{n} \underset{\alpha\beta}{\Gamma}{}^k_{jm} \underset{\alpha}{G}{}^i_k \underset{\beta}{\omega}{}^m. \qquad (3.3.30)$$

Differentiating (3.3.13) and using (3.3.30), we find that

$$d\underset{\alpha\beta}{\Gamma}{}^i_{jk} = \sum_{\gamma=1}^{n} \Big(\underset{\alpha\beta\gamma}{\Gamma}{}^i_{jkl} + \underset{\alpha\beta}{\Gamma}{}^i_{mk} \underset{\alpha\gamma}{\Gamma}{}^m_{jl} + \underset{\alpha\beta}{\Gamma}{}^i_{jm} \underset{\beta\gamma}{\Gamma}{}^m_{kl} \Big) \underset{\gamma}{\omega}{}^l, \quad \alpha \neq \beta, \qquad (3.3.31)$$

where

$$\underset{\alpha\beta\gamma}{\Gamma}{}^i_{jkl} = -\frac{\partial^3 F^i}{\partial u^p_\alpha \partial u^s_\beta \partial u^t_\gamma} \underset{\alpha}{G}{}^p_j \underset{\beta}{G}{}^s_k \underset{\gamma}{G}{}^t_l. \qquad (3.3.32)$$

Now let us substitute formulas (3.3.26), (3.3.31) into (3.3.25) and (3.3.24), (1.2.33) into (1.2.29). Comparison of the formulas obtained after these substitutions gives

$$\begin{aligned} \underset{\alpha\beta}{b}{}^i_{jkl} = -\frac{1}{2n(n-1)} \sum_{\gamma,\delta=1}^{n} \Big(& \underset{\gamma\delta\beta}{\Gamma}{}^i_{jkl} + \underset{\gamma\delta}{\Gamma}{}^i_{mk} \underset{\gamma\beta}{\Gamma}{}^m_{jl} + \underset{\gamma\delta}{\Gamma}{}^i_{jm} \underset{\delta\beta}{\Gamma}{}^m_{kl} \\ & - \underset{\gamma\delta\alpha}{\Gamma}{}^i_{jlk} - \underset{\gamma\delta}{\Gamma}{}^i_{ml} \underset{\gamma\alpha}{\Gamma}{}^m_{jk} - \underset{\gamma\delta}{\Gamma}{}^i_{jm} \underset{\delta\alpha}{\Gamma}{}^m_{lk} \Big) - \lambda^m_{j[k} \lambda^i_{|m|l]}, \quad \gamma \neq \delta. \end{aligned} \qquad (3.3.33)$$

When $u^i_\alpha = 0$, from the expansion (3.2.6) we find

$$\underset{\alpha}{F}{}^i_j \Big|_{u^m_\delta = 0} \underset{\alpha}{G}{}^i_j \Big|_{u^m_\delta = 0} = \delta^i_j, \quad \frac{\partial^2 F^i}{\partial u^i_\alpha \partial u^j_\beta} \Big|_{u^m_\delta = 0} = \underset{\alpha\beta}{\lambda}{}^i_{jk},$$

$$\frac{\partial^3 F^i}{\partial u^j_\alpha \partial u^k_\beta \partial u^l_\gamma} \Big|_{u^m_\delta = 0} = \underset{\alpha\beta\gamma}{\lambda}{}^i_{jkl}. \qquad (3.3.34)$$

Equations (3.3.20), (3.3.13), (3.3.32), taken for $u^i_\alpha = 0$, and (3.3.34) give

$$\left.\underset{\alpha\beta}{\Gamma}{}^i_{jk}\right|_{u^m_\delta=0} = -\underset{\alpha\beta}{\lambda}{}^i_{jk}, \quad \left.\underset{\alpha\beta\lambda}{\Gamma}{}^i_{jkl}\right|_{u^m_\delta=0} = -\underset{\alpha\beta\gamma}{\lambda}{}^i_{jkl},$$
$$\left.\lambda^i_{jk}\right|_{u^m_\delta=0} = -\frac{1}{n(n-1)}\sum_{\alpha,\beta=1}^{n}\underset{\alpha\beta}{\lambda}{}^i_{jk}. \tag{3.3.35}$$

Now the formulas (3.3.5) and (3.3.6) follow from (3.3.33) and (3.3.35). ■

3.4 The Relations between the Torsion Tensors and Alternators of Parastrophic Coordinate n-Quasigroups

We will now explain how the torsion tensors, and in the case $n = 3$ the coefficients α^i_{jk} of the alternators of the $(n+1)!$ mutually parastrophic coordinate n-quasigroups of a web $W(n+1,n,r)$ are related.

Theorem 3.4.1 *Under the passage from the coordinate n-quasigroup $q_{12\ldots n}$ to a parastrophic coordinate n-quasigroup the components of the torsion tensor of the web $W(n+1,n,r)$ undergo the following linear transformation:*

$$\left.\begin{array}{l}\sigma\, \underset{\alpha\beta}{a}{}^i_{jk} = \underset{\sigma(\alpha)\sigma(\beta)}{a}{}^i_{jk}, \quad h_\alpha(\underset{\alpha\beta}{a}{}^i_{jk}) = \underset{\alpha\beta}{a}{}^i_{jk},\\ h_\alpha(\underset{\hat{\alpha}\hat{\beta}}{a}{}^i_{jk}) = \underset{\hat{\alpha}\hat{\beta}}{a}{}^i_{jk} + \underset{\hat{\alpha}\alpha}{a}{}^i_{jk} + \underset{\alpha\hat{\beta}}{a}{}^i_{jk},\end{array}\right\} \tag{3.4.1}$$

where σ is a permutation of $1,\ldots,n$ *and h_α is a parastrophy defined in* (3.1.11) *and taken for* $\{\xi_1,\ldots,\xi_n\} = \{1,\ldots,n\}$:

$$h_\alpha q_{1,\ldots,\alpha-1,\,\alpha,\,\alpha+1,\ldots,n} = q_{1,\ldots\,,\alpha-1,\,n+1\,,\alpha+1,\ldots,n}. \tag{3.4.2}$$

In addition, under this passage the coefficients of the alternator of the ternary loop $L(a_1,a_2,a_3)$ of $W(4,3,r)$ undergo the following changes:

$$\sigma\alpha^i_{jk} = (-1)^{[\sigma]}\alpha^i_{jk}, \quad \underset{\alpha}{h}(\alpha^i_{jk}) = \underset{\beta\gamma}{a}{}^i_{[jk]}, \quad \beta,\gamma \neq \alpha, \tag{3.4.3}$$

where $(-1)^{[\sigma]} = 1$ for even σ and $(-1)^{[\sigma]} = -1$ for odd σ; α,β,γ, being distinct and forming an even permutation of the indices 1,2,3.

Proof. Let us first determine how the quantities $\underset{\alpha\beta}{\lambda}{}^i_{jk}$ in the expansions (3.2.6) vary under the passage from $q_{1,2,\ldots,n}$ to $q_{\sigma(1,2,\ldots,n)}$ and to $h_\alpha q_{1,2,\ldots,n}$. For $q_{\sigma(1,2,\ldots,n)}$ the expansion (3.2.6) assumes the form

$$u^i_{n+1} = \sum_\alpha u^i_\alpha + \frac{1}{2}\sum_{\alpha,\beta}\underset{\alpha\beta}{\lambda}{}^i_{jk} u^j_{\sigma(\alpha)} u^k_{\sigma(\beta)} + o(\rho^2). \tag{3.4.4}$$

Comparing (3.2.6) and (3.4.4), we see that

$$\tilde{\lambda}_{\alpha\beta}{}^{i}_{jk} = \sigma(\lambda_{\alpha\beta}{}^{i}_{jk}) = \lambda_{\sigma(\alpha)\sigma(\beta)}{}^{i}_{jk}. \tag{3.4.5}$$

From (3.4.5) and (3.3.4) we find

$$\sigma\, a_{\alpha\beta}{}^{i}_{jk} = a_{\sigma(\alpha)\sigma(\beta)}{}^{i}_{jk}. \tag{3.4.6}$$

We now determine the law of variation of $\lambda_{\alpha\beta}{}^{i}_{jk}$ under passage from $q_{1\,2\,\dots\,n}$ to $h_\alpha q_{1\,2\,\dots\,n}$. For $q_{1\,\dots\,\alpha-1\;n+1\;\alpha+1\,\dots\,n}$ from the expansion (3.4.4) it follows that

$$\begin{aligned} u^i_\alpha &= -\sum_{\hat{\alpha}\neq\alpha} u^i_{\hat{\alpha}} + u^i_{n+1} - \sum_{\hat{\alpha},\hat{\beta}} \lambda_{\hat{\alpha}\hat{\beta}}{}^{i}_{jk} u^i_{\hat{\alpha}} u^k_{\hat{\beta}} \\ &\quad + \sum_{\hat{\beta}\neq\alpha} \lambda_{\hat{\beta}\alpha}{}^{i}_{jk} u^i_{\hat{\beta}} \Big(\sum_{\hat{\gamma}} u^k_{\hat{\gamma}} - u^k_{n+1}\Big) - o(\rho^2). \end{aligned} \tag{3.4.7}$$

Let us make the isotopic transformation

$$\bar{u}^i_{n+1} = u^i_{n+1}, \quad \bar{u}^i_\alpha = u^i_\alpha, \quad \bar{u}^i_{\hat{\alpha}} = -u^i_{\hat{\alpha}} + \lambda_{\hat{\alpha}\alpha}{}^{i}_{jk} u^j_{\hat{\alpha}} u^k_{\hat{\alpha}}. \tag{3.4.8}$$

Using (3.4.8), we can write (3.4.7) in the form

$$\begin{aligned} \bar{u}^i_\alpha &= \sum_{\xi\neq\alpha} \bar{u}^i_\xi - \sum_{\hat{\alpha},\hat{\beta}} \lambda_{\hat{\alpha}\hat{\beta}}{}^{i}_{jk} \bar{u}^j_{\hat{\alpha}} \bar{u}^k_{\hat{\beta}} \\ &\quad + \sum_{\hat{\gamma}\neq\alpha} \lambda_{\hat{\gamma}\alpha}{}^{i}_{jk} \bar{u}^j_{\hat{\gamma}} \Big(\bar{u}^k_{n+1} + \sum_{\hat{\gamma}\neq\alpha,\hat{\gamma}} \bar{u}^k_{\hat{\gamma}}\Big) - o(\rho^2). \end{aligned} \tag{3.4.9}$$

It follows from (3.4.9) and (3.2.6) that

$$h_\alpha(\lambda_{\alpha\beta}{}^{i}_{jk}) = \lambda_{\alpha\beta}{}^{i}_{jk}, \quad h_\alpha(\lambda_{\hat{\alpha}\hat{\beta}}{}^{i}_{jk}) = -\lambda_{\hat{\alpha}\hat{\beta}}{}^{i}_{jk} + \lambda_{\hat{\alpha}\alpha}{}^{i}_{jk} + \lambda_{\alpha\hat{\beta}}{}^{i}_{jk}, \tag{3.4.10}$$

where α is fixed and $\hat{\alpha},\ \hat{\beta} \neq \alpha$. Equations (3.4.10) and (3.3.4) imply

$$h_\alpha(a_{\alpha\beta}{}^{i}_{jk}) = a_{\alpha\beta}{}^{i}_{jk}, \quad h_\alpha(a_{\hat{\alpha}\hat{\beta}}{}^{i}_{jk}) = a_{\hat{\alpha}\hat{\beta}}{}^{i}_{jk} + a_{\hat{\alpha}\alpha}{}^{i}_{jk} + a_{\alpha\hat{\beta}}{}^{i}_{jk}. \tag{3.4.11}$$

Suppose now that $n = 3$. The coefficients α^i_{jk} of the alternator of the ternary loop $L(a_1, a_2, a_3)$ have the expressions (3.3.7). From (3.3.7) and (3.4.6) we find that under passage to the ternary quasigroup $q_{\sigma(1,2,3)}$ we have

$$\sigma\alpha^i_{jk} = (-1)^{[\sigma]}\alpha^i_{jk} \tag{3.4.12}$$

where $(-1)^{[\sigma]} = 1$ or -1 respectively for even and odd σ.

From (3.3.7) and (3.4.11) we find that under passage to the ternary quasigroup $h_\alpha q_{123}$ we have

$$h_\alpha(\alpha^i_{jk}) = -\underset{\gamma\beta}{a}{}^i_{[jk]}, \quad \beta, \gamma \neq \alpha, \tag{3.4.13}$$

where (α, β, γ) is an even permutation of the indices 1,2,3. ■

Note that formulas (3.3.7) and (3.4.3) mean that the components of the alternator of any of parastrophic coordinate quasigroup $q_{\xi\eta\zeta}$ of a web $W(4,3,r)$ are equal or opposite to the corresponding components of the torsion tensor of the 3-subweb $[\xi, \eta, \zeta]$ of the web $W(4,3,r)$.

Note also that equations (3.4.13) make clear the relation (3.2.15) among the alternators: the torsion tensor of the 3-subweb $[1,2,3]$ is equal to the sum of the torsion tensors of three other 3-subwebs of the web $W(4,3,r)$ (see also formulas (1.3.69)).

3.5 Canonical Expansions of the Equations of a Local Analytic n-Quasigroup

Assume that the functions $u^i_{n+1} = F^i(u_1^{j_1}, \ldots, u_n^{j_n})$ defining the local n-loop $L(a_1, \ldots, a_n)$ (see equations (3.2.3)) are analytic. Then in the neighbourhoods U_α these functions can be represented as convergent power series:

$$u^i_{n+1} = \sum_{s=0}^{\infty} \underset{(s)}{\Lambda}{}^i(u_1^{j_1}, \ldots, u_n^{j_n}), \tag{3.5.1}$$

where

$$\underset{(s)}{\Lambda}{}^i(u_1^{j_1}, \ldots, u_n^{j_n}) = \frac{1}{s!} \sum_{\substack{p_\alpha=0 \\ (p_1+\cdots+p_n=s)}}^{s} \binom{s}{p_1}\binom{s-p_1}{p_2} \cdots \binom{s-p_1-\ldots-p_{n-2}}{p_{n-1}} \cdot \underset{(1)^{p_1}\ldots(n)^{p_n}}{\lambda}{}^i_{j_1\ldots j_{p_1}\ldots k_1\ldots k_{p_n}} u_1^{j_1} \ldots u_1^{j_{p_1}} \ldots u_n^{k_1} \ldots u_n^{k_{p_n}}; \tag{3.5.2}$$

in (3.5.2) the coefficients are constant and, for example, $(1)^{p_1}$ means $\underbrace{11\ldots1}_{p_1}$.

Definition 3.5.1 A transformation of the form

$$\bar{u}^i_\xi = \varphi^i_\xi(u^j_\xi), \quad \xi = 1, \ldots, n+1; \quad i = 1, \ldots, r, \tag{3.5.3}$$

where the φ^i_ξ are invertible analytic functions such that $\varphi^i_\xi(0) = 0$, is said to be an *isotopic* transformation of the variables u^i_ξ.

The reason why we call transformations (3.5.3) isotopic is: when we apply (3.5.3) to the variables of the n-loop $L(a_1, \ldots, a_n)$, we will obtain a new n-loop which is, according to Definition 3.1.12, is isotopic to the $L(a_1, \ldots, a_n)$.

We will show that by means of the isotopic transformations (3.5.3) the expansions (3.5.1) can be reduced to a specific canonical form. In other words, we will find an isotope of our n-loop $L(a_1, \ldots, a_n)$ which has in some sense the simplest expansions (3.5.1).

Theorem 3.5.2 *By means of the isotopic transformations* (3.5.3), *the expansions* (3.5.1) *can be reduced to the following canonical form:*

$$u^i_{n+1} = \sum_\alpha u^i_\alpha + \sum_{s=2}^{\infty} \underset{(s)}{\Lambda}{}^i(u_1^{j_1}, \ldots, u_n^{j_n}), \tag{3.5.4}$$

where

$$\underset{(s)}{\Lambda}{}^i = \frac{1}{s!} \sum_{\substack{p_\alpha=0 \\ (p_1+\cdots+p_n=s)}}^{s-1} \binom{s}{p_1}\binom{s-p_1}{p_2}\cdots\binom{s-p_1-\ldots-p_{n-2}}{p_{n-1}} \cdot$$

$$\underset{(1)^{p_1}\ldots(n)^{p_n}}{\lambda}{}^i_{j_1\ldots j_{p_1}\ldots k_1\ldots k_{p_n}} u_1^{j_1}\ldots u_1^{j_{p_1}}\ldots u_n^{k_1}\ldots u_n^{k_{p_n}}, \tag{3.5.5}$$

and the coefficients of (3.5.5) *satisfy the relations*

$$\sum_{\substack{p_\alpha=0 \\ (p_1+\cdots+p_n=s)}}^{s-1} \binom{s}{p_1}\binom{s-p_1}{p_2}\cdots\binom{s-p_1-\ldots-p_{n-2}}{p_{n-1}} \underset{(1)^{p_1}\ldots(n)^{p_n}}{\lambda}{}^i_{(j_1\ldots j_{p_1}\ldots k_1\ldots k_{p_n})} = 0. \tag{3.5.6}$$

Proof. Using the isotopic transformations (3.5.3), we will consecutively simplify in the expansion (3.5.1) the terms of the order 1,2, ...

Step 0. Observe first that the isotopic transformation

$$\bar{u}^i_{n+1} = u^i_{n+1} - \underset{(n+1)}{\Lambda}{}^i, \quad \bar{u}^i_\alpha = u^i_\alpha \tag{3.5.7}$$

yields $\underset{(n+1)}{\bar{\Lambda}}{}^i = 0$. After this the expansions (3.5.1) become

$$u^i_{n+1} = \sum_\alpha \underset{\alpha}{\lambda}{}^i_j u^j_\alpha + \sum_{s=2}^{\infty} \underset{(s)}{\Lambda}{}^i \tag{3.5.8}$$

(we omit the bar over u^i_α and $\underset{(s)}{\Lambda}{}^i$).

Step 1. To simplify the first order terms $\underset{(1)}{\Lambda}{}^i = \sum_\alpha \underset{\alpha}{\lambda}{}^i_j u^j_\alpha$ in (3.5.8), we note that expansions (3.5.8) imply

$$\underset{\alpha}{\lambda}{}^i_j = \left.\frac{\partial u^i_{n+1}}{\partial u^j_\alpha}\right|_{u^k_\gamma=0}. \tag{3.5.9}$$

Equations (3.5.9) and conditions (3.2.4) yield

$$\det\left(\underset{\alpha}{\lambda}{}^i_j\right) \neq 0. \tag{3.5.10}$$

By virtue of (3.5.10), the transformation

$$\bar{u}^i_\alpha = \underset{\alpha}{\lambda}{}^i_j u^j_\alpha \tag{3.5.11}$$

is isotopic. After applying the isotopic transformation (3.5.11), the first order terms in (3.5.8) assume the form $\underset{(1)}{\Lambda}{}^i = \sum_\alpha u^i_\alpha$ (we omit the bar over the u^i_α), and the the expansions (3.5.8) become

$$u^i_{n+1} = \sum_\alpha u^i_\alpha + \sum_{s=2}^{\infty} \underset{(s)}{\Lambda}{}^i. \tag{3.5.12}$$

We omit the bar over u^i_α and $\underset{(s)}{\Lambda}{}^i$, $s > 1$, which is obtained from similar terms in (3.5.8) by substitution (3.5.11).

Note that this much of the canonisation of expansions (3.5.1) has already been carried out in Section 3.2: from the requirements (3.2.5) we obtained the expansions (3.2.6), in which was achieved exactly the canonisation to this point.

Step 2. To prove formulas (3.5.5)–(3.5.6) for any degree $s \geq 2$ of the polynomials $\underset{(s)}{\Lambda}{}^i$, we will use the induction on s. To simplify the second order terms, we first employ an isotopic transformation

$$\bar{u}^i_{n+1} = u^i_{n+1} + \frac{1}{2}\gamma^i_{jk} u^i_{n+1} u^j_{n+1}, \quad \gamma^i_{jk} = \gamma^i_{kj}, \tag{3.5.13}$$

where, for the present, the coefficients γ^i_{jk} are arbitrary. After transformation (3.5.13) the expansion (3.5.12) will have the form

$$u^i_{n+1} = \sum_\alpha u^i_\alpha + \frac{1}{2}\left[\sum_\alpha \left(\underset{\alpha\alpha}{\lambda}{}^i_{jk} + \gamma^i_{jk}\right) u^j_\alpha u^k_\alpha + 2\sum_{(\alpha,\beta)} \left(\underset{\alpha\beta}{\lambda}{}^i_{jk} + \gamma^i_{jk}\right) u^j_\alpha u^k_\beta\right] + \sum_{s=3}^{\infty} \underset{(s)}{\Lambda}{}^i. \tag{3.5.14}$$

In (3.5.14) and in what follows the symbol (α, β) under the summation sign means that the summation is carried out over all combinations of indices $1, 2, \ldots, n$, taken two at a time. We now choose

$$\gamma^i_{jk} = -\frac{1}{\binom{n}{2}} \sum_{(\varepsilon,\mu)} \underset{\varepsilon\mu}{\lambda}{}^i_{(jk)}. \tag{3.5.15}$$

In addition, we perform the isotopic transformation

$$\bar{u}^i_\alpha = u^i_\alpha + \frac{1}{2}(\underset{\alpha\alpha}{\lambda}{}^i_{jk} + \gamma^i_{jk}) u^j_\alpha u^k_\alpha, \tag{3.5.16}$$

which does not change the terms $\underset{(1)}{\Lambda}{}^i$. As a result of (3.5.15) and (3.5.16), the expansions (3.5.14) assume the form

$$\bar{u}^i_{n+1} = \sum_\alpha \bar{u}^i_\alpha + \sum_{(\alpha,\beta)} \bar{\lambda}^{\,i}_{\alpha\beta jk} \bar{u}^j_\alpha \bar{u}^k_\beta + \sum_{s=3}^\infty \underset{(s)}{\Lambda}{}^i, \tag{3.5.17}$$

where

$$\bar{\lambda}^{\,i}_{\alpha\beta jk} = \lambda^{\,i}_{\alpha\beta jk} - \frac{1}{\binom{n}{2}} \sum_{(\varepsilon,\mu)} \lambda^{\,i}_{\varepsilon\mu(jk)}. \tag{3.5.18}$$

It follows from (3.5.18) that

$$\bar{\lambda}^{\,i}_{\alpha\beta[jk]} = \lambda^{\,i}_{\alpha\beta[jk]}, \quad \sum_{(\varepsilon,\mu)} \bar{\lambda}^{\,i}_{\varepsilon\mu(jk)} = 0. \tag{3.5.19}$$

Thus, equation (3.5.17) implies that

$$\underset{(2)}{\Lambda}{}^i = \sum_{(\alpha,\beta)} \lambda^{\,i}_{\alpha\beta jk} u^j_\alpha u^k_\beta, \tag{3.5.20}$$

where

$$\sum_{(\varepsilon,\mu)} \lambda^{\,i}_{\varepsilon\mu(jk)} = 0. \tag{3.5.21}$$

For simplicity, we omit the bar over u and λ.

Note that the transformations (3.5.13) and (3.5.16) which we used were isotopic: since in both cases $\left.\frac{\partial u^i_{n+1}}{\partial u^j_\alpha}\right|_{u^k_\gamma=0} = \delta^i_j$, $\det\left(\frac{\partial u^i_{n+1}}{\partial u^j_\alpha}\right) = 1$ at the point $u^k_\gamma = 0$ and therefore $\det\left(\frac{\partial u^i_{n+1}}{\partial u^j_\alpha}\right) \neq 0$ in a sufficiently small neighbourhood of the point $u^k_\gamma = 0$.

Step 3. To make our next (inductive) step more clear, let us go one step further. To simplify the third order terms, we perform an isotopic transformation

$$\bar{u}^i_{n+1} = u^i_{n+1} + \frac{1}{6}\gamma^i_{jkl} u^j_{n+1} u^k_{n+1} u^l_{n+1}, \tag{3.5.22}$$

where the γ^i_{jkl} are symmetric with respect to any pair of lower indices. From (3.5.22) and (3.5.16) we obtain

$$\begin{aligned} u^i_{n+1} = {} & \sum_\alpha u^i_\alpha + \sum_{(\alpha,\beta)} \lambda^{\,i}_{\alpha\beta jk} u^j_\alpha u^k_\beta + \frac{1}{6}\Big[\sum_\alpha \Big(\lambda^{\,i}_{\alpha\alpha\alpha jkl} + \gamma^i_{jkl}\Big) u^j_\alpha u^k_\alpha u^l_\alpha \\ & + 3 \sum_{\substack{\alpha,\beta \\ (\alpha\neq\beta)}} \Big(\lambda^{\,i}_{\alpha\beta\beta jkl} + \gamma^i_{jkl}\Big) u^j_\alpha u^k_\beta u^l_\beta + 6 \sum_{\substack{\alpha,\beta,\gamma \\ (\gamma\neq\alpha,\beta;\alpha\neq\beta)}} \Big(\lambda^{\,i}_{\alpha\beta\gamma jkl} + \gamma^i_{jkl}\Big) u^j_\alpha u^k_\beta u^l_\gamma\Big]. \end{aligned} \tag{3.5.23}$$

We choose

$$\gamma^i_{jkl} = -\frac{1}{A}\Big(3\sum_{\alpha,\beta} \lambda^{\,i}_{\alpha\beta\beta(jkl)} + 6\sum_{\alpha,\beta,\gamma} \lambda^{\,i}_{\alpha\beta\gamma(jkl)}\Big), \tag{3.5.24}$$

where

$$A = \sum_{\substack{p_\alpha=0 \\ (p_1+\dots+p_n=3)}}^{2} \binom{3}{p_1}\binom{3-p_1}{p_2}\dots\binom{3-p_1-\dots-p_{n-2}}{p_{n-1}}.$$

In addition, we perform the isotopic transformation of the form

$$\bar{u}^i_\alpha = u^i_\alpha + \frac{1}{6}(\underset{\alpha\alpha\alpha}{\lambda}{}^i_{jkl} + \gamma^i_{jkl})u^j_\alpha u^k_\alpha u^l_\alpha, \tag{3.5.25}$$

which does not change the terms $\underset{(1)}{\Lambda}{}^i$ and $\underset{(2)}{\Lambda}{}^i$.

As a result of (3.5.24) and (3.5.25), the third order terms assume the form

$$\underset{(3)}{\Lambda}{}^i = \frac{1}{2}\sum_{\alpha,\beta} \underset{\alpha\beta\beta}{\lambda}{}^i_{jkl} u^j_\alpha u^k_\beta u^l_\beta + \sum_{\alpha,\beta,\gamma} \underset{\alpha\beta\gamma}{\lambda}{}^i_{jkl} u^j_\alpha u^k_\beta u^l_\gamma, \tag{3.5.26}$$

where

$$2\sum_{\alpha,\beta,\gamma} \underset{\alpha\beta\gamma}{\lambda}{}^i_{(jkl)} + \sum_{\alpha,\beta} \underset{\alpha\beta\beta}{\lambda}{}^i_{(jkl)} = 0. \tag{3.5.27}$$

In (3.5.26) and (3.5.27) we again omit the bar over u and λ.

Step 4. Suppose now that

$$\underset{(m-1)}{\Lambda}{}^i = \frac{1}{(m-1)!} \sum_{\substack{p_\alpha=0 \\ (p_1+\dots+p_n=m-1)}}^{m-2} \binom{m-1}{p_1}\binom{m-1-p_1}{p_2}\dots\binom{m-1-p_1\dots-p_{n-2}}{p_{n-1}} \cdot \underset{(1)^{p_1}(2)^{p_2}\dots(n)^{p_n}}{\lambda}{}^i_{j_1\dots j_{r_1}k_1\dots k_{r_2}\dots l_1\dots l_{r_n}} u^{j_1}_1\dots u^{j_{r_1}}_1 u^{k_1}_2\dots u^{k_{r_2}}_2\dots u^{l_1}_n\dots u^{l_{r_n}}_n$$

and, in addition, that

$$\sum_{\substack{p_\alpha=0 \\ (p_1+\dots+p_n=m-1)}}^{m-1} \binom{m-1}{p_1}\binom{m-1-p_1}{p_2}\dots\binom{m-1-p_1-\dots-p_{n-2}}{p_{n-1}} \cdot \underset{(1)^{p_1}\dots(n)^{p_n}}{\lambda}{}^i_{(j_1\dots j_{p_1}\dots l_1\dots l_{p_n})} = 0.$$

Note that $\underset{(m-1)}{\Lambda}{}^i$ contains no terms in which all the numbers $p_1,\dots,p_n$ but one are simultaneously zero since then the non-zero number would have to be $m-1$, which is impossible, since the summation extends only to $m-2$.

Next we perform an isotopic transformation

$$\bar{u}^i_{n+1} = u^i_{n+1} + \frac{1}{m!}\gamma^i_{j_1\dots j_m} u^{j_1}_{n+1}\dots u^{j_m}_{n+1},$$

where the $\gamma^i_{j_1\dots j_m}$ are symmetric with respect to all lower indices. Such a transformation does not alter the terms $\underset{(s)}{\Lambda}{}^i$ for $s \leq m-1$. For the mth order terms we obtain

$$\underset{(m)}{\bar{\Lambda}}{}^{i} = \frac{1}{m!}\Big[\sum_{\substack{p_\alpha=0\\(p_1+\ldots+p_n=m)}}^{m} \binom{m}{p_1}\binom{m-p_1}{p_2}\cdots\binom{m-p_1-\ldots-p_{n-2}}{p_{n-1}} \cdot$$
$$\cdot\big(\underset{(1)^{p_1}\ldots(n)^{p_n}}{\lambda}{}^{i}_{(j_1\ldots j_{p_1}\ldots l_1\ldots l_{p_n})} + \gamma^i_{j_1\ldots j_{p_1}\ldots l_1\ldots l_{p_n})}\big)u_1^{j_1}\ldots u_1^{j_{p_1}}\ldots u_n^{l_1}\ldots u_n^{l_{p_n}}\Big]. \tag{3.5.28}$$

We choose the $\gamma^i_{j_1\ldots j_m}$ so that

$$\sum_{\substack{p_\alpha=0\\(p_1+\ldots+p_n=m)}}^{m-1} \binom{m}{p_1}\binom{m-p_1}{p_2}\cdots\binom{m-p_1-\ldots-p_{n-2}}{p_{n-1}} \cdot$$
$$\cdot\Big(\underset{(1)^{p_1}\ldots(n)^{p_n}}{\lambda}{}^{i}_{(j_1\ldots j_{p_1}\ldots l_1\ldots l_{p_n})} + \gamma^i_{j_1\ldots j_{p_1}\ldots l_1\ldots l_{p_n})}\Big) = 0.$$

For this we must put

$$\gamma^i_{(j_1\ldots j_{p_1}\ldots l_1\ldots l_{p_n})} = -\frac{1}{A}\sum_{\substack{p_\alpha=0\\(p_1+\ldots+p_n=m)}}^{m-1} \binom{m}{p_1}\binom{m-p_1}{p_2}\cdots\binom{m-p_1-\ldots-p_{n-2}}{p_{n-1}} \lambda^i_{(j_1\ldots j_{p_1}\ldots l_1\ldots l_{p_n})},$$

where

$$A = \sum_{\substack{p_\alpha=0\\(p_1+\ldots+p_n=m)}}^{m-1} \binom{m}{p_1}\binom{m-p_1}{p_2}\cdots\binom{m-p_1-\ldots-p_{n-2}}{p_{n-1}}.$$

We also perform the isotopic transformation

$$\bar{u}^i_\alpha = u^i_\alpha + \frac{1}{m!}\underset{(1)^0\ldots(\alpha)^m\ldots(n)^0}{\lambda}{}^{i}_{j_1\ldots j_n} u_\alpha^{j_1}\ldots u_\alpha^{j_m}.$$

Then

$$\underset{(m)}{\Lambda}{}^{i} = \frac{1}{m!}\sum_{\substack{p_\alpha=0\\(p_1+\ldots+p_n=m)}}^{m-1} \binom{m}{p_1}\binom{m-p_1}{p_2}\cdots\binom{m-p_1-\ldots-p_{n-2}}{p_{n-1}} \cdot$$
$$\cdot\underset{(1)^{p_1}\ldots(n)^{p_n}}{\lambda}{}^{i}_{j_1\ldots j_{p_1}\ldots l_1\ldots l_{p_n}} u_1^{j_1}\ldots u_1^{j_{p_1}}\ldots u_n^{l_1}\ldots u_n^{l_{p_n}}, \tag{3.5.29}$$

where

$$\sum_{\substack{p_\alpha=0\\(p_1+\ldots+p_n=m)}} \binom{m}{p_1}\binom{m-p_1}{p_2}\cdots\binom{m-p_1-\ldots-p_{n-2}}{p_{n-1}} \cdot$$
$$\cdot\underset{(1)^{p_1}\ldots(n)^{p_n}}{\lambda}{}^{i}_{(j_1\ldots j_{l_1}\ldots l_1\ldots l_{l_n})} = 0. \tag{3.5.30}$$

As with $\underset{(m-1)}{\Lambda}{}^i$, the expression for $\underset{(m)}{\Lambda}{}^i$ contains no terms in which all the numbers $p_1, \ldots, p_n$ but one are simultaneously zero, since then the non-zero number would have to be m, which is impossible, since the summation extends only to $m-1$.

In (3.5.29) and (3.5.30) we again omit the bar over u and λ.

Since we proved formulas (3.5.5)–(3.5.6) for $s = 2$ (Step 2) and for $s = m$ provided that they are true for $s = m - 1$ (Step 4), they hold for any $s \geq 2$. ■

This inductive proof shows that the expansions (3.5.1) can be reduced to the form (3.5.4) where the $\underset{(s)}{\Lambda}{}^i$ are defined by (3.5.5) and the coefficients of $\underset{(s)}{\Lambda}{}^i$ satisfy (3.5.6).

Corollary 3.5.3 *The expansions* (3.5.4) *define an n-loop with the unit* $e(0, \ldots, 0)$.

Proof. In fact, if $u^i_{\hat{\alpha}} = 0, \ \hat{\alpha} \neq \alpha$, then $u^i_{n+1} = u^i_\alpha$. ■

Corollary 3.5.4 *In the loop defined by the expansions* (3.5.4) *the following identity holds:*

$$F^i(u^{j_1}, \ldots, u^{j_n}) = nu^i. \tag{3.5.31}$$

Proof. To prove this, one must substitute $u^i_\alpha = u^i$ into expansions (3.5.4) and, using relations (3.5.6), confirm that the terms $\underset{(s)}{\Lambda}{}^i, \ s \geq 2$, vanish. ■

Remark 3.5.5 The expansions (3.5.4)-(3.5.6) differ from the expansions (3.2.3) since their coefficients satisfy relations (3.5.7).

Theorem 3.5.2 can be generalised in the following way.

Theorem 3.5.6 *The expansions* (3.5.1) *can be reduced to the form* (3.5.4) *where* $\underset{(s)}{\Lambda}{}^i$ *are expressed by* (3.5.5) *and their coefficients satisfy the relations :*

$$\sum_{\substack{p_\alpha=0 \\ (p_1+\ldots+p_n=m)}}^{s} \binom{s}{p_1}\binom{s-p_1}{p_2}\cdots\binom{s-p_1-\ldots-p_{n-2}}{p_{n-1}} \cdot \underset{(1)^{p_1}\ldots(n)^{p_n}}{\lambda}{}^i_{(j_1\ldots j_{p_1}\ldots k_1\ldots k_{p_n})}\sigma_2^{p_2}\ldots\sigma_n^{p_n} = 0 \tag{3.5.32}$$

provided that $\sigma_2, \ldots, \sigma_n$ *are fixed real numbers for which*

$$(\sigma_2 + \ldots \sigma_n)^s - \sigma_2^s - \ldots - \sigma_n^s - 1 \neq 0. \tag{3.5.33}$$

Proof. To prove this, we should modify the proof of Theorem 3.5.2 as follows. We substitute $u^i_{\hat{\alpha}} = \sigma^{p_n}_{\hat{\alpha}} u^i_1$, $\hat{\alpha} = 2, \ldots, n$, into (3.5.28) and find $\gamma^i_{j_1 \ldots j_m}$ from the condition

$$\sum_{\substack{p_\alpha = 0 \\ (p_1 + \ldots + p_n = m)}}^{m-1} \binom{m}{p_1} \binom{m - p_1}{p_2} \cdots \binom{m - p_1 - \ldots - p_{n-2}}{p_{n-1}} \cdot$$
$$\cdot \Big(\underset{(1)^{p_1} \ldots (n)^{p_n}}{\lambda}{}^{i}_{(j_1 \ldots l_{p_n})} + \gamma^i_{(j_1 \ldots l_{p_n})} \Big) \sigma_2^{p_2} \ldots \sigma_n^{p_n} = 0. \tag{3.5.34}$$

Then instead of (3.5.30) we obtain the condition (3.5.32). In order to be able to choose $\gamma^i_{j_1 \ldots j_m}$ using the method indicated above, it is necessary that its coefficient in (3.5.34) be different from zero, i.e,

$$\begin{aligned} \varphi(\sigma_2, \ldots \sigma_n) &= \sum_{\substack{p_\alpha = 0 \\ (p_1 + \ldots + p_n > 0)}}^{m-1} \binom{m}{p_1} \binom{m - p_1}{p_2} \cdots \binom{m - p_1 - \ldots - p_{n-2}}{p_{n-1}} \sigma_2^{p_2} \ldots \sigma_n^{p_n} \\ &= (\sigma_2 + \ldots + \sigma_n + 1)^m - \sigma_2^m - \ldots - \sigma_n^m - 1 \neq 0. \ \blacksquare \end{aligned} \tag{3.5.35}$$

Remark 3.5.7 In this more general canonisation we still have Corollary 3.5.2 and instead of the relations (3.5.31) we will get the more general relations

$$F^i(u^{j_1}, \sigma_2 u^{j_2}, \ldots, \sigma_n u^{j_n}) = (1 + \sigma_2 + \ldots + \sigma_n) u^i. \tag{3.5.36}$$

Theorem 3.5.2 is obtained from Theorem 3.5.6 when $\sigma_2 = \ldots = \sigma_n = 1$.

Definition 3.5.8 The expansions (3.5.4) discussed in Theorem 3.5.6 are called *canonical expansions*, and the variables occuring in them are called *canonical variables.*

In the sequel we will use the expansions (3.5.4)-(3.5.6).

Corollary 3.5.9 *In canonical coordinates the torsion tensor and the connection forms* ω^i_j *of a web* $W(n+1, n, r)$ *at the point* $u^l_\gamma = 0$ *have respectively the following expressions:*

$$\underset{\alpha\beta}{a}{}^i_{jk} \Big|_{u^l_\gamma = 0} = - \underset{\alpha\beta}{\lambda}{}^i_{jk}, \tag{3.5.37}$$

$$\omega^i_j \Big|_{u^l_\gamma = 0} = 0. \tag{3.5.38}$$

Proof. It follows from (3.3.4),(3.3.24), and (3.5.21). ■

Equations (3.5.37) and (3.5.38) show that the canonical variables play the role of normal geodesic coordinates (see [He 78]).

The canonical expansions we have obtained are changed under the general isotopic transformations (3.5.3). However, they are unique in the sense that they remain unchanged under concordant coordinate transformations

$$u^i_\xi = A^i_{j'} \bar{u}^{j'}_\xi. \tag{3.5.39}$$

Theorem 3.5.10 *The canonical expansions* (3.5.4) – (3.5.6) *preserve their form if and only if the variables* u^i_ξ *are transformed by the formulas* (3.5.39). *In addition the coefficients of* (3.5.4) *are transformed according to the tensor law.*

Proof. After a transformation (3.5.39) the expansions (3.5.4) become

$$A^i_{j'}\bar{u}^{j'}_{n+1} = \sum_\alpha A^i_{j'}\bar{u}^{j'}_\alpha + \sum_{s=2}^\infty \underset{(s)}{\Lambda}{}^i, \tag{3.5.40}$$

where

$$\underset{(s)}{\Lambda}{}^i = \frac{1}{s!} \sum_{\substack{p_\alpha=0 \\ (p_1+\ldots+p_n=s)}}^{s-1} \binom{s}{p_1}\binom{s-p_1}{p_2}\cdots\binom{s-p_1-\ldots-p_{n-2}}{p_{n-1}}.$$

$$\cdot \underset{(1)^{p_1}\ldots(n)^{p_n}}{\lambda}{}^i_{j_1\ldots j_{p_1}\ldots k_1\ldots k_{p_n}} A^{j_1}_{j'_1}\bar{u}^{j'_1}_1 \ldots A^{j_{p_1}}_{j'_{p_1}}\bar{u}^{j'_{p_1}}_1 \ldots A^{k_1}_{k'_1}\bar{u}^{k'_1}_n \ldots A^{k_{p_n}}_{k'_{p_n}}\bar{u}^{k'_{p_n}}_n. \tag{3.5.41}$$

Since det $(A^i_{j'}) \neq 0$, $(\tilde{A}^{i'}_j)$, the inverse matrix to $(A^i_{j'})$, exists,i.e., $\tilde{A}^{i'}_i A^i_{k'} = \delta^{i'}_{k'}$. Contracting both members of (3.5.40) within $\tilde{A}^{i'}_i$, we obtain

$$\bar{u}^{i'}_{n+1} = \sum_\alpha \bar{u}^{i'}_\alpha + \sum_{s=2}^\infty \underset{(s)}{\bar{\Lambda}}{}^{i'}, \tag{3.5.42}$$

where

$$\underset{(s)}{\bar{\Lambda}}{}^{i'} = \frac{1}{s!} \sum_{\substack{p_\alpha=0 \\ (p_1+\ldots+p_n=s)}}^{s-1} \binom{s}{p_1}\binom{s-p_1}{p_2}\cdots\binom{s-p_1-\ldots-p_{n-2}}{p_{n-1}}.$$

$$\cdot \underset{(1)^{p_1}\ldots(n)^{p_n}}{\bar{\lambda}}{}^{i'}_{j'_1\ldots j'_{p_1}\ldots k'_1\ldots k'_{p_n}} \bar{u}^{j'_1}_1 \ldots \bar{u}^{j'_{p_1}}_1 \ldots \bar{u}^{k'_1}_n \ldots \bar{u}^{k'_{p_n}}_n; \tag{3.5.43}$$

$$\underset{(1)^{p_1}\ldots(n)^{p_n}}{\bar{\lambda}}{}^{i'}_{j'_1\ldots j'_{p_1}\ldots k'_1\ldots k'_{p_n}} = \tilde{A}^{i'}_i A^{j_1}_{j'_1} \ldots A^{j_1}_{k'_1} \ldots A^{k_{p_n}}_{k'_{p_n}} \underset{(1)^{p_1}\ldots(n)^{p_n}}{\lambda}{}^i_{j_1\ldots j_{p_1}\ldots k_1\ldots k_{p_n}}. \tag{3.5.44}$$

We can see from this that equations (3.5.42) have the same form as (3.5.4) and that the coefficients of (3.5.42) are transformed by the tensor law (3.5.44) under the transformations (3.5.39).It is easy to prove that the coefficients of (3.5.42) satisfy relations of the form (3.5.6).

Conversely, suppose that the canonical expansions are preserved under certain transformations of the variables. We will show that such transformations have the form (3.5.39). Indeed, suppose they have the more general form

$$u^i_\alpha = \underset{\alpha}{A}{}^i_j\bar{u}^j_\alpha + o(\rho), \quad \bar{u}^i_{n+1} = \underset{n+1}{\tilde{A}}{}^i_j u^j_{n+1} + o(\rho). \tag{3.5.45}$$

Then from (3.5.45) we see that

$$\bar{u}^i_{n+1} = \underset{n+1}{\tilde{A}}{}^i_j u^j_{n+1} + o(\rho) = \underset{n+1}{\tilde{A}}{}^i_j \sum_\alpha u^j_\alpha + o(\rho) = \underset{n+1}{\tilde{A}}{}^i_j \sum_\alpha \underset{\alpha}{A}{}^j_k \bar{u}^k_\alpha + o(\rho). \tag{3.5.46}$$

It follows from (3.5.46) that the first order terms will have the form as in (3.5.4) if and only if

$$\underset{n+1}{\tilde{A}}{}^{i}_{j} \underset{\alpha}{A}{}^{j}_{k} = \delta^{i}_{k},$$

i.e., if the matrices $(\underset{\alpha}{A}{}^{j}_{k})$ are inverse to $(\underset{n+1}{\tilde{A}}{}^{i}_{j})$. Hence,

$$\underset{n+1}{A}{}^{i}_{j} = \underset{\alpha}{A}{}^{i}_{j} = A^{i}_{j}. \tag{3.5.47}$$

By (3.5.47), we can write the transformations (3.5.45) in the form

$$\left. \begin{aligned} u^{i}_{\alpha} &= \bar{u}^{i}_{\alpha} + \tfrac{1}{2} \underset{\alpha}{A}{}^{i}_{jk} \bar{u}^{j}_{\alpha} \bar{u}^{k}_{\alpha} + o(\rho^{2}), \\ \bar{u}^{i}_{n+1} &= u^{i}_{n+1} + \tfrac{1}{2} \underset{n+1}{\tilde{A}}{}^{i}_{jk} u^{j}_{n+1} u^{k}_{n+1} + o(\rho^{2}). \end{aligned} \right\} \tag{3.5.48}$$

From (3.5.48) we see that

$$\bar{u}^{i}_{n+1} = \sum_{\alpha} u^{i}_{\alpha} + \frac{1}{2} \sum_{\alpha} (\underset{\alpha}{A}{}^{i}_{jk} + \underset{n+1}{\tilde{A}}{}^{i}_{jk}) \bar{u}^{j}_{\alpha} \bar{u}^{k}_{\alpha} + \sum_{(\alpha,\beta)} (\underset{\alpha\beta}{\lambda}{}^{i}_{jk} + \underset{n+1}{\tilde{A}}{}^{i}_{jk}) \bar{u}^{j}_{\alpha} \bar{u}^{k}_{\beta} + o(\rho^{2}). \tag{3.5.49}$$

It follows from (3.5.49) that for the canonical expansion to be preserved we must have

$$\underset{\alpha}{A}{}^{i}_{jk} + \underset{n+1}{\tilde{A}}{}^{i}_{jk} = 0, \tag{3.5.50}$$

$$\sum_{(\alpha,\beta)} \underset{\alpha\beta}{\lambda}{}^{i}_{jk} + \frac{1}{\binom{n}{2}} \underset{n+1}{\tilde{A}}{}^{i}_{(jk)} = 0. \tag{3.5.51}$$

By virtue of (3.5.21), we obtain from (3.5.51) that $\underset{n+1}{\tilde{A}}{}^{i}_{jk} = 0$. Hence, in view of (3.5.50), we have $\underset{\alpha}{A}{}^{i}_{jk} = 0$.

Applying the method of mathematical induction and taking into account (3.5.5) and (3.5.47), we obtain that in the transformations of variables preserving the canonical expansions there can be no terms of second or higher orders, and they must have the form (3.5.39). ■

3.6 The One-Parameter n-Subquasigroups of a Differentiable n-Quasigroup

Suppose that in the manifolds $X_{\xi}, \xi = 1, \ldots, n+1$, there are given submanifolds $\tilde{X}_{\xi}$ of dimension $\tilde{r} < r$. We write the equations of the submanifolds $X_{\alpha}, \alpha = 1, \ldots, n$, in the form

$$x^{i}_{\alpha} = x^{i}_{\alpha}(u^{\hat{k}}_{\alpha}), \quad \hat{k} = 1, \ldots \tilde{r}. \tag{3.6.1}$$

By virtue of (3.6.1), the equations (3.2.1) of the n-quasigroup become

$$x^{i}_{n+1} = f^{i}[x^{j_1}_{1}(u^{\hat{k}_1}_{1}), \ldots, x^{j_n}_{n}(u^{\hat{k}_n}_{n})], \quad \hat{k}_1, \ldots, \hat{k}_n = 1, \ldots, \tilde{r}. \tag{3.6.2}$$

For the mapping $\tilde{f} : \tilde{X}_1 \times \ldots \times \tilde{X}_n \to \tilde{X}_{n+1}$ to define an n-quasigroup, the dimension of the manifold $\tilde{X}_{n+1}$ generated by the point x^i_{n+1} must be equal to $\tilde{r}$, i.e., we have

$$x^i_{n+1} = x^i_{n+1}(u^{\hat{k}_{n+1}}_{n+1}), \quad \hat{k}_{n+1} = 1, \ldots \tilde{r} \tag{3.6.3}$$

where

$$u^{\hat{k}_{n+1}}_{n+1} = p^{\hat{k}_{n+1}}(u^{\hat{k}_1}_1, \ldots u^{\hat{k}_n}_n). \tag{3.6.4}$$

Definition 3.6.1 An n-quasigroup $\tilde{f}$ defined by the equations (3.6.3) and (3.6.4) is called an *r-parameter subquasigroup* of the original n-quasigroup f.

Consider the case of one-parameter subquasigroup: $\tilde{r} = 1$. Then (3.6.2) gives

$$x^i_{n+1} = f^i[x^{j_1}_1(u_1), \ldots, x^{j_n}_n(u_n)] = x^i_{n+1}(u_{n+1}) \tag{3.6.5}$$

where

$$u_{n+1} = \phi(u_1, \ldots, u_n). \tag{3.6.6}$$

From (3.6.5) and (3.6.6) we see that

$$\frac{\partial f_i}{\partial u_\alpha} = \frac{\partial x^i_{n+1}}{\partial u_{n+1}} \frac{\partial u_{n+1}}{\partial u_\alpha}, \quad \alpha = 1, \ldots, n. \tag{3.6.7}$$

It follows from (3.6.7) that

$$\underset{\beta}{\lambda} \frac{\partial f_i}{\partial u_\alpha} - \underset{\alpha}{\lambda} \frac{\partial f^i}{\partial u_\beta} = 0, \tag{3.6.8}$$

where $\underset{\alpha}{\lambda} = \rho \dfrac{\partial u_{n+1}}{\partial u_\alpha}$ and ρ is an analytic function of $u_1, \ldots, u_n$. The system (3.6.8) contains $(n-1)r$ independent equations. It is easy to see that (3.6.8) implies (3.6.7).

We proved the following theorem.

Theorem 3.6.2 *For the equations* (3.6.5)–(3.6.6) *to define in the local n-quasigroup* (3.2.1) *a one-parameter subquasigroup, it is necessary and sufficient that the equations* (3.6.8) *be satisfied.* ■

Definition 3.6.3 The equations (3.6.8) are called the *differential equations of the one-parameter n-quasigroups.*

The next theorem gives conditions under which there exists in an arbitrarily given direction a one-parameter n-subloop or a one-parameter subgroup.

Theorem 3.6.4 *For there to exist in a principal isotope $L(a_1, \ldots, a_n)$ of an n-quasigroup Q_r a one-parameter n-subloop Q_1 or a one-parameter n-subgroup G_1 for each direction emanating from $e = a_{n+1} = f(a_1, \ldots, a_n)$, it is necessary and sufficient that one of the following conditions be satisfied:*

(i) *the coefficients of at least one of the canonical expansions* (3.5.4)–(3.5.6) *satisfy the relations: for an n-subloop* Q_1

$$\underset{(1)^{p_1}\dots(n)^{p_n}}{\lambda}{}^{i}{}_{(j_1\dots j_{p_1}\dots k_1\dots k_{p_n})} = \delta^i_{(j_1}\underset{(1)^{p_1}\dots(n)^{p_n}}{\lambda}{}_{j_2\dots j_{p_1}\dots k_1\dots k_{p_n})} \tag{3.6.9}$$

and for an n-subgroup G_1

$$\underset{(1)^{p_1}\dots(n)^{p_n}}{\lambda}{}^{i}{}_{(j_1\dots j_{p_1}\dots k_1\dots k_{p_n})} = 0; \tag{3.6.10}$$

(ii) *in a neighbourhood of* $e = a_{n+1}$ *the equations* (1.9.23) *for an n-subloop* Q_1 *and* (1.9.53) *for an n-subgroup* G_1 *obtain, i,e., the web* $W(n+1,n,r)$ *corresponding to* Q *is transversally geodesic or* $(2n+2)$*-hedral.*

Proof. Suppose the equations of the n-loop $L(a_1,\dots,a_n)$ have been reduced to the canonical form (3.5.4)–(3.5.6). Recall that for $m \geq 2$ we had

$$\underset{(m)}{\Lambda}{}^i(0,\dots,0,u_\alpha^{j_\alpha},0,\dots,0) = \underset{(m)}{\Lambda}{}^i(u_1^{j_1},u_1^{j_2},\dots,u_1^{j_n}) = 0. \tag{3.6.11}$$

In the manifolds X_α we assign curves passing through the points a_α. Let

$$u_\alpha^i = \sum_{s=1}^{\infty}\frac{1}{s!}\underset{\alpha}{a}{}_s^i x_\alpha^s \tag{3.6.12}$$

be the equations of the curve u_α. Here the $\underset{\alpha}{a}{}_1^i$ are the coordinates of the tangent vector to u_α at a_α.

Substituting (3.6.12) into the canonical expansions, we obtain

$$u_{n+1}^i = \sum_\alpha \underset{\alpha}{a}{}_1^i x_\alpha + \frac{1}{2}\sum_\alpha \underset{\alpha}{a}{}_2^i x_\alpha^2 + \sum_{(\alpha,\beta)}\underset{\alpha\beta}{\lambda}{}^i_{jk}\underset{\alpha}{a}{}_1^j\underset{\alpha}{a}{}_1^k x_\alpha x_\beta + o(\rho^2). \tag{3.6.13}$$

From (3.6.13) we see that

$$\frac{\partial u_{n+1}^i}{\partial x_\alpha} = \underset{\alpha}{a}{}_1^i + \underset{\alpha}{a}{}_2^i x_\alpha + \sum_{\beta\neq\alpha}\underset{\alpha\beta}{\lambda}{}^i_{jk}\underset{\alpha}{a}{}_1^j\underset{\beta}{a}{}_1^k x_\beta + o(\rho^2). \tag{3.6.14}$$

As we already know from Theorem 3.6.2, the tangent vectors to the curves u_α must satisfy equations of the form

$$\frac{\partial u_{n+1}^i}{\partial x_{\hat\alpha}} = \sigma_{\hat\alpha\alpha}\frac{\partial u_{n+1}^i}{\partial x_\alpha}, \quad \hat\alpha \neq \alpha, \quad \alpha \text{ fixed}, \tag{3.6.15}$$

where

$$\begin{aligned}\sigma_{\hat\alpha\alpha} &= \frac{\partial x_{n+1}}{\partial x_{\hat\alpha}}\Big/\frac{\partial x_{n+1}}{\partial x_\alpha} = \underset{\hat\alpha\alpha}{\sigma}{}_{n+1} + \sum_\beta \underset{\hat\alpha\alpha}{\sigma}{}_\beta x_\beta \\ &+\frac{1}{2}\sum_\beta \underset{\hat\alpha\alpha}{\sigma}{}_{\beta\beta}x_\beta^2 + \sum_{(\beta,\gamma)}\underset{\hat\alpha\alpha}{\sigma}{}_{\beta\gamma}x_\beta x_\gamma + o(\rho^2).\end{aligned} \tag{3.6.16}$$

From (3.6.14), (3.6.15), and (3.6.16), by equating the zero order terms, we obtain

$$\underset{\hat\alpha}{a}{}^i_1 = \underset{\hat\alpha}{\sigma}{}_{n+1} \underset{\hat\alpha}{a}{}^i_\alpha. \tag{3.6.17}$$

We can put

$$\tilde{x}_\alpha = x_\alpha, \quad \tilde{x}_{\hat\alpha} = \underset{\hat\alpha}{\sigma}{}_{n+1} x_{\hat\alpha}. \tag{3.6.18}$$

Then $\underset{\hat\alpha}{\sigma}{}_{n+1}$ in (3.6.18) are equal to 1, and the relations (3.6.17) assume the form

$$\underset{\alpha}{a}{}_i = \underset{\hat\alpha}{a}{}^i_1 = a^i_1. \tag{3.6.19}$$

Again from (3.6.14), (3.6.15), and (3.6.16), by equating the first order terms and using (3.6.19) and the fact that $\underset{\hat\alpha}{\sigma}{}_{n+1} = 1$, we obtain

$$\begin{cases} \underset{\hat\alpha}{a}{}^i_2 = \underset{\hat\alpha\alpha}{\sigma}{}_{\hat\alpha} a^i_1 + \underset{\alpha\hat\alpha}{\lambda}{}^i_{jk} a^j_1 a^k_1, \\ \underset{\alpha}{a}{}^i_2 = -\underset{\hat\alpha\alpha}{\sigma}{}_{\alpha} a^i_1 + \underset{\hat\alpha\alpha}{\lambda}{}^i_{jk} a^j_1 a^k_1. \end{cases} \tag{3.6.20}$$

$$(\underset{\hat\alpha\tilde\beta}{\lambda}{}^i_{jk} - \underset{\alpha\tilde\beta}{\lambda}{}^i_{jk}) a^j_1 a^k_1 = \underset{\hat\alpha\alpha}{\sigma}{}_{\tilde\beta} a^i_1, \quad \tilde\beta \neq \hat\alpha, \alpha. \tag{3.6.21}$$

If instead of α we fix another index $\beta \neq \alpha$, then we obtain an equation similar to the first equation of (3.6.20):

$$\underset{\hat\beta}{a}{}^i_2 = \underset{\hat\beta\beta}{\sigma}{}_{\hat\beta} a^i_1 + \underset{\beta\hat\beta}{\lambda}{}^i_{jk} a^j_1 a^k_1. \tag{3.6.22}$$

Let us take $\hat\gamma \neq \alpha, \beta$. Then (3.6.20) and (3.6.22) yield

$$(\underset{\alpha\hat\gamma}{\lambda}{}^i_{jk} - \underset{\beta\hat\gamma}{\lambda}{}^i_{jk}) a^j_1 a^k_1 = (\underset{\hat\gamma\beta}{\sigma}{}_{\hat\gamma} - \underset{\hat\gamma\alpha}{\sigma}{}_{\hat\gamma}) a^i_1. \tag{3.6.23}$$

In addition, the condition (3.5.21) implies

$$\sum_{\varepsilon,\mu} \underset{\varepsilon\mu}{\lambda}{}^i_{jk} a^j_1 a^k_1 = 0. \tag{3.6.24}$$

For $\underset{\varepsilon\mu}{\lambda}{}^i_{jk}$ we have the system consisting of equations (3.6.24), (3.6.23), and (3.6.21). Equations (3.6.21) and (3.6.23) and their consequences have in their left members all possible differences $(\underset{\varepsilon\mu}{\lambda}{}^i_{jk} - \underset{\rho\delta}{\lambda}{}^i_{jk}) a^j_1 a^k_1$. It is obvious, that for each fixed i there are $\binom{n}{2} - 1$ independent equations, among these equations, for example, those of them which contain in their left members the terms $(\underset{\alpha\beta}{\lambda}{}^i_{jk} - \underset{\rho\delta}{\lambda}{}^i_{jk}) a^j_1 a^k_1$ where α, β are fixed and ρ, δ taken any values such that $(\rho, \delta) \neq (\alpha, \beta)$.

The determinant of the system, for each fixed i arising from these equations and the equation (3.6.24), is of order $\binom{n}{2}$ and, up to its sign, can be written in the form:

$$\Delta = \begin{vmatrix} 1 & 1 & 1 & \dots & 1 \\ 1 & -1 & 0 & \dots & 0 \\ 1 & 0 & -1 & \dots & 0 \\ \dots & \dots & \dots & \dots & \dots \\ \dots & \dots & \dots & \dots & \dots \\ 1 & 0 & 0 & \dots & 1 \end{vmatrix}.$$

If we add to the first row of Δ all other rows, we obtain

$$\Delta = \begin{vmatrix} \binom{n}{2} & 0 & \dots & 0 \\ 1 & -1 & \dots & 0 \\ \dots & \dots & \dots & \dots \\ \dots & \dots & \dots & \dots \\ 1 & 0 & \dots & -1 \end{vmatrix} = (-1)^{\binom{n}{2}-1}\binom{n}{2} \neq 0.$$

Thus, for each fixed i the system has a unique solution. We can write it in the form

$$\underset{\alpha\beta}{\lambda}{}^{i}_{jk} a^j_1 a^k_1 = \underset{\alpha\beta}{\lambda} a^i_1, \tag{3.6.25}$$

where $\underset{\alpha\beta}{\lambda}$ are the completely determined function of $\underset{\hat{\alpha}\alpha}{\sigma}{}_{\hat{\alpha}}$, $\alpha = 1, \dots, n$. Equations (3.6.25) mean that the vector a^i_1 is not arbitrary. For it to be arbitrary, i.e., for any tangent vector to define a one-parameter n-subloop, it is necessary and sufficient that equations (3.6.25) be satisfied identically.

If we write $\underset{\alpha\beta}{\lambda}$ in the form $\underset{\alpha\beta}{\lambda} = \underset{\alpha\beta}{\lambda}{}_k a^k_1$, then (3.6.25) has the form

$$(\underset{\alpha\beta}{\lambda}{}^{i}_{jk} - \delta^i_j \underset{\alpha\beta}{\lambda}{}_k) a^j_1 a^k_1 = 0. \tag{3.6.26}$$

The equation (3.6.26) is an identity with respect to a^i_1 , if and only if

$$\underset{\alpha\beta}{\lambda}{}^{i}_{(jk)} = \delta^i_{(j} \underset{\alpha\beta}{\lambda}{}_{k)}. \tag{3.6.27}$$

It follows from (3.6.27) and (3.5.37), that at the point $u^l_\gamma = 0$ we have

$$\underset{\alpha\beta}{a}{}^{i}_{(jk)} = \delta^i_{(j} \underset{\alpha\beta}{a}{}_{k)}, \tag{3.6.28}$$

where $\underset{\alpha\beta}{a}{}_k = -\underset{\alpha\beta}{\lambda}{}_k$.

In view of (3.6.19), (3.6.20), and (3.6.27), equations (3.6.13) assume the form

$$u^i_{n+1} = a^i_1 \sum_\alpha x_\alpha + \frac{1}{2} a^i_1 \sum_\alpha l_\alpha x^2_\alpha + \sum_{(\alpha,\beta)} \underset{\alpha\beta}{\lambda}{}^{i}_{(jk)} a^j_1 a^k_1 x_\alpha x_\beta + o(\rho^2), \tag{3.6.29}$$

where the l_α are completely determined functions of $\underset{\alpha\beta}{\lambda}{}_k$ and therefore of $\underset{\hat{\alpha}\alpha}{\sigma}{}_{\hat{\alpha}}$. Further, performing the isotopic transformation

$$\bar{x}_\alpha = x_\alpha + \frac{1}{2} l_\alpha x_\alpha^2,$$

we reduce the quantities $\underset{\alpha}{a}{}^i_2$ to zero. After this, by (3.6.25), equations (3.6.29) assume the form

$$u^i_{n+1} = a^i_1 \sum_\alpha x_\alpha + a^i_1 \sum_{(\alpha,\beta)} \underset{\alpha\beta}{\lambda} x_\alpha x_\beta + o(\rho^2). \tag{3.6.30}$$

(we omit the bar over x_α) and equations (3.6.12) become

$$u^i_\alpha = a^i_1 x_\alpha + \sum_{s=3}^{\infty} \frac{1}{s!} \underset{\alpha}{a}{}^i_s x^s_\alpha. \tag{3.6.31}$$

Similarly, we can reduce to zero the quantities $\underset{\alpha}{a}{}^i_3, \underset{\alpha}{a}{}^i_4, \ldots$. For the existence of a one-parameter n-subloop in any direction a^i_1 emanating from the point $e = a_{n+1}$, we obtain conditions (3.6.9) analogous to (3.6.27).

The expansions (3.6.29) will have the form

$$u^i_\alpha = a^i_1 x_\alpha. \tag{3.6.32}$$

Conversely, if (3.6.9) and (3.6.32) hold, then any direction a^i_1 defines a one-parameter subloop in the n-loop $L(a_1, \ldots, a_n)$. Indeed, under these conditions we have

$$u^i_{n+1} = a^i_1 \sum_\alpha x_\alpha + \sum_{m=2}^{\infty} \underset{(m)}{\Lambda}{}^i, \tag{3.6.33}$$

where

$$\underset{(m)}{\Lambda}{}^i = \frac{1}{m!} \sum_{\substack{p_\alpha = 0 \\ (p_1+\ldots+p_n=m)}}^{m-1} \binom{m}{p_1}\binom{m-p_1}{p_2} \cdots \binom{m-p_1-\ldots-p_{n-2}}{p_{n-1}} \cdot$$

$$\cdot \underset{(1)^{p_1}\ldots(n)^{p_n}}{\lambda}{}^i_{(j_1\ldots j_m)} a^{j_1}_1 \ldots a^{j_m}_1 x^{p_1}_1 \ldots x^{p_n}_n = a^i_1 \underset{(m)}{\Lambda},$$

$$\underset{(m)}{\Lambda} = \frac{1}{m!} \sum_{\substack{p_\alpha = 0 \\ (p_1+\ldots+p_n=m)}}^{m-1} \binom{m}{p_1}\binom{m-p_1}{p_2} \cdots \binom{m-p_1-\ldots-p_{n-2}}{p_{n-1}} \cdot$$

$$\cdot \underset{(1)^{p_1}\ldots(n)^{p_n}}{\lambda}{}_{(j_2\ldots j_m)} a^{j_2}_1 \ldots a^{j_m}_1 x^{p_1}_1 \ldots x^{p_n}_n.$$

Now put

$$x_{n+1} = \sum_\alpha x_\alpha + \sum_{m=2}^{\infty} \underset{(m)}{\Lambda}. \tag{3.6.34}$$

Then the expansions (3.6.32) are reduced to the form $u^i_{n+1} = a^i_1 x_{n+1}$. Thus we have obtained a one-parameter n-subloop (3.6.34) in which the x_α are canonical variables.

Note that such a one-parameter n-subloop is an n-group if the condition (3.6.10) is satisfied. In particular, when $m = 2$, the equality (3.6.10) yields $\underset{\alpha\beta}{\lambda}{}^i_{(jk)} = 0$, hence

$$\underset{\alpha\beta}{a}{}^i_{(jk)} = 0. \tag{3.6.35}$$

To conclude the proof of the part (i), note that we have proved (3.6.28) and (3.6.35) at the point $u^i_\alpha = 0$, but under conditions (3.6.9) and (3.6.10) these equations are also true in a neighbourhood of this point. To see this, expand the tensors $\underset{\alpha\beta}{a}{}^i_{(jk)} - \delta^i_{(j}\underset{\alpha\beta}{a}{}_{k)}$ and $\underset{\alpha\beta}{a}{}_{(jk)}$ into Taylor series and use (3.6.9) and (3.6.10), respectively, to verify that all coefficients of these expansions are zeros. Note also that differentiating equations (3.6.28) and (3.6.35), we can get respectively (3.6.9) and (3.6.10).

To prove the part (ii), note that conditions (3.6.28) and (3.6.10) coincide with equations (1.9.23) and (1.9.53). According to our considerations in Section 1.9, it means that the web $W(n+1, n, r)$, corresponding to the n-loop under consideration, will be respectively transversally geodesic or $(2n+2)$-hedral. ■

Remark 3.6.5 Under the hypotheses of Theorem 3.6.4, the coefficients of any of the canonical expansions of Q_r satisfy (3.6.9) or (3.6.10).

We conclude this section by the following theorem giving a necessary and sufficient condition for all the canonical expansions to coincide.

Theorem 3.6.6 *For all the canonical expansions of the n-quasigroup Q_r introduced in Theorem 3.5.6 to agree, it is necessary and sufficient that the web $W(n+1, n, r)$ corresponding to Q_r be $(2n+2)$-hedral.*

Proof. We observe that if (3.6.10) holds, then the conditions (3.5.32) are satisfied identically and

$$F^i(u^{j_1}, \sigma_2 u^{j_2}, \ldots, \sigma_n u^{j_n}) = (1 + \sigma_2 + \ldots \sigma_n) u^i,$$

i.e., the conditions (3.5.36) are satisfied in Q_r for any $\sigma_2, \ldots, \sigma_n$ such that $(1 + \sigma_2 + \ldots + \sigma_n)^m - \sigma_2^m - \ldots - \sigma_n^m - 1 \neq 0$. In this case all the canonical expansions introduced in Theorem 3.5.6 agree with each other. The converse is obvious. ■

3.7 Comtrans Algebras

In this section we will show that in the tangent bundle of a coordinate n-loop $L(a_1, \ldots, a_n)$ there exists a certain algebraic structure generalising for analytic n-loops the concept of the Lie, Mal'cev, and Akivis algebras known respectively for the real analytic (in particular Lie) groups, real analytic Moufang loops, and arbitrary real analytic loops. We will also prove an analogue of Lie's Third Fundamental Theorem for n-loops.

3.7.1 Preliminaries

We will consider a local analytic n-loop $L(a_1, \dots, a_n)$ over $\mathbf{R}$ given by an r-tuple

$$F = (F^1, \dots, F^r) \tag{3.7.1}$$

where $F^i, i = 1, \dots, r$, are power series (3.2.6) satisfying (3.2.5) and (3.2.7). Thus we have

$$\begin{aligned} F^i &= \sum_{\alpha=1}^{n} u^i_\alpha + \sum_{(\alpha,\beta)} \lambda^{\ i}_{\alpha\beta\ jk} u^j_\alpha u^k_\beta + \frac{1}{2} \sum_{(\alpha,\beta)} \lambda^{\ i}_{\alpha\beta\beta\ jkl} u^j_\alpha u^k_\beta u^l_\beta \\ &+ \sum_{(\alpha,\beta,\gamma)} \lambda^{\ \ i}_{\alpha\beta\gamma\ jkl} u^j_\alpha u^k_\beta u^l_\gamma + o(\rho^3), \end{aligned} \tag{3.7.2}$$

and (3.2.5) holds:

$$F^i(0, \dots, 0, u^{j\alpha}_\alpha, 0, \dots, 0) = u^i_\alpha. \tag{3.7.3}$$

Definition 3.7.1 The *m-th degree chunk* F_m of the loop (3.7.1) is the n-loop $F_m(F^i_m, \dots, F^r_m)$ with

$$F^i_m = \underset{(1)}{\Lambda}{}^i + \underset{(2)}{\Lambda}{}^i + \cdots \underset{(m)}{\Lambda}{}^i \tag{3.7.4}$$

where $\underset{(1)}{\Lambda}{}^i, \dots, \underset{(m)}{\Lambda}{}^i$ are homogeneous terms of (3.7.2) of degrees $1, \dots, m$ respectively.

Note that F^i_m is an element of the polynomial ring $\mathbf{R}\ [u^1_1, \dots, u^r_1; \dots; u^1_n, \dots, u^r_n]$. In the sequel we will use only cubic, quadratic, and linear chunks: F_1, F_2, and F_3.

We will outline here the results known in the (binary) case $n = 2$.

Definition 3.7.2 An *Akivis algebra* over $\mathbf{R}$ is a real vector space A equipped with a bilinear operation $[x, y]$ (the (binary) *commutator*) and a trilinear operation (x, y, z) (the *associator*) such that the commutator is *anticommutative*:

$$[x, y] + [y, x] = 0, \tag{3.7.5}$$

and the *Akivis identity* (generalized Jacobi identity)

$$\begin{aligned} &[[x, y], z] + [[y, z,], x] + [[z, x], y] \\ = \ &(x, y, z) + (y, z, x) + (z, x, y) \\ - \ &(y, x, z) - (z, y, x) - (x, z, y) \end{aligned} \tag{3.7.6}$$

is satisfied.

The relationship between Akivis algebras and r-dimensional (binary) loops F over $\mathbf{R}$ may be summarised as follows for the purposes of this section. The quadratic chunk F_2 of F determines a commutator operation

$$[x_1, x_2] = F_2(x_1, x_2) - F_1(x_2, x_1) \tag{3.7.7}$$

on $\mathbf{R}^r$. The cubic chunk F_3 of F determines an associator operation

$$(x_1, x_2, x_3) = F_3(F_3(x_1, x_2), x_3) - F_3(x_1, F_3(x_2, x_3)) \tag{3.7.8}$$

on $\mathbf{R}^r$.

Theorem 3.7.3 (see [HS 86a], 2.7). *If F is an r-dimensional loop over $\mathbf{R}$, then the commutator* (3.7.7) *and the associator* (3.7.8) *form an Akivis algebra*

$$(\mathbf{R}^r, [\ ,\ ], (\ ,\ ,\ )) \tag{3.7.9}$$

on $\mathbf{R}$. ■

Theorem 3.7.4 (see [HS 86a, 2.8). *Let r be a positive integer. Suppose given an Akivis algebra over* $\mathbf{R}$

$$(\mathbf{R}^r, [\ ,\ ], (\ ,\ ,\ )). \tag{3.7.10}$$

Then there is an r-dimensional loop F over $\mathbf{R}$ such that its Akivis algebra (3.7.9) *as determined by Theorem* 3.7.3 *is the given Akivis algebra* (3.7.10). ■

3.7.2 Comtrans Structures

Suppose now that $n = 3$, i.e., we consider the ternary case. As we know in the case of a ternary loop we have a ternary operation (multiplication)

$$X \times X \times X \to X; \quad (x, y, z) \to xyz. \tag{3.7.11}$$

The multiplication (3.7.11) leads to the definition of two related ternary operations from X to X, the commutator and translator of the given multiplication.

Definition 3.7.5 The *commutator* $[x, y, z]$ of x, y, z is

$$[x, y, z] = xyz - yxz, \tag{3.7.12}$$

and the *translator* $< x, y, z >$ of x, y, z is

$$< x, y, z >= xyz - yzx. \tag{3.7.13}$$

It follows immediately from Definition 3.7.5 that the commutator is *left alternative* in the sense that it satisfies the identity

$$[x, y, z] + [y, x, z] = 0. \tag{3.7.14}$$

and the translator satisfies the *Jacobi identity*:

$$< x, y, z > + < y, z, x > + < z, x, y >= 0. \tag{3.7.15}$$

Finally, the commutator and translator together satisfy the *comtrans identity*

$$[x, y, z] + [z, y, x] =< x, y, z > + < z, y, x > . \tag{3.7.16}$$

Proposition 3.7.6 (i) *The product* (3.7.11) *is symmetric (in the sense of being invariant under permutations of the direct factors in* $X \times X \times X$*) if and only if the commutator and translator are identically zero.*

(ii) *The classical Jacobi identity for the binary commutator* $[x, y] = xy - yx$ *of an associative bilinear product* $X \times X \to X; (x, y) \to xy$ *is a special case of the current Jacobi identity* (3.7.15).

(iii) *The alternator* $\{x, y, z\}$ *introduced in Section* 3.2 *is expressed in terms of the commutator as follows:*

$$\{x, y, z\} = [x, y, z] + [y, z, x] + [z, x, y], \tag{3.7.17}$$

and the commutator and the translator of the alternator are

$$[\{x, y, z\}] = 2\{x, y, z\}, \quad < \{x, y, z\} >= 0. \tag{3.7.18}$$

Proof. (i) and (iii) follow from the corresponding definitions by straightforward calculations. As to (ii), it is obtained by observing that the repeated binary commutator $[[x, y], z]$ is the translator for the ternary multiplication

$$X \times X \times X \to X; \quad (x, y, z) \to -zxy - yxz. \tag{3.7.19}$$

Note that the commutator for this multiplication agrees with the translator, so that the comtrans identity is trivial here. Note also that the comtrans identity for the operation $\{x, y, z\}$ can be easily derived from (3.7.19) by applying the identity $\{x, y, z\} + \{z, y, x\} = 0$. ■

The properties (3.7.14)–(3.7.16) motivate the following definition.

Definition 3.7.7 A *comtrans structure* over $\mathbf{R}$ is a real vector space X with two ternary operations from X to X, known as the *commutator* $[x, y, z]$ and the *translator* $< x, y, z >$ respectively, such that the commutator is left alternative (3.7.14), the translator satisfies the Jacobi identity (3.7.15), and the two together satisfy the comtrans identity (3.7.16). The comtrans structure X is called a *comtrans algebra* if the commutator and translator are trilinear.

In a comtrans structure over $\mathbf{R}$ a new operation can be introduced.

Definition 3.7.8 The *bogus product* $/x, y, z/$ is

$$/x,y,z/ = ([x,y,z] + [y,z,x] + [z,x,y] + 2 < x,y,z > - 2 < z,x,y >). \quad (3.7.20)$$

The commutator $/x,y,z/ - /y,x,z/$ of the bogus product is called the *bogus commutator*; its translator $/x,y,z/ - /y,z,x/$ is called the *bogus translator*.

Proposition 3.7.9 *In a comtrans structure over* **R***:*
(i) *The bogus commutator agrees with the commutator.*
(ii) *The bogus translator agrees with the translator.*

Proof. (i) The bogus commutator is one-sixth of

$$\begin{array}{rl} & [x,y,z] + [y,z,x] + [z,x,y] + 2 < x,y,z > - 2 < z,x,y > \\ - & [y,x,z] - [x,z,y] - [z,y,x] - 2 < y,x,z > + 2 < z,y,x > . \end{array}$$

By the comtrans identity, the translator terms may be replaced by correspon ding commutator terms. By left alternativity, each commutator is a multiple of $[x,y,z]$, $[y,z,x]$, or $[z,x,y]$. One obtains

$$\begin{array}{rl} & [x,y,z] + [y,z,x] + [z,x,y] + 2[x,y,z] - 2[z,x,y] \\ + & [x,y,z] + [z,x,y] + [y,z,x] + 2[x,y,z] - 2[y,z,x], \end{array}$$

which cancels to $6[x,y,z]$.

(ii) The bogus translator is one-sixth of

$$\begin{array}{rl} & [x,y,z] + [y,z,x] + [z,x,y] + 2 < x,y,z > - 2 < z,x,y > \\ - & [y,z,x] - [z,x,y] - [x,y,z] - 2 < y,z,x > + 2 < x,y,z > . \end{array}$$

The commutator terms cancel immediately, while the translator terms reduce to $6 < x,y,z >$ by the Jacobi identity. ■

3.7.3 Masking

Let F be an r-dimensional ternary loop over **R**. The cubic chunk F_3 defines an operation

$$\mathbf{R}^r \times \mathbf{R}^r \times \mathbf{R}^r \to \mathbf{R}^r; \quad (x_1, x_2, x_3) \to F_3(x_1, x_2, x_3). \quad (3.7.21)$$

However, it is obvious that the commutator and translator of this operation are not multilinear. One possible way to see this is to see that in general F_3 contains terms with $x_1^2 x_2$, $x_2^2 x_3$, etc. To overcome this problem, the technique of *masking* is used.

Definition 3.7.10 The *masks* of the ternary loop F are the binary loops

$$\begin{aligned} F\{1\}(u_2, u_3) &= F(0, u_2, u_3), \\ F\{2\}(u_1, u_3) &= F(u_1, 0, u_3), \\ F\{3\}(u_1, u_2) &= F(u_1, u_2, 0), \end{aligned} \quad (3.7.22)$$

known respectively as the 1-*mask*, the 2-*mask*, and the 3-*mask*; in (3.7.22) u_α, $\alpha = 1,2,3$, means an r-tuple $u_\alpha = (u^1_\alpha, \ldots, u^r_\alpha)$. The *masked version* $M = M(F)$ of the ternary loop F is then defined by

$$M = F\{1\} + F\{2\} + F\{3\} - F. \tag{3.7.23}$$

It is easy to see that if F is a coordinate ternary loop of a web $W(4,3,r)$, then its α-masks $F\{\alpha\}, \alpha = 1,2,3$, are coordinate binary loops of 3-subwebs [2,3,4], [1,3,4], and [1,2,4].

Proposition 3.7.11 *Let M be the masked version of the ternary loop F. Then*

(i) *M is a ternary loop of the type* (3.7.1)–(3.7.3).

(ii) *The cubic chunk M_3 of M is the masked version $M(F_3)$ of the cubic chunk F_3 of F.*

(iii) *The commutator and translator of the ternary operation M_3 on* **R** *are trilinear.*

Proof. (i) If F^i are defined by (3.7.1), then

$$F\{\alpha\}^i = u^2_\beta + u^i_\gamma + \underset{\beta\gamma}{\lambda}{}^i_{jk} u^j_\beta u^k_\gamma + \frac{1}{2}(\underset{\beta\gamma\gamma}{\lambda}{}^i_{jkl} u^j_\beta u^k_\gamma u^l_\gamma + \underset{\gamma\beta\beta}{\lambda}{}^i_{jkl} u^j_\gamma u^k_\beta u^l_\beta) + o(\rho^3),\ \alpha = 1,2,3. \tag{3.7.24}$$

Next by (3.7.23), (3.7.24), and (3.7.2)

$$M^i = u^i_1 + u^i_2 + u^i_3 - \underset{123}{\lambda}{}^i_{jkl} u^j_1 u^k_2 u^l_3 + o(\rho^3), \tag{3.7.25}$$

so that M is a ternary loop.

(ii) By (3.7.25) the cubic chunk M_3 of M is

$$M^i_3 = u^i_1 + u^i_2 + u^i_3 - \underset{123}{\lambda}{}^i_{jkl} u^j_1 u^k_2 u^l_3. \tag{3.7.26}$$

The cubic chunk F_3 of F is

$$F^i_3 = u^i_1 + u^i_2 + u^i_3 + \sum_{(\alpha,\beta)} \underset{\alpha\beta}{\lambda}{}^i_{jk} u^j_\alpha u^k_\beta + \frac{1}{2} \sum_{(\alpha,\beta)} \underset{\alpha\beta\beta}{\lambda}{}^i_{jkl} u^j_\alpha u^k_\beta u^l_\gamma + \underset{123}{\lambda}{}^i_{jkl} u^j_1 u^k_2 u^l_3. \tag{3.7.27}$$

Since by (3.7.24)

$$F_3\{\alpha\} = u^i_\beta + u^i_\gamma + \underset{\beta\gamma}{\lambda}{}^i_{jk} u^j_\beta u^k_\gamma + \frac{1}{2}(\underset{\beta\gamma\gamma}{\lambda}{}^i_{jkl} u^j_\beta u^k_\gamma u^l_\gamma + \underset{\gamma\beta\beta}{\lambda}{}^i_{jkl} u^j_\gamma u^k_\beta u^l_\beta), \tag{3.7.28}$$

it follows that

$$M_3 = F_3\{1\} + F_3\{2\} + F_3\{3\} - F = M\{F_3\}, \tag{3.7.29}$$

as required.

(iii) By (3.7.26),

$$M_3^i(c_1^j, c_2^k, c_3^l) = c_1^i + c_2^i + c_3^i - \lambda^i_{jkl} c_1^j c_2^k c_3^l, \tag{3.7.30}$$

where $\lambda^i_{jkl} = \underset{123}{\lambda}{}^i_{jkl}$. Thus the i-th component of the commutator is

$$[c_1^j, c_1^k, c_3^l]^i = (\lambda^i_{kjl} - \lambda^i_{jkl}) c_1^j c_2^k c_3^l, \tag{3.7.31}$$

while the i-th component of the translator is

$$< c_1^j, c_2^k, c_3^l >^i = (\lambda^i_{klj} - \lambda^i_{jkl}) c_1^j c_2^k c_3^l. \tag{3.7.32}$$

Both of them are trilinear.. ■

Corollary 3.7.12 *The commutator and translator of $M_3 = M(F_3)$ determine a comtrans algebra on* $\mathbf{R}$.

Definition 3.7.13 The comtrans algebra of Corollary 3.7.12 is called the *comtrans algebra* of the ternary loop F.

3.7.4 Lie's Third Fundamental Theorems for Analytic 3-Loops

Let F be an r-dimensional ternary loop over $\mathbf{R}$. Each mask of F is a binary loop, and so determines an Akivis algebra on $\mathbf{R}$ by Theorem 3.7.3. Denote the Akivis algebra of the α-mask by $(\mathbf{R}, [\ ,\ ]_\alpha, (\ ,\ ,\ )_\alpha)$ for $\alpha = 1, 2, 3$. The formal analogue of the direct part of Lie's Third Fundamental Theorem for ternary loops may then be stated as follows, summarising the above results.

Theorem 3.7.14 *An r-dimensional ternary loop F over R determines an algebra structure*

$$(\mathbf{R}, [\ ,\ ]_\alpha, (\ ,\ ,\ )_\alpha, [\ ,\ ], <\ ,\ ,\ >) \tag{3.7.33}$$

on $\mathbf{R}^r$, $1 \le \alpha \le 3$, *comprising its comtrans algebra* $(\mathbf{R}, [\ ,\ ], <\ ,\ ,\ >)$ *and the three Akivis algebras* $(\mathbf{R}, [\ ,\ ]_\alpha, (\ ,\ ,\ )_\alpha)$ *of its α-masks.* ■

The converse of Theorem 3.7.14 gives the analogue of the converse part of Lie's Third Fundamental Theorem for ternary loops.

Theorem 3.7.15 *Let r be a positive integer. Suppose given Akivis algebras*

$$(\mathbf{R}, [\ ,\ ]_\alpha, (\ ,\ ,\ )_\alpha) \tag{3.7.34}$$

for $\alpha = 1, 2, 3$ and a comtrans algebra

$$(\mathbf{R}, [\ ,\ ], <\ ,\ ,\ >). \tag{3.7.35}$$

Then there is an r-dimensional ternary loop F over $\mathbf{R}$ *such that* (3.7.35) *is its comtrans algebra and such that the Akivis algebras* (3.7.34) *are the respective Akivis algebras of its α-masks for $\alpha = 1, 2, 3$.*

Proof. Use notation as in the proof of Proposition 3.7.11. By Theorem 3.7.4, the Akivis algebras (3.7.34) determine r-dimensional loops $G^i_\alpha(u_\beta, \ u_\gamma) = (3.7.19)$, $\alpha = 1, 2, 3$, whose Akivis algebras are again the Akivis algebras (3.7.34). Now consider the commutator and translator in (3.7.35). They determine a trilinear bogus product

$$\mathbf{R}^r \times \mathbf{R}^r \times \mathbf{R}^r \to \mathbf{R}^r; \ \ (c_1, c_2, c_3) \to /c_1, c_2, c_3/ \tag{3.7.36}$$

according to (3.7.20), say

$$/c^j_1, c^k_2, c^l_3/^i = -\lambda^i_{jkl} c^j_1 c^k_2 c^l_3. \tag{3.7.37}$$

By Proposition 3.7.9, the components of the commutator and translator in (3.7.35) are then given by (3.7.31) and (3.7.32) respectively. We define now the r-dimensional ternary loop F by

$$F^i = u^i_1 + u^i_2 + u^i_3 + \sum_{(\alpha,\beta)} {}_{\alpha\beta}\lambda^i_{jk} u^j_\alpha u^k_\beta + \frac{1}{2} \sum_{(\alpha,\beta)} {}_{\alpha\beta\beta}\lambda^i_{jkl} u^j_\alpha u^k_\beta u^l_\beta + \sum_{(\alpha,\beta,\gamma)} {}_{\alpha\beta\gamma}\lambda^i_{jkl} u^j_\alpha u^k_\beta u^l_\gamma. \tag{3.7.38}$$

Since F is equal to its cubic chunk F_3 , the α-masks are $G^i_1, \ \ G^i_2, \ \ G^i_3$ respectively, having (3.7.34) as their Akivis algebras. The masked version of F is given by (3.7.26) and (3.7.29), having (3.7.31) and (3.7.32) as its commutator and translator. The comtrans algebra of F is thus the given (3.7.35). ■

3.7.5 General Case of Analytic n-Loops

Let F be an n-ary loop. For each $(n-2)$-element subset σ of $\{1, 2, \ldots, n\}$ with complement $\{\beta, \gamma\}$, the *σ-mask* of F is the binary loop

$$F\sigma(u_\beta, u_\gamma) = F(0, \ldots, u_\beta, \ldots, u_\gamma, \ldots, 0) \tag{3.7.39}$$

obtained by setting $u_\delta = 0$ for δ in σ. For each $(n-3)$-element subset τ of $\{1, 2, \ldots, n\}$ with complement $\{\beta, \gamma, \delta\}$, the *$\tau$-mask* of F is the ternary loop

$$F\tau(u_\beta, u_\gamma, u_\delta) = F(0, \ldots, u_\beta, \ldots, u_\gamma, \ldots, u_\delta, \ldots, 0). \tag{3.7.40}$$

obtained by setting $u_\varepsilon = 0$ for ε in τ. Note that for a given τ, the three α-masks of F with $\tau \subset \sigma$ are the masks of the ternary loop $F\tau$ in the sense of (3.7.22). Then the n-loop F determine $\binom{n}{2}$ Akivis algebras and $\binom{n}{3}$ comtrans algebras. The Akivis algebras come from the σ-masks, and the comtrans algebras come from the masked versions of the τ-masks. Conversely, given $\binom{n}{2}$ Akivis algebras and $\binom{n}{3}$ comtrans algebras, indexed appropriately by $(n-2)$-element subsets σ and $(n-3)$-element subsets τ of $\{1, \ldots, n\}$, one may build (the cubic chunk of) an n-loop yielding these given Akivis and comtrans algebras according to the above construction.

NOTES

3.1–3.2. The results are due to the author (see [G 75a, 75b]) where for simplicity the case $n = 3$, i.e., a web $W(4,3,r)$ was considered).

3.3. The components of the torsion tensor were calculated by the author [G 75a, 75b] and those of the curvature tensor by Gerasimenko [Ge 84b].

3.4–3.6. The sections are from [G 75b].

3.7. In modern language, the direct part of Lie's Third Fundamental Theorem (see [Li 1893], p. 396) states that each local real analytic (in particular Lie) group determines a Lie algebra in its tangent space at the identity element. Furthermore, according to the converse part of Lie's Third Fundamental Theorem, to each finite dimensional real Lie algebra, there exists a corresponding real analytic group, determining the given Lie algebra in its tangent space at the identity. Mal'cev [Ma 55] extended Lie's Third Theorem to correspondences between local real analytic Moufang loops and Mal'cev algebras. Akivis [Ak 76] found an analogue of Lie and Mal'cev algebras, W-algebras, in the tangent space to the identity of an arbitrary real analytic loop. Hofmann and Strambach named W-algebras as Akivis algebras and they proved Lie's Third Theorem for a real analytic loop (see [HS 86a] or [CPS 88], Chapter).

The author (see [CPS 88], Problem X.3.9] or [Sc 84], p. 16) has raised the problem of finding an algebraic construction in the tangent bundle of the coordinate n-loop of a web $W(n+1,n,r), n > 2$, similar to the construction of the Akivis algebra for webs $W(3,2,r)$ or binary loops. Smith [Sm 88] found such a construction and proved Lie's Third Theorem for an analytic n-loop. His results are presented in this section. Note that we consider analytic n-loops over **R** while Smith studied them in more general setting: over a commutative ring **R** or (in the converse part of Lie's Third Fundamental Theorem) over a field of characteristic prime to 6.

Smith's construction has two salient features. The first is the specification of two ternary analogues of the binary commutator. He constructed the left alternative commutator (3.5.12) and the translator (3.5.13) satisfying the Jacobi identity (3.5.15). Together the commutator and translator satisfy a new identity called the "comtrans" identity (3.5.16). A ternary algebra in which all three indicated identities are satisfied is called a "comtrans algebra". This algebra is an analogue of Lie, Mal'cev and Akivis algebras. The second salient feature, which has no analogue in the binary cases, is that multilinear operations do not arise directly from n-ary loops with $n > 2$. They are extracted from ternary loops using the technique known as "masking". Three binary loops which are the "masks" of the ternary loop, are coordinate binary loops of 3-subwebs [1,2,4], [1,3,4], and [2,3,4] of a web $W(4,3,r)$. Trilinear operations forming a comtrans algebra then appear once the effects of the masks are removed.

In [HS 86a] and [Sm 87] one can find more detail when the Baker-Campbell-Hausdorff formula can or cannot be used for the proof of Lie's Third Fundamental Theorem.

Chapter 4

Special Classes of Multicodimensional $(n+1)$-Webs

4.1 Reducible $(n+1)$-Webs

In Chapter 1 we studied some special classes of multicodimensional $(n+1)$-webs: parallelisable, transversally geodesic, isoclinic, $(2n+2)$-hedral, hexagonal, etc. They were defined using a pure differential geometry approach. In Chapter 3 we introduced a concept of an l.d. n-quasigroup and established close relations between $(n+1)$-webs and l.d. n-quasigroups. This relationship allows us to introduce a few new interesting special classes of $(n+1)$-webs and find new properties of some of the special classes which were studied in Chapter 1.

Definition 4.1.1 An l.d. n-quasigroup, $n > 2$, is said to be *reducible* if its n-ary operation can be reduced to an $(h+1)$-ary and an $(n-h)$-ary operation, $1 \leq h < n$, i.e., if

$$u^i_{n+1} = F^i(u^{j_1}_1, \ldots, u^{j_n}_n) = F^i(u^{j_1}_{\alpha_1}, \ldots, g^k(u^{j_{h+1}}_{\alpha_{h+1}}, \ldots, u^{j_n}_{\alpha_n}), \ldots, u^{j_h}_{\alpha_h}). \qquad (4.1.1)$$

A web $W(n+1, n, r)$ is called *reducible* if at least one of its coordinate n-quasigroups is reducible.

Note that in (4.1.1) and in (3.2.4) the functions on the right-hand side are distinct: the number of variables on which they depend are different. However, to be able to use, without confusion, our formulas from Section 3.3 (for example, formulae (3.3.9), (3.3.11), etc.), it is convenient to denote all of them by F^i.

Remark 4.1.2 Since reducibility of an n-quasigroup is preserved under passage to an inverse operation (see [Be 72], p.18), any parastroph of a reducible n-quasigroup is reducible. Thus, in Definition 4.1.1 of a reducible $(n+1)$-web we claim reducibility of only one of its coordinate n-quasigroups.

Remark 4.1.3 Since the property of an n-quasigroup to be reducible is invariant under isotopy (see [Be 72], p. 108), all webs $W(n+1, n, r)$, which are equivalent to a reducible web $W(n+1, n, r)$ of type (4.1.1), are reducible of the same type.

We will give now a necessary and sufficient condition of reducibility (4.1.1) in terms of integrability of diagonal distributions of the web $W(n+1, n, r)$ which we studied in Section 1.8.

Theorem 4.1.4 *For a web* $W(n+1, n, r)$ *to be reducible of type* (4.1.1), *it is necessary and sufficient that the diagonal distribution* $\{\alpha_1, \ldots, \alpha_h\}$ *be integrable.*

Proof. *Sufficiency*. Let an $(n-1)r$-dimensional distribution $\{\alpha_1, \ldots, \alpha_h\}$ defined by equations (1.8.1) or (1.8.2) be integrable. As was shown in Section 1.8, this will be the case if and only if conditions (1.8.3) hold. Using (1.8.3), we can write the structure equations (1.2.16) in the form:

$$d\underset{\alpha_s}{\omega}^i = \underset{\alpha_s}{\omega}^j \wedge (\omega_j^i + \sum_{\beta_{s_1} \neq \alpha_s} \underset{\alpha_s \beta_{s_1}}{a}{}^i_{jk} \underset{\beta_{s_1}}{\omega}^k) + \sum_{t=h+1}^{n} \underset{\alpha_s \beta_t}{a}{}^i_{jk} \underset{\alpha_s}{\omega}^j \wedge \theta^k, \tag{4.1.2}$$

$$d\underset{\beta_t}{\omega}^i = \underset{\beta_t}{\omega}^j \wedge (\omega_j^i + \sum_{s=1}^{h} \underset{\beta_t \alpha_s}{a}{}^i_{jk} \underset{\alpha_s}{\omega}^k) + \sum_{\delta_{t_1} \neq \beta_t} \underset{\beta_t \delta_{t_1}}{a}{}^i_{jk} \underset{\beta_t}{\omega}^j \wedge \underset{\delta_{t_1}}{\omega}^k, \tag{4.1.3}$$

where

$$\theta^k = \sum_{t=h+1}^{n} \underset{\beta_t}{\omega}^k, \quad s, s_1 = 1, \ldots, h; \;\; t, t_1 = h+1, \ldots, n.$$

Equations (4.1.3) allow us to write the following structure equations for forms θ^i :

$$d\theta^i = \theta^j \wedge \omega_j^i + \sum_{s=1}^{h} \underset{\beta_t \alpha_s}{a}{}^i_{jk} \theta^j \wedge \underset{\alpha_s}{\omega}^k. \tag{4.1.4}$$

It follows from equations (4.1.4) that the system of forms θ^i is completely integrable. Therefore, the systems of equations

$$\underset{\beta_t}{\omega}^i = 0, \quad t = h+1, \ldots, n, \qquad \theta^i \equiv \sum_{t=h+1}^{n} \underset{\beta_t}{\omega}^i = 0 \tag{4.1.5}$$

defines, on the $(n-h)r$-dimensional manifold, a web $W(n-h+1, n-h, r)$. Integration of the systems (4.1.5) gives

$$v^k = g^k(u_{\alpha_{h+1}}^{j_{h+1}}, \ldots, u_{\alpha_n}^{j_n}). \tag{4.1.6}$$

Equations (4.1.2) and (4.1.6) show that the system of forms

$$-\underset{n+1}{\omega}^i = \theta^i + \sum_{s=1}^{h} \underset{\alpha_s}{\omega}^i \tag{4.1.7}$$

is completely integrable since

$$d\underset{n+1}{\omega}{}^i = \underset{n+1}{\omega}{}^j \wedge \omega_j^i.$$

Therefore, the systems of equations

$$\underset{\alpha_s}{\omega}{}^i = 0, \ \ s = 1, \dots, h; \ \ \theta^i = 0, \ \ \underset{n+1}{\omega}{}^i = 0 \tag{4.1.8}$$

defines, on the $(h+1)r$-dimensional manifold, a web $W(h+2, h+1, r)$.

Integration of (4.1.8) gives

$$u^i_{n+1} = F^i(u^{j_1}_{\alpha_1}, \dots, v^k, \dots, u^{j_n}_{\alpha_n}). \tag{4.1.9}$$

From (4.1.6) and (4.1.9) we obtain (4.1.1). This proves sufficiency.

Necessity. Since we now assume the coordinate n-quasigroup to be reducible, equations (4.1.1) hold. We need to prove that they imply equations (1.8.3) or that the equations

$$\underset{\alpha_s\beta_t}{\Gamma}{}^i_{jk} = \underset{\alpha_s\gamma_t}{\Gamma}{}^i_{jk}, \tag{4.1.10}$$

which are equivalent to (1.8.3) by virtue of (3.3.4) and (3.3.23) hold. From (4.1.1) we find

$$du^i_{n+1} = \sum_s \underset{\alpha_s}{F}{}^i_j du^j_{\alpha_s} + \frac{\partial f^i}{\partial g^m} \sum_t \frac{\partial g^m}{\partial u^j_{\beta_t}} du^j_{\beta_t}. \tag{4.1.11}$$

Therefore,

$$\underset{\alpha_s}{\omega}{}^i = \underset{\alpha_s}{F}{}^i_j du^j_{\alpha_s}, \tag{4.1.12}$$

$$\underset{\beta_t}{\omega}{}^i = \frac{\partial F^i}{\partial g^m} \frac{\partial g^m}{\partial u^j_{\beta_t}} du^j_{\beta_t}. \tag{4.1.13}$$

It follows from (4.1.13) that

$$\underset{\beta_t}{F}{}^j_i = \frac{\partial F^i}{\partial g^m} \frac{\partial g^m}{\partial u^j_{\beta_t}}. \tag{4.1.14}$$

Equations (4.1.14) mean that the matrix $(\underset{\beta_t}{F}{}^i_j)$ is the product of two matrices. Therefore the matrix $(\underset{\beta_t}{G}{}^i_j)$ inverse to it will be the product of the matrices (H^j_k) and $(\underset{\beta_t}{\varphi}{}^k_i)$, inverse to the matrices $(\frac{\partial F^i}{\partial g^m})$ and $(\frac{\partial g^m}{\partial u^j_{\beta_t}})$ but taken in the reverse order:

$$\underset{\beta_t}{G}{}^i_j = \underset{\beta_t}{\varphi}{}^i_m H^m_j. \tag{4.1.15}$$

Differentiating (4.1.12) and using (3.3.11), we obtain

$$\begin{aligned} d\underset{\alpha_s}{\omega}{}^i &= \sum_{s_1 \neq s} \frac{\partial^2 F^i}{\partial u^j_{\alpha_s} \partial u^k_{\beta_{s_1}}} \underset{\beta_{s_1}}{G}{}^k_l \underset{\alpha_s}{G}{}^j_m \underset{\beta_{s_1}}{\omega}{}^l \wedge \underset{\alpha_s}{\omega}{}^m \\ &+ \frac{\partial^2 F^i}{\partial u^j_{\alpha_s} \partial g^k} \sum_t \frac{\partial g^k}{\partial u^l_{\beta_t}} \underset{\beta_t}{G}{}^l_q \underset{\alpha_s}{G}{}^j_m \underset{\beta_t}{\omega}{}^q \wedge \underset{\alpha_s}{\omega}{}^m. \end{aligned} \tag{4.1.16}$$

Let us apply the results obtained to the case $n = 3$, i.e., to the case of a 4-web $W(4,3,r)$ and a ternary quasigroup. In this case the closure conditions (4.1.21) can be only one of the following types:

$$f(x_1, y_2, z_1) = f(x_1, y_1, z_2) \Longrightarrow f(x_2, y_2, z_1) = f(x_2, y_1, z_2), \tag{4.1.23}$$

$$f(x_1, y_1, z_2) = f(x_2, y_1, z_1) \Longrightarrow f(x_1, y_2, z_2) = f(x_2, y_2, z_1), \tag{4.1.24}$$

$$f(x_2, y_1, z_1) = f(x_1, y_2, z_1) \Longrightarrow f(x_2, y_1, z_2) = f(x_1, y_2, z_2). \tag{4.1.25}$$

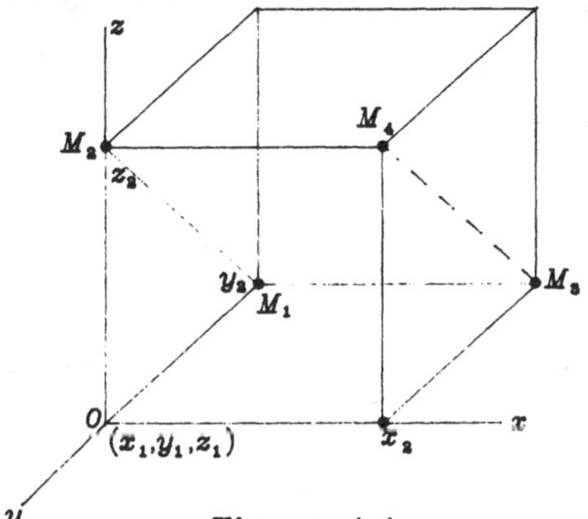

Figure 4.1

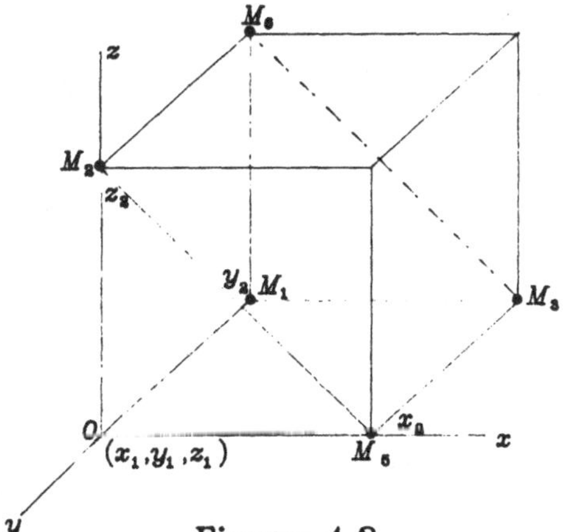

Figure 4.2

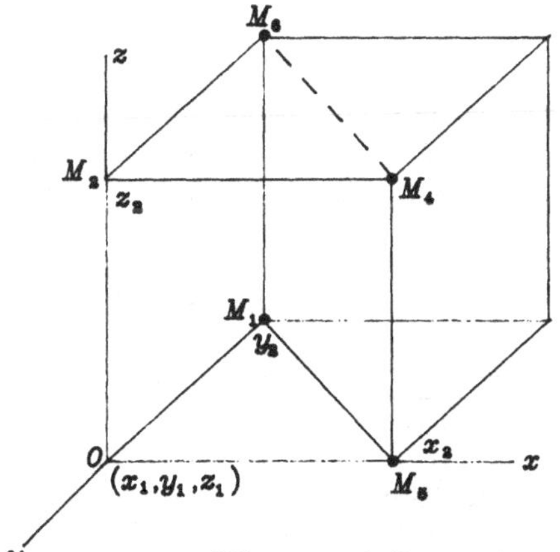

Figure 4.3

These closure conditions coincide with conditions $(D_1), (D_2)$, and (D_3) studied by Rado [R 60].

We will explain now the geometric meaning of these conditions. Suppose that leaves of the first three foliations of a web $W(4,3,r)$ are planes parallel to the coordinate planes of some Cartesian coordinate system. Then the points $(x_i, y_j, z_k) \in \mathbf{R}^{3r}$ in conditions (4.1.23), (4.1.24), and (4.1.25) are the points of intersections of the planes x_i =const., y_j =const., and z_k =const.,$i, j, k = 1, 2, 3$. The geometric meaning of conditions (4.1.23), (4.1.24), and (4.1.25) can be seen from Figures 4.1, 4.2, and 4.3.

For example, the condition (4.1.23) means that if points M_1 and M_2 lie on a leaf of the foliation λ_4, then the points M_3 and M_4 lie on another leaf of λ_4.

When $n = 3$, the conditional identities (4.1.22) have one of the following forms:

$$F(x,e,y) = F(x,y,e), \tag{4.1.26}$$

$$F(e,x,y) = F(y,x,e), \tag{4.1.27}$$

$$F(e,x,y) = F(x,e,y), \tag{4.1.28}$$

and the condition (4.1.1) in this case can take only one of the following forms:

$$F(x,y,z) = x \circ (y \cdot z), \tag{4.1.29}$$

$$F(x,y,z) = y \circ (x \cdot z), \tag{4.1.30}$$

$$F(x,y,z) = (x \cdot y) \circ z. \tag{4.1.31}$$

Finally, conditions (1.8.3) can be also only of three different types:

$$\underset{12}{a}{}^i_{jk} = \underset{13}{a}{}^i_{jk}, \tag{4.1.32}$$

$$\underset{23}{a}{}^i_{jk} = \underset{21}{a}{}^i_{jk}, \tag{4.1.33}$$

$$\underset{31}{a}{}^i_{jk} = \underset{32}{a}{}^i_{jk}. \tag{4.1.34}$$

Four-webs $W(4.3.r)$ satisfying these conditions are said to be 1-, 2-, or 3-*reducible*, respectively. It follows from our considerations that the four conditions (4.1.23), (4.1.26), (4.1.29), and (4.1.32) are equivalent to each other. The two other quadruples of conditions are also equivalent to each other.

Remark 4.1.9 When $r = 1$ and $n = 3$, in the case (4.1.23) the equation (4.1.19) assumes the form

$$\frac{\partial^2 F}{\partial u_1 \partial u_2} \frac{\partial F}{\partial u_3} = \frac{\partial^2 F}{\partial u_1 \partial u_3} \frac{\partial F}{\partial u_2}, \tag{4.1.35}$$

since in this case $\underset{\alpha}{G}{}^1_1 = \frac{1}{\frac{\partial F}{\partial u_\alpha}}$. Equation (4.1.35) is necessary and sufficient for a C^2-function $F(u_1, u_2, u_3)$ of three variables to be a superposition of two functions of two variables each, i.e., to be of the form

$$F(u_1, u_2, u_3) = \varphi[u_1, g(u_2, u_3)]. \tag{4.1.36}$$

Equation (4.1.35) giving a criterion for representability of a function $F(u_1, u_2, u_3)$ in the form (4.1.36), was first obtained by E. Goursat [Gou 99]. These equations are usually discussed in courses in nomography.

4.2 Multiple Reducible and Completely Reducible $(n+1)$-Webs

We now consider the case when an n-ary operation in a coordinate n-quasigroup of a web $W(n+1, n, r)$ can be reduced to several operations with fewer arguments.

Definition 4.2.1 We will say that an n-quasigroup is *multiple reducible* if its n-ary operation is reduced to s operations, $2 < s \leq n-1$, with fewer arguments. A web $W(n+1, n, r)$ is *multiple reducible* if one of its coordinate n-quasigroups is multiple reducible. If $s = n-1$, then an n-quasigroup and a web $W(n+1, n, r)$ are called *completely reducible.*

We will illustrate our ideas using examples from [Be 72] and [BS 66].

Let $n - 12$ and suppose that, in a coordinate-free notation, one of coordinate 12-quasigroups of a web $W(13, 12, r)$ has an equation:

$$u_{13} = A_1(A_2(u_1, u_2, u_3), A_4(u_4, A_5(u_5, u_6)), u_7, u_8, A_6(A_7(u_9, u_{10}, u_{11}), u_{12})). \quad (4.2.1)$$

We can put equation (4.2.1) in correspondence with the *tree* in Figure 4.4.

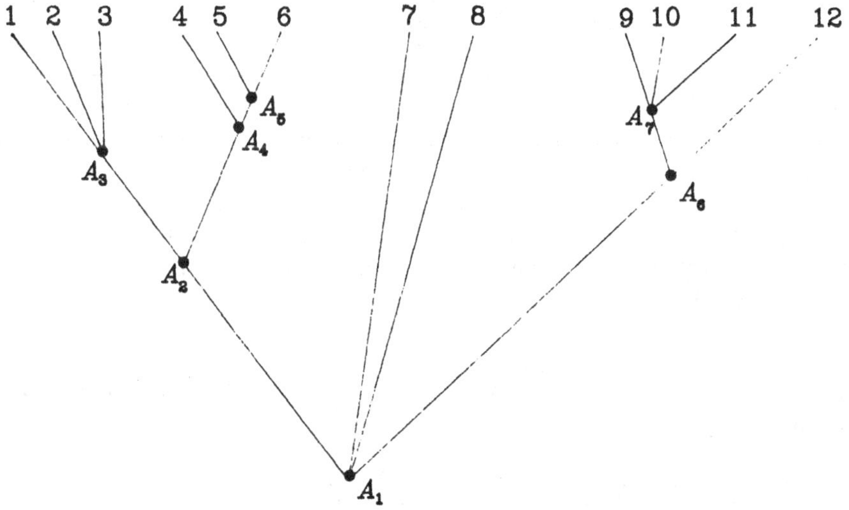

Figure 4.4

To the free elements there correspond the *vertices* of the tree while the operations correspond to the *nodes*. The node corresponding to the first operation is called the *root*. Each node is the root of some *subtree*. From each node *branches* emanate. For some node A_i , let the leftmost branch terminate at the vertex with ordinal number k_i while the the rightmost branch at the vertex with ordinal number l_i . We call the numbers k_i and l_i the *left* and the *right coordinates* of operation A_i. In our example (4.2.1) we have the coordinates indicated in Table 4.1:

i	1	2	3	4	5	6	7
k_i	1	1	1	4	5	9	9
l_i	12	6	3	6	6	12	11

Table 4.1: Coordinates of Operations A_i from (4.2.1)

Now, using Theorem 4.1.4, we can easily give the invariant characteristic of the web $W(13, 12, r)$ for which the 12-ary operation of the corresponding coordinate quasigroup has the form (4.2.1). Such a web is characterised by the following relations for the components of its torsion tensor:

$$\left\{\begin{array}{l} \underset{\lambda 1}{a}{}^i_{jk} = \underset{\lambda 2}{a}{}^i_{jk} = \underset{\lambda 3}{a}{}^i_{jk}, \quad \lambda \neq 1,2,3; \\ \underset{\lambda 5}{a}{}^i_{jk} = \underset{\lambda 6}{a}{}^i_{jk}, \quad \lambda \neq 5,6; \\ \underset{\lambda 4}{a}{}^i_{jk} = \underset{\lambda 6}{a}{}^i_{jk}, \quad \lambda \neq 4,5,6; \\ \underset{\lambda 1}{a}{}^i_{jk} = \underset{\lambda 6}{a}{}^i_{jk}, \quad \lambda \neq 1,\dots,6; \\ \underset{\lambda 9}{a}{}^i_{jk} = \underset{\lambda 10}{a}{}^i_{jk} = \underset{\lambda 11}{a}{}^i_{jk}, \quad \lambda \neq 9,10,11; \\ \underset{\lambda 9}{a}{}^i_{jk} = \underset{\lambda 12}{a}{}^i_{jk}, \quad \lambda \neq 9,10,11,12. \end{array}\right. \tag{4.2.2}$$

The rule according to which the equations (4.1.2) were written is as follows. To each operation there corresponds equations the number of which is one less than the operation's number of arguments. The second of the indices underneath a letter occuring in these equations, as representative of a component of the torsion tensor, runs through the indices of the free elements occurring in the subtree corresponding to the given operation (for a binary operation this index assumes the values of the operation's left and right coordinates). The first of these indices assumes all values distinct from the indices of the free elements occuring in the given subtree (therefore no equation corresponds to the root A_1).

Equations (4.2.2) mean, respectively, the complete integrability of the following systems of 1-forms:

$$\begin{array}{rcl} \underset{1}{\omega}{}^i + \underset{2}{\omega}{}^i + \underset{3}{\omega}{}^i &=& \underset{123}{\theta}{}^i, \quad \underset{5}{\omega}{}^i + \underset{6}{\omega}{}^i = \underset{56}{\theta}{}^i, \quad \underset{4}{\omega}{}^i + \underset{56}{\theta}{}^i = \underset{456}{\theta}{}^i, \\ \underset{123}{\theta}{}^i + \underset{456}{\theta}{}^i &=& \underset{1-6}{\theta}{}^i, \quad \underset{9}{\omega}{}^i + \underset{10}{\omega}{}^i + \underset{11}{\omega}{}^i = \underset{9-11}{\theta}{}^i, \quad \underset{9-11}{\theta}{}^i + \underset{12}{\omega}{}^i = \underset{9-12}{\theta}{}^i. \end{array}$$

This complete integrability means the integrability of certain completely defined $(n-1)r$-dimensional diagonal distributions of a web $W(n+1,n,r)$.

The chosen example shows how one can obtain an invariant characteristic for a multiple reducible web $W(n+1,n,r)$.

Note that if the free elements are first subjected to some substitution, then, in condition (4.2.2) it is necessary to perform this substitution on all the indices underneath the letters involved.

We will study now completely reducible webs $W(n+1,n,r)$. It is obvious that there are many different kinds of completely reducible $(n+1)$-webs. It is easy to see (see for example [H 75]) that the number of methods by which the sequence $u_1, u_2, \ldots, u_n$ may be obtained by means of binary nonassociative products equals $(2n-2)!/[n!(n-1)!]$ (for $n = 3,4,5,6,7$ we obtain, respectively, $2,5,14,42$,and 132).

In order to distinguish these methods, we agree to speak of complete reducibility of type $\sigma = \{k_1, \ldots, k_{n-1}\}$, where k is the left coordinate of binary operation A_i. In the case of complete reducibility, the right coordinates are uniquely determined by the left ones ([Be 72], [BS 66]).

The procedure for obtaining invariant characteristics of completely reducible webs $W(n+1,n,r)$ is exactly the same as that for multiple reducible webs. For example, if $n=8$ and we have a complete reducibility of type

$$u_9 = A_1(A_2(u_1, A_3(A_4(u_2,u_3), A_5(u_4,u_5))), A_6(A_7(u_6,u_7),u_8)), \tag{4.2.3}$$

then the tree of Figure 4.5 corresponds to it:

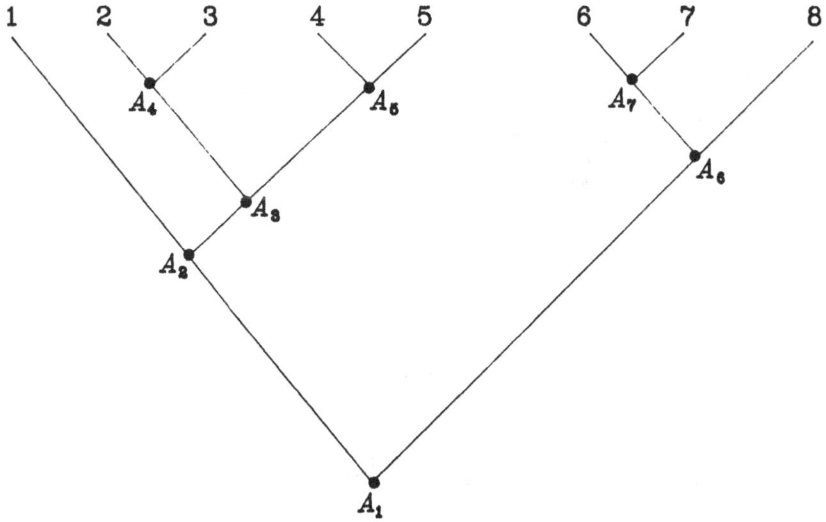

Figure 4.5

and operations A_i have the coordinates indicated in Table 4.2:

i	1	2	3	4	5	6	7
k_i	1	1	2	2	4	6	6
l_i	9	5	5	3	5	8	7

Table 4.2: Coordinates of Operation A_i from (4.2.3)

Therefore, in the case under consideration we have complete reducibility of type $\sigma = \{1,1,2,2,4,6,6\}$. According to the rule described for multiple reducible webs, such a completely reducible web $W(9,8,r)$ is characterised by the following relations between components of its torsion tensor:

$$\left\{\begin{array}{ll} \underset{\lambda 2}{a}{}^i_{jk} = \underset{\lambda 3}{a}{}^i_{jk}, \ \lambda \neq 2,3, & \underset{\lambda 1}{a}{}^i_{jk} = \underset{\lambda 5}{a}{}^i_{jk}, \ \lambda \neq 1,\dots,5; \\ \underset{\lambda 4}{a}{}^i_{jk} = \underset{\lambda 5}{a}{}^i_{jk}, \ \lambda \neq 4,5, & \underset{\lambda 6}{a}{}^i_{jk} = \underset{\lambda 7}{a}{}^i_{jk}, \ \lambda \neq 6,7; \\ \underset{\lambda 2}{a}{}^i_{jk} = \underset{\lambda 5}{a}{}^i_{jk}, \ \lambda \neq 2,3,4,5, & \underset{\lambda 6}{a}{}^i_{jk} = \underset{\lambda 8}{a}{}^i_{jk}, \ \lambda \neq 6,7,8. \end{array}\right. \tag{4.2.4}$$

In the case that the complete reducibility has the type $\sigma_0 = \{1,1,\dots,1\}$, the corresponding tree is on Figure 4.6:

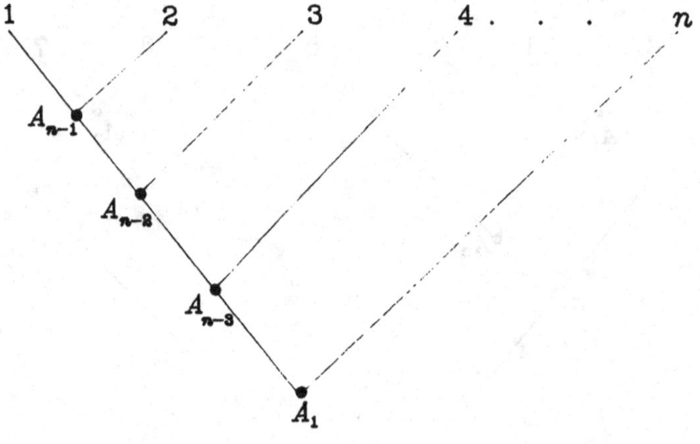

Figure 4.6

and the conditions on the torsion tensor of such a web are

$$\underset{31}{a}{}^i_{jk} = \underset{32}{a}{}^i_{jk}, \ \underset{41}{a}{}^i_{jk} = \underset{42}{a}{}^i_{jk} = \underset{43}{a}{}^i_{jk}, \dots, \underset{n1}{a}{}^i_{jk} = \underset{n2}{a}{}^i_{jk} = \dots = \underset{n.n-1}{a}{}^i_{jk}. \tag{4.2.5}$$

If, however, for the same type of complete reducibility, the free elements are first subjected to the substitution $\binom{1\ldots n}{\alpha_1\ldots\alpha_n}$, we will then have

$$\underset{\alpha_3\alpha_1}{a}{}^i_{jk} = \underset{\alpha_3\alpha_2}{a}{}^i_{jk}, \quad \underset{\alpha_4\alpha_1}{a}{}^i_{jk} = \underset{\alpha_4\alpha_2}{a}{}^i_{jk} = \underset{\alpha_4\alpha_3}{a}{}^i_{jk}, \ldots, \underset{\alpha_n\alpha_1}{a}{}^i_{jk} = \underset{\alpha_n\alpha_2}{a}{}^i_{jk} = \ldots = \underset{\alpha_n\alpha_{n-1}}{a}{}^i_{jk}. \tag{4.2.6}$$

4.3 Group $(n+1)$-Webs

If an n-quasigroup is an n-group with unit e, then, as proved by Hosszu [Ho 63] and Gluskin [Glu 64] (see also [Be 72]), n-ary operation in it reduces to $n-1$ operations of one and the same binary group operation:

$$F(u_1, u_2, \ldots, u_n) = u_1 \circ u_2 \circ \ldots \circ u_n,$$

where the operation $\circ$ is defined by

$$x \circ y = F(x, y, \underbrace{e, \ldots, e}_{n-2}).$$

Therefore, according to considerations of Section 4.2, an n-group with unit is completely reducible by many ways, for example, it is σ_0-reducible and σ_n-reducible, where $\sigma_0 = \{1, \ldots, 1\}$ and $\sigma_n = \{1, 2, \ldots, n-1\}$. For a specified order of elements $u_1, \ldots, u_n$ it is always possible, as we have mentioned in Section 4.2, to have $(2n-2)!/[n!(n-1)!]$ types of complete reducibility. The same number of types of complete reducibility will be for each of $n!$ permutations of elements $u_1, \ldots, u_n$.

Definition 4.3.1 We shall call a web $W(n+1, n, r)$ a *group web* if at least one of its coordinate n-quasigroup is an l.d. n-group.

According to Definition 3.1.3, an l.d. n-loop $L(a_1, \ldots, a_n)$ associated with any point $(a_1, \ldots, a_n)$ of a group web $W(n+1, n, r)$ will be an l.d. n-group with unit.

We now find a necessary and sufficient condition for a web $W(n+1, n, r)$ to be a group $(n+1)$-web. For definiteness we will assume that elements $u_1, \ldots, u_n$ are given in their natural order.

Theorem 4.3.2 *For a web $W(n+1, n, r)$ to be a group $(n+1)$-web, with the natural ordering of foliations $\lambda_1, \ldots, \lambda_n$ of the web, it is necesssary and sufficient that its torsion tensor satisfy the relations*

$$\underset{\alpha\beta}{a}{}^i_{jk} = a^i_{jk}, \quad a^i_{(jk)} = 0, \quad \alpha < \beta. \tag{4.3.1}$$

A coordinate n-quasigroup of such a web is a direct product of n mutually isomorphic r-parameter Lie groups.

Proof. *Necessity.* Let a coordinate n-quasigroup of a web $W(n+1, n, r)$ be an l.d. n-group. Then, its n-ary operation is (λ, μ)-associative, i.e.. for any $\lambda, \mu = 1, \ldots, n,\ \lambda \neq \mu$ and any $u_1, \ldots, u_n$ there holds the equation

$$\begin{aligned} & F(u_1, \ldots, u_{\lambda-1}, F(u_\lambda, \ldots, u_{n+\lambda-1}), u_{n+\lambda}, \ldots, u_{2n-1}) \\ = \ & F(u_1, \ldots, u_{\mu-1}, F(u_\mu, \ldots, u_{n+\mu-1}), u_{n+\mu}, \ldots, u_{2n-1}). \end{aligned} \tag{4.3.2}$$

Using (3.2.6), we find that (λ, μ)-associativity (4.3.2) implies

$$\underset{\alpha\beta}{\lambda}{}^i_{jk} = \lambda^i_{jk}, \quad \alpha < \beta. \tag{4.3.3}$$

Conditions (4.3.1) follow from (3.3.4) and (4.3.3).

We note that conditions (4.3.1) are obtained from (4.3.2) for any fixed λ and μ, i.e., each (λ, μ)-associativity implies conditions (4.3.1).

Sufficiency. We now show that conditions (4.3.1) are sufficient for a web $W(n+1, n, r)$ to be a group one. We consider the new co-basis, made up of the forms $\underset{11}{\theta}{}^i, \underset{12}{\theta}{}^i, \ldots, \underset{1n}{\theta}{}^i$ defined by equations

$$\underset{1\alpha}{\theta}{}^i = -\sum_{\beta=1}^{\alpha} \underset{\beta}{\omega}{}^i, \quad \alpha = 1, \ldots, n. \tag{4.3.4}$$

Note that $\underset{1n}{\theta}{}^i = \underset{n+1}{\omega}{}^i$. For these forms, by virtue of (1.2.16) and (4.3.1), we obtain the following structure equations:

$$d\underset{1\alpha}{\theta}{}^i = \underset{1\alpha}{\theta}{}^j \wedge \theta^i_j + a^i_{jk} \underset{1\alpha}{\theta}{}^j \wedge \underset{1\alpha}{\theta}{}^k, \tag{4.3.5}$$

where $\theta^i_j = \omega^i_j - a^i_{jk} \underset{1n}{\theta}{}^k$. Equations (4.3.5) show that the systems

$$\underset{1\alpha}{\theta}{}^i = 0, \quad \underset{1,\alpha+1}{\theta}{}^i = 0, \quad \underset{1\alpha}{\theta}{}^i - \underset{1,\alpha+1}{\theta}{}^i = 0, \quad \alpha = 1, \ldots, n-1,$$

are completely integrable and define, respectively on the $(2r)$-dimensional manifolds $\bigcap_{\xi \neq \alpha, \alpha+1, \alpha+2} \underset{\xi}{V}$, $\alpha = 1, \ldots, n-2$, $\bigcap_{\alpha \neq n-1, n-2} V_\alpha$ three-webs $W(3, 2, r)$ which have one and the same torsion tensor a^i_{jk} and the same connection forms θ^i_j. The closed form equations of these three-webs will have the form

$$\left\{ \begin{array}{l} v^i_1 = v^i_1(u^j_1, u^k_2), \\ v^i_2 = v^i_2(v^j_1, u^k_3), \\ \cdots\cdots\cdots\cdots\cdots\cdots\cdots \\ v^i_{n-2} = v^i_{n-2}(v^j_{n-3}, u^k_{n-1}), \\ v^i_{n+1} = u^i_{n+1}(v^j_{n-2}, u^k_n). \end{array} \right. \tag{4.3.6}$$

The curvature tensors $\underset{\alpha}{b}{}^i_{jkl}$ of these three-webs, according to (1.2.47), are determined from the equations

$$d\theta^i_j - \theta^k_j \wedge \theta^i_k = \underset{\alpha}{b}{}^i_{jkl} \underset{1\alpha}{\theta}{}^k \wedge \underset{1,\alpha+1}{\theta}{}^l \quad \alpha = 1, \ldots, n-1. \tag{4.3.7}$$

It follows from (4.3.7), by virtue of the independence of the products $\underset{1\alpha}{\theta}{}^k \wedge \underset{1,\alpha+1}{\theta}{}^l$ where $n > 2$, that the curvature forms and, therefore, also the curvature tensors, of all these three-webs are equal to zero: $\underset{\alpha}{b}{}^i_{jkl} = 0$. Consequently, all such three-webs are group three-webs (see Definition 1.2.8) and their coordinate binary quasigroups will be binary groups. These $n-1$ binary groups are mutually isomorphic , since they have one and the same torsion tensor a^i_{jk}. Therefore, by virtue of equations (4.3.6), one can assume that the n-ary operation reduces to an $(n-1)$-fold binary operation of one and the same group.

Since $d\theta^i_j - \theta^k_j \wedge \theta^i_k = 0$, there is an absolute parallelism, and one can turn to a familly of parallel moving frames in which $\theta^i_j = 0$. Then, by exterior differentiation of the structure equations (4.3.5), we find that $da^i_{jk} = 0$, i.e., the components of the torsion tensor will be constants. The structure equations (4.3.5) of such a web take the form

$$d\underset{1\alpha}{\theta}{}^i = a^i_{jk} \underset{1\alpha}{\theta}{}^j \wedge \underset{1\alpha}{\theta}{}^k. \tag{4.3.8}$$

An l.d. n-group with the structure equations (4.3.8) is the direct product $G_1 \times G_2 \times \ldots \times G_n$ of n r-parameter Lie groups G_α, which are mutually isomorphic and defined by invariant forms $\underset{1\alpha}{\theta}{}^i$, $\alpha = 1, \ldots, n$.

The converse is obvious: each such an l.d. n-group defines a group $(n+1)$-web. ■

Corollary 4.3.3 *For a web $W(n+1, n, r)$ to be a group $(n+1)$-web, with the natural ordering of foliations $\lambda_1, \ldots, \lambda_n$ of the web, it is necesssary and sufficient that in its coordinate n-loop $L(a_1, \ldots, a_n)$ the following identities be held:*

$$\begin{aligned} & F(x, y, e, \ldots, e) = && F(x, e, y, e, \ldots, e) = \ldots = F(x, e, \ldots, e, y) \\ = \; & F(e, x, y, e, \ldots, e) = && F(e, x, e, y, e, \ldots, e) = \ldots = F(e, x, e, \ldots, e, y) \\ = \; & \ldots = F(e, \ldots, e, x, y). && \end{aligned} \tag{4.3.9}$$

Proof. It follows from (3.2.6) and (3.3.4) that identities (4.3.9) are equivalent to conditions (4.3.1). ■

As we have already mentioned, a group web $W(n+1, n, r)$ is completely reducible in many ways, for example, it is σ_0-reducible where $\sigma_0 = \{1, \ldots, 1\}$. The converse is false: σ_0-reducibility is characterised by conditions (4.2.5) which do not imply conditions (4.3.1). The following corollary shows what should be added to σ_0-reducibility to obtain a new criterion for a group $(n+1)$-web.

Corollary 4.3.4 *For a web $W(n+1,n,r)$ to be a group $(n+1)$-web, with the natural ordering of foliations $\lambda_1,\ldots,\lambda_n$ of the web, it is necessary and sufficient that it be simultaneously completely reducible of type $\sigma_0 = \{1,\ldots,1\}$ and reducible of type*

$$u^i_{n+1} = \varphi^i(u_1^{j_1}, g^k(u_2^{j_2},\ldots,u_n^{j_n})). \tag{4.3.10}$$

Proof. Conditions (4.2.5), in conjunction with conditions

$$\underset{1\alpha}{a}{}^i_{jk} = \underset{1\beta}{a}{}^i_{jk}, \quad \alpha \neq \beta, \tag{4.3.11}$$

are equivalent to conditions (4.3.1). Equations (4.3.11) characterise the webs $W(n+1,n,r)$ with the integrable diagonal distribution $\{1\}$. By Theorem 4.3.2, such webs are reducible of type (4.3.10). ■

Note that Corollary 4.3.4 gives one of the possible criteria for a web $W(n+1,n,r)$ to be a group one. Others can be obtained by complementing the various types of reducibility and complete reducibility following from (4.3.1) with the corresponding requirements of certain reducibility.

When $n = 3$, by Corollary 4.3.4, the conditions for a web $W(n+1,n,r)$ to be a group one, are reduced to any two of conditions (4.1.23), (4.1.24), (4.1.25) or conditions equivalent to them.

If we take, for example, the conditions (4.1.24) and (4.1.25), it is easy to see that they are equivalent to any of the following conditions:

$$\begin{aligned} &f(x_2,y_1,z_1) = f(x_1,y_2,z_1), \quad f(x_1,y_1,z_3) = f(x_3,y_1,z_1) \\ \Longrightarrow\quad &f(x_2,y_1,z_3) = f(x_3,y_2,z_1) = f(x_1,y_2,z_3), \end{aligned} \tag{4.3.12}$$

$$F(e,x,y) = F(x,e,y) = F(y,x,e), \tag{4.3.13}$$

$$f(x,y,z) = y \circ x \circ z, \tag{4.3.14}$$

$$\underset{23}{a}{}^i_{jk} = \underset{13}{a}{}^i_{jk} = \underset{21}{a}{}^i_{jk}. \tag{4.3.15}$$

Note that in (4.3.14) the operation $\circ$ is associative. Similarly, if conditions (4.1.23)–(4.1.25) or ((4.1.23)–(4.1.24) hold, we have respectively

$$\begin{aligned} &f(x_2,y_1,z_1) = f(x_1,y_2,z_1), \quad f(x_1,y_3,z_1) = f(x_1,y_1,z_3) \\ \Longrightarrow\quad &f(x_2,y_3,z_1) = f(x_2,y_1,z_3) = f(x_1,y_2,z_3). \end{aligned} \tag{4.3.16}$$

$$F(e,x,y) = F(x,e,y) = F(x,y,e), \tag{4.3.17}$$

$$f(x,y,z) = x \circ y \circ z, \tag{4.3.18}$$

$$\underset{23}{a}{}^i_{jk} = \underset{13}{a}{}^i_{jk} = \underset{12}{a}{}^i_{jk}, \tag{4.3.19}$$

and

$$\begin{aligned} &f(x_1,y_2,z_1) = f(x_1,y_1,z_2), \quad f(x_3,y_1,z_1) = f(x_1,y_1,z_3) \\ \Longrightarrow\quad &f(x_3,y_2,z_1) = f(x_3,y_1,z_2) = f(x_1,y_2,z_3), \end{aligned} \tag{4.3.20}$$

$$F(e,x,y)=F(y,e,x)=F(y,x,e), \tag{4.3.21}$$

$$f(x,y,z)=x\circ z\circ y, \tag{4.3.22}$$

$$\underset{23}{a}{}^{i}_{jk}=\underset{31}{a}{}^{i}_{jk}=\underset{21}{a}{}^{i}_{jk}. \tag{4.3.23}$$

where an operation $\circ$ in (4.3.18) and (4.3.22) is a group operation. Conditions (4.3.16)–(4.3.19) correspond to the natural order $\lambda_1,\ldots,\lambda_n$ of the foliations of $W(4,3,r)$.

These three types of conditions coincide with conditions $(E_1),(E_2)$, and (E_3) obtained by F. Rado [R 60].

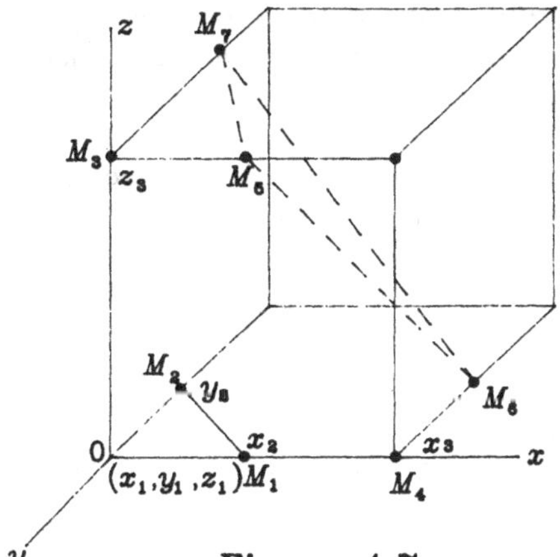

Figure 4.7

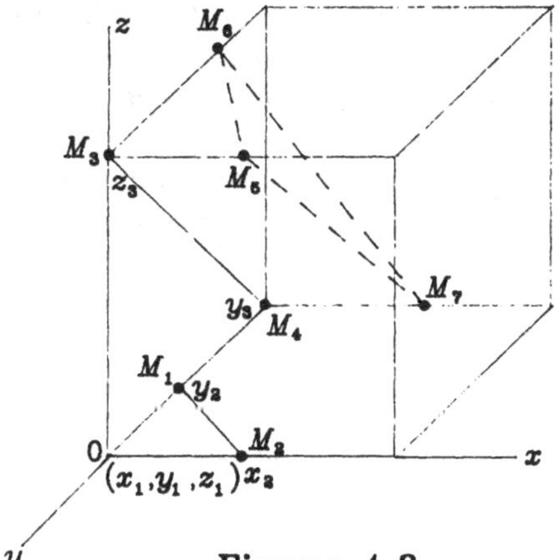

Figure 4.8

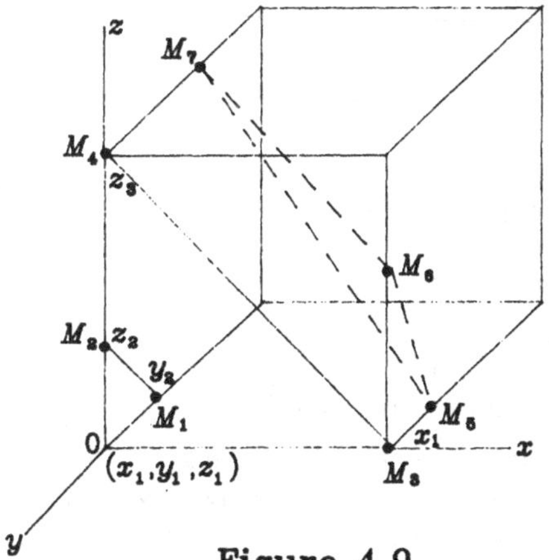

Figure 4.9

The geometric meaning of the conditions (4.3.12), (4.3.16), and (4.3.20) can be seen from Figures 4.7, 4.8, and 4.9, respectively. For example, the conditions (4.3.16) mean that if the points M_1, M_2 and M_3, M_4 belong to two leaves of the foliation λ_4, then points M_5, M_6, and M_7 also belong to a leaf of the foliation (see Figure 4.7).

We now find conditions under which a group web $W(n+1, n, r)$ will be parallelisable (see Section 1.5).

Theorem 4.3.5 *For a web $W(n+1, n, r)$ to be parallelisable, it is necessary and sufficient that it be a group one with the natural ordering of the foliations of the web and that any of its 4-subwebs $[n+1, \alpha, \beta, \gamma]$, where α, β, and γ are fixed and $\gamma < \alpha < \beta$, be α-reducible. With this, both the binary group by means of which there was defined the operation in the n-group, serving as a coordinate n-group for the parallelisable $(n+1)$-web, and the n-group itself, are Abelian.*

Proof. It is obvious that for parallelisability of a web $W(n+1, n, r)$ it is necessary and sufficient that to conditions (4.3.1) one adds a condition of the form

$$\underset{\alpha\beta}{a}{}^{i}_{jk} = \underset{\alpha\gamma}{a}{}^{i}_{jk}, \quad \gamma < \alpha < \beta. \tag{4.3.24}$$

where α, β and γ are fixed, since then conditions (4.3.1) and (4.3.24) give

$$a^{i}_{jk} = \underset{\gamma\alpha}{a}{}^{i}_{jk}, \quad a^{i}_{jk} = \underset{\alpha\beta}{a}{}^{i}_{jk} = \underset{\alpha\gamma}{a}{}^{i}_{jk} = \underset{\gamma\alpha}{a}{}^{i}_{kj}$$

and therefore

$$a^{i}_{jk} = 0, \tag{4.3.25}$$

i.e., the torsion tensor of the web vanishes, which, by Theorem 1.5.2, characterises parallelisable webs $PW(n+1, n, r)$.

From condition (4.3.25) we find

$$d\underset{\alpha}{\omega}{}^{i} = \underset{\alpha}{\omega}{}^{j} \wedge \omega^{i}_{j}, \qquad d\omega^{i}_{j} = \omega^{k}_{j} \wedge \omega^{i}_{k}.$$

Since, just as for any group web $W(n+1, n, r)$, the forms ω^i_j can be brought to zero (see the proof of Theorem 4.3.2), we find that

$$\underset{\alpha}{\omega}{}^{i} = 0.$$

These equations show that both, the binary group by means of which the operation was defined in the n-group, and the n-group itself, are Abelian. ■

It is easy to establish the geometrical meaning of relations (4.3.23). They are necessary and sufficient for the 4-subweb $[n+1, \alpha, \beta, \gamma]$ of the $W(n+1, n, r)$ defined by the system

$$\underset{\varepsilon}{\omega}{}^{i} = 0, \quad \varepsilon \neq \alpha, \beta, \gamma,$$

be α-reducible (see Section 4.1), i.e., for the ternary operation in its coordinate quasigroup be reduced in the following way to two binary operations:

$$u^i_{n+1} = \varphi^i(u^j_\alpha, v^k(u^l_\beta, u^m_\gamma)).$$

Corollary 4.3.6 *For a web $W(n+1,n,r)$ to be parallelisable, it is necessary and sufficient that it be a group web with natural ordering of the foliations of the web and that one of the following conditions be held:*

(i) *The web $W(n+1,n,r)$ be $\hat{\alpha}$-reducible, i.e., the operation in its coordinate n-quasigroup has the form*

$$u^i_{n+1} = f^i(u^{j_{\hat\alpha}}_{\hat\alpha}, g^k(u^{j_1}_1, \ldots, u^{j_{\hat\alpha-1}}_{\hat\alpha-1}, u^{j_{\hat\alpha+\nu}}_{\hat\alpha+1}))$$

at least for one $\hat{\alpha} \neq 1$.

(ii) *The web $W(n+1,n,r)$ be a group one for some ordering of the foliations of the web different from the natural ordering, i.e., it would be doubly a group one.*

Proof. It follows from the fact that both the condition (i) and the condition (ii) include conditions (4.3.23). ■

Remark 4.3.7 It is easy to understand that condition (ii) can be changed as follows: for a web $W(n+1,n,r)$ to be parallelisable, it is necessary and sufficient that it be doubly a group web, i.e., that its coordinate n-quasigroups corresponding to two different ordering of the foliations of the web be l.d. n-groups.

In case, when $n = 3$, conditions of parallelisability of a group $(n+1)$-web (Theorem 4.3.5 and Corollary 4.3.6) are reduced to the simultaneous realisation of all three types of reducibility: (4.1.23), (4.1.24), and (4.1.25).

These three conditions are equivalent to each of the following relations:

$$\begin{aligned} & f(x_2,y_1,z_1) = f(x_1,y_2,z_1), \quad f(x_3,y_1,z_1) = f(x_1,y_3,z_1) = f(x_1,y_1,z_3) \\ \Longrightarrow \quad & f(x_2,y_3,z_1) = f(x_3,y_2,z_1) = f(x_2,y_1,z_3) = f(x_1,y_2,z_3), \end{aligned} \tag{4.3.26}$$

$$F(x,y,e) = F(y,x,e) = F(x,e,y) = F(e,x,y), \tag{4.3.27}$$

$$f(x,y,z) = x \circ y \circ z, \tag{4.3.28}$$

$$\underset{\alpha\beta}{a}{}^i_{jk} = 0, \tag{4.3.29}$$

where the operation $\circ$ in (4.3.28) is a binary operation of a certain commutative group.

The geometrical meaning of the conditions (4.3.26) is seen from Figure 4.10: if the points M_1, M_2 and M_3, M_4, M_5 belong to two leaves of the foliation λ_4, then the points M_6, M_7, M_8, M_9 belong to a leaf of the same foliation λ_4.

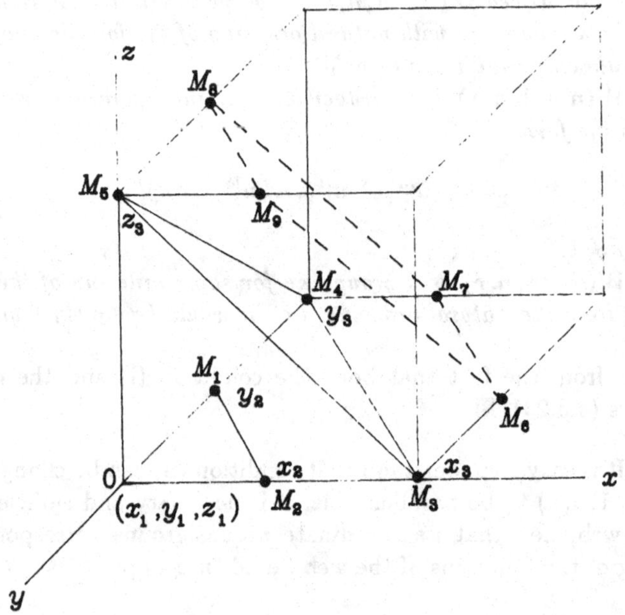

Figure 4.10

4.4 $(2n+2)$-Hedral $(n+1)$-Webs

Among the special classes of webs $W(n+1, n, r)$, a central position is occupied by the octahedral webs $W(4,3,r)$ (see [Bl 55] for $r=1$ and [G 75a] for $r>1$).

We will study now $(2n+2)$-hedral webs $W(n+1,n,r), n \geq 3, r \geq 2$, generalising octahedral webs $W(4,3,r)$ and will discuss octahedral four-webs as a particular case of $(2n+2)$-hedral $(n+1)$-webs.

Definition 4.4.1 We will call a web $W(n+1,n,r)$ $(2n+2)$-*hedral of type* t, $1 \leq t \leq n$, if, in its coordinate n-quasigroup, the following closure condition is met:

$$\begin{aligned} & f(x_2, y_1, z_1, \ldots, u_1) = f(x_1, y_2, z_1, \ldots, u_1) = \ldots = f(x_1, y_1, z_1, \ldots, u_2) \\ \Longrightarrow\; & f(\underbrace{x_2, y_2, \ldots, v_2}_{t}; \underbrace{\omega_1, \ldots, u_1}_{n-t}) = f(x_{\alpha_1}, y_{\alpha_2}, \ldots, v_{\alpha_t}, \omega_{\alpha_{t+1}}, \ldots, u_{\alpha_n}), \end{aligned} \tag{4.4.1}$$

where $\alpha_1, \ldots, \alpha_n$ is any permutation of the sequence, $\underbrace{2, \ldots, 2}_{t}\underbrace{1, \ldots, 1}_{n-t}$.

Kramareva [Kr 73] proved that the condition (4.4.1) is equivalent to the following identity in an l.d. n-loop $L(a_1, \ldots, a_n)$:

$$F(\underbrace{x, \ldots, x}_{t}, \underbrace{e, \ldots, e}_{n-t}) = F(\underbrace{e, \ldots, e}_{i_1}, \underbrace{x, \ldots, x}_{s_1}, \ldots, \underbrace{e, \ldots, e}_{i_k}, \underbrace{x, \ldots, x}_{s_k}), \tag{4.4.2}$$

where $i_m = 0, 1, \ldots, n-t$; $s_m = 0, 1, \ldots, t$; $m = 1, \ldots, k; i_1 + i_2 + \ldots + i_k + s_1 + s_2 + \ldots + s_k = n$; $s_1 + \ldots + s_k = t$

Obviously, we are interested only in the case $2 \leq t \leq n-1$ (if $t = 1$ or n, we have identities).

Theorem 4.4.2 *For a web $W(n+1, n, r)$ to be $(2n+2)$-hedral of type 2, it is necessary and sufficient that its torsion tensor satisfies the condition*

$$\underset{\alpha\beta}{a}{}^{i}_{(jk)} = 0. \tag{4.4.3}$$

If a web $W(n+1, n, r)$ is $(2n+2)$-hedral of type $t, 3 \leq t \leq n-2$, then it will also be $(2n+2)$-hedral of type 2.

Proof. Suppose first that $t = 2$. Then, we find from (4.4.2) and (3.2.6) that

$$\underset{\alpha\beta}{\lambda}{}^{i}_{(jk)} = \lambda^{i}_{jk}, \quad \lambda^{i}_{[jk]} = 0. \tag{4.4.4}$$

Equations (4.4.4) and (3.3.4) imply (4.4.3).

Let us prove the converse assertion: when conditions (4.4.3) are met, the closure conditions (4.4.1) hold with $t = 2$. We shall denote a point by the ordinal number of the leaves of the distinct foliations passing through it: leaves of the foliation λ_1 will be denoted by $1, 1'$, that of λ_2 by $2, 2'$, etc.

We consider an arbitrary $(2n+2)$-hedral figure formed by $n + \binom{n}{2}$ points:

$$\begin{array}{l}
[2, 3, \ldots, n, n+1], \; [2', 3', \ldots, n, n+1], \; [2, 3', 4', \ldots, n, n+1], \ldots, \\
[2, 3, \ldots, n', (n+1)'], \; [2, 3, \ldots, n, (n+1)'], \; [2', 3, \ldots, n-2, (n-1)', n, n+1] \\
[2, 3', \ldots, n-2, (n-1)', n+1], \; \ldots, [2, 3, \ldots, (n-2)', (n-1)', n, n+1], \\
[2, 3, \ldots, (n-1)', n, (n+1)'], [2', 3', 4, \ldots, n-2, (n-1)', n, n+1], \\
[2', 3, 4', \ldots, n-2, (n-1)', n, n+1], \ldots, [2, 3, 4, \ldots, (n-1)', n', (n+1)'].
\end{array}$$

We pass through all these points the required totally geodesic $(n-1)$-dimensional surfaces. Since all leaves of the web are totally geodesic, the $(n-1)$-dimensional surfaces constructed above will lie on the corresponding $(n-1)r$-dimensional leaves of the web. The constructed figure will belong to one transverally geodesic n-dimensional surface. By virtue of condition (4.4.3) and Theorem 1.9.18, the web $W(n+1, n, 1)$ on the latter will be parallelisable and, therefore, the last $\binom{n}{2}$ of the

$n + \binom{n}{2}$ points occuring in the $(2n+2)$-hedral figure will belong to a leaf $1'$ of the foliation λ_1.

Now, let $3 \leq t \leq n-2$. In this case we find from (4.4.2), (3.2.6), (3.3.4), (1.2.17), and (1.2.24) that, for fixed i, j, and k, the $\binom{n}{2}$ quantities $\underset{\alpha\beta}{a}{}^{i}_{jk}$ satisfy the system consisting of the $\binom{n}{t} \geq \binom{n}{2}$ linear homogeneous equations. Since the rank of the matrix of the coefficients of this system is equal to $\binom{n}{2}$, we again arrive at conditions (4.4.3). ■

Corollary 4.4.3 *Each group web $W(n+1, n, r)$ is $(2n+2)$-hedral of type 2.*

Proof. This follows from conditions (4.4.3),(4.3.1), and (1.2.17). ■

Corollary 4.4.4 *If a web $W(n+1, n, 1)$ is $(2n+2)$-hedral of type $t, 2 \leq t \leq n-2$, then it is parallelisable.*

Proof. It follows from equation (4.4.3). ■

Remark 4.4.5 The equations (4.4.3) coincide with (1.9.57). Now it is becoming clear why in Section 1.9, we called such webs $(2n+2)$-hedral. Theorem 1.9.18 gives another geometric characteristic of $(2n+2)$-hedral webs $W(n+1, n, r)$.

Suppose now that $n = 3$. In this case conditions (4.4.1) become

$$\begin{aligned} & f(x_2, y_1, z_1) = f(x_1, y_2, z_1) = f(x_1, y_1, z_2) \\ \Longrightarrow \quad & f(x_1, y_2, z_2) = f(x_2, y_1, z_2) = f(x_2, y_2, z_1), \end{aligned} \tag{4.4.5}$$

and the relations (4.4.2) assume the form

$$F(x, x, e) = F(x, e, x) = F(e, x, x). \tag{4.4.6}$$

Geometrically conditions (4.4.5) can be explained as follows. If the points M, M_1, M_3 lie on a leaf of the foliation λ_4, then the points M_2, M_4, M_5 also lie on a leaf of λ_4 (see Figure 4.11).

The points M, M_1, M_2, M_3, M_4, and M_5 are the vertices of an octahedron formed by leaves of the four-web. This is the reason why a web $W(4, 3, r)$ satisfying the condition (4.4.3) is called octahedral. Thus, condition (4.4.3) means that if seven triples of the six points $M, M_1, M_2, M_3, M_4, M_5$ lie on leaves of the four-web, then an eighth triple of points also lies on a leaf of the four-web.

The construction described above for a $(2n+2)$-hedral web $W(n+1, n, r)$ in the case $n = 3$ can be described as follows. Through an arbitrary point $M \in X^{3r}$ we pass leaves of the four foliations of the four-web. On the intersection of two of them we take an arbitrary point M and pass through it two other leaves of the four-web (two already pass through). They meet the r-dimensional intersections of the leaves

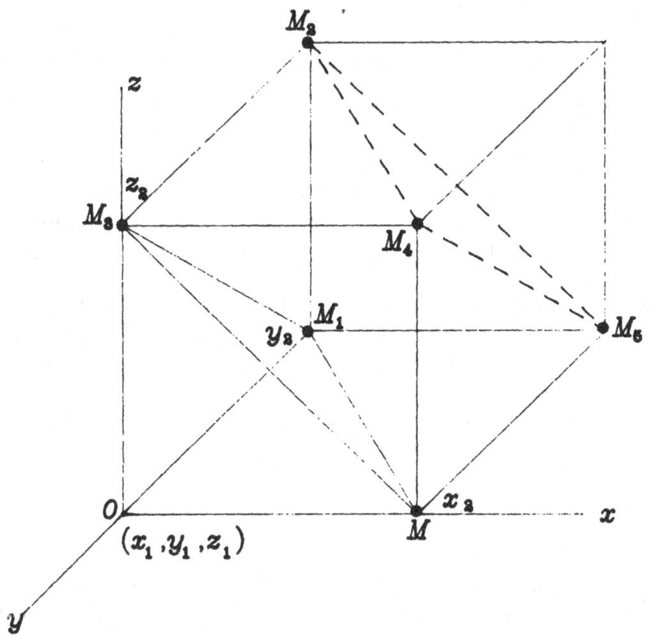

Figure 4.11

of the four-web passing through M in the points M_2 and M_3. Through M_3 pass the fourth leaf of the four-web (three already pass through). It intersects the r-dimensional intersections of the leaves of the four-web passing through M and M_1 in the points M_5 and M_4. Condition (4.4.3) confirms that the points M_2, M_4, M_5 must lie on a leaf of the four-web, namely a leaf in the same foliation containing the leaf passing through M, M_1, M_3.

Now consider an arbitrary octahedral figure $MM_1M_2M_3M_4M_5$. Pass through the points M, M_1, M_2; M, M_1, M_3; M, M_2, M_5; M, M_3, M_5; M_1, M_3, M_4; M_1, M_2, M_4; M_3, M_5, M_4 totally geodesic two-dimensional surfaces. Since the leaves of the four-web are totally geodesic, these two-dimensional surfaces will lie in $(2r)$-dimensional leaves of the four-web.

Now suppose that (4.4.3) holds. Then, for such a four-web, by Theorem 1.9.18, the constructed figure $MM_1M_2M_3M_4M_5$ lies on a transversally geodesic surface. Since the web $W(4,3,1)$ on it is parallelisable, the points M_2, M_4, M_5 lie on a leaf of the four-web.

4.5 Bol $(n+1)$-Webs

4.5.1 Definition and Properties of Bol and Moufang $(n+1)$-Webs

Definition 4.5.1 A web $W(n+1,n,r)$ is said to be a *Bol web* $B_\xi(n+1,n,r)$ if all its 3-subwebs $[\xi,\eta,\zeta]$, where ξ is a fixed and η,ζ are arbitrary indices, $\xi,\eta,\zeta = 1,\ldots,n+1$; $\eta,\zeta \neq \xi$, are Bol three-webs of type (B_m) (see Definition 1.2.8). The base of the foliation λ_ξ is called the *base* of the Bol web $B_\xi(n+1,n,r)$.

Note that if $n=2$, a three-web $B_\xi(n+1,n,r)$ is a Bol web of type (B_m).

In this section we will consider only Bol webs $B_{n+1}(n+1,n,r)$. However, all our considerations will be valid for Bol webs $B_\alpha(n+1,n,r), \alpha = 1,\ldots,n$.

Theorem 4.5.2 *A web $W(n+1,n,r)$ is a Bol web $B_{n+1}(n+1,n,r)$ if and only if on it the condition*

$$\underset{\alpha\beta}{\bar{b}}{}^{i}_{jkl} = 0 \tag{4.5.1}$$

holds for all $\alpha,\beta = 1,\ldots,n$.

Proof. It follows from Definition 4.5.1 and the fact that, according to Table 1.1 in Section 1.2, a Bol three-web $[n+1,\alpha,\beta]$ is characterised in the connection $\bar{\gamma}_{\alpha\beta}$ (see Section 1.3) by conditions (4.5.1). ■

Theorem 4.5.3 *A web $B_{n+1}(n+1,n,r)$ is hexagonal.*

Proof. By Definition 4.5.1, all 3-subwebs $[n+1,\alpha,\beta]$ of a web $B_{n+1}(n+1,n,r)$ are Bol three-webs of type (B_m), and, therefore, they are hexagonal since, according to Table 1.1 in Section 1.2, any Bol web $W(3,2,r)$ is hexagonal. Then, by Theorem 1.10.4, the web $B_{n+1}(n+1,n,r)$ is hexagonal. ■

Corollary 4.5.4 *If a web $W(n+1,n,r)$ is a Bol web $B_{n+1}(n+1,n,r)$, then it is transversally geodesic.*

Proof. This follows from Theorems 1.10.2 and 4.5.3. ■

Definition 4.5.5 A web $W(n+1,n,r)$ is called a *Bol web* $B_{\xi_{n+1},\xi_1,\ldots,\xi_m}(n+1,n,r)$, where ξ_s are certain fixed indices, $1 \leq \xi_s \leq n+1$; $s = 1,\ldots,m,n+1$; $0 \leq m \leq n$, if it is a Bol web $B_{\xi_s}(n+1,n,r)$ for all ξ_s. If $m=n$, then a web $B_{\xi_{n+1},\xi_1,\ldots,\xi_m}(n+1,n,r)$ is called a *Moufang web* and denoted by $M(n+1,n,r)$.

Note that a web $M(3,2,r)$ is an ordinary Moufang three-web (see Table 1.1 in Section 1.2).

Note also that in the notation $B_{\xi_{n+1},\xi_1,\ldots,\xi_m}(n+1,n,r)$ the order of indices $\xi_{n+1},\xi_1,\ldots,$ is immaterial.

Now we will find different criteria for a web $W(n+1,n,r)$ to be a Moufang web $M(n+1,n,r)$ or a Bol web $B_{n+1}(n+1,n,r)$.

Theorem 4.5.6 *A web $B_{\xi_1,\ldots,\xi_n}(n+1,n,r)$ is a Moufang web $M(n+1,n,r)$.*

Proof. For $n = 2$, i.e., for three-webs $W(3,2,r)$, it follows from [Ac 65], Theorem 2.1, or [Be 67], p. 201, where it is proved that if a $W(3,2,r)$ is a web $B_{\xi_1,\xi_2}(3,2,r)$, then it is a $B_{\xi_3}(3,2,r)$, and, therefore, it is also a Moufang web $M(3,2,r)$.

Let now $n > 2$. Let us, take, for example, a web $B_{12\ldots n}(n+1,n,r)$ and prove that it is also a web $B_{n+1}(n+1,n,r)$. For $B_{12\ldots n}(n+1,n,r)$ all its 3-subwebs $[\alpha,\xi,\eta], \alpha = 1,\ldots,n$, are Bol webs $B_\alpha(3,2,r)$. Consider any 3-subweb $[n+1,\alpha,\beta]$. One can consider it as the 3-subweb $[\alpha, n+1,\beta]$ or $[\beta,\alpha,n+1]$. These 3-subwebs are Bol webs $B_\alpha(3,2,r)$ and $B_\beta(3,2,r)$, respectively. Thus, the 3-subweb $[n+1,\alpha,\beta]$ is a Bol web simultaneously of both types,(B_l) and (B_r), and therefore it is a Bol web of type (B_m), i.e., a web $B_3(3,2,r)$. Using Definition 4.5.1, we obtain that the web $W(n+1,n,r)$ under consideration is a Bol web $B_{n+1}(n+1,n,r)$, and consequently a Moufang web $M(n+1,n,r)$. ■

Theorem 4.5.7 *A web $W(n+1,n,r)$ is a Moufang web if and only if all its 3-subwebs are Moufang webs $M(3,2,r)$.*

Proof. *Necessity.* Suppose that a web $W(n+1,n,r)$ is a Moufang web $M(n+1,n,r)$. Then from its definition and Definition 4.5.1 we obtain that each of its 3-subwebs is simultaneously a Bol web of types $(B_l),(B_r)$, and (B_m), i.e., it is a Moufang three-web.

Sufficiency. Suppose now that any 3-subweb of a web $W(n+1,n,r)$ is a Moufang web. Take any index $\xi_{n+1}, 1 \leq \xi_{n+1} \leq n+1$, and consider all 3-subwebs $[\xi_{n+1},\xi_1,\xi_2]$ where $\xi_1,\xi_2 = 1,\ldots,n+1$; ξ_1,ξ_2 and ξ_{n+1} are distinct. By condition of the theorem, all of them are Moufang three-webs and therefore, they are Bol webs $B_{\xi_{n+1}}(3,2,r)$. Then, by Definition 4.5.1, the web $W(n+1,n,r)$ is a Bol web $B_{\xi_{n+1}}(n+1,n,r)$. Since this is true for any $\xi_{n+1}, 1 \leq \xi_{n+1} \leq n+1$, then it follows from Definition 4.5.5 that the web $W(n+1,n,r)$ is a Moufang web $M(n+1,n,r)$. ■

Corollary 4.5.8 *For a web $W(n+1,n,r)$ to be a Moufang web $M(n+1,n,r)$, it is necessary and sufficient that all 3-subwebs of the form $[\alpha,\xi,\eta]$ where $\alpha = 1,\ldots,n; \xi,\eta = 1,\ldots,n+1$, be Bol three-webs of type (B_m).*

Proof. This follows from Theorems 4.5.6 and 4.5.7. ■

Theorem 4.5.9 *A web $W(n+1,n,r)$ is a Bol web $B_{n+1}(n+1,n,r)$ if and only if each of its $(m+1)$-subwebs $W(m+1,m,r)$, $1 < m < n$, of the form $[n+1,\alpha_1,\ldots,\alpha_m]$ is a Bol web $B_{n+1}(m+1,m,r)$.*

Proof. *Necessity.* Suppose that a web $W(n+1,n,r)$ is a Bol web $B_{n+1}(n+1,n,r)$. Then all its 3-subwebs $[n+1,\alpha,\beta]$ are Bol webs of type (B_m). Consider any $(m+1)$-subweb of the form $[n+1,\alpha_1,\ldots,\alpha_m]$. Since all its 3-subwebs $[n+1,\alpha_s,\alpha_t]$, $s,t = 1,\ldots,m$, are at the same time 3-subwebs of the whole web $W(n+1,n,r)$, then they are Bol three-webs of type (B_m). Thus, an $(m+1)$-subweb $W(m+1,m,r)$ is a Bol web $B_{n+1}(m+1,m,r)$.

Sufficiency. Suppose that each $(m+1)$-subweb $W(m+1,m,r)$ of the form $[n+1,\alpha_1,\ldots,\alpha_m]$ of a web $W(n+1,n,r)$ is a Bol web $B_{n+1}(m+1,m,r)$. Each of its 3-subwebs $[n+1,\alpha,\beta]$ is a 3-subweb of an $(m+1)$-subweb $[n+1,\alpha,\beta,\ldots]$. By the condition of the theorem, the $(m+1)$-subweb $[n+1,\alpha,\beta,\ldots]$ is a Bol web $B_{n+1}(m+1,m,r)$. Therefore, the 3-subweb $[n+1,\alpha,\beta]$ is a Bol three-web of type (B_m) and consequently the web $W(n+1,n,r)$ is a Bol web $B_{n+1}(n+1,n,r)$. ■

Using the same technique, we can prove a similar criterion for an $(n+1)$-web to be a Moufang $(n+1)$-web:

Theorem 4.5.10 *For a web $W(n+1,n,r)$ to be a Moufang web $M(n+1,n,r)$, it is necessary and sufficient that all its $(m+1)$-subwebs be Moufang webs $M(m+1,m,r)$.* ■

Proposition 4.5.11 *If a web $W(n+1,n,r)$ is a Bol web $B_{\xi_{n+1},\xi_1,\ldots,\xi_m}(n+1,n,r)$, then all its $(m+2)$-subwebs of the form $[\xi_{n+1},\xi_1,\ldots,\xi_m,\xi_{m+1}]$ are Moufang webs $M(m+2,m+1,r)$, $1<m<n;\xi_{m+1}=1,\ldots,n+1;\ \ \xi_{m+1}\neq\xi_1,\ \ s=1,\ldots,m,n+1$.*

Proof. Consider any $(m+2)$-subweb of the form $[\xi_{n+1},\xi_1,\ldots,\xi_{m+1}]$. Since a web $W(n+1,n,r)$ is a Bol web $B_{\xi_{n+1},\xi_1,\ldots,\xi_m}(n+1,n,r)$, then all its 3-subwebs of the form $[\xi_s,\eta,\zeta],\eta,\zeta=\xi_{n+1},\xi_1,\ldots,\xi_{m+1}$, are Bol three-webs of type (B_m). Since the same 3-subwebs are 3-subwebs of the $(m+2)$-subweb $[\xi_{n+1},\xi_1,\ldots,\xi_{m+1}]$, then this $(m+2)$-subweb is a Bol web $B_{\xi_{n+1},\xi_1,\ldots,\xi_m}(m+2,m+1,r)$. Finally, applying to this $(m+2)$-subweb Theorem 4.5.6, we conclude then it is a Moufang web $M(m+2,m+1,r)$. ■

4.5.2 The Bol Closure Conditions

Consider two different leaves of the foliation λ_{n+1} of a web $W(n+1,n,r)$. We will denote them by a^0_{n+1} and a^1_{n+1}. Take in the leaf a^0_{n+1} a point X. Through X there pass n leaves x_α of the foliations λ_α, $x_\alpha\in\lambda_\alpha$, $\alpha=1,\ldots,n$. Take any $n-1$ of them. Their intersection $\bigcap\limits_{\substack{\alpha=1\\(\alpha\neq\beta)}}^{n} x_\alpha$ has only one common point with the leaf a^1_{n+1}. Denote it by X_β:

$$X_\beta = a^1_{n+1}\cap\Big(\bigcap_{\substack{\alpha=1\\(\alpha\neq\beta)}}^{n} x_\alpha\Big). \tag{4.5.2}$$

We obtain n such points in the leaf a^1_{n+1}. Every point X_β uniquely determines a leaf y_β of the foliation λ_β passing through it. There will be n such leaves, they belong to different foliations, and, therefore, they have only one common point :

$$X=\bigcap_{\alpha=1}^{n} y_\alpha. \tag{4.5.3}$$

The point $\bar{X}$ uniquely determines a leaf $\bar{a}_{n+1}$ of the foliation λ_{n+1} (see Figure 4.12 which corresponds to the case $n = 3, r = 1$).

Definition 4.5.12 A figure formed by the leaves $a^0_{n+1}, a^1_{n+1}, x_\alpha, y_\alpha, \bar{a}_{n+1}$ and points of their intersection is said to be a *Bol n-parallelotope* and denoted by $B^{n+1}_{n+1}(X)$. The points X and $\bar{X}$ are called its *base* and *summit*, and the leaves a^0_{n+1} and a^1_{n+1} are called its *basis leaves.*

Consider the following intersection of the leaves x_α and y_α of a Bol n-parallelotope $B^{n+1}_{n+1}(X)$:

$$\left(\bigcap_{s=1}^{m} y_{\alpha_s}\right) \cap \left(\bigcap_{t=m+1}^{n} x_{\alpha_t}\right), \quad 1 < m < n,$$

Since n leaves of different foliations participate in this intersection, then it follows from Definition 1.1.1 taken for $d = n+1$ that this intersection is a point. Denote it by $X_{\alpha_1 \ldots \alpha_m}$:

$$X_{\alpha_1 \ldots \alpha_m} = \left(\bigcap_{s=1}^{m} y_{\alpha_s}\right) \cap \left(\bigcap_{t=m+1}^{n} x_{\alpha_t}\right). \tag{4.5.4}$$

Note that in the notation $X_{\alpha_1 \ldots \alpha_m}$ all indices are distinct and their order is immaterial. For $m = 0$, from (4.5.4) we obtain

$$X = \bigcap_{\alpha=1}^{n} x_\alpha,$$

for $m = 1$ equality (4.5.4) becomes

$$X_\beta = y_\beta \cap \left(\bigcap_{\substack{\alpha=1 \\ (\alpha \neq \beta)}}^{n} x_\alpha\right),$$

and for $m = n$ we have

$$X_{12\ldots n} = \bigcap_{\alpha=1}^{n} y_\beta = \bar{X}.$$

We will assume in the future that $0 \leq m \leq n$. Figure 4.12 shows the points $X_{\alpha_1 \ldots \alpha_m}$ in the case $n = 3, r = 1$.

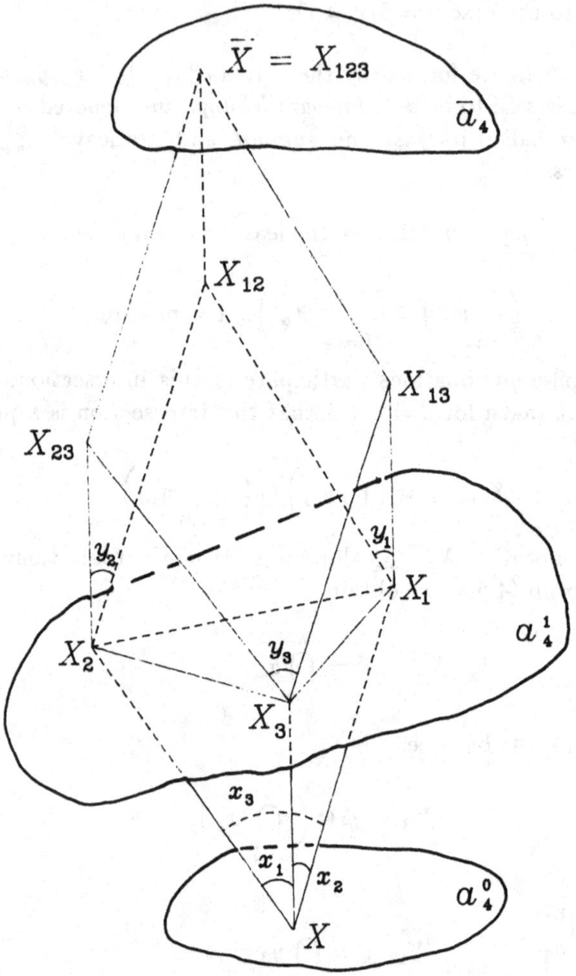

Figure 4.12

Definition 4.5.13 The point $X_{\alpha_1 \ldots \alpha_m}, 0 \leq m \leq n$, is said to be the *vertex of height* m of an n-parallelotope $B^{n+1}_{n+1}(X)$.

It is easy to see that the number of vertices of height m is equal to $\binom{n}{m}$, the vertex of height 0 is the base X of $B^{n+1}_{n+1}(X)$, and the vertex of height n is the summit $\bar{X}$ of $B^{n+1}_{n+1}(X)$.

Consider the intersection $\bigcap_{t=m+1}^{n} x_{\alpha_t}$ of the leaves x_{α_t}, $t = m+1, \ldots, n$, of the figure $B_{n+1}^{n+1}(X)$.

Lemma 4.5.14 *A Bol n-parallelotope $B_{n+1}^{n+1}(X)$ of a web $W(n+1,n,r)$ intersects the submanifold $\bigcap_{t=m+1}^{n} x_{\alpha_t}$ along a Bol m-parallelotope $B_{n+1}^{n+1}(X)$ of the $(m+1)$-subweb $[n+1, \alpha_1, \ldots, \alpha_m]$ of this web.*

Proof. The foliations $\lambda_{n+1}, \lambda_{\alpha_1}, \ldots, \lambda_{\alpha_m}$ cut on the intersection $\bigcap_{t=m+1}^{n} x_{\alpha_t}$ an $(m+1)$-subweb $[n+1, \alpha_1, \ldots, \alpha_m]$ of the web $W(n+1,n,r)$, and the leaves a_{n+1}^0, a_{n+1}^1, $x_{\alpha_s}, y_{\alpha_s}$, $s = 1, \ldots, m$, cut on it a Bol m-parallelotope $B_{n+1}^{n+1}(X)$ of the $(m+1)$-subweb $[n+1, \alpha_1, \ldots, \alpha_m]$ with basis leaves

$$a_{n+1}^0 \cap \Big(\bigcap_{t=m+1}^{n} x_{\alpha_t}\Big), \quad a_{n+1}^1 \cap \Big(\bigcap_{t=m+1}^{n} x_{\alpha_t}\Big),$$

base X and summit $X_{\alpha_1 \ldots \alpha_m}$. To the same figure $B_{n+1}^{n+1}(X)$ there belong all vertices of height less than m numbered by indices $\alpha_1, \ldots, \alpha_m$ and only by these indices. This completes the proof. ■

Consider now any leaf a_{n+1}^0 of the foliation λ_{n+1} and let $a_\alpha, \alpha = 1, \ldots, n$, be the leaves of the foliations λ_α. Suppose that the point of intersection of the a_α does not lie in a_{n+1}^0, i.e.,

$$a_{n+1}^0 \cap \Big(\bigcap_{\alpha=1}^{n} a_\alpha\Big) = \emptyset.$$

Exclude a leaf a_β, β fixed, $1 \leq \beta \leq n$, from the set of the leaves a_α. The remaining leaves have a unique common point with the leaf a_{n+1}. Denote it by X^β:

$$X^\beta = a_{n+1}^0 \cap \Big(\bigcap_{\substack{\alpha=1 \\ (\alpha \neq \beta)}}^{n} a_\alpha\Big)$$

(see Figure 4.13).

Definition 4.5.15 A figure cut in the leaf a_{n+1}^0 by the leaves a_α, $\alpha = 1, \ldots, n$, is said to be a *simplex* S_{n+1}, the intersections $a_\alpha \cap a_{n+1}^0$ are called the *faces* of the simplex S_{n+1} and the points X^β are called its *vertices*.

Through the vertex X^β of the simplex S_{n+1} in addition to the leaves defining X^β , there passes the leaf x_β of the foliation λ_β. Therefore, the point X^β can be represented in the form:

$$X^\beta = x_\beta \cap \Big(\bigcap_{\substack{\alpha=1 \\ (\alpha \neq \beta)}}^{n} a_\alpha\Big) \tag{4.5.5}$$

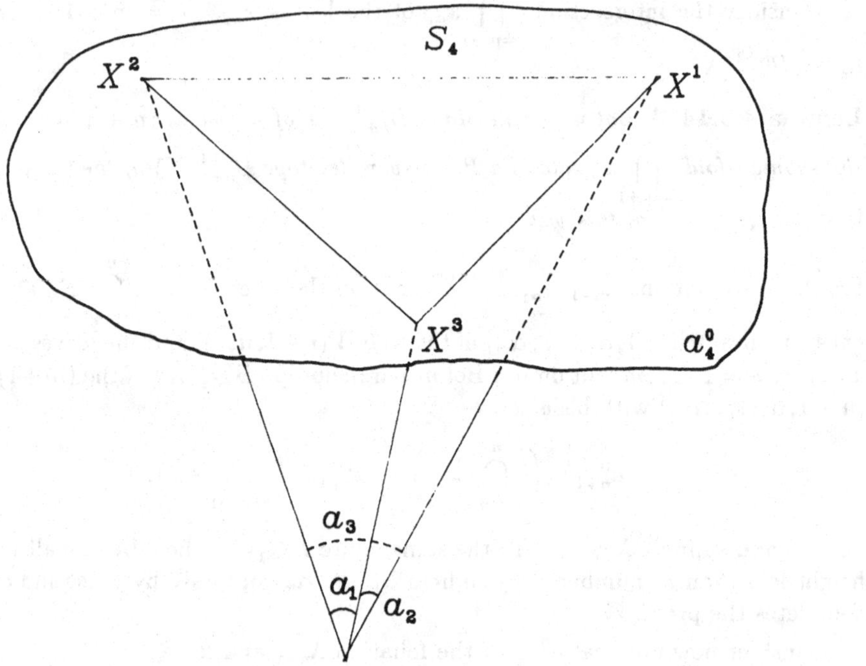

Figure 4.13

Lemma 4.5.16 *A simplex S_{n+1} in the leaf a^0_{n+1} is uniquely determined if we are given a point X in a^0_{n+1} and any leaf a_α of the foliation λ_β, β fixed, which does not pass through the point X.*

Proof. In fact, suppose that a_α, $\alpha = 1, \dots, \beta - 1, \beta + 1, \dots, n$, are leaves of the foliations λ_α passing through the point X. Since a_β does not pass through X, the point of intersection of the leaves a_α, $\alpha = 1, \dots, n$, does not lie in the leaf a^0_{n+1}. Therefore, the leaves $a_\alpha, \alpha = 1, \dots, n$, cut in the leaf a_{n+1} a simplex S_{n+1}, and the point X is a vertex X^β of the simplex S_{n+1}. ■

Let a^1_{n+1} be a leaf of the foliation λ_{n+1} sufficiently close to the leaf a^0_{n+1} but different from it. For each vertex X^α, $\alpha = 1, \dots, n$, of an arbitrary simplex S_{n+1} lying in a^0_{n+1}, we construct a Bol n-parallelotope $B^{n+1}_{n+1}(X^\alpha)$ with the basis leaves a^0_{n+1} and a^1_{n+1}. Let $\bar{X}^\alpha$ be its summit.

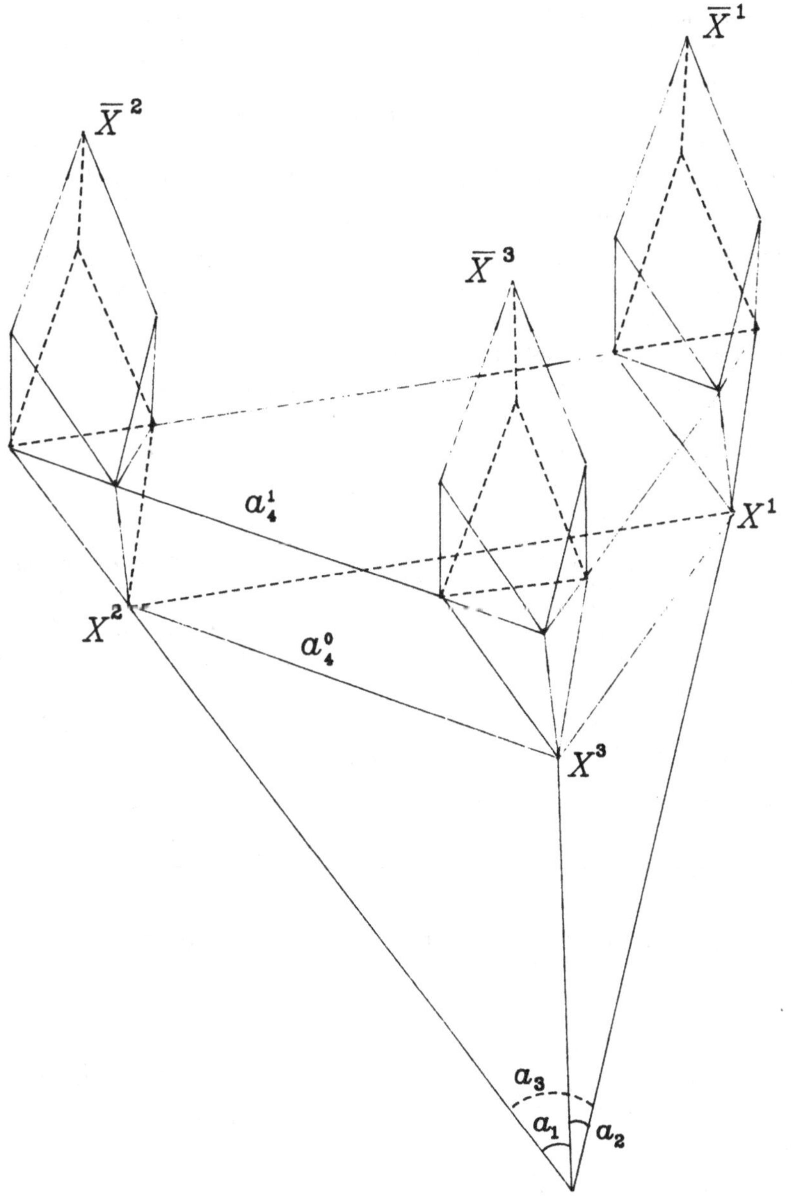

Figure 4.14

Definition 4.5.17 A figure formed by the leaves a^0_{n+1}, a^1_{n+1}, the simplex S_{n+1} in a^0_{n+1}, the Bol n-parallelotepes $B^{n+1}_{n+1}(X^\alpha)$ with the basis leaves a^0_{n+1}, a^1_{n+1} and the bases located in the vertices X^α of the symplex S_{n+1}, is said to be a *Bol figure* $B^{n+1}_{n+1}(S_{n+1})$ (see Figure 4.14 corresponding to the case $n=3, r=1$). A Bol figure $B^{n+1}_{n+1}(S_{n+1})$ is called *closed* if all the summits $\bar{X}^\alpha$ of the Bol n-parallelotepes $B^{n+1}_{n+1}(X^\alpha), \alpha = 1, \ldots, n$, lie in one and the same leaf of the foliation λ_{n+1}. We will say that a *Bol closure condition* (B^{n+1}_{n+1}) holds on a web $W(n+1, n, r)$ if all sufficiently small Bol figures $B^{n+1}_{n+1}(S_{n+1})$ are closed on it.

Since the foliations λ_ξ which form a web $W(n+1, n, r)$ are interchangeable , we can define the *Bol closure condition* (B^{n+1}_α), $\alpha = 1, \ldots, n$, using exactly the same way we have used to define the Bol closure condition (B^{n+1}_{n+1}).

Note that for $n = 2$ the Bol closure conditions (B^3_ξ), $\xi = 1, 2, 3$, agree with Bol closure conditions $(B_m), (B_r)$, and (B_l) which have been introduced in Definition 1.2.7.

There is another way to define Bol closure conditions. For this consider again two distinct leaves a^0_{n+1} and a^1_{n+1} of the foliation λ_{n+1}. Take in a^0_{n+1} two distinct points X and $Y: X, Y \in a^0_{n+1}, \;\; X \neq Y$.

Definition 4.5.18 A figure formed by the leaves a^0_{n+1}, a^1_{n+1} and the Bol n-parallelotopes $B^{n+1}_{n+1}(X)$ and $B^{n+1}_{n+1}(Y)$ constructed on them is said to be a *Bol figure* $B^{n+1}_{n+1}(X, Y)$ (see Figure 4.15 where again the case $n = 3, r = 1$ is presented). A Bol figure $B^{n+1}_{n+1}(X, Y)$ is called *closed* if the summits $\bar{X}$ and $\bar{Y}$ of the Bol n-parallelotopes $B^{n+1}_{n+1}(X)$ and $B^{n+1}_{n+1}(Y)$ lie in one and the same leaf $\bar{a}_{n+1}$ of the foliation λ_{n+1}. We will say that on a web $W(n+1, n, r)$ the *Bol closure condition* $(B^{n+1}_{n+1})'$ holds if on it all sufficiently small Bol figures $B^{n+1}_{n+1}(X, Y)$ are closed.

Theorem 4.5.19 *The Bol closure conditions* (B^{n+1}_{n+1}) *and* $(B^{n+1}_{n+1})'$ *are equivalent.*

Proof. First note that if $n = 2$, these two closure conditions coincide. Let us prove their equivalence for $n > 2$.

a) Suppose that on a web $W(n+1, n, r)$ the closure condition (B^{n+1}_{n+1}) holds. Let us consider a sufficiently small Bol figure $B^{n+1}_{n+1}(X, Y)$ with basis leaves a^0_{n+1} and a^1_{n+1} where

$$X, Y \in a^0_{n+1}. \tag{4.5.6}$$

Construct in a^0_{n+1} a sequence of simplices S^α_{n+1}, $\alpha = 1, \ldots, k$, $1 \leq k \leq n-1$, such that the point X is a vertex of the simplex S^1_{n+1}, Y is a vertex of the simplex S^k_{n+1}, and the simplices S^α_{n+1} and $S^{\alpha+1}_{n+1}$, $\alpha = 1, \ldots, k-1$, have a common vertex. Denote by $X^{\alpha\beta}$, $\alpha = 1, \ldots, n$; $\beta = 1, \ldots, k$, a vertex of the simplex S^β_{n+1} which does not belong to a face of S^β_{n+1} cut by a leaf of the foliation λ_α.

Suppose that b_γ and c_γ are the leaves of the foliation $\lambda_\gamma, \gamma = 1, \ldots, n$, passing through the points X and Y, respectively:

$$X = \bigcap_{\gamma=1}^{n} b_\gamma, \;\; Y = \bigcap_{\gamma=1}^{n} c_\gamma. \tag{4.5.7}$$

Since the points X and Y are distinct, then at least one of the leaves b_γ is different from the corresponding c_γ. Without loss of generality, we assume that $b_1 \neq c_1$. Therefore, the point X does not lie in the leaf c_1 and, by Lemma 4.5.16, the leaf c_1 and the point x determine a simplex in the leaf a^0_{n+1}. Denote it by S^1_{n+1}. Its faces are cut in a^0_{n+1} by the leaves $c_1, b_2, \ldots, b_n$. In addition, it follows from (4.5.6) and (4.5.7) and from the definition of the vertex X^{11} that

$$X^{11} = a^0_{n+1} \cap \left(\bigcap_{\gamma=2}^{n} b_\gamma \right) = X. \tag{4.5.8}$$

Suppose that we have constructed a simplex S^δ_{n+1} cut in the leaf a^0_{n+1} by the leaves $c_1, \ldots, c_\alpha, b_{\alpha+1}, \ldots, b_n$, $1 \leq \alpha \leq n-1$, $\delta \leq \alpha$. Consider its vertex:

$$X^{\alpha+1,\delta} = a^0_{n+1} \cap \left(\bigcap_{\gamma=1}^{\alpha} c_\gamma \right) \cap \left(\bigcap_{\gamma=\alpha+2}^{n} b_\gamma \right). \tag{4.5.9}$$

There are only two possibilities:

i) The leaves $c_{\alpha+1}, c_{\alpha+2}, \ldots, c_\beta, \alpha \leq \beta \leq n-2$, pass through the point $X^{\alpha+1,\delta}$ and this point does not belong to the leaf $c_{\beta+1}$. Then, by virtue of (4.5.9), we obtain that

$$X^{\alpha+1,\delta} = a^0_{n+1} \cap \left(\bigcap_{\gamma=1}^{\beta} c_\gamma \right) \cap \left(\bigcap_{\gamma=\beta+2}^{n} b_\gamma \right). \tag{4.5.10}$$

Since the point $X^{\alpha+1,\delta}$ does not lie in the leaf $c_{\beta+1}$, then, by Lemma 4.5.16, this point and the leaf $c_{\beta+1}$ determine in a^0_{n+1} a simplex $S^{\delta+1}_{n+1}$. It follows from (4.5.10) that the faces of the simplex $S^{\delta+1}_{n+1}$ are cut in a^0_{n+1} by the leaves $c_1, \ldots, c_{\beta+1}, b_{\beta+2}, \ldots, b_n$, and the point $X^{\alpha+1,\delta}$ is its vertex $X^{\beta+1,\delta+1}$, i.e., the simplices S^δ_{n+1} and $S^{\delta+1}_{n+1}$ have a common point.

ii) The leaves $c_{\alpha+1}, \ldots, c_{n+1}$ pass through the point $X^{\alpha+1,\delta}$. Then from (4.5.6), (4.5.7), and (4.5.9) we obtain that

$$X^{\alpha+1,\delta} = a^0_{n+1} \cap \left(\bigcap_{\gamma=1}^{n-1} c_\gamma \right) = Y. \tag{4.5.11}$$

Therefore, the point Y is a vertex of the simplex S^δ_{n+1} and in this case we have completed the construction of desired sequence of simplices.

Thus, using the construction described above no more than $n-1$ times, we will either reach the case ii) at some step or we will obtain the simplex S^k_{n+1} which is cut in a^0_{n+1} by the leaves $c_1, \ldots, c_{n-1}, b_n$. In the latter case, by virtue of (4.5.6) and

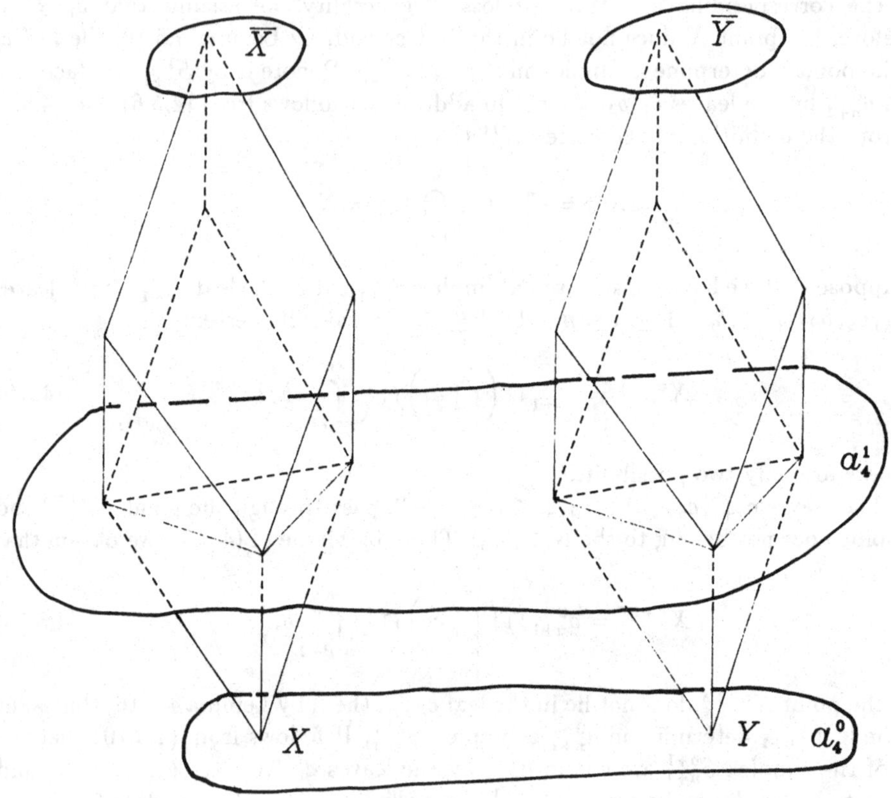

Figure 4.15

(4.5.7), the vertex X^{nk} of the simplex S_{n+1}^k coincides with the point Y :

$$X^{nk} = a_{n+1}^0 \cap \left(\bigcap_{\gamma=1}^{n-1} c_\gamma\right) = Y. \tag{4.5.12}$$

According to our construction, the simplices S_{n+1}^α and $S_{n+1}^{\alpha+1}$, $\alpha = 1, \ldots, k-1$, have a common vertex. Therefore, we have completed the construction of the desired sequence of simplices for all possible cases.

For each simplex S^α_{n+1}, $\alpha = 1, \dots, k$, we construct on the basis leaves a^0_{n+1} and a^1_{n+1} of the figure $B^{n+1}_{n+1}(X, Y)$ a Bol figure $B^{n+1}_{n+1}(S^\alpha_{n+1})$. Since the Bol closure condition (B^{n+1}_{n+1}) is met, all vertices of this figure belong to one and the same leaf b^α_{n+1} of the foliation λ_{n+1}. In addition, from (4.5.8), (4.5.11), and (4.5.12) we obtain that

$$\bar{X} \in b^1_{n+1}, \quad \bar{Y} \in b^k_{n+1}. \tag{4.5.13}$$

Consider the simplices S^α_{n+1} and $S^{\alpha+1}_{n+1}$. According to our construction, they have a common vertex. Therefore, the Bol figures $B^{n+1}_{n+1}(S^\alpha_{n+1})$ and $B^{n+1}_{n+1}(S^{\alpha+1}_{n+1})$ constructed on these simplices also have a common vertex. Then, from the definition of an $(n+1)$-web we obtain that

$$b^\alpha_{n+1} = b^{\alpha+1}_{n+1}. \tag{4.5.14}$$

The relations (4.5.13) and (4.5.14) imply that the summits $\bar{X}$ and $\bar{Y}$ of the Bol figure $B^{n+1}_{n+1}(X, Y)$ lie in one and the same leaf of the foliation λ_{n+1} and therefore on the web $W(n+1, n, r)$ the Bol closure condition $(B^{n+1}_{n+1})'$ holds.

b) Suppose that on a web $W(n+1, n, r)$ the closure condition $(B^{n+1}_{n+1})'$ holds. Consider a sufficiently small Bol figure $B^{n+1}_{n+1}(S_{n+1})$ with basis leaves a^0_{n+1} and a^1_{n+1}. It follows from the condition $(B^{n+1}_{n+1})'$ that all Bol figures $B^{n+1}_{n+1}(X^1, X^\alpha)$, $\alpha = 2, \dots, n$, constructed on the same basis leaves, are closed; here $X^1, \dots, X^n$ are vertices of the simplex S_{n+1}. Thus, all vertices $\bar{X}^\alpha$ of the figure $B^{n+1}_{n+1}(S_{n+1})$ lie in a leaf $\bar{a}_{n+1}$ of the foliation λ_{n+1} passing through the point $\bar{X}^1$. Then from Definition 4.5.17 we obtain that the Bol figure $B^{n+1}_{n+1}(S_{n+1})$ is closed and, therefore, the closure condition (B^{n+1}_{n+1}) holds on the web $W(n+1, n, r)$.

Thus, we completed the proof of equivalence of Bol closure conditions (B^{n+1}_{n+1}) and $(B^{n+1}_{n+1})'$. ■

4.5.3 A Geometric Characteristic of Bol $(n+1)$-Webs

Lemma 4.5.20 *If a web $W(n+1, n, r)$ is a Bol web $B_{n+1}(n+1, n, r)$, then all vertices of height 2 of any its Bol n-parallelotope $B^{n+1}_{n+1}(n+1, n, r)$ lie in one and the same leaf of the foliation λ_{n+1}.*

Proof. If $n = 2$, there is only one vertex of height 2, and the statement of the lemma is trivial.

Suppose that $n > 2$. Consider any sufficiently small Bol n-parallelotope $B^{n+1}_{n+1}(X)$ with basis leaves a^0_{n+1}, a^1_{n+1}, and the base $X \in a^0_{n+1}$:

$$X = \bigcap_{\alpha=1}^{n} x_\alpha;$$

here a_α is a leaf of the foliation λ_α, $\alpha = 1, \dots, n$, passing through the point X. Consider further a $(2r)$-dimensional submanifold $M_{\alpha\beta}(X)$ formed by an intersection of $n-2$ leaves :

$$M_{\alpha\beta}(X) = \bigcap_{\substack{\gamma=1 \\ (\gamma\neq\alpha,\beta)}}^{n} x_\gamma.$$

On $M_{\alpha\beta}(X)$ the leaves of the foliations $\lambda_{n+1}, \lambda_\alpha, \lambda_\beta$ cut a 3-subweb $[n+1, \alpha, \beta]$ which is the Bol web $B_{n+1}(3,2,r)$. It follows from Lemma 4.5.14 that the Bol n-parallelotope $B^{n+1}_{n+1}(X)$ intersects $M_{\alpha\beta}(X)$ in a Bol 2-parallelotope $B^3_{n+1}(X)$ of the 3-subweb $[n+1, \alpha, \beta]$. The base of $B^3_{n+1}(X)$ is the point X, its vertices of height 1 are the points X_α, X_β and the vertex of height 2 is the point $X_{\alpha\beta}$ (see (4.5.4)).

Choose in addition $n-2$ leaves $x_\gamma, \gamma \neq \delta, \varepsilon$. In their intersection a 3-subweb $[n+1, \delta, \varepsilon]$ is cut, and this 3-subweb is also a Bol web $B_{n+1}(3,2,r)$. The figure $B^{n+1}_{n+1}(X)$ intersects $M_{\delta\varepsilon}(X)$ in a Bol 2-parallelotope $B^3_{n+1}(X)$ of the 3-subweb $[n+1, \delta, \varepsilon]$ with the base X and the summit $X_{\delta\varepsilon}$. The points $X_{\alpha\beta}$ and $X_{\delta\varepsilon}$ uniquely determine the leaves $a^2_{n+1}(X_{\alpha\beta})$ and $a^2_{n+1}(X_{\delta\varepsilon})$ of the foliation λ_{n+1} passing through them. Let us prove that $a^2_{n+1}(X_{\alpha\beta}) = a^2_{n+1}(X_{\delta\varepsilon})$. Consider two cases:

a) One of the indices of the point $X_{\alpha\beta}$ coincides with one of those of the point $X_{\delta\varepsilon}$. Suppose, for example, that $\alpha = \delta$. In this case the intersection of the submanifolds $M_{\alpha\beta}(X)$ and $M_{\delta\varepsilon}(X)$ is an r-dimensional manifold: $M_\alpha(X) = \bigcap_{\substack{\gamma=1 \\ (\gamma\neq\alpha)}}^{n} x_\gamma$. Since the 3-subweb $[n+1, \alpha, \beta]$ is a Bol web $B_{n+1}(3,2,r)$, then it is hexagonal. Complete the Bol 2-parallelotope $B^3_{n+1}(X)$ of the 3-subweb $[n+1, \alpha, \beta]$ to a hexagonal figure (see Figure 1.8) with the centre at the point X_α (see Figure 4.16).

The points $X_\alpha, X, X_\beta, X_{\alpha\beta}, X'_{\alpha\beta}, X'_\beta$,and X' belong to this hexagonal figure. Since, by Theorem 4.5.3, the web $B_{n+1}(3,2,r)$ is hexagonal, the hexagonal figure is closed, and the points $X_{\alpha\beta}$ and $X'_{\alpha\beta}$ lie in one and the same leaf of the 3-subweb $[n+1, \alpha, \beta]$, and, therefore, $X'_{\alpha\beta} \in a^2_{n+1}(X_{\alpha\beta})$.

It has been proved by Fedorova [F 78] (see also [AGe 86]) that the points X, X_α, and $X'_{\alpha\beta}$ lie on one and the same geodesic line (in connection $\bar{\gamma}_{\alpha\beta}$), and the point X_α is the middle point of the geodesic segment with the ends X and $X'_{\alpha\beta}$.

Complete the Bol 2-parallelotope $B^3_{n+1}(X)$ of the 3-subweb $[n+1, \alpha, \varepsilon]$ to a hexagonal figure with the centre at the point X_α. Using similar considerations for this hexagonal figure, we obtain that the points X, X_α, and $X'_{\alpha\varepsilon}$ lie on one and the same geodesic line (in the connection $\bar{\gamma}_{\alpha\varepsilon}$), and the point X_α is the middle point of the geodesic segment with the ends X and $X'_{\alpha\varepsilon}$, and $X'_{\alpha\varepsilon} \in a^2_{n+1}(X_{\alpha\varepsilon})$ (see Figure 4.15).

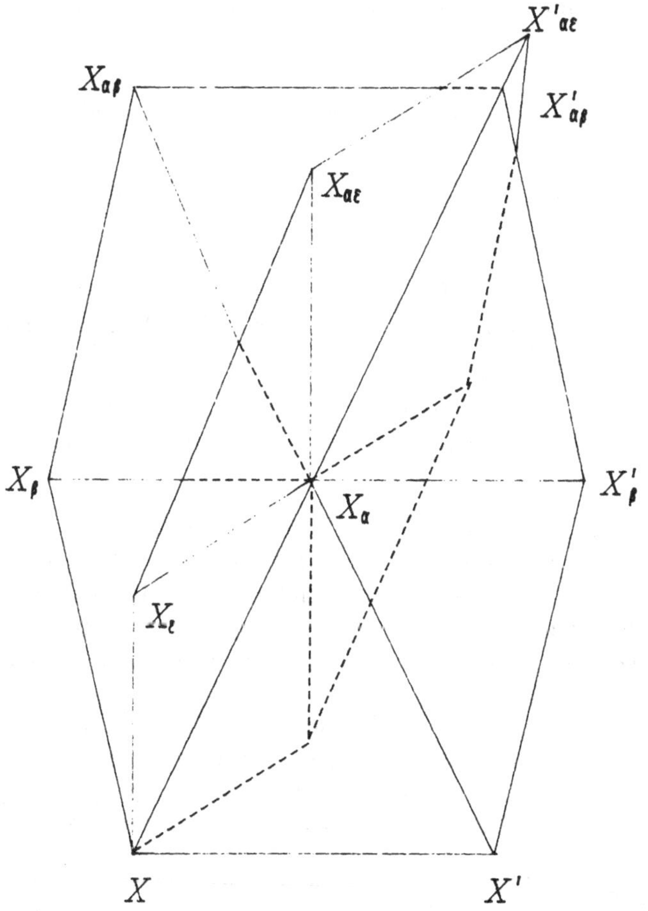

Figure 4.16

Two sufficiently close points X and X_α determine in the connection $\bar{\gamma}_{\alpha\beta}$ a unique geodesic line passing through them. According to Theorem 1.9.15, this line will be also geodesic in the connection $\bar{\gamma}_{\alpha\beta}$. Moreover, by Theorem 3.9.16, the geodesic parameters of this geodesic line agree in the connections $\bar{\gamma}_{\alpha\beta}$ and $\bar{\gamma}_{\alpha\varepsilon}$. Therefore, the points $X'_{\alpha\beta}$ and $X'_{\alpha\varepsilon}$ lie on one and the same geodesic line and coincide since they are equidistant from the points X and X_α. Then we have

$$a^2_{n+1}(X_{\alpha\beta}) \cap a^2_{n+1}(X_{\alpha\varepsilon}) \neq \emptyset,$$

and from the definition of an $(n+1)$-web we obtain that

$$a^2_{n+1}(X_{\alpha\beta}) = a^2_{n+1}(X_{\alpha\varepsilon}).$$

b) Suppose now that all indices $\alpha, \beta, \delta, \varepsilon$ are distinct. Consider the point $X_{\alpha\varepsilon}$. According to a), we have

$$a^2_{n+1}(X_{\alpha\beta}) = a^2_{n+1}(X_{\alpha\varepsilon}), \quad a^2_{n+1}(X_{\alpha\varepsilon}) = a^2_{n+1}(X_{\delta\varepsilon}),$$

and, therefore,

$$a^2_{n+1}(X_{\alpha\beta}) = a^2_{n+1}(X_{\delta\varepsilon}).$$

Thus, all the vertices of height 2 of the figure $B^{n+1}_{n+1}(X)$ lie in one and the same leaf a^2_{n+1} of the foliation λ_{n+1} . ■

Lemma 4.5.21 *Any two sufficiently close points X and Y of a web $W(n+1, n, r)$ can be connected by means of r-dimensional submanifolds which are intersections of leaves of the web.*

Proof. We will prove the lemma by induction on n. Suppose first that $n = 2$. Then leaves of the web are of dimension r. Consider a leaf a_1 of the foliation λ_1 passing through the point X and a leaf a_2 of the foliation λ_2 passing through the point Y. Since the leaves a_1 and a_2 are from different foliations, their intersection is a point, and therefore in this case the points X and Y can be connected by means of r-dimensional submanifolds.

Suppose now that the lemma is valid for all $k < n$. We prove it for $k = n$. Denote by a_{n+1} a leaf of the foliation λ_{n+1} passing through the point X and by a_α a leaf of the foliation $\lambda_\alpha, \alpha = 1, \ldots, n$, passing through the point Y. An intersection of any $n-1$ leaves a_α : $M_\beta(Y) = \bigcap_{\alpha\neq\beta} a_\alpha$, is an r-dimensional submanifold of the web $W(n+1, n, r)$. Since the leaves $a_{n+1}, a_\alpha, \alpha = 1, \ldots n; \alpha \neq \beta$, are from different foliations and their number is equal to n, they have a unique common point Z :

$$Z = a_{n+1} \cap \Big(\bigcap_{\substack{\alpha=1 \\ (\alpha\neq\beta)}}^{n} a_\alpha \Big) = a_{n+1} \cap M_\beta(Y).$$

By the construction, the points Y and Z are joined by an r-dimensional submanifold $M_\beta(Y)$ of the web $W(n+1, n, r)$. The points X and Z lie in the same leaf a_{n+1} of this web. In a_{n+1} the leaves of the foliations $\lambda_\alpha, \alpha = 1, \ldots, n$, cut an n-subweb $[1, 2, \ldots, n]$ of the web $W(n+1, n, r)$. By the inductive assumption, the lemma is valid for this subweb, and, therefore, the points X and Z can be connected by r-dimensional submanifolds which are cut in a_{n+1} by the leaves of the foliations λ_α, $\alpha = 1, \ldots, n$. Thus, the points X and Y can be joined by r-dimensional intersections of the leaves of the web $W(n+1, n, r)$. ■

Note that in order to connect any two sufficiently close points of the web $W(n+1, n, r)$ there is required not more than n such r-dimensional submanifolds.

Lemma 4.5.22 *If a web $W(n+1,n,r)$ is a Bol web $B_{n+1}(n+1,n,r)$, then for each its Bol n-parallelotope $B_{n+1}^{n+1}(X)$ the leaf a_{n+1}^2 does not depend on the choice of a point X and depends only on the choice of the basis leaves a_{n+1}^0 and a_{n+1}^1 of this Bol n-parallelotope.*

Proof. For $n=2$ the lemma follows from the definition of a web $B_3(3,2,r)$. Consider the case $n>2$. Let a_{n+1}^0, a_{n+1}^1 be the basis leaves of a Bol n-parallelotope $B_{n+1}^{n+1}(X)$ and let Y be any point in the leaf a_{n+1}^0 sufficiently close to X. We will prove that all vertices of height 2 of the Bol n-parallelotope $B_{n+1}^{n+1}(Y)$ constructed on the basis leaves a_{n+1}^0 and a_{n+1}^1 of $B_{n+1}^{n+1}(X)$ belong to the leaf a_{n+1}^2 of the Bol n-parallelotope $B_{n+1}^{n+1}(X)$. It follows from Lemma 4.5.21 that it is sufficient to consider the case when X and Y lie in one and the same r-dimensional submanifold cut in the leaf a_{n+1}^0 by the leaves of the foliations λ_α, $\alpha=1,\ldots,n$, of the web $W(n+1,n,r)$.

Suppose that the points X and Y lie in an r-dimensional intersection of the leaf a_{n+1}^0 and the submanifold $M_{\alpha\beta}=\bigcap\limits_{\gamma\neq\alpha,\beta} a_\gamma$:

$$X, Y \in a_{n+1}^0 \cap M_{\alpha\beta}.$$

It follows from Lemma 4.5.14 that the Bol n-parallelotopes $B_{n+1}^{n+1}(X)$ and $B_{n+1}^{n+1}(Y)$ intersect $M_{\alpha\beta}$ in Bol 2-parallelotopes $B_{n+1}^3(X)$ with the vertices $X, X_\alpha, X_\beta, X_{\alpha\beta}$ and $B_{n+1}^3(Y)$ with vertices $Y, Y_\alpha, Y_\beta, Y_{\alpha\beta}$ respectively. In addition, we have

$$X_\alpha, X_\beta, Y_\alpha, Y_\beta \in a_{n+1}^1 \cap M_{\alpha\beta},$$

$$X_{\alpha\beta} \in a_{n+1}^2 \cap M_{\alpha\beta}$$

(see Figure 4.17).

Since the 3-subweb $[n+1,\alpha,\beta]$ of the web $B_{n+1}(n+1,n,r)$ cut in $M_{\alpha\beta}$ is a Bol web $B_{n+1}(3,2,r)$, then $Y_{\alpha\beta}\in a_{n+1}^2\cap M_{\alpha\beta}$,i.e., on the three-web $[n+1,\alpha,\beta]$ the Bol figure (B_m) is closed. Consequently, by Lemma 4.5.21, we obtain that all the vertices of height 2 of the Bol n-parallelotope $B_{n+1}^{n+1}(Y)$ lie in the leaf a_{n+1}^2 of the Bol n-parallelotope $B_{n+1}^{n+1}(X)$. ■

Theorem 4.5.23 *If a web $W(n+1,n,r)$ is a Bol web $B_{n+1}(n+1,n,r)$, then all the vertices of height m, m fixed, of any its Bol n-parallelotopes $B_{n+1}^{n+1}(X)$ lie in one and the same leaf a_{n+1}^m of the foliation λ_{n+1} and the leaf a_{n+1}^m is uniquely determined by the choice of the leaves a_{n+1}^0 and a_{n+1}^1 and does not depend on the choice of a point X in the leaf a_{n+1}^0 where a_{n+1}^0 and a_{n+1}^1 are the basis leaves of the Bol n-parallelotope $B_{n+1}^{n+1}(X)$ and X is its base.*

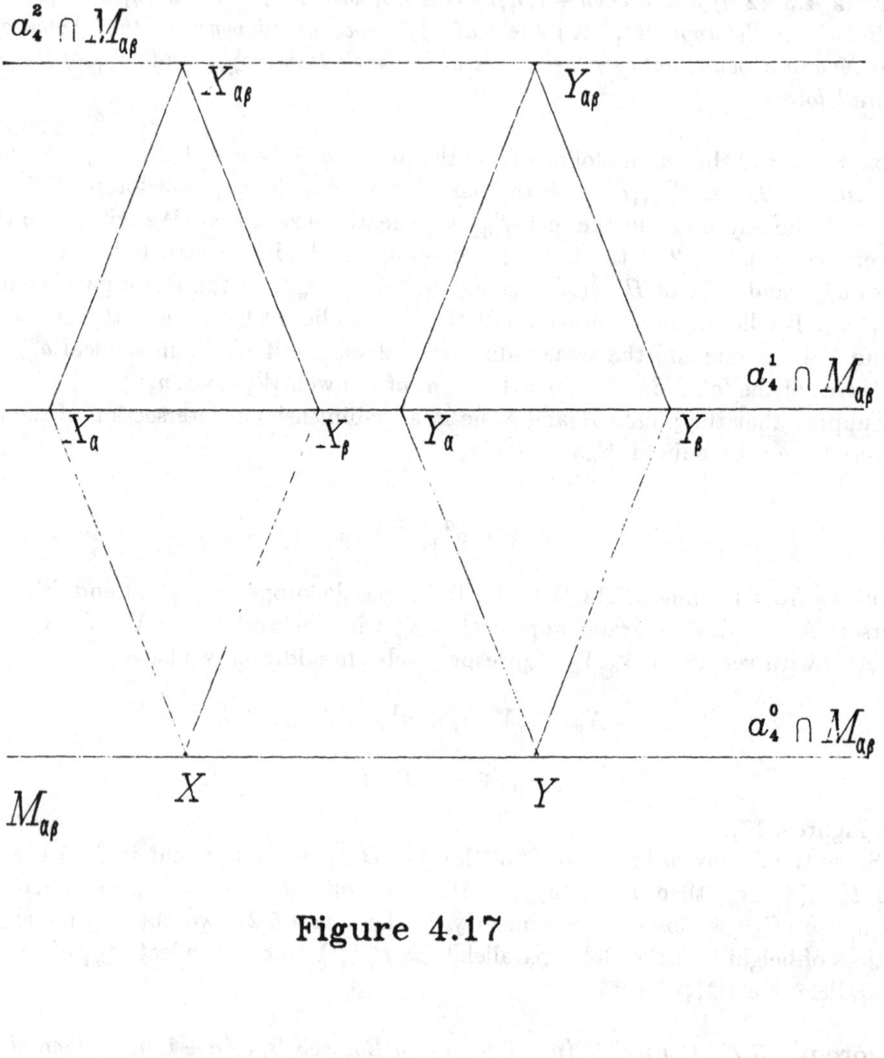

Figure 4.17

Proof. We will carry out the proof using induction on m. For $m = 0$ the theorem's statement is obvious. For $m = 1$ it follows from the definition of the Bol n-parallelotope $B_{n+1}^{n+1}(X)$. If $m = 2$, the statement was proved in Lemma 4.5.22.

Suppose now that the theorem is valid for all vertices of height $l, l \leq m+1; m > 0$. We prove that it is valid for vertices of height $m+2$ of the Bol n-parallelotope $B_{n+1}^{n+1}(X)$ provided that $m + 2 \leq n$.

Choose any two vertices $X_{\alpha_1 \ldots \alpha_{m+2}}$ and $X_{\beta_1 \ldots \beta_{m+2}}$ of height $m + 2$ of the Bol n-parallelotope $B_{n+1}^{n+1}(X)$. Consider the Bol n-parallelotope $B_{n+1}^{n+1}(X_{\alpha_1 \ldots \alpha_m})$ with the basis leaves a_{n+1}^m, a_{n+1}^{n+1} and the base $X_{\alpha_1 \ldots \alpha_m} \in a_{n+1}^m$. The point $X_{\alpha_1 \ldots \alpha_{m+2}}$ will be a vertex of height $m+2$ of this n-parallelotope. Let a_{n+1}^{n+1} be a leaf of the foliation λ_{n+1} passing through this point. Similarly the point $x_{\beta_1\ldots\beta_m}$ will be a vertex of height 2 of the n-parallelotope $B_{n+1}^{n+1}(X_{\beta_1\ldots\beta_m})$ with the same basis leaves. From Lemma 4.5.22 we obtain that $X_{\beta_1\ldots\beta_m} \in a_{n+1}^{m+2}$, and the leaf a_{n+1}^{m+2} depends only on the choice of the leaves a_{n+1}^{m} and a_{n+1}^{m+1}. According to the induction hypothesis, the leaves a_{n+1}^{m} and a_{n+1}^{m+1} are uniquely determined by the choice of the leaves a_{n+1}^{0} and a_{n+1}^{1}. Therefore, the leaf a_{n+1}^{m+2} also depends only on their choice. Thus, all the vertices of height $m+2$ of the Bol n-parallelotope $B_{n+1}^{n+1}(X)$ lie in one and the same leaf a_{n+1}^{m+2} of the foliation λ_{n+1} and this leaf is uniquely determined by the choice of the leaves a_{n+1}^{0} and a_{n+1}^{1} and does not depend on the choice of the point X in the leaf a_{n+1}^{0}. ■

Definition 4.5.24 A *vertex of height* m of a Bol figure $B_{n+1}^{m+1}(S_{n+1})$ is a vertex of height m of a Bol n-parallelotope $B_{n+1}^{m+1}(X_\alpha)$, $\alpha = 1, \ldots, n$, which is a part of the Bol figure $B_{n+1}^{n+1}(S_{n+1})$. A *vertex of height* m of a Bol figure $B_{n+1}^{n+1}(X, Y)$ is a vertex of height m of the Bol n-parallelotope $B_{n+1}^{n+1}(X)$ or $B_{n+1}^{n+1}(Y)$ forming $B_{n+1}^{n+1}(X, Y)$.

Corollary 4.5.25 *If a web $W(n+1, n, r)$ is a Bol web $B_{n+1}(n+1, n, r)$, then all the vertices of height m, m fixed, of any its Bol figure $B_{n+1}^{n+1}(S_{n+1})$ (or $B_{n+1}^{n+1}(X, Y)$) lie in one and the same leaf a_{n+1}^{m} of the foliation λ_{n+1} and the leaf a_{n+1}^{m} is uniquely determined by a choice of the leaves a_{n+1}^{0} and a_{n+1}^{1} of the figure $B_{n+1}^{n+1}(S_{n+1})$ (or respectively $B_{n+1}^{n+1}(X, Y)$).*

Proof. It follows from Theorem 4.5.23 and Definition 4.5.24. ■

Corollary 4.5.26 *If a web $W(n+1, n, r)$ is a Bol web $B_{n+1}(n+1, n, r)$, then on it the Bol closure conditions (B_{n+1}^{n+1}) and $(B_{n+1}^{n+1})'$ hold.*

Proof. It follows from Theorem 4.5.19 and Corollary 4.5.25 if one takes $m = n$. ■

Theorem 4.5.27 *If a web $W(n+1, n, r)$ is a Bol web $B_{n+1}(n+1, n, r)$ then all the vertices of any its Bol n-parallelotopes $B_{n+1}^{n+1}(X)$ lie in one and the same transversally geodesic surface of the web.*

Proof. Consider the vertices X_α, $\alpha = 1, \ldots, n$, of height 1 of the Bol n-parallelotope $B_{n+1}^{n+1}(X)$. The points X_α correspond to one another in the mapping φ_{n+1} established among the r-dimensional intersections $M_\beta = \bigcap_{\alpha \neq \beta} x_\alpha$ by the foliation λ_{n+1} (see Section 1.2), since they are obtained as a result of the intersection of the submanifolds $M_\beta, \beta = 1, \ldots, n$, with the leaf a_{n+1}^{1} of the foliation (see the beginning of the subsection 2 of this section and Section 1.2). The points X and X_α, α fixed, define in the affine connection γ_{n+1} (see Section 1.3) a unique geodesic line passing through them. The lines corresponding to this geodesic line in the correspondence φ_{n+1} will also be geodesic lines (Proposition 1.9.1), and they pass through the points $X_\beta, \beta = 1, \ldots, n; \beta \neq \alpha$. Since the web $B_{n+1}(n+1, n, r)$ is transversally geodesic (see Corollary 4.5.4), there exists an n-dimensional transversally geodesic surface V^n defined by the geodesic lines corresponding to one another. Moreover, these geodesic lines belong to the surface V^n, and therefore, V^n contains the points $X, X_\alpha, \alpha = 1, \ldots, n$.

By Lemmas 4.5.20, 4.5.22, Theorem 4.5.23, and the properties of transversally geodesic surfaces of a web $W(3,2,r)$, we obtain that any vertex $X_{\alpha_1 \ldots \alpha_{m+2}}$ of height $m+2$ of a Bol n-parallelotope $B^{n+1}_{n+1}(X)$ lies on a two-dimensional surface V^2 defined by the points $X_{\alpha_1 \ldots \alpha_m}, X_{\alpha_1 \ldots \alpha_m \alpha_{m+1}}$ and $X_{\alpha_1 \ldots \alpha_m \alpha_{m+2}}$. In addition, the surface V^2 is the intersection of V^n and a manifold carrying the 3-subweb $[n+1, \alpha_{m+1}, \alpha_{m+2}]$,i.e., V^2 is a transversally geodesic surface of this subweb. The geodecic lines joining the points $X_{\alpha_1 \ldots \alpha_m}$ and $X_{\alpha_1 \ldots \alpha_m \alpha_{m+1}}, X_{\alpha_1 \ldots \alpha_m}$ and $X_{\alpha_1 \ldots \alpha_m \alpha_{m+2}}, X_{\alpha_1 \ldots \alpha_m \alpha_{m+1}}$ and $X_{\alpha_1 \ldots \alpha_{m+2}}$, $X_{\alpha_1 \ldots \alpha_m \alpha_{m+2}}$ and $X_{\alpha_1 \ldots \alpha_{m+2}}$ lie on the surface V^2 and therefore they belong to V^n. Thus, any vertex of height $m+2$ of the Bol n-parallelotope $B^{n+1}_{n+1}(X)$ belongs to the surface V^n. Since the theorem is valid for $m = 0, 1$, then, by induction, we obtain that it is valid for any $m+2, m+2 \leq n$, i.e., all the vertices of the Bol n-parallelotope $B^{n+1}_{n+1}(X)$ lie in one and the same transversally geodesic surface of the web. ■

4.5.4 An Analytic Characteristic of the Bol Closure Condition (B^{n+1}_{n+1})

Consider a sufficiently small Bol figure $B^{n+1}_{n+1}(S_{n+1})$ with basis leaves a^0_{n+1}, a^1_{n+1} and vertices X^α of a simplex S_{n+1} which are located in the leaf a^0_{n+1}:

$$X^\alpha \in a^0_{n+1}, \quad \alpha = 1, \ldots, n, \tag{4.5.15}$$

and

$$X^\alpha = x_\alpha \cap \Big(\bigcap_{\substack{\beta=1 \\ (\beta \neq \alpha)}}^{n} a_\beta \Big) \tag{4.5.16}$$

(see Definition 4.5.15 and formula (4.5.5)). Denote by X^α_β the vertices of height 1 of the Bol n-parallelotope $B^{n+1}_{n+1}(X^\alpha)$ which are located in the leaf a^1_{n+1}:

$$X^\alpha_\beta \in a^1_{n+1}, \quad \alpha, \beta = 1, \ldots, n. \tag{4.5.17}$$

Using (4.5.17), from (4.5.2) we obtain

$$X^\alpha_\beta = a^1 \cap x_\alpha \cap \Big(\bigcap_{\substack{\gamma=1 \\ (\gamma \neq \alpha, \beta)}}^{n} a_\gamma \Big), \quad \alpha \neq \beta, \ \alpha, \beta = 1, \ldots, n, \tag{4.5.18}$$

$$X^\alpha_\alpha = a^1 \cap \Big(\bigcap_{\substack{\gamma=1 \\ (\gamma \neq \alpha)}}^{n} a_\gamma \Big). \tag{4.5.19}$$

Let y_β^α be a leaf of the foliation λ_β passing through the point X_β^α and $\bar{X}^\alpha$ be the summit of the Bol n-parallelotope $B_{n+1}^{n+1}(X^\alpha)$. Then from (4.5.16) we obtain

$$\bar{X}^\alpha = \bigcap_{\beta=1}^{n} y_\beta^\alpha. \tag{4.5.20}$$

Since $X_\beta^\alpha \in y_\beta^\alpha$, then by means of (4.5.18) and (4.5.19) we have

$$X_\beta^\alpha = y_\beta^\alpha \cap x_\alpha \cap \Big(\bigcap_{\substack{\gamma=1 \\ (\gamma\neq\alpha,\beta)}}^{n} a_\gamma \Big), \quad \alpha \neq \beta, \;\; \alpha, \beta = 1, \ldots, n, \tag{4.5.21}$$

$$X_\alpha^\alpha = y_\alpha^\alpha \cap \Big(\bigcap_{\substack{\gamma=1 \\ (\gamma\neq\alpha)}}^{n} a_\gamma \Big). \tag{4.5.22}$$

Consider a local coordinate n-quasigroup f of a web $W(n+1, n, r)$ (see Section 3.1). Using (4.5.16), we can write the condition of incidence of the points X^α and the leaf a_{n+1}^0 in the form

$$f(x_1, a_2, \ldots, a_n) = f(a_1, x_2, a_3, \ldots, n) = \ldots = f(a_1, \ldots, a_{n-1}, x_n). \tag{4.5.23}$$

Since the points X_β^α are determined by the formulae (4.5.21) and (4.5.22), the conditions (4.5.13) can be written in terms of the operation f in the form:

$$\begin{aligned} & f(y_1^1, a_2, \ldots, a_n) = f(x_1, y_2^1, a_3, \ldots, a_n) = \ldots = f(x_1, a_2, \ldots, a_{n-1}, y_n^1) \\ = \;& f(y_1^2, x_2, a_3, \ldots, a_n) = f(a_1, y_2^2, a_3, \ldots, a_n) = \ldots = f(a_1, x_2, a_3, \ldots, a_{n-1}, y_n^2) \\ = \;& f(y_1^n, a_2, \ldots, a_{n-1}, x_n) = f(a_1, y_2^n, a_3, \ldots, a_{n-1}, x_n) = \ldots = f(a_1, \ldots, a_{n-1}, y_n^n). \end{aligned} \tag{4.5.24}$$

On the web $W(n+1, n, r)$ the Bol closure condition (B_{n+1}^{n+1}) holds if and only if all the vertices $\bar{X}^\alpha$ lie in one and the same leaf $\bar{a}_{n+1}$ of the foliation λ_{n+1} :

$$\bar{X}^\alpha \in \bar{a}_{n+1}, \quad \alpha = 1, \ldots, n. \tag{4.5.25}$$

Since the points $\bar{X}^\alpha$ are determined by formulas (4.5.20), then, by means of the operation f, condition (4.5.25) can be written as follows:

$$f(y_1^1, y_2^1, \ldots, y_n^1) = f(y_1^2, y_2^2, \ldots, y_n^2) = \ldots = f(y_1^n, y_2^n, \ldots, y_n^n). \tag{4.5.26}$$

Definition 4.5.28 If in an n-quasigroup f the identities (4.5.23) and (4.5.24) imply the identities (4.5.25), we will say that in this n-quasigroup the *Bol conditional identities* B_{n+1} hold.

Our previous considerations show that the following theorem is valid.

Theorem 4.5.29 *On a web $W(n+1, n, r)$ the Bol closure condition (B^{n+1}_{n+1}) holds if and only if in a coordinate n-quasigroup f determined by the web the Bol conditional identities B_{n+1} hold.* ■

Consider now leaves a_α cutting in the leaf a^0_{n+1} a simplex S_{n+1}. There are n leaves a_α and they belong to different foliations λ_α. Therefore, they have a unique common point A. The point A uniquely determines a leaf a_{n+1} of the foliation λ_{n+1} passing through it. Therefore, $A = \bigcap_{\alpha=1}^{n} a_\alpha, A \in a_{n+1}$. Using the operation f, we can write this condition in the form:

$$a_{n+1} = f(a_1, a_2, \ldots, a_n). \tag{4.5.27}$$

Consider a coordinate n-loop $L(A)$ associated with the point A (see Section 3.1). The operation in $L(A)$ is given by the following formula (see (3.1.8):

$$\begin{aligned} w_{n+1} &= F(w_1, w_2, \ldots, w_n) = f(z_1, z_2, \ldots, z_n) \\ &= f(f_1(w_1, a_2, \ldots, a_n), f_2(a_1, w_2, a_3, \ldots, a_n), \ldots, f_n(a_1, \ldots, a_{n-1}, w_n)), \end{aligned}$$

where f_α is the αth inverse operation for f and

$$z_\alpha = f_\alpha(a_1, \ldots, a_{\alpha-1}, w_\alpha, a_{\alpha+1}, \ldots, a_n).$$

The unit of the n-loop $L(A)$ is the leaf a_{n+1}. Since

$$w_\alpha = f(a_1, \ldots, a_{\alpha-1}, z_\alpha, a_{\alpha+1}, \ldots, a_n), \tag{4.5.28}$$

then for $z_\alpha = a_\alpha$ from (4.5.27) and (4.5.28) we obtain $w_\alpha = a_{n+1} = e$. Using this, we will write the identities (4.5.23), (4.5.24), and (4.5.26) in terms of the operation F in the n-loop $L(A)$. Denote

$$\begin{aligned} f(a_1, \ldots, a_{\alpha-1}, x_\alpha, a_{\alpha+1}, \ldots, a_n) &= u_\alpha, \\ f(a_1, \ldots, a_{\alpha-1}, y^\beta_\alpha, a_{\alpha+1}, \ldots, a_n) &= v^\beta_\alpha. \end{aligned}$$

Then relations (4.5.23), (4.5.24), and (4.5.26) can be written respectively in the form:

$$u_1 = u_2 = \ldots = u_n, \tag{4.5.29}$$

$$\begin{aligned} v^1_1 &= F(u_1, v^1_2, e, \ldots, e) = \ldots = F(u_1, e, \ldots, e, v^1_n) \\ &= F(v^2_1, u_2, e, \ldots, e) = v^2_2 = \ldots = F(e, u_2, e, \ldots, e, v^2_n) = \ldots \\ &= F(v^n_1, e, \ldots, u_n) = F(e, v^n_2, e, \ldots, u_n) = \ldots = v^n_n, \end{aligned} \tag{4.5.30}$$

$$F(v^1_1, v^1_2, \ldots, v^1_n) = F(v^2_1, v^2_2, \ldots, v^2_n) = \ldots = F(v^n_1, v^n_2, \ldots, v^n_n). \tag{4.5.31}$$

It follows from (4.5.29) and (4.5.30) that u_α and v^α_α do not depend on the index α. We thus denote u_α by u and v^α_α by v. Using the operations F_α inverse for F, from (4.5.30) we can express $v^\beta_\alpha, \alpha \neq \beta$ as follows:

$$v^\beta_\alpha = F_\alpha(e, \ldots, e, v, e, \ldots, e, u, e, \ldots, e), \quad \alpha \neq \beta, \tag{4.5.32}$$

where v is in αth place, u is in βth place, and the unit e is in all remaining places. Denote this expression by $F^{\alpha\beta}_\alpha(v, u)$. Substituting (4.5.32) in more equations (4.5.31) which we have not already used, we obtain

$$\begin{aligned} & F(v, F^{21}_2(v,u), F^{31}_3(v,u), \ldots, F^{n1}_n(v,u)) \\ = \; & F(F^{12}_1(v,u), v, F^{32}_3(v,u), \ldots, F^{n2}_n(v,u)) = \ldots \\ = \; & F(F^{1n}_1(v,u), F^{2n}_2(v,u), \ldots, F^{n-1,n}_{n-1}(v,u), v). \end{aligned} \tag{4.5.33}$$

Therefore, we have proved that a closure of the Bol figure $B^{n+1}_{n+1}(S_{n+1})$ is equivalent to the identities (4.5.33) in the n-loop $L(A)$ associated with this figure, i.e. the following theorem holds :

Theorem 4.5.30 *On a web $W(n+1, n, r)$ the Bol closure condition (B^{n+1}_{n+1}) holds if and only if in any l.d. n-loop $L(A)$ associated with the web the identity* (4.5.33) *holds.* ∎

Let

$$\phi_\delta = F(F^{1\delta}_1(v,u), \ldots, F^{\delta-1,\delta}_{\delta-1}(v,u), \ldots, F^{n\delta}_n(v,u)) \tag{4.5.34}$$

be a common δth term of the identity (4.5.33). To get its Taylor expansion, in addition to the expansions (3.2.6), we need to have similar expansions for the inverse operations F_α. By the definition of inverse operations, we have

$$F(u_1, \ldots, u_{\alpha-1}, F_\alpha(u_1, \ldots, u_{\alpha-1}, u_{n+1}, u_{\alpha+1}, \ldots, u_n), u_{\alpha+1}, \ldots, u_n) = u_{n+1}. \tag{4.5.35}$$

We will look for the Taylor expansions of the functions F_α in the form

$$\begin{aligned} \underset{\alpha}{u}^i &= F^i_\alpha(\underset{1}{u}^{j_1}, \ldots, \underset{\alpha-1}{u}^{j_{\alpha-1}}, \underset{n+1}{u}^{j_{n+1}}, \underset{\alpha+1}{u}^{j_{\alpha+1}}, \ldots, \underset{n}{u}^{j_n}) \\ &= \overset{\alpha}{\lambda}{}^i + \sum_{\xi=1}^{n+1} \underset{\xi}{\overset{\alpha}{\lambda}}{}^i_j \underset{\xi}{u}^j + \frac{1}{2} \sum_{\xi,\eta=1}^{n+1} \underset{\xi\eta}{\overset{\alpha}{\lambda}}{}^i_{jk} \underset{\xi}{u}^j \underset{\eta}{u}^k + \frac{1}{6} \sum_{\xi,\eta,\zeta}^{n+1} \underset{\xi\eta\zeta}{\overset{\alpha}{\lambda}}{}^i_{jkl} \underset{\xi}{u}^j \underset{\eta}{u}^k \underset{\zeta}{u}^l + o(\rho^3), \end{aligned} \tag{4.5.36}$$

where $\xi, \eta, \zeta \neq \alpha; \rho = \max |\underset{\xi}{u}^i|, |\dfrac{o(t)}{t}| \to 0$, for $t \to 0$,

$$\underset{\xi\eta}{\overset{\alpha}{\lambda}}{}^i_{jk} = \underset{\eta\xi}{\overset{\alpha}{\lambda}}{}^i_{kj}, \quad \underset{\xi\eta\zeta}{\overset{\alpha}{\lambda}}{}^i_{jkl} = \underset{\sigma(\xi\eta\zeta)}{\overset{\alpha}{\lambda}}{}^i_{\sigma(jkl)}, \tag{4.5.37}$$

and σ is an arbitrary substitution of the indices. Substituting (3.2.6) and (4.5.36) into (4.5.35) and using (3.2.7) and (4.5.37), we obtain

$$
\begin{aligned}
\overset{\alpha}{\lambda}{}^i &= 0; \quad \underset{n+1}{\overset{\alpha}{\lambda}}{}^i_j = \delta^i_j; \quad \underset{\beta}{\overset{\alpha}{\lambda}}{}^i_j = -\delta^i_j;\\
\underset{n+1,n+1}{\overset{\alpha}{\lambda}}{}^i_{jk} &= 0; \quad \underset{\beta,n+1}{\overset{\alpha}{\lambda}}{}^i_{jk} = -\underset{\beta\alpha}{\lambda}{}^i_{jk}, \ \underset{\beta\gamma}{\overset{\alpha}{\lambda}}{}^i_{jk} = -\underset{\beta\gamma}{\lambda}{}^i_{jk} + \underset{\beta\alpha}{\lambda}{}^i_{jk} + \underset{\alpha\gamma}{\lambda}{}^i_{jk};\\
\underset{n+1,n+1,n+1}{\overset{\alpha}{\lambda}}{}^i_{jkl} &= 0; \quad \underset{\beta,n+1,n+1}{\overset{\alpha}{\lambda}}{}^i_{jkl} = -\underset{\beta\alpha\alpha}{\lambda}{}^i_{jkl};\\
\underset{\beta,\gamma,n+1}{\overset{\alpha}{\lambda}}{}^i_{jkl} &= -\underset{\beta\gamma\alpha}{\lambda}{}^i_{jkl} + \underset{\beta\alpha\alpha}{\lambda}{}^i_{jkl} + \underset{\alpha\gamma\alpha}{\lambda}{}^i_{jkl} + \underset{\beta\alpha}{\lambda}{}^i_{jm}\underset{\gamma\alpha}{\lambda}{}^m_{kl} + \underset{\gamma\alpha}{\lambda}{}^i_{km}\underset{\beta\alpha}{\lambda}{}^m_{jl};\\
\underset{\beta\gamma\delta}{\overset{\alpha}{\lambda}}{}^i_{jkl} &= -\underset{\beta\gamma\delta}{\lambda}{}^i_{jkl} + \underset{\beta\gamma\alpha}{\lambda}{}^i_{jkl} + \underset{\beta\alpha\delta}{\lambda}{}^i_{jkl} + \underset{\alpha\gamma\delta}{\lambda}{}^i_{jkl}\\
&\quad - \underset{\beta\alpha\alpha}{\lambda}{}^i_{jkl} - \underset{\alpha\gamma\alpha}{\lambda}{}^i_{jkl} - \underset{\alpha\alpha\delta}{\lambda}{}^i_{jkl} - \underset{\beta\alpha}{\lambda}{}^i_{jm}\big(\underset{\gamma\alpha}{\lambda}{}^m_{kl} + \underset{\alpha\delta}{\lambda}{}^m_{kl} - \underset{\gamma\delta}{\lambda}{}^m_{kl}\big)\\
&\quad - \underset{\gamma\alpha}{\lambda}{}^i_{km}\big(\underset{\beta\alpha}{\lambda}{}^m_{jl} + \underset{\alpha\delta}{\lambda}{}^m_{jl} - \underset{\beta\delta}{\lambda}{}^m_{jl}\big) - \underset{\delta\alpha}{\lambda}{}^i_{lm}\big(\underset{\beta\alpha}{\lambda}{}^m_{jk} + \underset{\alpha\gamma}{\lambda}{}^m_{jk} - \underset{\beta\gamma}{\lambda}{}^m_{jk}\big),
\end{aligned}
\tag{4.5.38}
$$

where $\beta, \gamma, \delta \neq \alpha$, $\beta, \gamma, \delta = 1, \dots, n$.

Now we will substitute (3.2.6) and (4.5.36) into (4.5.34). Using (3.2.7), (4.5.37), and (4.5.38), we obtain

$$
\begin{aligned}
\phi^i_\delta &= v^i + \sum_{\substack{\alpha=1\\(\alpha\neq\delta)}}^{n} \Big[v^i - u^i - \underset{\delta\alpha}{\lambda}{}^i_{jk}u^j v^k + \underset{\delta\alpha}{\lambda}{}^i_{(jk)}u^j v^k - \frac{1}{2}\underset{\delta\alpha\alpha}{\lambda}{}^i_{jkl}u^j v^k v^l\\
&\quad + \Big(\underset{\delta\alpha}{\lambda}{}^i_{(j|m|}\underset{\delta\alpha}{\lambda}{}^m_{k)l} + \underset{\delta\alpha\alpha}{\lambda}{}^i_{(jk)l} - \frac{1}{2}\underset{\delta\delta\alpha}{\lambda}{}^i_{jkl}\Big)u^j u^k v^l + \Big(\frac{1}{2}\underset{\delta\delta\alpha}{\lambda}{}^i_{(jkl)} - \frac{1}{2}\underset{\delta\alpha\alpha}{\lambda}{}^i_{(jkl)} - \underset{\delta\alpha}{\lambda}{}^i_{(j|m|}\underset{\delta\alpha}{\lambda}{}^m_{kl)}\Big)u^j u^k u^l\Big]\\
&\quad + \frac{1}{2}\sum_{\substack{\alpha,\beta=1\\(\alpha,\beta\neq\delta)}}^{n} \underset{\alpha\beta}{\lambda}{}^i_{jk}(v^j - u^j - \underset{\delta\alpha}{\lambda}{}^j_{ql}u^q v^l + \underset{\delta\alpha}{\lambda}{}^j_{(ql)}u^q v^l)(v^k - u^k - \underset{\delta\beta}{\lambda}{}^k_{mp}u^m v^p + \underset{\delta\beta}{\lambda}{}^k_{(mp)}u^m u^p)\\
&\quad + \sum_{\substack{\alpha=1\\(\alpha\neq\delta)}}^{n} \underset{\alpha\delta}{\lambda}{}^i_{jk}(v^j - u^j - \underset{\delta\alpha}{\lambda}{}^j_{ql}u^q v^l + \underset{\delta\alpha}{\lambda}{}^j_{(ql)}u^q u^l)v^k + \frac{1}{6}\sum_{\substack{\alpha,\beta,\gamma=1\\(\alpha,\beta,\gamma\neq\delta)}}^{n} \underset{\alpha\beta\gamma}{\lambda}{}^i_{jkl}(v^j - u^j)(v^k - u^k)(v^l -\\
&\quad + \frac{1}{2}\sum_{\alpha,\beta=1}^{n} \underset{\alpha\beta\delta}{\lambda}{}^i_{jkl}(v^j - u^j)(v^k - u^k)v^l + \frac{1}{2}\sum_{\substack{\alpha=1\\(\alpha\neq\delta)}}^{n} \underset{\alpha\delta\delta}{\lambda}{}^i_{jkl}(v^j - u^j)u^k v^l + o(\rho^3)\\
&= nv^i - (n-1)u^i + \frac{1}{2}\sum_{\alpha,\beta=1}^{n} \underset{\alpha\beta}{\lambda}{}^i_{(jk)}u^j u^k - \sum_{\alpha,\beta=1}^{n} \underset{\alpha\beta}{\lambda}{}^i_{(jk)}u^j v^k\\
&\quad + \frac{1}{2}\sum_{\alpha,\beta=1}^{n} \underset{\alpha\beta}{\lambda}{}^i_{(jk)}v^j v^k + \frac{1}{6}\sum_{\alpha,\beta,\gamma=1}^{n} \underset{\alpha\beta\gamma}{\lambda}{}^i_{(jkl)}v^j v^k v^l\\
&\quad + \Big[-\frac{1}{6}\sum_{\alpha,\beta,\gamma=1}^{n} \underset{\alpha\beta\gamma}{\lambda}{}^i_{(jkl)} + \frac{1}{2}\sum_{\alpha,\beta=1}^{n} \underset{\alpha\beta\delta}{\lambda}{}^i_{(jkl)} - \frac{1}{2}\sum_{\alpha=1}^{n} \underset{\alpha\alpha\delta}{\lambda}{}^i_{(jkl)} - \sum_{\alpha,\beta=1}^{n} \underset{\alpha\beta}{\lambda}{}^i_{(j|m|}\underset{\delta\beta}{\lambda}{}^m_{kl)}\Big]u^j u^k u^l\\
&\quad + \Big[\frac{1}{2}\sum_{\alpha,\beta,\gamma=1}^{n} \underset{\alpha\beta\gamma}{\lambda}{}^i_{(jk)l} - \sum_{\beta,\gamma=1}^{n} \underset{\delta\beta\gamma}{\lambda}{}^i_{(jk)l} + \sum_{\alpha=1}^{n} \underset{\delta\alpha\alpha}{\lambda}{}^i_{(jk)l} + \sum_{\alpha,\beta=1}^{n} \Big(\underset{\alpha\beta}{\lambda}{}^i_{lm}\underset{\delta\beta}{\lambda}{}^m_{(jk)} + \underset{\beta\alpha}{\lambda}{}^i_{(j|m|}\underset{\delta\alpha}{\lambda}{}^m_{k)l}\Big)\Big]u^j u^k v^l
\end{aligned}
$$

$$
+\Big[-\frac{1}{2}\sum_{\alpha,\beta,\gamma=1}^{n}\underset{\alpha\beta\gamma}{\lambda}{}^{i}_{j(kl)}+\frac{1}{2}\sum_{\alpha,\beta=1}^{n}\underset{\delta\alpha\beta}{\lambda}{}^{i}_{j(kl)}-\frac{1}{2}\sum_{\alpha=1}^{n}\underset{\delta\alpha\alpha}{\lambda}{}^{i}_{j(kl)}
$$

$$
-\frac{1}{2}\sum_{\alpha,\beta=1}^{n}\Big(\underset{\alpha\beta}{\lambda}{}^{i}_{m(k}\underset{\delta\alpha}{\lambda}{}^{m}_{|j|l)}+\underset{\beta\alpha}{\lambda}{}^{i}_{m(k}\underset{\delta\beta}{\lambda}{}^{m}_{|j|l)}\Big)\Big]u^{j}v^{k}v^{l}+o(\rho^{3}). \tag{4.5.39}
$$

If on a web $W(n+1,n,r)$ the Bol closure condition (B^{n+1}_{n+1}) holds, then in the n-loop $L(A)$ the identities (4.5.33) are satisfied and consequently the expression (4.5.34) of ϕ_δ must not depend on δ. Equating the expressions (4.5.39) written for δ and ε, $\delta\neq\varepsilon$, and using the fact that the relation obtained is an identity, we arrive to the following conditions:

$$
\sum_{\alpha,\beta=1}^{n}\Big(\underset{\alpha\beta}{\lambda}{}^{i}_{(jkl)}-2\underset{\alpha\beta}{\lambda}{}^{i}_{(j|m|}\underset{\delta\beta}{\lambda}{}^{m}_{kl)}\Big)=\sum_{\alpha,\beta=1}^{n}\Big(\underset{\alpha\beta\varepsilon}{\lambda}{}^{i}_{(jkl)}-2\underset{\alpha\beta}{\lambda}{}^{i}_{(j|m|}\underset{\varepsilon\beta}{\lambda}{}^{m}_{kl)}\Big),\quad \alpha\neq\beta, \tag{4.5.40}
$$

$$
\begin{aligned}
&\textstyle\sum_{\alpha,\beta=1}^{n}\Big(\underset{\delta\alpha\beta}{\lambda}{}^{i}_{(jk)l}-\underset{\alpha\beta}{\lambda}{}^{i}_{lm}\underset{\delta\beta}{\lambda}{}^{m}_{(jk)}-\underset{\beta\alpha}{\lambda}{}^{i}_{(j|m|}\underset{\delta\alpha}{\lambda}{}^{m}_{k)l}\Big)\\
=\;&\textstyle\sum_{\alpha,\beta=1}^{n}\Big(\underset{\varepsilon\alpha\beta}{\lambda}{}^{i}_{(jk)l}-\underset{\alpha\beta}{\lambda}{}^{i}_{lm}\underset{\varepsilon\beta}{\lambda}{}^{m}_{(jk)}-\underset{\beta\alpha}{\lambda}{}^{i}_{(j|m|}\underset{\varepsilon\alpha}{\lambda}{}^{m}_{k)l}\Big),\quad \alpha\neq\beta,
\end{aligned} \tag{4.5.41}
$$

$$
\begin{aligned}
&\textstyle\sum_{\alpha,\beta=1}^{n}\Big(\underset{\delta\alpha\beta}{\lambda}{}^{i}_{j(kl)}-\underset{\alpha\beta}{\lambda}{}^{i}_{m(k}\underset{\delta\alpha}{\lambda}{}^{m}_{|j|l)}-\underset{\beta\alpha}{\lambda}{}^{i}_{m(k}\underset{\delta\beta}{\lambda}{}^{m}_{|j|l)}\Big)\\
=\;&\textstyle\sum_{\alpha,\beta=1}^{n}\Big(\underset{\varepsilon\alpha\beta}{\lambda}{}^{i}_{j(kl)}-\underset{\alpha\beta}{\lambda}{}^{i}_{m(k}\underset{\varepsilon\alpha}{\lambda}{}^{m}_{|j|l)}-\underset{\beta\alpha}{\lambda}{}^{i}_{m(k}\underset{\varepsilon\beta}{\lambda}{}^{m}_{|j|l)}\Big),\quad \alpha\neq\beta.
\end{aligned} \tag{4.5.42}
$$

Applying the equations (3.2.7) and making a substitution of indices, we reduce relations (4.5.40)–(4.5.42) to the form:

$$
\sum_{\gamma,\delta=1}^{n}\underset{\gamma\delta\alpha}{\mu}{}^{i}_{(jkl)}=\sum_{\gamma,\delta=1}^{n}\underset{\gamma\delta\beta}{\mu}{}^{i}_{(jkl)},\quad \gamma\neq\delta, \tag{4.5.43}
$$

$$
\sum_{\gamma,\delta=1}^{n}\underset{\gamma\delta\alpha}{\mu}{}^{i}_{j(kl)}=\sum_{\gamma,\delta=1}^{n}\underset{\gamma\delta\beta}{\mu}{}^{i}_{j(kl)},\quad \gamma\neq\delta, \tag{4.5.44}
$$

$$
\sum_{\gamma,\delta=1}^{n}\underset{\gamma\delta\alpha}{\mu}{}^{i}_{(jk)l}=\sum_{\gamma,\delta=1}^{n}\underset{\gamma\delta\beta}{\mu}{}^{i}_{(jk)l)},\quad \gamma\neq\delta, \tag{4.5.45}
$$

where $\underset{\gamma\delta\alpha}{\mu}{}^{i}_{jkl}$ is determined by formulas (3.3.6). Note that by virtue of (3.2.7) and (3.3.6), for any α we have

$$
\underset{\gamma\delta\alpha}{\mu}{}^{i}_{jkl}=\underset{\delta\gamma\alpha}{\mu}{}^{i}_{kjl},\quad \gamma\neq\delta.
$$

But then

$$
\sum_{\gamma,\delta=1}^{n}\underset{\gamma\delta\alpha}{\mu}{}^{i}_{jkl}=\sum_{\substack{\gamma,\delta=1\\(\gamma<\delta)}}^{n}\Big(\underset{\gamma\delta\alpha}{\mu}{}^{i}_{jkl}+\underset{\delta\gamma\alpha}{\mu}{}^{i}_{jkl}\Big)
$$

$$= \sum_{\substack{\gamma,\delta=1\\(\gamma<\delta)}}^{n} \left[\frac{1}{2}\left(\underset{\gamma\delta\alpha}{\mu}{}^{i}_{jkl} + \underset{\gamma\delta\alpha}{\mu}{}^{i}_{kjl}\right) + \frac{1}{2}\left(\underset{\delta\gamma\alpha}{\mu}{}^{i}_{kjl} + \underset{\delta\gamma\alpha}{\mu}{}^{i}_{jkl}\right)\right]$$

$$= \sum_{\substack{\gamma,\delta=1\\(\gamma<\delta)}}^{n} \left(\underset{\gamma\delta\alpha}{\mu}{}^{i}_{(jk)l} + \underset{\delta\gamma\alpha}{\mu}{}^{i}_{(jk)l}\right) = \sum_{\gamma,\delta=1}^{n} \underset{\gamma\delta\alpha}{\mu}{}^{i}_{(jk)l}, \quad \gamma \neq \delta,$$

i.e.,

$$\sum_{\gamma,\delta=1}^{n} \underset{\gamma\delta\alpha}{\mu}{}^{i}_{jkl} = \sum_{\gamma,\delta=1}^{n} \underset{\gamma\delta\alpha}{\mu}{}^{i}_{(jk)l}, \quad \gamma \neq \delta. \tag{4.5.46}$$

Using (4.5.46), equation (4.5.45) can be rewritten in the form

$$\sum_{\gamma,\delta=1}^{n} \underset{\gamma\delta\alpha}{\mu}{}^{i}_{jkl} = \sum_{\gamma,\delta=1}^{n} \underset{\gamma\delta\beta}{\mu}{}^{i}_{jkl}, \quad \gamma \neq \delta. \tag{4.5.47}$$

It follows from this that the relations (4.5.43)-(4.5.44) are consequences of (4.5.47). This proves the following theorem:

Theorem 4.5.31 *If on a web $W(n+1,n,r)$ the Bol closure condition (B^{n+1}_{n+1}) holds, then in each its l.d. n-loop $L(A)$ the identity* (4.5.47) *holds for the coefficients of the expansions* (3.2.6) *of the closed form equations of the n-loop.* ■

The next theorem expresses the condition (4.5.47) in terms of the curvature tensor of the web $W(n+1,n,r)$.

Theorem 4.5.32 *If on a web $W(n+1,n,r)$ the Bol closure condition (B^{n+1}_{n+1}) holds, then*

$$\underset{\alpha\beta}{b}{}^{i}_{jkl} = R^{i}_{jkl}, \quad R^{i}_{j(kl)} = 0, \quad \alpha,\beta = 1,\dots,n, \tag{4.5.48}$$

where $\underset{\alpha\beta}{b}{}^{i}_{jkl}$ is the curvature tensor of the web.

Proof. First we will calculate the difference $\underset{\alpha\beta}{b}{}^{i}_{jkl} - \underset{\alpha\alpha}{b}{}^{i}_{jkl}$. Using (3.3.5), we obtain for this difference:

$$\underset{\alpha\beta}{b}{}^{i}_{jkl} - \underset{\alpha\alpha}{b}{}^{i}_{jkl} = \frac{1}{2n(n-1)} \sum_{\substack{\gamma,\delta=1\\(\gamma\neq\delta)}}^{n} \left(\underset{\gamma\delta\beta}{\mu}{}^{i}_{jkl} - \underset{\gamma\delta\alpha}{\mu}{}^{i}_{jkl}\right). \tag{4.5.49}$$

The relations (4.5.47) show that the right member of (4.5.49) vanishes, and therefore

$$\underset{\alpha\beta}{b}{}^{i}_{jkl} = \underset{\alpha\alpha}{b}{}^{i}_{jkl}, \quad \alpha,\beta = 1,\dots,n. \tag{4.5.50}$$

From (4.5.50), by means of (1.2.33), we obtain

$$\underset{\alpha\alpha}{b}{}^{i}_{jkl} = \underset{\beta\beta}{b}{}^{i}_{jkl}, \quad \alpha,\beta = 1,\dots,n. \tag{4.5.51}$$

Using (4.5.50) and (4.5.51), we find that

$$\underset{\alpha\beta}{b}{}^{i}_{jkl} = \underset{\gamma\delta}{b}{}^{i}_{jkl} \tag{4.5.52}$$

for any $\alpha, \beta, \gamma, \delta = 1, \ldots, n$. This means that the curvature tensor does not depend on indices α and β. Denote it by R^i_{jkl}. Then the relation (4.5.52) is written in the form:

$$\underset{\alpha\beta}{b}{}^{i}_{jkl} = R^i_{jkl}. \tag{4.5.53}$$

From (4.5.53) we obtain

$$\underset{\alpha\beta}{b}{}^{i}_{jkl} = \underset{\beta\alpha}{b}{}^{i}_{jkl}.$$

Using (1.2.33), we write this equation in the form

$$\underset{\alpha\beta}{b}{}^{i}_{j(kl)} = 0.$$

Then it follows from (4.5.53) that

$$R^i_{j(kl)} = 0. \quad \blacksquare$$

Corollary 4.5.33 *If a web $W(n+1,n,r)$ is a Bol web $B_{n+1}(n+1,n,r)$, then its curvature tensor satisfies the relations* (4.5.48).

Proof. This follows from Theorem 4.5.32 and Corollary 4.5.26. ■

Corollary 4.5.34 *If on a web $W(n+1,n,r)$ the Bol closure condition (B^{n+1}_{n+1}) holds, then the projection of the affine connection γ_{n+1} defined by the web on the base of the foliation λ_{n+1} is a torsion-free affine connection.*

Proof. In fact,the relations (4.5.48) imply that the structure equations (1.2.29) become

$$d\omega^i_j = \omega^k_j \wedge \omega^i_k + R^i_{jkl} \underset{n+1}{\omega}{}^k \wedge \underset{n+1}{\omega}{}^l. \tag{4.5.54}$$

The statement of the theorem follows now from (4.5.54) and (1.2.18). ■

Corollary 4.5.35 *If a web $W(n+1,n,r)$ is the Bol web $B_{n+1}(n+1,n,r)$, then the projection of the affine connection γ_{n+1} defined by the web on the base of the foliation λ_{n+1} is a torsion-free affine connection.*

Proof. It follows from Corollaries 4.5.35 and 4.5.26. ■

Proposition 4.5.36 *If a web $W(n+1,n,r)$ is the Bol web $B_{n+1}(n+1,n,r)$, then the following condition is satisfied:*

$$\underset{\alpha\beta\alpha}{s}{}^{i}_{jkl} = \underset{\alpha\beta\beta}{s}{}^{i}_{jkl}, \quad \alpha, \beta = 1, \ldots, n, \quad \alpha \neq \beta, \tag{4.5.55}$$

where $\underset{\alpha\beta\alpha}{s}{}^{i}_{jkl}$ and $\underset{\alpha\beta\beta}{s}{}^{i}_{jkl}$ are defined by (1.3.48).

Proof. The equations (4.5.55) follow from Theorem 4.5.2, Corollary 4.5.33, and relations (1.3.53). ■

Theorem 4.5.37 *If on a web $W(n+1,n,r)$ the Bol closure condition (B_{n+1}^{n+1}) and relations* (4.5.55) *hold, then the web is the Bol web $B_{n+1}(n+1,n,r)$.*

Proof. Suppose that on a web $W(n+1,n,r)$ the Bol closure condition (B_{n+1}^{n+1}) holds and the relations (4.5.55) are valid. Then, by Theorem 4.5.32, from (1.3.53) we obtain that

$$\bar{b}_{\alpha\beta}{}^{i}{}_{l(jk)} = 0, \quad \alpha, \beta = 1, \ldots, n, \quad \alpha \neq \beta.$$

From Theorem 4.5.2, this means that the web $W(n+1,n,r)$ is the Bol web $B_{n+1}(n+1,n,r)$. ■

NOTES

4.1–4.4. The results of these sections are due to the author (see [G 76] for $n \geq 3$ and [G 75a] for $n = 3$).

Reducibility of type (4.1.1) of n-quasigroups from the algebraic point of view was studied in [BS 66] (see also [Be 72] ans [S 65a, 65b]). The results obtained there refer to the case when $\alpha_{h+1}, \ldots, \alpha_n$ are successive subscripts from the set $1, 2, \ldots, n$. Theorem 4.1.4 corresponds to the situation when these successive subscripts have undergone some permutation. Such reducibility of n-quasigroups was studied from the algebraic point of view in [S 65b].

Examples of Section 4.2 are taken from [Be 72] and [BS 66] where they were considered only from the algebraic point of view.

4.5. This section is from [Ge 85b] (see also [AGe 86]).

Chapter 5

Realisation of Multicodimensional $(n+1)$-Webs

5.1 Grassmann $(n+1)$-Webs

5.1.1 Basic Definitions

In this chapter we will consider two projective realisations of multicodimensional webs $W(n+1,n,r)$: Grassmann $(n+1)$-webs and their special classes (algebraic, reducible, group, Bol, $(2n+2)$-hedral, parallelisable, etc.) and a diagonal web $W(4,3,r)$ formed by four pencils of multidimensional planes in a projective space.

We have already given the definition of a Grassmann web $GW(d,n,r)$ and an algebraic web $AW(d,n,r)$ (see Definitions 2.6.1. and 2.6.2). In this chapter we will consider only Grassmann and algebraic $(n+1)$-webs , i.e., we will assume that $d=n+1$.

It is worth to add to Definitions 2.6.1 and 2.6.2 that we will use the same term "Grassmann $(n+1)$-web" or "Algebraic $(n+1)$-web" for an $(n+1)$-web on $G(n-1,r+n-1)$ as well as for an $(n+1)$-web in P^{r+n-1} whose image is $GW(n+1,n,r)$ or respectively $AW(n+1,n,r)$. So, for example, for the second version of the Grassmann $(n+1)$-web which we have just described, we will say that the leaves of the foliation λ_ξ, $\xi=1,\dots,n,n+1$, are bundles of $(n-1)$-planes P^{n-1} with vertices located on the surfaces $U_1,\dots,U_n,U_{n+1}$ respectively.

To clarify what is a Grassmann web in P^{r+n-1} and what is its corresponding coordinate l.d. n-quasigroup, suppose that P_0^{n-1} be an $(n-1)$-plane in P^{r+n-1} which intersects the surfaces U_ξ, $\xi=1,\dots,n,n+1$, respectively in points M_ξ^0. Then any $(n-1)$-plane P^{n-1} passing through points $M_\alpha^0, \alpha=1,..,n$, from sufficiently small neighborhoods of the points M_α^0, intersects the surface U_{n+1} at the point M_{n+1}.

Therefore a mapping

$$h: U_1 \times U_2 \times \ldots \times U_n \rightarrow U_{n+1}$$

is defined, and this mapping is differentiable and invertible with respect to each argument. Thus this mapping is an l.d. n-quasigroup. The Grassmann web $GW(n+1,n,r)$ which we just defined in P^{r+n-1} corresponds to this l.d. n-quasigroup.

In what follows we will consider a domain D of the manifold of all $(n-1)$-planes P^{n-1} which consists of $(n-1)$-planes intersecting each of the surfaces U_ξ, $\xi = 1, \ldots, n+1$, at a single point M_ξ. We will also exclude from D $(n-1)$-planes for which any n of their points M_ξ of intersection with U_ξ are not in general position, i.e., they belong to an $(n-2)$-plane.

Definition 5.1.1 A domain D thus described is said to be a *domain of regularity* of a Grassmann web $GW(n+1,n,r)$.

We will define now Grassmann $(n-k+1)$-subwebs, $1 \leq k \leq n-2$, of a Grassmann web $GW(n+1,n,r)$ and l.d. $(n-k)$-quasigroups corresponding to them.

First let $k = 1$. Let M_1 be a point on the surface U_1. Consider the cones $V_2^1, \ldots, V_n^1, V_{n+1}^1$ with the common vertex M whose directrices are correspondingly the surfaces $U_2, \ldots, U_n, U_{n+1}$. The $(n-1)$-planes passing through M_1 define a differentiable and invertible mapping

$$h_1 : V_2^1 \times \ldots \times V_n^1 \to V_{n+1}^1$$

which is an l d. $(n-1)$-quasigroup. There is a web $W_1(n, n-1, r)$ associated with the $(n-1)$-quasigroup h_1 and formed by bundles of $(n-1)$-planes passing through generators $M_1M_2, \ldots, M_1M_n$, M_1M_{n+1} of the cones $V_2^1, \ldots, V_n^1, V_{n+1}^1$. It is natural to call this web an *n-subweb* of $GW(n+1,n,r)$. Similar l.d. n-quasigroups h_ξ, $\xi \neq 1$, and n-subwebs $W_\xi(n, n-1, r)$ are determined by points of each of the surfaces $U_2, \ldots, U_n, U_{n+1}$.

We will generalise this procedure to define $(n-k+1)$-subwebs of $GW(n+1,n,r)$ and corresponding l.d. $(n-k)$-quasigroups. Take the $(k-1)$-plane defined by the points $M_1, \ldots, M_k$, $1 \leq k \leq n-2$. Consider the cones $V_{k+1}^{1\ldots k}, \ldots, V_n^{1\ldots k}, V_{n+1}^{1\ldots k}$ with the common $(k-1)$-dimensional vertex δ whose directrices correspondingly are the surfaces $U_{k+1}, \ldots, U_n, U_{n+1}$. The $(n-1)$-planes passing through the $(k-1)$-plane $\delta = \{M_1, \ldots, M_k\}$ define a mapping

$$h_{1\ldots k} : V_{k+1}^{1\ldots k} \times \ldots \times V_n^{1\ldots k} \to V_{n+1}^{1\ldots k}$$

which is differentiable and invertible with respect to each argument. Therefore this mapping is an l.d. $(n-k)$-quasigroup. A web $W_{1\ldots k}(n-k+1, n-k, r)$ is associated to this quasigroup $h_{1\ldots k}$. This web is an $(n-k+1)$-*subweb* of $GW(n+1,n,r)$. It is formed by bundles of $(n-1)$-planes which pass through k-dimensional generators of the cones $V_{k+1}^{1\ldots k}, \ldots, V_n^{1\ldots k}, V_{n+1}^{1\ldots k}$ and are defined by the $(k-1)$-plane δ and the points $M_{k+1}, \ldots, M_n, M_{n+1}$. There are $\binom{n+1}{k}$ such l.d. $(n-k)$-quasigroups and $(n-k+1)$-subwebs. In particular, there are $\binom{n+1}{3}$ binary quasigroups $h_{\xi_1\ldots\xi_{n-2}}$ and 3-subwebs $W_{\xi_1\ldots\xi_{n-2}}(3,2,r)$ corresponding to them.

If we project a subweb $W_{\xi_1\ldots\xi_k}(n-k+1, n-k, r)$ from the $(k-1)$-plane δ onto an arbitrary fixed $(n+r-k-1)$-plane σ, which does not have common points with δ, we obtain on σ a Grassmann web $GW_{\xi_1\ldots\xi_k}(n-k+1, n-k, r)$ which is equivalent to the subweb $W_{\xi_1\ldots\xi_k}$ and generated by $n-k+1$ surfaces of dimension r.

In particular, if $k = n-2$ or $k = n-3$, we obtain respectively on a $(r+1)$- or $(r+2)$-plane a Grassmann three- or four-web generated by three or four surfaces of respective codimensions 1 or 2. Such Grassmann webs $GW(3,2,r)$ and $GW(4,3,r)$ were studied in [Ak 73] and [AG 74].

5.1.2 The Structure Equations of Projective Space

Now we give the structure equations for P^{r+n-1}, as this provides the model space for the study of the projective differential geometry of a family of $(n-1)$-planes in P^{r+n-1}. The index ranges

$$\begin{gathered} 1 \leq u, v, w \leq n+r, \\ 1 \leq \xi, \eta, \zeta \leq n+1, \\ 1 \leq \alpha, \beta, \gamma \leq n, \\ n+1 \leq i, j, k \leq n+r, \end{gathered}$$

will be used. As early, we shall work with real case. However, all our considerations can be carried over with the same notation to the complex case.

Definition 5.1.2 A *moving frame* in P^{r+n-1} is given by

$$F = \{A_1, \ldots, A_{n+r}\} = \{A_u\}$$

where the A_u form a basis in $\mathbf{R}^{r+n}$.

We will denote the manifold of all such frames by $\mathcal{F}(\mathbf{R}^{r+n})$. It can be identified with $\mathbf{GL}(n+r)$.

We shall frequently abuse notation and denote by X the point in P^{r+n-1} defined by the vector X from $\mathbf{R}^{r+n} - \{0\}$. Geometrically each frame $F = \{A_u\}$ defines a coordinate simplex in P^{r+n-1}. Any point $X \in P^{r+n-1}$ can be represented as $X = x^u A_u$ and the x^u are homogeneous projective coordinates of X with respect to $\{A_u\}$. A projective coordinate system is uniquely defined in P^{r+n-1} by the vertices A_u of the simplex F and the unit point $E(1, \ldots, 1)$ of this simplex which has all its coordinates equal to 1 in F.

If we study a family of $(n-1)$-planes P^{n-1} in P^{r+n-1}, it is convenient to place the points A_α of F in P^{n-1}. So,

$$P^{n-1} = \{A_1, \ldots, A_n\}.$$

Using this, we can say that there is a projection

$$\pi : \mathcal{F}(P^{r+n-1}) \to P^{r+n-1},$$

given by

$$\pi(F) = P^{n-1} = \{A_\alpha\},$$

and the fibre $\pi^{-1}(P^{n-1})$ consists of all frames

$$\tilde{F} = \{\tilde{A}_1, \ldots, \tilde{A}_{n+1}\}$$

where

$$\tilde{A}_\alpha = c_\alpha^\beta A_\beta, \quad \tilde{A}_i = c_i^u A_u, \quad i = n+1, \ldots, n+r, \tag{5.1.1}$$

and the matrices $c = (c_\alpha^\beta)$ and $C = (c_u^v)$ are non-singular:

$$\det(c) \neq 0, \ \det(C) \neq 0. \tag{5.1.2}$$

We can abbreviate equation (5.1.1) by writing F as a column vector and (c_v^u) as a matrix C. In this notation (5.1.1) becomes

$$\tilde{F} = C \cdot F. \tag{5.1.3}$$

It follows from (5.1.1) that

$$\tilde{A}_1 \wedge \ldots \wedge \tilde{A}_n = \det(c) A_1 \wedge \ldots \wedge A_n.$$

Thus, up to the homogeneity factor det (c), the fibre $\pi^{-1}(P^{n-1})$ over P^{n-1} consists of all the coordinate simplices whose coordinate $(n-1)$-plane, defined by the first n vertices, is P^{n-1}.

The frame vertices A_u may be considered as vector-valued functions

$$A_u : \mathcal{F}(P^{r+n-1}) \to \mathbf{R}^{r+n}$$

Expanding the exterior derivative dA_u at $F \in \mathcal{F}(P^{r+n-1})$ in terms of the basis determined by the coordinate simplex $\{A_u\}$, we obtain

$$dA_u = \theta_u^v A_v. \tag{5.1.4}$$

Equations (5.1.4) are the *equations of infinitesimal displacement* of the moving frame F and the matrix $\Theta = (\theta_u^v)$ gives the coefficients of this displacement. It is the *Maurer-Cartan matrix* on $\mathbf{GL}(n+1)$ (see [KN 63]). We can abbreviate equations (5.1.4) as follows:

$$dF = \Theta \cdot F. \tag{5.1.5}$$

Under a change of frame (5.1.3) the form Θ becomes

$$\tilde{\Theta} = dC \cdot C^{-1} + C\Theta C^{-1}. \tag{5.1.6}$$

In coordinate form equations (5.1.6) are

$$\left\{\begin{array}{lcl} \tilde{\theta}_\alpha^\beta & = & dc_\alpha^\gamma \tilde{c}_\gamma^\beta + c_\alpha^\gamma \theta_\gamma^\delta \tilde{c}_\delta^\beta, \\ \tilde{\theta}_\alpha^i & = & c_\alpha^\gamma \theta_\gamma^j \tilde{c}_j^i, \\ \tilde{\theta}_i^u & = & dc_i^v \tilde{c}_v^u + c_i^v \theta_v^w \tilde{c}_w^u, \end{array}\right. \tag{5.1.7}$$

where $i, j = n+1, \dots, n+r$, and $C^{-1} = (\tilde{c}_v^u)$ is the inverse matrix of $C = (c_v^u)$

Taking the exterior derivative of (5.1.5), we obtain the integrability conditions, or *Maurer-Cartan equation* (see [KN 63]):

$$d\Theta = \Theta \wedge \Theta. \tag{5.1.8}$$

In coordinate form equations (5.1.8) can be written as

$$d\theta_u^v = \theta_u^w \wedge \theta_w^v. \tag{5.1.9}$$

They represent the *structure equations* of the space P^{r+n-1}.

If we set

$$\theta_\alpha^i = \psi_\alpha^{i-n}, \tag{5.1.10}$$

then the 1-forms

$$\psi_1^1, \dots, \psi_1^r; \ \dots; \ \psi_n^1, \dots, \psi_n^r = \{\psi_\alpha^{i-n}\} \tag{5.1.11}$$

are horizontal in the fibering $\pi : \mathcal{F}(P^{r+n-1}) = P^{r+n-1}$, i.e., they vanish on the leaves $\pi^{-1}(P^{n-1})$(see [CDD 82], p.373). The horizontality of the forms (5.1.11) can be seen from the second equation of (5.1.7).

5.1.3 Specialisation of Moving Frames

In order to simplify our calculations, we will specialise our moving frames $\{A_u\}$. We have already made one step in this direction: the points A_α were placed in an $(n-1)$-plane P^{n-1}.

i) Since the points M_ξ in which an $(n-1)$-plane P^{n-1} intersects the surfaces U_ξ, $\xi = 1, \dots, n+1$, are in general position, we can place the points M_α, $\alpha = 1, \dots, n$, at the vertices A_α of a moving frame $\{A_u\}$ and normalise these vertices in such a way that the point M_{n+1} coincides with the unit point of the simplex $\{A_\alpha\}$ in P^{n-1}, i.e.,

$$M_{n+1} = \sum_{\alpha=1}^{n} A_\alpha \stackrel{\text{def}}{=} A_0.$$

Since the $(n-1)$-plane P^{n-1} does not belong to the tangent planes $T_{A_\alpha}(U_\alpha)$, the 1-forms θ_α^i, α fixed, can be taken as basis forms of the projectivised cotangent space $PT_{A_\alpha}(U_\alpha)$. Then, since when the plane P^{n-1} is fixed, the points M_α are also fixed, we have

$$\theta_\alpha^\beta = \lambda_{\alpha i}^\beta \theta_\alpha^i, \quad \beta \neq \alpha. \tag{5.1.12}$$

Now equations (5.1.4) and (5.1.12) give

$$dA_\alpha = \theta_\alpha^\alpha A_\alpha + \theta_\alpha^i \Big(A_i + \sum_{\beta \neq \alpha} \lambda_{\alpha i}^\beta A_\beta \Big). \tag{5.1.13}$$

In (5.1.13) and in the following the summation over the indices α, β, γ must be performed only if there is the summation sign.

ii) We now place the points A_i, $i = n+1, \dots, n+r$, in the plane $T_{M_{n+1}}(U_{n+1})$. Then

$$dA_0 = \theta A_0 + A_i \sum_\alpha \theta^i_\alpha, \tag{5.1.14}$$

where

$$\theta = \sum_\beta \theta^\alpha_\beta. \tag{5.1.15}$$

In the frame constructed above each of the systems

$$\theta^i_1 = 0, \dots, \theta^i_n = 0, \qquad \sum_\alpha \theta^i_\alpha = 0$$

is completely integrable. They define foliations of the bundles of $(n-1)$-planes forming a Grassmann web $GW(n+1, n, r)$. The forms θ^i_α are the basis forms of $GW(n+1, n, r)$.

Taking the exterior derivatives of (5.1.12) and applying the Cartan lemma, we obtain

$$\nabla\lambda^\beta_{\alpha i} + \lambda^\beta_{\alpha i}\lambda^\beta_{\alpha j}\theta^j_0 + \sum_{\gamma\neq\alpha,\beta}(\lambda^\beta_{\alpha i} - \lambda^\gamma_{\alpha i})(\lambda^\beta_{\alpha j} - \lambda^\beta_{\gamma j})\theta^j_\gamma + \theta^\beta_i = \lambda^\beta_{\alpha ij}\theta^j_\alpha, \tag{5.1.16}$$

where α, β and γ are distinct and

$$\lambda^\beta_{\alpha ij} = \lambda^\beta_{\alpha ji}, \quad \nabla\lambda^\beta_{\alpha i} = d\lambda^\beta_{\alpha i} - \lambda^\beta_{\alpha j}\phi^j_i;$$

$$\phi^j_i = \theta^j_j - \delta^j_i\theta, \tag{5.1.17}$$

$$\theta^i_0 = -\sum_\alpha \theta^i_\alpha. \tag{5.1.18}$$

iii) Let us consider the points

$$M_i = A_i + \lambda_i A_0, \tag{5.1.19}$$

where

$$\lambda_i = \frac{1}{2\binom{n}{2}} \sum_{\substack{\alpha,\beta \\ (\alpha\neq\beta)}} \lambda^\beta_{\alpha i}. \tag{5.1.20}$$

Denote by δ the differentiation symbol and by π^v_u the values of forms θ^v_u when the plane P^{n-1} is fixed (or $\theta^i_\alpha = 0$). Then differentiating (5.1.18) and using (5.1.4), (5.1.14), (5.1.12), and (5.1.20), we obtain

$$\delta M_i = \pi^j_i M_j. \tag{5.1.21}$$

Equation (5.1.21) shows that the $(r-1)$-plane $\mu^{r-1} = \{M_1, \dots, M_r\}$ spanned by the points M_i is invariant. We locate the points A_i in μ^{r-1}. By (5.1.19), in this case the quantities λ_i vanish and (5.1.20) gives

$$\sum_{\substack{\alpha,\beta \\ (\alpha\neq\beta)}} \lambda^\beta_{\alpha i} = 0. \tag{5.1.22}$$

Exterior differentiation of (5.1.12) gives

$$(\theta_i^\beta - \theta_i^\alpha) \wedge \theta_0^i = 0. \tag{5.1.23}$$

It follows from (5.1.23) that we can put

$$\theta_i^\alpha = \theta_i^0 + p_{ij}^\alpha \theta_0^j \tag{5.1.24}$$

where

$$p_{ij}^\alpha = p_{ji}^\alpha, \qquad \sum_\alpha p_{ij}^\alpha = 0,$$

and

$$\theta_i^0 = \frac{1}{n} \sum_\alpha \theta_\alpha^i. \tag{5.1.25}$$

On the other hand, it follows from (5.1.22) that for the form θ_i^0 defined by (5.1.25) we have $\pi_i^0 = 0$, i.e., the forms θ_i^0 are expressed in terms of the basis forms θ_α^i only:

$$\theta_i^0 = \sum_\beta q_{ij}^\beta \theta_\beta^j. \tag{5.1.26}$$

Equations (5.1.24) and (5.1.26) imply that

$$\theta_i^\alpha = \sum_\beta \left(q_{ij}^\beta - p_{ij}^\alpha\right) \theta_\beta^j. \tag{5.1.27}$$

Note that equations (5.1.16), (5.1.18), and (5.1.27) show that $\nabla_\delta \lambda_{\alpha i}^\beta = 0$ where $\nabla_\delta \lambda_{\alpha i}^\beta = \delta \lambda_{\alpha i}^\beta - \lambda_{\alpha j}^\beta \phi_i^j(\delta)$. This means that the quantities $\lambda_{\alpha i}^\beta$ form a $(0,1)$-tensor for any fixed α and β.

Finally, taking exterior derivatives of (5.1.22) and using (5.1.16) and (5.1.27), by means of the linear independence of the forms θ_α^i we get the following equation:

$$\binom{n}{2} q_{ij}^\alpha = \sum_{\beta \neq \alpha} \lambda_{\alpha ij}^\beta + \sum_{\beta, \gamma} \left[\lambda_{\gamma i}^\beta \lambda_{\gamma j}^\beta - \left(\lambda_{\gamma i}^\beta - \lambda_{\gamma i}^\alpha\right)\left(\lambda_{\gamma j}^\beta - \lambda_{\alpha j}^\beta\right)\right]. \tag{5.1.28}$$

5.1.4 The Structure Equations and the Fundamental Tensors of a Grassmann $(n+1)$-Web

We can now find the structure equations of a Grassmann web $GW(n+1,n,r)$ and its torsion and curvature tensors. Using (5.1.9), (5.1.12), and (5.1.15), we get the first set of the structure equations of the Grassmann web $GW(n+1,n,r)$:

$$d\theta_\alpha^i = \theta_\alpha^j \wedge \phi_j^i + \sum_{\beta \neq \alpha} \left(\delta_k^i \lambda_{\alpha j}^\beta + \delta_j^i \lambda_{\beta k}^\alpha\right) \theta_\alpha^j \wedge \theta_\beta^k, \tag{5.1.29}$$

where the forms ϕ^i_j are determined by (5.1.17). Comparison of (5.1.29) and (1.2.16) shows that the *torsion tensor* of $GW(n+1, n, r)$ is

$$\underset{\alpha\beta}{a}{}^i_{jk} = \delta^i_k \lambda^\beta_{\alpha j} + \delta^i_j \lambda^\alpha_{\beta k}. \tag{5.1.30}$$

In order to get the second set of the structure equations of $GW(n+1, n, r)$, we must take exterior derivatives of (5.1.17). Since equations (5.1.9), (5.1.15), (5.1.18), and (5.1.25) imply that

$$d\theta^i_j - \theta^k_j \wedge \theta^i_k = \textstyle\sum_\alpha \theta^\alpha_j \wedge \theta^i_\alpha,$$
$$d\theta = -\theta^k_0 \wedge \theta^0_k,$$

then, by means of (5.1.17), (5.1.24), and (5.1.26), we obtain the second set of the structure equations of $GW(n+1, n, r)$:

$$d\phi^i_j - \phi^k_j \wedge \phi^i_k = \sum_{\alpha,\beta} (\delta^i_l q^\alpha_{jk} + \delta^i_j q^\alpha_{lk} + \delta^i_k p^\alpha_{jl}) \theta^j_\alpha \wedge \theta^k_\beta. \tag{5.1.31}$$

Comparison of (5.1.31) and (1.2.29) shows that the *curvature tensor* of $GW(n+1, n, r)$ is

$$\underset{\alpha\beta}{b}{}^i_{jkl} = \frac{1}{2}(\delta^i_l q^\alpha_{jk} + \delta^i_j q^\alpha_{lk} + \delta^i_k p^\alpha_{jl}) - \frac{1}{2}(\delta^i_k q^\beta_{jl} + \delta^i_j q^\beta_{kl} + \delta^i_l p^\beta_{jk}). \tag{5.1.32}$$

Equations (5.1.32) and (5.1.28) show that this tensor is expressed in terms of the quantities p_{ij} only.

5.1.5 Transversally Geodesic and Isoclinic Surfaces of a Grassmann $(n+1)$-Web

Equations (5.1.30) and Theorem 1.11.12 prove that a Grassmann web $GW(n+1, n, r)$ is both transversally geodesic and isoclinic.

Transversally geodesic n-surfaces V^n of $GW(n+1, n, r)$ are determined by the equations $\theta^i_\alpha = \xi^i \theta_\alpha$ where the ξ^i are coordinates of a transversal n-vector which is tangent to V^n (see (1.9.4) and (1.9.3)). In addition, by (1.9.3), we have

$$d\xi^i + \xi^j \phi^i_j = \phi \xi^i. \tag{5.1.33}$$

From (5.1.4), (5.1.17), and (5.1.33) it follows that

$$\begin{aligned} dA_\alpha &= \theta_\alpha(\xi^i A_i) + \sum_\beta \theta^\beta_\alpha A_\beta, \\ d(\xi^i A_i) &= (\phi + \theta)(\xi^i A_i) + \xi^i \sum_\alpha \theta^\alpha_i A_\alpha. \end{aligned} \tag{5.1.34}$$

Equations (5.1.34) imply that the n-plane $\sigma^n = \{A_1, \ldots, A_n, \xi^i A_i\}$ defined by the points A_α and $\xi^i A_i$ is fixed. *The $(n-1)$-planes P^{n-1} lying in σ^n and intersecting all the surfaces U_ξ form the transversally geodesic surface V^n of $GW(n+1, n, r)$.*

The plane σ^n intersects each surface U_ξ along a curve u_ξ. In σ^n there arises an $(n+1)$-web formed by $n+1$ bundles of $(n-1)$-planes with centres on the curves u_ξ. This web is $W(n+1, n, 1)$ and it is dual to a web formed by $n+1$ one-parameter families of hyperplanes in a projective space P^n of dimension n. In the case when hyperplanes of each family form a pencil with an $(n-2)$-dimensional axis such web was studied by Bartsch [Ba 53].

Through any $(n-1)$-plane $P^{n-1} = \{A_1, \ldots, A_n\}$ of $GW(n+1, n, r)$ there passes an $(r-1)$-parameter family of n-planes carrying its transversally geodesic subwebs $W(n+1, n, 1)$. If we project a 3-subweb $W_{\xi_1, \ldots, \xi_{n-2}}(3, 2, r)$ of such subweb $W(n+1, n, 1)$ from an $(n-3)$-plane $\{M_{\xi_1}, \ldots, M_{\xi_{n-2}}\}$ on an $(r+1)$-plane in general position, then on the latter we obtain a so called *alignment chart* (see [Ac 65]). In the case $n = 2$, $r = 1$ the alignment chart consists of three curves corresponding to x-,y-, and z-values, and these x-, y-, and z-values belong together if the respective x-, y-, and z-points are on one straight line (see Figure 5.1). This alignment chart is dual to a web $W(3, 2, 1)$ formed by three pencils of straight lines.

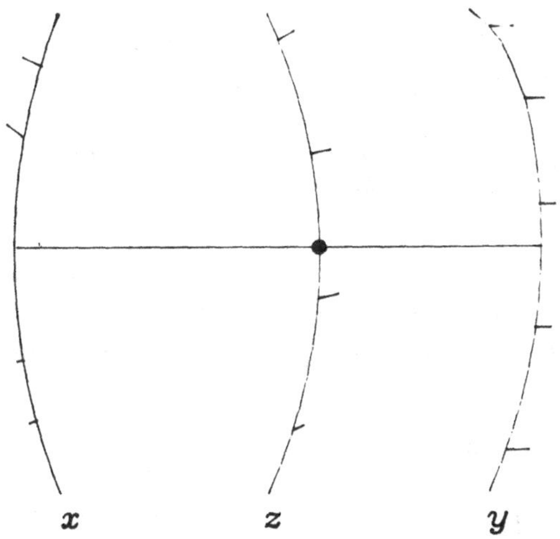

Figure 5.1

Thus, to three-webs which are cut on an intersection of any $n-2$ of the $(n-1)$-dimensional leaves of different foliations of the subweb $W(n+1, n, 1)$ by the remaining three its foliations, there correspond three-webs formed by pencils of $(n-1)$-planes whose $(n-2)$-dimensional axes are determined by the corresponding points of the curves u_ξ. This configuration being projected on an $(r-1)$-plane of general position gives an alignment chart.

Isoclinic r-surfaces V^r of $GW(n+1,n,r)$ are determined by the systems $\theta^i_j = \xi_\alpha \theta^i_0$ where ξ_α are coordinates of an isocline r-vector tangent to V^r (see (1.11.4) and (1.11.6)). In addition, by (1.11.2) and (1.11.17), we have

$$\sum_\alpha \xi_\alpha = -1, \quad d\xi_\alpha = \sum_{\beta\neq\alpha} \underset{\alpha\beta}{b}{}_j \xi_\beta \theta^j_0 \qquad (5.1.35)$$

Since for $GW(n+1,n,r)$, $\underset{\alpha\beta}{b}{}_j = \lambda^\beta_{\alpha j} - \lambda^\alpha_{\beta j}$, we obtain from (5.1.35) that

$$d\xi_\alpha = \sum_{\beta\neq\alpha} (\lambda^\beta_{\alpha j} - \lambda^\alpha_{\beta j})\xi_\beta \theta^j_0. \qquad (5.1.36)$$

Consider an $(n-2)$-plane μ^{n-2} determined by the points

$$M_\alpha = A_\alpha + \xi_\alpha A_0. \qquad (5.1.37)$$

Equations (5.1.36) and (5.1.37) imply that

$$dM_\alpha = \sum_\beta \theta^\beta_\alpha M_\beta. \qquad (5.1.38)$$

Equation (5.1.38) means that the plane μ^{n-2} is fixed. This plane is of dimension $n-2$ since by means of the first relation (5.1.35) the points M_α are linearly dependent.

Thus, *isoclinic surfaces V^r are bundles of $(n-1)$-planes P^{n-1} with $(n-2)$-dimensional centres. Each $(n-1)$-plane $\{A_1, \ldots, A_n\}$ defines an $(n-1)$-parameter family of isoclinic r-surfaces V^r.*

5.1.6 The Hexagonality Tensor of a Grassmann $(n+1)$-Web and the 2nd Fundamental Forms of Surfaces U_ξ

Let us find further the symmetric part of the curvature tensors of the 3-subwebs $W_{\xi_1\ldots\xi_{n-2}}(3,2,r)$. From (1.2.30), by means of (5.1.30) and (5.1.16) we find that

$$\underset{\alpha\beta\alpha}{a}{}^i_{jkl} = \delta^i_k(\lambda^\beta_{\alpha jl} + p^\beta_{jl} - q^\alpha_{jl}) - \delta^i_j(q^\alpha_{kl} - p^\alpha_{kl} + \lambda^\beta_{\alpha l}\lambda^\alpha_{\beta k}) - \delta^i_l \lambda^\alpha_{\beta k}\lambda^\beta_{\alpha j}. \qquad (5.1.39)$$

From (1.3.25), (5.1.39), (5.1.30), and (5.1.32) we find an expression of the symmetric part of the curvature tensor $\underset{\alpha\beta}{\bar{b}}{}^i_{jkl}$ of a 3-subweb $[n+1,\alpha,\beta]$ of $GW(n+1,n,r)$ in the affine connection γ_{n+1} which is at the same time is the hexagonality tensor of $GW(n+1,n,r)$ (see (1.10.2)):

$$\underset{\alpha\beta}{h}{}^i_{jkl} = \underset{\alpha\beta}{\bar{b}}{}^i_{(jkl)} = \delta^i_{(k}(p^\alpha_{jl)} - p^\beta_{jl)} + \lambda^\beta_{|\alpha|jl)} - \lambda^\alpha_{|\beta|jl)}). \qquad (5.1.40)$$

Let us find the 2nd fundamental forms of the r-surfaces U_{xi}. Note that the number of such forms for each r-surface is equal to the codimension of U_ξ, i.e., to $n-1$. Using (5.1.4), (5.1.12), (5.1.13), (5.1.16), and (5.1.27) we find that for the surfaces $U_\alpha = (A_\alpha)$ and $U_{n+1} = (A_{n+1})$ we have

$$\begin{cases} d^2 A_\alpha \equiv \sum_{\beta\neq\alpha} \lambda^\beta_{\alpha ij}\theta^i_\alpha\theta^j_\alpha A_\beta & \bmod A_\alpha,\ A_i + \sum_{\gamma\neq\alpha}\lambda^\gamma_\alpha A_\gamma; \\ d^2 A_0 \equiv \sum_{\beta\neq\alpha}(p^\alpha_{ij} - p^\beta_{ij})\theta^i_0\theta^j_0 A_\beta & \bmod A_0, A_i. \end{cases} \tag{5.1.41}$$

Relations (5.1.41) give the following structure of the *2nd fundamental forms* of surfaces U_α and U_{n+1}:

$$\begin{cases} \Phi^\beta_\alpha = \lambda^\beta_{\alpha ij}\theta^i_\alpha\theta^j_\alpha & (\beta\neq\alpha), \\ \Phi^\beta_{n+1} = (p^\alpha_{ij} - p^\beta_{ij})\theta^i_0\theta^j_0 & (\beta\neq\alpha). \end{cases} \tag{5.1.42}$$

fixed. The tensors $\lambda^\beta_{\alpha ij}$, $\beta\neq\alpha$ and $p^\alpha_{ij} - p^\beta_{ij}$, α fixed, in (5.1.42) are the *2nd fundamental tensors* of surfaces U_α and U_{n+1}.

The following two theorems summarise some of the results on a Grassmann web $GW(n+1,n,r)$ which we have proved in this section:

Theorem 5.1.3 *A Grassmann web $GW(n+1,n,r)$ is transversally geodesic and isoclinic.* ■

Theorem 5.1.4 *Transversally geodesic n-surfaces V^n of a Grassmann web $GW(n+1,n,r)$ are formed by $(n-1)$-planes P^{n-1} lying in fixed n-planes σ^n. Each such plane σ^n carries a web $W(n+1,n,1)$ formed by $n+1$ families of bundles of $(n-1)$-planes whose centres belong to the curves $u_\xi = \sigma^n \cap U_\xi$. The isoclinic r-surfaces of $GW(n+1,n,r)$ are bundles of $(n-1)$-planes with $(n-2)$-dimensional centres. Each $(n-1)$-plane P^{n-1} defines an $(r-1)$-parameter family of transversally geodesic n-surfaces and an $(n-1)$-parameter family of isoclinic r-surfaces.* ■

Theorem 5.1.3 admits an inversion:

Theorem 5.1.5 *A transversally geodesic and isoclinic web $W(n+1,n,r)$, $n>2$, $r>2$, is Grassmannisable, i.e., equivalent to a Grassmann web $GW(n+1,n,r)$.*

Proof. In fact, by Theorem 1.11.12, the torsion tensor of such a web, and only such a web, has the form (1.11.29)–(1.11.30). We showed in this section that the torsion tensor of a web $GW(n+1,n,r)$ has exactly the same structure (see (5.1.30)). As we can see from Theorem 1.2.6 and formulae (1.2.26), for $n>2$ a web $W(n+1,n,r)$ is determined by its torsion tensor. This completes our proof. ■

Note that Theorem 2.6.4 contains another proof of Theorem 1.5.5. Note also that Theorems 1.5.3 and 1.5.5 are equivalent to Theorem 2.6.4.

In the next section we will give the third proof of Theorem 2.6.4. This proof will be based on an estabishing an equivalence between a transversally geodesic and isoclinic web $W(n+1,n,r)$ and a Grassmann web $GW(n+1,n,r)$ by constructing a local diffeomorfism of domains D and $\tilde{D}$ of these two webs.

5.2 The Grassmannisation Theorem for Multicodimensional $(n+1)$-Webs

We will give now a new proof of Theorem 2.6.4 which will be of a pure differentially geometric nature. First let us reformulate Theorem 2.6.4.

Theorem 5.2.1 *A web $W(n+1, n, r)$, $n > 2$, $r > 2$, given in an open domain D of a differentiable manifold X^{nr} is Grassmannisable, i.e., equivalent to a Grassmann web $GW(n+1, n, r)$ given in an open domain $\tilde{D}$ of a projective space P^{r+n-1}, if and only if it is both transversally geodesic and isoclinic.*

Proof. The *necessity* has been proved in Theorems 5.1.3 and 2.6.4 and the *sufficiency* in Theorems 5.1.4 and 2.6.4.

We will give another proof of both the necessity and sufficiency by constructing a local diffeomorphism $\psi : D \to \tilde{D}$ from a domain D of a transversally geodesic and isoclinic web $W(n+1, n, r)$ to a domain $\tilde{D}$ of a Grassmann web $GW(n+1, n, r)$.

Consider now a mapping of D into $\tilde{D}$ given by the equations

$$\underset{\alpha}{\omega}^{\tilde{i}} = \theta^i_\alpha, \quad \tilde{i} = i - n = 1, \ldots, r. \tag{5.2.1}$$

In Chapters 1–4 the range for indices $i, j, k, \ldots$ was $1, \ldots, r$, while in Sections 5.1–5.6 it is $n+1, \ldots, n+r$. To make our previous notation consistent with that of this chapter, we denote $i - n$ by $\tilde{i}$, $i = n+1, \ldots, n+r$. So equation (5.2.1) means that $\underset{\alpha}{\omega}^{i-n} = \theta^i_\alpha$.

The mapping (5.2.1) transfers a point $X \in D$ defined by the completely integrable system $\underset{\alpha}{\omega}^{\tilde{i}} = 0$ into an $(n-1)$-plane P^{n-1} defined by the completely integrable system $\theta^i_\alpha = 0$. It follows from (5.1.13) and (5.1.14) that the forms θ^i_α, α fixed, and $\sum_\alpha \theta^i_\alpha$ are the basis forms of the cotangent spaces $PT^*_{A_\alpha}(U_\alpha)$ and $PT^*_{A_0}(U_{n+1})$. Along $V_\alpha \subset X_\alpha$ we have $\underset{\alpha}{\omega}^{\tilde{i}} = 0$, α fixed. Then from (5.2.1) it follows that $\theta^i_\alpha = 0$. This gives a point A_α on U_α. Therefore a leaf F_α of the foliation X_α is transferred by ψ into the bundle of $(n-1)$-planes with the vertex $A_\alpha \in U_\alpha$. In the same way one can prove that a leaf F_{n+1} of the foliation X_{n+1} is transferred by ψ into the bundle of $(n-1)$-planes with the vertex $A_0 \in U_{n+1}$. If we take $n+1$ leaves , $F_\alpha, \alpha = 1, \ldots, n$, and F_{n+1}, through a point $X \in D$, their images in P^{r+n-1} are $n+1$ bundles of $(n-1)$-planes whose vertices are the points of U_α and U_{n+1} belonging to an $(n-1)$-plane P^{n-1}. Thus $\psi(X_\xi) = \tilde{X}_\xi$, $\xi = 1, \ldots, n, n+1$.

We continue to study the mapping ψ. Taking the exterior derivatives of (5.2.1) by means of (1.2.16), (1.11.29), (5.1.9), and (5.1.12) and applying the Cartan lemma, we obtain

$$\theta^i_j - \omega^{\tilde{i}}_j - \delta^i_j \theta^\alpha_\alpha + \sum_{\beta \neq \alpha} (\lambda^\beta_{\alpha j} - \underset{\alpha\beta}{\lambda}{}_j) \theta_\beta - \delta^i_j \underset{\beta\alpha}{\lambda}{}_k \theta^k_\beta = \underset{\alpha}{\rho}{}^i_{jk} \theta^k_\alpha, \tag{5.2.2}$$

where $\underset{\alpha}{p}{}^i_{jk} = \underset{\alpha}{p}{}^i_{kj}$. It follows from (5.2.2) that

$$\begin{aligned} &\delta^i_j\Big(\theta^\alpha_\alpha - \theta^\beta_\beta - \sum_{\gamma\neq\beta} \underset{\gamma\beta}{\lambda}{}_k \theta^k_\gamma + \sum_{\gamma\neq\alpha} \underset{\gamma\alpha}{\lambda}{}_k \theta^k_\gamma\Big) \\ + \quad &\sum_{\gamma\neq\beta}(\lambda^\gamma_{\beta j} - \underset{\beta\gamma}{\lambda}{}_j)\theta^i_\gamma - \sum_{\gamma\neq\alpha}(\lambda^\gamma_{\alpha j} - \underset{\alpha\gamma}{\lambda}{}_j)\theta^i_\gamma = \underset{\beta}{\rho}{}^i_{jk}\theta^k_\beta - \underset{\alpha}{\rho}{}^i_{jk}\theta^k_\alpha. \end{aligned} \tag{5.2.3}$$

Equation (5.2.3) shows that

$$\theta^\alpha_\alpha - \theta^\beta_\beta - \sum_{\gamma\neq\beta} \underset{\gamma\beta}{\lambda}{}_k \theta^k_\beta - \sum_{\gamma\neq\alpha} \underset{\gamma\alpha}{\lambda}{}_k \theta^k_\gamma = \sum_\gamma \underset{\gamma}{\rho}{}_k \theta^k_\beta. \tag{5.2.4}$$

It follows from (5.2.3) and (5.2.4) that

$$\underset{\beta}{\rho}{}^i_{jk} = \delta^i_j \underset{\beta}{\rho}{}_k - \delta^i_k(\lambda^\beta_{\alpha j} - \underset{\alpha\beta}{\lambda}{}_j), \tag{5.2.5}$$

$$\sum_{\gamma\neq\alpha,\beta} \underset{\gamma}{\rho}{}^i_{jk}\theta^k_\gamma = 0. \tag{5.2.6}$$

Since $r > 2$, we can take $i \neq k$ in (5.2.5) and obtain

$$\underset{\beta}{\rho}{}^i_{jk} = \delta^i_j \underset{\beta}{\rho}{}_k, \tag{5.2.7}$$

$$\lambda^\beta_{\alpha j} = \underset{\alpha\beta}{\lambda}{}_j. \tag{5.2.8}$$

Since $\underset{\beta}{\rho}{}^i_{[jk]} = 0$ and $r > 2$, equation (5.2.8) gives

$$\underset{\beta}{\rho}{}_k = 0. \tag{5.2.9}$$

It follows from (5.2.5), (5.2.7), and (5.2.9) that

$$\underset{\beta}{\rho}{}^i_{jk} = 0. \tag{5.2.10}$$

Equation (5.2.10) shows that equations (5.2.8) are satisfied identically.

Using (5.2.1) and (5.2.10), we can rewrite equation (5.2.2) in the form

$$\theta^i_j - \omega^{\bar{i}}_j - \delta^i_j\Big(\theta^\alpha_\alpha + \sum_{\gamma\neq\alpha} \lambda^\alpha_{\gamma k}\theta^k_\gamma\Big) = 0. \tag{5.2.11}$$

Equation (5.2.11) implies

$$\theta^\alpha_\alpha - \theta^\beta_\beta = \sum_{\gamma\neq\beta} \lambda^\beta_{\gamma k}\theta^k_\gamma - \sum_{\gamma\neq\alpha} \lambda^\alpha_{\gamma k}\theta^k_\gamma, \quad \alpha \neq \beta. \tag{5.2.12}$$

Equation (5.2.12) allows us to put

$$\theta^\alpha_\alpha = \underset{\alpha\beta}{\tilde\theta} - \sum_{\gamma\neq\alpha} \lambda^\alpha_{\gamma k}\theta^k_\gamma, \tag{5.2.13}$$

where $\underset{\alpha\beta}{\tilde\theta} = \underset{\beta\alpha}{\tilde\theta}$. Since $n > 2$, we obtain from (5.2.13) that

$$\underset{\alpha\beta}{\tilde\theta} = \underset{\alpha\gamma}{\tilde\theta} = \underset{\gamma\alpha}{\tilde\theta} = \underset{\gamma\delta}{\tilde\theta} = \tilde\theta. \tag{5.2.14}$$

where α, β and γ are distinct indices. Equations (5.2.13) and (5.2.14) give

$$\theta^\alpha_\alpha + \sum_{\gamma\neq\alpha} \lambda^\alpha_{\gamma k}\theta^k_\gamma = \tilde\theta. \tag{5.2.15}$$

It follows from (5.2.11) and (5.2.15) that

$$\theta^i_j - \omega^{\bar i}_{\bar j} = \delta^i_j\tilde\theta. \tag{5.2.16}$$

Let us take the sum of all equations (5.1.15):

$$\sum_\alpha\Big(\theta^\alpha_\alpha + \sum_{\gamma\neq\alpha} \lambda^\alpha_{\gamma k}\theta^k_\gamma\Big) = n\theta. \tag{5.2.17}$$

Equations (5.2.15) and (5.2.17) give that $n\tilde\theta = n\theta$ and therefore

$$\tilde\theta = \theta. \tag{5.2.18}$$

Using (5.2.18), we can rewrite (5.2.15) and (5.2.16) in the form

$$\theta^\alpha_\alpha = \theta - \sum_{\gamma\neq\alpha} \lambda^\alpha_{\gamma k}\theta^k_\gamma, \tag{5.2.19}$$

$$\omega^{\bar i}_{\bar j} = \theta^i_j - \delta^i_j\theta. \tag{5.2.20}$$

Comparison of (5.1.17) and (5.2.20) shows that

$$\omega^{\bar i}_{\bar j} = \phi^i_j. \tag{5.2.21}$$

Taking the exterior derivative of (5.1.15) by means of (5.1.9) and (5.1.18), we obtain

$$d\theta = -\theta^i_0 \wedge \theta^\alpha_i + \sum_{\beta,\gamma}\theta^\gamma_\beta \wedge \theta^\alpha_\gamma. \tag{5.2.22}$$

If we add equations (5.2.22) corresponding to different α and use the equality $\sum_\alpha \theta^i_\alpha = n\theta^0_i$ (see (5.1.25)), we will have

$$nd\theta = -\theta^i_0 \wedge n\theta^0_i$$

or

$$d\theta = -\theta_0^i \wedge \theta_i^0. \tag{5.2.23}$$

The exterior differentiation of (5.2.19) also gives (5.2.23), the exterior differentiation of (5.2.23) leads to the identity, and the exterior differentiation of (5.2.20) (or (5.2.21)) brings us to the coincidence of the curvature tensor $\underset{\alpha\beta}{b}{}^{\tilde{i}}_{\tilde{j}\tilde{k}\tilde{l}}$ of the $W(n+1,n,r)$ and the curvature tensor $\underset{\alpha\beta}{b}{}^{i}_{jkl}$ of the $GW(n+1,n,r)$ determined by (5.1.32).

Therefore the mapping ψ is determined by equations (5.1.1), (5.1.19), and (5.1.20) where the exterior differential of θ is given by (5.2.23). It is easy to check that this system is completely integrable. Because of this, ψ is a local diffeomorphism of D and $\tilde{D}$. Since we have already proved that $\psi(X_\xi) = \tilde{X}_\xi$ we can now say that the transversally geodesic web $W(n+1,n,r)$ and the Grassmann web $GW(n+1,n,r)$ are equivalent to each other. ■

Note that this proof gives the coincidence of the torsion tensors of $W(n+1,n,r)$ and $GW(n+1,n,r)$ (see (1.11.29), (5.1.30), and (5.2.7)), their curvature tensors (see our proof above), and the forms of the affine connections induced by these webs (see (5.2.21)).

5.3 Reducible Grassmann $(n+1)$-Webs

Suppose that a Grassmann web $GW(n+1,n,r)$ is reducible. The following theorem characterises such Grassmann webs:

Theorem 5.3.1 *In a domain of reqularity of a Grassmann web $GW(n+1,n,r)$ the following three statements are equivalent:*

(i) *A web $GW(n+1,n,r)$ is reducible of type*

$$u_{n+1}^i = f^i(u_{\alpha_1}^{j_1}, \ldots, g^k(u_{\alpha_{\rho+1}}^{j_{\rho+1}}, \ldots, u_{\alpha_n}^{j_n}), \ldots, u_{\alpha_\rho}^{j_\rho}). \tag{5.3.1}$$

(ii) *The tensor $\lambda_{\alpha i}^{\beta}$ satisfies the conditions*

$$\lambda_{\alpha_s i}^{\beta_t} = \lambda_{\alpha_s i}, \quad \lambda_{\beta_t i}^{\alpha_s} = \lambda_i^{\alpha_s}; \quad s = 1, \ldots, \rho; \quad t = \rho+1, \ldots, n. \tag{5.3.2}$$

(iii) *The surfaces U_ξ generating the $GW(n+1,n,r)$ belong to two fixed planes: U_{n+1} and $U_{\alpha_s}, s = 1, \ldots, \rho$, to the $(r+\rho)$-plane*

$$\{A_0, A_{\alpha_1}, \ldots, A_{\alpha_\rho}, A_{n+1}, \ldots, A_{n+r}\},$$

and U_{β_t}, $t = \rho+1, \ldots, n$, to the $(n-\rho+r-1)$-plane

$$\{A_{\alpha_{\rho+1}}, \ldots, A_{\alpha_n}, A_{n+1} + \sum_s \lambda_1^{\alpha_s} A_{\alpha_s}, \ldots, A_{n+r} + \sum_s \lambda_r^{\alpha_s} A_{\alpha_s}\}.$$

Proof. Suppose that a web $GW(n+1,n,r)$ is reducible of type (5.3.1). By Theorem 4.1.4, this is equivalent to the fact that the distribution $\{\alpha_1,\ldots,\alpha_\rho\}$ is integrable, or analytically, it is equivalent to conditions (1.8.3) for the torsion tensor $\underset{\alpha\beta}{a}{}^i_{jk}$ of the web:

$$\underset{\alpha_s\beta_t}{a}{}^i_{jk} = \underset{\alpha_s\gamma_t}{a}{}^i_{jk}, \quad \beta_t \neq \gamma_t, \quad s=1,\ldots,\rho; \quad t=\rho+1,\ldots,n. \tag{5.3.3}$$

Since for a Grassmann web the torsion tensor $\underset{\alpha\beta}{a}{}^i_{jk}$ has the structure (5.1.30), it follows from this that relations (5.3.3) are equivalent to the equations (5.3.2). This proves an equivalence of (i) and (ii).

Now we will prove an equivalence of (ii) and (iii). Suppose that (5.3.2) holds. Differentiating (5.3.2) and using (5.1.16), (5.1.18), (5.1.27), and using the linear independence of the forms θ^i_α, we obtain that

$$p^{\beta_t}_{ij} = p^{\gamma_t}_{ij}, \quad \lambda^{\beta_t}_{\alpha_s ij} = \lambda^{\gamma_t}_{\alpha_s ij}, \quad \lambda^{\alpha_s}_{\beta_t ij} = \lambda^{\alpha_s}_{\gamma_t ij}. \tag{5.3.4}$$

By virtue of (5.3.2), equations (5.1.13) can be written in the form

$$\begin{cases} dA_{\alpha_s} = \theta^{\alpha_s}_{\alpha_s} A_{\alpha_s} + \theta^i_{\alpha_s}\Big(A_i + \lambda_{\alpha_s i}\sum\limits_t A_{\beta_t} + \sum\limits_{s_1\neq s}\lambda^{\beta_{s_1}}_{\alpha_s i} A_{\beta_{s_1}}\Big), \\ dA_{\beta_t} = \theta^{\beta_t}_{\beta_t} A_{\beta_t} + \theta^i_{\beta_t}\Big(A_i + \sum\limits_s \lambda^{\alpha_s}_i A_{\alpha_s} + \sum\limits_{t_1\neq t}\lambda^{\gamma_{t_1}}_{\beta_t i} A_{\gamma_{t_1}}\Big). \end{cases} \tag{5.3.5}$$

From equations (5.3.5) and the definition of the point A_0 we obtain the following relations:

$$dA_{\alpha_s} \wedge A_{\alpha_1} \wedge \ldots \wedge A_{\alpha_\rho} = \theta^i_{\alpha_s}(A_i + \lambda_{\alpha_s i} A_0). \tag{5.3.6}$$

By virtue of (5.3.4), equations (5.1.4) and (5.3.5) show that the $(r+\rho)$-plane

$$\{A_0, A_{\alpha_s}, \ldots, A_{\alpha_\rho}, A_{n+1}, \ldots, A_{n+r}\}$$

and the $(n-\rho+r-1)$-plane

$$\{A_{\alpha_{\rho+1}}, \ldots, A_{\alpha_n}, A_{n+1} + \sum_s \lambda^{\alpha_s}_{n+1} A_{\alpha_s}, \ldots, A_{n+r} + \sum_s \lambda^s_{n+r} A_{\alpha_s}\}$$

are fixed. In addition, it follows from (5.1.14) and (5.3.6) that:

a) An $(r-1)$-plane spanned by the points $A_i + \lambda_{\alpha_s i} A_0$ lies in the tangent r-plane of the surface U_{n+1} at the point A_0 and in the tangent subspace of the family of $(\rho-1)$-planes $\{A_{\alpha_1}, \ldots, A_{\alpha_s}\}$, and

b) An $(r-1)$ plane spanned by the points

$$A_i + \sum_s \lambda^{\alpha_s}_i A_{\alpha_s} + \sum_{t_1\neq t} \lambda^{\gamma_{t_1}}_{\beta_t i} A_{\gamma_{t_1}}$$

belongs to the tangent subspace of the family of $(n-\rho-1)$-planes $\{A_{\alpha_{\rho+1}}, \ldots, A_{\alpha_n}\}$.

It follows from a) and b) that the r-surfaces U_{n+1} and U_{α_s}, $s = 1, \ldots, \rho$, and U_{β_t}, $t = \rho+1, \ldots, n$, belong to the indicated fixed $(r+\rho)$- and $(n-\rho+r-1)$-plane respectively.

Conversely, equations (5.1.13) imply

$$\begin{cases} dA_{\alpha_s} \wedge A_{\alpha_1} \wedge \ldots \wedge A_{\alpha_\rho} \wedge A_0 \wedge A_{n+1} \wedge \ldots \wedge A_{n+r} \\ = \theta^i_{\alpha_s} \sum\limits_{t \neq \rho+1} (\lambda^{\beta_{\rho+1}}_{\alpha_s i} - \lambda^{\beta_t}_{\alpha_s i}) A_{\beta_t}, \\ dA_{\beta_t} \wedge A_{\beta_{\rho+1}} \wedge \ldots \wedge A_{\beta_n} = \theta^i_{\beta_t}(A_i + \sum\limits_s \lambda^{\alpha_s}_{\beta_t i} A_{\alpha_s}). \end{cases} \qquad (5.3.7)$$

The relations (5.3.7) show that if there is the imbedding indicated above of the surfaces U_ξ into the $(r+\rho)$- and $(n-\rho+r-1)$-planes, then conditions (5.3.2) hold and these two planes are fixed. Thus, the statements (ii) and (iii) are equivalent. ■

The equations (5.3.4) themselves mean a linear dependence of the 2nd fundamental forms of the surfaces U_ξ determined by (5.1.42). Note in conclusion that the two fixed planes indicated above intersect each other along the r-plane δ spanned by the points

$$A_0 - \sum_s A_{\alpha_s} = \sum_t A_{\beta_t} \quad \text{and} \quad A_i + \sum_s \lambda^{\alpha_s}_i A_{\alpha_s}.$$

Our proof implies the following corollary:

Corollary 5.3.2 *For a Grassmann web $GW(n+1, n, r)$, which is reducible of type (5.3.1), the two fixed hyperplanes indicated in Theorem 5.3.1 allow to perform an n-ary operation in a corresponding l.d. n-quasigroup in the following way: the point A_i which is the common point of the surface U_{n+1} and the $(n-1)$-plane $\{A_1, \ldots, A_n\}$ can be obtained first by finding the point of intersection $\sum_t A_{\beta_t} = A_0 - \sum_s A_{\alpha_s}$ of the $(n-\rho-1)$-plane $\{A_{\alpha_{\rho+1}}, \ldots, A_{\alpha_n}\}$ and the r-plane δ and then by finding the point A_0 as the common point of the surface U_{n+1} and the ρ-plane $\{A_{\alpha_1}, \ldots, A_{\alpha_s}, \sum_t A_{\beta_t}\}$.* ■

The construction described in Corollary 5.3.2 generalises for the case of a system of r equations in nr variables a composite alignment chart with a straight mute scale which corresponds to the case $n = 3, r = 1$ and is well-known in nomography (see [Gl 61], pp. 168-174).

Theorem 5.3.1 allows us to obtain a characterisation of multiple reducible Grassmann webs $GW(n+1, n, r)$. To illustrate this, we consider the following example.

Example 5.3.3 Suppose, that $n = 6$, i.e., we have a Grassmann web $GW(7, 6, r)$ given in P^{r+5}. Suppose further that this web is reducible of type

$$u^i_7 = f^i(u^{j_1}_1, u^{j_2}_2, \varphi^l(u^{j_3}_3, \psi^m(u^{j_4}_4, u^{j_5}_5, u^{j_6}_6))).$$

Then, by Theorem 5.3.1, part (ii), for such a 7-web, and only for it, the tensor $\lambda^{\beta}_{\alpha i}$ satisfies the following conditions:

$$\lambda^3_{1i} = \lambda^4_{1i} = \lambda^5_{1i} = \lambda^6_{1i}, \quad \lambda^1_{3i} = \lambda^1_{4i} = \lambda^1_{5i} = \lambda^1_{6i}, \quad \lambda^4_{3i} = \lambda^5_{3i} = \lambda^6_{3i},$$
$$\lambda^3_{2i} = \lambda^4_{2i} = \lambda^5_{2i} = \lambda^6_{2i}, \quad \lambda^2_{3i} = \lambda^2_{4i} = \lambda^2_{5i} = \lambda^2_{6i}, \quad \lambda^2_{4i} = \lambda^2_{5i} = \lambda^2_{6i}.$$

Next, by Theorem 5.3.1, part (iii), for such a 7-web, and only for it, the r-surfaces U_7, U_1, U_2 belong to a fixed $(r+2)$-plane Q^{r+2} and the surfaces U_3, U_4, U_5, U_6 belong to a fixed $(r+3)$-plane S^{r+3}. In addition, the surface U_3 lies in a fixed $(r+1)$-plane $G^{r+1} \subset S^{r+3}$ and surfaces U_4, U_5, U_6 lie in a fixed $(r+2)$-plane $H^{r+2} \subset S^{r+3}$ where $G^{r+1} \supset T_r$, $H^{r+1} \supset T^r$ and $T^r = Q^{r+2} \cap S^{r+3}$.

In the same manner, one can obtain a characterisation of completely reducible Grassmann webs $GW(n+1, n, r)$.

Example 5.3.4 Suppose that $n = 4$, i.e., we have a Grassmann web $GW(5, 4, r)$ in P^{r+3}, and this web is completely reducible of type

$$u^i_5 = f^i(u^{j_1}_1, g^k(u^{j_2}_2, h^l(u^{j_3}_3, u^{j_4}_4))).$$

According to Theorem 5.3.1, for such the $GW(5, 4, r)$, and only for it, the tensor $\lambda^{\beta}_{\alpha i}$ satisfies the relations

$$\lambda^2_{1i} = \lambda^3_{1i} = \lambda^4_{1i}, \quad \lambda^1_{2i} = \lambda^1_{3i} = \lambda^1_{4i}, \quad \lambda^3_{2i} = \lambda^4_{2i}, \quad \lambda^2_{3i} = \lambda^2_{4i},$$

and the surfaces U_5, U_1 belong to a fixed $(r+1)$-plane Q^{r+1}, the surfaces U_2, U_3, U_4 lie in a fixed $(r+2)$-plane R^{r+2}; in addition the U_2 belong to a fixed $(r+1)$-plane $G^{r+1} \subset R^{r+2}$ and U_3 and U_4 lie in a fixed $(r+1)$-plane $H^{r+1} \subset R^{r+2}$, where $G^{r+1} \supset T^r, H^{r+1} \supset T^r$ and $T^r = Q^{r+1} \cap R^{r+2}$.

Note in conclusion that in the case $n = 3$ there exist only three types of reducibility. This leads to the reducibilities of the types (4.1.29), (4.1.30), and (4.1.31) which we have indicated for the case $n = 3$ in Section 4.1.

5.4 Algebraic, Bol Algebraic and Reducible Algebraic $(n+1)$-Webs

5.4.1 General Algebraic $(n+1)$-Webs

We will study now the algebraic $AW(n+1, n, r)$. They were defined as the Grassmann webs $GW(n+1, n, r)$ for which the generating surfaces U_ξ belong to an algebraic r-dimensional surface V^r_{n+1} of degree $n+1$ (see Definition 2.6.2 for $d = n+1$).

First we shall consider the general algebraic webs $AW(n+1, n, r)$ and then the algebraic webs $AW(n+1, n, r)$ whose generating surface V^r_{n+1} decomposes into some components.

The following theorem gives an analytical criterion for a Grassmann web $GW(n+1, n, r)$ to be algebraic.

Theorem 5.4.1 *In a domain of its regularity, a Grassmann web $GW(n+1,n,r)$ is algebraic if and only if the 2nd fundamental tensors of the surfaces U_ξ generating $GW(n+1,n,r)$ are connected by the relations*

$$\lambda^{\beta}_{\alpha ij} - \lambda^{\alpha}_{\beta ij} + p^{\alpha}_{ij} - p^{\beta}_{ij} = 0. \tag{5.4.1}$$

Proof. It follows from Theorem 2.6.9 that a Grassmann web $GW(n+1,n,r)$ is algebraic if and only if it is hexagonal. On the other hand, by Corollary 1.10.6, it is hexagonal if and only if its hexagonality tensor $\underset{\alpha\beta}{h}{}^{i}_{jkl}$ vanishes. For a $GW(n+1,n,r)$ this tensor is expressed by (5.1.40), and its vanishing is equivalent to the relations (5.4.1). ■

5.4.2 Bol Algebraic $(n+1)$-Webs

Next we shall study multicodimensional Bol algebraic $(n+1)$-webs. First we will find under what condition a Bol $(n+1)$-web is algebraisable and after this we will establish a characteristic of Bol algebraic $(n+1)$-webs in terms of the structure of a generating surface V^r_{n+1}.

Theorem 5.4.2 *If a Bol web $B_\xi(n+1,n,r)$ is isoclinic, it is algebraisable.*

Proof. To fix our ideas, let us consider a Bol web $B_{n+1}(n+1,n,r)$. By Theorem 4.5.3 and Corollary 4.5.4, this web is hexagonal and transversally geodesic. According to Theorem 2.6.9, this means that the web $B_{n+1}(n+1,n,r)$ is algebraisable. ■

Since a Bol algebraisable $(n+1)$-web is equivalent to a Bol algebraic $(n+1)$-web, we will consider the latter.

Theorem 5.4.3 *An algebraic web $AW(n+1,n,r)$ is a Bol web $B_{n+1}(n+1,n,r)$ if and only if its generating surface V^r_{n+1} decomposes into an r-dimensional algebraic surface V^r_n of degree n and an r-plane P^r.*

Proof. *Necessity.* Let a web $AW(n+1,n,r)$ be a Bol web $B_{n+1}(n+1,n,r)$. Consider a 3-subweb $[n+1,\alpha,\beta]$ of the $AW(n+1,n,r)$. It is generated by $n-2$ points of general position taken respectively on the $n-2$ r-surfaces $U_\gamma, \gamma = 1,\dots,n;$ $\gamma \neq \alpha,\beta$. As we saw before, if we project the surface V^r_{n+1} from the centre Z^{n-1} spanned by the points indicated onto a fixed complimentary $(r+1)$-plane P^{r+1}, we obtain in P^{r+1} a cubic hypersurface V^r_3. On the Grassmannian $G(1,r+1)$ of straight lines of P^{r+1} this hypersurface generates a three-web which is equivalent to the 3-subweb $[n+1,\alpha,\beta]$. By the definition of a Bol $(n+1)$-web $B_{n+1}(n+1,n,r)$, a 3-subweb $[n+1,\alpha,\beta]$ is a Bol three-web. Therefore, the three-web being obtained in P^{r+1} is also a Bol three-web. According to Theorem 3 in [Ak 73], the cubic hypersurface V^r_3 generating this Bol three-web in P^{r+1} decomposes into a hyperquadric V^r_2 and a hyperplane P^r. It follows from this that the r-surface U_{n+1} generating the foliation

λ_{n+1} belongs to the r-plane P^r, i.e., the generating surface V^r_{n+1} decomposes into an r-plane P^r and an algebraic r-dimensional surface V^r_n of degree n.

Sufficiency. Let a web $AW(n+1,n,r)$ is generated by a surface V^r_{n+1} decomposing into an r-plane P^r and a surface V^r_n. To fix our ideas, we will assume that $U_{n+1} \subset P^r$ and $U_\alpha \subset V^r_n$, $\alpha = 1, \ldots, n$.

Consider a 3-subweb $[n+1, \alpha, \beta]$ of the $AW(n+1,n,r)$ determined by $n-2$ points of general position taken on V^r_n. Projecting the surface $V^r_{n+1} = V^r_n \cup P^r$ from the centre Z^{n-3} spanned by the $n-2$ points indicated above on a fixed complimentary $(r+1)$-plane P^{r+1}, we obtain in P^{r+1} a cubic hypersurface V^r_3 decomposing into a hyperquadric V^r_2 and a hyperplane P^r. Applying again Theorem 3 from [Ak 73], we conclude that the V^r_3 generates in P^{r+1} an algebraic Bol three-web $AW(3,2,r)$. Therefore, the 3-subweb $[n+1,\alpha,\beta]$ which is equivalent to the three-web in P^{r+1} is a Bol web. By Definition 4.5.1, it implies that the web $AW(n+1,n,r)$ being considered is a Bol web $B_{n+1}(n+1,n,r)$. ■

Corollary 5.4.4 *An algebraic web $AW(n+1,n,r)$ is the Bol web $B_\xi(n+1,n,r)$ if and only if the conditions* (5.4.1) *hold and the 2nd fundamental tensors of the surface U_{n+1} vanish.*

Proof. This follows from Theorems 5.4.1 and 5.4.3 and the fact that the surface U_ξ is an r-plane. ■

Corollary 5.4.5 *An algebraic web $AW(n+1,n,r)$ is a Bol web $B_{\xi_1 \ldots \xi_m}(n+1,n,r)$, $1 \leq m \leq n$, if and only if one of the following conditions is satisfied:*

(i) *Its generating surface V^r_{n+1} decomposes into m r-planes P^r_s, $s = 1, \ldots, m$, and an r-dimensional algebraic surface V^r_{n-m+1} of degree $n-m+1$.*

(ii) *Conditions* (5.4.1) *are met and the 2nd fundamental tensors of the surfaces $U_{\xi_1}, \ldots, U_{\xi_m}$ vanish.*

Corollary 5.4.6 *An algebraic web $AW(n+1,n,r)$ is a Moufang web $M(n+1,n,r)$ if and only if one of the following conditions is satisfied:*

(i) *Its generating surface V^r_{n+1} decomposes into $n+1$ r-planes.*

(ii) *The 2nd fundamental tensors of all the surfaces U_ξ vanish.*

Proof. The last two corollaries follow from Definition 4.5.5, Theorem 5.4.3, and Corollary 5.4.4. Note that for a Moufang web $AW(n+1,n,r)$ the conditions (5.4.1) are identically satisfied since all the 2nd fundamental tensors of the surfaces U_ξ are equal to zero. ■

5.4.3 Reducible Algebraic $(n+1)$-Webs

Now we consider different cases when a web $AW(n+1,n,r)$ is reducible. First we will find different tests for reducibility of a web $AW(n+1,n,r)$.

Theorem 5.4.7 *In a domain of regularity of an algebraic web $AW(n+1,n,r)$, the following four conditions are equivalent:*

(i) *The web $AW(n+1,n,r)$ is reducible of type* (5.3.1) *where $\rho = \sigma - 1$.*

(ii) *The surface V^r_{n+1} generating the $AW(n+1,n,r)$ decomposes into two algebraic r-surfaces V^r_σ and V^r_μ of degree σ and μ respectively, $\sigma + \mu = n+1, \sigma, \mu \geq 2$, and the surfaces U_{n+1}, $U_{\alpha_s} \subset V^r_\sigma$, $s = 1, \ldots, \sigma - 1$, and $U_{\beta_t} \subset V^r_\mu$, $t = \sigma, \ldots, n$.*

(iii) *The tensor $\lambda^\alpha_{\beta i}$ satisfies the conditions* (5.3.2), *where $\rho = \sigma - 1$ and the 2nd fundamental tensors of the surfaces U_ξ are connected by the relations* (5.4.1).

(iv) *Among $\binom{n+1}{3}$ 3-subwebs $AW_{\xi_1 \ldots \xi_{n-2}}(3,2,r)$ of $AW(n+1,n,r)$ there are $\mu\binom{\sigma}{2} + \sigma\binom{\mu}{2}$, $\sigma + \mu = n+1$, Bol webs $AW(3,2,r)$, and the remaining 3-subwebs are algebraic webs $AW(3,2,r)$ of general type.*

Proof. First we prove that the statements (ii) and (iv) are equivalent. Suppose that an algebraic surface V^r_{n+1} generating an $AW(n+1,n,r)$ decomposes and has two components:

$$V^r_{n+1} = V^r_\sigma \cup V^r_\mu, \quad \sigma + \mu = n+1, \quad \sigma, \mu \geq 2.$$

Then among its $\binom{n+1}{3}$ algebraic 3-subwebs there will be a certain number of Bol three-webs. These Bol 3-subwebs are obtained if the vertex of the cone associated with a 3-subweb is defined by $\sigma - 1$ points of V^r_σ and $\mu - 2$ points of V^r_μ or by $\mu - 1$ points of V^r_μ and $\sigma - 2$ points of V^r_σ. For all such cases, and only for them, when we project our configuration onto a fixed $(r+1)$-plane, we obtain an r-dimensional cubic surface decomposing into an r-plane and an r-dimensional quadric. It follows from this (see [Ak 73]) that the corresponding 3-subweb $AW_{\xi_1 \ldots \xi_{n-2}}(3,2,r)$ is a Bol three-web. It is obvious that there are $\mu\binom{\sigma}{2} + \sigma\binom{\mu}{2}$ such Bol 3-subwebs. In the remaining cases a projection of our configuration on a fixed $(r+1)$-plane gives a general r-dimensional cubic surface, and the corresponding 3-subwebs are general algebraic webs $AW(3,2,r)$.

Conversely, if the 3-subwebs $AW_{\xi_1 \ldots \xi_{n-2}}(3,2,r)$ of an algebraic web $AW(n+1,n,r)$ are of the types indicated in (iv), then $V^r_{n+1} = V^r_\sigma \cup V^r_\mu$.

To prove that the statements (i) and (ii) are equivalent, we note that, according to the classical result, the algebraic surfaces V^r_σ and V^r_μ lie in an $(r+\sigma-1)$-plane $Q^{r+\sigma-1}$ and an $(r+\mu-1)$-plane $R^{r+\mu-1}$ respectively (see [Ber 24], p. 213). These two planes intersect each other in an r-plane T^r. The surface $V^r_{\sigma+1} = V^r_\sigma \cup T^r$ in $Q^{r+\sigma-1}$ and the surface $V^r_{\mu+1} = V^r_\mu \cup T^r$ in $R^{r+\mu-1}$ generate algebraic webs $AW(\sigma+1,\sigma,r)$ and $AW(\mu+1,\mu,r)$, respectively. The generating algebraic surfaces $V^r_{\sigma+1}$ and $V^r_{\mu+1}$ consist respectively of the surfaces $U_{\xi_1}, \ldots, U_{\xi_{\sigma+1}}$, T^r and $U_{\xi_{\sigma+2}}, \ldots, U_{\xi_{n+1}}$. Since, according to (ii), the surface $U_{n+1} \subset V^r_\sigma$, by Theorem 5.3.1, the web $AW(n+1,n,r)$ is reducible of type (5.3.1) where $\rho = \sigma - 1$. The converse is also true: if a web $AW(n+1,n,r)$ is reducible of type (5.3.1) where $\rho = \sigma - 1$, then there exists a surface V^r_{n+1} generating $AW(n+1,n,r)$ and $V^r_{n+1} = V^r_\sigma \cup V^r_\mu$, $\sigma + \mu = n+1$.

To complete the proof, note that the statements (i) and (iii) are equivalent since hexagonality is equivalent to the relations (5.4.1) among the 2nd fundamental tensors of the surfaces U_ξ and the reducibility (5.3.1) is equivalent to the conditions (5.3.2), $\rho = \sigma - 1$ for the tensor $\lambda^{\beta}_{\alpha i}$. ■

5.4.4 Multiple Reducible Algebraic $(n+1)$-Webs

Next we will study different tests for multiple reducibility of a web $AW(n+1, n, r)$.

Theorem 5.4.8 *In a domain of regularity of an algebraic web $AW(n+1, n, r)$, the following four conditions are equivalent:*

(i) *The web $AW(n+1, n, r)$ is reducible of type*

$$u^i_{n+1} = f^i(u^{j_1}_{\alpha_1}, \ldots, g^k(u^{j_\sigma}_{\alpha_\sigma}, \ldots, \varphi^l(u^{j_{\sigma+\mu-1}}_{\alpha_{\sigma+\mu-1}}, \ldots, u^{j_n}_{\alpha_n}), \ldots, u^{j_{\sigma+\mu-2}}_{\alpha_{\sigma+\mu-2}}), \ldots, u^{j_{\sigma-1}}_{\alpha_{\sigma-1}}), \quad (5.4.2)$$

or of type

$$u^i_{n+1} = f^i(u^{j_1}_{\alpha_1}, \ldots, g^k(u^{j_{\sigma+\mu-1}}_{\alpha_{\sigma+\mu-1}}, \ldots, \varphi^l(u^{j_\sigma}_{\alpha_\sigma}, \ldots, u^{j_{\sigma+\mu-2}}_{\alpha_{\sigma+\mu-2}}), \ldots, u^{j_n}_{\alpha_n}), \ldots, u^{j_{\sigma-1}}_{\alpha_{\sigma-1}}). \quad (5.4.3)$$

(ii) *The surface V^r_{n+1} generating the $AW(n+1, n, r)$ decomposes into algebraic r-surfaces V^r_σ, V^r_μ, and V^r_τ of degree σ, μ and τ respectively, $\sigma + \mu + \tau = n + 1$; $\sigma, \mu, \tau \geq 2$, and the surfaces $U_{n+1}, U_{\alpha_{s_1}} \subset V^r_\sigma$, $s_1 = 1, \ldots, \sigma - 1$; $U_{\beta_{s_2}} \subset V^r_\mu$, $s_2 = \sigma, \ldots, \sigma + \mu - 1$; and $U_{\gamma_{s_3}} \subset V^r_\tau, s_3 = \sigma + \mu, \ldots, n$.*

(iii) *The tensor $\lambda^{\beta}_{\alpha i}$ satisfies the conditions*

$$\begin{cases} \lambda^{\beta_t}_{\alpha_s i} = \lambda^{\gamma_t}_{\alpha_s i}, \quad \lambda^{\alpha_s}_{\beta_t i} = \lambda^{\alpha_s}_{\gamma_t i}, \\ \lambda^{\beta_q}_{\alpha_p i} = \lambda^{\gamma_q}_{\alpha_p i} = \lambda^{\beta_q}_{\gamma_p i}, \quad \lambda^{\alpha_p}_{\beta_q i} = \lambda^{\alpha_p}_{\gamma_q i} = \lambda^{\gamma_p}_{\beta_q i}, \end{cases} \quad (5.4.4)$$

where

$$s = 1, \ldots \sigma - 1; \quad t = \sigma, \ldots, n; \quad p = \sigma, \ldots, \sigma + \mu - 2; \quad q = \sigma + \mu - 1, \ldots, n,$$

and the 2nd fundamental tensors of the surfaces U_ξ are connected by the relations (5.4.1).

(iv) *Among $\binom{n+1}{3}$ 3-subwebs $AW_{\xi_1 \ldots \xi_{n-2}}(3, 2, r)$ of $AW(n+1, n, r)$ there are $\sigma\mu\tau$ group webs $AW(3, 2, r)$, $\binom{\tau}{2}(\sigma+\mu) + \binom{\sigma}{2}(\mu+\tau) + \binom{\mu}{2}(\tau+\sigma)$ Bol three-webs $AW(3, 2, r)$, and the remaining 3-subwebs are algebraic webs $AW(3, 2, r)$ of general type.*

Proof. First let us prove that (ii) is equivalent to (iv). Suppose that the generating surface V^r_{n+1} decomposes and has more than two components. We can always assume that there are only three such components combining in one, if necessary, all the components except the first two:

$$V^r_{n+1} = V^r_\sigma \cup V^r_\mu \cup V^r_\tau, \quad \sigma + \mu + \tau = n + 1, \quad \sigma, \mu, \tau \geq 2.$$

Then among 3-subwebs $AW_{\xi_1 \ldots \xi_{n-2}}(3,2,r)$ there will be general algebraic webs $AW(3,2,r)$, Bol webs $AW(3,2,r)$, and group webs $AW(3,2,r)$. We obtain a group 3-subweb if the vertex of the cone associated with the 3-subweb is defined by $\sigma - 1$ points of $V_\sigma^r, \mu - 1$ points of V_μ^r and $\tau - 1$ points of V_τ^r. In this case, when we project our configuration onto a fixed $(r+1)$-plane, we get an r-dimensional cubic surface decomposing into three r-planes in general position. This means (see [Ak 73]) that the corresponding three-web is a group web $AW(3,2,r)$.

The way to get Bol 3-subwebs $AW(3,2,r)$ is the same as in the proof of Theorem 5.4.2. A count similar to one in Theorem 5.4.2 shows that in this case we have $\binom{\sigma}{2}(\tau+\mu) + \binom{\mu}{2}(\tau+\sigma) + \binom{\tau}{2}(\sigma+\mu)$ Bol 3-subwebs $AW(3,2,r)$ and $\sigma\mu\tau$ group 3-subwebs $AW(3,2,r)$; the remaining 3-subwebs are algebraic webs $AW(3,2,r)$ of general type.

Conversely, if the 3-subwebs $AW_{\xi_1 \ldots \xi_{n-2}}(3,2,r)$ of an $AW(n+1,n,r)$ are of the types indicated in (iv), then

$$V_{n+1}^r = V_\sigma^r \cup V_\mu^r \cup V_\tau^r. \tag{5.4.5}$$

Next let us prove that (i) is equivalent to (ii). If we have (ii), then the decomposition (5.4.5) holds. The surface V_σ^r lies in $(r+\sigma-1)$-plane $Q^{r+\sigma-1}$ and the surface $V_\mu^r \cup V_\tau^r$ lies in $(r+\mu+\tau-1)$-plane $R^{r+\mu+\tau-1}$. These two planes intersect each other at an r-plane T^r. If we assume, $U_{n+1} \subset V_\sigma^r$, then, by Theorem 5.4.2, we have in this case a reducibility of type

$$u_{n+1}^i = f^i(u_{\alpha_1}^{j_1}, \ldots, g^k(u_{\alpha_\sigma}^{j_\sigma}, \ldots, u_{\alpha_n}^{j_n}), \ldots, u_{\alpha_{\sigma-1}}^{j_{\sigma-1}}).$$

Further, in the plane $R^{r+\mu+\tau-1}$ we have a web $AW(\mu+\tau+1, \mu+\tau, r)$, and the surface generating it in this plane consists of the r-plane T^r and the surfaces V_μ^r and V_τ^r. The algebraic r-surface $T^r \cup V_\mu^r$ is of degree $\mu+1$ and lies in an $(r+\mu)$-plane $S^{r+\mu}$ and the r-surface V_τ^r belong to an $(r+\tau-1)$-plane $\tilde{S}^{r+\tau-1}$ where

$$S^{r+\mu} \cap \tilde{S}^{r+\tau-1} = \tilde{T}^r.$$

By Theorem 5.4.2, the web $AW(\mu+\tau+1, \mu+\tau, r)$ being considered is reducible of type

$$v^k = g^k(u_{\alpha_\sigma}^{j_\sigma}, \ldots, \varphi^l(u_{\alpha_{\sigma+\mu-1}}^{j_{\sigma+\mu-1}}, \ldots, u_{\alpha_n}^{j_n}), \ldots, u_{\alpha_{\sigma+\mu-2}}^{j_{\sigma+\mu-2}}),$$

and the whole web $AW(n+1,n,r)$ is reducible of type (5.4.2). If in the last consideration we interchange the surfaces V_μ^r and V_τ^r , we obtain that the web $AW(n+1,n,r)$ is reducible of type (5.4.3). In addition, it is clear that the reducibility (5.4.2) implies the reducibility (5.4.3) and conversely. So we have proved that (ii) implies (i).

It is obvious that (i) implies (ii): if a web $AW(n+1,n,r)$ is reducible of type (5.4.2), then we have the decomposition (5.4.5).

To complete the proof, note that the statements (i) and (iii) are equivalent since the hexagonality is equivalent to the relations (5.4.1) among the 2nd fundamental tensors of the surfaces U_ξ and the reducibility (5.3.2) is equivalent to the conditions (5.4.4) for the tensor $\lambda^\alpha_{\beta i}$. ■

As we saw, with an algebraic web $AW(n+1,n,r)$ which is reducible of type (5.4.2) there are associated three webs: the $AW(\sigma+1,\sigma,r)$ in the plane $Q^{r+\sigma-1}$, the $AW(\mu+2,\mu+1,r)$ in the plane $S^{r+\mu}$, and the $AW(\tau+1,\tau,r)$ in the plane $\tilde{S}^{r+\tau-1}$. They are generated by the components $V^r_\sigma, V^r_\mu, V^r_\tau$ and the r-planes $T^r, \tilde{T}^r$: the first web is generated by the surfaces $U_{n+1}, U_{\alpha_1}, \ldots, U_{\alpha_{\sigma-1}}$ and the plane T^r, the second one by the plane T^r, the surfaces $U_{\alpha_\sigma}, \ldots, U_{\alpha_{\sigma+\mu-1}}$ and the plane $\tilde{T}^r$, and the third one by the r-plane $\tilde{T}^r$ and the surfaces $U_{\alpha_{\sigma+\mu}}, \ldots, U_{\alpha_n}$. One can obtain similar results departing from the reducibility (5.4.3).

The following example illustrates our results.

Example 5.4.9 Let us consider an algebraic 6-web $AW(6,5,r)$ in P^{r+4} for which the generating r-surface V^r_σ decomposes into three r-dimensional quadrics : $V^r_6 = V^r_2 \cup \tilde{V}^r_2 \cup \tilde{\tilde{V}}^r_2$, i.e., $\sigma=\mu=\tau=2$. Suppose that

$$U_6, U_1 \subset V^r_2, \quad U_2, U_3 \subset \tilde{V}^r_2, \quad U_4, U_5 \subset \tilde{\tilde{V}}^r_2.$$

Such a web is reducible of type

$$u^i_6 = f^i(u^{j_1}_1, q^k(u^{j_2}_2, u^{j_3}_3, \varphi^l(u^{j_4}_4, u^{j_5}_5))),$$

or of type

$$u^i_6 = f^i(u^{j_1}_1, q^k(u^{j_4}_4, u^{j_5}_5, \varphi^l(u^{j_2}_2, u^{j_3}_3))).$$

For such a web the equations (5.4.1) hold where $\alpha, \beta = 1, \ldots, 5$, and its tensor $\lambda^\beta_{\alpha i}$ satisfies the equations (5.4.4):

$$\begin{cases} \lambda^2_{1i} = \lambda^3_{1i} = \lambda^4_{1i} = \lambda^5_{1i}, & \lambda^4_{2i} = \lambda^5_{2i} = \lambda^4_{3i} = \lambda^5_{3i}, \\ \lambda^1_{2i} = \lambda^1_{3i} = \lambda^1_{4i} = \lambda^1_{5i}, & \lambda^2_{4i} = \lambda^2_{5i} = \lambda^3_{4i} = \lambda^3_{5i}. \end{cases} \tag{5.4.6}$$

From (5.1.13), (5.1.14), and (5.4.6) it follows that the planes Q^{r+1}, R^{r+3}, S^{r+2}, $\tilde{S}^{r+2}$, T^r, and $\tilde{T}^r$ are defined as follows:

$$\begin{aligned} Q_{r+1} &= \{A_0, A_1, A_{n+1}, \ldots, A_{n+r}\}, \\ R_{r+3} &= \{A_2, A_3, A_4, A_5, A_{n+1} + \lambda^1_{2,n+1}A_1, \ldots, A_{n+r} + \lambda^1_{2,n+r}A_1\}, \\ S_{r+2} &= \{A_2, A_3, A_4 + A_5, A_{n+1} + \lambda^1_{2,n+1}A_1, \ldots, A_{n+r} + \lambda^1_{2,n+r}A_1\}, \\ \tilde{S}_{r+2} &= \{A_4, A_5, A_{n+1} + \lambda^1_{2,n+1}A_1 + \lambda^2_{4,n+1}(A_2+A_3), \ldots, \\ &\qquad A_{n+r} + \lambda^1_{2,n+r}A_1 + \lambda^2_{4,n+r}(A_2+A_3)\}, \\ S_{r+2} \cap \tilde{S}_{r+1} = \tilde{T}_r &= \{A_4 + A_5, A_{n+1} + \lambda^1_{2,n+1}A_1 + \lambda^2_{4,n+1}(A_2+A_3), \ldots, \end{aligned}$$

$$\begin{aligned} & A_{n+r} + \lambda^1_{2,n+r} A_1 + \lambda^2_{4,n+r}(A_2 + A_3)\} \\ = & \{A_4 + A_5, A_{n+1} + (\lambda^1_{2,n+1} - \lambda^2_{4,n+1})A_1 + \lambda^2_{4,n+1} A_0, \ldots, \\ & A_{n+r} + (\lambda^1_{2,n+r} - \lambda^2_{4,n+r})A_1 + \lambda^2_{4,n+r} A_0\}, \\ Q_{r+1} \cap R_{r+3} = T_r = & \{A_{n+r} + \lambda^1_{2,n+1} A_1, \ldots, A_{n+r} + \lambda^1_{2,n+r} A_1, A_2 + A_3 + A_4 + A_5\}. \end{aligned}$$

Note that in Theorem 5.4.6 we assumed that there are no r-planes among the components of the generating surface V^r_{n+1}. As to the webs discussed in Theorem 5.4.7, r-planes may be components of the V^r_r and as follows from Theorems 5.4.3 and Corollary 5.4.5, in this case the web $AW(n+1, n, r)$ will be a certain Bol $(n+1)$-web $B_{\xi_1 \ldots \xi_m}(n+1, n, r)$, $1 \leq m \leq n$.

5.4.5 Reducible Algebraic Four-Webs

To see possible geometric pictures if we allow r-planes to be among components of V^r_{n+1}, in this subsection we consider the case $n = 3$, i.e., a web $AW(4,3,r)$ in P^{r+2}. The web $AW(4,3,r)$ is interesting not only as an illustration but also since it represents an algebraic 4-web, next to an algebraic 3-web and since in this case we can renumerate all possible cases.

For an $AW(4,3,r)$ in P^{r+2} there is only one case corresponding to Theorem 5.4.6: $V^r_4 = V^r_2 \cup \tilde{V}^r_2$. Consequently all 4 3-subwebs of such $AW(4,3,r)$ are Bol webs $AW(3,2,r)$. However, by Theorem 5.4.3, the whole web $AW(4,3,r)$ is not a Bol $(n+1)$-web since none of the components of V^r_4 is an r-plane. This looks contradictory: all 3-subwebs are Bol webs and the whole 4-web is not a Bol web. However, it is easy to understand the reason of this. According to Definition 4.5.1, for a web $W(n+1, n, r)$ to be, for example, a Bol web $B_4(4,3,r)$, all its 3-subwebs $[4, \alpha, \beta]$ should be Bol three-webs of the same kind: (B_l), or (B_r), or (B_m). It means that when we project V^r_4 onto a fixed $(r+1)$-plane P^{r+1}, in P^{r+1} from a point taken on the surface U_γ, $\gamma \neq \alpha, \beta$, we must obtain a cubic hypersurface decomposing into an r-plane P^r and a hyperquadric and the projection of U_4 must belong to P^r. It will be the case when U_α and U_β belong to the same component V^r_2 or $\tilde{V}^r_2$ of V^r_4. So, this will be true only for one 3-subweb of type $[4, \alpha, \beta]$. For the two others, the projection of U_4 will belong to a hyperquadric $\tilde{\tilde{V}}^r_2$.

Let us consider now all the different cases when the generating surface V^r_4 decomposes and among its components there are one or two r-planes. In these cases

$$V^r_4 = V^r_1 \cup V^r_3 \quad \text{or} \quad V^r_4 = V^r_1 \cup \tilde{V}^r_1 \cup V^r_2.$$

The 2nd fundamental tensors of surfaces U_ξ are connected by relations (5.4.1) and correspondingly by one or two of the following conditions:

$$\begin{cases} \lambda^2_{1ij} = \lambda^3_{1ij} = 0, \\ \lambda^1_{2ij} = \lambda^3_{2ij} = 0, \\ \lambda^1_{3ij} = \lambda^2_{3ij} = 0, \\ p^1_{ij} = p^2_{ij} = p^3_{ij} = p_{ij}. \end{cases} \tag{5.4.7}$$

Each of conditions (5.4.7) implies the vanishing of one of the 2nd fundamental forms (5.1.42). In addition, in the second case one of the following two conditions holds:

$$\lambda^{\beta}_{\alpha i} = \lambda^{\gamma}_{\alpha i}, \quad \lambda^{\alpha}_{\beta i} = \lambda^{\alpha}_{\gamma i}, \quad \alpha, \beta, \gamma \text{ fixed.} \tag{5.4.8}$$

In the first case 3 and in the second case 2 out of 4 3-subwebs of $AW(4,3,r)$ are Bol webs $AW(3,2,r)$. The remaining 3-subweb in the first case is a general $AW(3,2,r)$ and the two remaining 3-subwebs in the second case are group webs $AW(3,2,r)$.

To prove this, we project three surfaces U_ξ onto a fixed hyperplane P^{r+1} from a fixed point of the fourth surface U_ξ. Then, in the first case in P^{r+1} we obtain a general cubic hypersurface if the centre of projection belongs to V_1^r and a cubic hypersurface decomposing into a hyperquadric and a hyperplane if the centre of projection belongs to V_3^r. In the second case we obtain in P^{r+1} a hyperplane and a hyperquadric if the centre projection is on V_1^r or $\tilde{V}_1^r$, and 3 hyperplanes of general position if the centre of projection belongs to V_2^r. The type of projections being obtained gives the type of 3-subwebs indicated above.

Note that, by Theorem 5.4.3 and Corollary 5.4.5, a web $AW(4,3,r)$ generated by V_4^r is a Bol web of type $B_\xi(4,3,r)$ and $B_{\xi_1\xi_2}(4,3,r)$ respectively.

Note also that in the second case two r-planes that are parts of V_4^r in general do not lie in an $(r+1)$-plane and because of this, the corresponding $AW(4,3,r)$ is not reducible. If V_1^r and $\tilde{V}_1^r$ belong to an $(r+1)$-plane, i.e., they have an $(r-1)$-dimensional intersection, then the corresponding web $AW(4,3,r)$ is reducible. Both the conditions (5.4.8) are satisfied for this web, and in contrast to the case $V_4^r = V_2^r \cup \tilde{V}_2^r$, only two of its 3-subwebs are Bol webs $AW(3,2,r)$ while two others are group webs $AW(3,2,r)$.

5.4.6 Completely Reducible Algebraic $(n+1)$-Webs

We will study now completely reducible algebraic $(n+1)$-webs $AW(n+1,n,r)$. Without loss of generality, we assume that an algebraic web $AW(n+1,n,r)$ is completely reducible of type

$$u^i_{n+1} = \overset{(1)}{f}{}^i(u_1^{j_1}, \overset{(2)}{f}{}^k(u_2^{j_2}, \ldots, \overset{(n-1)}{f}{}^l(u_{n-1}^{j_{n-1}}, u_n^{j_n}\underbrace{)\ldots))}_{n-1}. \tag{5.4.9}$$

For such a web, in addition to the conditions (5.4.1) of algebraisability, according to the results of Section 4.2, and equations (5.1.30), we have the relations

$$\left\{\begin{array}{l} \lambda^2_{1i} = \lambda^3_{1i} = \ldots = \lambda^n_{1i}, \quad \lambda^1_{2i} = \lambda^1_{3i} = \ldots = \lambda^1_{ni}, \\ \lambda^3_{2i} = \ldots = \lambda^n_{2i}, \quad \lambda^2_{3i} = \ldots = \lambda^2_{ni}, \\ \ldots\ldots\ldots\ldots\ldots\ldots\ldots\ldots\ldots\ldots\ldots\ldots \\ \lambda^{n-1}_{n-2,i} = \lambda^n_{n-2,i}, \quad \lambda^{n-2}_{n-1,i} = \lambda^{n-2}_{ni}. \end{array}\right. \tag{5.4.10}$$

Equations (5.4.10), (5.4.1), and (5.1.16) imply

$$\begin{cases} p^2_{ij} = p^3_{ij} = \ldots = p^n_{ij}, \\ \lambda^2_{1ij} = \lambda^3_{1ij} = \ldots = \lambda^n_{1ij} = p^1_{ij} - p^2_{ij}, \\ \lambda^3_{2ij} = \ldots = \lambda^n_{2ij} = 0, \\ \ldots\ldots\ldots\ldots\ldots\ldots\ldots\ldots \\ \lambda^{n-1}_{n-2,ij} = \lambda^n_{n-2,ij} = 0, \\ \lambda^n_{n-1,ij} = \lambda^{n-1}_{nij}, \\ \lambda^1_{2ij} = \lambda^1_{3ij} = \ldots = \lambda^1_{nij} = 0, \\ \lambda^2_{3ij} = \ldots = \lambda^2_{nij} = 0, \\ \ldots\ldots\ldots\ldots\ldots\ldots\ldots\ldots \\ \lambda^{n-2}_{n-1,ij} = \lambda^{n-2}_{nij} = 0. \end{cases} \tag{5.4.11}$$

Equations (5.4.11) and (5.1.42) show that the 2nd fundamental forms of the surfaces U_ξ are

$$\begin{array}{ll} \Phi^\beta_{n+1} = (p^1_{ij} - p^2_{ij})\theta^i_0\theta^j_0\,; & \\ \Phi^\beta_1 = (p^1_{ij} - p^2_{ij})\theta^i_1\theta^j_1\,; & \\ \Phi^\beta_2 = 0, \ \ \beta \neq 2; & \Phi^\beta_3 = 0, \ \ \beta \neq 3, \ldots, \Phi^\beta_{n-2} = 0, \ \ \beta \neq n-2; \\ \Phi^\beta_{n-1} = 0, \ \ \beta \neq n-1, n; & \Phi^n_{n-1} = \lambda^n_{n-1,ij}\theta^i_{n-1}\theta^j_{n-1}, \\ \Phi^\beta_n = 0, \ \ \beta \neq n-1, n; & \Phi^{n-1}_n = \lambda^n_{n-1,ij}\theta^i_n\theta^j_n. \end{array}$$

This structure of the 2nd fundamental forms means that the surfaces $U_2, \ldots, U_{n-2}$ are r-planes, and that each of the pairs U_{n+1}, U_1 and U_{n-1}, U_n belongs to its own an r-dimensional quadric.

The r-planes $U_2, \ldots, U_{n-2}$ which are components of V^r_{n+1} are not in general position with respect to the two r-dimensional quadrics $U_{n+1} \cap U_1$ and $U_{n+1} \cap U_n$. It follows from (5.1.13), (5.1.14), and (5.4.10) that the r-quadrics being indicated lie in the $(r+1)$-planes Q^{r+1} and $\tilde{Q}^{r+1}$ spanned by the points A_0, A_1, A_i and A_{n-1}, A_n, $A_i + \sum_{\alpha=1}^{n-2} \lambda^\alpha_{\alpha+1,i} A_\alpha$, and that the r-planes $U_2, \ldots, U_{n-2}$ are spanned respectively by the points:

$$\begin{array}{ll} A_2, & A_i + \lambda^3_{2i}(A_3 + \ldots + A_n); \\ A_3, & A_i + \lambda^2_{3i}A_2 + \lambda^4_{3i}(A_4 + \ldots + A_n); \\ \ldots\ldots & \ldots\ldots\ldots\ldots\ldots\ldots \\ A_{n-2}, & A_i + \sum\limits_{\alpha=1}^{n-2} \lambda^\alpha_{\alpha+1,i}A_\alpha + \lambda^{n-1}_{n-2,i}(A_{n-1} + A_n). \end{array}$$

It follows from this that the r-planes U_2 and U_{n-2} intersect the $(r+1)$-planes Q^{r-1} and $\tilde{Q}^{r-1}$ at $(r-1)$-planes spanned by the points

$$A_i + (\lambda^1_{2i} - \lambda^3_{2i})A_i + \lambda^3_{2i}A_0 \ \text{ and } \ A_i + \sum_{\alpha=1}^{n-2} \lambda^\alpha_{\alpha+1,i}A_\alpha + \lambda^{n-1}_{n-2,i}(A_{n-1} + A_n).$$

For a completely reducible algebraic $AW(n+1, n, r)$ the $n-2$ r-planes being obtained above can be invariantly defined. These r-planes are r-planes which are components of the generating surfaces of each of the webs associated with the completely reducible $AW(n+1,n,r)$. In the case of complete reducibility (5.4.9) the first such r-plane $\overset{(2)}{f}$ is obtained as an intersection of the $(r+1)$-plane Q^{r+1} and an $(r+n-2)$-plane, which contains the r-planes $U_2,\ldots,U_{n-2}$ and the r-quadric $U_{n-1}\cup U_n$. Since the latter is determined by the points $A_2,\ldots,A_n, A_i+\lambda^1_{2i}A_1$, then the r-plane $\overset{(2)}{f}$ is determined by the points $A_i+\lambda^1_{2i}A_1$ and $\sum_{\alpha=2}^n A_\alpha$.

The next r-plane $\overset{(3)}{f}$ is obtained as an intersection of an $(r+1)$-plane containing the r-planes $\overset{(2)}{f}$ and U_2, and an $(r+n-3)$-plane which contains the r-planes $U_3,\ldots,U_{n-2}$ and the r-quadric $U_{n+1}\cup U_n$. These two planes are spanned respectively by the points

$$A_2, A_i+\lambda^1_{2i}A_1, \ \sum_{\alpha=3}^n A_\alpha \quad \text{and} \quad A_3, A_4,\ldots,A_n, \ \ A_i+\lambda^1_{2i}\ A_1+\lambda^2_{3i}A_2,$$

and therefore the r-plane $\overset{(3)}{f}$ is spanned by the points

$$A_i+\lambda^1_{2i}A_1+\lambda^2_{3i}A_2 \quad \text{and} \quad \sum_{\alpha=3}^n A_\alpha.$$

In the same manner, we find that the r-planes $\overset{(4)}{f}, \overset{(5)}{f},\ldots, \overset{(n-1)}{f}$ are spanned respectively by the points

$$\begin{array}{ll} A_i+\sum\limits_{\alpha=2}^{4}\lambda^{\alpha-1}_{\alpha i}A_{\alpha-1}, & \sum\limits_{\alpha=5}^{n}A_\alpha; \\ A_i+\sum\limits_{\alpha=2}^{5}\lambda^{\alpha-1}_{\alpha i}A_{\alpha-1}, & \sum\limits_{\alpha=6}^{n}A_\alpha; \\ \dots\dots\dots & \dots\dots\dots \\ A_i+\sum\limits_{\alpha=2}^{n-2}\lambda^{\alpha-1}_{\alpha i}A_{\alpha-1}, & \sum\limits_{\alpha=n-1}^{n}A_\alpha. \end{array}$$

Now, for the algebraic web $AW(n+1,n,r)$ which is completely reducible of type (5.4.9), we can describe a procedure for getting the point $A_0\in U_{n+1}$ corresponding to the points $A_\alpha\in U_\alpha$:

$$A_{n-1}A_n\cap \overset{(n-1)}{f} = \overset{(n-1)}{F}, \quad \overset{(n-1)}{F}\ A_{n-2}\cap \overset{(n-2)}{f} = \overset{(n-2)}{F}, \quad \overset{(n-2)}{F}\ A_{n-3}\cap \overset{(n-3)}{f} = \overset{(n-3)}{F},$$
$$\ldots, \overset{(3)}{F}\ A_2\cap \overset{(2)}{f} = \overset{(2)}{F}, \quad \overset{(2)}{F}\ A_1\cap U_{n+1} = A_0.$$

Conversely, if the generating surface V^r_{n+1} consists of two r-quadrics $U_{n+1}\cup U_1, U_{n-2}\cup U_n$ and $n-3$ r-planes $U_2,\ldots,U_{n-2}$ where the r-planes U_2 and U_{n-1} intersect respectively the $(r+1)$-planes Q^{r+1} and $\tilde{Q}^{r+1}$ which contain $U_{n+1}\cup U_1$ and

$U_{n-1} \cup U_n$, at $(r-1)$-planes, then equations (5.4.10) and (5.4.11) hold and the complete reducibility (5.4.9) takes place.

The following theorem combines some of the results obtained in this subsection.

Theorem 5.4.10 *In a domain of regularity of an algebraic web $AW(n+1, n, r)$ the following three statements are equivalent:*

(i) *A web $AW(n+1, n, r)$ is completely reducible of type* (5.4.1) *or some other type.*

(ii) *The generating surface V^r_{n+1} of the $AW(n+1, n, r)$ decomposes into two r-dimensional quadrics and $n-3$ r-planes. Among these r-planes there are two r-planes such that one of them intersects an $(r+1)$-plane containing one r-quadric and the second one intersects another $(r+1)$-plane containing the second r-quadric.*

(iii) *The tensor $\lambda^\beta_{\alpha i}$ and the 2nd fundamental tensors of the surfaces U_ξ satisfy relations* (5.4.10) *and* (5.4.11) *in the case of complete reducibility* (5.4.9) *or similar conditions in the case of complete reducibility of some other type.* ■

To illustrate a structure of a completely reducible algebraic web $AW(n+1, n, r)$ for which the complete reducibility has a type different from the type (5.4.9), we consider an example.

Example 5.4.11 Suppose an algebraic 5-web $AW(5, 4, r)$ is given in P^{r+3} and it is completely reducible of type

$$u^i_5 = f^i(g^k(u^{j_1}_1, u^{j_2}_2), \varphi^l(u^{j_3}_3, u^{j_4}_4)). \tag{5.4.12}$$

From Section 4.2 and equations (5.1.30) we get the following relations for the tensor $\lambda^\beta_{\alpha i}$ of such a 5-web:

$$\lambda^3_{1i} = \lambda^4_{1i} = \lambda^3_{2i} = \lambda^4_{2i}, \quad \lambda^1_{3i} = \lambda^2_{3i} = \lambda^1_{4i} = \lambda^2_{4i}. \tag{5.4.13}$$

Equations (5.4.13), (5.1.16), and (5.4.1) give

$$\begin{cases} \lambda^3_{1ij} = \lambda^4_{1ij} = \lambda^3_{2ij} = \lambda^4_{2ij} = \lambda^1_{3ij} = \lambda^2_{3ij} = \lambda^1_{4ij} = \lambda^2_{4ij} = 0, \\ p^1_{ij} = p^2_{ij} = p^3_{ij} = p^4_{ij}, \quad \lambda^2_{1ij} = \lambda^1_{2ij}, \quad \lambda^4_{3ij} = \lambda^3_{4ij}. \end{cases} \tag{5.4.14}$$

From (5.4.14) it follows that the 2nd fundamental forms of U_ξ have the following structure:

$$\begin{aligned} \Phi^2_5 &= \Phi^3_5 = \Phi^4_5 = 0, \\ \Phi^3_1 &= \Phi^4_1 = \Phi^3_2 = \Phi^4_2 = 0, \quad \Phi^2_1 = \lambda^2_{1ij}\theta^i_1\theta^j_1, \quad \Phi^1_2 = \lambda^1_{2ij}\theta^i_2\theta^j_2, \\ \Phi^1_3 &= \Phi^2_3 = \Phi^1_4 = \Phi^2_4 = 0, \quad \Phi^4_3 = \lambda^4_{3ij}\theta^i_3\theta^j_3, \quad \Phi^3_4 = \lambda^4_{3ij}\theta^i_4\theta^j_4. \end{aligned}$$

This structure and equations (5.4.12) mean that the generating surface V^r_5 decomposes into an r-plane U_5 and two r-quadrics containing correspondingly the r-surfaces U_1, U_2 and U_3, U_4.

Let us find the r-planes g and φ which are invariantly connected with the $AW(5,4,r)$. For this we note that the r-plane U_5 and the r-quadric $U_1 \cup U_2$ lie in an $(r+2)$-plane Q^{r+2} spanned by the points $A_1, A_2, A_3 + A_4, A_i$, and that the r-quadric $U_3 \cup U_4$ lies in the $(r+1)$-plane R^{r+1} spanned by the points $A_3, A_4, A_i + \lambda^1_{3i}(A_1 + A_2)$. From this we find that the r-plane $\varphi = Q^{r+2} \cap R^{r+1}$ is spanned by the points $A_3 + A_4$ and $A_i + \lambda^1_{3i}(A_1 + A_2)$.

Using the same way, we find that the $(r+2)$- and $(r+1)$-planes $\tilde{Q}^{r+2}$ and $\tilde{R}^{r+1}$ where $U_5, U_3 \cup U_4$ and $U_1 \cup U_2$ are located, are spanned respectively by the points

$$A_3, A_4, A_1 + A_2, A_i \quad \text{and} \quad A_1, A_2, A_i + \lambda^3_{1i}(A_3 + A_4),$$

and their intersection $g = \tilde{Q}^{r+2} \cap \tilde{R}^{r+1}$ is spanned by the points

$$A_1 + A_2, \qquad A_i + \lambda^3_{1i}(A_3 + A_4).$$

In this case the procedure for getting the point $A_0 \in U_5$ provided that the points $A_\alpha \in U_\alpha$ are given, is as follows. Construct the straight lines A_3A_4 and A_1A_2. They intersect the r-planes φ and g at the points $A_3 + A_4$ and $A_1 + A_2$ respectively. The straight line joining the last two points intersects the r-plane U_5 at the point $A_0 = A_1 + A_2 + A_3 + A_4$.

As we already noted, in the case $n = 3$, there is only one type of reducible algebraic 4-webs $AW(4,3,r)$ (see subsection 5). It is obvious that every reducible algebraic web $AW(4,3,r)$ is completely reducible. Since the surfaces U_1, U_2, U_3, and U_4 must pairwise form the components V_2^r and $\tilde{V}_2^r$, such a reducibility can be realised by three different ways (compare the end of Section 5.3).

5.5 Moufang Algebraic $(n+1)$-Webs

Suppose first that we have a Grassmann web $GW(n+1,n,r)$ given in P^{r+n-1} by $n+1$ r-surfaces U_ξ. Note that if all the surfaces U_ξ are r-planes, then, since they belong to P^{r+n-1}, they determine a surface V^r_{n+1} (according to [Ber 24], V^r_{n+1} always belong to a P^{r+n}) and our web $GW(n+1,n,r)$ is algebraic. Conversely, if the generating surface of an algebraic web $AW(n+1,n,r)$ decomposes into r-planes, then the surfaces U_ξ are 4 r-planes belonging to these components of V^r_{n+1}. So, in this case we have an algebraic web $AW(n+1,n,r)$ whose generating surface V^r_{n+1} decomposes into $n+1$ r-planes. By Corollary 5.4.6, such a web $AW(n+1,n,r)$ is a Moufang web $M(n+1,n,r)$.

In this section we will study Moufang algebraic webs $M(n+1,n,r)$. First we consider the case when the r-planes U_ξ are in general position in P^{r+n-1}.

Theorem 5.5.1 *In a domain of regularity of a Grassmann web $GW(n+1,n,r)$ the following three statements are equivalent:*

(i) *The surfaces U_ξ generating a Grassmann web $GW(n+1,n,r)$ are r-planes in general position in P^{r+n-1}.*

(ii) *The 2nd fundamental tensors of all surfaces U_ξ vanish:*

$$\lambda^{\beta}_{\alpha ij} = 0, \quad p^{\alpha}_{ij} = p^{\beta}_{ij}, \quad \alpha, \beta = 1, \ldots, n, \quad \alpha \neq \beta \tag{5.5.1}$$

(iii) *All 3-subwebs $W_{\xi_1 \ldots \xi_{n-2}}(3,2,r)$ of a $GW(n+1,n,r)$ are group three-webs.*

Proof. The equivalence of (i) and (ii) is obvious. The equivalence of (i) and (iii) follows from the fact that the projection of V^r_{n+1} from an $(n-3)$-dimensional centre Z^{n-3} spanned by $n-2$ points taken by one from any $n-2$ surfaces U_ξ onto a fixed $(r+1)$-plane P^{r+1} , gives in P^{r+1} three hyperplanes in general position if and only if the surfaces U_ξ are r-planes in general position. This means (see [Ak 73]) that in P^{r+1} we have a group three-web $AW(3,2,r)$. This completes our proof. Note that the equivalence of (ii) and (iii) can be proved by showing that the conditions (5.5.1) are equivalent to the vanishing of the curvature tensors of all 3-subwebs $W_{\xi_1 \ldots \xi_{n-2}}(3,2,r)$ which characterises group three-webs (see Table 1.1 in Section 1.2). ■

We will consider now some special classes of Moufang webs $M(n+1,n,r)$ characterised by the fact that the $n+1$ r-planes into which the generating surface V^r_{n+1} of a Moufang web $M(n+1,n,r)$ decomposes are not in general position. We will single out some possible cases and indicate their characteristics.

1) r-planes $U_{\xi_1}, \ldots, U_{\xi_p}, p \leq n$, have a common $(r-1)$-plane. If the indicated r-planes are $U_{\alpha_1}, \ldots, U_{\alpha_p}$, and only in this case, we have

$$\lambda^{\alpha_\delta}_{\alpha_\rho i} = \lambda^{\alpha_\delta}_{i}, \quad \lambda^{\alpha_\delta}_{\alpha_\rho i} = \lambda^{\alpha_\sigma}_{\alpha_\tau i}, \quad \rho, \ \sigma, \ \tau = 1, \ldots, p; \ \delta = p+1, \ldots, n. \tag{5.5.2}$$

If the indicated r-planes are $U_{n+1}, U_{\alpha_1}, \ldots, U_{\alpha_{p-1}}$, and only in this case, we have

$$\lambda^{\beta}_{\alpha_\rho i} = \lambda_i, \quad \rho = 1, \ldots p-1; \quad \beta = 1, \ldots, n; \quad \beta \neq \alpha_\rho. \tag{5.5.3}$$

If $p \geq 3$, then $\binom{p}{3}$ of 3-subwebs $W_{\xi_1 \ldots \xi_{n-2}}(3,2,r)$ are parallelisable and the remaining 3-subwebs are group three-webs. We will not repeat the projection procedure that should be used for the proof. However, it is worth to note that, according to [Ak 73], if the generating surface in P^{r+1} decomposes into three hyperplanes belonging to a pencil of hyperplanes with an $(r-1)$-dimensional vertex, and only in this case, the corresponding three-web in P^{r+1} will be parallelisable.

2) r-planes $U_{n+1}, U_{\alpha_1}, \ldots, U_{\alpha_{p-1}}$ have a common $(r-1)$-plane and the remaining r-planes $U_{\alpha_p}, \ldots, U_{\alpha_n}$ have a common $(r-1)$-plane which is different from the first one. In this case, and only in this case, we have the following analytical conditions:

$$\begin{aligned} \lambda^{\beta}_{\alpha_\rho i} &= \lambda_i, \quad \rho = 1, \ldots, p-1; \quad \beta = 1, \ldots, n; \quad \beta \neq \alpha_\rho, \\ \lambda^{\alpha_p}_{\alpha_h i} &= \lambda^{\alpha_p}_{i}, \quad \lambda^{\alpha_g}_{\alpha_h i} = \lambda^{\alpha_g}_{\alpha_f i}, \quad f, g, h = p, \ldots, n, \end{aligned} \tag{5.5.4}$$

part of which, of course, agrees with (5.5.3). Conditions (5.5.4) imply, in particular, that the $(n+1)$-web under consideration is reducible of type

$$u^i_{n+1} = f^i(u^{j_1}_{\alpha_1}, \ldots, g^k(u^{j_p}_{\alpha_p}, \ldots, u^{j_n}_{\alpha_n}), \ldots, u^{j_{p-1}}_{\alpha_{p-1}}). \tag{5.5.5}$$

Moreover, in the case $n = 3, p = 2$, and only in this case, the reducibility (5.5.5) is not only necessary but also sufficient for the webs being considered.

3) r-planes $U_1, U_2, \ldots, U_n, U_{n+1}$ form a cycle where each of r-planes intersects the following one in an $(r-1)$-plane (U_1 follows U_{n+1}), and all $n+1$ such $(r-1)$-planes are mutually distinct. Analytically, this case is characterised by the conditions:

$$\lambda^\beta_{\alpha i} = -\lambda^\alpha_{\beta i} = \lambda_i, \quad \alpha < \beta. \tag{5.5.6}$$

It is easy to see that equations (5.1.30) show that conditions (4.3.1) are equivalent to conditions (5.5.6), i.e., conditions (5.5.6) hold for group Grassmann web $GW(n+1, n, r)$, and only for such webs. Since we consider the case when the surfaces U_ξ are r-planes, the web under consideration is algebraic. So, a group algebraic web $AW(n+1, n, r)$ is a special case of a Moufang web $M(n+1, n, r)$.

The $(r-1)$-planes of intersections of U_ξ are spanned respectively by the points

$$A_i + \lambda_i(A_1 + \ldots + A_n); \quad A_i + \lambda_i(-A_1 + A_2 + \ldots + A_n),$$

$$A_i + \lambda_i(-A_1 - A_2 + A_3 + \ldots + A_n); \ldots, A_i + \lambda_i(-A_1 - A_2 - \ldots - A_n).$$

The group algebraic web $AW(n+1, n, r)$ can be also obtained if one takes an algebraic web $AW(n+1, n, r)$ which is completely reducible of type (5.4.9) and reducible of type

$$u^i_{n+1} = f^i(g^k(u^{j_1}_1, \ldots, u^{j_{n-1}}_{n-1}), u^{j_n}_n), \tag{5.5.7}$$

since (5.4.9) and (5.5.7) implies the equalities

$$\lambda^\alpha_{ni} = \lambda^\beta_{ni}, \quad \lambda^n_{\alpha i} = \lambda^n_{\beta i}, \quad \alpha \neq \beta, \tag{5.5.8}$$

and (5.5.8), (5.4.10), and (5.1.22) lead to (5.5.1). The converse is trivial.

Because of the importance of this class of $(n+1)$-webs, we will outline all results we have obtained for this class in the following theorem.

Theorem 5.5.2 *In a domain of reqularity of an algebraic web $AW(n+1, n, r)$ the following four statements:*

(i) *The web $AW(n+1, n, r)$ is a group web.*

(ii) *The surfaces U_ξ generating the $AW(n+1, n, r)$ are r-planes and the r-planes $U_1, U_2, \ldots, U_n, U_{n+1}$ form a cycle where each r-plane intersects the neighboring ones in $(r-1)$-planes.*

(iii) *The 2nd fundamental forms of all U_ξ vanish and the conditions* (5.5.6) *hold.*

(iv) *A web $AW(n+1, n, r)$ is completely reducible of type* (5.4.9) *and reducible of type* (5.5.7). ■

4) Suppose now that all the $n+1$ r-planes U_ξ generating a Moufang algebraic web $M(n+1,n,r)$ have a common $(r-1)$-plane. In this case, and only in this case, we have

$$\lambda^{\beta}_{\alpha i} = 0. \tag{5.5.9}$$

It is clear from (5.1.13) and (5.1.14) that this $(r-1)$ plane is spanned by the points A_i . Moreover, the form (5.1.30) of the torsion tensor of a Grassmann web show that condition (5.5.9) is equivalent to the vanishing of the torsion tensor $\underset{\alpha\beta}{a}{}^{i}_{jk} : \underset{\alpha\beta}{a}{}^{i}_{jk} = 0$. By Theorem 1.5.2, this means that (5.5.9) is equivalent to the parallelisability of the web under consideration.

We proved the following theorem.

Theorem 5.5.3 *In a domain of regularity of a Grassmann web $GW(n+1,n,r)$ the following three statements are equivalent:*

(i) *The web $GW(n+1,n,r)$ is parallelisable.*

(ii) *The surfaces U_ξ generating the $GW(n+1,n,r)$ are r-planes passing through an $(r-1)$-plane.*

(iii) *The tensor $\lambda^{\beta}_{\alpha i}$ vanishes.* ■

Remark 5.5.4 Note that we can get the parallelisable Grassmann web $GW(n+1,n,r)$ from a group algebraic web $AW(n+1,n,r)$ by claiming that at least two of the $(r-1)$-planes of intersections of U_ξ coincide. Another way to get the parallelisable Grassmann web $GW(n+1,n,r)$ is to combine some of special positions of the r-planes which we discussed in 1) and 2). For example, a Grassmann web $GW(n+1,n,r)$ is parallelisable if and only if the r-planes $U_{n+1}, U_\alpha, \alpha \neq \alpha_p, p$ fixed, pass through an $(r-1)$-plane and the same is true for the r-planes $U_{n+1}, U_\alpha, \alpha \neq \alpha_q, q \neq p, q$ fixed, or for all r-planes U_α.

We conclude this section by indicating in the case $n=3$ the most interesting cases of Moufang Grassmann webs $GW(4,3,r)$. As we mentioned in the beginning of this section, they automatically are Moufang algebraic webs $AW(n+1,n,r)$.

a) Each of the pairs of r-planes U_β, U_γ and U_α, U_4 has a common $(r-1)$-plane. In this, and only in this case, the Moufang Grassmann web $GW(4,3,r)$ is α-reducible: $u^i_4 = f^i(u^j_\alpha, g^k(u^l_\beta, u^m_\gamma))$.

b) Not only each of the pairs U_β, U_γ and U_α, U_4 but also each of the pairs U_α, U_γ and U_β, U_4 have a common $(r-1)$-plane. In this case, and only in this case,a Grassmann web is algebraisable and reducible of two different types (see Section 4.3) and therefore it will be a group 4-web.

The configuration being described corresponds to the general case of a group Grassmann web $GW(4,3,r)$. In fact the r-planes U_4, U_1, U_2, U_3 taken in this order gives the situation corresponding to the case 3.

c) Three out of four r-planes U_ξ have a common $(r-1)$-plane. In this case, and only in this case, one of the 3-subwebs $W_{\xi_1\xi_2\xi_3}(3,2,r)$ of a Moufang Grassmann web $GW(4,3,r)$ is parallelisable and three others are group three-webs.

d) All four r-planes U_ξ have a common $(r-1)$-plane. In this case, and only in this case, the Moufang Grassmann web $GW(n+1,n,r)$ is parallelisable.

5.6 $(2n+2)$-Hedral Grassmann $(n+1)$-Webs

In this section we will study $(2n+2)$-hedral Grassmann webs $GW(n+1,n,r)$. We will prove that they are always algebraic, and we will describe a structure of their generating surfaces V^r_{n+1}.

Theorem 5.6.1 *In a domain of regularity of a Grassmann web $GW(n+1,n,r)$ the following three statements are equivalent:*

(i) *A Grassmann web $GW(n+1,n,r)$ is $(2n+2)$-hedral.*

(ii) *The tensor $\lambda^\beta_{\alpha i}$ of a $GW(n+1,n,r)$ satisfies the conditions*

$$\lambda^\beta_{\alpha i} = \lambda^\alpha_\beta = 0. \qquad (5.6.1)$$

(iii) *The surfaces U_ξ generating a $GW(n+1,n,r)$ belong to one and the same algebraic r-dimensional surface of degree 2^{n-1} which is the intersection of $n-1$ hyperquadrics of the space P^{r+n-1}.*

Proof. First we will prove that (i) and (ii) are equivalent. Suppose that a Grassmann web $GW(n+1,n,r)$ is $(2n+2)$-hedral. According to Theorem 4.4.2, a $(2n+2)$-hedral web is distinguished by the condition $\underset{\alpha\beta}{a}{}^i_{(jk)} = 0$ for its torsion tensor. Since for a Grassmann web $GW(n+1,n,r)$ the torsion tensor has a structure (5.1.30), the condition $\underset{\alpha\beta}{a}{}^i_{(jk)} = 0$ is equivalent to the relations (5.6.1). Note that if we differentiate (5.6.1) by means of (5.1.16), then by virtue of the linear independence of the 1-forms θ^i_α, we obtain as one of consequences the algebraisability conditions (5.4.1). So, $(2n+2)$-hedral Grassmann webs $GW(n+1,n,r)$ are always algebraic.

Next we will prove that (i) and (iii) are equivalent. Suppose that we have a Grassmann $(2n+2)$-hedral web $GW(n+1,n,r)$. By Theorem 1.9.18, its transversally geodesic subwebs $GW(n+1,n,1)$ cut on fixed n-planes σ (see Theorem 5.1.4), are also $(2n+2)$-hedral. According to [Ba 51a], a $(2n+2)$-hedral Grassmann web $GW(n+1,n,1)$ is dual to a $(2n+2)$-hedral $(n+1)$-web formed by $(n-1)$-planes that are common for a $(\binom{n}{2}) - 2)$-parameter system of hypersurfaces of the second class. Therefore, the $(2n+2)$-hedral Grassmann web $GW(n+1,n,r)$ is generated by an algebraic curve of degree 2 that is common for $n-1$ hyperquadrics. The lines $u_\xi = \sigma \cap U_\xi$ belong to this curve (see Section 5.1.5). Since the n-plane σ is an arbitrary n-plane, the surfaces U_ξ themselves belong to one and the same r-dimensional algebraic surface of degree 2^{n-1} which is the intersection of $n-1$ hyperquadrics. ■

Note that $2^{n-1} \geq n+1$, and equality holds only if $n=3$, i.e., for an octahedral web $GW(4,3,r)$ an r-dimensional algebraic surface of order 4, which is obtained as the intersection of two hyperquadrics, coincides with the surface V_4^r generating the 4-web and consisting of four sheets U_ξ.

If $n > 3$, the generating surface V_{n+1}^r for the $(n+1)$-web consisting of $n+1$ sheets U_ξ is a part of an algebraic r-surface of order 2^{n-1} which generates a $(2n+2)$-hedral $(n+1)$-web.

Theorem 5.6.1 generalises the Graf-Sauer theorem for octahedral 4-webs formed by 2-planes in a three-dimensional projective space (see Theorem 2.6.6). Note that in the case $r = 1, n > 1$ a generalisation has been made by Bartsch [Ba 53] and in the case $n = 3, r \geq 1$ by Akivis and Goldberg [AG 74].

Now we will be able to find another criterion for a Grassmann web $GW(n+1, n, r)$ to be a group $(n+1)$-web.

Theorem 5.6.2 *For a Grassmann web $GW(n+1, n, r)$ to be a group $(n+1)$-web, it is necessary and sufficient that it be $(2n+2)$-hedral and that all its 3-subwebs $W_{\xi_1 \ldots \xi_{n-2}}(3, 2, r)$ be group three-webs.*

Proof. *Necessity.* From conditions (5.5.6) which with some other conditions characterises a group web $GW(n+1, n, r)$, follow equations (5.6.1) characterising $(2n+2)$-hedral webs $GW(n+1, n, r)$ as well as conditions (5.5.1) meaning that all 3-subwebs $W_{\xi_1 \ldots \xi_{n-2}}(3, 2, r)$ are group three-webs.

Sufficiency. We must prove that conditions (5.5.1) and (5.6.1) imply conditions (5.5.6) or similar conditions which with (5.5.1) give that the web $GW(n+1, n, r)$ is a group $(n+1)$-web. In fact, differentiating (5.6.1) and using (5.1.16), (5.1.27), and (5.5.1), we first find that

$$\overset{1}{q}_{ij} = \overset{2}{q}_{ij} = \ldots = \overset{n}{q}_{ij} = q_{ij}, \tag{5.6.2}$$

and secondly, taking into account (5.6.2), we find that for any mutually distinct α, β, γ we have

$$\lambda_{\alpha i}^{\beta} \lambda_{\alpha j}^{\beta} = q_{ij} - p_{ij}, \tag{5.6.3}$$

$$\lambda_{\alpha i}^{\beta}(\lambda_{\gamma j}^{\alpha} + \lambda_{\beta j}^{\gamma}) + \lambda_{\beta i}^{\gamma}(\lambda_{\alpha j}^{\beta} + \lambda_{\gamma j}^{\alpha}) + \lambda_{\gamma i}^{\alpha}(\lambda_{\beta j}^{\gamma} + \lambda_{\alpha j}^{\beta}) = 2(p_{ij} - q_{ij}). \tag{5.6.4}$$

Equations (5.6.3) and (5.6.4) show that

$$(\lambda_{\alpha i}^{\beta} + \lambda_{\beta i}^{\gamma} + \lambda_{\gamma i}^{\alpha})(\lambda_{\alpha j}^{\beta} + \lambda_{\beta j}^{\gamma} + \lambda_{\gamma j}^{\alpha}) = q_{ij} - p_{ij}. \tag{5.6.5}$$

It follows from (5.6.3) and (5.6.5) that if $\lambda_{\alpha i}^{\beta} = 0$ for some fixed α, β, then all $\lambda_{\varepsilon i}^{\delta} = 0$, i.e., the web $GW(n+1, n, r)$ is parallelisable, but a parallelisable web $GW(n+1, n, r)$ is a group $(n+1)$-web.

Suppose now that $\lambda_{\alpha i}^{\beta} \neq 0$ for any α and β. Then, for any fixed α, β, γ, from (5.6.3) and (5.6.5) we obtain that for $i = j$ we have

$$\lambda_{\beta i}^{\gamma} = \varepsilon_1 \lambda_{\alpha i}^{\beta}, \quad \lambda_{\gamma i}^{\alpha} = \varepsilon_2 \lambda_{\alpha i}^{\beta}, \quad \varepsilon_1 = \pm 1, \quad \varepsilon_2 = \pm 1. \tag{5.6.6}$$

It follows from (5.6.6), (5.6.3), and (5.6.5) that $(1+\varepsilon_1+\varepsilon_2)^2 = 1$. This gives that either $\varepsilon_1+\varepsilon_2 = 0$ or $\varepsilon_1+\varepsilon_2 = -2$. As result, we have three possibilities:

$$-\lambda^{\beta}_{\alpha i} = \lambda^{\gamma}_{\beta i} = \lambda^{\alpha}_{\gamma i}, \quad \lambda^{\beta}_{\alpha i} = -\lambda^{\gamma}_{\beta i} = \lambda^{\alpha}_{\gamma i}, \quad \lambda^{\beta}_{\alpha i} = \lambda^{\gamma}_{\beta i} = -\lambda^{\alpha}_{\gamma i}. \tag{5.6.7}$$

Each of the conditions (5.6.7) means that the 4-subweb $[n+1, \alpha, \beta, \gamma]$ is a group 4-web for a certain numbering of its foliations. It is obvious that then the whole web $GW(n+1, n, r)$ is a group web. Note that depending on which one of three

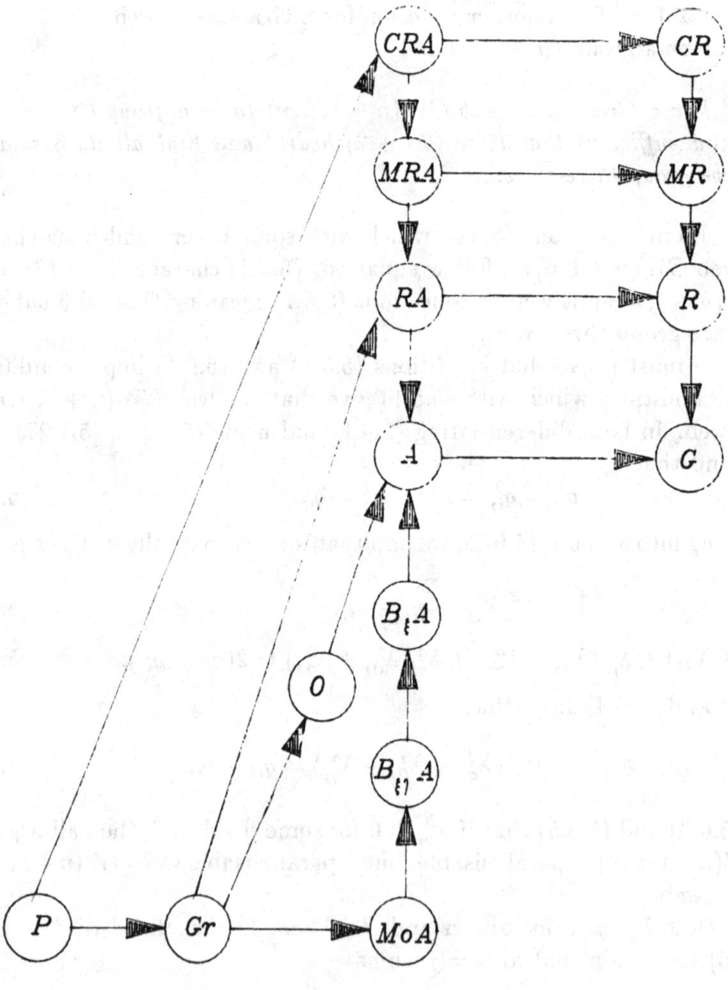

Figure 5.2

possible numbering of the foliations of $[n+1, \alpha, \beta, \gamma]$ we have, we will get group webs $GW(n+1, n, r)$ corresponding to different ordering of its foliations.

In conclusion we will indicate a scheme of inclusions of the different classes of Grassmann $(n+1)$-webs which we studied in this chapter (see Figure 5.2):

On Figure 5.2 the letters G, R, M, C, A, B, Mo, Gr, O, P abbreviate respectively the words Grassmann, reducible, multiple, completely, algebraic, Bol, Moufang, group,$(2n+2)$-hedral, and parallelisable. In the case $n = 3$, i.e, for Grassmann webs $GW(4, 3, r)$ the scheme indicated on Figure 5.2 can be seen on Figure 5.3 :

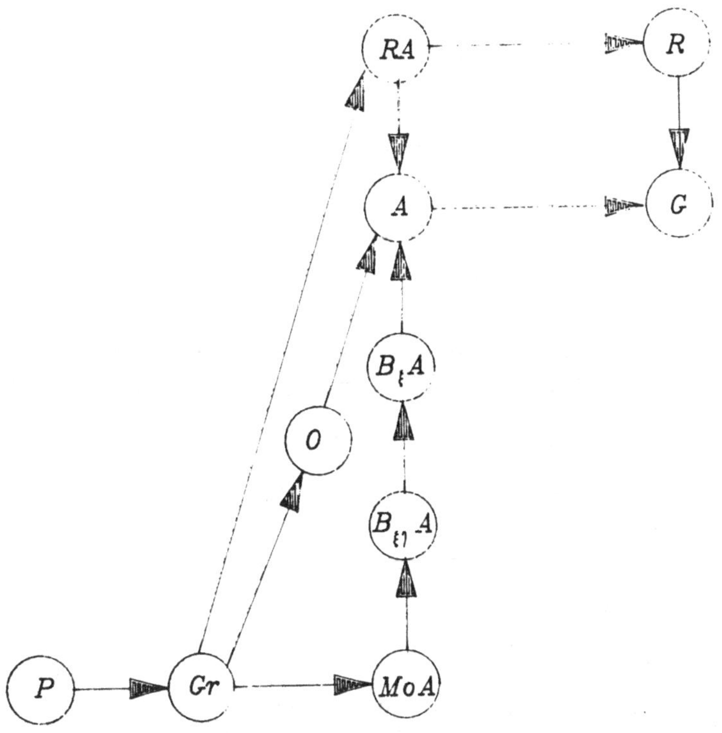

Figure 5.3

5.7 The Fundamental Equations of a Diagonal 4-Web Formed by Four Pencils of $(2r)$-Planes in P^{3r}

We will consider now another projective realisation of multidimensional webs $W(n+1,n,r)$.

Suppose that in a projective space P^{3r} of dimension $3r$ a 4-web is given and this 4-web is formed by four r-parametric foliations of $(2r)$-planes.

In this section and the next the index range

$$0 \leq u,v,w \leq 3r,$$
$$1 \leq i_1,j_1,k_1 \leq r,$$
$$r+1 \leq i_2,j_2,k_2 \leq 2r,$$
$$2r+1 \leq i_3,j_3,k_3 \leq 3r,$$
$$1 \leq \alpha,\beta,\gamma \leq 3,$$
$$1 \leq \xi,\eta,\zeta \leq 4$$

will be used.

With each point of P^{3r} we associate a moving frame $F=\{A_0,A_{i_1},A_{i_2},A_{i_3}\}$. The equations of infinitesimal displacement of the moving frame F are (cf. (5.1.4)):

$$dA_u=\omega_u^v A_v,\quad u,v=0,1,\dots 3r, \tag{5.7.1}$$

where the ω_u^v are 1-forms satisfying the structure (Maurer–Cartan) equations of the space P^{3r} (cf. (5.1.9)):

$$d\omega_u^v=\omega_u^w\wedge\omega_w^v. \tag{5.7.2}$$

Suppose that a $(2r)$-plane of the foliation λ_1 is spanned by the points A_0,A_{i_2},A_{i_3}, that of the foliation λ_2 by the points A_0,A_{i_1},A_{i_3} , and that of the foliation λ_3 by the points A_0,A_{i_2},A_{i_2}. Then, by virtue of (5.7.1), the foliations λ_1,λ_2, and λ_3 are defined respectively by the following systems of Pfaffian equations: $\omega^{i_1}=0,\omega^{i_2}=0,\omega^{i_3}=0$, where $\omega^{i_\alpha}=\omega_0^{i_\alpha}$.

We can always assume (cf.(1.2.5)) that the foliation λ_4 is defined by the system

$$\omega^{i_4}\equiv-(\omega^{i_1}+\omega^{i_2}+\omega^{i_3})=0. \tag{5.7.3}$$

By virtue of (5.7.1), equations (5.7.3) mean that an r-plane of the λ_4 is determined by the points $A_0,A_{i_2}-A_{i_1},A_{i_3}-A_{i_1}$.

The leaves of the foliations λ_1,λ_2, and λ_3 are $(2r)$-planes if and only if

$$\begin{cases} dA_{i_\beta}\wedge A_0\wedge A_{1_\beta}\wedge\dots\wedge A_{r_\beta}\wedge A_{1_\gamma}\wedge\dots\wedge A_{r_\gamma}\equiv 0 \mod \omega^{l_\alpha}, \\ dA_{i_\gamma}\wedge A_0\wedge A_{1_\beta}\wedge\dots\wedge A_{r_\beta}\wedge A_{1_\gamma}\wedge\dots\wedge A_{r_\gamma}\equiv 0 \mod \omega^{l_\alpha}, \end{cases} \tag{5.7.4}$$

where $\beta, \gamma \neq \alpha$. It follows from (5.7.4) and (5.7.1) that

$$\omega_{i_\alpha}^{j_\beta} = a_{i_\alpha k_\beta}^{j_\beta} \omega^{k_\beta}. \tag{5.7.5}$$

Similarly the leaves of the foliation λ_4 are $(2r)$-planes if and only if we have

$$\begin{aligned} d(A_{i_\beta} - A_{i_\alpha}) \wedge A_0 \wedge (A_{1_\beta} - A_{1_\alpha}) \wedge \ldots \wedge (A_{r_\beta} - A_{r_\alpha}) \\ \wedge (A_{1_\gamma} - A_{1_\alpha}) \wedge \ldots \wedge (A_{r_\gamma} - A_{r_\alpha}) \equiv \bmod \omega^{l_4}, \quad \beta, \gamma \neq \alpha, \end{aligned} \tag{5.7.6}$$

and by virtue of (5.7.1) equations (5.7.6) are equivalent to

$$\omega_{i_\alpha}^{j_\alpha} - \omega_{i_\beta}^{j_\beta} + \omega_{i_\alpha}^{j_\beta} + \omega_{i_\alpha}^{j_\gamma} - \omega_{i_\beta}^{j_\alpha} - \omega_{i_\beta}^{j_\gamma} = \underset{\alpha\beta}{m}{}^j_{ik}(\omega^{k_1} + \omega^{k_2} + \omega^{k_3}), \tag{5.7.7}$$

where $\underset{\alpha\beta}{m}{}^j_{ik} = -\underset{\beta\alpha}{m}{}^j_{ik}$. Using (5.7.6), we can rewrite equations (5.7.7) in the form

$$\omega_{i_\alpha}^{j_\alpha} - \omega_{i_\beta}^{j_\beta} = (\underset{\alpha\beta}{m}{}^j_{ik} + a_{i_\beta k_\alpha}^{j_\alpha})\omega^{k_\alpha} + (\underset{\alpha\beta}{m}{}^j_{ik} - a_{i_\alpha k_\beta}^{j_\beta})\omega^{k_\beta} + (\underset{\alpha\beta}{m}{}^j_{ik} - a_{i_\alpha k_\gamma}^{j_\gamma} + a_{i_\beta k_\gamma}^{j_\gamma})\omega^{k_\gamma}, \tag{5.7.8}$$

where

$$\underset{\alpha\beta}{m}{}^j_{ik} + \underset{\beta\gamma}{m}{}^j_{ik} + \underset{\gamma\alpha}{m}{}^j_{ik} = 0. \tag{5.7.9}$$

Equations (5.7.8) allow us to put

$$\omega_{i_\alpha}^{j_\alpha} = \tilde{\theta}_i^j + \delta_i^j \omega_0^0 + \sum_{\beta\neq\alpha} (\underset{\alpha\beta}{m}{}^j_{ik} - a_{i_\alpha k_\beta}^{j_\beta})\omega^{k_\beta}. \tag{5.7.10}$$

Equations (5.7.10) show that the structure equation (5.7.2) for $u = 0$ becomes

$$d\omega^{i_\alpha} = \omega^{j_\alpha} \wedge \tilde{\theta}_j^i + \sum_{\beta\neq\alpha} (\underset{\alpha\beta}{m}{}^i_{jk} - a_{j_\alpha k_\beta}^{i_\beta} - a_{k_\beta j_\alpha}^{i_\alpha})\omega^{j_\alpha} \wedge \omega^{k_\beta} \tag{5.7.11}$$

where we used the notation $\omega_0^{i_\alpha} = \omega^{i_\alpha}$. Equations (5.7.11) are the structure equations of the 4-web under consideration. From (5.7.3) and (5.7.11) we find that

$$d\omega^{i_4} = \omega^{j_4} \wedge (\tilde{\theta}_j^i + 2\underset{12}{m}{}^i_{(jk)}\omega^{k_2} + 2\underset{13}{m}{}^i_{(jk)}\omega^{k_3}). \tag{5.7.12}$$

In order that the form of the equations of (5.7.11) agrees with that of the structure equations (1.2.16) of the general web $W(4, 3, r)$, we put

$$\omega_j^i = \tilde{\theta}_j^i + \frac{2}{3} \sum_{\gamma,\beta(\gamma\neq\beta)} \underset{\gamma\beta}{m}{}^i_{(jk)}\omega^{k_\beta}. \tag{5.7.13}$$

By virtue of (5.7.13), equations (5.7.11) become

$$d\omega^{i_\alpha} = \omega^{j_\alpha} \wedge \theta_j^i + \sum_{\beta\neq\alpha} (\underset{\alpha\beta}{m}{}^i_{jk} - a_{j_\alpha k_\beta}^{i_\beta} - a_{k_\beta j_\alpha}^{i_\alpha} - \frac{2}{3}\sum_{\gamma\neq\beta} \underset{\gamma\beta}{m}{}^i_{(jk)})\omega^{j_\alpha} \wedge \omega^{k_\beta}. \tag{5.7.14}$$

Now equations (5.7.12) have the form

$$d\omega^{i_4} = \omega^{j_4} \wedge \theta^i_j. \tag{5.7.15}$$

The form of equations (5.7.14) and (5.7.15) agrees with that of (1.2.16) and (1.2.18). It follows from (5.7.14) that the torsion tensor of the 4-web being considered has the form:

$$\underset{\alpha\beta}{a}{}^i_{jk} = \underset{\alpha\beta}{m}{}^i_{jk} - a^{i_\beta}_{j_\alpha k_\beta} - a^{i_\alpha}_{k_\beta j_\alpha} - \frac{2}{3}\sum_{\gamma\neq\beta} \underset{\gamma\beta}{m}{}^i_{(jk)}. \tag{5.7.16}$$

Equations (5.7.9) and (5.7.16) show that the components of the torsion tensor satisfy conditions (1.2.17). In addition, according to (1.2.24), we must have

$$\sum_{\alpha,\beta(\alpha\neq\beta)} \underset{\alpha\beta}{a}{}^i_{jk} = 0. \tag{5.7.17}$$

By (5.7.16) and (5.7.9), condition (5.7.17) gives

$$\sum_{\alpha,\beta(\alpha\neq\beta)} \left(a^{i_\alpha}_{j_\beta k_\alpha} + a^{i_\alpha}_{k_\beta j_\alpha}\right) = 0. \tag{5.7.18}$$

Each of the foliations λ_ξ of the 4-web under consideration is an r-parameter family of $(2r)$-planes in the space P^{3r}. Therefore, locally through each point of P^{3r} there passes a unique $(2r)$-plane of each of the foliations. Such families of planes are called *congruences* (see [Gei 67]). So each foliation λ_ξ is a congruence of $(2r)$-planes in P^{3r}.

Definition 5.7.1 A point $G = x^0 A_0 + x^{i_\alpha} A_{i_\alpha} + x^{i_\beta} A_{i_\beta}$ is said to be a *focus* of an $(2r)$-plane $\pi^{2r}_\gamma = \{A_0, A_{i_\alpha}, A_{i_\beta}\}$ if $dG \in \pi^{2r}_\gamma$ when we have a displacement in a certain direction. A locus of foci is called a *focal surface* of the congruence of $(2r)$-planes.

According to (5.7.1) and (5.7.13), we will get a focus if and only if

$$\left(x^0\delta^j_k + x^{i_\alpha} a^{j_\gamma}_{i_\alpha k_\gamma} + x^{i_\beta} a^{j_\gamma}_{i_\beta k_\gamma}\right)\omega^{k_\gamma} = 0. \tag{5.7.19}$$

The system (5.7.19) is homogeneous with respect to ω^{k_γ}. It consists of r equations in r unknowns and has a non-trivial solutions if and only if the determinant of the matrix of its coefficients vanishes:

$$\det\left(\delta^j_k x^0 + a^{j_\gamma}_{i_\alpha k_\gamma} x^{i_\alpha} + a^{j_\gamma}_{i_\beta k_\gamma} x^{i_\beta}\right) = 0. \tag{5.7.20}$$

Equation (5.7.20) is of order r with respect to $x^0, x^{i_\alpha}, x^{i_\beta}$. Therefore, the focal surface of the plane π^{2r}_γ is a $(2r-1)$-dimensional algebraic surface of degree r.

Similar considerations show that the point

$$\Phi = x^0 A_0 + x^{i_2}(A_{i_2} - A_{i_1}) + x^{i_3}(A_{i_3} - A_{i_1})$$

is a focus of the plane $\pi_4^{2r} = \{A_0, A_{i_2} - A_{i_1}, A_{i_3} - A_{i_1}\}$ if and only if

$$\det\,(\delta_k^j x^0 - \underset{12}{m}{}^j_{ik} x^{i_2} - \underset{13}{m}{}^j_{ik} x^{i_3}) = 0. \tag{5.7.21}$$

Equation (5.7.21) is an equation of the focal surface of the plane π_4^{2r}. It shows that the focal surface of π_4^{2r} is also a $(2r-1)$-dimensional algebraic surface of degree r. Thus, we can conclude that the geometric objects $a^{k_\beta}_{i_\alpha j_\beta}$ and $\underset{\alpha\beta}{m}{}^i_{jk}$ determine the focal surfaces on $(2r)$-planes of the 4-web being studied.

Let us find focal surfaces of r-planes of each of the six families of r-planes which are common for any two $(2r)$-planes of different foliations λ_ξ passing through a point of P^{3r}. Each of these families of r-planes depends on $2r$ parameters and therefore is a congruence of r-planes in P^{3r}. The r-planes of the six families indicated above are spanned by the following points: the r-plane $\pi^r_{\beta\gamma}$ by the points A_0, A_{i_α} and the r-plane $\pi^r_{4\gamma}$ by the points $A_0, A_{i_\alpha} - A_{i_\beta}$. If the points

$$K = x^0 A_0 + x^{i_\alpha} A_{i_\alpha}, \quad L = x^0 A_0 + x^{i_\alpha}(A_{i_\alpha} - A_{i_\beta})$$

are foci of these two r-planes, then we have two systems

$$\begin{cases} (\delta_k^j x^0 + a^{j_\beta}_{i_\alpha k_\beta} x^{i_\alpha})\omega^{k_\beta} = 0, \\ (\delta_k^j x_0 + x^{i_\alpha} a^{j_\gamma}_{i_\alpha k_\gamma})\omega^{k_\gamma} = 0, \end{cases}$$

$$\begin{cases} (x^0 \delta_k^j + \underset{\alpha\beta}{m}{}^j_{ik} x^{i_\alpha})(\omega^{k_\alpha} + \omega^{k_\beta}) + x^{i_\beta}(\underset{\alpha\beta}{m}{}^j_{ik} - a^{j_\gamma}_{i_\alpha k_\gamma} + a^{j_\gamma}_{i_\beta k_\gamma})\omega^{k_\gamma} = 0, \\ \left[\delta_k^j x^0 + x^{i_\alpha}(a^{j_\gamma}_{i_\alpha k_\gamma} - a^{j_\gamma}_{i_\beta k_\gamma}\right]\omega^{k_\gamma} = 0. \end{cases} \tag{5.7.22}$$

Each of the systems (5.7.22) is a system of $2r$ homogeneous equations in $2r$ unknowns $\omega^{k_\beta}, \omega^{k_\gamma}$ and $\omega^{k_\alpha} + \omega^{k_\beta}, \omega^{k_\gamma}$. They have non-trivial solutions if and only if

$$\det\left(\delta_k^j x^0 + a^{j_\beta}_{i_\alpha k_\beta} x^{i_\alpha}\right) \cdot \det\left(\delta_k^j x^0 + a^{j_\gamma}_{i_\alpha k_\gamma} x^{i_\alpha}\right) = 0, \tag{5.7.23}$$

$$\det\left(\delta_k^j x^0 + \underset{\alpha\beta}{m}{}^j_{ik} x^{i_\alpha}\right) \cdot \det\left(\delta_k^j x^0 + (a^{k_\gamma}_{i_\alpha j_\gamma} - a^{k_\gamma}_{i_\beta j_\gamma}) x^{i_\alpha}\right) = 0. \tag{5.7.24}$$

Equations (5.7.23) and (5.7.24) show that the focal surfaces of the planes $\pi^r_{\beta\gamma}$ and $\pi^r_{4\gamma}$ are $(r-1)$-dimensional algebraic surfaces of degree $2r$. Each of them decomposes into two $(r-1)$-dimensional algebraic surfaces of degree r. It follows from (5.7.20), (5.7.21), (5.7.23), and (5.7.24) that the focal surfaces of the r-planes $\pi^r_{\beta\gamma}$ and $\pi^r_{4\gamma}$ are respectively the intersections of focal surfaces of the $(2r)$-planes π^{2r}_β and π^{2r}_γ with the r-plane $\pi^r_{\beta\gamma}$ and of the focal surfaces of the $(2r)$-planes π^{2r}_4 and π^{2r}_γ with the r-plane $\pi^r_{4\gamma}$.

Suppose now that each of the foliations λ_ξ of the 4-web is a pencil of $(2r)$-planes with a $(2r-1)$-dimensional axis. Let the axis of the foliation λ_{i_γ} be spanned by the points $\underset{\gamma}{M}_{i_\alpha} = A_{i_\alpha} - \underset{\gamma}{\lambda}_{i_\alpha} A_0$, $\gamma \neq \alpha$. Then this axis must be fixed. Since we have

$$d\underset{\gamma}{M}_{i_\alpha} = \Big(-d\underset{\gamma}{\lambda}_{i_\alpha} - \underset{\gamma}{\lambda}_{i_\alpha}\omega_0^0 - \underset{\gamma}{\lambda}_i \sum_{\beta\neq\gamma} \underset{\gamma}{\lambda}_{j_\beta}\omega^{j_\beta} + \sum_{\beta\neq\gamma} \underset{\gamma}{\lambda}_{j_\beta}\omega_{i_\alpha}^{j_\beta} + \omega_{i_\alpha}^0\Big)A_0$$
$$+\Big(\omega_{i_\alpha}^{j_\gamma} - \lambda_{i_\alpha}\omega^{j_\gamma}\Big)A_{j_\gamma} + \Big(\omega_{i_\alpha}^{j_\beta} - \underset{\gamma}{\lambda}_{i_\alpha}\omega^{j_\beta}\Big)\underset{\gamma}{M}_{j_\beta} + \Big(\omega_{i_\alpha}^{j_\alpha} - \underset{\gamma}{\lambda}_{i_\alpha}\omega^{j_\alpha}\Big)\underset{\gamma}{M}_{j_\alpha}, \qquad (5.7.25)$$

it follows from (5.7.25) that it will be fixed if and only if

$$a_{i_\alpha k_\gamma}^{j_\gamma} = \delta_k^j \underset{\gamma}{\lambda}_{i_\alpha}, \qquad (5.7.26)$$

$$\Delta\underset{\gamma}{\lambda}_{i_\alpha} \equiv d\underset{\gamma}{\lambda}_{i_\alpha} + \underset{\gamma}{\lambda}_{i_\alpha}\omega_0^0 - \omega_{i_\alpha}^0 - \sum_{\beta\neq\gamma} \underset{\gamma}{\lambda}_{j_\beta}\omega_{i_\alpha}^{j_\beta} + \underset{\gamma}{\lambda}_{i_\alpha} \sum_{\beta\neq\gamma} \underset{\gamma}{\lambda}_{j_\beta}\omega^{j_\beta} = 0. \qquad (5.7.27)$$

Conversely, if for a fixed i the equations (5.7.26) hold, then

$$\omega_{i_\alpha}^{j_\gamma} = \underset{\gamma}{\lambda}_{i_\alpha}\omega^{j_\gamma}. \qquad (5.7.28)$$

Taking exterior derivative of (5.7.28), we obtain

$$\Delta\underset{\gamma}{\lambda}_{i_\alpha} \wedge \omega^{j_\gamma} = 0. \qquad (5.7.29)$$

If $r > 1$, equations (5.7.29) imply $\Delta\underset{\gamma}{\lambda}_{i_\alpha} = 0$, i.e., equations (5.7.27) are valid and consequently the axis spanned by the points $\underset{\gamma}{M}_{i_\alpha}$ is fixed.

Note that if $r = 1$, equations (5.7.27) can be obtained by extending equations (5.7.8). Thus the conditions (5.7.26) are necessary and sufficient for the foliation λ_γ to be a pencil with a $(2r-1)$-dimensional axis.

Let us obtain similar conditions for the foliation λ_4. Suppose that the axis of this pencil is spanned by the points

$$\underset{\alpha\beta}{N}_i = A_{i_\alpha} - A_{i_\beta} + \underset{\alpha\beta}{\sigma}_i A_0,$$

where

$$\underset{\alpha\beta}{\sigma}_i = -\underset{\beta\alpha}{\sigma}_i, \qquad \underset{\alpha\beta}{\sigma}_i + \underset{\beta\gamma}{\sigma}_i + \underset{\gamma\alpha}{\sigma}_i = 0.$$

The plane spanned by the points $\underset{\alpha\beta}{N}_i$ is fixed if and only if

$$\underset{\alpha\beta}{m}_{ik}^j = \underset{\beta\alpha}{\sigma}_i \delta_k^j, \qquad (5.7.30)$$

$$
\begin{aligned}
\Delta \underset{\alpha\beta}{\sigma}{}_i &= d\underset{\alpha\beta}{\sigma}{}_i + \underset{\alpha\beta}{\sigma}{}_i \omega_0^0 - \underset{\alpha\beta}{\sigma}{}_i \sum_{\gamma\neq\beta} \underset{\gamma\beta}{\sigma}{}_j \omega^{j_\gamma} \\
&+ \omega^0_{i_\alpha} - \omega^0_{i_\beta} - \sum_{\gamma\neq\beta}\Big(\underset{\gamma\beta}{\sigma}{}_j \omega^{j_\gamma}_{i_\alpha} + \underset{\gamma\beta}{\sigma}{}_j \omega^{j_\gamma}_{i_\beta}\Big) = 0.
\end{aligned} \tag{5.7.31}
$$

The conditions (5.7.30) and (5.7.31) are necessary and sufficient for the foliation λ_4 to be a pencil with a $(2r-1)$-dimensional axis. Formulae (5.7.20) and (5.7.21) show that under conditions (5.7.26) and (5.7.30) the focal surfaces of the planes π^{2r}_α and π^{2r}_4 become r-fold $(2r-1)$-planes coinciding with the corresponding axes. The equations of these axes are

$$
x^{i_\alpha} = 0, \quad x^0 + \sum_{\beta\neq\alpha} x^{i_\beta} \underset{\alpha}{\lambda}{}_{i_\beta} = 0, \quad \alpha = 1,2,3; \tag{5.7.32}
$$

$$
x^{i_1} + x^{i_2} + x^{i_3} = 0, \quad x^0 - \sum_{\beta\neq\alpha} x^{i_\beta} \underset{\beta\alpha}{\sigma}{}_i = 0. \tag{5.7.33}
$$

In addition, note that by means of (5.7.26) the relations (5.7.18) take the form

$$
\sum_{\substack{\alpha,\beta \\ (\alpha\neq\beta)}} \underset{\beta}{\lambda}{}_{i_\alpha} = 0. \tag{5.7.34}
$$

The six points $\xi^i \underset{\beta}{M}{}_{i_\alpha}$ belong pairwise to three straight lines passing through the point A_0 and correspondingly through the points $\xi^i A_{i_\alpha}$. Therefore, these six points span a 3-plane. Equations (5.7.34) mean that the 3-plane spanned by the points $\xi^i \underset{\beta}{M}{}_{i_\alpha}$ and the 2-plane $\{\xi^i A_{i_1}, \xi^i A_{i_2}, \xi^i A_{i_3}\}$ have the common point

$$
2\xi^i (A_{i_1} + A_{i_2} + A_{i_3}) = \sum_{\substack{\alpha,\beta \\ (\alpha\neq\beta)}} \xi^i \underset{\beta}{M}{}_{i_\alpha}.
$$

Next, the planes π^{2r}_γ and π^{2r}_β intersect each other in the r-plane $\pi^r_{\gamma\beta}$ spanned by the points $A_0, A_{i_\alpha}, \alpha \neq \beta, \gamma$. The $(2r-1)$-dimensional axes of the pencils λ_γ and λ_β intersect the r-plane $\pi^r_{\gamma\beta}$ in the $(r-1)$-planes spanned respectively by the points $\underset{\gamma}{M}{}_{i_\alpha} = A_{i_\alpha} - \underset{\gamma}{\lambda}{}_{i_\alpha} A_0$ and $\underset{\beta}{M}{}_{i_\alpha} = A_{i_\alpha} - \underset{\beta}{\lambda}{}_{i_\alpha} A_0$. For each fixed i we locate the points A_{i_α} in such a way that

$$
\Big(A_0, A_{i_\alpha}; \underset{\beta}{M}{}_{i_\alpha} \underset{\gamma}{M}{}_{i_\alpha}\Big) = -1. \tag{5.7.35}
$$

Equation (5.7.35) implies

$$
\underset{\beta}{\lambda}{}_{i_\alpha} + \underset{\gamma}{\lambda}{}_{i_\alpha} = 0. \tag{5.7.36}
$$

By virtue of (5.7.36), equations (5.7.26) become

$$
\left\{
\begin{aligned}
\omega^{j_3}_{i_1} &= \lambda_{i_1}\omega^{j_3}, & \omega^{j_1}_{i_2} &= \lambda_{i_2}\omega^{j_1}, & \omega^{j_2}_{i_3} &= \lambda_{i_3}\omega^{j_2}, \\
\omega^{j_2}_{i_1} &= -\lambda_{i_1}\omega^{j_2}, & \omega^{j_3}_{i_2} &= -\lambda_{i_2}\omega^{j_3}, & \omega^{j_1}_{i_3} &= -\lambda_{i_3}\omega^{j_1},
\end{aligned}
\right. \tag{5.7.37}
$$

where for brevity we used the notations

$$\underset{3}{\lambda}_{i_1} = -\underset{2}{\lambda}_{i_1} = \lambda_{i_1}, \ \underset{1}{\lambda}_{i_2} = -\underset{3}{\lambda}_{i_2} = \lambda_{i_2}, \ \underset{2}{\lambda}_{i_3} = -\underset{1}{\lambda}_{i_3} = \lambda_{i_3}. \tag{5.7.38}$$

If we take exterior derivative of (5.7.37) and apply the Cartan lemma, we obtain (5.7.27). Equations (5.7.27) and (5.7.38) give

$$\omega_{i_\alpha}^0 = \lambda_{i_\alpha}\Big(\lambda_{j_\alpha}\omega^{j_\alpha} - \sum_{\beta\neq\alpha} \lambda_{j_\beta}\omega^{j_\beta}\Big), \tag{5.7.39}$$

$$\Delta\lambda_{i_\alpha} \equiv d\lambda_{i_\alpha} + \lambda_{i_\alpha}\omega_0^0 - \lambda_{j_\alpha}\omega_{i_\alpha}^{j_\alpha} + \lambda_{i_\alpha}\Big(\lambda_{j_\gamma}\omega^{j_\gamma} - \lambda_{j_\beta}\omega^{j_\beta}\Big) = 0 \tag{5.7.40}$$

where in (5.7.40) the indices α, β, γ are fixed and form an even permutation of the numbers 1,2,3. Exterior differentiation of (5.7.39) and (5.7.40) leads to identities.

By (5.7.26), (5.7.30), and (5.7.38), equations (5.7.8) can be written in the form

$$\omega_{i_\alpha}^{j_\alpha} - \omega_{i_\beta}^{j_\beta} = (\underset{\beta\alpha}{\sigma}{}_i + \lambda_{i_\beta})\omega^{j_\alpha} + (\underset{\beta\alpha}{\sigma}{}_i + \lambda_{i_\alpha})\omega^{j_\beta} + (\underset{\beta\alpha}{\sigma}{}_i - \lambda_{i_\alpha} - \lambda_{i_\beta})\omega^{j_\gamma}, \tag{5.7.41}$$

where α, β, γ are fixed mutually distinct indices from 1,2,3.

In addition, we note that by virtue of (5.7.38), the equations (5.7.34) will have the form

$$\lambda_{i_1} + \lambda_{i_2} + \lambda_{i_3} = 0. \tag{5.7.42}$$

The relations (5.7.42) mean that the 2-planes

$$\Big\{\xi^i \underset{2}{M}_{i_1}, \ \xi^i \underset{3}{M}_{i_2}, \xi^i \underset{1}{M}_{i_3}\Big\}, \Big\{\xi^i \underset{3}{M}_{i_1}, \ \xi^i \underset{1}{M}_{i_2}, \xi^i \underset{2}{M}_{i_3}\Big\}, \Big\{\xi^i A_{i_1}, \xi^i A_{i_2}, \xi^i A_{i_3}\Big\}$$

have common point

$$\xi^i\Big(A_{i_1} + A_{i_2} + A_{i_3}\Big) = \xi^i\Big(\underset{2}{M}_{i_1} + \underset{3}{M}_{i_2} + \underset{1}{M}_{i_3}\Big) = \xi^i\Big(\underset{3}{M}_{i_1} + \underset{1}{M}_{i_2} + \underset{2}{M}_{i_3}\Big).$$

On each of the 2-planes $\tau = \{\xi^i A_{i_1}, \xi^i A_{i_2}, \xi^i A_{i_3}\}$ the $(2r)$-planes of different foliations λ_ξ passing through a point A_0 cut a complete quadrilateral. The six r-planes of the intersections of these four $(2r)$-planes cut a 2-plane τ in the vertices of the quadrilateral. If we join opposite vertices by straight lines, we obtain the diagonal triangle of the quadrilateral. To its vertices there correspond three r-planes passing through the point A_0.

Definition 5.7.2 Three r-dimensional distributions defined by these r-planes are called *diagonal.* A 4-web formed by four pencils of $(2r)$-planes in P^{3r} is said to be *diagonal* if all three its diagonal distributions are integrable and have r-planes as their integral surfaces.

It is obvious that the the diagonal distributions are defined by the following systems of Pfaffian equations:

$$\omega^{i_\alpha} + \omega^{i_\gamma} = 0, \quad \omega^{i_\beta} + \omega^{i_\gamma} = 0, \quad \gamma \neq \alpha, \beta; \quad \alpha \neq \beta. \tag{5.7.43}$$

Introduce the notation:

$$\omega^{i_\alpha} = \omega^{i_\beta} = -\omega^{i_\gamma} = \omega^i. \tag{5.7.44}$$

Since

$$dA_0 \equiv \varphi^0 A_0 + \varphi^i (A_{i_\alpha} + A_{i_\beta} - A_{i_\gamma}) \mod \omega^{i_\alpha} + \omega^{i_\gamma}, \quad \omega^{i_\beta} + \omega^{i_\gamma}, \tag{5.7.45}$$

the tangent r-plane to the integral surface indicated above is spanned by the points A_0 and $A_{i_\alpha} + A_{i_\beta} - A_{i_\gamma}$. Using (5.7.41), (5.7.42), (5.7.44), and (5.7.44), we find

$$\begin{aligned} d^2 A_0 \quad \equiv \quad & \varphi^0 A_0 + \varphi^i (A_{i_\alpha} + A_{i_\beta} - A_{i_\gamma}) + \omega^i \omega^j \Big[\Big(\lambda_{i_\beta} - \underset{\alpha\gamma}{\sigma}{}_i\Big) A_{j_\alpha} \\ & + \Big(\underset{\gamma\beta}{\sigma}{}_i - \lambda_{i_\alpha}\Big) A_{j_\beta}\Big] \mod \omega^{i_\alpha} + \omega^{i_\gamma}, \quad \omega^{i_\beta} + \omega^{i_\gamma}. \end{aligned} \tag{5.7.46}$$

For such a surface to be an r-plane, it is necessary and sufficient that the point $d^2 A_0$ belongs to its tangent r-plane. It follows from (5.7.46) that this will be the case if and only if

$$\underset{\alpha\gamma}{\sigma}{}_i = \lambda_{i_\beta}, \quad \underset{\gamma\beta}{\sigma}{}_i = \lambda_{i_\alpha}. \tag{5.7.47}$$

Equations (5.7.47) mean that

$$\underset{13}{\sigma}{}_i = \lambda_{i_2}, \quad \underset{21}{\sigma}{}_i = \lambda_{i_3}, \quad \underset{32}{\sigma}{}_i = \lambda_{i_1}. \tag{5.7.48}$$

Thus, equations (5.7.48) are necessary and sufficient for a 4-web formed by four pencils of $(2r)$-planes to be diagonal.

By virtue of (5.7.38) and (5.7.48), equations (5.7.32) and (5.7.33) of the axes of the pencils become

$$\begin{cases} x^{i_1} = 0, & x^0 + \lambda_{i_2} x^{i_2} - \lambda_{i_3} x^{i_3} = 0, \\ x^{i_2} = 0, & x^0 - \lambda_{i_1} x^{i_1} + \lambda_{i_3} x^{i_3} = 0, \\ x^{i_3} = 0, & x^0 + \lambda_{i_1} x^{i_1} - \lambda_{i_2} x^{i_2} = 0, \end{cases} \tag{5.7.49}$$

$$x^{i_1} + x^{i_2} + x^{i_3} = 0, \quad x^0 - \lambda_{i_3} x^{i_2} + \lambda_{i_2} x^{i_3} = 0. \tag{5.7.50}$$

By (5.7.48) and (5.7.39), equations (5.7.31) are reduced to (5.7.40). By (5.7.48), equations (5.7.41) become

$$\omega_{i_\alpha}^{j_\alpha} - \omega_{i_\beta}^{j_\beta} = -\lambda_{i_\alpha} \omega^{j_\alpha} - \lambda_{i_\beta} \omega^{j_\beta} + 2\lambda_{i_\gamma} \omega^{j_\gamma}. \tag{5.7.51}$$

Exterior differentiation of (5.7.51) leads to the relations

$$\lambda_{i_\alpha} \lambda_{j_\beta} = \lambda_{i_\beta} \lambda_{j_\alpha}, \quad \alpha \neq \beta. \tag{5.7.52}$$

Relations (5.7.52) imply that

$$\lambda_{i_\alpha} = \lambda_\alpha \lambda_i, \quad \alpha = 1, 2, 3; \quad i = 1, \ldots, r, \tag{5.7.53}$$

where, by virtue of (5.7.42), the quantities λ_α are connected by the condition

$$\lambda_1 + \lambda_2 + \lambda_3 = 0. \tag{5.7.54}$$

Finally note that exterior differentiation of (5.7.52), by means of (5.7.40), leads to the identities.

Now we are able to prove the following theorem:

Theorem 5.7.3 *The system of equations defining a diagonal 4-web in P^{3r} consists of the Pfaffian equations* (5.7.37), (5.7.39), (5.7.40), (5.7.51) *and relations* (5.7.53), (5.7.54). *This system is completely integrable, and its solution depends on* $8r^2+3r+2$ *arbitrary constants.*

Proof. We have already proved that this system is completely integrable. To calculate the number of arbitrary constants, we must count the number of independent Pfaffian equations. There are $6r^2$ independent equations (5.7.37), $3r$ independent equations (5.7.39), by (5.7.53) and (5.7.54) only $2r$ independent equations among (5.7.40), and since $i_\alpha \neq j_\alpha$, $i_\beta \neq j_\beta$, $2r(r-1)$ independent equations among (5.7.51): $6r^2 + 3r + 2r + 2r(r-1) = 8r^2 + 3r + 2$. ■

5.8 The Geometry of Diagonal 4-Webs in P^{3r}

In this section we will study in detail a geometry of a diagonal 4-web whose definition and fundamental equations were given in Section 5.7

Theorem 5.8.1 *A diagonal 4-web in P^{3r}, $r > 2$, is transversally geodesic and isoclinic (and therefore Grassmannisable). Four $(2r-1)$-dimensional axes of four pencils of $(2r)$-planes of the web, lie on one and the same hypercone of second order. This hypercone has $(3r-4)$-dimensional vertex and its directrix is a hyperboloid of one sheet lying in a 3-plane which is in general position with its vertex. Each of $(3r-2)$-dimensional generators of two families of generators of the hypercone is a linear span of the vertex and a generator on the director hyperboloid. The four vertices belong to four $(3r-2)$-dimensional generators of one family of generators of the hypercone, and the anharmonic ratio of these four generators is equal to $-\lambda_2/\lambda_1$ (see* (5.7.53)*). Each of the axes intersects the vertex in a $(2r-3)$-plane.*

Proof. First note that, by virtue of conditions (5.7.26), (5.7.30), (5.7.38), and (5.7.48), equations (5.7.16) take the form

$$\underset{\alpha\beta}{a}{}^{i}_{jk} = \frac{2}{3}\Big[\Big(\lambda_{j_\alpha} - \lambda_{j_\beta}\Big)\delta^i_k + \Big(\lambda_{k_\alpha} - \lambda_{k_\beta}\Big)\delta^i_j\Big], \tag{5.8.1}$$

or

$$\underset{\alpha\beta}{a}{}^{i}_{jk} = \delta^i_{(j}\,\underset{\alpha\beta}{a}{}_{k)}, \tag{5.8.2}$$

where

$$\underset{\alpha\beta}{a}{}_{k} = \frac{3}{4}\left(\lambda_{k_\alpha} - \lambda_{k_\beta}\right). \tag{5.8.3}$$

It follows from (5.8.2) that

$$\underset{\alpha\beta}{a}{}^{i}_{[jk]} = 0. \tag{5.8.4}$$

By Theorem 1.9.7, conditions (5.8.2) give that a diagonal 4-web in P^{3r} is transversally geodesic. By Theorem 1.11.4, conditions (5.8.4) mean that a diagonal 4-web is isoclinic and, moreover, by Definition 1.11.9, it is isoclinicly geodesic. By Theorem 2.6.4, it follows from this that a diagonal 4-web in P^{3r} is Grassmannisable provided that $r > 2$. Note that for isoclinicly geodesic 4-webs the condition $\underset{\alpha\beta}{a}{}^{i}_{(jk)} = 0$ of octahedrality is equivalent to the condition $\underset{\alpha\beta}{a}{}^{i}_{jk} = 0$ of parallelisability.

Now we will prove that the axes of four pencils of $(2r)$-planes of a diagonal 4-web lie on a hypercone of second order.

In our moving frame $\{A_u\}$ an equation of a general hyperquadric Q has the form:

$$a_{uv}x^u x^v = 0, \quad u, v = 0, 1, \ldots, 3r. \tag{5.8.5}$$

where the coefficients a_{uv} are symmetric in u and v. The coefficients a_{uv} in (5.8.5) are defined up to a common factor. Moreover, they depend upon Q and upon the choice of a frame $\{A_u\}$. If Q is fixed, we will find how the coefficients a_{uv} change when the frame $\{A_u\}$ is changed infinitesimally.

Consider first the case of a point $M = x^u A_u$ in P^{3r} with homogeneous coordinates x^u with respect to the frame $\{A_u\}$. The point M in P^{3r} will remain fixed under an infinitesimal change of the frame $\{A_u\}$ as long as

$$dM = \lambda M, \tag{5.8.6}$$

where λ is some 1-form. Equations (5.7.1) and (5.8.6), using the linear independence of the A_u, imply that

$$dx^u + x^v \omega_v^u = \lambda x^u. \tag{5.8.7}$$

Conditions (5.8.7) are necessary and sufficient for a point M to be fixed.

Suppose next that M lies on the hypersurface Q. Then x^u satisfy equation (5.8.5). Q remains invariant under an infinitesimal change of the frame $\{A_u\}$, so in particular equations (5.8.7) hold. Note that the polynomial in (5.8.5) generates the ideal of Q, so we have

$$d(a_{uv}x^u x^v) = \theta(a_{uv}x^u x^v) \tag{5.8.8}$$

where θ is a 1-form. To find the law of change of the coefficients a_{uv} of (5.8.5), we must differentiate (5.8.5). Using (5.8.7) and (5.8.8), this differentiation leads to

$$\nabla a_{uv}x^u x^v = \theta(a_{uv}x^u x^v), \tag{5.8.9}$$

where

$$\nabla a_{uv} = da_{uv} - a_{uw}\omega_v^w - a_{wv}\omega_u^w,$$

and since M is an arbitrary point of Q, (5.8.9) implies

$$\nabla a_{uv} = \theta a_{uv}. \tag{5.8.10}$$

The conditions (5.8.10) are necessary and sufficient for a hyperquadric (5.8.5) to be fixed.

We will find a hyperquadric passing through three axes (5.7.49) and will prove that it passes also through the fourth axis (5.7.50). We will assume that $\lambda_\alpha \neq 0$. All the cases when one or more λ_α vanishes, will be considered later on.

It is easy to see that the axes (5.7.49) belong to Q if and only if

$$a_{i_\alpha 0} = 0, \quad a_{i_\alpha k_\alpha} = -a_{00}\lambda_{i_\alpha}\lambda_{k_\alpha}, \quad a_{i_\alpha k_\beta} = a_{00}\lambda_{i_\alpha}\lambda_{k_\beta}, \quad \beta \neq \alpha. \tag{5.8.11}$$

By virtue of (5.8.11), equation (5.8.5) of Q becomes

$$(x^0)^2 - \sum_\alpha \lambda_{i_\alpha}\lambda_{k_\alpha}x^{i_\alpha}x^{k_\alpha} + 2\sum_{\substack{\alpha,\beta\\(\alpha<\beta)}} \lambda_{i_\alpha}\lambda_{k_\beta}x^{i_\alpha}x^{k_\beta} = 0. \tag{5.8.12}$$

Since in (5.8.12) $a_{00} = 1$, we find from one of conditions (5.8.10) that

$$\theta = -2\omega_0^0. \tag{5.8.13}$$

The remaining equations (5.8.10) will be identically satisfied by means of (5.8.11), (5.8.13), (5.7.37), (5.7.39), and (5.7.40).

To prove that the fourth vertex (5.7.50) belongs to the hyperquadric (5.8.12), it is convenient to write its equations (5.7.50) in a parametric form:

$$x^{i_2} = u^i, \quad x^{i_3} = v^i, \quad x^{i_1} = -(u^i + v^i), \quad x^0 = -\lambda_{i_2}v^i + \lambda_{i_3}u^i. \tag{5.8.14}$$

Substituting the coordinates x^u from (5.8.14) into (5.8.12), we convince ourselves that equation (5.8.12) will be satisfied identically.

To study the hyperquadric Q defined by equation (5.8.12), we will make a coordinate change putting

$$x^0 = y^0, \quad \lambda_{i_1}x^{i_1} = y^1, \quad \lambda_{i_2}x^{i_2} = y^2, \quad \lambda_{i_3}x^{i_3} = y^3 \tag{5.8.15}$$

and not changing the remaining coordinates. After the change (5.8.15) equation (5.8.12) becomes

$$(y^0)^2 - (y^1)^2 - (y^2)^2 - (y^3)^2 + 2(y^1y^2 + y^2y^3 + y^3y^1) = 0. \tag{5.8.16}$$

Equation (5.8.16) contains four variables out of $3r+1$. Therefore, if $r = 1$, the equation (5.8.16) is an equation of a non-degenerate hyperquadric and, if $r > 1$, this equation is an equation of a hypercone of second order.

The hypercone has a $(3r-4)$-dimensional vertex determined by equations $y^0 = y^1 = y^2 = y^3 = 0$, or

$$x^0 = 0, \quad \lambda_{i_1} x^{i_1} = 0, \quad \lambda_{i_2} x^{i_2} = 0, \quad \lambda_{i_3} x^{i_3} = 0. \tag{5.8.17}$$

A quadric (5.8.16) lying in the plane $y^0 y^1 y^2 y^3$ is a directrix of the hypercone.

Let us reduce the left member of (5.8.16) to the sum of squares. For this we make the following change of coordinates:

$$y^0 = z^0, \quad y^1 - y^2 - y^3 = z^1, \quad y^2 + y^3 = z^2, \quad y^2 - y^3 = z^3. \tag{5.8.18}$$

Solving (5.8.18) with respect to y^0, y^1, y^2, and y^3 ,we obtain

$$z^0 = y^0, \quad \frac{1}{2}(z^2 + z^3) = y^2, \quad \frac{1}{2}(z^2 - z^3) = y^3, \quad z^1 + z^2 = y^1. \tag{5.8.19}$$

Substituting y's from (5.8.19) into (5.8.16), we see that in new coordinates the equation (5.8.16) of our hypercone becomes

$$(z^0)^2 - (z^1)^2 - (z^2)^2 - (z^3)^2 = 0. \tag{5.8.20}$$

A surface determined in the 3-plane y^0, y^1, y^2, y^3 by equation (5.8.20) is a hyperboloid of one sheet. It carries two families of rectilinear generators. The hypercone also carries two families of plane generators. Each of them is of dimension $3r-2$ and is determined by the $(3r-4)$-dimensional vertex (5.8.17) and one-dimensional generator of the hyperboloid (5.8.20).

Next we will prove that four $(2r-1)$-dimensional axes of the pencils of $(2r)$-planes forming the 4-web belong to four $(3r-2)$-dimensional generators of one family of the hypercone. For this we shall write the equations of generators of this family in the form:

$$z^0 + z^1 = k(z^3 + z^2), \quad k(z^0 - z^1) = z^3 - z^2. \tag{5.8.21}$$

Returning to the original coordinates x^u, we can write equation (5.8.21) in the form:

$$\begin{cases} x^0 + \lambda_{i_1} x^{i_1} - \lambda_{i_2} x^{i_2} - \lambda_{i_3} x^{i_3} = 2k\lambda_{i_2} x^{i_2}, \\ k(x^0 - \lambda_{i_1} x^{i_1} + \lambda_{i_2} x^{i_2} + \lambda_{i_3} x^{i_3}) = -2\lambda_{i_3} x^{i_3}. \end{cases} \tag{5.8.22}$$

Note that the equations (5.8.22) of $(3r-2)$-dimensional generators corresponding to $k = -1, \infty, 0, \lambda_3/\lambda_2$ will be satisfied identically if we put into them the relations (5.7.49) and (5.7.50) connecting the coordinates of four axes. It can be easily checked that each of the four axes intersects the vertex in a $(2r-3)$-plane.

We calculate the anharmonic ratio for four generators containing four axes taken them in the order indicated above. To be able to perform the calculation, we will take a generator of another family of the hyperboloid (5.8.20) and find the coordinates of its intersection points with four generators indicated above. As a result, we obtain that the anharmonic ratio is equal to $-\lambda_2/\lambda_1$. Note that by means of (5.7.40), for the quantity $\lambda_2/\lambda_1 = \lambda_{i_2}/\lambda_{i_1}$ we have $d(\lambda_2/\lambda_1) = 0$, i.e., this quantity is an absolute invariant. ■

Note that Theorem 5.8.1 does not exclude the case $r = 1$. We only should keep in mind that as usual the dimension -1 corresponds to the empty set.

The next theorem gives a form and a location of the diagonal families of r-planes (see Definition 5.7.2).

Theorem 5.8.2 *The r-planes of each of the three diagonal foliations of r-planes of the diagonal 4-web intersect in an $(r-1)$-planes one and the same (for a given family) pair of $(3r-2)$-dimensional generators of the hypercone where the axes of four pencils of $(2r)$-planes of a 4-web are located. Moreover, these three pairs of generators belong to the same family of generators as the generators containing the axes. The four generators containing the axes can be splitted into two pairs by three ways. The two pairs of each splitting are simultaneously harmonically divided by the pair of generators corresponding to one of the diagonal families.*

Proof. As we already mentioned in Section 5.7, the diagonal r-plane is spanned by the points A_0 and $A_{i_\alpha} + A_{i_\beta} - A_{i_\gamma}$ and therefore it has the following equations in the moving frame $\{A_u\}$:

$$x^{i_\alpha} + x^{i_\gamma} = 0, \quad x^{i_\beta} + x^{i_\alpha} = 0. \tag{5.8.23}$$

The intersection of the diagonal r-plane (5.8.23) with the hypercone (5.8.12) is determined by the equations (5.8.23) and

$$x^0 = \pm 2\sqrt{\lambda_{i_\alpha}\lambda_{j_\beta} x^{i_\alpha} x^{j_\beta}}. \tag{5.8.24}$$

Thus, a diagonal r-plane intersects the hypercone in two $(r-1)$-planes determined by the equations (5.8.23) and (5.8.24).

An easy calculation shows that each of these $(r-1)$-planes is incident exactly with one generator of the same family to which the four axes belong. This pair of generators is obtained from (5.8.22) for the following values of k: if $\alpha = 2, \beta = 3$, then $k = \pm\sqrt{-\lambda_3/\lambda_2}$; if $\alpha = 3, \beta = 1$, then $k = (\lambda_3 \pm \sqrt{-\lambda_3\lambda_2})/\lambda_2$, and if $\alpha = 1, \beta = 2$, then $k = -1 \pm \sqrt{-\lambda_1/\lambda_2}$.

Let

$$M = x^0 A_0 + \sum_\alpha x^{i_\alpha} A_{i_\alpha}$$

be an arbitrary point in P^{3r}. If it is fixed, then we have (5.8.6) or in the coordinate form (5.8.7). If we write the equations (5.8.7) separately for x^0 and x^{i_α}, we obtain

$$\begin{cases} dx^0 + x^0\omega_0^0 + \sum\limits_\alpha x^{i_\alpha}\omega_{i_\alpha}^0 = \lambda x^0, \\ dx^{i_\alpha} + x^0\omega^{i_\alpha} + \sum\limits_\beta x^{i_\beta}\omega_{j_\beta}^{i_\alpha} = \lambda x^{i_\alpha}. \end{cases} \tag{5.8.25}$$

Using equations (5.8.25), it is easy to show that the plane generators determined by equations (5.8.22) are fixed. Since the hypercone is fixed, the plane generators of the second family are transfered to each other.

Since a diagonal r-plane intersects a pair of generators of the family (5.8.22) and these generators are fixed under an ininitesimal displacement, the r-planes of each of three diagonal families of r-planes intersect one and the same (for a given family) pair of generators in $(r-1)$-planes.

Thus, in addition to the four generators containing the axes, three more pairs of generators are distinguished in the same family. The quadruple of generators containing the axes can be split into two pairs by three different ways. A pair of generators corresponding to one diagonal family harmonically intersects both pairs of one of such splittings. More precisely, the pair for which $k = \pm\sqrt{-\lambda_3/\lambda_2}$ harmonically divides the pairs for which $k = 0, \infty$ and $k = -1, \lambda_3/\lambda_2$; the pair for which $k = -1 \pm \sqrt{-\lambda_1/\lambda_2}$ harmonically divides the pairs for which $k = \infty, -1$ and $k = 0, \lambda_3/\lambda_2$ and finally the pair for which $k = (\lambda_3 \pm \sqrt{-\lambda_3\lambda_1})/\lambda_2$ harmonically divides the pairs for which $k = \infty, \lambda_3/\lambda_2$ and $k = 0, -1$. ■

Remark 5.8.3 Since all the quantities $\lambda_1, \ \lambda_2, \ \lambda_3 \neq 0$ and they satisfy condition (5.7.54), there are two of them of the same sign and the third one will be of the opposite sign. Therefore one of the products $\lambda_\alpha\lambda_\beta, \alpha \neq \beta$ is negative while two others are positive. As a consequence, the generators of one of the pairs corresponding to the diagonal families are imaginary while the generators of the two others are real.

The following two theorems describe diagonal 4-webs for which one or two of the quantities λ_α vanishes.

Theorem 5.8.4 *For a diagonal 4-web in P^{3r} the following six conditions are equivalent:*

(i) $\lambda_\alpha = 0, \alpha$ *fixed.*

(ii) *A diagonal 4-web is α-reducible.*

(iii) *The hypercone (5.8.12) containing four axes of the pencils of a diagonal 4-web decomposes into a pair of hyperplanes*

$$x^0 \pm \lambda_{i_\beta}(x^{i_\beta} + x^{i_\gamma}) = 0.$$

(iv) *The axes of the pencils λ_β and λ_γ lie in the hyperplane $x^0 - \lambda_{i_\beta}(x^{i_\beta} + x^{i_\gamma}) = 0$ and the axes of the pencils λ_α and λ_4 lie in the hyperplane $x^0 + \lambda_{i_\beta}(x^{i_\beta} + x^{i_\gamma}) = 0$.*

(v) *The axes of the pencils λ_β and λ_γ have a common $(r-1)$-plane $x^0 = x^{i_\beta} = x^{i_\gamma} = 0$ and the axes of the pencils λ_α and λ_4 have a common $(r-1)$-plane*

(vi) *The axes of the pencils λ_β and λ_γ, λ_α and λ_4 intersect an r-plane of intersection of corresponding $(2r)$-planes of these pencils in one and the same $(r-1)$-plane.*

Proof. Suppose that

$$\lambda_\alpha = 0, \quad \alpha \text{ fixed.} \tag{5.8.26}$$

Then by means of (5.8.26) and (5.7.54), we have

$$\lambda_\beta = -\lambda_\gamma. \tag{5.8.27}$$

By (5.8.26) and (5.8.27), equations (5.7.49) and (5.7.50) of the axes will have the form:

$$\begin{aligned} x^{i_\alpha} &= 0, & x^0 + \lambda_{i_\beta}(x^{i_\beta} + x^{i_\gamma}) &= 0, \\ x^{i_\beta} &= 0, & x^0 - \lambda_{i_\beta} x^{i_\gamma} &= 0, \\ x^{i_\gamma} &= 0, & x^0 - \lambda_{i_\beta} x^{i_\beta} &= 0, \\ x^{i_\alpha} + x^{i_\beta} + x^{i_\gamma} &= 0 & x^0 - \lambda_{i_\beta} x^{i_\alpha} &= 0. \end{aligned} \tag{5.8.28}$$

It is easy to see from (5.8.28) that the axes of the pencils λ_β and λ_γ have a common $(r-1)$-plane $x^0 = x^{i_\beta} = x^{i_\gamma} = 0$ and these two axes belong to the hyperplane $x^0 - \lambda_{i_\beta}(x^{i_\beta} + x^{i_\gamma}) = 0$. Similarly, the axes of the pencils λ_α and λ_4 have a common $(r-1)$-plane $x^0 = x^{i_\alpha} = x^{i_\beta} + x^{i_\gamma} = 0$ and lie in the hyperplane $x^0 + \lambda_{i_\beta}(x^{i_\beta} + x^{i_\gamma}) = 0$. In this case

$$\underset{\beta}{M}{}_{i_\alpha} = \underset{\gamma}{M}{}_{i_\alpha} = A_{i_\alpha}, \;\; \underset{\alpha}{M}{}_{i_\beta} - \underset{\alpha}{M}{}_{i_\gamma} = \underset{\beta\alpha}{N}{}_i - \underset{\gamma\alpha}{N}{}_i = A_{i_\beta} - A_{i_\gamma}. \tag{5.8.29}$$

Equations (5.8.29) prove the coincidence of the $(r-1)$-planes in which the axes of the two pairs of pencils indicated above intersect an r-plane of intersection of the corresponding $(2r)$-planes of the pencils.

In the case (5.8.26) the hypercone (5.8.12) decomposes into two hyperplanes mentioned above. The hyperplanes themselves have a common $(3r-2)$-plane π whose equations are

$$x^0 = 0, \;\; \lambda_{i_\beta}(x^{i_\beta} + x^{i_\gamma}) = 0. \tag{5.8.30}$$

The $(3r-2)$-plane π determined by (5.8.30) intersects the axes of the pencils λ_β and λ_γ in $(2r-2)$-planes defined by equations

$$x^{i_\beta} = x^0 = \lambda_{i_\beta} x^{i_\gamma} = 0, \;\; x^{i_\gamma} = x^0 = \lambda_{i_\beta} x^{i_\beta} = 0. \tag{5.8.31}$$

The $(2r-2)$-planes (5.8.31) have a common an $(r-1)$-plane with equations $x^0 = x^{i_\beta} = x^{i_\gamma} = 0$. Similarly, the $(3r-2)$-plane (5.8.30) intersects the axes of the pencils λ_α and λ_4 in a $(2r-2)$-planes defined by the equations

$$x^0 = x^{i_\alpha} = \lambda_{i_\beta}(x^{i_\beta} + x^{i_\gamma}) = 0, \;\; x^0 = \lambda_{i_\beta} x^{i_\alpha} = x^{i_\alpha} + x^{i_\beta} + x^{i_\gamma} = 0, \tag{5.8.32}$$

and the $(2r-2)$-planes (5.8.32) intersect each other in an $(r-1)$-plane with equations $x^0 = x^{i_\alpha} = x^{i_\beta} + x^{i_\gamma} = 0$. We proved that (i) implies (iii)–(vi). The converse statements can be easily checked.

Note finally that, by (5.8.2), condition (5.8.26) is equivalent to the condition

$$\underset{\alpha\beta}{a}{}^{i}_{jk} = \underset{\alpha\gamma}{a}{}^{i}_{jk}. \tag{5.8.33}$$

Condition (5.8.33) coincide with one of conditions (4.1.32)–(4.1.34) and mean that a diagonal 4-web is α-reducible. ■

Theorem 5.8.5 *For a diagonal* 4*-web in* P^{3r} *the following six conditions are equivalent:*

(i) $\lambda_1 = \lambda_2 = \lambda_3 = 0$.

(ii) *A diagonal* 4*-web is parallelisable.*

(iii) *The hypercone* (5.8.12) *decomposes in the* 2*-fold hyperplane* $x^0 = 0$.

(iv) *All four axes of the pencils of* $(2r)$*-planes which form a diagonal* 4*-web mutually intersect each other in* $(r-1)$*-planes.*

(v) *All four axes of the pencils of* $(2r)$*-planes which form a diagonal* 4*-web belong to the hyperplane* $x^0 = 0$.

(vi) *The axes of any two pencils which form a diagonal* 4*-web intersect the common* r*-plane of corresponding* $(2r)$*-planes of these two pencils in an* $(r-1)$*-plane.*

Proof. Suppose that

$$\lambda_\alpha = \lambda_\beta = 0. \tag{5.8.34}$$

Then, by (5.7.54) and (5.8.34), we have

$$\lambda_\gamma = 0. \tag{5.8.35}$$

Equations (5.8.34) and (5.8.35) imply that

$$\underset{\beta}{M}_{i_\alpha} = \underset{\gamma}{M}_{i_\alpha} = A_{i_\alpha}, \tag{5.8.36}$$

and conversely. Therefore, in the case $\lambda_1 = \lambda_2 = \lambda_3 = 0$, and only in this case, for a diagonal 4-web, we have the properties described in (iii), (iv), (v), and (vi). Finally, by (5.8.2), conditions (5.8.34) and (5.8.35) are equivalent to the condition $\underset{\alpha\beta}{a}{}^{i}_{jk} = 0$. By Theorem 1.5.2, this means that the diagonal 4-web is parallelisable. ■

In conclusion we will prove the following theorem:

Theorem 5.8.6 *Through each point of any of four axes of a diagonal* 4*-web there passes a* $(2r-1)$*-plane intersecting the other three axes in* $(2r-1)$*-planes.*

Proof. Suppose first that $\lambda_\alpha \neq 0, \alpha = 1, 2, 3$. Let us take, for example, an arbitrary point K belonging to the axis of the pencil λ_α and not belonging to the vertex (5.8.17) of the hypercone (5.8.12). Take through K a 3-plane σ not having common points with the vertex (5.8.17). The hypercone (5.8.12) intersects σ along a quadric F_2. Through K there pass two rectilinear generators of the quadric F_2. One of them together with the vertex (5.8.17) gives that generator of the hypercone where the axis of the pencil is located. Take another generator of the quadric F_2 passing through the point K and draw through it the $(3r-2)$-dimensional generator of the hypercone. Consider in the latter a $(2r-3)$-plane determined by the common $(2r-1)$-plane of the vertex (5.8.17) and the axis of the pencil λ_α and by the second generator of the quadric F_2 passing through the point K. The $(2r-1)$-plane intersects the axes of the pencils $\lambda_\beta, \lambda_\gamma$, and λ_4 in $(r-1)$-planes. Each of these three $(r-1)$-planes is determined by the common point of the second generator of F_2 passing through K and that generator of F_2 which defines the axis of the pencil λ_β (or λ_γ, or λ_4) and by a common $(r-2)$-plane of the axes of the pencils λ_α and λ_β(or λ_γ, or λ_4).

Suppose further that a point K lies in the vertex of the hypercone. Then any $(2r-1)$-plane passing through the point K and intersecting the vertex in a $(2r-2)$-plane, intersects the other three axes in $(2r-2)$-planes.

Consider next the case when $\lambda_\alpha = 0, \alpha$ fixed. Suppose first that a point K lies in the axis of the pencil λ_α and $K \notin \pi$. In this case the desired $(2r-1)$-plane is determined by the point K and the common $(2r-2)$-plane of π and the axis of the pencil λ_β (or λ_γ). If $K \in \pi$, then as the desired $(2r-1)$-plane, one can take any $(2r-1)$-plane passing through the common $(2r-2)$-plane of π and the axes of the pencil λ_γ and through any point of the hyperplane where the axes of the pencils λ_β and λ_γ are located.

Finally in the case $\lambda_1 = \lambda_2 = \lambda_3 = 0$ all four axes of the pencils lie in the hyperplane $x^0 = 0$ and consequently as the desired $(2r-1)$-plane, one can take any $(2r-1)$-plane passing through a point in one of the axes and lying in the hyperplane $x^0 = 0$. ■

NOTES

5.1, 5.3–5.6. These sections, except Section 5.4.2, are from the author's paper[G 75e]. Section 5.4.2 is due to Gerasimenko [Ge 85b] (see also [AGe 86]). Note that in [Ge 85b] and in [AGe 86] wrongly indicated that an isoclinic Moufang web $M(n+1,n,r)$ is a group one. According to Corollary 5.4.6, the surfaces U_ξ generating algebraic $M(n+1,n,r)$ are r-planes in general position while for algebraic $M(n+1,n,r)$ to be a group one, these r-planes are in special position indicated in Theorem 5.5.2.

5.2. In this section we have followed the exposition in [G 82b].

5.7–5.8. The results of these two sections are due to the author [G 75f].

Chapter 6

Applications of the Theory of $(n+1)$-Webs

6.1 The Application of the Theory of $(n+1)$-Webs to the Theory of Point Correspondences of $n+1$ Projective Lines

In this chapter we consider the applications of the theory of $(n+1)$-webs which we have developed in Chapters 1 – 4 to the theory of point correspondences among $n+1$ projective lines (Section 6.1) and $n+1$ projective spaces (Section 6.2) and to the theory of holomorphic mappings between polyhedral domains (Section 6.3).

We will start with applications to the theory of point correspondences among $n+1$ projective lines.

6.1.1 The Fundamental Equations

We consider $n+1$ projective lines P_ξ, $\xi, \eta, \zeta = 1, \ldots, n+1$, and a point correspondence $C : P_1 \times P_2 \times \ldots \times P_n \rightarrow P_{n+1}$ among them. Suppose that a non-homogeneous coordinate x_ξ is introduced on the line P_ξ and a point O_ξ is the origin of the coordinate system on P_ξ. If an equation

$$f(x_1, \ldots, x_{n+1}) = 0 \tag{6.1.1}$$

is satisfied for the $(n+1)$-tuple $(O_1, \ldots, O_{n+1})$, i.e.,

$$f(0, \ldots, 0) = 0, \tag{6.1.2}$$

and, in addition,

$$\frac{\partial f}{\partial x_\xi} = 0, \quad \xi = 1, \ldots, n+1, \tag{6.1.3}$$

then equation (6.1.1) is uniquely solvable in a neighbourhood of the points O_ξ with respect to each of the variables x_ξ.

Definition 6.1.1 The equation (6.1.1) together with the conditions (6.1.2) and (6.1.3) defines a correspondence among the $n+1$ lines P_ξ which is *regular at the* $(n+1)$*-tuple* $(O_1, \ldots, O_{n+1})$.

Here and in what follows we shall assume that that the function $f(x_1, \ldots, x_{n+1})$ is analytic or sufficiently smooth.

Suppose that $\underset{\xi}{M}_0$ are corresponding points of the lines P_ξ. We will associate with the points $\underset{\xi}{M}_0$ projective moving frames $F_\xi = \{\underset{\xi}{M}_0, \underset{\xi}{M}_0\}$ of the lines P_ξ . The equations of infinitesimal displacements of the moving frames F_ξ are:

$$\begin{cases} d\underset{\xi}{M}_0 = \underset{\xi}{\omega}{}_0^0 \underset{\xi}{M}_0 + \underset{\xi}{\omega}{}_0^1 \underset{\xi}{M}{}_1^0, \\ d\underset{\xi}{M}{}_1^0 = \underset{\xi}{\omega}{}_1^0 \underset{\xi}{M}_0 + \underset{\xi}{\omega}{}_1^1 \underset{\xi}{M}_1, \end{cases} \tag{6.1.4}$$

where $\underset{\xi}{\omega}{}_v^u$, $u, v, w = 0, 1$, are 1-forms satisfying the structure equations of the projective lines P_ξ :

$$d\underset{\xi}{\omega}{}_u^v = \underset{\xi}{\omega}{}_u^w \wedge \underset{\xi}{\omega}{}_w^v \tag{6.1.5}$$

(cf. equations (5.1.4) and (5.1.9)).

In the chosen system of moving frames the equation of the correspondence C is written as a linear relation among forms $\underset{\xi}{\omega}{}_0^1$:

$$\sum_\xi h_\xi \underset{\xi}{\omega}{}_0^1 = 0 \tag{6.1.6}$$

If we denote $h_\xi \underset{\xi}{\omega}{}_0^1$ by $\underset{\xi}{\tilde{\omega}}{}_0^1$, then $\underset{\xi}{\omega}{}_0^1 = \underset{\xi}{\tilde{\omega}}{}_0^1 / h_\xi$. Equations (6.1.4) give

$$d\underset{\xi}{M}_0 = \underset{\xi}{\omega}{}_0^0 \underset{\xi}{M}_0 + \underset{\xi}{\tilde{\omega}}{}_0^1 \underset{\xi}{M}_1 / h_\xi.$$

By normalising the points $\underset{\xi}{M}_1$: $\underset{\xi}{\tilde{M}}_1 = \underset{\xi}{M}_1 / h_\xi$, we reduce relation (6.1.6) to the form:

$$\sum_\xi \underset{\xi}{\omega}{}_0^1 = 0. \tag{6.1.7}$$

In (6.1.7) we suppress a wave over ω . Equation (6.1.7) is a *fundamental equation* of the correspondence we study.

Taking the exterior derivative of (6.1.7) by means of (6.1.5) and substituting $\underset{n+1}{\omega}{}_0^1$ from (6.1.7) into the quadratic equation obtained, we have

$$\sum_{\alpha=1}^{n} \left(\underset{\alpha}{\omega} - \underset{n+1}{\omega} \right) \wedge \underset{\alpha}{\omega}{}_0^1 = 0 \tag{6.1.8}$$

where $\underset{\xi}{\omega} = \underset{\xi}{\omega}{}_1^1 - \underset{\xi}{\omega}{}_0^0$. Application of the Cartan lemma to (6.1.8) implies

$$\underset{\alpha}{\omega} - \underset{n+1}{\omega} = \sum_{\beta} \underset{\alpha\beta}{\lambda}\, \underset{\beta}{\omega}{}_0^1 \tag{6.1.9}$$

where $\underset{\alpha\beta}{\lambda} = \underset{\beta\alpha}{\lambda}$. After exterior differentiation of (6.1.9), by using (6.1.5), (6.1.7), and (6.1.9), we obtain

$$\sum_{\beta}\left[\nabla \underset{\alpha\beta}{\lambda} - \underset{\alpha\beta}{\delta}(\underset{\alpha}{\omega}{}_1^0 + \underset{\beta}{\omega}{}_1^0) - 2\underset{n+1}{\omega}{}_1^0 - \sum_{\gamma} \underset{\alpha\beta}{\lambda}(\underset{\alpha\gamma}{\lambda} + \underset{\beta\gamma}{\lambda})\underset{\gamma}{\omega}{}_0^1\right] \wedge \underset{\beta}{\omega}{}_0^1 = 0 \tag{6.1.10}$$

where $\nabla \underset{\alpha\beta}{\lambda} = d\underset{\alpha\beta}{\lambda} - \underset{\alpha\beta}{\lambda}\,\underset{n+1}{\omega}$. By Cartan's lemma, we obtain from (6.1.10)

$$\nabla \underset{\alpha\beta}{\lambda} = \underset{\alpha\beta}{\delta}(\underset{\alpha}{\omega}{}_1^0 + \underset{\beta}{\omega}{}_1^0) + 2\underset{n+1}{\omega}{}_1^0 + \sum_{\gamma}\left[\underset{\alpha\beta\gamma}{\lambda} + \underset{\alpha\beta}{\lambda}(\underset{\alpha\gamma}{\lambda} + \underset{\beta\gamma}{\lambda})\right]\underset{\gamma}{\omega}{}_1^1 \tag{6.1.11}$$

where the quantities $\underset{\alpha\beta\gamma}{\lambda}$ are symmetric over all indices, and $\alpha, \beta, \gamma = 1, \ldots, n$.

The quantities $\underset{\alpha\beta}{\lambda}$ and $\underset{\alpha\beta\gamma}{\lambda}$ in (6.1.9) and (6.1.11), as well as similar objects arising if one makes the consecutive prolongations of (6.1.11) define the differential geometry of the correspondence C. The objects $\underset{\alpha\beta}{\lambda}$ belonging to the second differential neighbourhood of C allow us to assign an invariant equipment on the lines P_ξ of the correspondence C and to relate invariant moving frames to the correspondence C.

To show this, first note that the point $\underset{\xi}{M}_0$ was already chosen as an invariant point. Analytically this is confirmed by the equation $\delta\underset{\xi}{M}_0 = \underset{\xi}{\pi}_0\underset{\xi}{M}_0$ which follows from (6.1.4) and where, as usually δ replaces d when the point $\underset{\xi}{M}_0$ is fixed, i.e., $\underset{\xi}{\omega}_0 = 0$, and $\underset{\xi}{\pi}{}_v^u = \underset{\xi}{\omega}{}_v^u(\delta)$. To find another invariant point

$$\underset{\xi}{A}_0 = \underset{\xi}{M}_0 - \underset{\xi}{x}\,\underset{\xi}{M}_0 \tag{6.1.12}$$

on the line P_ξ , we differentiate (6.1.12) keeping $\underset{\xi}{\omega}{}_0^1 = 0$ and using (6.1.4), (6.1.12), and (6.1.9). As result, we have

$$\delta\underset{\xi}{A}_1 \wedge \underset{\xi}{A}_1 = (\underset{\xi}{\pi}{}_1^0 - \nabla_\delta \underset{\xi}{x})\underset{\xi}{M}_0 \tag{6.1.13}$$

where $\nabla_\delta \underset{\xi}{x} = \delta\underset{\xi}{x} - \underset{\xi}{x}\,\underset{\xi n+1}{\pi}$. The relation (6.1.13) shows that the point $\underset{\xi}{A}_1$ is invariant if and only if

$$\nabla_\delta \underset{\xi}{x} = \underset{\xi}{\pi}{}_1^0. \tag{6.1.14}$$

So, to find an invariant point $\underset{\xi}{A_1}$, we should construct a quantity $\underset{\xi}{x}$ satisfying (6.1.14). Equations (6.1.9) and (6.1.11) allow us to check that the quantities

$$\underset{n+1}{l} = \frac{1}{2n(n-1)} \sum_{\alpha,\beta(\alpha\neq\beta)} \underset{\alpha\beta}{\lambda}, \quad \underset{\alpha}{l} = \frac{1}{2}\underset{\alpha\alpha}{\lambda} - \underset{n+1}{l} \tag{6.1.15}$$

satisfy (6.1.14). If we set

$$\underset{\xi}{A_0} = \underset{\xi}{M_0}, \quad \underset{\xi}{A_1} = \underset{\xi}{M_1} - \underset{\xi}{l}\underset{\xi}{M_0} \tag{6.1.16}$$

where the $\underset{\xi}{l}$ are determined by (6.1.15), then the moving frames $\{\underset{\xi}{A_0}, \underset{\xi}{A_0}\}$ of the lines P_ξ are invariant moving frames defined by the correspondence C.

For the frame $\{\underset{\xi}{A_0}, \underset{\xi}{A_1}\}$ we have the following equations of infinitesimal displacement:

$$d\underset{\xi}{A_u} = \underset{\xi}{\theta^v_u}\underset{\xi}{A_v}. \tag{6.1.17}$$

If we differentiate (6.1.16), substitute $\underset{\xi}{M_u}$ from (6.1.16) into obtained equations and compare them with (6.1.17), we will have the following relations between components $\underset{\xi}{\theta^v_u}$ of infinitesimal displacement of the invariant frames and the analogous components of arbitrary frames:

$$\begin{cases} \underset{\xi}{\theta^1_0} = \underset{\xi}{\omega^1_0}, & \underset{\xi}{\theta^0_0} = \underset{\xi}{\omega^0_0} + \underset{\xi}{l}\underset{\xi}{\omega^1_0}, \\ \underset{\xi}{\theta^0_1} = \underset{\xi}{\omega^0_1} - d\underset{\xi}{l} + \underset{\xi}{l}(\underset{\xi}{\omega} - \underset{\xi}{l}\underset{\xi}{\omega^1_0}), & \underset{\xi}{\theta^1_1} = \underset{\xi}{\omega^1_1} - \underset{\xi}{l}\underset{\xi}{\omega^1_0}. \end{cases} \tag{6.1.18}$$

Using (6.1.18), we can write equations (6.1.7) and (6.1.9) for the invariant frames:

$$\sum_\xi \underset{\xi}{\theta^1_0} = 0, \quad \underset{\alpha}{\theta} - \underset{n+1}{\theta} = \sum_\beta \underset{\alpha\beta}{a}\underset{\beta}{\theta^1_0}. \tag{6.1.19}$$

where $\underset{\xi}{\theta} = \underset{\xi}{\theta^1_1} - \underset{\xi}{\theta^0_0}$, $\underset{\alpha\beta}{a} = \underset{\alpha\beta}{\lambda} - \underset{\alpha\beta}{\delta}(\underset{\alpha}{l} + \underset{\beta}{l}) - 2\underset{n+1}{l}$. We proved the following theorem.

Theorem 6.1.2 *A point correspondence C among $n+1$ projective lines determines an invariant equipment of these lines in the second differential neighbourhood of C. The equations of C with respect to the invariant frames have the form* (6.1.19). ■

The quantities $\underset{\alpha\beta}{a}$ are expressed in terms of $\underset{\alpha\beta}{\lambda}$ and therefore belong to the second differential neighbourhood of the correspondence C. As we will see below, they are invariants. Since the $\underset{\alpha\beta}{\lambda}$ are symmetric and the $\underset{\xi}{l}$ are expressed by (6.1.15), the quantities $\underset{\alpha\beta}{a}$ satisfy the relations

$$\underset{\alpha\alpha}{a} = 0, \quad \underset{\alpha\beta}{a} = \underset{\beta\alpha}{a}, \quad \sum_{\alpha,\beta(\alpha\neq\beta)} \underset{\alpha\beta}{a} = 0. \tag{6.1.20}$$

As follows from (6.1.20), the total number of independent invariants $\underset{\alpha\beta}{a}$ is equal to $n(n-1)/2-1$.

For the invariant frames the 1-forms are $\underset{\xi}{\theta_1^0}$ expressed in terms of $\underset{\xi}{\theta_0^1}$ since the point $\underset{\xi}{A_1}$ is fixed when $\underset{\xi}{A_0} = \underset{\xi}{M_0}$ is fixed. We therefore set

$$\underset{\xi}{\theta_1^0} = \sum_\alpha \underset{\xi\alpha}{b}\, \underset{\alpha}{\theta_0^1}. \tag{6.1.21}$$

The quantities $\underset{\alpha\beta}{a}$ and $\underset{\xi\alpha}{b}$ satisfy additional relations. To find these, first note that by virtue of (6.1.19), (6.1.20), and (6.1.21) we have the following equations for $\underset{\alpha\beta}{a}$:

$$\nabla \underset{\alpha\beta}{a} = \sum_\gamma \left[\underset{\alpha\beta\gamma}{a} + \underset{\alpha\beta}{a}\big(\underset{\alpha\gamma}{a} + \underset{\beta\gamma}{a}\big)\right] \underset{\gamma}{\theta_0^1} \tag{6.1.22}$$

where $\nabla \underset{\alpha\beta}{a} = d\underset{\alpha\beta}{a} - \underset{\alpha\beta n+1}{a\ \theta}$ and

$$\underset{\alpha\beta\gamma}{a} = \underset{\beta\alpha\gamma}{a}, \quad \underset{\alpha\alpha\gamma}{a} = 0, \quad \sum_{\alpha,\beta(\alpha\neq\beta)} \left[\underset{\alpha\beta\gamma}{a} + \underset{\alpha\beta}{a}\big(\underset{\alpha\gamma}{a} + \underset{\beta\gamma}{a}\big)\right] = 0. \tag{6.1.23}$$

Next, by exterior differentiation of (6.1.11) and substitution of the values of $\underset{\xi}{\theta_1^0}, \nabla \underset{\alpha\beta}{a}$ from (6.1.21) and (6.1.22) into the equations obtained, by virtue of the linear independence of $\underset{\alpha}{\theta_0^1}$, we deduce to the following equations

$$\underset{\alpha\beta\gamma}{a} - \underset{\alpha\gamma\beta}{a} - 2\big(\underset{n+1,\alpha}{b} - \underset{n+1,\beta}{b}\big) - \underset{\alpha\beta}{\delta}\big(\underset{\alpha\gamma}{b} + \underset{\beta\gamma}{b}\big) + \underset{\alpha\gamma}{\delta}\big(\underset{\alpha\beta}{b} + \underset{\gamma\beta}{b}\big) = 0. \tag{6.1.24}$$

The equations (6.1.24) can be simplified. Suppose that $n \geq 3$ (the case $n=2$, i.e., the point correspondence among three projective lines was considered in [Bo 82]). Then two mutual combinations of the values α, β and γ are possible. In the first case the indices α, β and γ are distinct. Then from (6.1.24) we have

$$2\big(\underset{n+1,\gamma}{b} - \underset{n+1,\beta}{b}\big) = \underset{\alpha\beta\gamma}{a} - \underset{\alpha\gamma\beta}{a}, \quad \alpha \neq \beta. \tag{6.1.25}$$

In the second case two indices are identical. Suppose that $\alpha = \gamma \neq \beta$ (the case $\beta = \gamma \neq \alpha$ is equivalent to this since the expression in (6.1.24) is skew-symmetric with respect to β and γ). Then frcm (6.1.24) we have

$$2\underset{\alpha\beta}{b} = 2\big(\underset{n+1,\alpha}{b} - \underset{n+1,\beta}{b}\big) - \underset{\alpha\beta\gamma}{a}, \quad \alpha \neq \beta. \tag{6.1.26}$$

The relations (6.1.25) and (6.1.26) are equivalent to (6.1.24) for $n \geq 3$.

6.1.2 Correspondences Among $n+1$ Projective Lines and One-Codimensional $(n+1)$-Webs

Under the conditions we have assumed the correspondence C among the $n+1$ projective lines P_ξ is a local differentiable n-quasigroup (see Definition 3.1.1). As we know (see Section 3.1), to each n-quasigroup there is associated an $(n+1)$-web. In our case this web can be constructed as follows. Consider the Cartesian product $P_1 \times \ldots \times P_{n+1}$. Then in this product the correspondence C corresponds to an n-dimensional manifold. The fixation of the points of the line P_ξ generates a foliation of codimension one on this manifold. The set of all $n+1$ foliations of codimension one defines a web $W(n+1, n, 1)$ on the manifold.

It is convenient to write the structure equations of the $W(n+1, n, 1)$ in the system of invariant frames of the lines P_ξ . Using (6.1.19), (6.1.21), and (6.1.5), we have

$$\begin{cases} \sum\limits_\xi \underset{\xi}{\theta}{}_0^1 = 0, \quad d\underset{\alpha}{\theta}{}_0^1 = \underset{\alpha}{\theta}{}_0^1 \wedge \underset{n+1}{\theta} + \sum\limits_\beta \underset{\alpha\beta\alpha}{a} \underset{\beta}{\theta}{}_0^1 \wedge \underset{\beta}{\theta}{}_0^1, \\ d\theta = \sum\limits_{\substack{\alpha,\beta \\ (\alpha<\beta)}} 2(\underset{n+1,\alpha}{b} - \underset{n+1,\beta}{b}) \underset{\alpha}{\theta}{}_0^1 \wedge \underset{\beta}{\theta}{}_0^1. \end{cases} \tag{6.1.27}$$

Comparison of (6.1.27) and (1.2.16)–(1.2.29) shows that the invariants $\underset{\alpha\beta}{a}$ and $2(\underset{n+1,\alpha}{b} - \underset{n+1,\beta}{b})$ of the correspondence C form the torsion and curvature tensors of the web $W(n+1, n, 1)$ associated with C. If $n > 2$, it folllows from (6.1.25), that the curvature tensor of $W(n+1, n, 1)$ is expressed in terms of the torsion tensor and its Pfaffian derivatives. This agrees with the equations (1.2.26) obtained for a general $W(n+1, n, r), r > 1$.

6.1.3 Parallelisable Correspondences

Definition 6.1.3 A correspondence C is said to be *parallelisable* if the web $W(n+1, n, 1)$ associated with C is parallelisable.

Theorem 6.1.4 *For a correspondence C among $n+1$ projective lines P_ξ, $n > 2$, to be parallelisable, it is necessary and sufficient that the invariant projective equipment of the lines P_ξ which is determined by C, on each line depends only on the choice of the points of this line.*

Proof. Since $n > 2$, by Theorem 1.5.2 and Definition 3.1.3, the correspondence C is parallelisable if and only if its torsion tensor vanishes: $\underset{\alpha\beta}{a} = 0$. If we substitute $\underset{\alpha\beta}{a} = 0$ into (6.1.11) and use the relations (6.1.25) and (6.1.26), we obtain the equations of a parallelisable correspondence C in the form:

$$\sum_\xi \underset{\xi}{\theta}{}_0^1 = 0, \quad \underset{\alpha}{\theta} = \underset{n+1}{\theta}, \quad \underset{\alpha}{\theta}{}_1^0 = \underset{\alpha\alpha}{b} \underset{\alpha}{\theta}{}_0^1, \quad \underset{n+1}{\theta}{}_1^0 = -\underset{n+1,1}{b} \underset{n+1}{\theta}{}_0^1. \tag{6.1.28}$$

In the general case the invariant equipment of the lines P_ξ depends on the choice of the points $\underset{\xi}{A}_0$ and $\underset{\xi}{A}_1$. If C is parallelisable, then, as follows from (6.1.28),

$$d\underset{\xi}{A}_1 \equiv \underset{\xi}{\theta}{}^1_0 \underset{\xi}{A}_1 \mod \underset{\xi}{\theta}{}^1_0,$$

and the invariant frame of the line P_ξ is completely defined by the choice of the point $\underset{\xi}{A}_0 \in P_\xi$ only.

Conversely, suppose that the invariant frame $\{\underset{\xi}{A}_0, \underset{\xi}{A}_1\}$ of each line P_ξ, ξ fixed, is defined by the choice of only the point $\underset{\xi}{A}_0$. Then $d\underset{\xi}{A}_0 = \sigma \underset{\xi}{A}_0 \mod \underset{\xi}{\theta}{}^1_0$ where σ is an 1-form. By the equations $d\underset{\xi}{A}_1 = \underset{\xi}{\theta}{}^0_1 \underset{\xi}{A}_0 + \underset{\xi}{\theta}{}^1_0 \underset{\xi}{A}_1$ and (6.1.21), we have $\underset{n+1,\alpha}{b} = \underset{n+1,\beta}{b}$, $\underset{\alpha\beta}{b} = 0, \;\; \alpha \neq \beta$.

We substitute these values into (6.1.25) and (6.1.26). Then we obtain $\underset{\alpha\beta\alpha}{a} = 0$. Further, setting $\gamma = \alpha$ in (6.1.23), we finally obtain $\underset{\alpha\beta}{a} = 0$. Therefore, the correspondence C is parallelisable.

The equations (6.1.28) of a parallelisable correspondence admit integration in a closed form. In fact, since, by (6.1.27) and (6.1.28), $d\underset{n+1}{\theta} = d\underset{\xi}{\theta}{}^0_0 = 0$, the forms $\underset{n+1}{\theta}, \underset{\xi}{\theta}{}^0_0$ in this case are total differentials. Set: $\underset{n+1}{\theta} = d\ln a, \; \underset{\xi}{\theta}{}^0_0 = d\ln \underset{\xi}{a}$. Then, choosing appropriate a and $\underset{\xi}{a}$, we reduce the forms $\underset{n+1}{\theta}, \underset{\xi}{\theta}{}^0_0$ to 0. Equations (6.1.28) show that in addition we have $\underset{\alpha}{\theta} = 0$ and $\underset{\xi}{\theta}{}^1_0 = 0$. Since $\underset{\xi}{\theta}{}^0_0 = \underset{\xi}{\theta}{}^1_0 = 0$, we have $d\underset{\xi}{\theta}{}^0_0 = 0$, i.e., the form $\underset{\xi}{\theta}{}^1_0$ is a total differential. Setting $\underset{\xi}{\theta}{}^1_0 = du_\xi$ and substituting it into (6.1.28), we obtain

$$\sum_\xi du_\xi = 0, \;\; \underset{\xi}{\theta} = 0, \;\; \underset{\xi}{\theta}{}^0_1 = \underset{\xi}{b} du_\xi, \tag{6.1.29}$$

where $\underset{\xi}{b} = \underset{\xi}{b}(u_\xi)$ since $d\underset{\xi}{\theta}{}^0_1 = 0$ and the form $\underset{\xi}{\theta}{}^0_1$ is also a total differential. Substituting these values into the equations of infinitesimal displacement of the frame $\{\underset{\xi}{A}_0, \underset{\xi}{A}_0\}$, we find

$$d\underset{\xi}{A}_0 = du_\xi \underset{\xi}{A}_1, \;\; d\underset{\xi}{A}_1 = \underset{\xi}{b}(u) du_\xi \underset{\xi}{A}_0. \tag{6.1.30}$$

For a fixed value of ξ equation (6.1.30) leads to the differential equation

$$\frac{d^2 \underset{\xi}{A}_0}{(du_\xi)^2} = b_\xi(u_\xi) \underset{\xi}{A}_0. \tag{6.1.31}$$

The general solution of (6.1.30) has the form

$$\underset{\xi}{A}_0 = x_{0\xi}(u_\xi) \underset{\xi}{A}{}^0_0 + x_{1\xi}(u_\xi) \underset{\xi}{A}{}^0_0, \tag{6.1.32}$$

where the $x_{0\xi}(u_\xi)$, $x_{1\xi}(u_\xi)$, are two linearly independent solutions of the scalar equation $d^2y/(du_\xi)^2 = b_\xi(u_\xi)y_\xi$ and $\underset{\xi}{A}{}_0^0$, $\underset{\xi}{A}{}_0^0$ are fixed points of the line P_ξ.

The functions $x_{0\xi}$, $x_{1\xi}$ are homogeneous coordinates of the point $\underset{\xi}{A_0}$. We introduce the following non-homogeneous coordinates : $z_\xi = x_{0\xi}/x_{1\xi} = \psi_\xi(u_\xi)$. Since the solutions $x_{0\xi}, x_{1\xi}$ are independent, the functions $z_\xi = \psi_\xi(u_\xi)$ are invertible. Suppose $u_\xi = \tilde{\psi}_\xi(z_\xi)$. Then by (6.1.29) we obtain the equations of the correspondence C in a closed form

$$\sum_\xi \tilde{\psi}(z_\xi) = 0,$$

or in homogeneous coordinates

$$\sum_\xi \tilde{\psi}\left(\frac{x_{0\xi}}{x_{1\xi}}\right) = 0.$$

This not only proves the existence of parallelisable correspondences but also explicitly shows that such correspondences depend on $n+1$ arbitrary functions $\tilde{\psi}_\xi$ of one variable. Of course, we will arrive to the same number of arbitrary functions if we prove that the system (6.1.28) is in involution.

6.1.4 Hexagonal Correspondences

A point correspondence among $n+1$ projective lines generates $\binom{n+1}{3}$ families of correspondences among three projective lines. They arise if one fixes the points on $n-2$ lines. Each of these correspondences is a binary quasigroup with which a three-web $W(3,2,1)$ is associated.

The equations of the $\binom{n+1}{3}$ families of correspondences $C_{n+1,\alpha\beta} : P_\alpha \times P_\beta \to P_{n+1}$ are written in the form: $\underset{\gamma}{\theta}{}_0^1 = 0$ where $\gamma \neq \alpha, \beta$ and α, β are fixed, which implies that $\underset{n+1}{\theta}{}_0^1 + \underset{\alpha}{\theta}{}_0^1 + \underset{\beta}{\theta}{}_0^1 = 0$. Substituting these values into (6.1.27), we obtain

$$d\underset{\alpha}{\theta}{}_0^1 = \underset{\alpha}{\theta}{}_0^1 \wedge \theta, \quad d\underset{\beta}{\theta}{}_0^1 = \underset{\beta}{\theta}{}_0^1 \wedge \theta \tag{6.1.33}$$

where $\theta = \underset{n+1}{\theta} + \underset{\alpha\beta}{a}(\underset{\alpha}{\theta}{}_0^1 + \underset{\beta}{\theta}{}_0^1)$ is the form of the affine connection related to the associated three-web. Setting

$$d\theta = \underset{n+1,\alpha\beta}{\kappa}\,\underset{\alpha}{\theta}{}_0^1 \wedge \underset{\beta}{\theta}{}_0^1,$$

we find the curvature of the correspondence $C_{n+1,\alpha\beta}$ in the form

$$\underset{n+1,\alpha\beta}{\kappa} = 2(\underset{n+1,\beta}{b} - \underset{n+1,\alpha}{b}) - \underset{\alpha\beta\beta}{a} + \underset{\alpha\beta\alpha}{a} \tag{6.1.34}$$

(cf. (1.3.25)). Using the same technique as in Section 1.3, we find that the curvature of the correspondence $C_{\alpha\beta\gamma}$ is given by the formula

$$\underset{\alpha\beta\gamma}{\kappa} = \underset{\beta\gamma\gamma}{a} + \underset{\alpha\beta\beta}{a} + \underset{\gamma\alpha\alpha}{a} - \underset{\beta\gamma\beta}{a} - \underset{\alpha\beta\alpha}{a} - \underset{\gamma\alpha\gamma}{a}. \tag{6.1.35}$$

(cf. (1.3.79)). It follows from (6.1.34) and (6.1.35) that for any quadruple of correspondences $C_{n+1,\alpha\beta}, C_{n+1,\beta\gamma}, C_{n+1,\gamma\alpha}, C_{\alpha\beta\gamma}$ the relation

$$\underset{n+1,\alpha\beta}{\kappa} + \underset{n+1,\beta\gamma}{\kappa} + \underset{n+1,\gamma\alpha}{\kappa} + \underset{\alpha\beta\gamma}{\kappa} = 0 \tag{6.1.36}$$

holds (cf. (1.10.1)). The relation (6.1.36) is similar to the well-known theorem of Dubourdieu for webs $W(4,3,1)$ (see Section 1.10).

Definition 6.1.5 A subcorrespondence $C_{\xi\eta\zeta}$ is said to be *hexagonal* if it is curvature-free. A correspondence C is *hexagonal* if all its subcorrespondences $C_{\xi\eta\zeta}$ are hexagonal.

Theorem 6.1.6 *The following three statements are equivalent to each other:*

(i) *A correspondence C is hexagonal.*

(ii) *The conditions*

$$2\Big(\underset{n+1,\beta}{b} - \underset{n+1,\alpha}{b}\Big) - \underset{\beta\alpha\beta}{a} + \underset{\alpha\beta\alpha}{a} = 0 \tag{6.1.37}$$

hold.

(iii) *The invariant* 1*-form*

$$\tau = (n-1)\underset{n+1}{\theta} + \sum_{\alpha,\beta(\alpha<\beta)} \underset{\alpha\beta}{a}\,\underset{\beta}{\theta_0^1} \tag{6.1.38}$$

is a total differential.

Proof. The equivalence of (i) and (ii) follows from (6.1.34) and (6.1.36). To prove the equivalence of (ii) and (iii), note that, by means of (6.1.22) and (6.1.27),

$$d\tau = \sum_{\alpha,\beta(\alpha\neq\beta)} \Big[2\Big(\underset{n+1,\beta}{b} - \underset{n+1,\alpha}{b}\Big) - \underset{\beta\alpha\beta}{a} + \underset{\alpha\beta\alpha}{a}\Big] \underset{\alpha}{\theta_0^1} \wedge \underset{\beta}{\theta_0^1}. \tag{6.1.39}$$

It follows from (6.1.39) and (6.1.34) that $d\tau = 0$ if and only if (6.1.37) holds. This proves that (ii) and (iii) are equivalent. ■

It is obvious that parallelisable correspondences are always hexagonal. The converse is false: in [Bo 83] the existence of a hexagonal correspondence which is not parallelisable was proved.

6.1.5 The Godeaux Homography

Definition 6.1.7 A correspondence $C : P_1 \times \ldots \times P_n \to P_{n+1}$ is said to be a *Godeaux homography* if for fixation of any set of $n-1$ points $\underset{\xi}{A}_0 \in P_\xi$ the correspondence $C_{\eta\zeta} : P_\eta \to P_\zeta$ between the remaining lines P_η and P_ζ is projective.

The equations of the correspondence $C_{n+1,\alpha}$ is obtained from (6.1.19) in the form:

$$\underset{n+1}{\theta}{}_0^1 + \underset{\alpha}{\theta}{}_0^1 = 0, \quad \underset{\alpha}{\theta} - \underset{n+1}{\theta} = 0. \tag{6.1.40}$$

A projective mapping K is *tangent* to $C_{n+1,\alpha}$ if K and $C_{n+1,\alpha}$ have a tangency of order one at any pair of corresponding points $\underset{n+1}{A}{}_0, \underset{\alpha}{A}{}_0$ i.e. if for any curve of class $C^s, s \geq 1$, passing through the point $\underset{n+1}{A}$ in P_{n+1} its images in P_α under the mappings K and $C_{n+1,\alpha}$ have an analytic tangency at least of the first order. The correspondence $C_{n+1,\alpha}$ is projective if it coincides with a projective mapping K tangent to it. An arbitrary projective mapping K tangent to $C_{n+1,\alpha}$ at the pair of corresponding points $\underset{n+1}{A}{}_0, \underset{\alpha}{A}{}_0$, is determined by the relations

$$K \underset{n+1}{A}{}_0 = \underset{\alpha}{A}{}_0, \quad K(\underset{n+1}{A}{}_1 + l \underset{n+1}{A}{}_0) = -\underset{\alpha}{A}{}_1 \tag{6.1.41}$$

where l is a parameter. Suppose that K does not depend on the choice of corresponding points. Then

$$dK \underset{n+1}{A}{}_0 = \sigma \underset{\alpha}{A}{}_0, \quad dK(\underset{n+1}{A}{}_1 + l \underset{n+1}{A}{}_0) = -\sigma \underset{\alpha}{A}{}_1 \tag{6.1.42}$$

where σ is a suitable 1-form. Developing (6.1.42), we arrive to the following relations:

$$\sigma = \underset{1}{\theta}{}_0^0 - \underset{n+1}{\theta}{}_0^0, \quad l = 0, \quad \underset{n+1}{\theta}{}_1^0 + \underset{\alpha}{\theta}{}_1^0 = 0. \tag{6.1.43}$$

Hence, using (6.1.21), we obtain that projectivity conditions (6.1.43) for the correspondence $C_{n+1,\alpha}$ are equivalent to

$$\underset{n+1,\alpha}{b} + \underset{\alpha\alpha}{b} = 0. \tag{6.1.44}$$

Analogous computations show that projectivity conditions for the correspondence $C_{\alpha\beta}$ are:

$$\underset{\alpha\alpha}{b} - \underset{\beta\beta}{b} - \underset{\alpha\beta}{b} + \underset{\beta\alpha}{b} + \underset{\alpha\beta\alpha}{a} - \underset{\beta\alpha\beta}{a} = 0. \tag{6.1.45}$$

Theorem 6.1.8 *A Godeaux homography among $n+1$ projective lines is a hexagonal correspondence.*

Proof. In fact, if we substitute the values $\underset{\alpha\alpha}{b}, \underset{\alpha\beta}{b}$ from (6.1.44) and (6.1.26) into (6.1.45), we obtain (6.1.37). ∎

6.1.6 Parallelisable Godeaux Homographies

By equations (6.1.28) and relations (6.1.44), the equations of this class of correspondences take the form:

$$\sum_{\xi} \underset{\xi}{\theta}{}_0^1 = 0, \quad \underset{\alpha}{\theta} = \underset{n+1}{\theta}, \quad \underset{\xi}{\theta}{}_1^0 = b\underset{\xi}{\theta}{}_0^1, \quad db - 2b\underset{n+1}{\theta} = 0. \tag{6.1.46}$$

It is easy to check that the system (6.1.46) is closed. Therefore, it is completely integrable and its solution depends on $2n+3$ constants (the number of independent equations in (6.1.46)).

6.2 The Application of the Theory of $(n+1)$-Webs to the Theory of Point Correspondences of $n+1$ Projective Spaces

We will continue to study applications of the theory of $(n+1)$-webs and consider in this section an application to the theory of point correspondences among $n+1$ projective r-dimensional spaces. The case when $r = 1$ and $n > 2$, was considered in Section 6.1. In this section we will study the most general case $n > 2$, $r > 1$.

6.2.1 The Fundamental Equations

We consider $n+1$ r-dimensional projective spaces $P_\xi, \xi, \eta, \zeta = 1, \ldots, n+1$, and a point correspondence $C : P_1 \times P_2 \times \ldots P_n \to P_{n+1}$ among them.

Suppose that the $\underset{\xi}{\overline{M}}{}_0$ are corresponding points of the spaces P_ξ. The correspondence C generates $\binom{n+1}{2}$ point correspondences $\underset{\xi\eta}{T} : P_\xi \to P_\eta$ arising if one fixes corresponding points $\underset{\zeta}{\overline{M}}{}_0$ where $\zeta \neq \xi, \eta$. We shall assume that the mappings $\underset{\xi\eta}{T}$ are differentiable and invertible for each pair of the spaces P_ξ, P_η provided that corresponding points of the remaining spaces are fixed.

With each point $\underset{\xi}{\overline{M}}{}_0 \in P_\xi$ we associate a projective moving frame $F_\xi = \{\underset{\xi}{\overline{M}}{}_0, \underset{\xi}{\overline{M}}{}_i\}$ of the space P_ξ where $i, j, k = 1, \ldots, r$. The equations of infinitesimal displacement of the moving frame F_ξ are:

$$\begin{cases} d\underset{\xi}{\overline{M}}{}_0 = \underset{\xi}{\overline{\omega}}{}_0^0 \underset{\xi}{\overline{M}}{}_0 + \underset{\xi}{\overline{\omega}}{}_0^i \underset{\xi}{\overline{M}}{}_i, \\ d\underset{\xi}{\overline{M}}{}_j = \underset{\xi}{\overline{\omega}}{}_j^0 \underset{\xi}{\overline{M}}{}_0 + \underset{\xi}{\overline{\omega}}{}_j^i \underset{\xi}{\overline{M}}{}_i, \end{cases} \tag{6.2.1}$$

where $\overline{\omega}{}_v^u$, $u, v, w = 0, 1, \ldots, r$, are 1-forms satisfying the structure equations of the projective spaces P_ξ :

$$d\overline{\omega}{}_v^u = \overline{\omega}{}_v^w \wedge \overline{\omega}{}_w^u \tag{6.2.2}$$

(cf. equations (5.1.4) and (5.1.9)). The forms $\underset{\xi}{\overline{\omega}}{}^i_0$ determine transformations of the points $\underset{\xi}{\overline{M}}_0$ of the spaces P_ξ. Since the points $\underset{\xi}{\overline{M}}_0$ are connected by the correspondence C, the forms $\underset{\xi}{\overline{\omega}}{}^i_0$ must satisfy linear relations of the form:

$$\sum_\xi \underset{\xi}{t}{}^i_j \underset{\xi}{\overline{\omega}}{}^j_0 = 0. \tag{6.2.3}$$

For a fixed value of ξ the forms $\underset{\xi}{\omega}{}^j_0$ are linearly independent, so

$$\det(\underset{\xi}{t}{}^i_j) \neq 0. \tag{6.2.4}$$

Let us change our moving frames by putting $\underset{\xi}{\overline{M}}_0 = \underset{\xi}{M}_0$, $\underset{\xi}{\overline{M}}_i = \underset{\xi}{t}{}^j_i \underset{\xi}{M}_j$ and denote the components of the infinitesimal displacements of the frames $\{\underset{\xi}{M}_0, \underset{\xi}{M}_i\}$ by $\underset{\xi}{\omega}{}^u_v$. Then, since

$$d\underset{\xi}{M}_0 = \underset{\xi}{\overline{\omega}}{}^0_0 \underset{\xi}{M}_0 + \underset{\xi}{\omega}{}^j_0 \underset{\xi}{t}{}^i_j \underset{\xi}{M}_i,$$

we have $\underset{\xi}{\omega}{}^i_0 = \underset{\xi}{t}{}^i_j \underset{\xi}{\overline{\omega}}{}^j_0$ and equations (6.2.1) become

$$\sum_\xi \underset{\xi}{\omega}{}^i_0 = 0. \tag{6.2.5}$$

The equations of infinitesimal displacement of the moving frame $\{\underset{\xi}{M}_0, \underset{\xi}{M}_i\}$ have the form

$$d\underset{\xi}{M}_u = \underset{\xi}{\omega}{}^v_u \underset{\xi}{M}_v, \tag{6.2.6}$$

and the forms $\underset{\xi}{\omega}{}^v_u$ satisfy the structure equations

$$d\underset{\xi}{\omega}{}^v_u = \underset{\xi}{\omega}{}^w_u \wedge \underset{\xi}{\omega}{}^v_w. \tag{6.2.7}$$

We remind ourselves that a homography K is tangent to the mapping $\underset{\xi\eta}{T}$ if K and $\underset{\xi\eta}{T}$ have a tangency of order one at any pair of corresponding points $\underset{\xi}{M}_0, \underset{\eta}{M}_0$ i.e., if for any curve of class C^s, $s \geq 1$, passing through the point $\underset{\xi}{M}_0$ in P_ξ its images in P_η under mappings K and $\underset{\xi\eta}{T}$ have an analytic tangency at least of the first order (see Section 6.1). With each mapping $\underset{\xi\eta}{T}$ a bundle of tangent homographies (projective mappings) is connected. The equations (6.2.5) mean geometrically that the moving frames $\{\underset{\xi}{M}_0, \underset{\xi}{M}_i\}$ are chosen at the corresponding points $\underset{\xi}{M}_0$ in such a way that the lines $\underset{\xi}{M}_0\underset{\xi}{M}_i$ and $\underset{\eta}{M}_0 \underset{\eta}{M}_i$ of the spaces P_ξ and P_η correspond to each other under homographies tangent to the mappings $\underset{\xi\eta}{T}$.

The equations (6.2.5) are the *fundamental equations* of the correspondence we study. Let us find their differential consequences.

Taking the exterior derivative of (6.2.5) by means of (6.2.7) and substituting $\underset{n+1}{\omega}{}^i_0$ from (6.2.5) into the obtained quadratic equation, we have

$$\left(\underset{\alpha}{\Omega}{}^i_j - \underset{n+1}{\Omega}{}^i_j\right) \wedge \underset{\alpha}{\omega}{}^j_0 = 0, \quad \alpha, \beta, \gamma = 1, \dots, n, \tag{6.2.8}$$

where $\underset{\xi}{\Omega}{}^i_j = \underset{\xi}{\omega}{}^i_j - \delta^i_j \underset{\xi}{\omega}{}^0_0$. Application of the Cartan lemma to (6.2.8) gives

$$\underset{\alpha}{\Omega}{}^i_j - \underset{n+1}{\Omega}{}^i_j = \sum_\beta \underset{\alpha\beta}{\lambda}{}^i_{jk} \underset{\beta}{\omega}{}^k_0, \tag{6.2.9}$$

where $\underset{\alpha\beta}{\lambda}{}^i_{jk} = \underset{\beta\alpha}{\lambda}{}^i_{kj}$. After the exterior differentiation of (6.2.9), by using (6.2.7), (6.2.5), and (6.2.9), we obtain

$$\left[\nabla \underset{\alpha\beta}{\lambda}{}^i_{jk} - \underset{\alpha\beta}{\delta}\left(\delta^i_{(k} \underset{\alpha}{\omega}{}^0_{j)} + \delta^i_{(k} \underset{\beta}{\omega}{}^0_{j)}\right) - 2\delta^i_{(k} \underset{n+1}{\omega}{}^0_{j)}\right] \wedge \underset{\beta}{\omega}{}^k_0 = 0 \tag{6.2.10}$$

where $\nabla \underset{\alpha\beta}{\lambda}{}^i_{jk} = d \underset{\alpha\beta}{\lambda}{}^i_{jk} - \underset{\alpha\beta}{\lambda}{}^i_{mk} \underset{n+1}{\Omega}{}^m_j - \underset{\alpha\beta}{\lambda}{}^i_{jm} \underset{n+1}{\Omega}{}^m_k + \underset{\alpha\beta}{\lambda}{}^m_{jk} \underset{n+1}{\Omega}{}^i_m$. By Cartan's lemma, we obtain from (6.2.10) that

$$\nabla \underset{\alpha\beta}{\lambda}{}^i_{jk} = \underset{\alpha\beta}{\delta}\left(\delta^i_{(k} \underset{\alpha}{\omega}{}^0_{j)} + \delta^i_{(k} \underset{\beta}{\omega}{}^0_{j)}\right) + 2\delta^i_{(k} \underset{n+1}{\omega}{}^0_{j)} + \sum_\gamma \underset{\alpha\beta\gamma}{\lambda}{}^i_{jkl} \underset{\gamma}{\omega}{}^l \tag{6.2.11}$$

where the quantities $\underset{\alpha\beta\gamma}{\lambda}{}^i_{jkl}$ satisfy some additional relations.

We introduce the quantities

$$\underset{\alpha\beta}{b}{}^i_{jk} = \underset{\alpha\beta}{\lambda}{}^i_{jk} - \frac{1}{\binom{n}{2}} \sum_{\gamma \neq \delta} \underset{\gamma\delta}{\lambda}{}^i_{(jk)}. \tag{6.2.12}$$

It follows from the equations (6.2.11) that the $\underset{\alpha\beta}{b}{}^i_{jk}$ defined by (6.2.12) satisfy the equations

$$\nabla_\delta \underset{\alpha\beta}{b}{}^i_{jk} = 0 \tag{6.2.13}$$

where $\nabla_\delta \underset{\alpha\beta}{b}{}^i_{jk} = \delta \underset{\alpha\beta}{b}{}^i_{jk} - \underset{\alpha\beta}{b}{}^i_{mk} \underset{n+1}{\Omega}{}^m_j(\delta) - \underset{\alpha\beta}{b}{}^i_{jm} \underset{n+1}{\Omega}{}^m_k(\delta) + \underset{\alpha\beta}{b}{}^m_{jk} \underset{n+1}{\Omega}{}^i_m(\delta)$ and δ as usually replaces d when the point $\underset{\xi}{M}_0$ is fixed, i.e., $\underset{\xi}{\omega}{}^i_0 = 0$. Equation (6.2.13) means that the quantities $\underset{\alpha\beta}{b}{}^i_{jk}$ form a tensor in the second differential neighbourhood of the correspondence C. This tensor plays a fundamental role in the theory of point correspondences. We will discuss its geometrical meaning later on.

The system of moving frames $\{\underset{\xi}{M}_0, \underset{\xi}{M}_i\}$ in which the equations (6.2.5) of the correspondence C are written admits transformations of the form:

$$'\underset{\xi}{M}_0 = \underset{\xi}{M}_0, \quad '\underset{\xi}{M}_i = \underset{\xi}{M}_i - \underset{\xi}{t}_i \underset{\xi}{M}_0. \tag{6.2.14}$$

The transformations (6.2.14) preserve the form of the equations (6.2.5).

Denote by $'\omega_u^v$ and $'\underset{\alpha\beta}{\lambda}{}^i_{jk}$ the values of the forms ω_u^v and the functions $\underset{\alpha\beta}{\lambda}{}^i_{jk}$ in the moving frames $\{'\underset{\xi}{M}_0, '\underset{\xi}{M}_i\}$. If we differentiate (6.2.14) and substitute $\underset{\xi}{M}_u$ from (6.2.14) into the equations obtained by virtue of the linear independence of the points $'\underset{\xi}{M}_u$, we obtain the following expressions for $'\omega_u^v$ and $'\underset{\alpha\beta}{\lambda}{}^i_{jk}$:

$$\begin{cases} '\underset{\xi}{\omega}{}_0^i = \underset{\xi}{\omega}{}_0^i, \quad '\underset{\xi}{\omega}{}_0^0 = \underset{\xi}{\omega}{}_0^0 + \underset{\xi}{t}_i \underset{\xi}{\omega}{}_0^i, \quad '\underset{\xi}{\omega}{}_i^j = \underset{\xi}{\omega}{}_i^j - \underset{\xi}{t}_i \underset{\xi}{\omega}{}_0^j, \\ '\underset{\xi}{\omega}{}_i^0 = \underset{\xi}{\omega}{}_i^0 - d\underset{\xi}{t}_i + \underset{\xi}{t}_k \underset{\xi}{\Omega}{}_i^k - \underset{\xi}{t}_i \underset{\xi}{t}_j \underset{\xi}{\omega}{}_0^j, \end{cases} \tag{6.2.15}$$

$$'\underset{\alpha\beta}{\lambda}{}^i_{jk} = \underset{\alpha\beta}{\lambda}{}^i_{jk} - \underset{\alpha\beta}{\delta}\left(\delta^i_{(k}\underset{\alpha}{t}_{j)} + \delta^i_{(k}\underset{\beta}{t}_{j)}\right) - 2\delta^i_{(k}\underset{n+1}{t}_{j)}. \tag{6.2.16}$$

Using (6.2.15) and (6.2.16), it is easy to check that in addition to equations (6.2.5) the forms

$$\omega^i_j = \underset{n+1}{\Omega}{}^i_j - \frac{1}{\binom{n}{2}} \sum_{\alpha,\beta(\alpha\neq\beta)} \underset{\alpha\beta}{\lambda}{}^i_{jk} \underset{n+1}{\omega}{}^k_0 \tag{6.2.17}$$

are invariant under the transformations (6.2.14) of the moving frames.

Let us now find moving frames in the set of admissible moving frames $\{\underset{\xi}{M}_0, \underset{\xi}{M}_i\}$ which are invariantly connected with the correspondence C.

For this, first note that the point $\underset{\xi}{M}_0$ was already chosen as an invariant point. Analytically this is confirmed by the equation $\delta \underset{\xi}{M}_0 = \underset{\xi}{\pi}{}^0_0 \underset{\xi}{M}_0$ (where $\underset{\xi}{\pi}{}^u_v = \underset{\xi}{\omega}{}^u_v(\delta)$) which follows from (6.2.6). To find other invariant points

$$\underset{\xi}{A}_k = \underset{\xi}{M}_k - \underset{\xi}{x}_k \underset{\xi}{M}_0 \tag{6.2.18}$$

in the space P_ξ , we differentiate (6.2.18) keeping $\underset{\xi}{\omega}{}^i_0 = 0$ and using (6.2.6), (6.2.18), and (6.2.9). As a result, we have

$$\delta \underset{\xi}{A}_k \wedge \underset{\xi}{A}_1 \wedge \ldots \wedge \underset{\xi}{A}_r = (\underset{\xi}{\pi}{}^0_k - \nabla_\delta \underset{\xi}{x}_k)\underset{\xi}{M}_0 \tag{6.2.19}$$

where $\nabla_\delta \underset{\xi}{x}_k = \delta \underset{\xi}{x}_k - \underset{\xi}{x}_m \underset{n+1}{\Omega}{}^m_k(\delta)$. Relation (6.2.19) shows that the space spanned by the points $\underset{\xi}{A}_k$ is invariant if and only if

$$\nabla_\delta \underset{\xi}{x}_k = \underset{\xi}{\pi}{}^0_k. \tag{6.2.20}$$

So, to find points $\underset{\xi}{A}_i$ defining invariant $(r-1)$-planes $\{\underset{\xi}{A}_1, \dots, \underset{\xi}{A}_r\}$ of the correspondence C, we should construct a quantity $\underset{\xi}{x}_k$ satisfying (6.2.20). Equations (6.2.9) and (6.2.11) allow us to check that the quantities

$$\underset{n+1}{l}{}_k = \frac{1}{(r+1)\binom{n}{2}} \sum_{\alpha,\beta(\alpha<\beta)} \underset{\alpha\beta}{\lambda}{}^i_{(ik)}, \quad \underset{\alpha}{l}{}_k = \frac{1}{(r+1)}\left(\underset{\alpha\alpha}{\lambda}{}^i_{ik} - \frac{1}{\binom{n}{2}} \sum_{\alpha,\beta(\alpha<\beta)} \underset{\alpha\beta}{\lambda}{}^i_{(ik)}\right) \quad (6.2.21)$$

satisfy (6.2.20). If we set

$$\underset{\xi}{A}_0 = \underset{\xi}{M}_0, \quad \underset{\xi}{A}_k = \underset{\xi}{M}_k - \underset{\xi}{l}_k \underset{\xi}{M}_0. \quad (6.2.22)$$

where the $\underset{\xi}{l}_k$ are determined by (6.2.21), then the moving frames $\{\underset{\xi}{A}_0, \underset{\xi}{A}_k\}$ of the spaces P_ξ are *invariant moving frames* defined by the correspondence C.

For the frame $\{\underset{\xi}{A}_0, \underset{\xi}{A}_k\}$ we have the following equations of infinitesimal displacement:

$$d\underset{\xi}{A}_u = \underset{\xi}{\theta}{}^v_u \underset{\xi}{A}_v. \quad (6.2.23)$$

The relations between the components $\underset{\xi}{\theta}{}^v_u$ of the infinitesimal displacement of the invariant frames and the analogous components $\underset{\xi}{\omega}{}^v_u$ of arbitrary frames are similar to (6.2.15).

Using these relations, we can write the equations (6.2.5) and (6.2.9) of the correspondence C with respect to the invariant frames in the following form:

$$\sum_\xi \underset{\xi}{\theta}{}^i_0 = 0, \quad \underset{\alpha}{\Theta}{}^i_j - \underset{n+1}{\Theta}{}^i_j = \sum_\beta \underset{\alpha\beta}{a}{}^i_{jk} \underset{\beta}{\theta}{}^k_0. \quad (6.2.24)$$

where

$$\underset{\xi}{\Theta}{}^i_j = \underset{\xi}{\theta}{}^i_j - \delta^i_j \underset{\xi}{\theta}{}^0_0, \quad \underset{\alpha\beta}{a}{}^i_{jk} = \underset{\alpha\beta}{\lambda}{}^i_{jk} - \underset{\alpha\beta}{\delta}(\delta^i_{(k} \underset{\alpha}{l}_{j)} + \delta^i_{(k} \underset{\beta}{l}_{j)}) - 2\delta^i_{(k} \underset{n+1}{l}_{j)}.$$

In addition, by virtue of equations (6.2.20), the forms $\underset{\xi}{\theta}{}^0_i$ are expressed only in terms of the forms $\underset{\alpha}{\theta}{}^j_0$:

$$\underset{\xi}{\theta}{}^0_i = \sum_\alpha \underset{\xi\alpha}{b}_{ij} \underset{\alpha}{\theta}{}^j_0. \quad (6.2.25)$$

We proved the following theorem :

Theorem 6.2.1 *A point correspondence C among $n+1$ r-dimensional projective spaces determines an invariant equipment of these spaces in the second differential neighbourhood of C. The equations of C with respect to the invariant frames have the form* (6.2.24).

The quantities $\underset{\alpha\beta}{a}{}^i_{jk}$ in equations (6.2.24) are expressed in terms of $\underset{\alpha\beta}{\lambda}{}^i_{jk}$ and therefore belong to the second differential neighbourhood of the correspondence C.

Definition 6.2.2 We will call the quantities $\underset{\alpha\beta}{a}{}^i_{jk}$ the *fundamental tensors* of the correspondence C.

Since $\underset{\alpha\beta}{\lambda}{}^i_{jk} = \underset{\beta\alpha}{\lambda}{}^i_{kj}$ and $\underset{\xi}{l}_k$ are expressed by (6.2.21), the quantities $\underset{\alpha\beta}{a}{}^i_{jk}$ satisfy the conditions:

$$\sum_{(\alpha,\beta)} \underset{\alpha\beta}{a}{}^i_{(ik)} = 0, \quad \underset{\alpha\alpha}{a}{}^i_{(ik)} = 0, \quad \underset{\alpha\beta}{a}{}^i_{jk} = \underset{\beta\alpha}{a}{}^i_{kj}, \quad \alpha \neq \beta. \tag{6.2.26}$$

Construct a new operator ∇_Θ by means of the forms $\underset{n+1}{\Theta}{}^j_i$ and set

$$\nabla_\Theta \underset{\alpha\beta}{a}{}^i_{jk} = \sum_\gamma \Big(\underset{\alpha\beta\gamma}{a}{}^i_{jkl} + \underset{\alpha\beta}{a}{}^i_{mk} \underset{\gamma\alpha}{a}{}^m_{lj} + \underset{\alpha\beta}{a}{}^i_{jm} \underset{\beta\alpha}{a}{}^m_{kl} \Big) \underset{\gamma}{\theta}{}^l_0. \tag{6.2.27}$$

where $\nabla_\Theta \underset{\alpha\beta}{a}{}^i_{jk} = d\underset{\alpha\beta}{a}{}^i_{jk} - \underset{\alpha\beta}{a}{}^i_{mk}\Theta^m_j - \underset{\alpha\beta}{a}{}^i_{jm}\Theta^m_k + \underset{\alpha\beta}{a}{}^m_{jk}\Theta^i_m$.

The quantities $\underset{\alpha\beta\gamma}{a}{}^i_{jkl}$ form tensors in the third differential neighbourhood of the correspondence C. They satisfy some additional conditions. To find them, we should take the exterior derivative of (6.2.24), replace the forms $\underset{\xi}{\theta}{}^0_i$ and $\nabla_\Theta \underset{\alpha\beta}{a}{}^i_{jk}$ in the obtained exterior quadratic equations by their values from (6.2.25) and (6.2.27), and use the linear independence of the forms $\underset{\alpha}{\theta}{}^i_0$. This gives

$$\underset{\alpha\beta\gamma}{a}{}^i_{jkl} - \underset{\alpha\gamma\beta}{a}{}^i_{jlk} - \underset{\alpha\beta}{\delta}\Big(\delta^i_{(k}\underset{\alpha\gamma}{b}{}_{j)l} + \delta^i_{(k}\underset{\beta\gamma}{b}{}_{j)l}\Big) + \underset{\alpha\gamma}{\delta}\Big(\delta^i_{(l}\underset{\alpha\beta}{b}{}_{j)k} + \delta^i_{(l}\underset{\gamma\beta}{b}{}_{j)k}\Big)$$
$$-2\delta^i_{(k}\underset{n+1,\gamma}{b}{}_{j)l} + 2\delta^i_{(l}\underset{n+1,\beta}{b}{}_{j)l} = 0. \tag{6.2.28}$$

Apart from this, differentiation of (6.2.26) leads to the following additional conditions for the quantities $\underset{\alpha\beta\gamma}{a}{}^i_{jkl}$:

$$\left\{\begin{array}{l} \underset{\alpha\beta\gamma}{a}{}^i_{jkl} = \underset{\beta\alpha\gamma}{a}{}^i_{kjl}, \\ \underset{\alpha\alpha\gamma}{a}{}^i_{ikl} - \underset{\alpha\alpha}{a}{}^i_{mk}\underset{\gamma\alpha}{a}{}^m_{li} = 0, \\ \displaystyle\sum_{\alpha,\beta(\alpha\neq\beta)} \underset{\alpha\beta\gamma}{a}{}^i_{(lk)i} + \sum_{\alpha\neq\beta}\Big(\underset{\alpha\beta}{a}{}^i_{m(k}\underset{\gamma\alpha}{a}{}^m_{|l|i)} - \underset{\alpha\beta}{a}{}^i_{(l|m|}\underset{\beta\gamma}{a}{}^m_{k)l}\Big) = 0. \end{array}\right. \tag{6.2.29}$$

6.2.2 Correspondences Among $n+1$ Projective Spaces and Multicodimensional $(n+1)$-Webs

From algebraic point of view, the point correspondence C among $n+1$ r-dimensional projective spaces P_ξ is a local differentiable n-quasigroup (see Definition 3.1.1). As we know (see Section 3.1), to each n-quasigroup there is associated an $(n+1)$-web $W(n+1,n,r)$. In our case this web can be constructed as follows. Consider the Cartesian product $P_1 \times \ldots \times P_{n+1}$. In the space of this product to the correspondence

C there corresponds a smooth (nr)-dimensional manifold. To each space P_ξ there corresponds a foliation of codimension r on this manifold. The set of all $n+1$ foliations of codimension r forms a web $W(n+1,n,r)$ on the manifold.

We write out the structure equations of the $W(n+1,n,r)$ in the system of the invariant frames of the spaces P_ξ . Using (6.2.24), (6.2.25), and (6.2.7), we have

$$\begin{cases} \sum\limits_\xi \underset{\xi}{\theta}{}^i_0 = 0, \\ d\underset{\alpha}{\theta}{}^i_0 = \underset{\alpha}{\theta}{}^j_0 \wedge \omega^i_j + \sum\limits_{\alpha,\beta} \underset{\alpha\beta}{b}{}^i_{jk} \underset{\alpha}{\theta}{}^j_0 \wedge \underset{\beta}{\theta}{}^k_0, \\ d\omega^i_j - \omega^k_j \wedge \omega^l_k = \sum\limits_\alpha \underset{\alpha\alpha}{b}{}^i_{jkl} \underset{\alpha}{\theta}{}^k_0 \wedge \underset{\alpha}{\theta}{}^l_0 + \sum\limits_{\alpha,\beta} \underset{\alpha\beta}{b}{}^i_{jkl} \underset{\alpha}{\theta}{}^k_0 \wedge \underset{\beta}{\theta}{}^l_0. \end{cases} \tag{6.2.30}$$

The forms ω^i_j and the tensor $\underset{\alpha\beta}{b}{}^i_{jk}$ are the connection forms and the torsion tensor of the $W(n+1,n,r)$. They were introduced in (6.2.17) and (6.2.12). In particular, it follows from (6.2.12) and the expressions for the $\underset{\alpha\beta}{a}{}^i_{jk}$ that the torsion tensor is expressed in terms of the $\underset{\alpha\beta}{a}{}^i_{jk}$ as follows:

$$\underset{\alpha\beta}{b}{}^i_{jk} = \underset{\alpha\beta}{a}{}^i_{jk} - \frac{1}{\binom{n}{2}} \sum_{\gamma,\delta(\gamma<\delta)} \underset{\gamma\delta}{a}{}^i_{(jk)}, \quad \alpha \neq \beta. \tag{6.2.31}$$

Introduce the notation:

$$c^i_{jk} = \frac{1}{\binom{n}{2}} \sum_{\gamma,\delta(\gamma<\delta)} \underset{\gamma\delta}{a}{}^i_{(jk)} \tag{6.2.32}$$

and suppose that

$$\nabla_\Theta c^i_{jk} = \sum_\alpha \underset{\alpha}{c}{}^i_{jkl} \underset{\alpha}{\theta}{}^l_0. \tag{6.2.33}$$

Then, using (6.2.30), (6.2.32), and (6.2.33), we find the values of the components of the curvature tensor of the web $W(n+1,n,r)$:

$$\begin{cases} \underset{\alpha\alpha}{b}{}^i_{jkl} = \underset{\alpha}{c}{}^i_{j[kl]} + 2\delta^i_{(j} \underset{n+1\,\alpha}{b}{}_{l)k} - 2\delta^i_{(j} \underset{n+1\,\alpha}{b}{}_{k)l} + c^t_{j[l} c^i_{k]t}, \\ 2\underset{\alpha\beta}{b}{}^i_{jkl} = \underset{\beta}{c}{}^i_{jkl} - \underset{\alpha}{c}{}^i_{jlk} - 2\delta^i_{(j} \underset{\beta}{b}{}_{k)l} + 2\delta^i_{(j} \underset{\alpha}{b}{}_{l)k} + c^t_{jl} c^i_{kt} - c^t_{jk} c^i_{lt}. \end{cases} \tag{6.2.34}$$

As we know from Section 1.2, the curvature tensor satisfies some additional relations. To find these relations in our case, we put

$$\nabla_\omega \underset{\alpha\beta}{b}{}^i_{jk} = \sum_\gamma \Big(\underset{\alpha\beta\gamma}{b}{}^i_{jkl} + \underset{\alpha\beta}{b}{}^i_{mk} \underset{\gamma\alpha}{b}{}^m_{lj} + \underset{\alpha\beta}{b}{}^i_{jm} \underset{\beta\gamma}{b}{}^m_{kl} \Big) \underset{\gamma}{\theta}{}^l_0 \tag{6.2.35}$$

where $\nabla_\omega \underset{\alpha\beta}{b}{}^i_{jk} = d\,\underset{\alpha\beta}{b}{}^i_{jk} - \underset{\alpha\beta}{b}{}^i_{mk}\omega^m_j - \underset{\alpha\beta}{b}{}^i_{jm}\omega^m_k + \underset{\alpha\beta}{b}{}^m_{jk}\omega^i_m$. Using (6.2.35), (6.2.26), (6.2.28), and (6.2.29), after some calculations we find

$$\underset{\gamma\gamma}{b}{}^i_{jkl} = \underset{\alpha\gamma\gamma}{b}{}^i_{j[kl]}, \quad 2\underset{\alpha\beta}{b}{}^i_{jkl} = \underset{\gamma\alpha\beta}{b}{}^i_{jkl} - \underset{\beta\gamma\alpha}{b}{}^i_{ljk}, \quad 2\underset{\alpha\beta}{b}{}^i_{[jk]l} + \underset{\alpha\alpha}{b}{}^i_{ljk} = 0, \quad \underset{\alpha\beta}{b}{}^i_{[jkl]} = 0. \tag{6.2.36}$$

The relations (6.2.36) show, in particular, that for $n > 2$ the curvature tensor of the $W(n+1, n, r)$ is expressed in terms of the torsion tensor $\underset{\alpha\beta}{b}{}^i_{jk}$ and its Pfaffian derivatives (cf. Section 1.2).

6.2.3 Parallelisable Correspondences

In Chapters 1–3 we studied different special classes of multicodimensional $(n+1)$-webs and gave an invariant geometric characteristic to each of them. To each of these classes of $(n+1)$-webs there corresponds a certain class of point correspondences among the $n+1$ r-dimensional projective spaces P_ξ . The point correspondences in turn possess certain special properties which together with the properties of the associated webs can be taken as a basis for a classification.

We consider the geometric pecularities of certain classes of corrrespondences. We will start from a parallelisable correspondence C.

Definition 6.2.3 A correspondence C is said to be *parallelisable* if the web $W(n+1, n, r)$ associated with C is parallelisable.

Since $n > 2$, by Theorem 1.5.2 and Definition 3.1.3, the correspondence C is parallelisable if and only if its torsion tensor vanishes:

$$\underset{\alpha\beta}{b}{}^i_{jk} = 0. \tag{6.2.37}$$

In an arbitrary system of moving frames the tensor $\underset{\alpha\beta}{b}{}^i_{jk}$ is expressed by (6.2.12). We put

$$\lambda^i_{jk} = \frac{1}{\binom{n}{2}} \sum_{\gamma,\delta(\gamma\neq\delta)} \underset{\gamma\delta}{\lambda}{}^i_{(jk)}. \tag{6.2.38}$$

Then equations (6.2.5) and (6.2.9) for a parallelisable correspondence C take the form

$$\sum_\xi \underset{\xi}{\omega}{}^i_0 = 0, \quad \underset{\alpha}{\Omega}{}^i_j - \underset{n+1}{\Omega}{}^i_j = \underset{\alpha}{\lambda}{}^i_{jk}\underset{\alpha}{\omega}{}^k_0 - \lambda^i_{jk}\underset{n+1}{\omega}{}^k_0 \tag{6.2.39}$$

where $\underset{\alpha}{\lambda}{}^i_{jk} = \underset{\alpha\alpha}{\lambda}{}^i_{jk} - \lambda^i_{jk}$.

6.2.4 Godeaux Homographies

Let us consider again the mappings $\underset{\xi\eta}{T} : P_\xi \rightarrow P_\eta$ arising if one fixes the corresponding points $\underset{\zeta}{M}_0$, $\zeta \neq \xi, \eta$, of the spaces P_ξ .

Definition 6.2.4 A correspondence $C : P_1 \times \ldots \times P_n \rightarrow P_{n+1}$ is said to be a *Godeaux homography* if for any system of the corresponding points $\underset{\xi}{M}_0 \in P_\xi$ all the mappings $\underset{\xi\eta}{T}$ are projective.

Proposition 6.2.5 *The mapping $\underset{\xi\eta}{T} : P_\xi \to P_\eta$ is projective if and only if its fundamental tensor vanishes.*

Proof. Let us write out the equations of the mapping $\underset{\xi\eta}{T}$ in the invariant system of frames $\{\underset{\xi}{A_0}, \underset{\xi}{A_i}\}$. For the mapping $\underset{n+1,\alpha}{T}$ we find from (6.2.24) that

$$\underset{\alpha}{\theta}{}_0^i + \underset{n+1}{\theta}{}_0^i = 0, \quad \underset{\alpha}{\Theta}{}_j^i - \underset{n+1}{\Theta}{}_j^i = \underset{\alpha\alpha}{a}{}^i_{jk}\underset{\alpha}{\theta}{}_0^k. \tag{6.2.40}$$

It follows from (6.2.40) and Definition 6.2.2 that the quantities $\underset{\alpha\alpha}{a}{}^i_{jk}$ form the fundamental tensor of the mapping $\underset{n+1,\alpha}{T}$.

Similarly we find the equations of the mapping $\underset{\alpha\beta}{T}$. They have the form:

$$\underset{\alpha}{\theta}{}_0^i + \underset{\beta}{\theta}{}_0^i = 0, \quad \underset{\alpha}{\Theta}{}_j^i - \underset{\beta}{\Theta}{}_j^i = (\underset{\alpha\alpha}{a}{}^i_{jk} + \underset{\beta\beta}{a}{}^i_{jk} - 2\,\underset{\alpha\beta}{a}{}^i_{(jk)})\underset{\alpha}{\theta}{}_0^k. \tag{6.2.41}$$

Equations (6.2.41) and Definition 6.2.2 show that the fundamental tensor of the mapping $\underset{\alpha\beta}{T}$ is $\underset{\alpha\alpha}{a}{}^i_{jk} + \underset{\beta\beta}{a}{}^i_{jk} - 2\,\underset{\alpha\beta}{a}{}^i_{(jk)}$.

Suppose now that $\underset{\alpha\alpha}{a}{}^i_{jk} = 0$ for the point $\underset{\alpha}{M_0}$. Then it follows from (6.2.40) that $\underset{\alpha}{\Theta}{}_j^i - \underset{n+1}{\Theta}{}_j^i = 0$. Taking the exterior derivative of the latter equation and keeping in mind that we consider the case $r > 1$, we obtain $\underset{\alpha}{\theta}{}_j^0 + \underset{n+1}{\theta}{}_j^0 = 0$. A projective mapping

$$K\underset{n+1}{A_0} = \underset{\alpha}{A_0}, \quad K\underset{n+1}{A_i} = -\underset{n+1}{A_i}$$

which is tangent to the mapping $\underset{n+1,\alpha}{T}$ becomes constant, and therefore, it coincides with $\underset{n+1,\alpha}{T}$.

Conversely, let the mapping $\underset{n+1,\alpha}{T}$ be projective. Take the moving frames at the points $\underset{n+1}{M_0}$ and $\underset{\alpha}{M_0}$ in such a way that the points $\underset{n+1}{M_i}$ and $\underset{\alpha}{M_i}$ correspond to each other in this mapping. Then it is obvious that $\underset{\alpha\alpha}{a}{}^i_{jk} = 0$.

We have proved Proposition 6.2.5 for a mapping $\underset{n+1,\alpha}{T}$. Since the spaces P_ξ are interchangeable, Proposition 6.2.5 is valid for a mapping $\underset{\alpha\beta}{T}$. Of course, this can be checked directly. ■

Theorem 6.2.6 *A point correspondence C among $n+1$ r-dimensional projective spaces is a Godeaux homography if and only if the quantities $\underset{\alpha\beta}{a}{}^i_{jk}$ satisfy the following conditions:*

$$\underset{\alpha\alpha}{a}{}^i_{jk} = 0, \qquad \underset{\alpha\beta}{a}{}^i_{(jk)} = 0. \tag{6.2.42}$$

Proof. It follows from Definition 6.2.4 and Proposition 6.2.5 that a correspondence C is a Godeaux homography if and only if the fundamental tensors of all the mappings $\underset{\xi\eta}{T}$ vanish. According to the form of the fundamental tensors of $\underset{\xi\eta}{T}$ which we indicated above, this will take place if and only if the conditions (6.2.42) hold. ■

Corollary 6.2.7 *A web $W(n+1,n,r)$ associated with a Godeaux homography is $(2n+2)$-hedral of type* 2.

Proof. In fact, by Theorem 4.4.2, a web $W(n+1,n,r)$ is $(2n+2)$-hedral of type 2 if and only if its torsion tensor satisfies the condition $\underset{\alpha\beta}{b}{}^{i}_{(jk)} = 0$. The equations (6.2.31) and (6.2.42) show that this condition is satisfied for a Godeaux homography. ■

6.2.5 Parallelisable Godeaux Homographies

A parallelisable Godeaux homography is a particular case of a Godeaux homography. Such correspondences are distinguished from the whole set of point correspondences by the equations (6.2.37) and (6.2.42). By (6.2.31), these two equations are equivalent to the condition $\underset{\alpha\beta}{a}{}^{i}_{jk} = 0$.

Let us write out the equations of a parallelisable Godeaux correspondence. For this we substitute the values $\underset{\alpha\beta}{a}{}^{i}_{jk} = 0$ into equations (6.2.24). This gives

$$\sum_{\xi} \underset{\xi}{\theta}{}^{i}_{0} = 0, \quad \underset{\alpha}{\Theta}{}^{i}_{j} - \underset{n+1}{\Theta}{}^{i}_{j} = 0. \tag{6.2.43}$$

Taking the exterior derivative of (6.2.43) and using the fact that $n > 2$, $r > 1$, we will come to the additional equations:

$$\underset{\xi}{\theta}{}^{0}_{k} = 0. \tag{6.2.44}$$

It is easy to check that the system (6.2.43), (6.2.44) is closed. Therefore, it is completely integrable and its solution depends on $nr^2 + (n+1)r$ constants (the number of independent equations in (6.2.43), (6.2.44)).

The equations (6.2.43), (6.2.44) of a parallelisable Godeaux homography admit integration in closed form. In fact, in this case the forms $\underset{\xi}{\theta}{}^{0}_{0}$ and $\underset{n+1}{\Theta}{}^{i}_{j}$ are closed since $d\underset{\xi}{\theta}{}^{0}_{0} = d\Theta^{i}_{j} = 0$. Therefore these forms are total differentials: $\underset{\xi}{\theta}{}^{0}_{0} = d\underset{\xi}{a}$, $\underset{n+1}{\Theta}{}^{i}_{j} = da^{i}_{j}$. Taking appropriate values for $\underset{\xi}{a}$ and a^{i}_{j}, we reduce the forms $\underset{\xi}{\theta}{}^{0}_{0}$ and $\underset{n+1}{\Theta}{}^{i}_{j}$ to zero: $\underset{\xi}{\theta}{}^{0}_{0} = \underset{n+1}{\Theta}{}^{i}_{j} = 0$. Then, by (6.2.43), we have $\underset{\xi}{\Theta}{}^{i}_{j} = 0$ for all ξ. After this each of the forms $\underset{\xi}{\theta}{}^{i}_{0}$ is closed and therefore is a total differential: $\underset{\xi}{\theta}{}^{i}_{0} = d\underset{\xi}{u}{}^{i}$.

From equations (6.2.23) we find now that $d\underset{\xi}{A_0} = d\underset{\xi}{u^i}\underset{\xi}{A_i}$. The solution of the latter equations have the form: $\underset{\xi}{A_0} = \underset{\xi}{X^0}\underset{\xi}{A_0^0} + \underset{\xi}{X^i}\underset{\xi}{A_i^0}$ where $\underset{\xi}{A_0^0}$, $\underset{\xi}{A_i^0}$ are fixed points in the space P_ξ. The quantities $\underset{\xi}{X^u}$ are homogeneous coordinates of the point $\underset{\xi}{A_0}$. The non-homogeneous coordinates $\underset{\xi}{Z^i} = \underset{\xi}{X^i}/\underset{\xi}{X^0}$ of this point must be expressed linearly in terms of the variables $\underset{\xi}{u^i}$,i.e.,$\underset{\xi}{Z^i} = \underset{\xi}{X^i}/\underset{\xi}{X^0} = \underset{\xi}{\tilde{c}^i_j}\underset{\xi}{u^j}$ where $\det(\underset{\xi}{\tilde{c}^i_j}) \neq 0$. Solving these equations with respect to $\underset{\xi}{u^i}$, we have $\underset{\xi}{u^i} = \underset{\xi}{c^i_j}\underset{\xi}{X^j_0}/\underset{\xi}{X^0}$. Substitute these values into the fundamental equations of the correspondence. Integrating the equation obtained after this substitution and putting the constants of integration equal to zero, we obtain the closed form equations of a parallelisable Godeaux homography in the following form: $\sum_{\xi}\prod_{\hat{\eta}} \underset{\xi}{c^i_j}\underset{\xi}{X^j}\underset{\hat{\eta}}{X^0} = 0$ where $\hat{\eta} \neq \xi$.

6.2.6 Paratactical Correspondences

A point correspondence $C : P_1 \times \ldots \times P_n \to P_{n+1}$ generates $\binom{n+1}{3}$ families of subcorrespondences $\underset{\xi\eta\zeta}{C} : P_\xi \times P_\eta \to P_\zeta$ arising if one fixes $n-2$ corresponding points. To each of these subcorrespondences there corresponds a three-web $W(3,2,r)$. We write out the equations of the subcorrespondences $\underset{\xi\eta\zeta}{C}$ and also the structure equations of the three-webs associated with them.

First of all we consider a correspondence of the form $\underset{n+1,\alpha\beta}{C}$. Its equations are

$$\underset{\gamma}{\theta^i_0} = 0, \quad \gamma \neq \alpha, \beta, \quad \alpha, \beta \text{ fixed} \tag{6.2.45}$$

or, as follows from equations (6.2.24),

$$\underset{\alpha}{\theta^i_0} + \underset{\beta}{\theta^i_0} + \underset{n+1}{\theta^{\,i}_{\,0}} = 0. \tag{6.2.46}$$

If we substitute the values of $\underset{\alpha}{\theta^i_0}$, $\underset{n+1}{\theta^{\,i}_{\,0}}$ from (6.2.45), (6.2.46) into the structure equations (6.2.30) of the web $W(n+1,n,r)$ associated with the correspondence, after some transformations we obtain the structure equations of the three-web $W(3,2,r)$ associated with the correspondence $\underset{n+1,\alpha\beta}{C}$:

$$\begin{cases} d\underset{\alpha}{\theta^i_0} = \underset{\alpha}{\theta^j_0} \wedge \left(\underset{n+1}{\theta^{\,i}_{\,j}} + \underset{\alpha\beta}{a^{\,i}_{jk}}\underset{\beta}{\theta^k_0} + \underset{\alpha\beta}{a^{\,i}_{kj}}\underset{\alpha}{\theta^k_0}\right) + \underset{\alpha\beta}{a^{\,i}_{[jk]}}\underset{\alpha}{\theta^j_0} \wedge \underset{\alpha}{\theta^k_0}, \\ d\underset{\beta}{\theta^i_0} = \underset{\beta}{\theta^j_0} \wedge \left(\underset{n+1}{\theta^{\,i}_{\,j}} + \underset{\alpha\beta}{a^{\,i}_{jk}}\underset{\beta}{\theta^k_0} + \underset{\alpha\beta}{a^{\,i}_{kj}}\underset{\alpha}{\theta^k_0}\right) + \underset{\alpha\beta}{a^{\,i}_{[jk]}}\underset{\beta}{\theta^j_0} \wedge \underset{\beta}{\theta^k_0}. \end{cases} \tag{6.2.47}$$

It follows from (6.2.47) that the connection forms of the $W(3,2,r)$ are $\underset{n+1}{\theta^{\,i}_{\,j}} + \underset{\alpha\beta}{a^{\,i}_{jk}}\underset{\beta}{\theta^k_0} + \underset{\alpha\beta}{a^{\,i}_{kj}}\underset{\alpha}{\theta^k_0}$ and its torsion tensor is $\underset{\alpha\beta}{a^{\,i}_{[jk]}} = \underset{\alpha\beta}{b^{\,i}_{[jk]}}$. Similar computations show that the torsion tensor of the three-web associated with the subcorrespondence $\underset{\alpha\beta\gamma}{C}$ is $\underset{\alpha\beta}{a}{}^{i}_{[jk]} + \underset{\beta\gamma}{a}{}^{i}_{[jk]} + \underset{\gamma\alpha}{a}{}^{i}_{[jk]}$ (cf. (1.3.23) and (1.3.69)).

Definition 6.2.8 A subcorrespondence $\underset{\xi\eta\zeta}{C}$ is called *paratactical* if the web $W(3,2,r)$ associated with the $\underset{\xi\eta\zeta}{C}$ is *paratactical* (see Definition 1.6.1). A correspondence C is said to be *paratactical* if all its subcorrespondences $\underset{\xi\eta\zeta}{C}$ are paratactical.

Theorem 6.2.9 *A correspondence C among the $n+1$ projective r-dimensional spaces P_ξ is paratactical if and only if*

$$\underset{\alpha\beta}{a}{}^{i}_{[jk]} = 0. \tag{6.2.48}$$

Proof. This follows from Definition 6.2.8 and Theorem 1.6.2. ∎

Corollary 6.2.10 *A correspondence C is paratactical if and only if the $\binom{n}{2}$ subcorrespondences $\underset{\xi\eta\zeta}{C}$ where ξ is fixed are paratactical.*

Proof. For the correspondences $\underset{n+1,\alpha\beta}{C}$ this is obvious since the vanishing of the torsion tensors of the webs associated with the $\underset{n+1,\alpha\beta}{C}$ implies the vanishing of the torsion tensors of the three-webs associated with the remaining subcorrespondences. The validity of the assertion for other correspondences $\underset{\xi\eta\zeta}{C}$ where $\xi \neq n+1$ follows from interchangebility of the spaces P_ξ. ∎

Corollary 6.2.11 *A paratactical Godeaux homography among the $n+1$ projective r-dimensional spaces P_ξ coincides with a parallelisable Godeaux homography.*

Proof. This follows from Theorems 6.2.9 and 6.2.6 and equations (6.2.37) and (6.2.31). ∎

6.3 The Application of the Theory of $(n+1)$-Webs to the Theory of Holomorphic Mappings between Polyhedral Domains

6.3.1 Introductory Note

In this section we will present an application of the methods of web geometry to the theory of holomorphic mappings.

The development of the basic concepts of Complex Analysis which underly this approach would be beyond the objectives of this text. Therefore this presentation will be limited to a sketch of some passages, especially if they are straightforward exercises of Complex Analysis methods, or if they would require too much preparation.

In [Bau 82] the ideas presented here are developed more extensively. In particular, this gives the basic preliminaries and many additional examples.

The aim of this section is to outline the basic geometric ideas that have a common ground with the differential geometric theory of $(n+1)$-webs over the field **R** of real numbers. In particular, the approach chosen tries to imitate classic notions that can formally be carried through without discussing too far the complications that might be caused by the different field **C** and the attached holomorphic and meromorphic functions and mappings. For example, we continue to speak about $(n+1)$-webs $W(n+1, n, 1)$ and web structure preserving (holomorphic) mappings which also consistently and equivalently can be considered as the theory of special real $(n+1)$-webs $W(n+1, n, 2)$ where the leaves of the foliations of the web are of (real) codimension 2 and they satisfy additionally some harmonic partial differential equations (those for holomorphy of course), with a special class of web preserving morphisms (namely those where the component functions satisfy harmonic partial differential equations that ensure the holomorphy of the mapping).

Apparently this would be a significant complication which our approach somehow elegantly camouflages.

In addition, the meromorphic webs as presented here also link global arguments and concepts that are characteristic and powerful ideas of classic and modern Complex Analysis with web-theoretic considerations.

Last but not least this context should also acknowledge the works of Timoshenko (see [Ti 75, 77]) who extended the theory of three-webs to algebraic structures over fields other than the field **R** of real numbers.

6.3.2 Analytical Polyhedral Domains in $\mathbf{C}^n$, $n > 1$

We will consider the complex field **C**, and (open) domains of $\mathbf{C}^n$, $n > 1$, that are naturally isomorphic to a $(2n)$-dimensional real vector space with Euclidean geometry.

Holomorphic functions of several variables are functions that can be given locally (i.e., around each point x on a suitable open neighbourhood) as a convergent power series in their variables. *Meromorphic functions* are then defined via local quotients of locally defined holomorphic functions.

Holomorphic mappings between complex vector spaces or complex manifolds are defined locally in the usual manner as mappings whose components are holomorphic functions. For explicit definitions we refer to [Na 73], [GuR 65] or [Wh 72].

An *analytical polyhedron* is defined as an elementary Euclidean polyhedron. The latter is defined by linear functions. Consequently we can perceive an *analytical polyhedron* as a connected domain defined via analytical (i.e., holomorphic) functions.

Of course, the Euclidean polyhedrons in a $(2n)$-dimensional real vector space are naturally analytical polyhedrons in an n-dimensional complex vector space.

We shall give now more explicit definitions.

Definition 6.3.1 a) An *analytical polyhedron* P in $\mathbf{C}^n$, $n > 1$, is a compact subset in $\mathbf{C}^n$ such that there exists an open neighbourhood U of P and a finite number of holomorphic functions $f_1, \ldots, f_k$, so that the subset P is given as

$$P = \{z \in U : | f_1(z) | \leq 1, \ldots, | f_k(z) | \leq 1\}.$$

b) An *(analytical) polyhedral domain* A in $\mathbf{C}^n$, $n > 1$, is a connected component of the open interior of an analytical polyhedron P such that the following holds: the neighbourhood U of P and the functions $f_1, \ldots, f_k$ (see a)) can be chosen in such a way that for each $i = 1, \ldots, k$ the subset

$$\{z \in U : | f_i(z) | = 1\} \cap \mathrm{bd}\, A \neq 0$$

has topological dimension $2n-1$ (bd A designates the boundary of A).

c) Let A be an (analytical) polyhedral domain A in $\mathbf{C}^n$, $n > 1$. A non-constant function g which is holomorphic in an open neighbourhood U of A, is called a *partition function* of A if the following holds true: the boundary of A, bd A, contains a topologically $(2n-1)$-dimensional subset R, so that: $|g|\Big|_R = \text{constant}$.

The partition of A via g – defined as the partition given by the connected components of the sets $\{z \in A : g(z) = \text{constant}\}$ – is called a *characteristic partition* of the polyhedral domain A (cf. [Ri 64a]).

The characteristic partitions of (analytic) polyhedrons can be considered as (generalised) foliations that are induced in a natural way on analytical polyhedral domains and will lead to meromorphic webs on those domains if there is a sufficient number of "different" partition functions on A.

In order to clarify this, we will proceed to consider the 1-forms associated with the partition functions of an analytic polyhedral domain.

Definition 6.3.2 Let A be an (analytical) polyhedral domain A in $\mathbf{C}^n$, $n > 1$. A system $g_1, \ldots, g_s$ of partition functions of A is called a *partition representation* of the polyhedral domain if and only if two conditions are met :

i)

$$dg_i \wedge dg_j \neq 0 \quad \text{for all} \quad i \neq j.$$

ii) If g is an arbitrary chosen partition function of A, then there exists a j' with $1 < j' < s$, so that $dg \wedge dg_{j'} = 0$.

The number s is called the *number of tiers* of A.

Lemma 6.3.3 *Consider holomorphic functions $f_1, \ldots, f_k$ defined in an open non-empty domain $U \subset \mathbf{C}^n$, $n > 1$, with the following properties:*

i) $P = \{x \in U : | f_1(z) | \leq 1, \ldots, | f_k(z) | \leq 1\} \subset U$ *is a compact subset of* U *and the open set*

$$A = \{x \in U : | f_1(z) | < 1, \ldots, | f_k(z) | < 1\} \subset U$$

is connected (i.e., A is a domain).

ii) *For each $i = 1, \ldots, k$, let A define a non-empty set of* bd A:

$$A_i = \{x \in U : \ |f_i(z)| = 1, \ |f_j(z)| < 1 \text{ for } j \neq i\} \neq \emptyset,$$

and suppose that the holomorphic 1-forms df_i do not vanish identically on A_i : $df_i|_{A_i} = 0$. Then A defines an analytical polyhedral domain and there are numbers $0 < i_i < \ldots < i_s < k$ such that $(f_{i_1}, \ldots, f_{i_s})$ is a partition representation of this polyhedral domain A with number of tiers s.

Definition 6.3.4 A partition of a domain $D \subset \mathbf{C}^n$, $n > 1$, is said to be a *complex affine partition* of D if and only if there exists a complex linear mapping of rank 1: $A : \mathbf{C}^n \to \mathbf{C}$ such that the partition can be described as $\{z \in \mathbf{C}^n : A(z) = c_j\} \cap D$ with c_j varying in $U \subset \mathbf{C}$. In other words, the partition is described via a complex affine function A, and the partition elements are naturally parallel pieces of complex hyperplanes.

Example 6.3.5 a) A bounded convex Euclidean polyhedron defined in a $(2n)$-dimensional real vector space is an analytical polyhedron defined by the inequalities:

$$|\exp(a_{i_1} z_1 + \ldots + a_{i_n} z_n + b)| \leq 1, \quad i = 1, \ldots, k, \quad k \geq 2n + 1$$

for the real components $(x_1, y_1, \ldots, x_n, y_n)$ and $z_j = x_j + i y_j$ for $j = 1, \ldots, n$, where $a_{i_j}, b \in \mathbf{C}$ are appropriately chosen. The interior of such a polyhedron is an analytic polyhedral domain of $\mathbf{C}^n$. All the characteristic partitions are naturally parallel pieces of complex hyperplanes.

The number of tiers is at least n, and for simplices this can take on each value between $n + 1$ and $2n + 1$.

b) Analytical polyhedral domains defined by inequalities

$$| a_1 z_1 + \ldots + a_n z_n + b | < 1, \quad a_i, b \in \mathbf{C},$$

will be named *complex affine polyhedral domains* (1).

c) Consider a *real number* a with $1 < a < 2$. Then the sets

$$P^a = \{(z, w) \in \mathbf{C}^2 : | z | < 1, \quad | w | < 1, \quad | z + w + a z^2 w | < 1\}$$

define analytical polyhedral domains with number of tiers 3.

d) Consider natural numbers r and t. Then the sets

$$P_{r,t} = \{(z, w) \in \mathbf{C}^2 : | z | < 1, \quad | w | < 1, \quad | z^r + z^t w | < 1\}$$

define analytical polyhedral domains with number of tiers 3.

The number of tiers of a polyhedral domain is an essential invariant with respect to proper holomorphic mappings between such domains. It should be recalled the notion of proper mappings. A *proper mapping* f has the topological property that the inverse image $f^{-1}(K)$ is compact for all compact sets K of the image. Within the context of Complex Analysis the proper holomorphic mappings are of great importance: they generalise the concept of holomorphic covering mappings. Proper holomorphic mappings between domains of finite-dimensional complex vector spaces are nearly everywhere (i.e., with exception of thin analytical varieties) locally biholomorphic.

We present now the classical results that allow the use of the characteristic partitions of polyhedral domains in a web-theoretic way.

Theorem 6.3.6 (See [Ro 35] for the biholomorphic case and [RS 60] for the proper holomorphic case). *Consider analytical polyhedral domains A and A' in $\mathbf{C}^n$, $n > 1$, with number of tiers s and s' respectively. Let $F : A \to A'$ be a proper holomorphic mapping. Then the following holds true:*

a) *The characteristic partitions of A are naturally mapped onto those of A': each partition Z of A is associated with a partition Z' of A' via F, so that each element of Z is mapped onto an element of Z'. Different partitions of A are mapped onto different partitions of A'.*

b) *Let $(g_1, \ldots, g_s)$ and $(g_1', \ldots, g_{s'}')$ be partition representations of A and A'. Then*

(i) *the number of tiers is identical: $s = s'$* [Ri 64a]*; and*
(ii) *there is a permutation $\sigma \in S_s$, such that for all $j = 1, \ldots, s$, we have $dg_j(z) \wedge dg_{\sigma(j)}(F(z)) = 0$ if $z \in A$.*

Rothstein in [Ro 35] gave also the first applications $(^2)$.

Proposition 6.3.7 ([RS 60], p. 186, Satz 5). *Consider analytical polyhedral domains A and A' in $\mathbf{C}^n$, $n > 1$, whose partitions consist of pieces of parallel complex hyperplanes only. Let $F : A \to A'$ be a surjective holomorphic mapping that maps the partitions of A onto the partitions of A'. Then F is a restriction of an affine mapping, i.e. $F(z) = az + b$ with $z = (z_1, \ldots, z_n)$, $a \in \mathbf{C}^* = \mathbf{C} - \{0\}$, and $b \in \mathbf{C}^n$.*

This can be proved by a local differential argument by considering the partial derivations of the component functions of F, which indicates that this was only the special case of a more general theorem in a web-theoretic context. We will present this theorem later.

6.3.3 Meromorphic Webs in Domains of $\mathbf{C}^n$, $n > 1$

In order to adjust the theory of $(n+1)$-webs which we have constructed in Chapters 1–4 and allow us to exploit the properties of the naturally induced characteristic partitions with respect to proper holomorphic mappings between analytical polyhedral domains, we cannot confine ourselves to holomorphic functions and biholomorphic mappings. In particular, if we want to use the fruitful methods using 1-forms and the structure equations of $(n+1)$-webs $W(n+1,n,1)$, we naturally must consider 1-forms with meromorphic coefficients: this allows us to calculate nearly as conveniently as with diffeomorphisms and non-vanishing functions in the real case. We can work with the meromorphic functions as with numbers in a field.

Definition 6.3.8 A system $(\omega_1, \dots, \omega_{n+1})$ of meromorphic 1-forms shall be called a *meromorphic $(n+1)$-web* $W(n+1,n,1)$ in a domain $D \subset \mathbf{C}^n$, $n > 1$, if the following holds:

a) $\omega_\xi \wedge \omega_\eta \neq 0$ for all $\xi \neq \eta$;
b) $d\omega_\xi = \omega_\xi \wedge q_\xi$ for a suitable meromorphic 1-form q_ξ and for all $\xi = 1, \dots, n+1$;
c) $\omega_{\xi_1} \wedge \dots \wedge \omega_{\xi_n} \neq 0$ for all $1 \leq \xi_1 \leq \dots \leq \xi_n \leq n+1$;
d) $\omega_1 + \dots + \omega_{n+1} = 0$.

Example 6.3.9 a) Consider a domain $D \subset \mathbf{C}^2$, with coordinates (z,w) and let there be given a meromorphic function f with partial derivations f_z and f_w that are not identically zero (i.e., f does not depend solely on z or w). Then $(f_z dz,\ f_w dw,\ -df)$ defines a meromorphic three-web with $q_\xi = -f_{zw}(f_z dz + f_w dw) = q$; here, however, q is a simultaneous meromorphic 1-form satisfying the condition b) in Definition 6.3.8. Note that this simultaneous $q = q_\xi$ is symmetric, i.e., independent on the particular order of the foliations $X_1, \dots, X_{n+1}$. In general, in an n-dimensional space all q_ξ's can be found if one starts with a meromorphic function $f(z_1, \dots, z_n)$ with certain conditions on its partial derivatives (all the partial derivatives f_{z_α} not identically zero). However, a simultaneous 1-form $q = q_\xi$ (for all ξ and all orderings of $\{1, \dots, n+1\}$) does not exist in general.

b) Consider a domain $D \subset \mathbf{C}^2$, with coordinates (z,w). Let f_1, f_2, f_3 be meromorphic functions in D, that are not mutually dependent, i.e., $df_\xi \wedge df_\eta \neq 0$, if $\xi \neq \eta$. Then there are meromorphic functions a and b as well as a meromorphic 1-form q, so that $(a \cdot df_1,\ b \cdot df_2,\ df_3)$ is a 3-web $W(3,2,1)$ with $q_\xi = q$. This can be obtained using linear algebra to solve the equations

$$a(f_1)_z + b(f_2)_z = -(f_3)_z, \quad a(f_1)_w + b(f_2)_w = -(f_3)_w.$$

They can be solved since $df_\xi \wedge df_\eta \neq 0$. The 1-form q is obtained as a consequence of the integrability conditions $\binom{3}{}$.

An analogous procedure would succeed in the n-dimensional case.

c) Let a domain $D \subseteq \mathbf{C}^2$ with coordinates (z,w) be given. Consider again the web as in a) induced by a meromorphic function f with the derivatives $f_z,\ f_w \neq 0$. Obviously we can also consider the meromorphic function f^2 and its corresponding meromorphic 3-web $W(3,2,1)$ defined as in b). Then the 3-webs $(f_z dz,\ f_w dw,\ -df)$ and $(2f \cdot f_z dz,\ 2f \cdot f_w dw,\ -d(f^2))$ have topologically identical foliations since the defining functions as well as the 1-forms of the webs are analytically dependent, but these two webs have quite different 1-forms $q(f)$ and $q(f^2)$ respectively.

This is described by the following structure equations.

Proposition 6.3.10 *Consider a meromorphic $(n+1)$-web $W(n+1,n,1)$ in a domain $D \subset \mathbf{C}^n$, $n > 1$, denoted by $(\omega_1, \dots, \omega_{n+1})$. Then the following holds:*

a) *There exists a uniquely defined meromorphic 1-form q as well as uniquely defined meromorphic functions $T_{\alpha\beta}$ and $K_{\alpha\beta}$ in D such that:*

$$d\omega_{n+1} = \omega_{n+1} \wedge q, \tag{6.3.1}$$

$$d\omega_\alpha = \omega_\alpha \wedge q + T_{\alpha 1}\omega_\alpha \wedge \omega_1 + \dots + T_{\alpha n}\omega_\alpha \wedge \omega_n, \quad \alpha = 1, \dots, n, \tag{6.3.2}$$

$$T_{\alpha\alpha} = 0, \quad T_{\alpha\beta} = T_{\beta\alpha}, \quad \alpha \neq \beta, \quad \alpha, \beta = 1, \dots, n, \tag{6.3.3}$$

$$dq = \sum_{\alpha,\beta(\alpha<\beta)} K_{\alpha\beta}\omega_\alpha \wedge \omega_\beta. \tag{6.3.4}$$

(cf. equations (1.2.16), (1.2.17), and (1.2.24) in the case $r = 1$ that give equations (6.3.1)–(6.3.4) if one makes the following convenient change of notation: $\underset{\xi}{\omega^1} = \omega_\xi$, $\omega_1^1 = q$, $\underset{\alpha\beta}{a}{}^1_{11} = T_{\alpha\beta}$, $\underset{\alpha\beta}{b}{}^1_{111} = K_{\alpha\beta}$).

b) *Let $(\nu_1, \dots, \nu_{n+1})$ be another meromorphic $(n+1)$-web $W(n+1,n,1)$ in D with corresponding q', $T'_{\alpha\beta}$, $K'_{\alpha\beta}$ and the following property:*

$$\omega_\xi \wedge \nu_\xi = 0, \quad \xi = 1, \dots, n+1,$$

i.e., ω_ξ and ν_ξ are pairwise dependent 1-forms. Then there exists a meromorphic (not identically zero) function h in D such that

$$\nu_\alpha = h\omega_\alpha, \quad \alpha = 1, \dots, n, \tag{6.3.5}$$

$$\begin{aligned} q &= q' + d(\log h), \\ T_{\alpha\beta} &= hT'_{\alpha\beta}, \\ K_{\alpha\beta} &= hK'_{\alpha\beta}, \quad \alpha, \beta = 1, \dots, n; \ \alpha < \beta. \end{aligned}$$

This can be derived with the usual web-theoretic formalism, because we can operate with meromorphic functions formally as with rational numbers. The dependence of meromorphic 1-forms as in (6.3.5) leads to direct equations for the meromorphic function h in D.

Definition 6.3.11 a) Consider a meromorphic $(n+1)$-web $W(n+1,n,1)$ in a domain $D \subset \mathbf{C}^n$, $n > 1$, written as $W = (\omega_1, \dots, \omega_{n+1})$. The $W(n+1,n,1)$ is called an *$(n+1)$-web of type* (0) if $T_{\alpha\beta} = 0$ and $K_{\alpha\beta} = 0$ for all $\alpha, \beta = 1, \dots, n$.

b) Consider two meromorphic $(n+1)$-webs $W(n+1,n,1)$ in a domain $D \subset \mathbf{C}^n$, $n > 1$, say, $W = (\omega_1, \dots, \omega_{n+1})$ and $V = (\nu_1, \dots, \nu_{n+1})$. The web W is called *related* (or *similar*) to V, if

$$\omega_\xi \wedge \nu_\xi = 0, \quad \xi = 1, \dots, n+1.$$

Remark 6.3.12 (on "semi-invariants of various degrees"). Functions with similar behaviour to the functions $T_{\alpha\beta}, K_{\alpha\beta}$ are easily generated. Assume, for instance, that not all $T_{\alpha\beta}$, $K_{\alpha\beta}$ are identically zero. Then the equations

$$dT_{\alpha\beta} - T_{\alpha\beta}q = T_{\alpha\beta,1}\omega_1 + \ldots + T_{\alpha\beta,n}\omega_n, \tag{6.3.6}$$

$$dK_{\alpha\beta} - 2K_{\alpha\beta}q = K_{\alpha\beta,1}\omega_1 + \ldots + K_{\alpha\beta,n}\omega_n \tag{6.3.7}$$

will yield new functions so that

$$T_{\alpha\beta,\gamma} = h^2 T'_{\alpha\beta,\gamma}, \quad K_{\alpha\beta,\gamma} = h^3 K'_{\alpha\beta,\gamma}.$$

with notation as in Definition 6.3.10 b). It is worth noting that the exponent of h increases by 1. Obviously this process can be iterated further, yielding such "semi-invariants" of arbitrary high "degree". The functions $T_{\alpha\beta}$ and $K_{\alpha\beta}$ are examples of semi-invariants of degree 1 and 2 of the $(n+1)$-web $W(n+1, n, 1)$ – for semi-invariants see [B 54], pp. 2–15.

The same reasoning could be also applied to linear combinations of "semi-invariants" of the same degree, multiplication of semi-invariants (the degree of the product will be the sum of the degrees of the factors) as well as to quotients of two semi-invariants (with well-defined nominator) which for meromorphic functions is very reasonable (the degree of the resulting semi-invariant is then the difference of the individual ones). The latter case is especially interesting and could be applied without any difficulty in the global meromorphic framework: in this way we might arrive at semi-invariants of degree 0, i.e., to meromorphic functions that are "invariantly" attached to the class of meromorphic similar $(n+1)$-webs $W(n+1, n, 1)$ in a fixed domain D. We need only to have not identically vanishing $T_{\alpha\beta}$ or $K_{\alpha\beta}$ which can be easily constructed.

By generating other semi-invariants of higher degrees we can then consider a suitable quotient of not trivial semi-invariants in order to find an invariant.

It is therefore convenient to introduce the following definition.

Definition 6.3.13 Consider a meromorphic $(n+1)$-web $W(n+1, n, 1)$ in a domain $D \subset \mathbf{C}^n$, $n > 1$, written as $W = (\omega_1, \ldots, \omega_{n+1})$. The web W is called an *$(n+1)$-web of type (k)* with $0 < k < n+1$ if in D there exist "invariant" meromorphic functions $I_1, \ldots, I_k$, constructible as in Remark 6.3.12, such that

$$dI_1 \wedge \ldots \wedge dI_k \neq 0 \tag{6.3.8}$$

in D and such that for any other invariant J (constructible as in Remark 6.3.12) we have

$$dI_1 \wedge \ldots \wedge dI_k \wedge dJ = 0 \tag{6.3.9}$$

in D.

Example 6.3.14 For most purposes it is often very practical if we could assume that the web can be described by $n+1$ local functions (as in Example 6.3.9 b)) – a property that can be achieved with help of the integrability condition (Definition 6.3.8 b)) – in the meromorphic context at least outside of a thin analytical (singular) variety.

Also, it is reasonable – outside of a certain analytical variety – that the local web functions $f_1, \dots, f_{n+1}$ in a U allow the definition of $f = f_{n+1} \cdot (f_1, \dots, f_n)^{-1}$ as a holomorphic function, inducing a meromorphic $(n+1)$-web $W(n+1, n, 1)$ on $V = (f_1, \dots, f_n)(U) \subset \mathbf{C}^n$. This reduces most web-theoretic situations to Example 6.3.9 a). In Complex Analysis this manifold point of view is quite natural due to the identity theorem (or the continuity principle) that allows the confirmation of a holomorphic property (like vanishing or other identity equations) if it can be established in an open neighbourhood.

Consider the case $n = 2$ with $W = (\omega_1, \omega_2, \omega_3)$ in $D \subseteq \mathbf{C}^2$. Then there is no $T_{\alpha\beta}$, and only one function $K_{\alpha\beta} = K$. In this case the equations (6.3.2), (6.3.4), and (6.3.7) are as follows (see Example 6.3.9):

$$\begin{aligned} d\omega_\alpha &= \omega_\alpha \wedge q, \\ dq &= K\omega_1 \wedge \omega_2 = K\omega_2 \wedge \omega_3 = K\omega_3 \wedge \omega_1, \\ dK - 2Kq &= K_2\omega_1 + K_1\omega_2 = K_3\omega_2 + K_2\omega_3 = K_1\omega_3 + K_3\omega_1. \end{aligned}$$

Consequently, if W is considered locally given by meromorphic functions $f_1(z,w) = z$, $f_2(z,w) = w$ and $f_3(z,w) = f(z,w)$ with not identically vanishing f_z and f_w in an open $U \subset \mathbf{C}^2$, then we find (see [Bl 55], p.9 or [Bau 82], I.1.4)

$$\begin{aligned} K &= (\log\,(f_w/f_z)_{zw}/(f_z \cdot f_w), \\ K_1 &= -K_w/f_w - 2K f_{zw}/(f_z \cdot f_w), \\ K_2 &= K_z/f_z + 2K f_{zw}/(f_z \cdot f_w), \\ K_3 &= K_w/f_w - K_z/f_z \end{aligned}$$

and so on for $K_{11}, K_{12}, K_{21}, K_{22}, K_{111}, \dots$. It means that all these functions are rational expressions of meromorphic functions, and therefore, they are themselves meromorphic functions.

For determination of the type of the three-web we proceed as follows. If the three-web is not of type (0), i.e, if K is not identical zero, then using the well-known identity (see [BD 28], p. 214 or [Bau 82], I.1.9):

$$K_{12} - K_{21} = 2K^2,$$

we will easily find at least one non-constant invariant function

$$I_1 = K_1^2/K^3 \quad \text{or} \quad I_2 = K_2^2/K^3.$$

(see [Bau 82], I.1.9). Obviously we can always assume that this is the case for I_1 : otherwise, reorder the foliations.

It is a straightforward exercise that only by inspection of the invariants I_2, K_{11}/K^2 and K_{12}/K^2 we can find an additional invariant (if the web is of type (2))(see [Bau 82], II.2.3). As mentioned above, the locally established existence continues in the entire domain of the considered web.

Consider now three-webs in a domain of $\mathbf{C}^2$ defined by three functions f_1, f_2, f_3 as in Example 6.3.9 b). For

$$f_1(z,w) = z + w,\ f_2(z,w) = z^2 + w^2,\ f_3(z,w) = z \cdot w$$

we have a 3-web of type (0). For

$$f_1(z,w) = z,\ f_2(z,w) = z + w^3,\ f_3(z,w) = z \cdot w$$

we have a 3-web of type (1). For

$$f_1(z,w) = z,\ f_2(z,w) = w,\ f_3(z,w) = z + w + z^2 \cdot w$$

we have a 3-web of type (2).

Remark 6.3.15 Consider an $(n+1)$-web $W = (\omega_1, \ldots, \omega_{n+1})$ of type (0) in D. Then we can choose a suitable small open domain U in D, so that via the Frobenius integrability theorem we will find holomorphic functions $f_1, \ldots, f_{n+1}$ such that

$$\omega_\xi \wedge df_\xi = 0 \text{ and } df_\xi = a\omega_\xi$$

with a unique meromorphic function a in U for all $\xi = 1, \ldots, n+1$, and so that the identity $f_1 + \ldots + f_{n+1} = 0$ will hold in the whole of U. Of course, we can extend these functions as far as possible into the remaining parts of D. In other words, in the language of etale spaces one considers the connected continuation component of f_ξ in the sheaf of holomorphic functions (see [GR 84] and [GF 76]). However, according to the main paradigm of Complex Analysis, in general, we will not be able to accomplish this for the entire domain D (at best it can be only extended to a Riemannian manifold covering over a part of D because during the continuation of holomorphic functions one often has to avoid singularity sets where the function may wind around those for a finite or infinite number — the standard example is the well-known definition of the complex logarithm function — see [We 64]).

For example, consider the following $(n+1)$-web in $\mathbf{C}^n$, $n > 1$:

$$\begin{aligned} f_\alpha(z_1, \ldots, z_n) &= z_\alpha,\ \alpha = 1, \ldots, n, \\ f_{n+1}(z_1, \ldots, z_n) &= z_1 \ldots z_n. \end{aligned}$$

For this web, outside the analytical variety

$$\{(z_1, \ldots, z_n) \in \mathbf{C}^n,\ z_\alpha = 0 \quad \text{for at least one } \alpha\},$$

in a simply connected domain U, we will find locally well-defined meromorphic functions $F_\alpha(z_1, \ldots, z_n) = \log z_\alpha$, $\alpha = 1, \ldots, n$, $F_{n+1}(z_1, \ldots, z_n) = \log(z_1 \ldots z_n)$ satisfying the above described identity $F_1 + \ldots + F_{n+1} = 0$ in U.

Therefore it is reasonable to give the following

Definition 6.3.16 Consider an $(n+1)$-web $W = (\omega_1, \ldots, \omega_{n+1})$ of type (0) in a domain $D \subseteq \mathbf{C}^n$, $n > 1$. The web W is called *globally parallelisable* $(^4)$ if there exist meromorphic (global and well-defined) functions $f_1, \ldots, f_{n+1}$ in D such that the following holds:

$$f_1 + \ldots + f_{n+1} = 0, \tag{6.3.10}$$

$$\omega_\xi \wedge df_\xi = 0, \ \ \xi = 1, \ldots, n+1. \tag{6.3.11}$$

The following theorem generalises Rothstein's application which we discussed above.

Theorem 6.3.17 *Consider meromorphic $(n+1)$-webs $W(n+1, n, 1)$ $W = (\omega_1, \ldots, \omega_{n+1})$ and $V = (\nu_1, \ldots, \nu_{n+1})$ in a domain $D \subseteq \mathbf{C}^n$, $n > 1$, both of type (0) with global parallelisations $f_1, \ldots, f_{n+1}$ and $g_1, \ldots, g_{n+1}$ meromorphic in D. Suppose furthermore that the $(n+1)$-webs W and V are related in $D : \omega_\xi \wedge \nu_\xi = 0$, $\xi = 1, \ldots, n+1$. Then there exist a non-zero complex number a and n complex numbers b_α, $\alpha = 1, \ldots, n$, such that: $g_\alpha = af_\alpha + b_\alpha$, $\alpha = 1, \ldots, n$.*

Sketch of Proof (see Proposition 6.3.7, [Bau 82], II.1.3 or [RS 60], p. 186, Satz 15). The main idea of the proof is to show the assertion for an open part of D and then the continuity principle of Complex Analysis will automatically extend the assertion to the entire D.

a) *Local factorisation.* Similarly to the real case, related 1-forms imply that integrated functions are dependent and may be factorised. This fact will be used here. As the 1-forms ω_ξ and ν_ξ are related, we conclude (at least in a suitable open part of D where the 1-forms are regular) that the parallelisation functions f_ξ and g_ξ may be factorised: $g_\xi = h_\xi f_\xi$, $\xi = 1, \ldots, n+1$, for suitable holomorphic functions h_ξ in a certain domain in $\mathbf{C}$. This can be proved by applying Theorem 6.5F of [Wh 72] (p. 198) or Satz 0.8 of [Bau 82] .

b) *Differentiation and integration of the factorisation functions.* Consider now the well-defined function

$$H = h_1'(f_i) = \ldots = h_{n+1}'(f_{n+1})$$

where $h_\xi' = dh_\xi/dt$ in accordance with Definition 6.3.11, which can be derived locally with help of the (parallelisation) identities

$$0 = \sum_{\xi=1}^{n+1} dg_\xi = \sum_{\xi=1}^{n+1} h_\xi'(f_\xi) df_\xi; \ \ 0 = \sum_{\xi=1}^{n+1} df_\xi.$$

Now we can conclude that $dH = 0$, because the web is of type (0) and the web 1-forms $df_1, \ldots, df_n$ are independent as demanded in Definition 6.3.8 a) (see [Bau 82] for more explicite elaborations). Reversing these steps by integration gives $h_\xi : z \to az + b_\xi$,

$\xi = 1, \ldots, n+1$, for suitable chosen complex numbers $a \neq 0$ and b_ξ, $\xi = 1, \ldots, n+1$ $\left(^5\right)$. Note that $a \neq 0$ since g_ξ are not constants.

c) *Application of the continuity principle.* From b) we conclude that the assertion holds for an open part of D. According to the identity theorem (or the continuity principle of Complex Analysis, see [Wh 72], Proposition 0.3 D, p. 10), we may consider the identity $af_\xi + b_\xi - g_\xi = 0$, $\xi = 1, \ldots, n+1$, in a small open part of D which must then extend to the entire D. ■

Now we are ready to use the web-theoretic approach for construction of invariant meromorphic functions with respect to proper holomorphic mappings between analytical polyhedral domains.

6.3.4 Partition Webs Generated by Analytical Polyhedral Domains

Consider an analytical polyhedral domain $A \subset \mathbf{C}^n$, $n > 1$, with a representation system $g_1, \ldots, g_s, s > n$, of holomorphic functions in A (see Lemma 6.3.3).

First we will study the problem of existence of meromorphic webs that are induced by partition functions in A: what kind of m-webs $W(m, n, 1)$, $m = n+1, \ldots, s$, are introduced on a given analytical polyhedral domain A ?

It should be remarked (and this will be elaborated on later) that Definition 6.3.8 implies that $n+1$ partition functions may lead to $(n+1)!$ (formally) different $(n+1)$-webs arising by way of permutations of those partition functions (according to Definition 3.1.10, the coordinate n-quasigroups of such webs are parastrophic to each other).

Definition 6.3.18 Consider an analytical polyhedral domain $A \subset \mathbf{C}^n$, $n > 1$, with a partition representation $g_1, \ldots, g_s$, where s is the number of tiers of A. Then there exist non-negative numbers

$$s_2, \ldots, s_n, \; p_3, \ldots, p_n, \; r_1, \ldots, r_{s-n}$$

defined as follows:

(a) s_j is the number of j-tuples $(i_1, \ldots, i_j)$, $s_j \leq \binom{j}{s}$ so that $0 < i_1 < \ldots < i_j \leq s$ and in A

$$dg_{i_1} \wedge \ldots \wedge dg_{i_j} \neq 0, \quad j = 2, \ldots, n.$$

(b) p_k is the number of k-tuples $(i_1, \ldots, i_k)$ so that $0 < i_1 < \ldots < i_k \leq s$, and in A

$$dg_{i_1} \wedge \ldots \wedge dg_{i_k} = 0.$$

Further, for all permutations $\sigma \in S_k$ (permutation group of order k) we have in A

$$dg_{i_{\sigma(1)}} \wedge \ldots \wedge dg_{i_{\sigma(k-1)}} \neq 0$$

and

$$dg_{i_{\sigma(1)}} \wedge \ldots \wedge dg_{i_{\sigma(k-1)}} \wedge dg_{j_1} \wedge \ldots \wedge dg_{j_{n-k+1}} \neq 0$$

for suitable indices $j_1, \ldots, j_{n-k+1}$, where $k = 3, \ldots, n$.

(c) r_j is the number of $(j+n)$-tuples $(i_1, \ldots, i_{j+n})$ so that $0 < i_1 < \ldots < i_{j+n} \leq s$ and in A

$$dg_{i_{\sigma(1)}} \wedge \ldots \wedge dg_{i_{\sigma(n)}} \neq 0$$

for all permutations $\sigma \in S_{j+n}$ (of order $j+n$).

The system $S_A = (s_2, \ldots, s_n, \quad p_3, \ldots, p_n, \quad r_1, \ldots, r_{s-n})$ is independent of the choice of the partition representation and may be called the *web levels* of the polyhedral domain A.

Proposition 6.3.19 *Consider analytical polyhedral domains A and A' in $\mathbf{C}^n$, $n > 1$, and a proper holomorphic mapping $F : A \to A'$. Then the web level numbers of A and A' coincide: $S_A = S_{A'}$. In other words: S_A is an invariant of A with respect to proper holomorphic mappings onto polyhedral domains.*

Sketch of proof (see [Bau 82], III.1.4–1.5). According to Theorem 6.3.5, we can find partition representations $f_1, \ldots, f_s$ and $g_1, \ldots, g_s$ so that $dg_\alpha(F) \wedge df_\alpha = 0$ in the whole of A for all $\alpha = 1, \ldots, s$. By Theorem 6.5F, p. 198, from the book [Wh 72], there is an analytic (thin) variety Z in A so that in $A - Z$ we can locally factorise the functions $g_\alpha(F)$ and $f_\alpha : g_\alpha(F) = h_\alpha f_\alpha$ for suitable holomorphic h_α in a small open domain $U \subset A$, so that even the restriction $F\mid_\nu$ is biholomorphic and all the dh_α/dt do not vanish on $h_\alpha(f_\alpha(U))$. Then the assertion can be proved using locally the pull-back F^* and the identity theorem of Complex Analysis for meromorphic functions applied to all the involved functions and component functions (see [Wh 72], 0.3 D, p. 10). ■

Practically this proposition may be used to construct analytic polyhedral domains that cannot be mapped onto each other via proper holomorphic mappings.

Remark 6.3.20 a) The numbers r_j indicate how many $(j+n)$-webs $W(j+n, n, 1)$ exist in A that are induced via partition functions of A. For instance, there are $k = (n+1)!r$ partition $(n+1)$-webs $W(n+1, n, 1)$ in an analytical polyhedral domain A in $\mathbf{C}^n$, $n > 1$, with $s > n$. This can be seen in a straightforward constructive way with the help of Cramer's rule (as in [Bau 82], Lemma III.2.1, p. 72).

b) The numbers p_j indicate (if there are no partition $(n+1)$-webs $W(n+1, n, 1)$ in A), that it is reasonable to consider k-webs $W(k, k-1, 1)$, $k < n+1$.

c) The case $n = 2$ does not allow us to define any number p_j and therefore obviously the information for the invariant S_A for a polyhedral domain A reduces to $(s, \binom{s}{3}, \; r_2, \ldots, r_{s-2})$, the number r_1 being also redundant because the web condition (see Definition 6.3.8 c)) holds for $n = 2$ anyway. However, for higher $n > 2$ this condition gives an additional information.

Example 6.3.21 Consider the analytical polyhedral domains

$$\begin{aligned} A &= \{(z,w,u) \in \mathbf{C}^3 : |z| < 1, |w| < 1, |u| < 1, |z+w| < 1\}, \\ B &= \{(z,w,u) \in \mathbf{C}^3 : |z| < 1, |w| < 1, |u| < 1, |z+w+u| < 1\}. \end{aligned}$$

Then $S_A \neq S_B$ because $r_1 = 0$ for the domain A.

Therefore no proper holomorphic mappings can exist between A and B. Naturally the type of partition $(n+1)$-webs will yield also some distinctive information on analytical polyhedral domains whose characteristic partitions allow $(n+1)$-webs, i.e., $r_1 \neq 0$.

Definition 6.3.22 Consider an analytical polyhedral domain $A \subset \mathbf{C}^n$, $n > 1$. First, we will assume $r_1 \neq 0$ (see Definition 6.3.18). Then there exist $m = \binom{r_1}{n+1}$ different partition $(n+1)$-webs $W(n+1,n,1)$ in A and the system $T_A = (t_0, \ldots, t_n)$ is well-defined; here t_i is the number of partition $(n+1)$-webs of type (i) in A in the sense of Definition 6.3.11. In the "degenerate case" $r_1 = 0$ we can consider k-webs $W(k, k-1, 1)$ where $k < n+1$. Let $k(< n+1)$ be the highest number with $p_k \neq 0$ (see Definition 6.3.18). Then the system $T_A = (t_0, \ldots, t_k)$ where t_i is the number of partition k-webs of type (i) in A is well-defined. The system T_A is called the *type* of the analytical polyhedral domain A.

Remark 6.3.23 a) T_A is independent of the choice of the partition representation. It is trivial that $t_0 = m - t_1 - \ldots - t_n$ and respectively $t_0 = m - t - \ldots - t_k$ in the degenerate case.

For $n > 2$, as the polyhedral domain A in Example 6.3.23 or similar ones show, it is possible that T_A degenerates to $T_A = (t_0, \ldots, t_k)$. However, we can still attach web invariants to such domains.

b) A proposition analogous to Proposition 6.3.19 is valid: T_A is an invariant with respect to proper holomorphic mappings. This follows from Theorem 6.3.5 in a similar way to that in the proof of Proposition 6.3.19.

Example 6.3.24 a) Consider the polyhedral domain A from Example 6.3.21 and the following analytical polyhedral domains

$$\begin{aligned} C &= \{(z,w,u) \in \mathbf{C}^3 : |z| < 1, \ |w| < 1, \ |u| < 1, \ |2zwu| < 1\}, \\ D &= \{(z,w,u) \in \mathbf{C}^3 : |z| < 1, \ |w| < 1, \ |u| < 1, \ |z+w+z^2w| < 1\}. \end{aligned}$$

D, as well as A (see Remark 6.3.20) is of degenerate type, but for the type of the domains we find: $T_A = (1,0,0)$, $T_C = (1,0,0,0)$, $T_D = (0,0,1)$ (see Example 6.3.14). Therefore, there will be no proper holomorphic mappings among the polyhedral domains A, C and D.

b) The following example gives a non-degenerate case for $n = 2$:

$$\begin{aligned} E &= \{(z,w) \in \mathbf{C}^2 : |z| < 1, \ |w| < 1, \ |2zw| < 1, \ |z + z^2 w| < 1\}, \\ F &= \{(z,w) \in \mathbf{C}^2 : |z| < 1, \ |w| < 1, \ |z + w| < 1, \ |z + z^2 w| < 1\}. \end{aligned}$$

In this case we find that $T_E = (4,0,0) \neq (2,2,0) = T_F$, and we conclude that there exists no proper holomorphic mapping between the polyhedral domains E and F (see [Bau 82], III.2.10).

c) A look at the polyhedral domains B and C (see Examples 6.3.21 and 6.3.24 a) shows that they both induce 4-webs $W(4,3,1)$ of type (0) (see Definition 6.3.11). However, according to Definition 6.3.16, all 4-webs in B are obviously globally parallelisable. None of the 4-webs in C is globally parallelisable as already indicated in Remark 6.3.15. As a consequence of Theorem 6.3.17, we conclude:

(i) There exists no biholomorphic mapping between B and C (as this would carry over the global parallelisation functions of B onto C).

(ii) We infer immediately that there cannot exist a proper holomorphic mapping from C onto B.

d) The set

$$X = \{(z,w,u) \in \mathbf{C}^3 : |z| < 1, \ |w| < 1, \ |u| < 1, \ |z + w| < 1, \ |w + u| < 1\}$$

defines an analytical polyhedral domain with number of tiers 5. However, no 4-web $W(4,3,1)$ is induced in X by this polyhedral domain and $T_X = (2,0,0)$ obviously.

6.3.5 Partition Webs with Parallelisable Foliations

Polyhedral domains which give rise to partition $(n+1)$-webs $W(n+1,n,1)$ that possess global meromorphic parallelisation functions, have some significant properties with respect to proper holomorphic mappings. Locally they resemble nearly everywhere (i.e., with exception of an analytic (thin) variety ([6])) foliations consisting of pieces of parallel affine hyperplanes: they are locally (outside of a singularity variety) biholomorphic to $U \cap \{z \in \mathbf{C}^n : \sum_{\alpha=1}^n a_\alpha z_\alpha = \text{const} \in \mathbf{C}\}$ where U is open in $\mathbf{C}^n$.

The complex version of the Poincare's Lemma (see for example [K 34] or [Bau 82], 0.3, p.1) allows to find locally (outside the meromorphic singular set) parallelisation functions that can be patched along paths together with the help of Theorem 6.3.17. Thus, in simply connected domains we can construct global parallelisation functions. This process depends on the homological properties of the domain outside of the singular variety (see [Bau 82], II.1). But we will not pursue this direction here.

We will now consider analytical polyhedral domains with affine partitions which are defined as $\{\sum_\alpha a_\alpha z_\alpha = \ \text{const} \ \in \mathbf{C}\} \cap U$ where U is open in $\mathbf{C}^n$. It is easy to construct analytical polyhedral domains with affine parallel pieces as partition elements since the conditions of Lemma 6.3.3 can be easily satisfied and they can be described with normalised partition representations (see [Bau 82], IV.1.1).

Definition 6.3.25 Consider an $(n+1)\times(n+1)$ matrix $M=(a_{\xi\eta})$, $\xi,\eta=1,\ldots,$ $n+1$, of complex numbers such that all its $n\times n$ submatrices are of rank n. The affine functions

$$a_\xi(z_1,\ldots,z_n)=a_{\xi 1}z_1+\ldots+a_{\xi n}z_n+c_\xi$$

for arbitrary $c_1,\ldots,c_{n+1}$ define an *$(n+1)$-web $W(n+1,n,1)$ with parallelisations* that are given globally in $\mathbf{C}^n$ by affine functions. In particular, there exist non-zero complex numbers $A_1,\ldots,A_n$ and a complex number c such that:

$$A_1a_1+\ldots+A_na_n+a_{n+1}+c=0$$

in $\mathbf{C}^n$. We will denote such $(n+1)$-webs $W(n+1,n,1)$ by W.

Remark 6.3.26 Consider an $(n+1)$-web W with such an "affine" parallelisation as in Definition 6.3.25. For each $P\subset\mathbf{C}^n$ we find another global affine parallelisation

$$A_1(a_1-a_1(P)),\ldots,A_n(a_n-a_n(P)),\ \ a_{n+1}-a_{n+1}(P).$$

Every other parallelisation can be generated with a non-zero $a\neq 0$, and complex numbers $b_1,\ldots,b_n$ according to Theorem 6.3.16.

Definition 6.3.27 Consider a meromorphic $(n+1)$-web $W(n+1,n,1)$ in a domain U in $\mathbf{C}^n$ such that there exists an $(n+1)$-web W_M so that the following holds: $W_M\big|_U$ is similar to $W(n+1,n,1)$ (in the sense of Definition 6.3.11 b)). The web W is called an *affine $(n+1)$-web*.

Consider now affine $(n+1)$-webs $W(n+1,n,1)$ and $V(n+1,n,1)$ given respectively in domains A and B of $\mathbf{C}^n$ with global parallelisations $L_1,\ldots,L_{n+1}$ and $L_1',\ldots,L_{n+1}'$ respectively and the following condition:

(c) There exist constants $b_1,\ldots,b_n$ and a non-zero a such that:

$$(aL_1+b_1,\ldots,aL_n+b_n)(A)=(L_1',\ldots,L_n')(B).$$

Then there exists a surjective holomorphic (even affine) mapping $F:A\to B$ that maps all web foliations of $W(n+1,n,1)$ onto those of $V(n+1,n,1)$ and vice versa. This mapping is explicitly given by

$$F=(L_1',\ldots,L_n')^{-1}(a_1L_1+b_1,\ldots,a_nL_n+b_n):A\to B \qquad (6.3.12)$$

and is obviously affine (and therefore biholomorphic). One can consider this constructive principle modulo permutations of the foliations (parastrophies).

this is a necessary condition for existence of a web preserving F. This proposition relates the image sets $(L_1, \ldots, L_n)(A)$ and $(L'_1, \ldots, L'_n)(B)$ via affine mappings if and only if the foliations can be mapped onto each other with the web structure preserved. Obviously we may even conclude that (c) is sufficient if some local well-defined construction of F in the form (6.3.12) can be extended to the whole of A. This can be used to determine the existence of proper holomorphic mappings between polyhedral domains carrying affine partition webs or partition webs allowing global parallelisation functions.

Proposition 6.3.28 *Consider polyhedral domains A and B in $\mathbf{C}^n$, $n > 1$, inducing partition $(n+1)$-webs. Suppose further that there exists a proper holomorphic mapping $F : A \to B$. Then the following statements are true:*

a) *If an affine partition $(n+1)$-web of A is mapped via F onto an affine web of B, then F is an affine mapping.*

b) *If all partition $(n+1)$-webs of type (0) of A are affine and there exists one affine partition $(n+1)$-web in B, then the mapping F is affine. This is also valid if the roles of A and B are interchanged.*

c) *If all partition $(n+1)$-webs of type (0) on B have global holomorphic parallelisation functions and there exists one partition $(n+1)$-web in A with a global parallelisation $L_1, \ldots, L_{n+1}$ so that the mapping*

$$G = (L_1, \ldots, L_n)^{-1} : (L_1, \ldots, L_n)(A) \to A$$

is well-defined and holomorphic, then F is biholomorphic.

Proof. First note that this last assertion c) could be further extended for everywhere locally biholomorphic G if B is simply connected. All these assertions are consequences of the diagram on Figure 6.1, Theorem 6.3.17 and the condition (c) where $T(z_1, \ldots, z_n) = (az_1 + b_1, \ldots, az_n + b_n)$ for suitable $a \neq 0$ and $b_1, \ldots, b_n \in \mathbf{C}$. ■

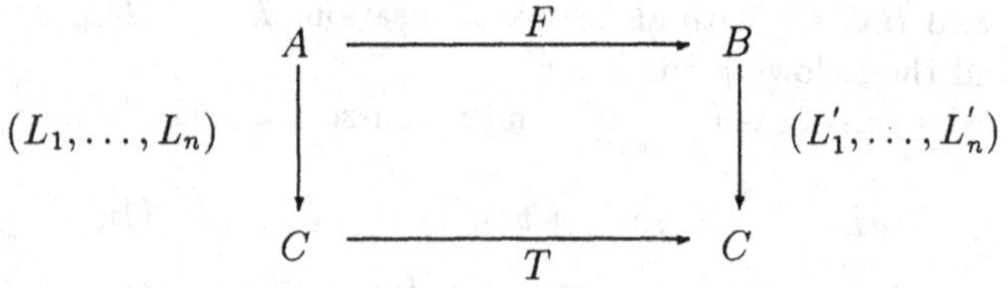

Figure 6.1

Proposition 6.3.28 is a generalisation of Satz 15 in [RS 60] (p. 186). Consequently the following criterion can be formulated:

Theorem 6.3.29 (Criterion for polyhedral domains with affine foliations.) *Consider analytical polyhedral domains A and B in $\mathbf{C}^n$, $n > 1$, so that all the partition $(n+1)$-webs of type (0) are affine webs. Consider also the image sets with respect to parallelisation functions in A and B notated as $A_1, \ldots, A_k$ and $B_1, \ldots, B_k$ respectively. If there is an A_i which is not similar to any of the B_j, then there exists no biholomorphic mapping $F : A \to B$.*

We will make some comments on this theorem:

a) For comparison the following parallelisation sets might be used: $A_i = L_i(A)$ and $B_j = L_j^{'}(B)$ for some parallelisation functions and consequently other subsystems belonging to a web as well:

$$A^1 = (L_1, \ldots, L_t)(A) \text{ and } B^1 = (L_1^{'}, \ldots, L_t^{'})(B).$$

"Similar" means that there exists an affine mapping $T : \mathbf{C}^t \to \mathbf{C}^t$ where

$$T(z_1, \ldots, z_t) = (az_1 + b_1, \ldots, az_t + b_t)$$

for suitable $a \neq 0$ and $b_1, \ldots, b_t$, so that $T(A_i) = B_j$ for at least one j.

b) The criterion could be extended if one technically demands that there is an 1-to-1 assignment between the A_i and B_j by similarity. The number of similar pairs of (A_i, B_j) is necessarily matching. For instance, for $A_i = L_i(A)$ and $B_j = L_j^{'}(B)$ (i.e., for image sets generated by using only one parallelisation function) there are practically $m = k/(n+1)$ disjunctive sets of $n+1$ indices and m affine mappings T_k (as in a)) so that for each B_j we find an A_i such that $B_j = T_r(A_i)$ for an appropriate $r(\leq m)$. Image sets generated by more than one partition function need more such technical description.

c) The above criterion can be constructively applied to proper holomorphic mappings if the word "affine" is replaced by the condition that there is at least one $(n+1)$-web $W(n+1, n, 1)$ with global parallelisation functions that allows a well-defined inversion $(L_1, \ldots, L_n)^{-1}$.

d) The above criterion can be applied negatively to conclude the non-existence of proper holomorphic mappings if

i) there are global parallelisations in at least one polyhedral domain, and
ii) there is at least one image set that is not similar to any such ones in the other polyhedral domain.

For example, consider the polyhedral domains B and C from Examples 6.3.21 and 6.3.24:

$$\begin{aligned} B &= \{(z, w, u) \in \mathbf{C}^3 : |z| < 1, \ |w| < 1, \ |u| < 1, \ |z + w + u| < 1\}, \\ C &= \{(z, w, u) \in \mathbf{C}^3 : |z| < 1, \ |w| < 1, \ |u| < 1, \ |2zwu| < 1\}. \end{aligned}$$

There is no proper holomorphic mapping $F : C \to B$ since, according to Remark 6.3.15, the domain does not allow global parallelisation. A mapping F, however, would carry those of B over to C. One may notice that the parallelisation image sets of C are unbounded since the logarithm is unlimited near the singularity set.

Now we will give some applications.

Theorem 6.3.30 *There exist no proper holomorphic mappings between convex Euclidean and complex affine polyhedral domains in* $\mathbf{C}^n$, $n > 1$, *if their number of tiers is greater than* n.

Sketch of Proof (see the proof in detail in [Bau 82], IV.1.9). Such domains were described in Examples 6.3.5. The idea of the proof is simple and geometric. It is based on the trivial observation that polygons and circular areas are not similar in the elementary Euclidean geometry of $\mathbf{R}^{2n}$. It is obvious that the Euclidean polyhedral domains induce parallelisation sets that are described by polygons in the real plane. However, complex affine polyhedral domains induce parallelisation sets where the boundary is constituted by circular pieces. Therefore, these sets are not similar: they cannot be transformed into each other by translations and homotheties or dilations or a combination of them.

It is evident that many such examples may be easily constructed by mixing Euclidean and complex inequalities in the definition of polyhedral domains (so that they cannot be mapped onto each other via a proper holomorphic or a biholomorphic mapping). The above criterion may also be used for investigation of automorphisms of polyhedral domains.

Definition 6.3.31 A domain A in $\mathbf{C}^n$, $n > 1$, is called *rigid* if there exists no proper holomorphic mapping of A onto itself.

Our approach can be used to construct such objects.

It should be remarked that for parallelisations the $(n+1)$st parallelisation set is the sum set of the other n image sets (the so called Minkowski sum of sets).

Example 6.3.32 a) Rothstein [Ro 35] gave the following example. The domains

$$\begin{aligned} A_t &= \{(z,w) \in \mathbf{C}^2 : 0 < \mathrm{Re}(z),\ \mathrm{Im}(z) < 1,\ 0 < \mathrm{Re}(w),\ \mathrm{Im}(w) < 2, \\ &\quad \mathrm{Re}(w + (1 - it/2)z + t - 3) < 0\} \end{aligned}$$

are rigid for $0 < t < 3$ whereas for $t = 0$ there are automorphisms. The set $\{A : 0 \le t < 3\}$ is a family of inequivalent domains with respect to proper holomorphic mappings. This can be verified by looking at the parallelisation sets: there are two rectangles and an octagon that is intersected by a half-plane parametrised by t. If

$t = 0$ and the intersection is undisturbed, the domain is symmetric and therefore there are some automorphisms.

This example can be easily generalised for $n > 2$. See also [Bau 82], III.2.10 for this classic and further similar examples.

b) Another example (see [Bau 82], IV.1.8):

$$\begin{aligned} H &= \{(z,w,b) \in \mathbf{C}^3, \ |z| < 1, \ |2w| < 1, \ |3z+3w+3b| < 1, \\ &\quad |z+w+5/6| < 1, \ 0 < \mathrm{Re}(b-5/6) < 1/6, \ 0 < \mathrm{Im}(b) < 1/6)\}; \\ H_b &= H \cap \mathbf{C}^2 \times \{b\}. \end{aligned}$$

H is rigid as domain in $\mathbf{C}^3$ and is fibred by the family H_b of the rigid polyhedral domains $H_b = (pr_z, pr_w)(H_b)$ in $\mathbf{C}^2$.

Now we will discuss a holomorphic deformation of a rigid domain with rigid domains. First we will give an example dealing with non-affine foliations.

Example 6.3.33 Consider the following analytical polyhedral domains:

$$P^c_{a,b} = \{(z,w) \in \mathbf{C}^2 : |z| < 1, \ |w| < 1, \ |az^2 + bw^2 + c| < 1\}$$

where $0 < a \le b \le 1 < a+b$ and $0 \le c < 1+a+b$ for suitable real numbers a, b, and c. For these polyhedral domains the following holds:

(i) $P^c_{a,b}$ is rigid if and only if $a \neq b$ and $c \neq 0$.

(ii) $(P^c_{a,b} : 0 \le a \le b < 1 < a+b; \ \ 0 \le c < 1+a+b)$ constitutes a family of polyhedral domains (with non-affine partitions) that cannot be mapped onto each other via proper holomorphic mappings.

Obviously there are parallelisation functions az^2, bw^2, and $az^2 + bw^2$, and each parallelisation set is an intersection of two circles. The conditions for a, b and c follow then consequently by excluding similar matching of those sets (see [Bau 82], IV.1.10).

Example 6.3.33 can be generalised for higher exponents and $n > 2$.

Example 6.3.34 Another class of domains that can be investigated with the affine partition $(n+1)$-webs are the simplest Euclidean polyhedrons: the simplices of $\mathbf{C}^n = \mathbf{R}^{2n}$ (see [Bau 82], IV.2). The number of tiers of a simplex domain in $\mathbf{C}^n$, $n > 1$, ranges from $n+1$ to $2n+1$ and their partitions are all affine. Therefore, we can find global parallelisations that can be normalised as in Proposition 6.3.28 a) and b). The parallelisation image sets are triangles in the real plane $\mathbf{R}^2$. They can be normalised as in Proposition 6.3.28 b) so that they have a common vertex in the origin 0.

The *simplest case* is the case when the number of tiers is $n+1$. In this case the parallelisation image set can be normalised so that

i) there is a common vertex in 0;

ii) there is a parallel line to the y-axis carrying two vertices of all but one triangle, and

iii) the two vertices of that other triangle are on the line that is symmetric to the first one with respect to the y-axis, and this other triangle is the convex hull (rotated with 180°) of the other ones.

The picture in the case $n = 2$ is presented on Figure 6.2:

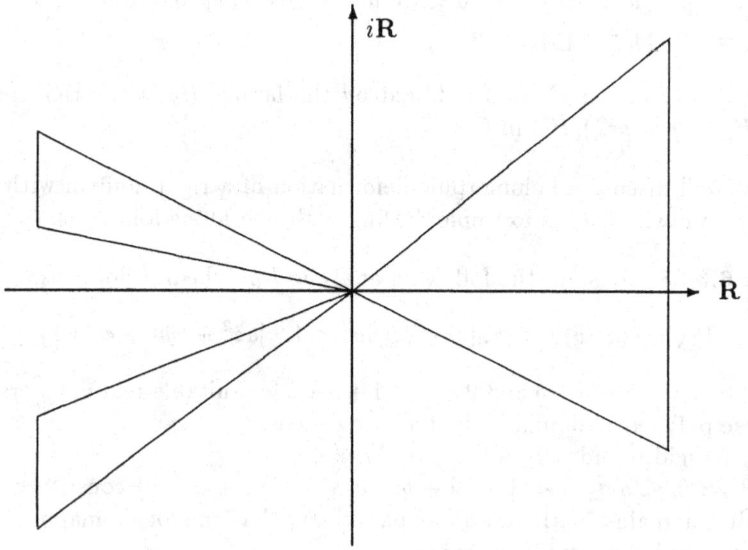

Figure 6.2

For $n > 2$ the picture is similar. There are n triangles on the left side and the other is on the right side being the convex hull of rotated ones (see Figure 6.3).

Generally (modulo affine mappings) this elementary geometric situation can also be described in terms of the convex hull of points $P_0, \ldots, P_{2n}$ as follows: let g_k be the complex line through P_k and P_{2k} for $k = 1, \ldots, n$. Then all the g_k meet in P_0. Consequently automorphisms must interchange those lines.

Further, in general all essentially different simplex domains can be described as follows. Consider the parameter set

$$\begin{aligned} A_n &= \{(a_1, \ldots, a_n) \in \mathbf{C}^n : \operatorname{Im}(a_j) > 0,\ 1 \le j \le n; \\ &\qquad 0 < |a_1| \le \ldots \le |a_n|\} \end{aligned}$$

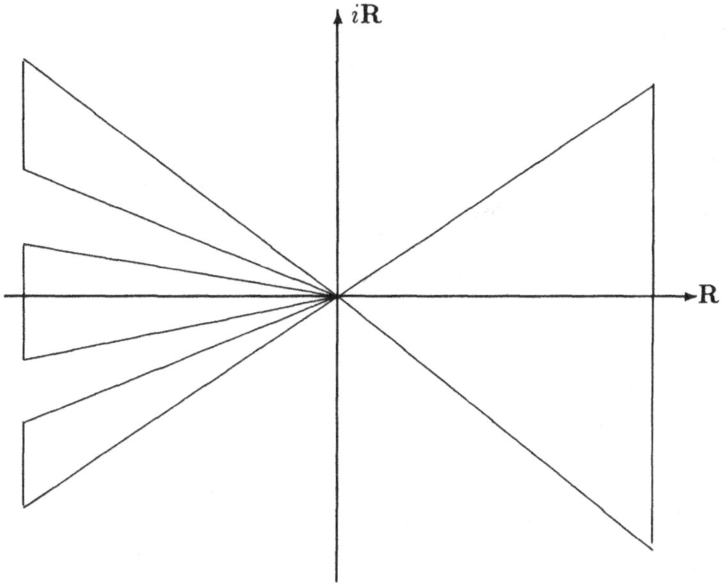

Figure 6.3

with the following property :

$$\text{if } |a_k| = |a_m| \text{ for some } k < m, \text{ then } \arg\,(a_k) < \arg\,(a_m).$$

Then every $a = (a_1, \ldots, a_n) \in A$ defines a simplex domain given by the vertex points: $P_0 = 0$; $P_k = e_k$, $P_{n+k} = a_k e_k$ for $k = 1, \ldots, n$, where $(e_1, \ldots, e_n)$ is the complex vector space basis of $\mathbf{C}^n$, $n > 1$.

Non-rigid simplex domains are those that allow a permutation $\sigma \in S_n$ so that $(a_1, \ldots, a_n) = (a_{\sigma(1)}, \ldots, a_{\sigma(k)})$ (see [Bau 82], IV.2.5).

In the *general case* if the number of tiers is greater than $n+1$, the situation becomes more complicated. For the sake of simplicity we consider here the case $n = 2$ only.

For $n = 2$ we can geometrically conclude the following. Consider a simplex domain S in $\mathbf{C}^2$ being the convex hull of the points $P_0, \ldots, P_4$. Then the following holds:

a) S has a number of tiers 3 if and only if:

i) There are two complex lines g and h (defined by the vertices) such that:

$$g \cap h \in \{P_0, \ldots, P_4\} \subseteq g \cup h.$$

ii) There is only one partition 3-web in S.

iii) All points lie on those two complex lines, and one of those points is the intersection of these lines.

b) S has a number of tiers 4 if and only if:

i) There are two complex lines g and h such that:

$$g \cap h \notin \{P_0, \ldots, P_4\} \subseteq g \cup h.$$

ii) There are exactly two partition 3-webs $W(3,2,1)$ in S.

iii) All points lie on those two complex lines but not in their intersection.

c) S has a number of tiers 5 if and only if:

i) There are no complex lines g and h such that:

$$\{P_0, \ldots, P_4\} \subseteq g \cup h,$$

ii) There are 10 partition 3-webs $W(3,2,1)$ in S.

iii) The points are in general position.

Obviously the 3-web concept becomes here rather cumbersome and it would be necessary to consider 4-webs and 5-webs (and subsequently $(n+k)$-webs in the general case). The Euclidean simplex domains of $\mathbf{C}^2$ fall into the following categories:

a) The foliations define a 3-web $W(3,2,1)$ as discussed above.

b) The foliations define a 4-web $W(4,2,1)$ (see Chapter 7 and [Bau 82], IV. 2.6 and IV. 2.8).

c) The foliations define a 5-web $W(5,2,1)$ (see [Bau 82], IV.2.10).

Consider now some examples.

Example 6.3.35 a) The points $(0,0),(1,0),(0,1),(i,0)$, and $(0,i/2)$ define a rigid simplex domain with only one partition 3-web.

b) The points $(1,0),(0,1),(i,0),(0,-1)$, and $(0,(1+i)/2)$ define a rigid domain with two 3-webs. The foliations define naturally a 4-web $W(4,2,1)$.

c) The points $(1,0),(0,1),(i,0),(0,i/2)$, and (i,i) define a simplex domain with number of tiers 5; similarly the points $(1,0),(0,1),(i,0),(0,i)$, and (i,i) define a simplex domain with 10 partition 3-webs that allows an automorphism: $(z,w) \rightarrow (w,z)$, $\binom{7}{}$.

6.3.6 Partition Webs with Invariant Functions

In the previous section a particular useful idea stemmed from the fact that polyhedral domains which induce partition $(n+1)$-webs with global meromorphic parallelisation functions produce characteristic sets in the plane which could indicate whether proper holomorphic mappings might exist or not. Those sets were defined up to similiar mappings of the real plane (according to Theorem 6.3.17 in a natural interpretation of complex numbers as translations and homotheties). Similarly the invariant functions as introduced in Remark 6.3.12 have such significant properties with respect to proper holomorphic mappings as well – they are even more practical as there are no affine transformations to be taken into account.

Remark 6.3.36 Consider analytical polyhedral domains P and P' in $\mathbf{C}^n$, $n > 1$, with partition $(n+1)$-webs $W = (\omega_1, \ldots, \omega_{n+1})$ in P and $W' = (\omega_1', \ldots, \omega_{n+1}')$ in P' that are related via a proper holomorphic mapping $F : P \to P'$, i.e., the pull-back web $F^*(W')$ and W are related as in Definition 6.3.11 b).

Assume that W and W' are not of type (0) so that there are invariants I for W and I' for W' that are defined similarly (see [Bau 82], I.1 and I.2). Then the following diagram holds:

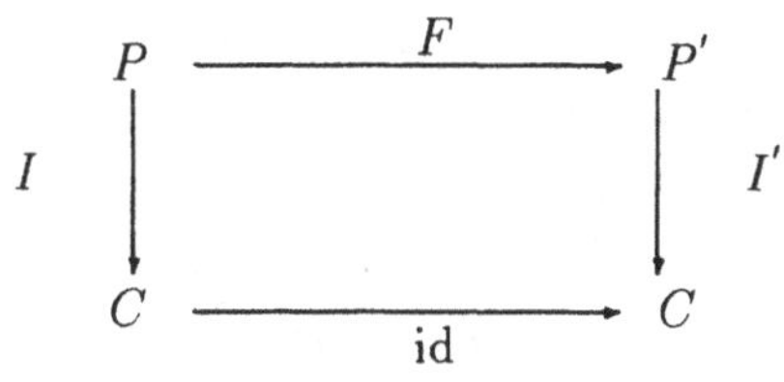

Figure 6.4

Consequently the set theoretic identity $I(P) = I'(P')$ is a necessary condition for the existence of a proper holomorphic mapping between P and P' that relates these partition $(n+1)$-webs W and W'. The diagram in Figure 6.4 is analogous to the diagram in Figure 6.1 for globally parallelisable $(n+1)$-webs $W(n+1, n, 1)$ but with "id" instead of an affine mapping T.

Of course, the above applies to a system of invariants as well. Assume that there are a number of invariant functions $I_1, \ldots, I_k$ of W and correspondingly defined invariant functions $I_1', \ldots, I_k'$ of W'. Then the diagram presented on Figure 6.5 holds, and accordingly for the image sets $(I_1, \ldots, I_k)(P) = (I_1', \ldots, I_k')(P')$.

It is noteworthy to record some further observations as well:

a) If for a given set of well-defined invariants of a web W the analogous invariants for W' can not be defined – say because a denominator function is identically zero – then this indicates that there is no proper holomorphic mapping that preserves the web structures. In particular, the identical vanishing of some semi-invariants is an invariant characteristic of the polyhedral domain.

b) An example for $n = 2$ might be constructed with the function

$$f(z, w) = e^z + \frac{e^z}{2w - z}$$

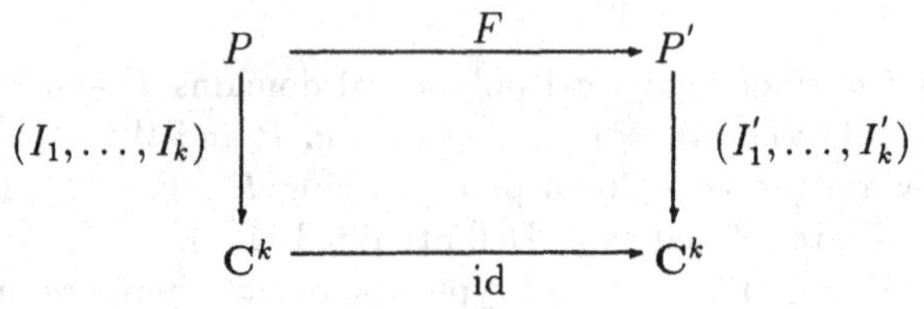

Figure 6.5

which implies the identity $K_3 = 0$ (see formula for K_3 in Example 6.3.14; cf. [BD 28], pp. 213-214 or [BB 38], pp. 160–161).

c) For the purpose of deciding whether there exist some proper holomorphic mappings between given polyhedral domains P and P', one should to check:

- whether the numbers of tiers of P and P' are equal;
- whether the web levels of P and P' (see Definition 6.3.18) coincide;
- whether the type distributions T_P and $T_{P'}$ of P and P' are identical;
- whether the numbers of webs with global parallelisation functions are compatible;
- whether image sets of parallelisation functions for P and P' are similar in the real plane (or combinations using parallelisation function as components) and a mapping can be developed via local inversions that can be "zoomed" to a surjective matching, i.e. adapted with appropriate $a \neq 0$; $b_\alpha, \alpha = 1, \ldots, n$;
- whether partition $(n+1)$-webs of fixed type $(k), k > 0$, of P and P' have coinciding image sets of invariant functions and local inversions exist and may locally allow to construct a web preserving surjective mapping (by patching together and continuity).

Consequently the following criterion can be formulated:

Theorem 6.3.37 *Consider analytical polyhedral domains A and B in $\mathbf{C}^n$, $n > 1$, with $r(>0)$ partition $(n+1)$-webs of type (k) for a fixed $k > 0$. Consider the image sets of the correspondingly defined invariants in A and B denoted by $A_1, \ldots, A_r$ and $B_1, \ldots, B_r$ respectively. If there is an A_i which does not coincide with any of the B_j, then there exists no proper holomorphic mapping $F : A \to B$ or $G : B \to A$.*

Remark 6.3.38 a) For comparison the following invariant sets might be used: $A_i = I_i(A)$ and $B_j = I'_j(B)$ for the correspondingly defined invariants and consequently for other subsystems belonging to a web as well: $A^1 = (I_1, \ldots, I_t)(A)$ and $B^1 = (I'_1, \ldots, I'_t)(B)$.

b) The criterion could be extended technically to require that there is an 1-to-1 assignment between the A_i and B_j via identity of the sets. The number of similar pairs of (A_i, B_j) has to match. For instance for $A_i = I_i(A), \ldots$ and $B_j = I'_j(B), \ldots$ (i.e., for image sets generated by using only one invariant function) in practice there must be r matching pairs $A_{j_1}, B_{j_1}; \ldots; A_{j_r}, B_{j_r}$ so that: $B_j = A_{j'}$ where $j' = \sigma(j)$ for all $j = 1, \ldots, r$ and for a certain permutation $\sigma \in S_r$.

Image sets generated by more than one invariant function require more such technical description.

c) The above criterion can be applied constructively to proper holomorphic mappings if there exist some partition $(n+1)$-webs of type (n) with n holomorphic invariant functions $I_1, \ldots, I_n$ in A and correspondingly defined $I'_1, \ldots, I'_n$ of the target domain B that allow a holomorphic inversion $(I'_1, \ldots, I'_n)^{-1}$ in $(I'_1, \ldots, I'_n)(B) = (I_1, \ldots, I_n)(A)$ (surjective!).

The holomorphic mapping produced in such a manner still has to be tested as to whether the partition webs are mapped onto each other – this translates into some necessary identities of the invariants projectively constructed with the $(I_j)_i$ (see [Bau 82], II.2).

It should be remarked that

- locally a construction of F may be started by finding coinciding sets of points in addition the conditions on $(I_j)_i$ are satisfied, generally resulting in a correspondence (a "meromorphic inversion"); and
- for partition $(n+1)$-webs of type (k) with $0 < k < n$ one should start with a possible constructive approach by looking for the appropriate conditions on I_j and $(I_j)_i$ and substituting for $n - k$ free variables by integration. (The practical idea is to look at the image range of the functions $I_j, \ldots$ so that $\mathbf{C} \supseteq I_j(P) \cap I'_{j'}(P') \neq \emptyset$.)

d) Above criterion can be applied negatively to conclude the non-existence of proper holomorphic mappings if

- there are invariant functions on at least one polyhedral domain; and
- there is at least one image set of an invariant function that does not coincide with any image sets of correspondingly defined invariant functions from the other polyhedral domain.

Remark 6.3.39 As we already remarked above, our definition of $(n+1)$-webs associates with each web $n+1$ partitions that allow the structure of $(n+1)$-webs to have $(n+1)!$ different formal descriptions depending on which of the $(n+1)!$ local differentiable coordinate n-quasigroup is chosen.

In dealing with (global) parallelisations this effects a permutation of the parallelisation functions and their image sets.

However, in dealing with invariant functions (in particular, if $n > 2$) the connection between invariants of "permuted" webs is more complicated and cumbersome. For example, consider the case of the permuted 3-webs (webs with parastrophic coordinate quasigroups according to the terminology of Chapter 3) in $\mathbf{C}$ as elaborated in the Table 6.1 (see [Bau 82], III.2.6), where

$$I_1 = K_1^2K^{-3}, \quad I_2 = K_2^2K^{-3}, \quad I = K_1K_2K^{-3}, \quad a \in \mathbf{C},$$
$$J_{11} = K_{11}K^{-2}, \quad J_{12} = K_{12}K^{-2}, \quad J_{21} = K_{21}K^{-2}, \quad J_{22} = K_{22}K^{-2}.$$

Remark 6.3.40 Another method to remedy the complications of permuted webs is to consider *symmetrised invariant functions.* This can be done with the help of the symmetric polynomials:

$$P_1(x_1, \dots, x_n) = x_1 + x_2 + \dots + x_n,$$
$$P_2(x_1, \dots, x_n) = x_1x_2 + x_2x_3 + \dots + x_nx_1,$$
$$\dots\dots\dots\dots\dots\dots\dots\dots\dots\dots\dots\dots$$
$$P_n(x_1, \dots, x_n) = x_1x_2 \dots x_n.$$

If invariant functions are substituted into these polynomials, then the resulting functions are invariant up to the sign which can be removed by multiplication with itself.(see [Bau 82], III.2.7).

Using the notations introduced in Example 6.3.14, we find that in the case $n = 2$ the lowest symmetrised invariants are

$$\begin{aligned} J_1 &= I^2(I_1 + I_2 + 2I), \\ J_2 &= (I_1 + I_2 + I)^2. \end{aligned}$$

We will give now some examples of analytical polyhedral domains in $\mathbf{C}^2$ with invariant functions. We will use the same notation as in Example 6.3.14 and in Table 6.1.

Example 6.3.41 The following examples are examples of families of non-equivalent analytical polyhedral domains of type (1) $\binom{8}{}$.

a) The functions

$$f_1(z, w) = z, \quad f_2(z, w) = w, \quad f_3(z, w) = z^r + z^s w$$

$\begin{pmatrix}123\\123\end{pmatrix}$	$\begin{pmatrix}123\\132\end{pmatrix}$	$\begin{pmatrix}123\\312\end{pmatrix}$	$\begin{pmatrix}123\\321\end{pmatrix}$	$\begin{pmatrix}123\\231\end{pmatrix}$	$\begin{pmatrix}123\\213\end{pmatrix}$
I_1	$-I_1$	$I_1 + I_2 + 2I$	$-I_1 - I_2 - 2I$	I_2	$-I_2$
I_2	$-I_1 - I_2 - 2I$	I_1	$-I_2$	$I_1 + I_2 + 2I$	$-I_1$
I	$I_1 - I$	$-I_1 - I$	$I_2 + I$	$-I_2 - I$	$-I$
$I_1 + aI$	$(a-1)I_1 + aI$	$(a-1)I_1 + I_2$ $+(2-a)I$	$(a-1)I_2 - I_1$ $+(a-2)I$	$(a-1)I_2 - aI$	$-I_2 - aI$
$I_2 + aI$	$(a-1)I_1 - I_2$ $+(a-2)I$	$(1-a)I_1 - aI$	$(a-1)I_2 + aI$	$I_1 + (1-a)I_2$ $+(2-a)I$	$-I_1 - aI$
$I_1 + aI_2$	$-(1+a)I_1$ $-aI_2 - 2aI$	$(1+a)I_1$ $+I_2 + 2I$	$-I_1 - (1+a)I_2$ $-2I$	$I_1 + (a+1)I_2$ $+2aI$	$-I_2 - 2aI_1$
J_{11}	$-J_{11}$	$J_{11} + J_{12}$ $+J_{21} + J_{22}$	$-J_{11} - J_{12}$ $-J_{21} - J_{22}$	J_{22}	$-J_{22}$
J_{12}	$J_{11} + J_{12}$	$-J_{11} - J_{21}$	$J_{12} - J_{22}$	$-J_{21} - J_{22}$	$-J_{21}$
J_{21}	$J_{11} + J_{21}$	$-J_{11} - J_{12}$	$J_{21} + J_{22}$	$-J_{12} - J_{22}$	$-J_{12}$
J_{22}	$-J_{11} - J_{12}$ $-J_{21} - J_{22}$	J_{11}	$-J_{22}$	$J_{11} + J_{12}$ $+J_{21} + J_{22}$	$-J_{11}$

Table 6.1: Invariants of "Permuted" Webs $W(3,2,1)$

define for all positive integers $r, s,\ r \neq s$, the following meromorphic 3-webs $W(3,2,1)$ of type (1) in $\mathbf{C}^2$:

$$W_{r,s} = (-(rz^{r-1} + sz^{s-1}w)dz, \quad -z^s dw, \quad d(z^r + z^s w)).$$

In particular, the first semi-invariant functions are:

$$
\begin{aligned}
K &= -r(s-r)\frac{z^{r-3}}{(rz^{r-1}+sz^{s-1}w)^3}, \\
K &= -rs^2(s-r)\frac{z^{r-4}}{(rz^{r-1}+sz^{s-1}w)^4}, \\
K &= -rs(s-r)^2\frac{(2rz^{2r-5}-sz^{r+s-5}w)^2}{(rz^{r-1}+sz^{s-1}w)^5},
\end{aligned}
$$

and invariants $I_1 = K_1^2K^{-3}$, $I_2 = K_2^2K^{-3}$, $I = K_1K_2K^{-3}$ are then given explicitly as

$$
\begin{aligned}
I_1 &= \frac{s^4}{r(r-s)}(r+sz^{s-r}w), \\
I_2 &= s^2(r-s)\frac{(2r-sz^{s-r}w)^2}{r(r+sz^{s-r}w)}, \\
I &= -2s^3+\frac{s^4}{r}z^{s-r}w.
\end{aligned}
$$

Then the invariants

$$
\begin{aligned}
J_0 &= I_1+\frac{s}{s-r}I = \frac{3s^4}{r-s} \quad \text{(i.e. constant)}, \\
J &= rs^4(I+2s^3) = z^{s-r}w
\end{aligned}
$$

are particularly interesting.

Since the $W_{r,s}$ define 3-webs of type (1), all other invariants are analytically dependent on J.

b) Consider a rational number $q = r/s$ with $0 < q < 1$, where r, s are relatively prime integers. Then the following well-defined simply connected domains

$$P_q = \{(z,w) \in \mathbf{C}^2 : |z| < 1,\ \ |w| < 1,\ \ |z^r + z^s w| < 1\}$$

define analytical polyhedral domains with partition 3-webs (i.e. number of tiers equal to 3), and the partition 3-web is of type (1). In particular, the $\dot{P}_q$ belong to the class of so called "H.Cartan domains" or rotating domains ($(1, r-s)$-domains) ([9]) because there exist automorphisms (for all $\sigma \in [0,2)$)

$$f_\sigma : (z,w) \to ze^{i\sigma},\ \ we^{i(r-s)\sigma}.$$

The conditions for Lemma 6.3.3 can be verified straightforwardly (see [Bau 82], IV.3). This confirms that P_q are in fact analytical polyhedral domains.

c) Consider a rational number $q = r/s$ with $0 < q < 1/2$ where r, s are again relatively prime integers. Then $(P_q : q \in \mathbf{Q} \cap (0, 1/2))$, where $\mathbf{Q}$ is the field of rational numbers, defines a family

$$P_q = \{(z, w) \in \mathbf{C}^2 : |z| < 1, \quad |w| < 1, \quad |z^r + z^s w| < 1\}$$

of analytical polyhedral H.Cartan domains of type (1) with a number of tiers 3 that are non-nequivalent via proper holomorphic mappings, i.e., they cannot be mapped proper holomorphically onto each other. To see this consider the invariants J and J_0 as they were defined above in a). When checking whether there exists for given $q = q'$ a holomorphic mapping, we see that J_0 yields constant values, a unique one in the form $I_1 + aI$, and J is bounded for fixed q.

Inspecting the possibility of permutations of the partitions, with the help of J, it can be concluded by looking at Table 6.1 that almost all permutations $\sigma \in S_3$ result in an unbounded J with exception of $\binom{123}{132}$ which by inspection of J_0 can be excluded (see [Bau 82], IV.3).

d) Consider the 3-tiered analytical polyhedral domains of type (1)

$$P_q^{a,b} = \{(z, w) \in \mathbf{C}^2 : |z| < a, \quad |w| < b, \quad |z^r + z^s w| < 1\}$$

where $(a, b) \in \{(x, y) \in \mathbf{R}^2 : |x^{r-s} - x^{-s}| < y < x^{r-s} + x^{-s}; \quad 0 < x, y\}$ for a fixed q (a rational number between 0 and 1/2). This is also a continuous family of non-inequivalent polyhedral domains with respect to proper holomorphic mappings (see [Bau 82], IV.3]).

The bounded invariant image domains for the invariant J are different.

e) Now we will consider some examples of rigid polyhedral domains of type (1). First consider for $0 < a < 1$ the simply connected polyhedral domains:

$$P_q^a = \{(z, w) \in \mathbf{C}^2, \quad |z - a| < 1, \quad |w| < b, \quad |z^r + z^s w| < 1\}.$$

For a fixed $0 < q < 1/2$ they define a family of rigid non-equivalent polyhedral domains with respect to proper holomorphic mappings which approximate for $a \to 0$ a non-rigid domain (cf. [Bau 82], IV.3). This can be investigated by considering the image sets of J which for various values of a differ already when intersected by $\mathbf{R}$, i.e., $J(P_q^a) \cap \mathbf{R} =]-1-a, 0]$.

f) Further examples for 3-webs of type (2) can result from the following functions:

$$f_3(z, w) = A(z^r w^t + z^s w^l)$$

yielding $(l - t, r - s)$- domains, or

$$f_3(z, w) = z + w^2 + 2zw + zw^2.$$

yielding another type (2) web $W(3, 2, 1)$.

Example 6.3.42 We will now give examples of polyhedral domains with partition 3-webs of type (2).

a) The functions

$$f_1(z,w)=z,\ \ f_2(z,w)=w,\ \ f_3(z,w)=z+w+az^2w$$

define for each non-zero complex number a the following meromorphic 3-webs of type (1) on $\mathbf{C}^2$:

$$W_a=(-(1+azw)dz,\ \ -(1+az^2)dw,\ \ d(z+w+az^2w)).$$

In particular, the first semi-invariant functions are (see [Bau 82], IV.3.6):

$$\begin{aligned} K &= \frac{-2a}{(1+az^2)(1+2azw)^3}, \\ K_1 &= \frac{4a^2z}{(1+az^2)^2(1+2azw)^4}, \\ K_2 &= \frac{4a^2(z-3w-az^2w)}{(1+az^2)^2(1+2azw)^5} \end{aligned}$$

and invariants $I_1=K_1^2K^{-3}, I_2=K_2^2K^{-3}, I=K_1K_2K^{-3}$ are then given explicitly as

$$\begin{aligned} I_1 &= \frac{-2az^2(1+2azw)}{(1+az^2)}, \\ I_2 &= \frac{-2a(z-3w-az^2w)}{(1+az^2)(1+2azw)}, \\ I &= \frac{-2az(z-3w-az^2w)}{(1+az^2)}. \end{aligned}$$

In particular, the following special invariants

$$\begin{aligned} J &= I_1-I=6azw, \\ J_1 &= I_1+2I=6az\frac{2w-z}{1+az^2} \end{aligned}$$

satisfy $dJ\wedge dJ_1\neq 0$, i.e., J and J_1 are two independent invariant functions and therefore W_a is of type (2) for any non-zero a.

b) For all $a, 1<a<2$, the simply connected domains P_a are well-defined and induce a partition 3-web of type (2) so that the system $(P_a : 0<a<2)$ defines a family of rigid analytical polyhedral domains that cannot be mapped onto each other with respect to proper holomorphic mappings. Again the conditions of Lemma

6.3.3 can be directly verified and the invariant J is bounded: Table 6.1 implies for the permuted 3-webs that the corresponding invariants will be unbounded (having meromorphic singularities at $(ia, 0)$). This implies that only the identical permutation is allowed and the image set $J(P)$ is definitely characteristic. It is therefore sufficient to consider for various a the intersection with positive real numbers: $J(P_a) \cap \mathbf{R} = [0, 6a)$ in order to yield inequivalence for proper holomorphic mappings (see [Bau 82], IV.3).

c) Similarly for $k > 0$ the functions

$$f_1(z,w) = z, \quad f_2(z,w) = w, \quad f_3(z,w) = z + w + az^k w$$

define for each non-zero complex number a the following meromorphic 3-web $W(3,2,1)$ of type (1) in $\mathbf{C}^2$:

$$W_a^k = (-(1 + akz^{k-1}w)dz, \quad -(1 + az^k)dw, \quad d(z + w + az^k w)).$$

In particular, the first semi-invariant functions are:

$$\begin{aligned}
K &= \frac{-ak(k-1)az^{k-2}}{(1 + akz^{k-1})(1 + az^k w)}, \\
K_1 &= \frac{a^2k^2(k-1)z^{2k-3}}{(1 + akz^{k-1}w)^4(1 + az^k w)^2}, \quad \text{etc} \\
I_1 &= -\frac{1 + akz^{k-1}w}{1 + az^k} akz^k, \\
I &= -\frac{k - 2 + 2a(k-1)z^k + a^2k(2k-1)z^{k-1}w + k(k-1)z^{2k-1}w}{(k-1)(1 + az^k)},
\end{aligned}$$

(see [Bau 82], IV.3). They have a bounded invariant

$$J = I_1 - I = \frac{k - 2 - k(2k-1)az^{k-1}w}{k-1} = \frac{k-2}{k-1} + k\frac{2k-1}{1-k}az^{k-1}w.$$

Therefore, constructions similar to the above can be applied.

More formulae for the other first invariants can be found in [Bau 82], IV 3.8 b) ([10]).

NOTES

6.1–6.2. All results here are due to Bolodurin [Bo 84, 85]. In our exposition we follow these two papers.

6.3. All results of this section are from the Baumann Doctoral Dissertation [Bau 82]. He selected the material essential for this section and kindly provided a draft of the exposition.

One characteristic phenomenon of Complex Analysis is the role of boundary behaviour for the existence of functions or mappings. A well-known fact is that a holomorphic function can be defined only with the help of its values on the boundary of its domain.

A variation on this theme would that the characteristics of the boundary of a domain have some decisive impact on the existence of functions of its interior or mappings onto other domains.

Of course, the topological structure of the boundary is of immense importance. One recent successful attempt used differential geometric invariants constructed on strictly pseudoconvex domains, i.e., domains with very smooth but non-analytic boundaries (see [CM 74]).

An approach that is closer to the web-theoretic considerations would be to deal with fibrations or foliations that are induced in a natural manner in the domains. For instance this could be achieved if there are some functions defined in a domain D that are invariantly related to the automorphic or isomorphic transformations within an appropriately defined category.

A web-theoretic morphism would naturally require that the characteristic fibrations or foliations are mapped onto characteristic ones.

A typical complex analytic idea would be that this needs only to be true for a small (open or "essential") part of the domain. (An "essential" domain is a sequence of pieces of holomorphic hypersurfaces on which the assertions hold). For instance, a classic result of complex analysis asserts that a biholomorphic mapping would extent to parts of the boundary that are "fibered" by analytic smooth curves [Ro 35]. Parts of those "small" analytic curves would be mapped on curves that "fiber" the boundary of the image. Therefore, we could consider a category of domains with heavily fibered boundaries (say, by families of analytic hypersurfaces) together with a suitable morphism that should preserve the "fibered" structure of an essential part of the boundary.

A further remark is as follows. Apart from just taking biholomorphic and automorphic mappings, the context of Complex Analysis provides a more interesting candidate: the proper holomorphic mappings, i.e., the generalised covering mappings (the problem of inversion of those leads to the multivalued, i.e., partly locally invertible mappings – the correspondences). (see [Ste 58]).

Let us outline the general idea. A web-theoretic category within Complex Analysis is exemplified by domains in $\mathbf{C}^n$, $n > 1$, whose boundaries contain "essential" parts that are defined by families of holomorphic hypersurfaces, the morphisms between these domains being the proper holomorphic mappings (boundary preserving) or even the surjective holomorphic correspondences. The necessary theorem relating this was proved for the biholomorphic case in [Ro 35], later generalised in [RS 60] and extended to correspondences in [Ri 64a, 64b]. This context is useful in clarifying the theory of holomorphic (nearly everywhere non-degenerate) mappings between analytic polyhedral domains or such domains that have some polyhedral boundary parts.

(1) Examples a) and b) are examples of polyhedrons with parallel (complex) affine foliations, i.e. foliations that can be described via $U \cap \{\sum_{\alpha=1}^{n} a_\alpha z_\alpha = c \in \mathbf{C}\}$ with appropriate $U \subseteq \mathbf{C}^n$, $a_\alpha \in \mathbf{C}$ (not degenerate, of course).

(2) Rothstein gave the first application of this theorem in [Ro 35] and a slight generalisation of his example is given in [RS 60], p. 186, Satz 15. The example in [Bau 82] (see

IV.1.4) is based on the Rothstein's classic example. The Example 6.3.32 a) is a special case of the latter example which can be formulated more generally.

([3]) The linear system for the unknown meromorphic functions a and b in the text can be obtained by writing the equation $a \cdot df_1 + b \cdot df_2 + df_3 = 0$ in the component form. The linear system is soluble since $df_\xi \wedge df_\eta \neq 0$ for $\xi \neq \eta$.

To obtain q, introduce the notation: $\omega_1 = a \cdot df_1,\ \ \omega_2 = b \cdot df_2,\ \ \omega_3 = df_3$. Then

$$\begin{aligned} d\omega_1 &= \omega_1 \wedge p_2\omega_2, \\ d\omega_2 &= \omega_2 \wedge p_1\omega_1, \end{aligned}$$

where $p_2 = -(\partial a/\partial f_2)/(ab)$ and $p_1 = -(\partial a/\partial f_1)/(ab)$. It follows from this that $\omega_\xi = \omega_\xi \wedge q$ where $q = p_1\omega_1 + p_2\omega_2$. Since $\omega_1 = df_3$, we have $d\omega_3 = 0$. This implies $p_1 = p_2 = p$. Hence $q = p(\omega_1 + \omega_2)$.

([4]) We will clarify here the notion of global parallelisability in this meromorphic context and compare this notion with the notion of local parallelisability.

This notion brings up some essential differences with the real case. First of all, the complex approach is slightly more general since particular meromorphic 1-forms do not need to be regular everywhere. Secondly, we have to deal with the transition from local considerations to global situations.

If the meromorhic 1-forms are used, then we must take into account that they can be degenerate over certain analytic varieties (thin sets defined by analytic equations and of codimension less than or equal to $n-1$, where the 1-forms have lower rank) but in Complex Analysis it is still meaningful to deal with them in these sets of singularities. However, the integrability may cease to be meaningful (although there may exist extensions of the Frobenius theorem – see [K 34], p. 57 or [Mal 73], p. 163 or [Bau 82], 0.0.6).

If the parallelisation exists, this gives an additional non-trivial condition that continues over singularity varieties of the 1-forms defining the web (see [Bau 82, p. 28).

Only in exceptional cases global parallelisations exist that continue over singularities of the 1-forms defining the web (see [Bau 82], II.1).

The notion of local parallelisation for meromorphic webs can be given in various stages:

i) Our Definition 6.3.16 of webs of type (0) may be seen as the weakest approach. In general, this allows us only to find local parallelisations outside of the singularity set where all 1-forms are regular and can be integrated.

It is often possible to continue local integrated parallelisations from the regular part into the singularity set, but this does not always work (e.g. it is well-known that the logarithm function cannot be continued into the 0).

ii) A meromorphic web $W(n+1, n, 1)$ in a domain D of $\mathbf{C}^n$, $n > 1$, of type (0) should be called *locally parallelisable* if and only if for *all* x in D there exists a neighbourhood $U(x)$ in D such that there is a parallelisation of $W(n+1,n,1)\Big|_{U(x)}$ in $U(x)$.

This definition is in accordance with the real case in Section 1.5.

iii) To achieve a global parallelisation, the local ones have to be patched together along paths. In general, there may be obstructions to prevent a well-defined global parallelisation and continuations may wind finitely or infinitely around singularit y sets (varieties). Generally, the continuations will lead to covering manifolds with finite or infinite leaves where all the local parallelisations tie together.

The existence of global parallelisations (as they are defined in Definition 6.3.16) of given webs on given domains is a rather strong assumption depending on the particular web partitions and the topological properties of the domains. For example, if the underlying domain is simply connected and does not contain 0, then the logarithm may be well-defined in the domain. Some relevant homological conditions are discussed in [Bau 82], II.1. For further references on this globalisation framework see [GF 76] or [GRe 84].

(5) The part b) of the proof may also be found in [RS 60], p. 189, proof of Satz 15, in a slightly less general form (for the complex affine partitions only). Rothstein gave similar rather implicit arguments in [Ro 36].

(6) An "analytic " variety in a manifold D (or even more general "complex spaces") is a well-defined notion in Complex Analysis meaning a set described locally with holomorphic functions $f_1, \ldots, f_k$ in $U \subset D$ as $\{z \in U \subset D \, : \, f_1(z) = \ldots = f_k(z) = 0\}$.

(7) As the tiers of the Euclidean simplex polyhedrons in $\mathbf{C}^n$, $n > 1$, range from $n+1$ to $2n+1$, there are automatically n different categories of the Euclidean simplex polyhedrons that are disjoint relative to the proper holomorphic mappings (see [Bau 82], IV. 2.4).

Let $1 \leq m \leq n$. Then the Euclidean polyhedrons with the number of tiers equal to $n+m$ define a partition $(n+m)$-web $W(n+m, n, 1)$ that is characteristic for these simplex polyhedrons with respect to the proper holomorphic mappings.

The characteristic partitions are naturally given by the complex affine functions. The $(n+m)$-web has $n+m$ affine parallelisation functions $f_1, \ldots, f_{n+m}$ so that we have the following equations:

$$\begin{cases} \omega_1 + \omega_2 + \ldots + \omega_n + \omega_{n+1} = 0, \\ \omega_1 + a_{12}\omega_2 + \ldots + a_{1n}\omega_n + \omega_{n+2} = 0, \\ \ldots\ldots\ldots\ldots\ldots\ldots\ldots\ldots\ldots\ldots\ldots \\ \omega_1 + a_{m2}\omega_2 + \ldots + a_{mn}\omega_n + \omega_{n+m} = 0 \end{cases}$$

with constant complex coefficients $a_{\alpha\beta}$, $\alpha = 1, \ldots, m$; $\beta = 1, \ldots, n$, and the same relations for the parallelisation functions f_ξ . This may be then used to determine the equivalence and the automorphy properties between the Euclidean polyhedrons in a similar manner as in Figures 6.2 and 6.3.

This technique can be applied to simplify the consideration of automorphisms of the simplex domains with the number of tiers greater or equal than $n+2$. For instance a simplex domain P in $\mathbf{C}^2$ of 4 tiers induces the web $W(4,2,1)$. This affine web has the three essential parastrophies (the principal affinor is respectively equal to $-1/2, 1/2$ for them - see [G 82c]). Therefore we may expect at most the three non-trivial automorphisms, and these are realised by the simplex domain generated by the following points:$(1,0), (e^{\frac{2\pi}{3}i}, 0), (e^{\frac{4\pi}{3}i}, 0), (0,1), (0,a)$;
$a \in \mathbf{C} - \mathbf{R}$. This approach elegantly simplifies the lengthy considerations in [Bau 82], IV. 2.7–2.8.

(8) The fact that all the webs of Example 6.3.41 are of type (1) follows from [Bau 82], II.2.3. Further explicit formulae for other webs of type (1) are given in [Bau 82], IV. 3.8.e 1)–5), p. 150. Other criteria for the type (1) webs can be found in [Bau 82], Corollary 2.9 and Example II. 2.10.

(9) An open connected set $G \subset \mathbf{C}^2$ is called a *H. Cartan domain* or *(p,q)-rotating domain* if there exist relatively prime integers m and p such that for all real d the following

holds: all the mappings $T(p,m,d) : (z,w) \to ((z-z_o)e^{imd} + z_o, (w-w_o)e^{ipd} + w_o)$ define an automorphism of the domain G . Such domains were investigated in classical Complex Analysis (see [BT 70]).

(^{10}) Further examples of webs of type (0) can be found in [Bau 82], II. 1.10 and of types (1) and (2) in [Bau 82], II. 2.10, IV. 3.8 b) and e).

The singular varieties that arise during the process of constructing the meromorphic webs connected with the partition foliations of polyhedral domains partially depend on the specific choice of the 1-forms defining the web. However, the characteristic singularity varieties of the polyhedral domain, that are essential, have to be defined:

- where the integrability for parallelisation fails, and
- where the invariant functions are singular.

The Complex Analysis methods can deal with these singularities in specific cases (for instance, resolve them via proper holomorphic mappings) and improve the situation (for instance, via blow ups, see [Wh 72]).

Chapter 7

The Theory of Four-Webs $W(4,2,r)$

7.1 Differential Geometry of Four-Webs $W(4,2,r)$

7.1.1 Basic Notions and Equations

In Chapters 1-6 we have constructed the theory of webs $W(n+1,n,r)$. The next step should be a construction of the theory of webs $W(d,n,r)$, $d > n+1$. We have already defined them (Section 1.1) and discussed some problems connected with them (Section 2.6). However, a general theory of such webs has not yet been constructed. In this chapter we will construct such the theory for webs $W(4,2,r)$ and in the next chapter we will do it for webs $W(d,2,r)$.

So, suppose now that a web $W(4,2,r)$ is given in an open domain D of a differentiable manifold X^{2r} by four foliations λ_ξ, $\xi = 1,2,3,4$, of codimension r in general position (see Definition 1.1.1 for $d = 4, n = 2$).

Example 7.1.1 Consider a web $W(4,3,r)$ in X^{3r} and take a surface S of codimension r in X^{3r} which is in general position with respect to the web $W(4,3,r)$. The leaves of all four foliations cut on the surface S a web $W(4,2,r)$.

Let the foliations λ_ξ are given by the completely integrable systems of Pfaffian equations

$$\underset{\xi}{\bar{\omega}}^i = 0, \quad i = 1, \ldots r; \quad \xi = 1,2,3,4. \tag{7.1.1}$$

Suppose that the 1-forms $\underset{1}{\bar{\omega}}^i$ and $\underset{2}{\bar{\omega}}^i$ are basis forms of the manifold X^{2r}. Then the 1-forms $\underset{3}{\bar{\omega}}^i$ and $\underset{4}{\bar{\omega}}^i$ are expressed linearly in terms of $\underset{1}{\bar{\omega}}^i$ and $\underset{2}{\bar{\omega}}^i$:

$$-\underset{3}{\bar{\omega}}^i = p^i_j \underset{1}{\bar{\omega}}^j + q^i_j \underset{2}{\bar{\omega}}^j, \quad -\underset{4}{\bar{\omega}}^i = \bar{r}^i_j \underset{1}{\bar{\omega}}^j + \bar{s}^i_j \underset{2}{\bar{\omega}}^j. \tag{7.1.2}$$

Since any $2r$ forms $\underset{\xi}{\bar{\omega}}^i$, $\underset{\eta}{\bar{\omega}}^i$ can be taken as basis forms of X^{2r} , equations (7.1.2). are solvable with respect to any $\underset{\xi}{\bar{\omega}}^i$, $\underset{\eta}{\bar{\omega}}^i$, ξ, η fixed. Because of this, the matrices

(p^i_j), (q^i_j), $(\bar{r}^i_j)$, and $(\bar{s}^i_j)$ in (7.1.2) are non-singular:

$$\det(p^i_j) \neq 0, \quad \det(q^i_j) \neq 0, \quad \det(\bar{r}^i_j) \neq 0, \quad \det(\bar{s}^i_j) \neq 0. \tag{7.1.3}$$

We will change the basis forms $\underset{1}{\bar{\omega}}{}^i$ and $\underset{2}{\bar{\omega}}{}^i$ for $\underset{1}{\omega}{}^i$ and $\underset{2}{\omega}{}^i$ putting

$$\underset{1}{\omega}{}^i = p^i_j \underset{1}{\bar{\omega}}{}^j, \quad \underset{2}{\omega}{}^i = q^i_j \underset{2}{\bar{\omega}}{}^j. \tag{7.1.4}$$

By (7.1.4), equations (7.1.2) become

$$-\underset{3}{\bar{\omega}}{}^i = \underset{1}{\omega}{}^i + \underset{2}{\omega}{}^i, \quad -\underset{4}{\bar{\omega}}{}^i = r^i_j \underset{1}{\omega}{}^j + s^i_j \underset{2}{\omega}{}^j, \tag{7.1.5}$$

where $r^i_j = \bar{r}^i_k \tilde{p}^k_j$, $s^i_j = \bar{s}^i_k \tilde{q}^k_j$ and $(\tilde{p}^k_j)$, $(\tilde{q}^k_j)$ are the matrices inverse to the matrices (p^k_j), (q^k_j) respectively.

In addition to (7.1.3), suppose that

$$\underset{3}{\bar{\omega}}{}^i = \underset{3}{\omega}{}^i, \quad \underset{4}{\omega}{}^i = \tilde{s}^i_j \underset{4}{\bar{\omega}}{}^j, \quad \tilde{s}^i_k r^k_j = \lambda^i_j, \tag{7.1.6}$$

where $(\tilde{s}^i_j)$ is the inverse matrix of (s^i_j). By (7.1.6), equations (7.1.5) will have the form

$$-\underset{3}{\omega}{}^i = \underset{1}{\omega}{}^i + \underset{2}{\omega}{}^i, \quad -\underset{4}{\omega}{}^i = \lambda^i_j \underset{1}{\omega}{}^j + \underset{2}{\omega}{}^i, \tag{7.1.7}$$

Since the system (7.1.7) must be solvable with respect to $\underset{1}{\omega}{}^i$, $\underset{3}{\omega}{}^i$ and $\underset{1}{\omega}{}^i$, $\underset{2}{\omega}{}^i$, then

$$\det(\lambda^i_j) \neq 0, \quad \det(\delta^i_j - \lambda^i_j) \neq 0. \tag{7.1.8}$$

Under conditions (7.1.8), the system (7.1.7) is automatically solvable with respect to $\underset{1}{\omega}{}^i$, $\underset{4}{\omega}{}^i$; $\underset{2}{\omega}{}^i$, $\underset{3}{\omega}{}^i$; and $\underset{2}{\omega}{}^i$, $\underset{4}{\omega}{}^i$.

To find out a geometrical meaning of the specialisation we have performed, let us consider the tangent space $T_p(X^{2r})$ to the manifold X^{2r} at the point $p \in D$. Suppose that vectors e^i_1 and e^i_2 form its basis. If dx is a tangent vector to X^{2r} at p, we have

$$dx = \underset{1}{\omega}{}^i e^1_i + \underset{2}{\omega}{}^i e^2_i. \tag{7.1.9}$$

By (7.1.7), equality (7.1.9) can be also written as follows:

$$dx = -\underset{3}{\omega}{}^i e^1_i + \underset{2}{\omega}{}^i (e^2_i - e^1_i), \quad dx = -\underset{4}{\omega}{}^i e^2_i + \underset{1}{\omega}{}^i (e^1_i - \lambda^j_i e^2_j). \tag{7.1.10}$$

It follows from (7.1.9) and (7.1.10) that the vectors e^2_i, e^1_i, $e^3_i = e^2_i - e^1_i$, $e^4_i = e^1_i - \lambda^j_i e^2_j$ are tangent respectively to the leaves F_1, F_2, F_3, F_4 passing through the point p and defined by the systems

$$\underset{1}{\omega}{}^i = 0, \quad \underset{2}{\omega}{}^i = 0, \quad \underset{1}{\omega}{}^i + \underset{2}{\omega}{}^i = 0, \quad \lambda^i_j \underset{1}{\omega}{}^j + \underset{2}{\omega}{}^i = 0. \tag{7.1.11}$$

which are equivalent to the systems in (7.1.1).

It is obvious that concordant transformations of the 1-forms $\underset{\xi}{\omega}^i$ of the form

$$'\underset{\xi}{\omega}^i = A^i_j \underset{\xi}{\omega}^j, \tag{7.1.12}$$

and only such transformations, preserve equations (7.1.7). Under transformations (7.1.12) the matrix $\Lambda = (\lambda^i_j)$ is transformed according to the formula

$$'\Lambda = A^{-1}\Lambda A \tag{7.1.13}$$

where $A = (A^i_j)$. The formula (7.1.13) means that the quantities form (1,1)-tensor.

Definition 7.1.2 The tensor with components λ^i_j is called the *basis affinor* of a web $W(4,2,r)$.

7.1.2 The Geometrical Meaning of the Basis Affinor

There are two ways to establish a one-to-one correspondence between the leaves F_{ξ_1} and F_{ξ_2} of different foliations λ_{ξ_1} and λ_{ξ_2} in a neighbourhood of a point $p \in D$. These two ways differ only by the fact the leaves of what foliation, λ_{ξ_3} or λ_{ξ_4}, give corresponding points on F_{ξ_1} and F_{ξ_2} while cutting them. Analytically each of the correspondences mentioned above is defined by the equations of the foliation by means of which it is established.

Let Γ_{ξ_1} be a line on F_{ξ_1} and $\Gamma^{\xi_3}_{\xi_1\xi_2}$ be the corresponding line on F_{ξ_2} in the correspondence established by the foliation λ_{ξ_3}. We map the latter line back on F_{ξ_1} by means of λ_{ξ_4}. Then on F_{ξ_1} we obtain a line $\Gamma^{\xi_3\xi_4}_{\xi_1\xi_2\xi_1}$. Suppose that

$$\underset{\xi_1}{\omega}^i = 0, \;\; \underset{\xi_2}{\omega}^i = \xi^i dt; \;\; \underset{\xi_1}{\omega}^i = 0, \;\; \underset{\xi_2}{\omega}^i = \eta^i dt \tag{7.1.14}$$

are the equations of the lines Γ_{ξ_1} and $\Gamma^{\xi_3\xi_4}_{\xi_1\xi_2\xi_1}$. In (7.1.14) ξ^i and η^i are the coordinates of the vectors $\xi = \xi^i e^{\xi_1}_i$ and $\eta = \eta^i e^{\xi_1}_i$ tangent to the lines at the point p. Let us find an expression of the vector η in terms of the vector ξ. For the leaves F_1 and F_2 we have

$$\begin{aligned}
\Gamma_1 &: \underset{1}{\omega}^i = 0, \;\; \underset{2}{\omega}^i = \xi^i dt; & \Gamma^3_{12} &: \underset{2}{\omega}^i = 0, \;\; \underset{1}{\omega}^i = -\xi^i dt;\\
\Gamma^{34}_{121} &: \underset{1}{\omega}^i = 0, \;\; \underset{2}{\omega}^i = \lambda^i_j \xi^j dt; & \Gamma^4_{12} &: \underset{2}{\omega}^i = 0, \;\; \underset{1}{\omega} = -\tilde{\lambda}^i_j \xi^j dt;\\
& & \Gamma^{43}_{121} &: \underset{1}{\omega}^i = 0, \;\; \underset{2}{\omega}^i = \tilde{\lambda}^i_j \xi^j dt,
\end{aligned}$$

where $(\tilde{\lambda}^i_j)$ is the inverse matrix of (λ^i_j).

Similarly for the remaining five pairs of the corresponding lines we find:

$$
\begin{array}{ll}
\Gamma^{24}_{131} : \underset{1}{\omega}^i = 0, \ \underset{3}{\omega}^i = (\delta^i_j - \lambda^i_j)\xi^j dt \ ; & \Gamma^{42}_{131} : \underset{1}{\omega}^i = 0, \ \underset{3}{\omega}^i = \tilde{\tilde{\lambda}}{}^i_j \xi^i dt \ ; \\
\Gamma^{23}_{141} : \underset{1}{\omega}^i = 0, \ \underset{4}{\omega}^i = (\delta^i_j - \tilde{\lambda}^i_j)\xi^j dt \ ; & \Gamma^{32}_{141} : \underset{1}{\omega}^i = 0, \ \underset{4}{\omega} = -\lambda^i_j \tilde{\tilde{\lambda}}{}^j_k \xi^k dt \ ; \\
\Gamma^{21}_{343} : \underset{3}{\omega}^i = 0, \ \underset{4}{\omega}^i = \lambda^i_j \xi^j dt \ ; & \Gamma^{21}_{343} : \underset{3}{\omega}^i = 0, \ \underset{4}{\omega}^i = \tilde{\lambda}^i_j \xi^j dt \ ; \\
\Gamma^{13}_{242} : \underset{2}{\omega}^i = 0, \ \underset{4}{\omega}^i = (\delta^i_j - \lambda^i_j)\xi^j dt; & \Gamma^{31}_{242} : \underset{2}{\omega}^i = 0, \ \underset{4}{\omega}^i = \tilde{\tilde{\lambda}}{}^i_j \xi^j dt \ , \\
\Gamma^{14}_{232} : \underset{2}{\omega}^i = 0, \ \underset{3}{\omega}^i = (\delta^i_j - \tilde{\lambda}^i_j)\xi^j dt \ ; & \Gamma^{41}_{232} : \underset{2}{\omega}^i = 0, \ \underset{3}{\omega} = -\tilde{\tilde{\lambda}}{}^i_j \lambda^j_k \xi^k dt \ ,
\end{array}
$$

where $\tilde{\tilde{\lambda}}{}^i_j$ are components of the matrix $(E - \Lambda)^{-1}$ inverse to the matrix $E - \Lambda$.

Thus, if we denote by $[\xi, \eta]$ the composition of transformations by means of the foliations λ_ξ, λ_η, we obtain the following expression for the vector η in terms of the vector ξ :

$$
\left\{
\begin{array}{llll}
[3,4], \ [1,2]: & \eta = \Lambda\xi, & [4,3], \ [2,1]: & \tilde{\eta} = \Lambda^{-1}\xi, \\
[2,4], \ [1,3]: & \eta = (E - \Lambda)\xi, & [4,2], \ [3,1]: & \tilde{\eta} = (E - \Lambda)^{-1}\xi, \\
[2,3], \ [1,4]: & \eta = (E - \Lambda^{-1})\xi, & [3,2], \ [4,1]: & \tilde{\eta} = (E - \Lambda^{-1})^{-1}\xi.
\end{array}
\right. \tag{7.1.15}
$$

Hence the basis affinor Λ gives the law of transformation of a vector ξ tangent to a leaf $F_1(F_3)$ under transformation [3,4] ([1,2]). The transformations of vectors tangent to other leaves under other transformations $[\xi, \eta]$ are also constructed by using the affinor Λ .

7.1.3 Transversal Bivectors Associated with a 4-Web

A 4-web $W(4,2,r)$ contains four 3-subwebs $[\xi_1, \xi_2, \xi_3]$ formed by the foliations λ_{ξ_1}, λ_{ξ_2}, λ_{ξ_3}. The three vectors $h^a = h^i \underset{i}{e}^a$, $a = \xi_1, \xi_2, \xi_3$ are tangent to the leaves F_{ξ_1}, F_{ξ_2}, F_{ξ_3} passing through a point $p \in D$ and belonging to the same 2-plane. So, at each point there are associated with a 4-web $W(4,2,r)$ four bivectors $h^\xi \wedge h^\eta$. Each of them is a transversal bivector of a 3-subweb of $W(4,2,r)$ (cf. Definition 1.2.1).

It is easy to see from the expressions of the vectors $\underset{i}{e}^\xi$ and $\underset{i}{e}^\eta$ that these four transversal bivectors $h^\xi \wedge h^\eta$ coincide if and only if the vector h^i is an eigenvector of the matrix $\Lambda = (\lambda^i_j)$.

Hence at each point of the $W(4,2,r)$, in general, there exist not more than r bivectors common for all four 3-subwebs.

If there are more than r such common bivectors, then they exist for any direction h^i , i.e., an arbitrary nonzero vector h^i will be an eigenvector for the matrix Λ and consequently the matrix Λ is a scalar matrix: $\Lambda = \lambda E$. In this case the 3-subweb [1,2,3] is isoclinic (see Definition 1.11.5), i.e., there exists an one-parameter family of foliations λ_4 such that each of them together with the 3-subweb [1,2,3] form a 4-web $W(4,2,r)$.

The following theorem combines the results obtained in subsections 7.2 and 7.3.

Theorem 7.1.3 *The basis affinor of a web $W(4,2,r)$ gives the matrix of the transformation of a tangent vector ξ to a leaf $F_1(F_3)$ under the transformation* $[3,4]$ $([1,2])$. *At each point of a 4-web there exist not more than r bivectors common for all four 3-subwebs of the 4-web. If there are more than r such common bivectors, then such bivectors exist at any direction and each 3-subweb $[\xi_1,\xi_2,\xi_3]$ of the $W(4,2,r)$ is isoclinic. In this case there are one-parameter family of foliations λ_{ξ_4} each of which together with the 3-web $[\xi_1,\xi_2,\xi_3]$ form a 4-web $W(4,2,r)$.* ■

7.1.4 Permutability of Transformations $[\xi,\eta]$

Theorem 7.1.4 *The transformations* $[1,2]$(*or*$[3,4]$), $[1,3]$(*or*$[2,4]$), *and* $[1,4]$ (*or*$[2,3]$) *are permutational if and only if one of the following two conditions is satisfied:*

(i) *The matrix Λ of the basis affinor is scalar: $\Lambda = \lambda E$ where λ is equal correspondingly to* $-1, 2$, *and* $1/2$.

(ii) *The matrices $\Lambda, E-\Lambda$, and $E-\Lambda^{-1}$ are respectively involutory matrices.*

Proof. The relations (7.1.15) show that the transformation $[1,2]([3,4])$ is permutational if and only if

$$\Lambda^2 = E, \tag{7.1.16}$$

and the transformations $[1,3]$(or $[2,4]$), $[1,4]$(or $[2,3]$) are permutational if and only if we have respectively: $(E-\Lambda)^2 = E$ and $(E-\Lambda^{-2})^2 = E$. The last two relations and (7.1.16) mean that the matrices $\Lambda, E-\Lambda$, and $E-\Lambda^{-2}$ are involutory. The last two relations are equivalent respectively to

$$\Lambda = 2E, \tag{7.1.17}$$

$$\Lambda = E/2. \tag{7.1.18}$$

Note further that equation (7.1.16) implies that the eigenvalues of the matrix Λ are equal to ± 1. However, none of them can be equal to 1 since otherwise the second condition (7.1.8) will be violated. Hence all the eigenvalues are equal to -1, and equation (7.1.16) is equivalent to the equation

$$\Lambda = -E. \quad \blacksquare \tag{7.1.19}$$

Corollary 7.1.5 *If the 3-subweb $[1,2,3]$ is isoclinic, then only for three 4-webs in an one-parameter family of 4-webs generated by the $[1,2,3]$ two pairs out of six pairs of the transformations $[\xi,\eta]$ are permutational.*

Proof. This follows from the comparison of statements in Theorems 7.1.3 and 7.1.4. ■

The next theorem establishes a geometric meaning of the permutability of transformations $[\xi,\eta]$.

Theorem 7.1.6 *In an arbitrary hyperplane* P^{2r-1} *of the tangent space* $T_p(X^{2r})$ *at a point* $p \in D$ *the tangent* r*-planes* $T_p(F_\xi)$ *to the leaves* F_ξ *passing through* p *cut* $(r-1)$*-planes* P_ξ^{r-1}. *If one of the transformations* $[\eta, \zeta]$ *is permutational, then through any point of each of the planes* P_ξ^{r-1} *there passes a straight line intersecting the remaining three planes* P_ξ^{r-1} . *On each of such straight lines the planes* P_ξ^{r-1} *cut the points* A_ξ . *In the case of permutabilities indicated in Theorem* 7.1.4 *the following relations hold respectively:*

$$(A_1, A_2; A_3 A_4) = -1, \quad (A_1, A_3; A_2, A_4) = -1, \quad (A_1, A_4; A_2, A_3) = -1. \tag{7.1.20}$$

Proof. If $r = 1$, then the cross-ratio of the four curves $F_1,\ F_2,\ F_3,\ F_4$ of the 4-web passing through a point p is equal to λ (the only component of the basis affinor in the case $r = 1$). Since under condition (7.1.16), we have $\lambda^2 = 1$, then, by (7.1.8), only the case $\lambda = -1$ is possible, i.e., in this case the curves $F_1,\ F_2,\ F_3,\ F_4$ form a harmonic sheaf of curves. It is easy to check that in the case $r = 1$ the conditions (7.1.17) and (7.1.18) mean respectively that the curves $F_2,\ F_4$ are harmonically separated by the curves $F_1,\ F_3$ and the curves $F_2,\ F_3$ are harmonically separated by the curves $F_1,\ F_4$ since the cross-ratios of the curves indicated above are equal respectively to $1 - \lambda^2$ and $1 - \lambda^{-2}$.

In order to get a geometric characteristic in the case $r > 1$, we consider the tangent space $T_p(X^{2r})$ to X^{2r} at the point p. Let $T_\xi = T_p(F_\xi)$ be the tangent r-planes to the leaves F_ξ at p. As we saw early, the r-planes $T_1,\ T_2,\ T_3,\ T_4$ are determined respectively by the vectors $e_i^2,\ e_i^1,\ e_i^3 = e_i^2 - e_i^1,\ e_i^4 = e_i^1 - \lambda_j^i e_j^2$. Let P^{2r-1} be a hyperplane of $T_p(X^{2r})$ not passing through the point p. The planes T_ξ intersect P^{2r-1} in $(r-1)$-planes P_ξ^{r-1}. In the cases (7.1.19), (7.1.17), and (7.1.18) through any point of P_4^{r-1} there passes the only straight line intersecting P_1^{r-1}, P_2^{r-1}, and P_3^{r-1}. On each of such straight lines the planes $P_1^{r-1},\ P_2^{r-1},\ P_3^{r-1},\ P_4^{r-1}$ define harmonic points $A_1,\ A_2,\ A_3 = A_2 - A_1,\ A_4 = A_2 - \lambda A_1$ where λ is equal respectively to $-1, 2, 1/2$. In the case (7.1.19) the points $A_3,\ A_4$ are harmonically separated by the points $A_1,\ A_2$,i.e., $(A_1, A_2; A_3, A_4) = -1$; in the cases (7.1.17) and (7.1.18) we have respectively $(A_1, A_3; A_2, A_4) = -1$ and $(A_1, A_4; A_2, A_3) = -1$. ■

7.1.5 Fundamental Equations of a Web $W(4,2,r)$

Since for the web $[1,2,3]$ we have made the same specialisation as in Section 1.2, the equations (1.2.46), (1.2.47), (1.2.51), and (1.2.57) hold. Let us write these equations suppressing a bar over the letters b and ω :

$$d\underset{1}{\omega}^i = \underset{1}{\omega}^j \wedge \omega_j^i + a_{jk}^i \underset{1}{\omega}^j \wedge \underset{1}{\omega}^k, \quad d\underset{2}{\omega}^i = \underset{2}{\omega}^j \wedge \omega_j^i - a_{jk}^i \underset{2}{\omega}^j \wedge \underset{2}{\omega}^k, \tag{7.1.21}$$

$$d\omega_j^i - \omega_j^k \wedge \omega_k^i = b_{jkl}^i \underset{1}{\omega}^k \wedge \underset{2}{\omega}^l, \tag{7.1.22}$$

$$\nabla a_{jk}^i = b_{[j|l|k]}^i \underset{1}{\omega}^l + b_{[jk]l}^i \underset{2}{\omega}^l, \tag{7.1.23}$$

$$b^i_{|jkl|} = 2a^m_{[jk}a^i_{|m|l]}. \tag{7.1.24}$$

In these equations $\nabla a^i_{jk} = da^i_{jk} - a^i_{ik}\omega^i_j - a^i_{jl}\omega^l_k + a^l_{jk}\omega^i_l$, and $a^i_{jk} = -a^i_{kj}$. In addition to these equations, we will need the following equations which are obtained by taking exterior derivatives of (7.1.22) and (7.1.23):

$$\begin{cases} (\nabla b^i_{jkl} - b^l_{jpl}a^p_{km}\underset{1}{\omega}^m + b^i_{jkp}a^p_{lm}\underset{2}{\omega}^m)\underset{1}{\omega}^k \wedge \underset{2}{\omega}^l = 0, \\ (\nabla b^i_{[j|l|k]} - b^i_{[j|p|k]}a^p_{lm}\underset{1}{\omega}^m) \wedge \underset{1}{\omega}^l + (\nabla b^i_{[jk]l} + b^i_{[jk]p}a^p_{lm}\underset{2}{\omega}^m) \wedge \underset{2}{\omega}^l \\ = (a^i_{pj}b^p_{klm} - a^i_{pk}b^p_{jlm} + a^p_{jk}b^i_{plm})\underset{1}{\omega}^l \wedge \underset{2}{\omega}^m; \end{cases} \tag{7.1.25}$$

here $\nabla b^i_{jkl} = db^i_{jkl} - b^i_{mkl}\omega^m_j - b^i_{jml}\omega^m_k - b^i_{jkm}\omega^m_l + b^m_{jkl}\omega^i_m$.

We recall that the quantities a^i_{jk} and b^i_{jkl} are the torsion and curvature tensors of the 3-subweb [1, 2, 3].

The equations (7.1.7) and (7.1.21)–(7.1.25) give some of the fundamental equations of a web $W(4,2,r)$. To get the remaining equations, we must find the conditions of complete integrability of the system $\underset{4}{\omega}^i = 0$ defining the foliation λ_4. From (7.1.7) we have

$$d\underset{4}{\omega}^i = \underset{4}{\omega}^j \wedge (\omega^i_j + a^i_{jk}\underset{4}{\omega}^k + 2a^i_{jk}\lambda^k_m\underset{1}{\omega}^m) - \nabla\lambda^i_j \wedge \underset{1}{\omega}^j - (\lambda^i_m a^m_{kl} - a^i_{pq}\lambda^p_k\lambda^q_l)\underset{1}{\omega}^k \wedge \underset{1}{\omega}^l. \tag{7.1.26}$$

It follows from (7.1.26) and $\underset{4}{\omega}^i = 0$ that $\nabla\lambda^i_j = d\lambda^i_j - \lambda^i_k\omega^k_j + \lambda^k_j\omega^i_k$ are expressed in terms of the basis forms $\underset{1}{\omega}^i$ and $\underset{2}{\omega}^i$. Suppose that

$$\nabla\lambda^i_j = \lambda^i_{jk}\underset{1}{\omega}^k + \mu^i_{jk}\underset{2}{\omega}^k. \tag{7.1.27}$$

Substitute (7.1.27) and (7.1.7) into (7.1.26). Then, by linear independence of the forms $\underset{1}{\omega}^j \wedge \underset{1}{\omega}^k$, we obtain the condition of complete integrability of the system $\underset{4}{\omega}^i = 0$ in the form:

$$\lambda^i_{[jk]} - \mu^i_{[j|p|}\lambda^p_{k]} = \lambda^i_p a^p_{jk} + \lambda^p_{[k}\lambda^q_{j]}a^i_{pq}. \tag{7.1.28}$$

Under conditions (7.1.28), equations (7.1.26) take the form

$$d\underset{4}{\omega}^i = \underset{4}{\omega}^j \wedge (\omega^i_j + a^i_{jk}\underset{4}{\omega}^k + 2a^i_{jk}\lambda^k_m\underset{1}{\omega}^m + \mu^i_{kj}\underset{1}{\omega}^k).$$

Equations (7.1.27) show one more time that the quantities are components of a (1, 1)-tensor.

Proposition 7.1.7 *The conditions under which the tensor λ^i_j is covariantly constant in the canonical affine connection γ_{123} (see Section 1.3) on the foliations λ_1, λ_2, λ_3, λ_4 are respectively:*

$$\mu^i_{jk} = 0, \quad \lambda^i_{jk} = 0, \quad \lambda^i_{jk} = \mu^i_{jk}, \quad \lambda^i_{jk} = \mu^i_{jl}\lambda^l_k. \tag{7.1.29}$$

Proof. To prove this for the foliations λ_1, λ_2, λ_3, λ_4 we should substitute $\underset{1}{\omega}^i = 0$, $\underset{2}{\omega}^i = 0$, $\underset{2}{\omega}^i = -\underset{1}{\omega}^i$, $\underset{2}{\omega} = -\lambda^i_j\underset{1}{\omega}^j$ respectively into (7.1.27). ■

Proposition 7.1.8 *If the basis affinor λ_j^i is covariantly constant in the canonical affine connection γ_{123} on two of the foliations of a 4-web, it is covariantly constant in γ_{123} on the whole manifold X^{2r} .*

Proof. The sufficiency is obvious. The necessity of the assertion is trivial and follows directly from (7.1.27) for the pairs λ_1, λ_2; λ_1, λ_3; λ_1, λ_4; λ_2, λ_3; λ_2, λ_4. If λ_j^i is covariantly constant in γ_{123} on λ_3 and λ_4 , we obtain from (7.1.27) that $\lambda_{jk}^i = \mu_{jk}^i$, $\lambda_{jl}^i(\delta_k^l - \lambda_k^l) = 0$ and, by (7.1.8), we get $\lambda_{jk}^i = 0$ and consequently $\mu_{jk}^i = 0$. ∎

Taking exterior derivatives of (7.1.27), we obtain

$$\nabla\lambda_{jk}^i \wedge \underset{1}{\omega}^k \; + \; \nabla\mu_{jk}^i \wedge \underset{2}{\omega}^k + \lambda_{jk}^i a_{lm}^k \underset{1}{\omega}^l \wedge \underset{1}{\omega}^m - \mu_{jk}^i a_{lm}^k \underset{2}{\omega}^l \wedge \underset{2}{\omega}^m$$
$$+ (\lambda_k^i b_{jlm}^k - \lambda_j^k b_{klm}^i)\underset{1}{\omega}^l \wedge \underset{2}{\omega}^m = 0, \tag{7.1.30}$$

where $\nabla\lambda_{jk}^i = d\lambda_{jk}^i - \lambda_{lk}^i\omega_j^l - \lambda_{jl}^i\omega_k^l + \lambda_{jk}^l\omega_l^i$, $\nabla\mu_{jk}^i = d\mu_{jk}^i - \mu_{lk}^i\omega_j^l - \mu_{jl}^i\omega_k^l + \mu_{jk}^l\omega_l^i$. From (7.1.30) we find that

$$\nabla\lambda_{jk}^i = \lambda_{jkl}^i\underset{1}{\omega}^l + \beta_{jkl}^i\underset{2}{\omega}^l, \quad \nabla\mu_{jk}^i = a_{jkl}^i\underset{1}{\omega}^l + \mu_{jkl}^i\underset{2}{\omega}^l, \tag{7.1.31}$$

$$\alpha_{jml}^i - \beta_{jlm}^i + \lambda_k^i b_{jlm}^k = 0, \quad \lambda_{j[ml]}^i = \lambda_{jk}^i a_{ml}^k, \quad \mu_{j[ml]}^i = \mu_{jk}^i a_{lm}^k. \tag{7.1.32}$$

Equations (7.1.7), (7.1.21)–(7.1.25), (7.1.27), (7.1.30)–(7.1.32) are the *fundamental equations* of a web $W(4,2,r)$.

Now for a web $W(4,2,r)$ we are able to prove the fundamental theorem similar to Theorem 1.2.6 for a web $W(n+1,n,r)$.

Theorem 7.1.9 *Let there be given in an open domain D of an analytic manifold X^{2r} the 1-forms $\underset{1}{\omega}^i$, $\underset{2}{\omega}^i$ and tensors $\{\lambda_j^i\}$, $\{a_{jk}^i\}$ $(a_{jk}^i = -a_{kj}^i)$, $\{b_{jkl}^i\}$ satisfying relations (7.1.24), (7.1.28), Pfaffian equations (7.1.23), (7.1.27), the structure equations (7.1.21), (7.1.22) and the integrability conditions (7.1.25), (7.1.30). Then in D the systems of Pfaffian equations*

$$\underset{1}{\omega}^i = 0, \; \underset{2}{\omega}^i = 0, \; \underset{1}{\omega}^i + \underset{2}{\omega}^i = 0, \; \lambda_j^i\underset{1}{\omega}^i + \underset{2}{\omega}^i = 0$$

define up to an analytic transformation a unique web $W(4,2,r)$ for which the tensors indicated above are respectively the basis affinor and the torsion and curvature tensor of the 3-subweb $[1,2,3]$.

Proof. This follows from E. Cartan's theorem on equivalence of two systems of 1-forms (see [Ca 08]). ∎

7.1.6 The Affine Connections Associated with a Web $W(4,2,r)$

In Subsection 1.3.4 we studied the canonical affine connection induced by a three-web $W(3,2,r)$. For the 3-subweb $[1,2,3]$ we can deduce directly from the structure equations (7.1.21) and (7.1.22) that the connection forms of the canonical affine connection λ_{123} are

$$\omega_J^I = \begin{pmatrix} \omega_j^i & 0 \\ 0 & \omega_j^i \end{pmatrix}, \quad I,J,K = 1,\dots 2r, \tag{7.1.33}$$

and its torsion and curvature tensors have correspondingly the form:

$$R_{JK}^I = \left\{ \begin{pmatrix} a_{jk}^i & 0 \\ 0 & 0 \end{pmatrix}, \begin{pmatrix} 0 & 0 \\ 0 & -a_{jk} \end{pmatrix} \right\}, \tag{7.1.34}$$

$$R_{JKL}^I = \begin{pmatrix} A & 0 \\ 0 & A \end{pmatrix}, \quad \text{where } A = \begin{pmatrix} 0 & \frac{1}{2} b_{jkl}^i \\ -\frac{1}{2} b_{jkl}^i & 0 \end{pmatrix}. \tag{7.1.35}$$

Using the same method which we used in Theorem 1.3.2 for the connection γ_ξ, it is easy to show that the connections induced by the canonical connection γ_{123} on the leaves of λ_1 and λ_2 are connections with absolute parallelism (curvature-free connections), and that the components of the torsion tensors of the connections induced on F_1, F_2, F_3 are the same and coincide with the components of the torsion tensor of the 3-subweb $[1,2,3]$.

On a leaf $F_4 \subset \lambda_4$ determined by the system $\lambda_j^i \underset{1}{\omega^j} + \underset{2}{\omega^i} = 0$, we have

$$d\underset{1}{\omega^i} = \underset{1}{\omega^j} \wedge \omega_j^i + a_{jk}^i \underset{1}{\omega^j} \wedge \underset{1}{\omega^k}, \quad d\omega_j^i - \omega_j^k \wedge \omega_k^i = -b_{jkl}^i \lambda_m^l \underset{1}{\omega^k} \wedge \underset{1}{\omega^m}. \tag{7.1.36}$$

Equations (7.1.36) show that the torsion tensor of the connection induced by the canonical connection γ_{123} on the leaf F_4 coincides with the torsion tensor of the 3-subweb $[1,2,3]$, and the curvature tensor of this connection is obtained from the curvature tensor of the 3-subweb $[1,2,3]$ by a linear transformation by means of the basis affinor.

In general the equations of the geodesic lines of X^{2r} have the form:

$$d\omega^I + \omega^J \omega_J^I = \theta \omega^I. \tag{7.1.37}$$

where d is the symbol of the ordinary (not exterior) differentiation and θ is a 1-form. In the case when X^{2r} carries a web $W(4,2,r)$, using (7.1.33), we can write equations (7.1.37) in the form

$$d\underset{1}{\omega^i} + \underset{1}{\omega^j} \omega_j^i = \theta \underset{1}{\omega^i}, \quad d\underset{2}{\omega^i} + \underset{2}{\omega^j} \omega_j^i = \theta \underset{2}{\omega^i}. \tag{7.1.38}$$

It follows from equations (7.1.38) that the leaves of the foliations λ_1, λ_2, λ_3 determined by equations $\underset{1}{\omega^i} = 0$, $\underset{2}{\omega^i} = 0$, $\underset{1}{\omega^i} + \underset{2}{\omega^i} = 0$, are totally geodesic surfaces of X^{2r} in the connection γ_{123}. Let us find under what conditions the leaves of λ_4 are also totally geodesic in γ_{123}. Since for $F_4 \subset \lambda_4$ we have $\underset{2}{\omega}^i = -\lambda^i_j \underset{1}{\omega}^j$, from (1.7.38) it follows that F_4 is totally geodesic if and only if

$$\nabla \lambda^i_j \underset{1}{\omega}^j = 0. \tag{7.1.39}$$

By (7.1.28), condition (7.1.39) is equivalent to

$$(\lambda^i_{jk} - \mu^i_{jl}\lambda^l_k)\underset{1}{\omega}^j \underset{1}{\omega}^k = 0. \tag{7.1.40}$$

Equations (7.1.40) become identities if and only if

$$\lambda^i_{(jk)} = \lambda^l_{(j}\mu^i_{k)l}. \tag{7.1.41}$$

We studied the canonical affine connection γ_{123} induced by the 3-subweb $[1,2,3]$ where we distinguish the foliation λ_3. In general, there are 12 canonical affine connections $\gamma_{\xi_1\xi_2\xi_3}$ induced by the 3-subwebs $[\xi_1, \xi_2, \xi_3]$ where the foliation λ_{ξ_3} is distinguished. All these affine connections possess similar properties.

The following theorem combines the results obtained so far in this subsection.

Theorem 7.1.10 *A 4-web $W(4,2,r)$ possesses the following properties:*

(i) *It induces on the manifold X^{2r} 12 canonical affine connections $\gamma_{\xi_1\xi_2\xi_3}$ determined by the 3-subwebs $[\xi_1, \xi_2, \xi_3]$ and the foliation λ_{ξ_3} in them. The connection forms of γ_{123} have the form* (7.1.33), *and the torsion and curvature tensors of γ_{123} are determined by* (7.1.34), (7.1.35).

(ii) *The connection γ_{123} induces the affine connections on the leaves of the 4-web. The leaves of the foliations λ_1 and λ_2 are surfaces with absolute parallelism in all these connections, the torsion tensors of all leaf F_ξ coincide with the torsion tensor of the 3-subweb* [1,2,3], *the curvature tensor of a leaf F_4 coincides with the curvature tensor of the 3-subweb* [1,2,3], *and the curvature tensor of a leaf F_4 is obtained from the latter by a linear transformation by means of the basis affinor.*

(iii) *The leaves of the foliations λ_1, λ_2, λ_3 are totally geodesic in γ_{123} , and the leaves of λ_4 are totally geodesic in γ_{123} if and only if the basis affinor satisfies conditions* (7.1.41). ∎

7.1.7 Conditions of Geodesicity of Some Lines on the Leaves of a 4-web in the Canonical Affine Connection γ_{123}

In this subsection we will consider the canonical affine connection γ_{123} and leaves F_1, F_2, F_3, F_4 passing through a point $p \in D$. Consider a geodesic line on the leaf F_1 :

$$\underset{1}{\omega}^i = 0, \ \underset{2}{\omega}^i = \xi^i dt, \ d\xi^i + \xi^j \omega^i_j = \theta \xi^i. \tag{7.1.42}$$

On the leaf F_2 in the correspondence established by the foliation λ_3 to the line (7.1.42) there corresponds the line

$$\underset{2}{\omega}^i = 0, \quad \underset{1}{\omega}^i = -\xi^i dt, \tag{7.1.43}$$

and in the correspondence established by the foliation λ_4 to the same line (7.1.42) there corresponds the line

$$\underset{2}{\omega}^i = 0, \quad \underset{1}{\omega}^i = -\tilde{\lambda}^i_j \xi^j dt. \tag{7.1.44}$$

On the leaf F_3 in the correspondences established by λ_2 and λ_4 to the line (7.1.42) there correspond respectively the lines

$$\underset{3}{\omega}^i = 0, \quad \underset{1}{\omega}^i = \xi^i dt, \tag{7.1.45}$$

$$\underset{3}{\omega}^i = 0, \quad \underset{1}{\omega}^i = \tilde{\tilde{\lambda}}^i_j \xi^j dt. \tag{7.1.46}$$

Finally, on the leaf F_4 in the correspondences established by λ_2 and λ_3 to the line (7.1.42) there correspond the lines

$$\underset{4}{\omega}^i = 0, \quad \underset{1}{\omega}^i = \xi^i dt, \tag{7.1.47}$$

$$\underset{4}{\omega}^i = 0, \quad \underset{1}{\omega}^i = \tilde{\tilde{\lambda}}^i_j \lambda^j_k \xi^k dt. \tag{7.1.48}$$

Using (7.1.42), it is easy to show that the lines (7.1.43) and (7.1.45) are geodesic in γ_{123}. The line (7.1.47) is geodesic in γ_{123} if and only if the leaf F_4 is totally geodesic, i.e., if the conditions (7.1.41) are met.

To find under what conditions the lines (7.1.44), (7.1.46), and (7.1.48) are geodesic, we note that from

$$\lambda^i_k \tilde{\lambda}^k_j = \delta^i_j, \quad (\delta^i_k - \lambda^i_k)\tilde{\tilde{\lambda}}^k_j = \delta^i_j$$

it follows that

$$\nabla \tilde{\lambda}^i_j = -\tilde{\lambda}^k_j \nabla \lambda^l_k \tilde{\lambda}^i_l, \quad \nabla \tilde{\tilde{\lambda}}^i_j = -\tilde{\tilde{\lambda}}^k_j \nabla \lambda^l_k \tilde{\tilde{\lambda}}^i_l \tag{7.1.49}$$

where $\nabla \tilde{\lambda}^i_j = d\tilde{\lambda}^i_j - \tilde{\lambda}^i_k \omega^k_j + \tilde{\lambda}^k_j \omega^i_k$, $\nabla \tilde{\tilde{\lambda}}^i_j = d\tilde{\tilde{\lambda}}^i_j - \tilde{\tilde{\lambda}}^i_k \omega^k_j + \tilde{\tilde{\lambda}}^k_j \omega^i_k$. Further equations (7.1.44), (7.1.38), and (7.1.42) give

$$\nabla \tilde{\lambda}^i_j \xi^j = 0. \tag{7.1.50}$$

By virtue of (7.1.49), (7.1.44), and (7.1.27), it follows from (7.1.50) that

$$\tilde{\lambda}^m_k \tilde{\lambda}^l_j \lambda^i_{lm} \xi^j \xi^k = 0. \tag{7.1.51}$$

Equations (7.1.51) are satisfied identically if and only if

$$\lambda^i_{(jk)} = 0. \tag{7.1.52}$$

Similarly we can find that the conditions of geodesicity of the line (7.1.46) in γ_{123} are

$$\lambda^i_{(jk)} = \mu^i_{(jk)}, \tag{7.1.53}$$

and for the line (7.1.48) the conditions of geodesicity are (7.1.41) and

$$\lambda^m_{(j}(\lambda^i_{k)m} - \mu^i_{k)q}\lambda^q_m) = 0. \tag{7.1.54}$$

We proved the following theorem.

Theorem 7.1.11 *For the lines* (7.1.42)–(7.1.48) *the following statements are valid:*

(i) *In the connection γ_{123} to the geodesic line* (7.1.42) *on F_1 there corresponds the geodesic lines* (7.1.43) *on F_2 and* (7.1.45) *on F_3 if the correspondence is established by the foliations λ_2 and λ_3.*

(ii) *The lines* (7.1.43) *on F_2 and* (7.1.46) *on F_3 corresponding to the line* (7.1.42) *on F_1 in the correspondences established by the foliation λ_4 are geodesic in γ_{123} if and only if conditions* (7.1.52) *and* (7.1.53) *are satisfied.*

(iii) *The lines* (7.1.47) *and* (7.1.48) *on F_4 corresponding to the line* (7.1.42) *of F_1 in the correspondences established by λ_2 and λ_3 are geodesic in γ_{123} if and only if the respective conditions* (7.1.41) *and* (7.1.41), (7.1.54) *are satisfied.* ■

7.2 Special Classes of Webs $W(4,2,r)$

7.2.1 Parallelisable Webs $W(4,2,r)$

Definition 7.2.1 A web $W(4,2,r)$ is said to be *parallelisable* if it is equivalent to a parallel web $PW(4,2,r)$ of an affine space A^{2r} of dimension $2r$.

Theorem 7.2.2 *A web $W(4,2,r)$ is parallelisable if and only if its basis affinor is covariantly constant on X^{2r} in γ_{123} and its 3-subweb* $[1,2,3]$ *is parallelisable.*

Proof. *Necessity.* Suppose that in an affine space A^{2r} a parallel web $PW(4,2,r)$ is given and it is formed by four foliations λ_ξ of parallel r-planes F_ξ. With each point $p \in A^{2r}$ we associate a moving frame $\{p,\ e^1_i,\ e^2_i;\ i = 1,...,r\}$ such that its vectors $e^2_i;\ e^1_i;\ e^3_i = e^2_i - e^1_i;\ e^4_i = e^1_i - \lambda^j_i e^2_j$ belong respectively to the r-planes $F_1,\ F_2,\ F_3,\ F_4$ passing through the point p. Then, for any tangent vector dx from $T_p(A^{2r})$, we have

$$dx = \underset{1}{\omega}{}^i e^1_i + \underset{2}{\omega}{}^i e^2_i, \tag{7.2.1}$$

and equations of 4 foliations of the parallel web $W(4,2,r)$ have the form

$$\underset{1}{\omega}^i = 0, \quad \underset{2}{\omega}^i = 0, \quad \underset{1}{\omega}^i + \underset{2}{\omega}^i = 0, \quad \lambda^i_j \underset{1}{\omega}^j + \underset{2}{\omega}^i = 0. \tag{7.2.2}$$

Since the r-planes of the foliation λ_1 are parallel to each other and the same is true for the r-planes of λ_2, we have

$$de^1_i = \overset{1}{\omega}{}^j_i e^1_j, \quad de^2_i = \overset{2}{\omega}{}^j_i e^2_j. \tag{7.2.3}$$

By (7.1.3), for the vectors e^3_i and e^4_i we have

$$de^3_i = \overset{1}{\omega}{}^j_i e^3_j + (\overset{2}{\omega}{}^j_i - \overset{1}{\omega}{}^j_i) e^1_j, \quad de^4_i = \overset{1}{\omega}{}^j_i e^4_j - (d\lambda^j_i - \lambda^k_i \overset{2}{\omega}{}^j_k + \lambda^j_k \overset{1}{\omega}{}^k_i) e^2_j. \tag{7.2.4}$$

Since the r-planes of the foliations λ_3 and λ_4 are also parallel and the vectors e^3_j, e^1_j and e^4_j, e^2_j are linearly independent, it follows from (7.2.4) that

$$\overset{1}{\omega}{}^j_i = \overset{2}{\omega}{}^i_j = \omega^i_j, \quad \nabla\lambda^i_j \equiv d\lambda^i_j - \lambda^i_k \omega^k_j + \lambda^k_j \omega^i_k = 0. \tag{7.2.5}$$

Now the equations (7.2.3) take the form

$$de^1_i = \omega^j_i e^1_j, \quad de^2_i = \omega^j_i e^2_j. \tag{7.2.6}$$

Taking exterior derivatives of (7.2.1) and (7.2.6), we obtain the following structure equations of a parallel web $PW(4,2,r)$:

$$d\underset{1}{\omega}^i = \underset{1}{\omega}^j \wedge \omega^i_j, \quad d\underset{2}{\omega}^i = \underset{2}{\omega}^j \wedge \omega^i_j, \quad d\omega^i_j - \omega^k_j \wedge \omega^i_k = 0. \tag{7.2.7}$$

Comparison of equations (7.2.5) and (7.2.7) with the similar equations (7.1.21), (7.1.22), and (7.1.27) of a general web $W(4,2,r)$ shows that for a parallel web $PW(4,2,r)$ we have

$$a^i_{jk} = 0, \quad b^i_{jkl} = 0, \quad \lambda^i_{jk} = \mu^i_{jk} = 0. \tag{7.2.8}$$

Sufficiency. Let us show that the conditions (7.2.8) are sufficient for the parallelisability of a web $W(4,2,r)$. In fact, under conditions (7.2.8) the structure equations have the form (7.2.7) and $\nabla\lambda^i_j = 0$. But we have proved that a parallel web $PW(4,2,r)$ has such equations. Using Theorem 7.1.9, we conclude that our web $W(4,2,r)$ is parallelisable. ■

7.2.2 Webs $W(4,2,r)$ with Special 3-Subwebs

To study such webs, first we will find the forms of the canonical affine connection $\gamma_{\xi_1\xi_2\xi_3}$ and the torsion and curvature tensors of the 3-subwebs $[\xi_1, \xi_2, \xi_3]$. We remind ourselves that for the 3-subweb $[1,2,3]$ the connection forms and the tensors mentioned above are:

$$\underset{123}{\omega}{}^i_j = \omega^i_j, \quad \underset{123}{a}{}^i_{jk} = a^i_{jk}, \quad \underset{123}{b}{}^i_{jkl} = b^i_{jkl}. \tag{7.2.9}$$

Consider the 3-subweb $[1,2,4]$. We will change the basis forms $\underset{1}{\omega}^i$, $\underset{2}{\omega}^i$ and the forms $\underset{4}{\omega}^i$ according to the following formulae:

$$\underset{1}{\bar{\omega}}^i = \lambda^i_j \underset{1}{\omega}^j, \quad \underset{2}{\bar{\omega}}^i = \underset{2}{\omega}^i, \quad \underset{4}{\bar{\omega}}^i = \underset{4}{\omega}^i. \tag{7.2.10}$$

Then the second equation in (7.1.7) becomes

$$\underset{1}{\bar{\omega}}^i + \underset{2}{\bar{\omega}}^i + \underset{4}{\bar{\omega}}^i = 0. \tag{7.2.11}$$

Equation (7.2.11) shows that the forms $\underset{1}{\bar{\omega}}^i$, $\underset{2}{\bar{\omega}}^i$, $\underset{4}{\bar{\omega}}^i$ satisfy the relations which allowed us to derive the structure equations of [1,2,3] in the form (7.1.21).

Taking exterior derivatives of (7.2.10) and using (7.1.21), (7.1.27), and (7.2.10), we obtain

$$d\underset{1}{\bar{\omega}}^i = \underset{1}{\bar{\omega}}^j \wedge \underset{124}{\omega}{}^i_j + \underset{124}{a}{}^i_{jk} \underset{1}{\bar{\omega}}^j \wedge \underset{1}{\bar{\omega}}^k, \quad d\underset{2}{\bar{\omega}}^i = \underset{2}{\bar{\omega}}^j \wedge \underset{124}{\omega}{}^i_j - \underset{124}{a}{}^i_{jk} \underset{2}{\bar{\omega}}^j \wedge \underset{2}{\bar{\omega}}^k, \tag{7.2.12}$$

where

$$\underset{124}{\omega}{}^i_j = \omega^i_j - \mu^i_{mq} \tilde{\lambda}^m_j \underset{2}{\omega}^q, \tag{7.2.13}$$

$$\underset{124}{a}{}^i_{jk} = a^i_{jk} + \mu^i_{m[j} \tilde{\lambda}^m_{k]}. \tag{7.2.14}$$

are the connection forms of the canonical connection γ_{124} and the torsion tensor of the 3-subweb $[1,2,4]$.

The curvature tensor of the 3-subweb $[1,2,4]$ is found from the equation

$$d \underset{124}{\omega}{}^i_j - \underset{124}{\omega}{}^k_j \wedge \underset{124}{\omega}{}^i_k = \underset{124}{b}{}^i_{jkl} \underset{1}{\bar{\omega}}^k \wedge \underset{2}{\bar{\omega}}^l. \tag{7.2.15}$$

By virtue of (7.2.13), (7.2.12), (7.1.22), (7.1.31), (7.1.48), we find from (7.2.15) that

$$\underset{124}{b}{}^i_{jkl} = \tilde{\lambda}^p_k (b^i_{jpl} - \tilde{\lambda}^m_j \alpha^i_{mlp} + \tilde{\lambda}^r_j \tilde{\lambda}^m_s \lambda^s_{rp} \mu^i_{ml}). \tag{7.2.16}$$

For the 3-subweb $[2,3,4]$ we make the following change of forms:

$$\underset{2}{\bar{\omega}}^i = (\delta^i_j - \lambda^i_j) \underset{2}{\omega}^j, \quad \underset{3}{\bar{\omega}}^i = -\lambda^i_j \underset{3}{\omega}^j, \quad \underset{4}{\bar{\omega}}^i = \underset{4}{\omega}^i. \tag{7.2.17}$$

Then

$$\underset{2}{\bar{\omega}}^i + \underset{3}{\bar{\omega}}^i + \underset{4}{\bar{\omega}}^i = 0 \tag{7.2.18}$$

and

$$d\underset{2}{\bar{\omega}}^i = \underset{2}{\bar{\omega}}^j \wedge \underset{234}{\omega}{}^i_j + \underset{234}{a}{}^i_{jk} \underset{2}{\bar{\omega}}^j \wedge \underset{2}{\bar{\omega}}^k, \quad d\underset{3}{\bar{\omega}}^i = \underset{3}{\bar{\omega}}^j \wedge \underset{234}{\omega}{}^i_j - \underset{234}{a}{}^i_{jk} \underset{3}{\bar{\omega}}^j \wedge \underset{3}{\bar{\omega}}^k, \tag{7.2.19}$$

$$d \underset{234}{\omega}{}^i_j - \underset{234}{\omega}{}^k_j \wedge \underset{234}{\omega}{}^i_k = \underset{234}{b}{}^i_{jkl} \underset{2}{\bar{\omega}}^k \wedge \underset{3}{\bar{\omega}}^l, \tag{7.2.20}$$

where

$$\underset{234}{\omega}{}^i_j = \omega^i_j - \lambda^i_{mk}\tilde{\tilde{\lambda}}^m_j \underset{3}{\omega}{}^k + \tilde{\lambda}^m_j(\lambda^i_{mk} - \mu^i_{mk} - 2\lambda^i_p a^p_{mk})\underset{2}{\omega}{}^k, \tag{7.2.21}$$

$$\underset{234}{a}{}^i_{jk} = -(\lambda^i_m a^m_{qp} + \lambda^i_{pq})\tilde{\lambda}^q_{[j}\tilde{\lambda}^p_{k]} + \lambda^i_{qp}\tilde{\lambda}^p_{[k}\tilde{\tilde{\lambda}}^q_{j]}, \tag{7.2.22}$$

$$\begin{aligned}\underset{234}{b}{}^i_{jkl} = -\tilde{\tilde{\lambda}}^s_k\tilde{\lambda}^t_l\{& b^i_{jts} + \tilde{\tilde{\lambda}}^m_j(\lambda^i_{mts} - \beta^i_{mts}) + \tilde{\lambda}^m_j(\lambda^i_{mst} - \alpha^i_{mst}) \\ & + \lambda^i_{at}\tilde{\tilde{\lambda}}^a_p\tilde{\lambda}^r_j\tilde{\tilde{\lambda}}^q_r(\lambda^p_{qs} - \mu^p_{qs}) - 2\lambda^i_{qt}\tilde{\tilde{\lambda}}^q_p\tilde{\lambda}^m_j a^p_{ms} - 2\tilde{\lambda}^p_j\lambda^i_{mt}a^m_{ps} \\ & + \tilde{\lambda}^a_p\lambda^p_{qt}\tilde{\tilde{\lambda}}^q_b\tilde{\lambda}^b_j(\lambda^i_{as} - \mu^i_{as} - 2\lambda^i_m a^m_{as}) - 2\tilde{\lambda}^p_j\lambda^i_m b^m_{[p|t|s]}\}.\end{aligned} \tag{7.2.23}$$

Note that (7.1.48) implies

$$(\tilde{\tilde{\lambda}}^i_k + \tilde{\lambda}^i_k)\lambda^k_j = \tilde{\tilde{\lambda}}^i_j. \tag{7.2.24}$$

Contracting both members of (7.2.24) with λ^j_l, we get

$$\tilde{\tilde{\lambda}}^i_l + \tilde{\lambda}^i_l = \tilde{\tilde{\lambda}}^i_k\tilde{\lambda}^k_l. \tag{7.2.25}$$

The relations (7.2.25) were used to obtain (7.2.21).

Finally for the 3-subweb [3, 1, 4] we will make the following change of forms:

$$\underset{3}{\bar{\omega}}{}^i = -\underset{3}{\omega}{}^i, \quad \underset{1}{\bar{\omega}}{}^i = -(\delta^i_j - \lambda^i_j)\underset{1}{\omega}{}^j, \quad \underset{4}{\bar{\omega}}{}^i = \underset{4}{\omega}{}^i. \tag{7.2.26}$$

This gives

$$\begin{cases} d\underset{3}{\bar{\omega}}{}^i = \underset{3}{\bar{\omega}}{}^j \wedge \underset{314}{\omega}{}^i_j + \underset{314}{a}{}^i_{jk}\underset{3}{\bar{\omega}}{}^j \wedge \underset{1}{\bar{\omega}}{}^k, \\ d\underset{1}{\bar{\omega}}{}^i = \underset{1}{\bar{\omega}}{}^j \wedge \underset{314}{\omega}{}^i_j - \underset{314}{a}{}^i_{jk}\underset{3}{\bar{\omega}}{}^j \wedge \underset{1}{\bar{\omega}}{}^k, \\ d\underset{314}{\omega}{}^i_j - \underset{314}{\omega}{}^k_j \wedge \underset{314}{\omega}{}^i_k = \underset{314}{b}{}^i_{jkl}\underset{3}{\bar{\omega}}{}^k \wedge \underset{1}{\bar{\omega}}{}^l, \end{cases} \tag{7.2.27}$$

and

$$\underset{314}{\omega}{}^i_j = \omega^i_j + 2a^i_{jk}\underset{1}{\omega}{}^k - \mu^i_{mk}\tilde{\tilde{\lambda}}^m_j\underset{3}{\omega}{}^k, \tag{7.2.28}$$

$$\underset{314}{a}{}^i_{jk} = -a^i_{jk} + \mu^i_{m[j}\tilde{\tilde{\lambda}}^m_{k]}, \tag{7.2.29}$$

$$\begin{aligned}\underset{314}{b}{}^i_{jkl} = b^i_{pjk}\tilde{\tilde{\lambda}}^p_l + \tilde{\tilde{\lambda}}^p_j\tilde{\tilde{\lambda}}^q_l(& \alpha^i_{pkq} - \mu^i_{pkq} + \mu^i_{tk}\tilde{\tilde{\lambda}}^t_s\lambda^s_{pq} - \mu^i_{tk}\mu^s_{pq}\tilde{\tilde{\lambda}}^t_s \\ & -2\mu^i_{pt}a^t_{kq} + 2a^i_{tq}\mu^t_{pk}) - 2a^t_{jq}\mu^i_{pk}\tilde{\tilde{\lambda}}^q_l\tilde{\tilde{\lambda}}^p_t.\end{aligned} \tag{7.2.30}$$

Now we will distinguish certain classes of webs $W(4, 2, r)$ characterised by the vanishing of the torsion and curvature tensors of some of the 3-subwebs $[\xi_1, \xi_2, \xi_3]\$andthecova$

As we noted in Section 7.1, the conditions (7.1.29) are necessary and sufficient for the basis affinor $\Lambda = (\lambda^i_j)$ of the web $W(4,2,r)$ to be covariantly constant in the connection γ_{123} on the respective foliations $\lambda_1, \lambda_2, \lambda_3, \lambda_4$. Let us indicate some consequences of this in the cases when the basis affinor is covariantly constant in γ_{123} on λ_1 or λ_2.

In the first case $\mu^i_{jk} = 0$; by (7.1.31), it implies $\alpha^i_{jkl} = \mu^i_{jkl} = 0$. In this case, using (7.2.13), (7.2.14), (7.2.16), (7.2.29), (7.2.30), we obtain

$$\underset{124}{a}{}^i_{jk} = -\underset{314}{a}{}^i_{jk} = a^i_{jk}, \quad \underset{124}{b}{}^i_{jkl} = \tilde{\lambda}^p_k b^i_{jpl}, \quad \underset{314}{b}{}^i_{jlm} = (2b^i_{jkl} - b^i_{kjl})\tilde{\tilde{\lambda}}^k_m, \quad \underset{124}{\omega}{}^i_j = \omega^i_j, \tag{7.2.31}$$

where the last equation is necessary and sufficient for this class (it implies $\mu^i_{jk} = 0$).

In the second case $\lambda^i_{jk} = 0$; by (7.1.31), this implies $\lambda^i_{jkl} = \beta^i_{jkl} = 0$. In this case, using (7.2.22), (7.2.23), (7.2.13), (7.2.21), we obtain

$$\begin{cases} \underset{234}{a}{}^i_{jk} = -\lambda^i_m \tilde{\lambda}^q_j \tilde{\lambda}^p_k a^m_{qp}, \\ \underset{234}{b}{}^i_{jkl} = -\tilde{\tilde{\lambda}}^s_k \tilde{\lambda}^t_l (b^i_{jst} - \tilde{\lambda}^p_j \alpha^i_{pts} - 2\tilde{\lambda}^p_j \lambda^i_m b^m_{[p|s|t]}), \\ \underset{234}{\omega}{}^i_j = \omega^i_j - 2\lambda^i_m \tilde{\lambda}^k_j a^m_{kl} \underset{2}{\omega}{}^l. \end{cases} \tag{7.2.32}$$

Now we will study the cases when one or more of 3-subwebs of $W(4,2,r)$ are paratactical (see Definition 1.6.1). If the basis affinor is covariantly constant in the whole manifold X^{2r} in one of the canonical affine connections, then it is covariantly constant in X^{2r} in any other connection. Because of this, in such cases we will not indicate an affine connection.

Theorem 7.2.3 *The 3-subweb [1,2,3] is paratactical if and only if one of 3-subwebs, [1,2,4] or [3,1,4], is paratactical and the basis affinor is covariantly constant on the foliation λ_2 in the connection γ_{123}. The 3-subweb [1,2,3] is paratactical and the basis affinor is covariantly constant in the connection γ_{123} on λ_1, λ_2 or in the whole X^{2r} if and only if respectively*

$$\underset{314}{\omega}{}^i_j = \omega^i_j, \quad \underset{234}{\omega}{}^i_j = \underset{124}{\omega}{}^i_j, \quad \underset{124}{\omega}{}^i_j = \underset{234}{\omega}{}^i_j = \underset{314}{\omega}{}^i_j = \omega^i_j$$

In these cases the 3-subwebs [1,2,4], [3,1,4]; [2,3,4]; [1,2,4], [3,1,4], [2,3,4] are correspondingly paratactical.

Proof. The 3-subweb $[1,2,3]$ is paratactical if and only if $a^i_{jk} = 0$ (see Definition

1.6.1). The statements of the theorem follow from the following implications:

$$
\begin{aligned}
1)\quad a^i_{jk} &= \mu^i_{jk} = 0 \Longrightarrow \underset{124}{a}{}^i_{jk} = \underset{314}{a}{}^i_{jk} = 0,\\
&\Updownarrow\\
\underset{314}{\omega}{}^i_j &= \omega^i_j;\\
2)\quad \underset{124}{a}{}^i_{jk} &= \mu^i_{jk} = 0 \Longrightarrow a^i_{jk} = 0;\\
3)\quad \underset{314}{a}{}^i_{jk} &= \mu^i_{jk} = 0 \Longrightarrow a^i_{jk} = 0;\\
4)\quad a^i_{jk} &= \mu^i_{jk} = 0 \Longrightarrow \underset{234}{a}{}^i_{jk} = 0, \quad \underset{234}{a}{}^i_{jk} = \lambda^i_{jk} = 0 \Longrightarrow a^i_{jk} = 0;\\
&\Updownarrow\\
\underset{234}{\omega}{}^i_j &= \omega^i_j;\\
5)\quad a^i_{jk} &= \lambda^i_{jk} = \mu^i_{jk} = 0 \Longrightarrow \underset{124}{a}{}^i_{jk} = \underset{234}{a}{}^i_{jk} = \underset{314}{a}{}^i_{jk} = 0;\\
&\Updownarrow\\
\underset{124}{\omega}{}^i_j &= \underset{234}{\omega}{}^i_j = \underset{314}{\omega}{}^i_j = \omega^i_j. \quad \blacksquare
\end{aligned}
$$

Next we will discuss cases when one or more of 3-subwebs $[\xi_1, \xi_2, \xi_3]$ are group three-webs (see Table 1.1 at the end of Section 1.2).

Theorem 7.2.4 *The following two statements are valid for a web $W(4,2,r)$:*
(i) *If a 3-subweb $[1,\xi,\eta]$ (or $[2,\xi,\eta]$) is a group web and the basis affinor λ^i_j is covariantly constant in the canonical connection γ_{123} on the foliation $\lambda_1(\lambda_2)$, then the 3-subwebs $[1,\xi,\zeta]$, $[1,\eta,\zeta]$ ($[2,\xi,\zeta]$, $[2,\eta,\zeta]$) are also group webs.*
(ii) *If the basis affinor is covariantly constant in the whole X^{2r} and one of the 3-subwebs $[\xi,\eta,\zeta]$ is a group web, then the remaining 3-subwebs $[\xi,\eta,\zeta]$ are also group webs.*

Proof. As we noted in Table 1.1, the 3-subweb $[1,2,3]$ is a group web if and only if $b^i_{jkl} = 0$. Note further the following implications:

$$
\begin{aligned}
\mu^i_{jk} = 0, \quad \underset{1\xi\eta}{b}{}^i_{jkl} = 0 \quad &\Rightarrow \quad \underset{1\xi\zeta}{b}{}^i_{jkl} = \underset{1\eta\zeta}{b}{}^i_{jkl} = 0,\\
\lambda^i_{jk} = 0, \quad \underset{2\xi\eta}{b}{}^i_{jkl} = 0 \quad &\Rightarrow \quad \underset{2\xi\zeta}{b}{}^i_{jkl} = \underset{2\eta\zeta}{b}{}^i_{jkl} = 0,\\
\lambda^i_{jk} = \mu^i_{jk} = 0, \quad \underset{\xi\eta\zeta}{b}{}^i_{jkl} = 0 \quad &\Rightarrow \quad \underset{\xi\eta\zeta}{b}{}^i_{jkl} = \underset{\eta\zeta\rho}{b}{}^i_{jkl} = \underset{\xi\zeta\rho}{b}{}^i_{jkl} = 0.
\end{aligned}
$$

We will prove these implications for the cases:

$$
\begin{aligned}
a)\ \mu^i_{jk} = 0, \quad \underset{314}{b}{}^i_{jkl} = 0 &\Rightarrow \underset{123}{b}{}^i_{jkl} = \underset{124}{b}{}^i_{jkl} = 0,\\
b)\ \lambda^i_{jk} = 0, \quad \underset{234}{b}{}^i_{jkl} = 0 &\Rightarrow \underset{123}{b}{}^i_{jkl} = \underset{124}{b}{}^i_{jkl} = 0,\\
c)\ \lambda^i_{jk} = 0, \quad \underset{124}{b}{}^i_{jkl} = 0 &\Rightarrow \underset{123}{b}{}^i_{jkl} = \underset{234}{b}{}^i_{jkl} = 0.
\end{aligned}
$$

In the other cases the proof readily follows from (7.1.31), (7.1.32). In the cases indicated above we have:

a) In this case $\alpha^i_{jkl} = \mu^i_{jkl} = 0$ and hence from $\underset{314}{b}{}^i_{jkl} = 0$ it follows that $b^i_{jkl} = 0$, i.e., $\underset{123}{b}{}^i_{jkl} = 0$ and $\underset{124}{b}{}^i_{jkl} = 0$.

b) Now we have

$$\beta^i_{jkl} = \lambda^i_{jkl} = 0, \quad \alpha^i_{jml} = \lambda^k_j b^i_{klm} - \lambda^i_k b^k_{jlm}. \tag{7.2.33}$$

Hence from $\underset{234}{b}{}^i_{jkl} = 0$ it follows that

$$b^i_{jkl} = \tilde{\lambda}^p_j \alpha^i_{plk} + 2\tilde{\lambda}^p_j \lambda^i_m b^m_{[p|k|l]}. \tag{7.2.34}$$

Substituting α^i_{jml} from (7.2.33) into (7.2.34), we find that

$$\underset{123}{b}{}^i_{jkl} = \alpha^i_{jkl} = \underset{124}{b}{}^i_{jkl} = 0.$$

c) We still have (7.2.33). From $\underset{124}{b}{}^i_{jkl} = 0$ it follows

$$b^i_{jpl} - \lambda^m_j \alpha^i_{mlp}. \tag{7.2.35}$$

Substituting α^i_{jkl} from (7.2.33) into (7.2.35), we find that

$$\underset{123}{b}{}^i_{jkl} = \alpha^i_{jkl} = \underset{124}{b}{}^i_{jkl} = 0. \quad \blacksquare$$

7.2.3 Group Webs $W(4,2,r)$

A web $W(4,2,r)$ discussed in the part (ii) of Theorem 7.2.3 deserves a special study.

Definition 7.2.5 A web $W(4,2,r)$ whose all 3-subwebs are group webs and whose basis affinor is covariantly constant in the whole manifold X^{2r} is said to be a *group 4-web*.

As the part (ii) of Theorem 7.2.3 shows, for a web $W(4,2,r)$ to be a group 4-web, it is necessary and sufficient that its basis affinor be covariantly constant in the whole manifold X^{2r} and one of its 3-subwebs be a group web.

A group web $W(4,2,r)$ can be represented as follows. Take an r-parameter Lie group G with invariant forms $\underset{2}{\omega}{}^i$ satisfying the structure equations

$$d\underset{2}{\omega}{}^i = -a^i_{jk}\underset{2}{\omega}{}^j \wedge \underset{2}{\omega}{}^k \tag{7.2.36}$$

where the $-a^i_{jk}$ are the structure constants of G. The *anti-automorphism* $\varphi(g) = g^{-1}$,

$g \in G$, defines the group G^{-1} with invariant forms $\underset{1}{\omega}^i$ satisfying the structure equations

$$d\underset{1}{\omega}^i = a^i_{jk}\underset{1}{\omega}^j \wedge \underset{1}{\omega}^k. \tag{7.2.37}$$

The direct product $G \times G^{-1}$ determines a $(2r)$-parameter Lie group with the invariant forms $\underset{1}{\omega}^i$, $\underset{2}{\omega}^i$ satisfying the structure equations (7.2.36), (7.2.37).

Suppose that a field of constant affinor $\lambda^i_j (d\lambda^i_j = 0)$ be given on the $(2r)$-dimensional analytic group manifold of the group $G \times G^{-1}$. Then on this group manifold the completely integrable systems

$$\underset{1}{\omega}^i = 0, \ \ \underset{2}{\omega}^i = 0, \ \ \underset{1}{\omega}^i + \underset{2}{\omega}^i = 0, \ \ \lambda^i_j \underset{1}{\omega}^j + \underset{2}{\omega}^i = 0$$

determine four r-parameter foliations of codimension r. One leaf of each foliation, namely the leaf passing through the unit (e,e) of the group $G \times G^{-1}$ (e is the unit of the group G), is a subgroup of the group $G \times G^{-1}$, and the other leaves of this foliation are cosets with respect to this subgroup.

For the web $W(4,2,r)$ constructed above the torsion tensor coincides with the structure tensor of the group G, and the connection forms $\omega^i_j = 0$. By (7.1.22), it follows from this that $b^i_{jkl} = 0$.

Conversely, suppose that for some web $W(4,2,r)$ $b^i_{jkl} = \lambda^i_{jk} = \mu^i_{jk} = 0$. Then the equations (7.1.22) and (7.1.23) have the form

$$d\omega^i_j - \omega^k_j \wedge \omega^i_k = 0, \quad \nabla a^i_{jk} = 0. \tag{7.2.38}$$

The first group of equations (7.2.38) means that the manifold X^{2r} is a space with absolute parallelism. Because of this, we can take a frame at a point $p \in D \subset X^{2r}$ and move it parallel across D. In the family of moving frames which we obtained in the described manner we have $\omega^i_j = 0$. Then equations (7.1.22) give $da^i_{jk} = 0$, i.e., the components of the torsion tensor are constant. Consequently equations (7.1.21) coincide with the structure equations of the group $G \times G^{-1}$, and the relations (7.1.24) become the Jacobi identities for the constants of the group G. Finally, by $\lambda^i_{jk} = \mu^i_{jk} = 0$, relations (7.1.28) will take the form

$$\lambda^i_p a^p_{jk} + \lambda^p_{[k}\lambda^q_{j]} a^i_{pq} = 0. \tag{7.2.39}$$

We have proved the following theorem:

Theorem 7.2.6 *A web $W(4,2,r)$ is a group 4-web if and only if its 3-subweb* [1,2,3] *is a group 3-web and its ambient space X^{2r} is endowed with the field of a constant affinor λ^i_j connected with the torsion tensor of the three-web* [1,2,3] *by relations* (7.2.39).

■

7.2.4 Parallelisable Webs $W(4,2,r)$ (Continuation)

We will give here two new criteria for a web $W(4,2,r)$ to be parallelisable.

Proposition 7.2.7 *If the 3-subweb $[1,2,3]$ is parallelisable and the basis affinor is covariantly constant in the canonical affine connection γ_{123} on the foliation λ_1 (λ_2 or in the whole X^{2r}), then the 3-subwebs $[1,2,4]$, $[3,1,4]$ ($[2,3,4]$ or $[1,2,4]$, $[3,1,4]$, $[2,3,4]$) are also parallelisable. In the latter case the whole web $W(4,2,r)$ is parallelisable.*

Proof. The 3-subweb $[1,2,3]$ is parallelisable if and only if $a^i_{jk} = b^i_{jkl} = 0$ (see Table 1.1, Section 1.2.). The proposition follows from the following implications:

$$a^i_{jk} = b^i_{jkl} = \mu^i_{jk} = 0 \;\Rightarrow\; \underset{124}{a}{}^i_{jk} = \underset{314}{a}{}^i_{jk} = \underset{124}{b}{}^i_{jkl} = \underset{314}{b}{}^i_{jkl} = 0,$$
$$a^i_{jk} = b^i_{jkl} = \lambda^i_{jk} = 0 \;\Rightarrow\; \underset{234}{a}{}^i_{jk} = \underset{234}{b}{}^i_{jkl} = 0,$$
$$a^i_{jk} = b^i_{jkl} = \lambda^i_{jk} = \mu^i_{jk} = 0 \;\Rightarrow\; \underset{124}{a}{}^i_{jk} = \underset{314}{a}{}^i_{jk} = \underset{234}{a}{}^i_{jk} = \underset{124}{b}{}^i_{jkl} = \underset{314}{b}{}^i_{jkl} = \underset{234}{b}{}^i_{jkl} = 0. \blacksquare$$

Proposition 7.2.8 *A web $W(4,2,r)$ is parallelisable if and only if it is generated by the direct product of an r-parameter commutative Lie group G and the group G^{-1} whose elements are obtained from elements of G by means of the anti-automorphism $\varphi(g) = g^{-1}$, $g \in G$, and if its basis affinor is covariantly constant on the group manifold of the group $G \times G^{-1}$.*

Proof. If in our representation of a group 4-web the group G, and therefore G^{-1}, is commutative, then $a^i_{jk} = 0$. In this case the relations (7.2.39) are satisfied identically, and the web $W(4,2,r)$ is parallelisable.

Conversely, any parallelisable web $W(4,2,r)$ can be considered as a 4-web generated on the group manifold of $G \times G^{-1}$, where G is an r-parameter commutative Lie group, by the method indicated above, if in addition a field of a constant affinor is given on this manifold. ■

7.2.5 Webs $W(4,2,r)$ with a Group 3-Subweb

We will study here a web $W(4,2,r)$ for which the 3-subweb $[1,2,3]$ is a group three-web and which satisfies an additional property.

Theorem 7.2.9 *If the 3-subweb $[1,2,3]$ of a web $W(4,2,r)$ is a group three-web and the canonical connections γ_{123} and γ_{124} induced on X^{2r} by the 3-subwebs $[1,2,3]$ and $[1,2,4]$ have common geodesic lines, then the 3-subweb $[1,2,4]$ is a Moufang three-web.*

Proof. Let the 3-subweb $[1,2,3]$ be a group three-web generated by an r-parameter Lie group G with the structure constants $-a^i_{jk}$. Then its basis forms $\underset{1}{\omega}^i$ and $\underset{2}{\omega}^i$ satisfy equations (7.2.36) and (7.2.37). This means that for the web $[1,2,3]$ the connection forms $\omega^i_j = 0$ and consequently the curvature tensor $b^i_{jkl} = 0$. This implies that relations (7.1.24) become the Jacobi identities for the structure constants a^i_{jk} of the group G generating the group three-web $[1,2,3]$. In addition, it follows from this that the canonical connection γ_{123} is one of the two well-known Cartan curvature-free affine connections [Ca 27].

The canonical connection forms for the 3-subweb $[1,2,4]$ have the form (7.2.13) where $\omega^i_j = 0$. These forms and its basis forms defined by (7.2.10) satisfy the structure equations (7.2.12) and (7.2.15), and the torsion and curvature tensors of [1,2,4] are expressed by (7.2.14) and (7.2.16). The canonical connection γ_{124} is defined by the forms

$$\underset{124}{\omega}{}^I_J = \begin{pmatrix} \underset{124}{\omega}{}^i_j & 0 \\ 0 & \underset{124}{\omega}{}^i_j \end{pmatrix}, \qquad I, J = 1, \ldots, 2r.$$

Let us find under what conditions geodesic lines of X^{2r} in the connections γ_{123} and γ_{124} coincide. Since $\omega^i_j = 0$, equations (7.1.37) of geodesic lines of X^{2r} in the connection γ_{123} take the form

$$d\underset{1}{\omega}^i = \theta \underset{1}{\omega}^i, \quad d\underset{2}{\omega}^i = \theta \underset{2}{\omega}^i. \tag{7.2.40}$$

On the other hand, the equations of geodesic lines of X^{2r} in the connection γ_{124} are written in the form

$$d\theta^i + \theta^j \underset{124}{\omega}{}^i_j = \varphi\theta^i, \quad d\underset{2}{\omega} + \underset{2}{\omega}^j \underset{124}{\omega}{}^i_j = \varphi \underset{2}{\omega}^i \tag{7.2.41}$$

where, by (7.2.10), $\theta^i = \lambda^i_j \underset{1}{\omega}^j$ and φ is an 1-form. Substituting $\theta^i = \lambda^i_j \underset{1}{\omega}^j$ from (7.2.10) into (7.2.41) and using (7.2.13), (7.1.17), we rewrite the system (7.2.41) in the form

$$\lambda^i_{jk} \underset{1}{\omega}^j \underset{1}{\omega}^k = 0, \quad \mu^i_{mq} \tilde{\lambda}^m_j \underset{2}{\omega}^j \underset{2}{\omega}^q = 0. \tag{7.2.42}$$

It follows from (7.2.42) that the systems (7.2.40) and (7.2.41) have the same solutions if and only if the following two conditions are satisfied:

$$\mu^i_{m(q} \tilde{\lambda}^m_{j)} = 0, \tag{7.2.43}$$

$$\lambda^i_{(jk)} = 0. \tag{7.2.44}$$

Now we can prove that the 3-subweb $[1,2,4]$ is a Moufang three-web. According to Table 1.1 (Section 1.2), this will be the case if and only if

$$\underset{124}{b}{}^i_{jkl} = \underset{124}{b}{}^i_{[jkl]}. \tag{7.2.45}$$

To prove (7.2.45), first note that condition (7.2.43) is equivalent to

$$\mu^i_{(q|m|}\lambda^m_{j)} = 0. \qquad (7.2.46)$$

In fact, contracting (7.2.43) with $\lambda^j_k\lambda^q_p$, we get (7.2.46), and conversely, contracting (7.2.46) with $\tilde{\lambda}^q_p\tilde{\lambda}^q_k$, we get (7.2.43). By (7.2.44) and (7.2.46), equations (7.1.28) take the form

$$\lambda^i_{jk} = \mu^i_{jp}\lambda^p_k + \lambda^i_p a^p_{jk} + \lambda^p_k\lambda^q_j a^i_{pq}. \qquad (7.2.47)$$

By (7.2.47), equations (7.1.17) are written in the form

$$d\lambda^i_j = (\mu^i_{jp}\lambda^p_k + \lambda^i_p a^p_{jk} + \lambda^p_k\lambda^q_j a^i_{pq})\underset{1}{\omega}^k + \mu^i_{jk}\underset{2}{\omega}^k. \qquad (7.2.48)$$

Taking exterior derivative of (7.2.48) and applying the Cartan lemma to the obtained exterior quadratic equation, we obtain

$$d\mu^i_{jk} = \alpha^i_{jkl}\underset{1}{\omega}^l + \mu^i_{jkl}\underset{2}{\omega}^l, \qquad (7.2.49)$$

$$\mu^i_{j[kl]} = \mu^i_{jp}a^p_{kl}, \qquad (7.2.50)$$

$$\alpha^i_{jlk} = \lambda^p_k\mu^i_{jpl} + \mu^i_{jp}\mu^p_{kl} + a^p_{jk}\mu^i_{pl} + 2\mu^p_{[k|l|}\lambda^q_{j]}a^i_{pq}, \qquad (7.2.51)$$

$$\lambda^i_m a^m_{jp}a^p_{kl} + \lambda^q_j\lambda^m_{[k}\lambda^t_{l]}a^i_{pq}a^p_{mt} = 0. \qquad (7.2.52)$$

Note that to obtain (7.2.52), one should use the Jacobi identities for the structure constants a^i_{jk} .

On the other hand, differentiation of (7.2.46) by means of (7.2.49), (7.2.48) leads to relations:

$$\alpha^i_{(k|ml|}\lambda^m_{j)} + \mu^i_{(k|m|}\mu^m_{j)p}\lambda^p_l + \mu^i_{(k|m|}a^p_{j)l}\lambda^m_p - \mu^i_{(k|m|}\lambda^p_{j)}\lambda^q_l a^m_{pq} = 0, \qquad (7.2.53)$$

$$\mu^i_{(k|ml|}\lambda^m_{j)} + \mu^i_{(k|m|}\mu^m_{j)l} = 0. \qquad (7.2.54)$$

Using (7.2.51), (7.2.54), and (7.2.50), we exclude α^i_{jkl} and μ^i_{jkl} from (7.2.53). From this we deduce

$$\mu^i_{(j|m|}\lambda^p_{k)}a^m_{lp} + \mu^q_{(k|l|}\lambda^m_{j)}a^i_{qm} + \mu^i_{(j|p|}\mu^p_{k)l} = 0. \qquad (7.2.55)$$

Substituting $b^i_{jkl} = 0$ into (7.2.16), contracting it with λ^k_s and symmetrising the relations obtained with respect to indices j and l, we obtain

$$\lambda^k_s \underset{124}{b}{}^i_{(j|k|l)} = \tilde{\lambda}^m_{(j}(-\alpha^i_{|m|l)s} + \mu^i_{|r|l)}\tilde{\lambda}^r_q\lambda^q_{ms}). \qquad (7.2.56)$$

Let us show that the right member of (7.2.56) vanishes. In fact, using (7.2.47), we can rewrite equations (7.2.53) in the form

$$\alpha^i_{(k|lq|}\lambda^l_{j)} + \mu^i_{(k|l|}\lambda^l_{j)q} = 0. \qquad (7.2.57)$$

Contracting equations (7.2.57) with $\tilde{\lambda}^j_p\tilde{\lambda}^k_s$, we get

$$\tilde{\lambda}^m_{(j}\alpha^i_{k)mq} + \tilde{\lambda}^j_p\tilde{\lambda}^k_s\lambda^l_{(j|q|}\mu^i_{k)l} = 0. \tag{7.2.58}$$

Substituting (7.2.58) into (7.2.56) and using (7.2.54), we deduce

$$\lambda^k_s \underset{124}{b}{}^i_{(j|k|l)} = 0. \tag{7.2.59}$$

By (7.1.8), we obtain from (7.2.59) that

$$\underset{124}{b}{}^i_{(j|k|l)} = 0. \tag{7.2.60}$$

Next we contract (7.2.56) with λ^j_t and symmetrise the obtained relations with respect to the indices t and s. As a result, we get

$$\lambda^j_{(t}\lambda^k_{s)}\underset{124}{b}{}^i_{jkl} = -\alpha^i_{(t|l|s)} + \tilde{\lambda}^m_q\lambda^q_{ts}\mu^i_{ml}. \tag{7.2.61}$$

By (7.2.44), (7.2.51), and (7.2.54), it follows from (7.2.61) that

$$\lambda^j_{(t}\lambda^k_{s)}\underset{124}{b}{}^i_{jkl} = 0. \tag{7.2.62}$$

Contraction of (7.2.62) with $\tilde{\lambda}^t_p\tilde{\lambda}^s_q$ leads to

$$\underset{124}{b}{}^i_{jkl} = 0. \tag{7.2.63}$$

The relations (7.2.60) and (7.2.63) imply (7.2.45). ■

Suppose that $r = 2$, i.e., we consider a web $W(4,2,2)$.

Proposition 7.2.10 *A web $W(4,2,2)$ web satisfying the conditions of Theorem 7.2.8 is defined by the following closed form equations:*
$\lambda_1 : x^i = a^i = \text{const.}; \quad \lambda_2 : y^i = b^i = \text{const.};$
$\lambda_3 : z^i = c^i = \text{const.}; \quad \lambda_4 : u^i = d^i = \text{const.},$
where (x^1, x^2, y^1, y^2) are local coordinates of the manifold X^4 carrying the web $W(4,2,2)$ and

$$z^1 = x^1 + x^2y^1, \quad z^2 = x^2y^2, \tag{7.2.64}$$
$$u^1 = (k_1x^1 + k_2)/x^2 + y^1/x^2, \quad u^2 = y^2/x^2, \tag{7.2.65}$$

and $k_1 \neq 0$, k_2 are constants.

Proof. First note that since a four-dimensional Moufang loop is a group [Ku 70], then, by Theorem 7.2.8, not only the 3-subweb [1,2,3], but also the 3-subweb [1,2,4] is a group three-web.

Secondly, it is well-known [Ja 62] that there exists a unique non-commutative two-dimensional Lie algebra. Its structure tensor has the only non-zero component: $a^1_{12} = 1$. Hence the structure equations (7.2.36), (7.2.37) of the 3-subweb $[1,2,3]$ will have the form

$$\begin{cases} d\underset{1}{\omega}^1 = 2\underset{1}{\omega}^1 \wedge \underset{1}{\omega}^2, & d\underset{1}{\omega}^2 = 0, \\ d\underset{2}{\omega}^1 = -2\underset{2}{\omega}^1 \wedge \underset{2}{\omega}^2, & d\underset{2}{\omega}^2 = 0. \end{cases} \tag{7.2.66}$$

Integration of (7.2.66) gives us the invariant forms of the Lie group corresponding to the 3-subwebs $[1,2,3]$ and $[1,2,4]$:

$$\underset{1}{\omega}^1 = \frac{dx^1}{x^2}, \quad \underset{1}{\omega}^2 = \frac{1}{2} d \ln x^2; \tag{7.2.67}$$

$$\underset{2}{\omega}^1 = \bar{y}^2 d\bar{y}^1, \quad \underset{2}{\omega}^2 = \frac{1}{2} d \ln \bar{y}^2, \tag{7.2.68}$$

where $(x^i, \bar{y}^j)$ are local coordinates of a point of the manifold X^{2r} carrying the web $W(4,2,2)$. To give more details of how equations (7.2.67) and (7.2.68) are obtained from equations (7.2.66), note that equations (7.2.66) shows that the forms $\underset{1}{\omega}^1$ and $\underset{2}{\omega}^1$ are not total differentials, i.e., they have the form $\underset{1}{\omega}^1 = \alpha dx^1$, $\underset{2}{\omega}^1 = \beta d\bar{y}^1$, where the coefficient α depends not only on x^1 and the coefficient β depends not only on $\bar{y}^1$. Taking α and β as $1/x^2$ and $\bar{y}^2$ respectively, we arrive to the first equations of (7.2.67) and (7.2.68). As to the second equations of (7.2.67) and (7.2.68), they follow from the fact that, by (7.2.66), the forms $\underset{1}{\omega}^2$ and $\underset{2}{\omega}^2$ are total differentials of some functions. To satisfy (7.2.66), these functions may be taken as $(\ln x^2)/2$ and $(\ln \bar{y}^2)/2$.

Let us find the components λ^i_j of the basis affinor Λ. For this we substitute the values of the components of the torsion tensor of the web $[1,2,3]$ into equations (7.2.52). Taking consecutively $i = 2$ and $i = 1$, we obtain

$$\lambda^2_1 = 0, \quad \lambda^2_2 = -1. \tag{7.2.69}$$

Note that $\lambda^2_1 = 0$ and the inequalities (7.1.8) imply that $\lambda^1_1 \neq 0$ and $\lambda^2_2 \neq 1$. The latter inequality was used while getting $\lambda^2_2 = -1$.

Further, it follows from (7.1.27) and (7.2.69) that

$$\lambda^2_{jk} = 0, \quad \mu^2_{jk} = 0. \tag{7.2.70}$$

By (7.2.70), equations (7.2.46) take the form

$$\mu^1_{11} = 0, \quad \mu^1_{12} = \mu^1_{21}\lambda^1_1, \quad \mu^1_{22} = \mu^1_{21}\lambda^1_2. \tag{7.2.71}$$

Next, by (7.2.71), the relations (7.2.55) for $i = k = 1$ and $j = l = 2$ imply $\mu^1_{21} = 0$. This and the relations (7.2.71) mean that all

$$\mu^i_{jk} = 0. \tag{7.2.72}$$

We know from Proposition 7.1.7 that the equations (7.2.72) mean that the basis affinor is covariantly constant in the connection γ_{123} on the foliation λ_1. Let us find another geometric meaning of equations (7.2.72).

The group G is a coordinate quasigroup of the 3-subweb $[1,2,3]$. Denote by $'G$ a coordinate quasigroup of the 3-subweb $[1,2,4]$. The first group of equations (7.2.10): $\underset{1}{\bar{\omega}}^i = \lambda^i_j \underset{1}{\omega}^j$ defines a mapping F of the group G onto the group $'G$. According to (7.2.67), the invariant forms of G are expressed in terms of the variables x^1, x^2 only. Therefore, the mapping F is isotopic if and only if $\lambda^i_j = \lambda^i_j(x^1, x^2)$. It follows from equations (7.1.27) and (7.2.47) that this will be the case if and only if $\mu^i_{jk} = 0$. Therefore, conditions (7.2.72) mean that the groups G and $'G$ are isotopic. (Note that by Albert's theorem (see [Al 43] or [Be 67]) the isotopic groups are isomorphic.) Since G and $'G$ are coordinate groups of the 3-subwebs $[1,2,3]$ and $[1,2,4]$, then these 3-subwebs are equivalent.

By virtue of (7.2.72) and (7.2.47), the differential equations (7.1.27) take the form

$$d\lambda^i_j = (\lambda^i_p a^p_{jk} + \lambda^p_k \lambda^q_j a^i_{pq}) \underset{1}{\omega}^k. \tag{7.2.73}$$

Using (7.2.69), we can write equations (7.2.73) in the form

$$d\lambda^1_1 = \lambda^1_1 d\ln x^2, \quad d\lambda^1_2 = -2\lambda^1_1 dx^1/x^2. \tag{7.2.74}$$

Integrating (7.2.74), we find that

$$\lambda^1_1 = k_1 x^2, \quad \lambda^1_2 = -2k_1 x^1 - 2k_2, \tag{7.2.75}$$

where k_1 and k_2 are constants and $k_1 \neq 0$.

Let us find the closed form equations of the leaves of all the foliations of the web $W(4,2,2)$. By (7.2.67), the equations $\underset{1}{\omega}^i = 0$ are equivalent to $dx^i = 0$, $i = 1,2$. It follows from this that the equations of the leaves of λ_1 are $x^i = a^i = \text{const}$. Similarly we find from (7.2.68) that the equations of the leaves of λ_2 are $\bar{y}^i = \bar{b}^i = \text{const}$.

The leaves of λ_3 are determined by the equations $\underset{1}{\omega}^i + \underset{2}{\omega}^i = 0$ (see (7.1.7)) or

$$\frac{dx^1}{x^2} + \bar{y}^2 d\bar{y}^1 = 0, \quad d\ln x^2 + d\ln \bar{y}^2 = 0. \tag{7.2.76}$$

Integrating (7.2.76), we get

$$x^1 + c^2 \bar{y}^1 = c^1, \quad x^2 \bar{y}^2 = c_2 \tag{7.2.77}$$

where c_1, c_2 are constants.

The equations of the leaves of λ_4 are $\lambda^i_j \underset{1}{\omega}^j + \underset{2}{\omega}^i = 0$ (see (7.1.7)). For $i = 2$, by (7.2.69), this gives

$$\frac{\bar{y}^2}{x^2} = d^2, \tag{7.2.78}$$

where d^2 is a constant. Similarly for $i = 1$ after integration we get

$$\frac{k_1x^1 + k_2 + \bar{y}^1\bar{y}^2}{x^2} = d^1. \tag{7.2.79}$$

Excluding from (7.2.77), (7.2.78), (7.2.79) the local coordinates x^i, $\bar{y}^j$ and denoting $y^2 = \bar{y}^2$, $y^1 = y^2\bar{y}^1$, we find the equations of our web $W(4,2,2)$ in the form:

$$z^1 = x^1 + x^2y^1, \quad z^2 = x^2y^2, \tag{7.2.80}$$

$$u^1 = \frac{(k_1x^1 + k_2)}{x^2} + \frac{y^1}{x^2}, \quad u^2 = \frac{y^2}{x^2}. \quad \blacksquare \tag{7.2.81}$$

Corollary 7.2.11 *The canonical connection γ_{123} induced on X^4 by the 3-subweb $[1,2,4]$ is one of the Cartan curvature-free affine connections,*

Proof. This follows from (7.2.72) and (7.2.13) where $\omega_j^i = 0$. ■

Corollary 7.2.12 *The web $W(4,2,2)$ defined by (7.2.80) and (7.2.81) is not a group four-web.*

Proof. Excluding from (7.2.80) and (7.2.81) variables x^i , we get the equations of the 3-subweb $[2,3,4]$ in the form:

$$u^1 = \frac{z^1y^2 + y^1(k_1 - y^2(z^2)^2)}{k_1 - y^2}, \quad u^2 = z^2y^2. \tag{7.2.82}$$

It can be checked by inspection that a loop defined by (7.2.82) is not associative. Hence the 3-subweb, and consequently the web $W(4,2,2)$ are not a group 3-web and 4-web.

Note that Corollary 7.2.11 also follows from the fact that the non-zero components of the basis affinor of the web $W(4,2,2)$ under consideration satisfy the differential equation (7.2.75) , and therefore, in general the basis affinor is not covariantly constant in the whole X^4 which, by Definition 7.2.4, means that the $W(4,2,2)$ is not a group 4-web. ■

7.3 The Canonical Expansions of the Equations of a Pair of Orthogonal Quasigroups Associated with a Web $W(4,2,r)$

7.3.1 A Pair of Orthogonal Quasigroups Associated with a Web $W(4,2,r)$

Two binary quasigroups are connected with a web $W(4,2,r)$. A leaf $x^i = a^i$ $=$ const of the foliation λ_1 and a leaf $y^i = b^i =$ const of the foliation λ_2 define the point (a^j, b^k) on the manifold X^{2r} through which a leaf of λ_3 and a leaf of λ_4 pass. In other words, we have two mappings

$$A : X_1 \times X_2 \rightarrow X_3, \qquad B : X_1 \times X_2 \rightarrow X_4, \tag{7.3.1}$$

where X_ξ, $\xi = 1, 2, 3, 4$, are differentiable manifolds of the same dimension r, and the mappings A and B satisfy conditions similar to those for the mapping f in Definition 3.1.1. The quasigroups A and B defined by (7.3.1) have closed form equations:

$$z^i \doteq A^i(x^j, y^k), \quad i, j, k = 1, \ldots, r, \tag{7.3.2}$$

$$u^i = B^i(x^j, y^k), \quad i, j, k = 1, \ldots, r. \tag{7.3.3}$$

The quasigroups A and B are *orthogonal* [Be 71] because, by the definition of four-web, the system of equations

$$A^i(x^j, y^k) = \alpha^i, \qquad B^i(x^j, y^k) = \beta^i$$

has a unique solution (a^j, b^k). The constant parameters of the leaves of the foliations λ_1 and λ_2 passing through the point of intersection of the leaves $z^i = \alpha^i$ and $u^j = \beta^j$ of the foliations λ_3 and λ_4 give this solution.

7.3.2 The Canonical Expansions of the Equations of the Quasigroups A and B

In a neighbourhood of any point $p \in D \subset X^{2r}$ it is possible to introduce coordinates (x^j, y^k) in such a way that for the point p we will have $x^j = y^k = 0$. Then in this neghbourhood the right members of the equations (7.3.2) (and (7.3.3)) can be expanded in Taylor series. Using isotopic transformations $\bar{x}^i = \alpha^i(x^j)$, $\bar{y}^i = \beta^i(y^j)$, $\bar{z}^i = \gamma^i(z^j)$ of the variables x^i, y^i, z^i (see Definition 3.5.1) and the method outlined in Theorem 3.5.2, we can reduce these expansions to the following form (see [AS 71a]:

$$z^i = x^i + y^i + \alpha^i_{jk} x^j y^k + \frac{1}{2}(\alpha^i_{jkr+l} x^j x^k y^l + \alpha^i_{jr+kr+l} x^j y^k y^l) + o(\rho^2) \tag{7.3.4}$$

where $\rho = \max(|x^i|, |y^i|)$ and $\alpha^i_{(jk)} = 0$, $\alpha^i_{[jk]\, r+l} = 0$, $\alpha^i_{j[r+k\, r+l]} = 0$.

The expansions (7.3.4) are invariant only under concordant isomorphic transformations (cf. (3.5.39)):

$$x^i = A^i_j \bar{x}^j, \quad y^i = A^i_j \bar{y}^j, \quad z^i = A^i_j \bar{z}^j. \tag{7.3.5}$$

We write now the expansions of the right members of equations (7.3.3) in Taylor series:

$$u^i = \gamma^i + \gamma^i_j x^j + \gamma^i_{r+j} y^j + \tfrac{1}{2}\Big(\gamma^i_{jk} x^j x^k + 2\gamma^i_{j\, r+k} x^j y^k + \gamma^i_{r+j\, r+k} y^j y^k\Big)$$
$$+\tfrac{1}{6}\Big(\gamma^i_{jkl} x^j x^k x^l + 3\gamma^i_{jk\, r+l} x^j x^k y^l + 3\gamma^i_{j\, r+k\, r+l} x^j y^k y^l + \gamma^i_{r+j\, r+k\, r+l} y^j y^k y^l\Big) + o(\rho^3). \tag{7.3.6}$$

The matrices (γ^i_j) and (γ^i_{r+j}) in (7.3.6) are non-singular since the operation of the quasigroup B is invertible.

We perform consecutively the following isotopic transformations:

$$u^i_1 = u^i - \gamma^i, \quad u^i_2 = \tilde{\gamma}^i_{r+j} u^j_1,$$
$$u^i_3 = u^i_2 - \frac{1}{2}\gamma^i_{r+j\,r+k} u^j_2 u^k_2 - \frac{1}{6}\gamma^i_{r+j\,r+k\,r+l} u^j_2 u^k_2 u^l_2 + o(\rho^3),$$

where $(\tilde{\gamma}^i_{r+j})$ is the inverse matrix of (γ^i_{r+j}) and the variables x^i, y^j, and z^k are not changed. If we denote by $\lambda^i_j = \tilde{\gamma}^i_{r+k}\lambda^k_j$ and write u^i instead of u^i_3, then we reduce the equations (7.3.6) to the form:

$$u^i = \lambda^i_j x^j + y^i + \frac{1}{2}\left(\gamma^i_{jk} x^j x^k + 2\gamma^i_{j\,r+k} x^j y^k\right) + \frac{1}{6}\left(\gamma^i_{jkl} x^j x^k x^l + 3\gamma^i_{jk\,r+l} x^j x^k x^l + 3\gamma^i_{j\,r+k\,r+l} x^j y^k y^l\right) + o(\rho^3). \tag{7.3.7}$$

The equations (7.3.4) and (7.3.7) give

$$A^i(x^j, 0) = x^i, \quad A^i(0, y^k) = y^i, \quad B^i(0, y^k) = y^i. \tag{7.3.8}$$

It follows from (7.3.8) that the quasigroup A is a loop with the unit $e(e^i = 0)$, and the unit e of the loop A is the left unit of the quasigroup B.

The expansions (7.3.4) and (7.3.7) are preserved only under concordant isomorphic transformations of variables x^i, y^i, z^i, u^i:

$$x^i = A^i_j \bar{x}^j, \quad y^i = A^i_j \bar{y}^j, \quad z^i = A^i_j \bar{z}^j, \quad u^i = A^i_j \bar{u}^j. \tag{7.3.9}$$

Definition 7.3.1 The expansions (7.3.4) and (7.3.7) of the equations (7.3.2) and (7.3.3) of the quasigroups A and B are called the *canonical expansions.*

Note that they are *canonical* only for the terms up to the third order. This is enough for our purposes.

Using the method outlined in the proof of Theorem 3.3.1, we can prove that the torsion and curvature tensors of the 3-subweb [1, 2, 3] at the point $x^i = y^i = 0$ are expressed in terms of the coefficients of the expansions (7.3.4) by the following formulae (see [AS 71a]):

$$a^i_{jk} = -\alpha^i_{jk}, \tag{7.3.10}$$

$$b^i_{jkl} = \alpha^i_{k\,r+j\,r+l} - \alpha^i_{kj\,r+l} + \alpha^m_{jl}\alpha^i_{km} - \alpha^m_{kj}\alpha^i_{ml}. \tag{7.3.11}$$

We will find now the expressions of the torsion and curvature tensors of the 3-subweb [1, 2, 4] in terms of the coefficients of the expansion (7.3.7). For this, we first differentiate equations (7.3.2) and (7.3.3). It gives

$$dz^i = \underset{1}{A^i_j} dx^j + \underset{2}{A^i_j} dy^j, \tag{7.3.12}$$

$$du^i = \underset{1}{B}{}^i_j dx^j + \underset{2}{B}{}^i_j dy^j, \tag{7.3.13}$$

where

$$\underset{1}{A}{}^i_j = \frac{\partial A^i}{\partial x^j}, \quad \underset{2}{A}{}^i_j = \frac{\partial A^i}{\partial y^j}, \quad \underset{1}{B}{}^i_j = \frac{\partial B^i}{\partial x^j}, \quad \underset{2}{B}{}^i_j = \frac{\partial B^i}{\partial y^j}.$$

We denote further by $(\underset{\alpha}{\tilde{A}}{}^i_j)$ and $(\underset{\alpha}{\tilde{B}}{}^i_j)$ the inverse matrices of the matrices $(\underset{\alpha}{A}{}^i_j)$ and $(\underset{\alpha}{B}{}^i_j)$, $\alpha = 1, 2$, and set

$$\underset{3}{\omega}{}^i = -dz^i, \quad \underset{1}{\omega}{}^i = \underset{1}{A}{}^i_j dx^j, \quad \underset{2}{\omega}{}^i = \underset{2}{A}{}^i_j dy^j, \quad \underset{4}{\omega}{}^i = -\underset{2}{A}{}^i_m \underset{2}{\tilde{B}}{}^m_k du^k. \tag{7.3.14}$$

Then the equations (7.3.12) and (7.3.13) take the form (7.1.7) where

$$\lambda^i_j = \underset{2}{A}{}^i_p \underset{2}{\tilde{B}}{}^p_q \underset{1}{B}{}^q_s \underset{1}{\tilde{A}}{}^s_j. \tag{7.3.15}$$

It follows from (7.3.4) and (7.3.7) that at the point $x^i = y^i = 0$ we have

$$\begin{cases} \underset{1}{A}{}^i_j = \underset{2}{A}{}^i_j = \underset{2}{B}{}^i_j = \delta^i_j, \quad \underset{1}{B}{}^i_j = \lambda^i_j, \\ \dfrac{\partial^2 A^i}{\partial x^j \partial y^k} = \alpha^i_{jk}, \ldots, \dfrac{\partial^3 B^i}{\partial x^j \partial y^k \partial y^l} = \gamma^i_{j\,r+k\,r+l}. \end{cases} \tag{7.3.16}$$

We can see from (7.3.15) and (7.3.16) that at this point the components of the basis affinor λ^i_j of the web $W(4,2,r)$ coincide with the coefficients of x^j in equations (7.3.7) – this was the reason for our notation in (7.3.7).

The calculation of the torsion and curvature tensors of the 3-subweb [1, 2, 4] at the point $x^i = y^i = 0$ is similar to that of the 3-subweb [1, 2, 3] (see [AS 71a]) and gives the following formulae:

$$\underset{124}{a}{}^i_{jk} = \gamma^i_{l[r+j}\tilde{\lambda}^l_{k]}, \tag{7.3.17}$$

$$\begin{aligned} \underset{124}{b}{}^i_{jkl} &= \gamma^i_{m\,r+j\,r+l}\tilde{\lambda}^m_k - \gamma^i_{mn\,r+l}\tilde{\lambda}^m_j\tilde{\lambda}^n_k + \gamma^i_{t\,r+l}\gamma^p_{mn}\tilde{\lambda}^t_p\tilde{\lambda}^m_j\tilde{\lambda}^n_k \\ &+\gamma^m_{a\,r+l}\gamma^i_{b\,r+m}\tilde{\lambda}^a_j\tilde{\lambda}^b_k + \gamma^m_{a\,r+j}\gamma^i_{b\,r+l}\tilde{\lambda}^a_k\tilde{\tilde{\lambda}}{}^b_m, \end{aligned} \tag{7.3.18}$$

where $(\tilde{\lambda}^i_j)$ and $(\tilde{\tilde{\lambda}}{}^i_j)$ are the inverse matrices of the matrices (λ^i_j) and $(\delta^i_j - \lambda^i_j)$.

The following theorem combined all results obtained in this section.

Theorem 7.3.2 *In a neighbourhood of the point $x^i = y^i = 0$ the Taylor expansions of the right members of the equations (7.3.2) and (7.3.3) of the quasigroups A and B can be reduced with the help of isotopic transformations to the canonical form (7.3.4) and (7.3.7) which is preserved only under the concordant isomorphic transformations of the variables of the form (7.3.9). In addition, the quasigroup A is a loop and its unit $e(e^i = 0)$ is the left unit of the quasigroup B. The torsion and curvature tensors of the 3-subwebs [1, 2, 3] and [1, 2, 4] at the point $x^i = y^i = 0$ are expressed in terms of the coefficients of the canonical expansions (7.3.4) and (7.3.7) by the formulae (7.3.10), (7.3.11), and (7.3.17), (7.3.18).* ■

Later on we shall need expressions for the Pfaffian derivatives of the basis affinor λ^i_j at the point $x^i = y^i = 0$ in the canonical connection γ_{123} in terms of the coefficients of the canonical expansions (7.3.4) and (7.3.7). From (7.1.27), (7.1.31), (7.3.4), (7.3.7), (7.3.11), and (7.3.18) we find that at the point $x^i = y^i = 0$:

$$\lambda^i_{jk} = \gamma^i_{jk} - \lambda^q_j \gamma^i_{k\,r+q} + \lambda^i_p \alpha^p_{kj}, \quad \mu^i_{jk} = \gamma^i_{j\,r+k} - \lambda^p_j \alpha^i_{pk}, \tag{7.3.19}$$

$$\begin{cases} \alpha^i_{jkl} = \lambda^p_j (b^i_{plk} - \lambda^q_l \underset{124}{b}{}^i_{pqk}) + \tilde{\lambda}^t_s \mu^i_{tk} \gamma^s_{jl} + \mu^i_{pk}\alpha^p_{lj} - \lambda^p_j \lambda^b_m \gamma^m_{l\,r+p} \mu^i_{bk}, \\ \beta^i_{jkl} = \lambda^i_t b^t_{jkl} - \lambda^p_j \lambda^q_k \underset{124}{b}{}^i_{pql} - \lambda^t_{jk} \alpha^i_{tl} + \tilde{\lambda}^t_s \gamma^i_{t\,r+l} \lambda^s_{jk}, \\ \mu^i_{jkl} = \gamma^i_{j\,r+k\,r+l} - \lambda^t_j \alpha^i_{t\,r+k\,r+l} - \gamma^i_{s\,r+k}\alpha^s_{jl} + \lambda^t_j \alpha^i_{pk} \alpha^p_{tl} \\ \qquad + \mu^i_{jp}\alpha^p_{kl} + \mu^i_{pk}\alpha^p_{jl} - \mu^p_{jk}\alpha^i_{pl} - (\gamma^t_{j\,r+l} - \lambda^t_s \alpha^s_{jl})\alpha^i_{tk}. \end{cases} \tag{7.3.20}$$

In conclusion we shall prove the following lemma.

Lemma 7.3.3 *If the basis affinor λ^i_j of a web $W(4,2,r)$ is covariantly constant on the foliation λ_ξ in one of the canonical affine connections $\gamma_{\xi\eta\zeta}$, ξ fixed, then it is covariantly constant in all other canonical connections $\gamma_{\xi\eta\zeta}$.*

Proof. To prove the lemma, we must write 24 expressions $\overset{\xi\eta\zeta}{\nabla} \lambda^i_j$ using the formulae (7.2.13), (7.2.21), (7.2.28), (7.2.14), (7.2.22), (7.2.29), (7.2.16), (7.2.23), (7.2.30) and the fact that under parastrophies the connection forms and the torsion and curvature tensors of a three-web $[\xi, \eta, \zeta]$ undergo the following transformations (see [AS 71b]):

$$\begin{cases} \underset{\alpha\beta\gamma}{\omega}{}^i_j = \underset{\beta\alpha\gamma}{\omega}{}^i_j, \quad \underset{\beta\gamma\alpha}{\omega}{}^i_j = \underset{\alpha\beta\gamma}{\omega}{}^i_j - 2\underset{\alpha\beta\gamma}{a}{}^i_{jk}\underset{\beta}{\omega}{}^k, \quad \underset{\gamma\alpha\beta}{\omega}{}^i_j = \underset{\alpha\beta\gamma}{\omega}{}^i_j = -2\underset{\alpha\beta\gamma}{a}{}^i_{jk}\underset{\alpha}{\omega}{}^k; \\ \underset{\alpha\beta\gamma}{a}{}^i_{jk} = -\underset{\beta\alpha\gamma}{a}{}^i_{jk}, \quad \underset{\beta\gamma\alpha}{a}{}^i_{jk} = \underset{\gamma\alpha\beta}{a}{}^i_{jk} = \underset{\alpha\beta\gamma}{a}{}^i_{jk}; \\ \underset{\alpha\beta\gamma}{b}{}^i_{jkl} = -\underset{\beta\alpha\gamma}{b}{}^i_{jlk}, \quad \underset{\alpha\beta\gamma}{b}{}^i_{jkl} = \underset{\beta\gamma\alpha}{b}{}^i_{klj} = \underset{\gamma\alpha\beta}{b}{}^i_{ljk}. \end{cases} \tag{7.3.21}$$

Then we find that the necessary and sufficient conditions for the basis affinor λ^i_j to be covariantly constant on the foliation λ_ξ, $\xi = 1, 2, 3, 4$, in the connection $\gamma_{\xi\eta\zeta}$ do not depend on η and ζ and have respectively the form

$$\mu^i_{jk} = o, \quad \lambda^i_{jk} = 0, \quad A^i_{jk} = 0, \quad B^i_{jk} = 0 \tag{7.3.22}$$

where

$$\begin{aligned} A^i_{jk} &= \lambda^i_{jk} - \mu^i_{jk} - 2\lambda^i_p a^p_{jk} + 2\lambda^p_j a^i_{pk}, \\ B^i_{jk} &= \lambda^i_{kj} - \lambda^i_p \mu^p_{kj} - 2\lambda^i_p a^p_{kj} + 2\lambda^i_p \lambda^t_k a^p_{jt}. \end{aligned} \tag{7.3.23}$$

For example, for the connections γ_{342}, γ_{134}, γ_{412} we have

$$\begin{cases} \overset{342}{\nabla} \lambda^i_j = -\tilde{\tilde{\lambda}}^m_k B^i_{jm} \underset{3}{\omega}{}^k + \lambda^i_l \tilde{\lambda}^t_j \tilde{\tilde{\lambda}}^m_k A^l_{tm} \underset{4}{\omega}{}^k, \\ \overset{134}{\nabla} \lambda^i_j = A^i_{jk} \underset{1}{\omega}{}^k - (\delta^i_l - \lambda^i_l) \tilde{\tilde{\lambda}}^m_j \mu^l_{mk} \underset{3}{\omega}{}^k, \\ \overset{412}{\nabla} \lambda^i_j = B^i_{jk} \underset{1}{\omega}{}^k - \lambda^i_l \mu^l_{mk} \tilde{\lambda}^m_j \underset{4}{\omega}{}^k. \end{cases} \tag{7.3.24}$$

For these connections the lemma follows from (7.3.24). For some foliations λ_ξ we derive the conditions (7.3.22) directly from (7.3.24). For others we need to contract a condition derived from (7.3.24) with the appropriate matrix which is a product of the matrices (λ^i_j), $(\delta^i_j - \lambda^i_j)$ and their inverse matrices.

7.4 Webs $W(4,2,r)$ Satisfying the Desargues and Triangle Closure Conditions

7.4.1 Webs $W(4,2,r)$ Satisfying the Desargues Closure Condition D_1

Definition 7.4.1 We shall say that on a web $W(4,2,r)$ the *Desargues closure condition* D_1 is realised if at any point $P \in D$ the figure represented in Figure 7.1 is closed.

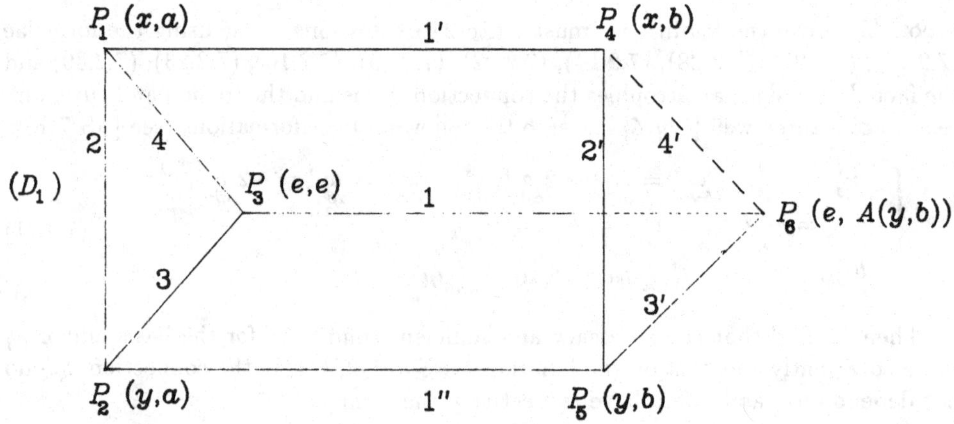

Figure 7.1

As usually, on Figure 7.1 segments represent the leaves of the web.

Theorem 7.4.2 *On a web* $W(4,2,r)$ *the figures* D_1 *are closed if and only if the transformation*

$$\bar{x}^i = \lambda^i_j x^j + \frac{1}{2}\gamma^i_{jk} x^j x^k + \frac{1}{6}\gamma^i_{jkl} x^j x^k x^l + \dots \tag{7.4.1}$$

gives an isotopy of the quasigroups A and B. In this case all 3-subwebs $[1,\xi,\eta]$ are group webs and the isotopy mentioned above is an isomorphism between A and B.

Proof. We introduce coordinates on the web in such a way that the summits have the coordinates indicated on Figure 7.1 In particular, the point P_3 has coordinates (e,e).

As $A(e,e)=0$, the condition D_1 can be written in the form

$$A(y,a)=0, \quad B(x,a)=0 \to B(x,b)=A(y,b). \tag{7.4.2}$$

From (7.4.2), (7.3.4), and (7.3.7) we obtain that

$$\begin{cases} \gamma^i_{j\,r+k} = \lambda^a_j \alpha^i_{ak}, \\ \gamma^i_{j\,r+k\,r+l} = \lambda^a_j \alpha^i_{a\,r+k\,r+l}, \\ \gamma^i_{jk\,r+l} = \lambda^a_j \lambda^b_k \alpha^i_{ab\,r+l} + \gamma^a_{jk}\alpha^i_{al}. \end{cases} \tag{7.4.3}$$

The equations (7.4.3) are necessary and sufficient for conditions D_1 to be realised.

It follows from (7.4.3) and (7.3.19) that at the point $x^i = y^i = 0$ we have

$$\mu^i_{jk} = 0, \tag{7.4.4}$$

i.e., the basis affinor is covariantly constant on the foliation λ_1 in the connection γ_{123}. It follows from (7.4.4) and (7.1.31) that at the point $x^i = y^i = 0$ we have

$$\mu^i_{jkl} = 0, \qquad \alpha^i_{jkl} = 0. \tag{7.4.5}$$

The equations (7.3.20), (7.4.4), and (7.4.5) give the equation

$$(\delta^q_l - \lambda^q_l) b^l_{iqk} = 0. \tag{7.4.6}$$

Using (7.1.8), we find from (7.4.6) that

$$b^i_{jkl} = 0. \tag{7.4.7}$$

The equations (7.3.10), (7.3.11), (7.3.17), and (7.3.18) show that

$$\underset{124}{a}{}^i_{jk} = a^i_{jk}, \tag{7.4.8}$$

$$\underset{124}{b}{}^i_{jkl} = b^i_{jkl}. \tag{7.4.9}$$

Further, it follows from (7.4.5), (7.4.6), (7.4.7), and (7.2.30) that

$$\underset{124}{b}{}^i_{jkl} = \underset{314}{b}{}^i_{jkl}. \tag{7.4.10}$$

The equations (7.4.7) and (7.4.10) mean that the 3-subwebs $[1,\xi,\eta]$ are group three-webs (see Table 1.1, Section 1.2).

The conditions (7.4.8) and (7.4.9) did not appear by chance. If we substitute the values of $\gamma^i_{j\,r+k}$, $\gamma^i_{jk\,r+l}$ and $\gamma^i_{j\,r+k\,r+l}$ from (7.4.3) into (7.3.7) and make the isotopic transformation (7.4.1), the expansions (7.3.7) will have the form

$$u^i = \bar{x}^i + y^i + \alpha^i_{jk}\bar{x}^j y^k + \frac{1}{2}(\alpha^i_{jk\,r+l}\bar{x}^j\bar{x}^k y^l + \alpha^i_{j\,r+k\,r+l}\bar{x}^j y^k y^l) + o(\rho^3). \tag{7.4.11}$$

The expansions (7.3.4) and (7.4.11) imply that the quasigroups A and B are isotopic. As both are groups, they are isomorphic (see [Al 43] or [Be 67]).

Thus, the relations (7.4.3) give:

(i) The covariant constancy of the basis affinor on the foliation λ_1 in the connection γ_{123},
(ii) The vanishing of the curvature tensors of the 3-subwebs $[1, \xi, \eta]$, and
(iii) The isomorphism of the groups A and B which is realised by the isotopy (7.4.1).

It is easy to show that the converse statement is also true: if the isotopic transformation (7.4.1) reduces to the isotopy of the quasigroups A and B, then the relations (7.4.3) hold and the figures D_1 are closed on the web $W(4,2,r)$. ■

Note that there is a criterion for any figure D_ξ which is similar to that in Theorem 7.4.2.

The next theorem gives another criterion for the closure of the figures D_1.

Theorem 7.4.3 *On a web $W(4,2,r)$ the figures D_1 are closed if and only if the basis affinor is covariantly constant on the foliation λ_1 in one of the canonical affine connections $\gamma_{1\xi\eta}$ and all the 3-subwebs $[1,\xi,\eta]$ are group webs.*

Proof. The *necessity* of this statement has been already proved. To proof its *sufficiency*, let us note that the covariant constancy of the basic affinor λ^i_j on the foliation λ_1 in one of the connections $\gamma_{1\xi\eta}$ implies (7.4.4) and hence the first part of conditions (7.4.3) and conditions (7.4.5). The second part of conditions (7.4.3) follows from (7.4.5). The third part of conditions (7.4.3) is obtained from (7.4.9), (7.3.18), and the first and second parts of (7.4.3). ■

We now establish one more criterion for the closure of the figures D_1.

Suppose that the 3-subwebs $[1,2,3]$, $[1,2,4]$, and $[3,1,4]$ are group webs. Then Table 1.1 of Section 1.2 and (7.2.16) and (7.2.30) give

$$\begin{cases} b^i_{jkl} = 0, \\ \alpha^i_{jlp} = \tilde{\lambda}^m_s \lambda^s_{jp} \mu^i_{ml}, \\ \mu^i_{jlm} = \tilde{\lambda}^t_s \lambda^s_{jm} \mu^i_{tl} + \tilde{\tilde{\lambda}}^t_q \mu^i_{tl} (\lambda^q_{jm} - \mu^q_{jm}) - 2\mu^i_{jt} a^t_{lm} \\ \qquad +2a^i_{tm} \mu^t_{jl} - 2a^t_{jm} \mu^i_{rl} \tilde{\tilde{\lambda}}^r_t + 2\lambda^s_j a^t_{sm} \mu^i_{rl} \tilde{\tilde{\lambda}}^r_t. \end{cases} \tag{7.4.12}$$

Substituting α^i_{jkl} and μ^i_{jkl} from (7.4.12) into the second equality of (7.1.31) and then applying exterior differentiation to it, we obtain

$$M^i_{bk} \Lambda^b_{jqp} \underset{1}{\omega^p} \wedge \underset{2}{\omega^q} = 0, \tag{7.4.13}$$

where

$$\begin{cases} M^i_{bk} = \tilde{\tilde{\lambda}}^m_c \lambda^c_b \mu^i_{mk} \\ \Lambda^b_{jpq} = \lambda^b_{jqp} - \tilde{\lambda}^t_s \lambda^s_{jp} \lambda^b_{tq} + \tilde{\tilde{\lambda}}^m_c \lambda^b_{mp} (A^c_{iq} + 2\lambda^c_l a^l_{jq} - 2a^c_{jq}) + 2\lambda^b_c \tilde{\lambda}^t_s \lambda^s_{jp} a^c_{tq}. \end{cases} \tag{7.4.14}$$

It follows from (7.4.13) that

$$M^i_{bk}\Lambda^b_{jqp} = 0. \tag{7.4.15}$$

If, at least for one pair of indices q and p, the rank of the matrix (Λ^b_{jqp}) is equal to r, then it follows from (7.4.15) that $M^i_{bk} = 0$. Hence, contracting (7.4.14) with the product $(\delta^p_j - \lambda^p_j)\tilde{\lambda}^b_p$, we obtain (7.4.4).

We have proved the following theorem.

Theorem 7.4.4 *The figures D_1 are closed on a web $W(4,2,r)$ if and only if the 3-subwebs $[1,\xi,\eta]$ are group webs and, at least for one pair of indices q and p, the rank of the matrix of the tensor $\{\Lambda^b_{jqp}\}$ is equal to r.* ■

Cases for which the rank of (Λ^b_{jqp}) is less than r are not studied here.

The next theorem proves the existence of webs $W(4,2,r)$ on which the figures D_1 are closed.

Theorem 7.4.5 *Webs $W(4,2,r)$ on which the figures D_1 are closed exist and depend on r functions of r variables.*

Proof. Webs $W(4,2,r)$ on which the figures D_1 are closed are defined by equations (7.4.4), (7.4.7), (7.1.21). For them the equations (7.1.22), (7.1.23), (7.1.24), (7.1.27), and (7.1.28) take the form

$$\begin{cases} d\omega^i_j - \omega^k_j \wedge \omega^i_k = 0, & \nabla\lambda^i_j = \lambda^i_{jk}\underset{1}{\omega}^k, \\ \lambda^i_{[jk]} = \lambda^i_p a^p_{jk} + \lambda^p_{[k}\lambda^q_{j]}a^i_{pq}, & a^i_{m[j}a^m_{kl]} = 0. \end{cases} \tag{7.4.16}$$

Taking the exterior derivative of (7.4.16), we get

$$[\nabla\lambda^i_{(jk)} + (\lambda^i_{jl}a^l_{ak} + T^i_{jka})\underset{1}{\omega}^a] \wedge \underset{1}{\omega}^k = 0 \tag{7.4.17}$$

where

$$T^i_{jka} = \lambda^i_{la}a^l_{jk} + a^i_{pq}(\lambda^q_j\lambda^p_{ka} - \lambda^q_k\lambda^p_{ja}).$$

Thus, we have that the number of the unknown functions $\nabla\lambda^i_{(jk)}$ is equal to $q = r(r+1)/2$. The consecutive characters (see [Ca 45]) are

$$s_1 = r^2, \quad s_2 = r(r-1), \quad s_3 = r(r-2), \ldots, s_r = r.$$

The Cartan number Q is equal to

$$Q = s_1 + 2s_2 + \ldots + rs_r = \frac{1}{6}r^2(r+1)(r+2).$$

The number N of arbitrary parameters on which the rth integral element depends, is equal to $N = r^2(r+1)(r+2)/6$. Because $Q = N$, the system is in involution, and its solution depends on r functions of r variables. ■

7.4.2 Group Webs $W(4,2,r)$ and Webs $W(4,2,r)$ Satisfying Two Desargues Closure Conditions, D_1 and D_2

Note first that for the figures D_2 the foliations λ_1 and λ_2 on Figure 7.1 must be transposed (see Figure 7.2):

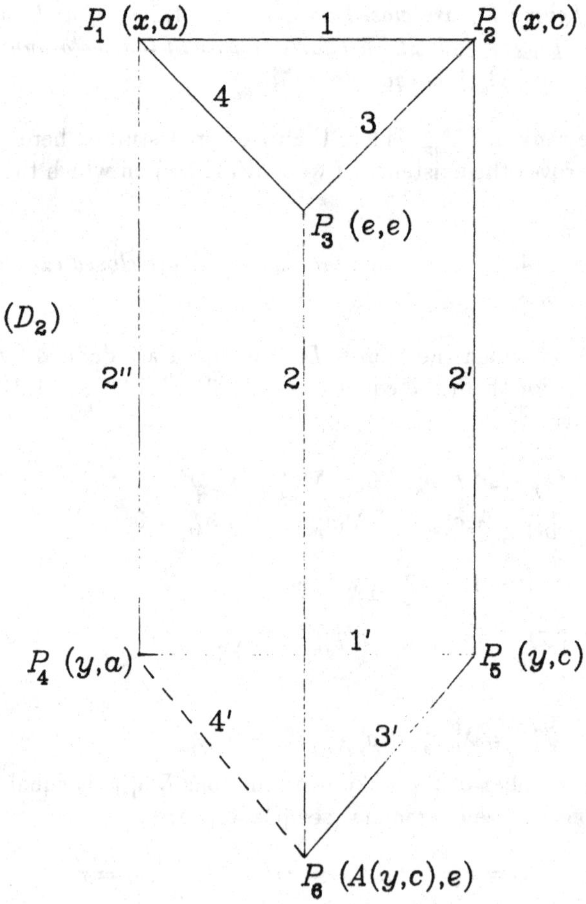

Figure 7.2

To D_2 there corresponds the following system of identities:

$$A(x,c)=0,\;\; B(x,a)=0 \rightarrow B(y,a)=B(A(y,c),0). \tag{7.4.18}$$

The relations (7.4.18), (7.4.4), and (7.4.7) imply

$$\begin{cases} \gamma^i_{j\,r+k} = (\lambda^i_a \alpha^a_{jp} + \gamma^i_{jp})\tilde{\lambda}^p_k, \\ \gamma^i_{jk\,r+l} = (\lambda^i_a \alpha^a_{jk\,r+p} + 2\gamma^i_{(j|b|}\alpha^b_{k)p} + \gamma^i_{jkp})\tilde{\lambda}^p_l, \\ \gamma^i_{j\,r+k\,r+l} = (\lambda^i_a \alpha^a_{j\,r+p\,r+q} + 2\gamma^i_{pb}\alpha^b_{jq} + \gamma^i_{jpq} - \gamma^i_{j\,r+a}\gamma^a_{pq})\tilde{\lambda}^p_{(j}\tilde{\lambda}^q_{k)}. \end{cases} \tag{7.4.19}$$

We obtain from (7.4.19) that

$$\lambda^i_{jk} = 0, \tag{7.4.20}$$

$$\lambda^i_{jkl} = \beta^i_{jkl} = 0, \tag{7.4.21}$$

$$b^i_{jkl} = \underset{124}{b}{}^i_{jkl} = \underset{234}{b}{}^i_{jkl} = 0. \tag{7.4.22}$$

Note that to obtain (7.4.22) we must use (7.1.8), (7.2.23), (7.4.21), and (7.3.20). Equations (7.4.20) mean that the basis affinor λ^i_j is covariantly constant on the foliation λ_2 in the connection γ_{123}.

Now let both the figures D_1 and the figures D_2 be closed on a web $W(4,2,r)$. Then we have (7.4.4) and (7.4.20), i.e., the affinor λ^i_j is covariantly constant on the whole X^{2r} in any affine connection, in particular, in the connections γ_{123} and γ_{124}. Moreover, by (7.4.10) and (7.4.47), all $\underset{\alpha\beta\gamma}{b}{}^i_{jkl} = 0$. Hence the web $W(4,2,r)$ is a group web (see Definition 7.2.4).

We will make an additional normalisation $A^i(x^j, x^k) = 2x^i$ (cf. (3.5.31) and [Ak 69b]). It gives

$$\alpha^i_{(jk\,r+l)} + \alpha^i_{(j\,r+k\,r+l)} = 0. \tag{7.4.23}$$

It follows from (7.4.23), (7.4.3), and (7.4.19) that

$$\gamma^i_{jkl} = 0, \tag{7.4.24}$$

$$\begin{cases} \lambda^i_a \alpha^a_{jk} = \lambda^a_j \lambda^b_k \alpha^i_{ab}, \\ \lambda^i_a \alpha^a_{jk\,r+l} = \lambda^a_j \lambda^b_k \lambda^c_l \alpha^i_{ab\,r+c}, \\ \lambda^i_a \alpha^a_{j\,r+k\,r+l} = \lambda^a_j \lambda^b_k \lambda^c_l \alpha^i_{a\,r+b\,r+c}. \end{cases} \tag{7.4.25}$$

By (7.4.3), (7.4.19), and (7.4.24), the expansions (7.3.7) take the form

$$\begin{aligned} u^i = \lambda^i_j x^j + y^i + \lambda^t_j \alpha^i_{tk} x^j y^k + \frac{1}{2}(\lambda^a_j \lambda^b_k \alpha^i_{ab\,r+l} x^j y^i \\ + \lambda^a_j \alpha^i_{a\,r+k\,r+l} x^j y^k y^l) + o(\rho^3) \end{aligned} \tag{7.4.26}$$

and, in addition, the equations (7.4.25) hold. We have proved

Theorem 7.4.6 *A web $W(4,2,r)$ is a group web if and only if on it both the figures D_1 and D_2 are closed. In addition, the expansions (7.3.7) of the equations of the quasigroup B have the form (7.4.26) and the isomorphism of the groups A and B is established by the transformation $\bar{x}^i = \lambda^i_j x^j$. In this case the coefficients of the expansions (7.3.4) satisfy the relations (7.4.25).* ■

Note that if in (7.3.7) we perform the admissible transformations

$$x^i = \lambda^i_j \bar{x}^j, \quad y^i = \lambda^i_j \bar{y}^j, \quad z^i = \lambda^i_j \bar{z}^j, \tag{7.4.27}$$

then, by (7.4.27), the relations (7.4.25) are equivalent to

$$\bar{\alpha}^i_{jk} = \alpha^i_{jk}, \quad \bar{\alpha}^i_{jk\, r+l} = \alpha^i_{jk\, r+l}, \quad \bar{\alpha}^i_{j\, r+k\, r+l} = \alpha^i_{j\, r+k\, r+l}. \tag{7.4.28}$$

We now consider a group web $W(4,2,2)$.

Proposition 7.4.7 *A group web $W(4,2,2)$ is parallelisable.*

Proof. Let us investigate the first of the equations (7.4.25) in our case when $r = 2$. The tensor α^i_{jk} has two essential components in this case: α^1_{12} and α^2_{12}. It follows from (7.4.25) that these components satisfy the system:

$$\begin{cases} (\Delta - \lambda^1_1)\alpha^1_{12} - \lambda^1_2 \alpha^2_{12} = 0, \\ -\lambda^2_1 \alpha^1_{12} + (\Delta - \lambda^2_2)\alpha^2_{12} = 0 \end{cases} \tag{7.4.29}$$

where $\Delta = \det(\lambda^i_j) = \lambda^1_1\lambda^2_2 - \lambda^2_1\lambda^1_2$. The determinant of the system (7.4.29) is equal to $\Delta \cdot \Delta_1$ where $\Delta_1 = \det(\delta^i_j - \lambda^i_j) = \Delta - \lambda^1_1 - \lambda^2_2 + 1$ and, by (7.1.8), this determinant is different from zero. Consequently the system (7.4.29) has only the trivial solution: $\alpha^1_{12} = \alpha^2_{12} = 0$. ■

The next theorem proves the existence of group webs $W(4,2,r)$.

Theorem 7.4.8 *A group web $W(4,2,r)$ exists and depends on $r^2(r+1)/2 - p$ constants where p is the number of independent equations of the system*

$$\lambda^i_p a^p_{jk} = \lambda^p_{[j} \lambda^q_{k]} a^i_{pq}, \quad a^i_{p[j} a^p_{kl]} = 0. \tag{7.4.30}$$

If $p = r^2(r-1)/2$, i.e., p is maximal, a group web $W(4,2,r)$ is parallelisable. This will be the case, for example, if $r = 2$, or if the matrix of the basis affinor is scalar: $\lambda^i_j = \delta^i_j \lambda$, $\lambda \neq 0, 1$.

Proof. A non-parallelisable group web $W(4,2,r)$ is defined by the the following system:

$$\nabla a^i_{jk} = 0, \quad \nabla \lambda^i_j = 0, \quad \lambda^i_p a^p_{jk} + \lambda^p_{[k} \lambda^q_{j]} a^i_{pq} = 0, \quad a^i_{m[j} a^m_{kl]} = 0. \tag{7.4.31}$$

The system (7.4.31) is closed under exterior differentiation. Hence it is completely integrable and determines the group webs $W(4,2,r)$ with the arbitrariness $r^2(r+1)/2 - p$ constants where p is the number of the independent relations (7.4.30). We have seen in the proof of Proposition 7.4.7 that for $r = 2$ the relations (7.4.30) give $a^i_{jk} = 0$ and the web $W(4,2,r)$ is parallelisable. In this case $p = r^2(r-1)/2$, and the number of independent constants will be equal to r^2 , i.e., to the number of the components of the basis affinor λ^i_j. ■

7.4.3 Webs $W(4,2,2)$ Satisfying the Desargues Closure Condition D_{12}

Definition 7.4.9 We shall say that on a web $W(4,2,r)$ the *Desargues closure condition* D_{12} is realised if at any point $P \in D$ the figure represented in Figure 7.3 is closed.

(D_{12})

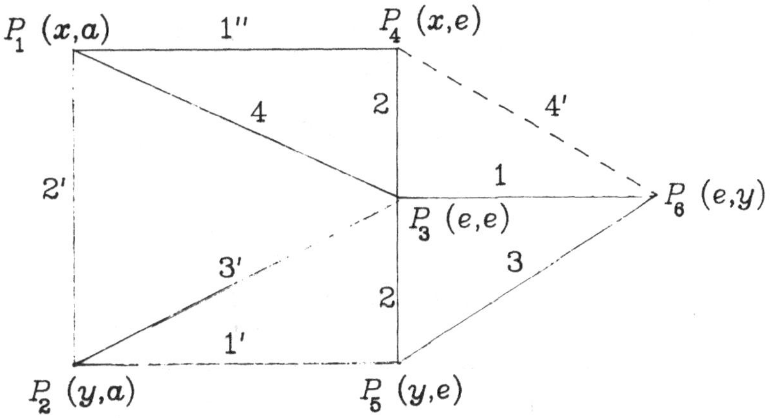

Figure 7.3

We will establish now a criterion for webs $W(4,2,r)$ on which the figures D_{12} are closed.

Theorem 7.4.10 *The figures D_{12} are closed on a web $W(4,2,r)$ if and only if the following condition is satisfied: the transformation $[3,4]$ maps a geodesic line of a leaf F_1 in the connection γ_{123} in a geodesic line of F_2 in the same connection and the hexagonality tensors of the 3-subwebs $[1,2,3]$ and $[1,2,4]$ are equal to each other.*

Proof. As it follows from Figure 7.3, if the figure D_{12} is closed, we have the following system of identities:

$$A(y,a) = 0, \quad B(x,a) = 0 \Longrightarrow B(x,0) = y. \tag{7.4.32}$$

The condition D_{12} is a special case of the condition D_1 and is obtained from it if $b = 0$ (compare (7.4.2)). Relations (7.4.32), (7.3.4), and (7.3.7) imply that (7.4.32) is valid if and only if

$$\lambda^a_{(j}\gamma^i_{k)r+a} = 0, \tag{7.4.33}$$

$$-\lambda^b_{(l}\gamma^i_{jk)r+b} + \lambda^a_{(k}\lambda^b_l\gamma^i_{j)\,r+a\,r+b} - \gamma^t_{(kl}\gamma^i_{j)\,r+l} + (\alpha^i_{ab\,r+c} - \alpha^i_{a\,r+b\,r+c})\lambda^a_{(j}\lambda^b_k\lambda^c_{l)} = 0. \tag{7.4.34}$$

By (7.3.19), conditions (7.4.33) are equivalent to

$$\lambda^l_{(j}\mu^i_{k)l} = 0, \tag{7.4.35}$$

and, by (7.3.12) and (7.3.18), conditions (7.4.34) are equivalent to

$$\underset{124}{b}{}^i_{(jkl)} = b^i_{(jkl)}. \tag{7.4.36}$$

The differentiation of (7.4.35) and use of (7.3.20), (7.3.12), and (7.3.18) give as one of consequences the equation

$$b^i_{(j|l|k)} = \lambda^q_l \underset{124}{b}{}^i_{(j|q|k)}. \tag{7.4.37}$$

It follows from (7.4.36) and (7.4.37) that

$$b^i_{(jkl)} = \tilde{\lambda}^a_k b^i_{(j|a|l)}. \tag{7.4.38}$$

Conversely, (7.4.36) follows from (7.4.38) and (7.4.37).

We now clarify a geometrical meaning of conditions (7.4.35). We saw in Section 7.1 that a line $\Gamma_1 : \underset{1}{\omega}{}^i = 0,\ \ \underset{2}{\omega}{}^i = \xi^i dt$ of a leaf F_1 is transferred into a line Γ^{34}_{121} : $\underset{1}{\omega}{}^i = 0,\ \underset{2}{\omega}{}^i = \lambda^i_j \xi^j dt$ by the transformation $[3,4]$. By (7.1.41), the line Γ_1 is geodesic in the connection γ_{123} if and only if $d\xi^i + \xi^j \omega^i_j = \theta \xi^i$. By (7.1.37), the line Γ^{34}_{121} is geodesic in the same connection γ_{123} if and only if $\nabla \lambda^i_j \xi^j = 0$ or $\mu^i_{jm} \lambda^m_k \xi^j \xi^k = 0$. The last equations are identities if and only if the conditions (7.4.35) are satisfied. ■

Corollary 7.4.11 *On a web $W(4,2,r)$ the figures D_{12} are closed if the transformation $[3,4]$ maps a geodesic line of a leaf F_1 in the connection γ_{123} in a geodessic line of F_1 in the same connection and the 3-subwebs $[1,2,3]$ and $[1,2,4]$ are Bol three-webs of type B_r (see Table 1.1, Section 1.2).*

Proof. The proof follows from the fact that the conditions (7.4.36), (7.4.37), and (7.4.38) are realised when the 3-subwebs $[1,2,3]$ and $[1,2,4]$ both are Bol three-webs of type B because then, by Table 1.1, Section 1.2, $\underset{123}{b}{}^i_{(j|k|l)} = \underset{124}{b}{}^i_{(j|k|l)} = 0$. ■

Similar theorems can be proved for other figures $D_{\xi\eta}$. For example, for the figure D_{21} we have

$$\lambda^i_{(jk)} = 0, \tag{7.4.39}$$

$$\begin{cases} \lambda^a_{(j}\lambda^b_k\lambda^c_{l)} \underset{124}{b}{}^i_{abc} = \lambda^i_t b^t_{(jkl)}, \\ \lambda^a_{(j}\lambda^b_{k)} \underset{124}{b}{}^i_{abl} = \lambda^i_t b^t_{(jk)l}. \end{cases} \tag{7.4.40}$$

The equations (7.4.39) mean that the transformation $[3,4]$ maps a geodesic line Γ_2 of a leaf F_2 in the connection γ_{123} into a geodesic line Γ^{34}_{212} of F_2 in the same connection. Equations (7.4.40) are realised if the 3-subwebs $[1,2,3]$ and $[1,2,4]$ both are Bol three-webs of type B_l (see Table 1.1, Section 1.2) since then $\underset{123}{b}{}^i_{(jk)l} = \underset{124}{b}{}^i_{(jk)l} = 0$.

7.4.4 Webs $W(4,2,r)$ Satisfying the Triangle Closure Conditions

Definition 7.4.12 We shall say that on a web $W(4,2,r)$ the *triangle closure condition* (Δ) of the type $\left(\begin{array}{cc|c} \xi_1 & \xi_2 & \xi_3 \\ \xi_1 & \xi_2 & \xi_4 \end{array}\right)$ is realised if on it the figure represented in Figure 7.4 is closed at each point.

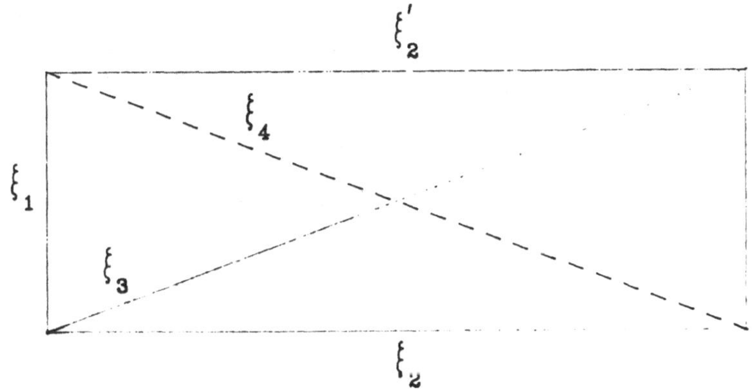

Figure 7.4

Theorem 7.4.13 *On a web $W(4,2,r)$ the figures (Δ) of the types*

$$\left(\begin{array}{cc|c} 1 & 2 & 3 \\ 1 & 2 & 4 \end{array}\right), \quad \left(\begin{array}{cc|c} 1 & 4 & 2 \\ 1 & 4 & 3 \end{array}\right), \quad \left(\begin{array}{cc|c} 1 & 3 & 2 \\ 1 & 3 & 4 \end{array}\right)$$

are closed if and only if the web $W(4,2,r)$ is parallelisable and its basis affinor is scalar: $\lambda^i_j = \delta^i_j \lambda$ where λ is equal respectively to -1, 2, 1/2.

Proof. We observe first the figure (Δ) of the type $\left(\begin{array}{cc|c} 1 & 2 & 3 \\ 1 & 2 & 4 \end{array}\right)$. Let the summits for it have the coordinates indicated on Figure 7.5.

Then $A(x,0) = A(0,y) \rightarrow B(x,y) = 0$. Since $A(x,0) = A(0,y)$ gives $x = y$, we have

$$B(x,x) = 0. \tag{7.4.41}$$

By (7.3.7), the identity (7.4.41) is equivalent to the equations

$$\lambda^i_j = -\delta^i_j, \tag{7.4.42}$$

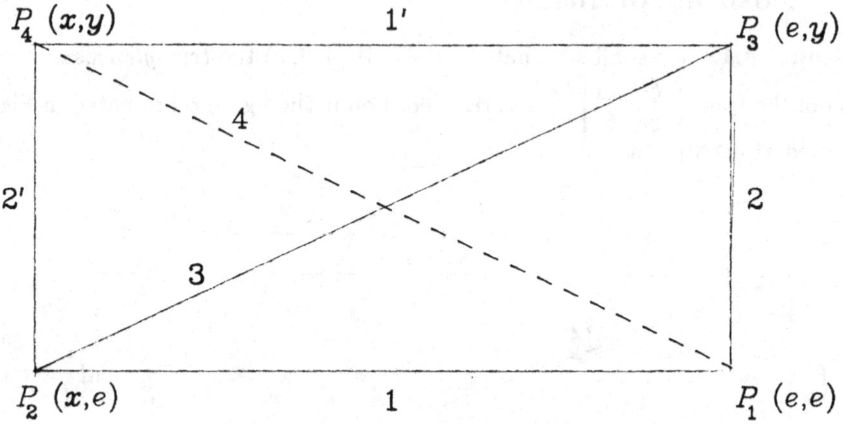

Figure 7.5

$$\gamma^i_{jk} = -2\gamma^i_{(j\,r+k)}, \tag{7.4.43}$$

$$\gamma^i_{jkl} = -3(\gamma^i_{(jk\,r+l)} + \gamma^i_{(j\,r+k\,r+l)}. \tag{7.4.44}$$

It follows from (7.4.42) that

$$\lambda^i_{jk} = \mu^i_{jk} = 0, \quad \lambda^i_{jkl} = \mu^i_{jkl} = \alpha^i_{jkl} = \beta^i_{jkl} = 0. \tag{7.4.45}$$

From (7.3.19), (7.4.42), and (7.4.45) we obtain

$$\gamma^i_{j\,r+k} + \alpha^i_{jk} = 0, \quad \gamma^i_{jk} + \gamma^i_{k\,r+j} = \alpha^i_{kj}. \tag{7.4.46}$$

It follows from (7.4.46) that $\gamma^i_{(jr+k)} = 0$ and consequently

$$\gamma^i_{jk} = 0, \quad \alpha^i_{jk} = 0, \quad \gamma^i_{j\,r+k} = 0. \tag{7.4.47}$$

Under the conditions (7.4.47), equations (7.4.43) are identically satisfied. From (7.3.10) and (7.3.17) we obtain

$$a^i_{jk} = 0, \qquad \underset{124}{a}{}^i_{jk} = 0. \tag{7.4.48}$$

By (7.4.48), it follows from (7.2.22) and (7.2.29) that

$$\underset{134}{a}{}^i_{jk} = \underset{234}{a}{}^i_{jk} = 0. \tag{7.4.49}$$

Formulae (7.2.16), (7.2.23), and (7.2.30) give

$$-\underset{124}{b}{}^{i}_{jkl} = 2\underset{314}{b}{}^{i}_{jkl} = 2\underset{234}{b}{}^{i}_{jkl} = b^{i}_{jkl}. \tag{7.4.50}$$

Equations (7.4.42), (7.4.45), and (7.3.19) imply

$$\begin{cases} -\alpha^{i}_{l\,r+j\,r+k} + \alpha^{i}_{jl\,r+k} + \gamma^{i}_{l\,r+j\,r+k} + \gamma^{i}_{jl\,r+k} = 0, \\ \gamma^{i}_{j\,r+k\,r+l} + \alpha^{i}_{j\,r+k\,r+l} = 0, \\ \gamma^{i}_{jkl} = \alpha^{i}_{kl\,r+j} - \gamma^{i}_{kl\,r+j}. \end{cases} \tag{7.4.51}$$

It follows from (7.4.51) that $\alpha^{i}_{jk\,r+l}$, $\gamma^{i}_{jk\,r+l}$, $\alpha^{i}_{j\,r+k\,r+l}$, $\gamma^{i}_{j\,r+k\,r+l}$ are symmetric with respect lower indices and

$$\gamma^{i}_{jkl} = -2(\gamma^{i}_{jk\,r+l} + \gamma^{i}_{j\,r+k\,r+l}). \tag{7.4.52}$$

The comparison of (7.4.44) and (7.4.52) gives

$$\gamma^{i}_{jk\,r+l} + \gamma^{i}_{j\,r+k\,r+l} = 0. \tag{7.4.53}$$

Then, by (7.4.51), we have

$$\alpha^{i}_{jk\,r+l} = \alpha^{i}_{j\,r+k\,r+l}, \tag{7.4.54}$$

$$\gamma^{i}_{jkl} = 0. \tag{7.4.55}$$

By (7.4.48), (7.4.54), (7.3.11, and (7.4.50), we obtain

$$\underset{\xi\eta\zeta}{b}{}^{i}_{jkl} = 0. \tag{7.4.56}$$

It follows from ((7.4.45), (7.4.48), (7.4.49), and (7.4.56) that all 3-subwebs $[\xi,\eta,\zeta]$ are parallelisable and the basis affinor of the web $W(4,2,r)$ is covariantly constant in X^{2r} in the connection γ_{123}. Hence, by Definition 7.2.1, the whole web $W(4,2,r)$ is parallelisable. ■

It can be easily demonstrated that conditions (7.4.42) are necessary and sufficient for the closure of the figures (Δ) of the types $\left(\begin{array}{cc|c} 3 & 4 & 1 \\ 3 & 4 & 2 \end{array}\right)$, $\left(\begin{array}{cc|c} 3 & 4 & 2 \\ 3 & 4 & 1 \end{array}\right)$, $\left(\begin{array}{cc|c} 1 & 2 & 4 \\ 1 & 2 & 3 \end{array}\right)$. For the closure of the figures (Δ) of the type $\left(\begin{array}{cc|c} 1 & 4 & 2 \\ 1 & 4 & 3 \end{array}\right)$ or $\left(\begin{array}{cc|c} 1 & 4 & 2 \\ 1 & 4 & 3 \end{array}\right)$, $\left(\begin{array}{cc|c} 2 & 3 & 1 \\ 2 & 3 & 4 \end{array}\right)$, $\left(\begin{array}{cc|c} 2 & 3 & 4 \\ 2 & 3 & 1 \end{array}\right)$ and $\left(\begin{array}{cc|c} 1 & 3 & 2 \\ 1 & 3 & 4 \end{array}\right)$ or $\left(\begin{array}{cc|c} 1 & 3 & 4 \\ 1 & 3 & 2 \end{array}\right)$, $\left(\begin{array}{cc|c} 2 & 4 & 1 \\ 2 & 4 & 3 \end{array}\right)$, $\left(\begin{array}{cc|c} 2 & 4 & 3 \\ 2 & 4 & 1 \end{array}\right)$ the conditions $\lambda^{i}_{j} = 2\delta^{i}_{j}$ and $\lambda^{i}_{j} = \delta^{i}_{j}/2$ are necessary and sufficient respectively. In each of these two cases it can be also proved that the web $W(4,2,r)$ is parallelisable.

Corollary 7.4.14 *The triangle figures (Δ) of the types mentioned in Theorem 7.4.13 are closed on a web $W(4,2,r)$ if and only if one of the closure condition D_ξ is realised on it and its basis affinor has the structure mentioned in Theorem 7.4.13.*

Proof. Evidently, if any of the closure conditions given in Theorem 7.4.13 is realised, then the condition D_1 is also realised. It can be easily seen that if the condition D_1 is realised and $\lambda^i_j = \delta^i_j \lambda$ where $\lambda = -1, 2$ or $1/2$, then we come to the three types of parallelisable webs mentioned in Theorem 7.4.13. ■

Remark 7.4.15 Note that the three types of webs which arose in Theorem 7.4.13 have also been discussed in Theorem 7.1.4. The latter theorem gives another characteristic property for these three types of webs by means of permutability of the transformations $[\xi, \eta]$.

We note further that these three types of parallelisable webs $W(4,2,r)$ can be obtained by joining to a parallelisable three-web $[1,2,3]$ a foliation λ_4 defined by the system $\lambda \underset{1}{\omega}^i + \underset{2}{\omega}^i = 0$ where λ is respectively equal to -1, 2, $1/2$.

The foliations of these three types of parallelisable webs are defined by the systems: $\underset{1}{\omega}^i = 0$, $\underset{2}{\omega}^i = 0$, $\underset{2}{\omega}^i = -\underset{1}{\omega}^i$, $\underset{2}{\omega}^i = -\lambda \underset{1}{\omega}^i$, $\lambda = -1,\ 2,\ 1/2$. Since $d\underset{1}{\omega}^i = d\underset{2}{\omega}^i = 0$, the forms $\underset{1}{\omega}^i$ and $\underset{2}{\omega}^i$ are total differentials: $\underset{1}{\omega}^i = dx^i_1$, $\underset{2}{\omega}^i = dx^i_2$. Substituting them into the equations of the foliations, we can easily integrate the latter ones and arrive to the following equations of the foliations λ_ξ of the 3 types of parallelisable webs indicated above:

$$
\begin{aligned}
x^i_1 &= c^i_1; \quad x^i_2 = c^i_2; \quad x^i_3 = -(x^i_1 + x^i_2) = c^i_3; \\
x^i_4 &= -(\lambda x^i_1 + x^i_2) = c^i_4; \quad \lambda = -1,\ 2,\ 1/2,
\end{aligned}
\tag{7.4.57}
$$

where c^i_ξ, $\xi = 1,2,3,4$, are constants.

Remark 7.4.16 In our considerations we proved the equations (7.4.4), (7.4.20), (7.4.35), (7.4.42) and others concerning quantities associated with a web $W(4,2,r)$, at the point $x^i = y^i = 0$. To prove that they are also valid at in a neighbourhood of this point, we must expand the tensors μ^i_{jk}, λ^i_{jk}, $\lambda^l_{(j}\mu^i_{k)l}$, $\lambda^i_j + \delta^i_j$ and others into Taylor series and use respectively (7.4.3); (7.4.19); (7.4.33)–(7.4.34); (7.4.47),(7.4.53)–(7.4.55) and similar equations to verify that all coefficients of these Taylor expansions are zeros (cf. the proof of the part (i) of Theorem 3.6.4).

7.4.5 Properties of Orthogonal Quasigroups Associated with Webs $W(4,2,r)$ Satisfying the Desargues and Triangle Closure Conditions

The Desargues closure conditions D_ξ, $D_{\xi\eta}$ and triangle closure conditions (Δ) of different types analysed in this section are equivalent to some identities in the pair of orthogonal quasigroups A and B associated with a web $W(4,2,r)$.

We now denote by $\cdot$ and $+$ the operations in A and B and by ${}^{-1}a$, a^{-1} and $-a$, $(a-)$ the left and right inverse elements of a in A and B respectively. Then, using

(7.4.2), (7.4.18), (7.4.32), (7.4.41), we obtain the identities equivalent to D_1, D_2, D_{12}, and (Δ) of the type $\left(\begin{array}{cc|c} 1 & 2 & 3 \\ 1 & 2 & 4 \end{array}\right)$. To write the identities equivalent to other closure conditions D_ξ, $D_{\xi\eta}$, and (Δ), we should first find conditional identities similar to (7.4.2), (7.4.18), (7.4.32), (7.4.41).

Recall that e is the unit of A and the left unit of B, i.e., $a \cdot e = e \cdot a = a$, $e + a = a$. We will list below some of the identities mentioned above:

(D_1) $\quad -a + b = {}^{-1}a \cdot b,$

(D_2) $\quad b + (a-) = b \cdot a + e.$

(D_3) $\quad b + {}^{-1}b \cdot (a \cdot b^{-1} + e) = a + b^{-1},$

(D_4) $\quad (a \cdot b) \cdot (a-)^{-1} + (a-) = (a + b) + e,$

(D_{12}) $\quad {}^{-1}a = -a + e,$

(D_{13}) $\quad a^{-1} + (-a)^{-1} = -a + (-a)^{-1},$

(D_{14}) $\quad a^{-1} + [({}^{-1}a)-] = -a + [({}^{-1}a)-],$

(D_{21}) $\quad (a-) = a^{-1} + e,$

(D_{23}) $\quad {}^{-1}(a-) + (a-) = {}^{-1}(a-) \cdot a^{-1} + e,$

(D_{24}) $\quad -(a^{-1}) + (a-) = [-(a^{-1})] \cdot a^{-1} + e.$

Although for closure conditions (Δ) it is sufficient to write three identities, we will write all 12 of them (3 equivalent quadruples) because they are of separate interest:

$$-a + a = e\,, \qquad \left(\begin{array}{cc|c} 1 & 1 & 3 \\ 2 & 2 & 3 \end{array}\right);$$

$$^{-1}a = a + e\,, \qquad \left(\begin{array}{cc|c} 1 & 2 & 4 \\ 1 & 2 & 3 \end{array}\right);$$

$$(-a) \cdot a^{-1} = {}^{-1}a + a\,, \qquad \left(\begin{array}{cc|c} 3 & 4 & 1 \\ 3 & 4 & 2 \end{array}\right);$$

$$a \cdot (a-) = a + a^{-1}\,, \qquad \left(\begin{array}{cc|c} 3 & 2 & 2 \\ 3 & 4 & 1 \end{array}\right);$$

$$(-a) \cdot a = {}^{-1}a\,, \qquad \left(\begin{array}{cc|c} 1 & 3 & 2 \\ 1 & 3 & 4 \end{array}\right);$$

$$^{-1}a + a = a + e, \qquad \left(\begin{array}{cc|c} 1 & 3 & 4 \\ 1 & 3 & 2 \end{array}\right);$$

$$a + a^{-1} = (a-), \qquad \left(\begin{array}{cc|c} 2 & 4 & 1 \\ 2 & 4 & 3 \end{array}\right);$$

$$a \cdot (a-) = a + e, \quad \left(\begin{array}{cc|c} 2 & 4 & 3 \\ 2 & 4 & 1 \end{array}\right);$$

$$a^{-1} = a \cdot (a-), \quad \left(\begin{array}{cc|c} 2 & 3 & 1 \\ 2 & 3 & 4 \end{array}\right);$$

$$a = a + {}^{-1}a, \quad \left(\begin{array}{cc|c} 2 & 3 & 4 \\ 2 & 3 & 1 \end{array}\right);$$

$${}^{-1}a + a = a + e, \quad \left(\begin{array}{cc|c} 1 & 4 & 2 \\ 1 & 4 & 3 \end{array}\right);$$

$$a = (-a) \cdot a + e, \quad \left(\begin{array}{cc|c} 1 & 4 & 3 \\ 1 & 4 & 2 \end{array}\right).$$

7.5 A Classification of Group Webs $W(4,2,3)$

A group web $W(4,2,r)$ was defined as a web $W(4,2,r)$ all of whose 3-subwebs are group webs and whose basis affinor is covariantly constant in the whole manifold X^{2r} (see Definition 7.2.4). In Theorem 7.2.3 we proved that for a web $W(4,2,r)$ to be a group web, it is necessary and sufficient that its basis affinor be covariantly constant in the whole X^{2r} and one of its 3-subwebs be a group web.

As we saw in Section 7.2, a group web $W(4,2,r)$ is determined by the Pfaffian equations (7.1.7), inequalities (7.1.8), and by the following system of equations (see (7.2.36), (7.2.37), (7.2.38) where $\omega_j^i = 0$, (7.2.39), (7.1.27), where $\omega_j^i = 0$, and (7.1.24) where $b^i_{jkl} = 0$):

$$d\underset{1}{\omega}^i = a^i_{jk}\underset{1}{\omega}^j \wedge \underset{1}{\omega}^k, \quad d\underset{2}{\omega}^i = -a^i_{jk}\underset{2}{\omega}^j \wedge \underset{2}{\omega}^k, \tag{7.5.1}$$

$$da^i_{jk} = 0, \quad d\lambda^i_j = 0, \tag{7.5.2}$$

$$\lambda^i_p a^p_{jk} + \lambda^p_{[k}\lambda^q_{j]} a^i_{pq} = 0, \tag{7.5.3}$$

$$a^m_{[jk} a^i_{|m|l]} = 0. \tag{7.5.4}$$

In Theorem 7.4.8 we have investigated the system (7.5.1)–(7.5.4), in particular relations (7.5.3), in detail and proved that the system is completely integrable and defines group webs $W(4,2,r)$ with the arbitrariness of $r^2(r+1)/2 - p$ constants where p is the number of independent relations (7.5.3)–(7.5.4). When p is maximal, i.e., $p = r^2(r-1)/2$ (as we saw, this will be the case when, for example, $r = 2$, or $\lambda^i_j = \delta^i_j \lambda$), it follows from (7.5.3) that $a^i_{jk} = 0$, and a group web $W(4,2,r)$ turns out to be a parallelisable web.

The question arises: *Is it possible for p not to be maximal, i.e., could there exist non-parallelisable group webs $W(4,2,r)$?*

In this section we will get a positive answer to this question giving a complete classification of group webs $W(4,2,3)$, i.e., for $r=3$.

The classification comes from a consideration of all possible types of the Jordan form of the matrix (λ^i_j) of the basis affinor of a web $W(4,2,3)$.

Case 1. Let $\lambda^i_j = \delta^i_j \lambda$. In this case we obtain from (7.5.3) that $(\lambda - \lambda^2) a^i_{jk} = 0$. Since, by (7.1.8), $\lambda \neq 0, 1$, we get in this case that all $a^i_{jk} = 0$, and, by Theorem 1.5.2, the web $W(4,2,3)$ is parallelisable.

Case 2. Let the matrix (λ^i_j) be diagonalisable and $\lambda_1, \lambda_2, \lambda_3$ be its distinct eigenvalues. Then (7.5.3) gives

$$(\lambda^i - \lambda_j \lambda_k) a^i_{jk} = 0. \qquad (7.5.5)$$

We have written the index i in λ as a superscript since there is no summation with respect to i.

Case 2a). Suppose that $\lambda^i \neq \lambda_j \lambda_k$ for different i, j, and k. Then (7.5.5) implies $a^i_{jk} = 0$, and the web $W(4,2,3)$ is parallelisable.

Case 2b). Suppose that $\lambda^i = \lambda_j \lambda_k$ for some different i, j, and k. Then not all a^i_{jk} must be equal to zero. Let, for example, $\lambda_1 = \lambda_2 \lambda_3$. Then $a^1_{23} \neq 0$. Equations (7.5.1) after the substitution $a^1_{23}\omega^2 \rightarrow \omega^2$ can be written in the form

$$\begin{cases} d\underset{1}{\omega}^1 = \underset{1}{\omega}^2 \wedge \underset{1}{\omega}^3, & d\underset{1}{\omega}^2 = d\underset{1}{\omega}^3 = 0, \\ d\underset{2}{\omega}^1 = -\underset{2}{\omega}^2 \wedge \underset{2}{\omega}^3, & d\underset{2}{\omega}^2 = d\underset{2}{\omega}^3 = 0. \end{cases}$$

In what follows in this section, we will write the structure equations of the group G^{-1} only and will omit the index under ω:

$$d\omega^1 = \omega^2 \wedge \omega^3, \qquad d\omega^2 = d\omega^3 = 0. \qquad (7.5.6)$$

Equations (7.5.6) coincide with the structure equations of the Lie group $\Gamma_{3,7}$ in [Va 69] where all Lie groups of three parameters are found. Note that in [Va 69] were also found invariant forms and parametric groups corresponding to the 3-parameter Lie groups.

Case 3. Let the matrix (λ^i_j) be diagonalisable but $\lambda_2 = \lambda_3 \neq \lambda_1$. Then, by (7.1.8), $\lambda_2 \neq 0$ and $\lambda_1 \neq 0, 1$, and therefore $\lambda_2 \neq \lambda_1 \lambda_2$. If $\lambda_1 = (\lambda_2)^2$, we again obtain a non-parallelisable group web $W(4,2,3)$ on the manifold $G \times G^{-1}$ where G^{-1} is the group $\Gamma_{3,7}$.

Case 4. Let the Jordan form of the matrix (λ^i_j) be

$$\begin{pmatrix} \lambda & 0 & 0 \\ 1 & \lambda & 0 \\ 0 & 1 & \lambda \end{pmatrix}.$$

In this case the system (7.5.3) has the form

$$\begin{cases} \lambda(2-\lambda)a^1_{12} + 2a^2_{12} = 0, \\ \lambda(2-\lambda)a^1_{13} + 2a^2_{13} - \lambda a^1_{12} = 0, \\ \lambda(2-\lambda)a^1_{23} + 2a^2_{23} - \lambda a^1_{13} - a^1_{12} = 0, \\ \lambda(2-\lambda)a^2_{12} + 2a^3_{12} = 0, \\ \lambda(2-\lambda)a^2_{13} + 2a^3_{13} - \lambda a^2_{12} = 0, \\ \lambda(2-\lambda)a^2_{23} + 2a^3_{23} - \lambda a^2_{13} - a^2_{12} = 0, \\ \lambda(2-\lambda)a^3_{12} = 0, \\ \lambda(2-\lambda)a^3_{13} - \lambda a^3_{12} = 0, \\ \lambda(2-\lambda)a^3_{23} - \lambda a^3_{13} - a^3_{12} = 0. \end{cases} \tag{7.5.7}$$

Case 4a). If $\lambda \neq 2$, the system (7.5.7) gives $a^i_{jk} = 0$, and we have a parallelisable web $W(4,2,3)$.

Case 4b). If $\lambda = 2$, the system (7.5.7) gives

$$\begin{cases} a^2_{12} = a^3_{12} = a^3_{13} = 0, \quad a^1_{13} = b, \\ a^3_{23} = a^2_{13} = a^1_{12} = c, \quad a^1_{23} = a. \end{cases} \tag{7.5.8}$$

The Jacobi identities (7.5.4) and conditions (7.5.8) imply now that $c(4b + c) = 0$. There are two possibilities:

Case 4b$_1$). $c = 0$. In this case the structure equations of G^{-1} have the form

$$d\omega^1 = a\omega^2 \wedge \omega^3 - b\omega^3 \wedge \omega^1, \quad d\omega^2 = \omega^3 \wedge \omega^1, \quad d\omega^3 = 0,$$

or after the substitution $b\omega^3 \to \omega^3$ and $a\omega^2/b \to \omega^2$

$$d\omega^1 = \omega^1 \wedge \omega^3 + \omega^2 \wedge \omega^3, \quad d\omega^2 = \omega^2 \wedge \omega^3, \quad d\omega^3 = 0. \tag{7.5.9}$$

Equations (7.5.9) are the structure equations of the group $\Gamma_{3,5}$ (see [V 69]).

Case 4b$_2$). $c = -4b$. Then the structure equations of the group G^{-1} have the form

$$\begin{cases} d\omega^1 = a\omega^2 \wedge \omega^3 - b\omega^3 \wedge \omega^1 - 4b\omega^1 \wedge \omega^2, \\ d\omega^2 = -b\omega^2 \wedge \omega^3 + 4b\omega^3 \wedge \omega^1, \\ d\omega^3 = -4b\omega^2 \wedge \omega^3. \end{cases} \tag{7.5.10}$$

The matrix of the structure constants of G^{-1} is

$$\begin{pmatrix} a & -b & -4b \\ -b & 4b & 0 \\ -4b & 0 & 0 \end{pmatrix}$$

It has the following principal minors:

$$\Delta_0 = 1, \quad \Delta_1 = a, \quad \Delta_2 = b(4a - b), \quad \Delta_3 = 64b^3.$$

If $a > 0,\ b > 0,\ 4a > b$ or $a < 0,\ b < 0,\ 4|a| > |b|$, equations (7.5.10) can be reduced to the form

$$d\omega^1 = \omega^2 \wedge \omega^3, \quad d\omega^2 = \omega^3 \wedge \omega^1, \quad d\omega^3 = \omega^1 \wedge \omega^2. \tag{7.5.11}$$

Equations (7.5.11) coincide with the structure equations of the group $\Gamma_{3,1}$ (see [Va 69]). After the substitution

$$\omega^1 = \omega_1^2, \quad \omega^2 = \omega_1^3, \quad \omega^3 = \omega_3^2$$

they are reduced to the structure equations of the three-dimensional orthogonal group $O(3)$.

In other cases equations (7.5.10) are reduced to the form

$$d\omega^1 = \omega^2 \wedge \omega^3, \quad d\omega^2 = \omega^3 \wedge \omega^1, \quad d\omega^3 = \omega^2 \wedge \omega^1. \tag{7.5.12}$$

Equations (7.5.12) are the structure equations of the group $\Gamma_{3,2}$ (see [Va 69]). It is shown in [Va 69] that in this case the group G^{-1} is the group of projective transformations on a straight line.

Case 5. Let the Jordan form of the matrix (Δ) be

$$\begin{pmatrix} \lambda & 0 & 0 \\ 1 & \lambda & 0 \\ 0 & 0 & \lambda \end{pmatrix}.$$

We again have two possibilities:

Case 5a). $\lambda \neq 2$. The system (7.5.3) gives $a^i_{jk} = 0$, i.e., we have a parallelisable web $W(4,2,3)$.

Case 5b). $\lambda = 2$. From (7.5.3) and (7.5.4) we find

$$a^2_{12} = a^2_{13} = a^3_{13} = 0, \quad a^2_{23} = a^1_{13} = a, \quad a^3_{23} = \alpha, \quad a^3_{12} = \beta, \quad a^1_{23} = b,$$

and therefore

$$a(\alpha + c) = 0, \quad a\beta = 0. \tag{7.5.13}$$

From (7.5.13) it follows that two variants are possible:

Case 5b$_1$). $a = 0$. Then the structure equations of the group G^{-1} have the form

$$\begin{cases} d\omega^1 = b\omega^2 \wedge \omega^3 + c\omega^1 \wedge \omega^2, \\ d\omega^2 = 0, \\ d\omega^3 = \alpha\omega^2 \wedge \omega^3 + \beta\omega^1 \wedge \omega^2, \end{cases} \tag{7.5.14}$$

and actually our problem reduces to the problem of classification of the Lie groups of two parameters. It was shown in [Va 69] that in this case the types of the reduced form of equations (7.5.14) depend on the type of the matrix $\begin{pmatrix} b & c \\ \alpha & \beta \end{pmatrix}$ and that the following reduced forms of (7.5.14) are possible:

$$\begin{cases} d\omega^1 = m\omega^1 \wedge \omega^2, \\ d\omega^2 = 0, \\ d\omega^3 = \omega^3 \wedge \omega^2; \end{cases} \qquad (\Gamma_{3,3})$$

$$\begin{cases} d\omega^1 = \lambda\omega^1 \wedge \omega^2 - \omega^3 \wedge \omega^2, \\ d\omega^2 = 0, \\ d\omega^3 = \omega^1 \wedge \omega^2 + \lambda\omega^3 \wedge \omega^2; \end{cases} \qquad (\Gamma_{3,4})$$

$$\begin{cases} d\omega^1 = \omega^1 \wedge \omega^2 + \omega^3 \wedge \omega^2, \\ d\omega^2 = 0, \\ d\omega^3 = \omega^3 \wedge \omega^2; \end{cases} \qquad (\Gamma_{3,5})$$

$$\begin{cases} d\omega^1 = \omega^1 \wedge \omega^2, \quad \omega^2 = d\omega^3 = 0; & (\Gamma_{3,6}) \\ d\omega^1 = \omega^1 \wedge \omega^2, \quad d\omega^2 = d\omega^3 = 0; & (\Gamma_{3,7}) \\ \qquad\qquad d\omega^1 = d\omega^2 = d\omega^3 = 0. & (\Gamma_{3,8}) \end{cases}$$

Case 5b$_2$). $\beta = 0$, $\alpha = -c$. Now the structure equations of G^{-1} are

$$\begin{cases} d\omega^1 = b\omega^2 \wedge \omega^3 - a\omega^3 \wedge \omega^1 + c\omega^1 \wedge \omega^2, \\ d\omega^2 = a\omega^2 \wedge \omega^3, \\ d\omega^3 = -c\omega^2 \wedge \omega^3. \end{cases}$$

After the substitution $c\omega^2 + a\omega^3 \to \omega^3$, $(b/a)\omega^2 \to \omega^2$ we obtain the structure equations of the group $\Gamma_{3,5}$.

Case 6. Finally, let the Jordan form of the matrix (λ^i_j) be

$$\begin{pmatrix} \lambda_1 & 0 & 0 \\ 0 & \lambda_2 & 0 \\ 0 & 1 & \lambda_2 \end{pmatrix}.$$

In this case the system (7.5.3) has the form

$$\begin{cases} \lambda_1(2-\lambda_2)a^1_{12} = 0, \\ \lambda_1(2-\lambda_2)a^1_{13} - \lambda_1 a^1_{12} = 0, \\ \left[2\lambda_1 - (\lambda_2)^2\right] a^1_{23} = 0, \\ \lambda_2(2-\lambda_1)a^2_{12} + 2a^3_{12} = 0, \\ \lambda_2(2-\lambda_2)a^2_{13} + 2a^3_{13} - \lambda_1 a^2_{12} = 0, \\ \lambda_2(2-\lambda_2)a^2_{23} + 2a^3_{23} = 0, \\ \lambda_2(2-\lambda_1)a^3_{12} = 0, \\ \lambda_2(2-\lambda_1)a^3_{13} - \lambda_1 a^3_{12} = 0, \\ \lambda_2(2-\lambda_2)a^3_{23} = 0. \end{cases} \qquad (7.5.15)$$

There are three possibilities:

6a) $\lambda_1 = 2$, $\lambda_2 \neq 2$; 6b) $\lambda_1 \neq 2$, $\lambda_2 = 2$; 6c) $\lambda_1 \neq 2$, $\lambda_2 \neq 2$.

Let us consider them consecutively.

Case 6a). $\lambda_1 = 2,\ \lambda_2 \neq 2$. From (7.5.15) it follows that

$$a^1_{12} = a^3_{12} = a^3_{23} = a^2_{23} = a^1_{13} = 0, \quad a^3_{13} = a^2_{12} = a, \quad a^2_{31} = c, \quad [(\lambda_2)^2 - 4]a^1_{23} = 0.$$

We have to distinguish three cases:

Case 6a$_1$). $\lambda_2 \neq -2$. Then $a^1_{23} = 0$. The structure equations of the group G^{-1} have the form

$$\begin{cases} d\omega^1 = 0, \\ d\omega^2 = c\omega^3 \wedge \omega^1 + a\omega^1 \wedge \omega^2, \\ d\omega^3 = -a\omega^3 \wedge \omega^1. \end{cases}$$

The substitution $-a\omega^1 \to \omega^1, \quad -c\omega^3/a \to \omega^3$ reduces these equations to the structure equations of the group $\Gamma_{3,5}$.

Case 6a$_2$). $\lambda_2 = -2$. Now the structure equations of G^{-1} are

$$\begin{cases} d\omega^1 = b\omega^2 \wedge \omega^3, \\ d\omega^2 = c\omega^3 \wedge \omega^1 + a\omega^1 \wedge \omega^2, \\ d\omega^3 = -a\omega^3 \wedge \omega^1. \end{cases}$$

and the Jacobi identities (7.5.4) give $ab = 0$. Hence, we have two possibilities:

Case 6a$_{21}$). $a = 0$. The structure equations

$$d\omega^1 = b\omega^2 \wedge \omega^3, \quad d\omega^2 = -c\omega^1 \wedge \omega^3, \quad d\omega^3 = 0$$

of the group G^{-1} can be reduced to the structure equations of the group $\Gamma_{3,4}$ where $\lambda = 0$, if we make the following substitution: $\omega^1 \to \sqrt{c}\omega^1, \quad \omega^2 \to \sqrt{b}\omega^2$, $\omega^3 \to \sqrt{bc}\omega^3$. Thus, we have the *group of the motions in the plane.*

Case 6a$_{22}$). $b = 0$. The structure equations

$$\begin{cases} d\omega^1 = 0, \\ d\omega^2 = c\omega^3 \wedge \omega^1 + a\omega^1 \wedge \omega^2, \\ d\omega^3 = -a\omega^3 \wedge \omega^1 \end{cases}$$

of the group G^{-1} can be reduced to the structure equations of the group $\Gamma_{3,5}$ by means of the substitution $-a\omega^1 \to \omega^1, \quad -(c/a)\omega^3 \to \omega^3$.

Case 6b). $\lambda_1 \neq 2, \quad \lambda_2 = 2$. In this case

$$a^1_{12} = a^1_{23} = a^3_{23} = a^3_{12} = a^3_{13} = a^2_{12} = a^2_{13} = 0, \quad a^1_{13} = a, \quad a^2_{23} = b,$$

and

$$d\omega^1 = a\omega^1 \wedge \omega^3, \quad d\omega^2 = b\omega^2 \wedge \omega^3, \quad d\omega^3 = 0.$$

We again obtain the group $\Gamma_{3,3}$.

Case 6c). $\lambda_2 \neq 2, \quad \lambda_2 \neq 2$. Now

$$a^1_{12} = a^1_{13} = a^2_{12} = a^2_{13} = a^2_{23} = a^3_{13} = a^3_{12} = a^3_{23} = 0, \quad [2\lambda_1 - (\lambda_2)^2]a^1_{23} = 0.$$

If $\lambda_1 \neq (\lambda_2)^2/2$, then all $a^i_{jk} = 0$. If $\lambda_1 = (\lambda_2)^2/2$, then $a^1_{23} \neq 0$ and we again obtain the group $\Gamma_{3,7}$.

In the following theorem we list all cases for which there exists a non-parallelisable group web $W(4,2,3)$.

Theorem 7.5.1 *On a six-dimensional differentiable manifold X^6 there exists a non-parallelisable group web $W(4,2,3)$ if and only if X^6 is a group manifold of the group $G \times G^{-1}$ where G^{-1} is the Lie group of three parameters and the basis affinor of the web has the Jordan form of a certain type. We indicate below the Jordan form of (λ^i_j) and a type of the group G^{-1}:*

1) $\begin{pmatrix} \lambda\mu & 0 & 0 \\ 0 & \lambda & 0 \\ 0 & 0 & \mu \end{pmatrix}$, $\lambda \neq \mu;\ \lambda, \mu \neq 0, 1;$ $(\Gamma_{3,7})$

2) $\begin{pmatrix} \lambda^2 & 0 & 0 \\ 0 & \lambda & 0 \\ 0 & 0 & \lambda \end{pmatrix}$, $\lambda \neq 0, 1;$ $(\Gamma_{3,7})$

3) $\begin{pmatrix} 2 & 0 & 0 \\ 1 & 2 & 0 \\ 0 & 1 & 2 \end{pmatrix}$, $(\Gamma_{3,5}, \Gamma_{3,1}, \Gamma_{3,2})$

4) $\begin{pmatrix} 2 & 0 & 0 \\ 1 & 2 & 0 \\ 0 & 0 & 2 \end{pmatrix}$, $(\Gamma_{3,3}, \Gamma_{3,4}, \Gamma_{3,5}, \Gamma_{3,6}, \Gamma_{3,7})$

5) $\begin{pmatrix} 2 & 0 & 0 \\ 0 & \lambda & 0 \\ 0 & 1 & \lambda \end{pmatrix}$, $\lambda \neq 2, -2, 0, 1;$ $(\Gamma_{3,5})$

6) $\begin{pmatrix} 2 & 0 & 0 \\ 0 & -2 & 0 \\ 0 & 1 & -2 \end{pmatrix}$, $(\Gamma_{3,4}, \Gamma_{3,5})$

7) $\begin{pmatrix} \lambda & 0 & 0 \\ 0 & 2 & 0 \\ 0 & 1 & 2 \end{pmatrix}$, $\lambda \neq 0, 1, 2;$ $(\Gamma_{3,3})$

8) $\begin{pmatrix} \frac{1}{2}\lambda^2 & 0 & 0 \\ 0 & \lambda & 0 \\ 0 & 1 & \lambda \end{pmatrix}$, $\lambda \neq 0, 1, 2, -2;$ $(\Gamma_{3,7})$ ■

7.6 Grassmann Webs $GW(4,2,r)$

7.6.1 Basic Notions

In Sections 7.6–7.8 we shall study Grassmann webs $GW(4,2,r)$ of codimension r on a differentiable manifold X^{2r} and their special classes. Such webs form a very important special class of webs $W(4,2,r)$. In addition, they give a realisation of webs $W(4,2,r)$

and play an important role in a description of the webs $W(4,2,r)$ of maximum rank (see Chapter 8).

We will begin with Grassmann webs $GW(4,2,r)$. The definition of Grassmann webs $GW(d,n,r)$ has been given in Section 2.6 (see Definition 2.6.1). So, to obtain the definition of webs $GW(4,2,r)$ we should take $d=4$ and $n=2$ in Definition 2.6.1.

We will use the same term "Grassmann 4-web" for a 4-web on $G(1,r+1)$ as well as for a 4-web in a projective space P^{r+1} whose image is $GW(4,2,r)$.

So, in the second interpretation of the Grassmann 4-web, we will say that the leaves of the foliations λ_ξ, $\xi=1,2,3,4$, are bundles of straight lines l with vertices located on the surfaces U_1, U_2, U_3, U_4 respectively.

To clarify what is a Grassmann web $GW(4,2,r)$ in P^{r+1} and what are its corresponding pair of l.d. quasigroups, suppose that l^0 be a straight line in P^{r+1} which intersects each of the hypersurfaces U_ξ respectively in points M_ξ^0 . Then any straight line l passing through points K and L from sufficiently small neighbourhoods of the points M_1^0 and M_2^0 intersects the hypersurfaces U_3 and U_4 at the points M and N.

Therefore we obtain two mappings:

$$q_{123}: U_1 \times U_2 \to U_3 \quad \text{and} \quad q_{124}: U_1 \times U_2 \to U_4.$$

They are differentiable and invertible and determine two l.d. quasigroups q_{123} and q_{124} which are orthogonal (see Section 7.3). The Grassmann web $GW(4,2,r)$ connected with the quasigroups q_{123} and q_{124} is formed on the $(2r)$-dimensional manifold of straight lines of P^{r+1} by four foliations of bundles of straight lines whose vertices are points of the hypersurfaces U_ξ.

Henceforth we will confine ourselves without special stipulation to the domain D of the manifold of straight lines of P^{r+1} cutting each of the hypersurfaces U_ξ at a single point. As in Definition 5.1.1, we will call such a domain D a *domain of regularity* of $GW(4,2,r)$.

7.6.2 Specialisation of Moving Frames

We associate the space P^{r+1} with a moving frame $\{A_u\}$, $u,v,w=0,1,\dots,r+1$. The equations of infinitesimal displacement of this frame have the form (cf. (5.1.4)):

$$dA_u = \omega_u^v A_v, \quad u,v = 0,1,\dots,r+1, \tag{7.6.1}$$

where the Pfaffian forms ω_u^v satisfy the structure equations of the projective space P^{r+1} (cf. ((5.1.9)):

$$d\omega_u^v = \omega_u^w \wedge \omega_w^v, \quad u,v,w = 0,1,\dots,r+1. \tag{7.6.2}$$

We consider now such moving frames connected with points $K \in U_1$ and $L \in U_2$ for which $K=A_0$, $L=A_{r+1}$ and the points A_i, $i=1,\dots,r$, are in the $(r-1)$-dimensional intersection of the planes $T_K(U_1)$ and $T_L(U_2)$ tangent to U_1 and U_2 at K and L.

Then the equations of U_1 and U_2 in this frame have the form

$$\omega_0^{r+1} = 0, \quad \omega_{r+1}^0 = 0, \tag{7.6.3}$$

and the forms ω_0^i and ω_{r+1}^i are the basis forms of the manifold under consideration. Let us find the equations of the hypersurfaces U_3 and U_4 generated by the points M and N. Since M and N lie on the straight line A_0A_{r+1}, we have $M = xA_0 + yA_{r+1}$ and $N = \lambda A_0 + \mu A_{r+1}$.

Differentiating these relations and using (7.6.1), we obtain

$$\begin{cases} dM = (dx + x\omega_0^0)A_0 + (dy + y\omega_{r+1}^{r+1})A_{r+1} + (x\omega_0^i + y\omega_{r+1}^i)A_i, \\ dN = (d\lambda + \lambda\omega_0^0)A_0 + (d\mu + \mu\omega_{r+1}^{r+1})A_{r+1} + (\lambda\omega_0^i + \mu\omega_{r+1}^i)A_i. \end{cases} \tag{7.6.4}$$

Since M and N are geometrically invariant, we have $\delta M = \pi M$, $\delta N = \phi N$ where δ is used for d when a straight line $l = A_0A_{r+1}$ is fixed, i.e., when $\omega_0^i = \omega_{r+1}^i = 0$. Therefore, we have

$$\begin{cases} \delta x + \pi_0^0 x = \pi x, \delta y + \pi_{r+1}^{r+1} y = \pi y, \\ \delta\lambda + \pi_0^0\lambda = \phi\lambda, \delta\mu + \pi_{r+1}^{r+1}\mu = \phi\mu, \end{cases} \tag{7.6.5}$$

where $\pi_u^v = \omega_u^v(\delta)$.

It is easy to see that we can reduce the coefficients x, y, and μ to one: $x = y = \mu = 1$. Let us show this, for example, for the coefficient x. Since $M \in U_3$ and $A_{r+1} \in U_2$, it is obvious that $x \neq 0$. We can put all $\pi_u^v = 0$, except π_0^0. The form $\pi - \pi_0^0$ will depend on one parameter only and can be considered as the total differential of this parameter: $\pi - \pi_0^0 = \delta t$. Hence the first equation of (7.6.5) can be written in the form

$$\delta \ln x = \delta t. \tag{7.6.6}$$

Integrating (7.6.6), we obtain $\ln x = t + C$. Taking $t = -C$, we find $\ln x = 0$ and consequently $x = 1$. After this we will have $\pi_0^0 = \pi$.

When, using the method indicated above for x, we reduce x, y, and μ to one, we will have

$$\pi_0^0 = \pi_{r+1}^{r+1} = \pi = \phi, \quad \delta\lambda = 0. \tag{7.6.7}$$

Equations (7.6.7) imply that the forms $\omega_0^0 - \omega_{r+1}^{r+1}$ and $d\lambda$ are expressed in terms of the basis forms ω_0^i, ω_{r+1}^i only. In addition, we have the following expressions for the points M and N generating the hypersurfaces U_3 and U_4 :

$$M = A_0 + A_{r+1}, \quad N = \lambda A_0 + A_{r+1}. \tag{7.6.8}$$

The expressions (7.6.4) for the differentials of these points will take the form:

$$\begin{cases} dM = \omega_{r+1}^{r+1} M + (\omega_0^0 - \omega_{r+1}^{r+1})A_0 + (\omega_0^i + \omega_{r+1}^i)A_i, \\ dN = \omega_{r+1}^{r+1} N + [d\lambda + \lambda(\omega_0^0 - \omega_{r+1}^{r+1})]A_0 + (\lambda\omega_0^i + \omega_{r+1}^i)A_i. \end{cases} \tag{7.6.9}$$

Since the points M and N generate r-dimensional surfaces and each of the sets of the forms $\omega_0^i + \omega_{r+1}^i$ and $\lambda\omega_0^0 + \omega_{r+1}^i$ is linearly independent, we have

$$\omega_0^0 - \omega_{r+1}^{r+1} = a_i(\omega_0^i + \omega_{r+1}^i), \quad d\lambda + \lambda(\omega_0^0 - \omega_{r+1}^{r+1}) = b_i(\lambda\omega_0^i + \omega_{r+1}^i). \tag{7.6.10}$$

Equations (7.6.10) imply that

$$d\lambda = \lambda(b_i - a_i)\omega_0^i + (b_i - \lambda a_i)\omega_{r+1}^i. \tag{7.6.11}$$

The first equation of (7.6.10) and the equation (7.6.11) are the differential equations of the hypersurfaces U_3 and U_4 in the adapted moving frame.

Using (7.6.10) and (7.6.11), we write (7.6.9) in the form

$$dM = \omega_{r+1}^{r+1} M + (\omega_0^i + \omega_{r+1}^i) M_i, \tag{7.6.12}$$
$$dN = \omega_{r+1}^{r+1} N + (\lambda\omega_0^i + \omega_{r+1}^i) N_i, \tag{7.6.13}$$

where $M_i = A_i + a_i A_0, \quad N_i = A_i + b_i A_0$.

As for the bundles of straight lines which are leaves of the web $GW(4,2,r)$, in the adapted frame they are defined by the systems

$$\omega_0^i = 0, \quad \omega_{r+1}^i = 0, \quad \omega_0^i + \omega_{r+1}^i = 0, \quad \lambda\omega_0^i + \omega_{r+1}^i = 0. \tag{7.6.14}$$

7.6.3 The Structure Equations, the Fundamental Tensors, and the Basis Affinor of a Grassmann Web $GW(4,2,r)$

The forms ω_0^i and ω_{r+1}^i are the basis forms of the $(2r)$-dimensional manifold of straight lines of the space P^{r+1}. Equations (7.6.10) show that

$$\omega_0^0 = \omega + a_i\omega_0^i, \quad \omega_{r+1}^{r+1} = \omega - a_i\omega_{r+1}^i, \tag{7.6.15}$$

where ω is an 1-form. Using equations (7.6.2) and (7.6.15), we find the first group of the structure equations of the web $GW(4,2,r)$:

$$d\omega_0^i = \omega_0^j \wedge \theta_j^i + a_j\delta_k^i\omega_0^j \wedge \omega_0^k, \quad d\omega_{r+1}^i = \omega_{r+1}^j \wedge \theta_j^i - a_j\delta_k^i\omega_{r+1}^j \wedge \omega_{r+1}^k, \tag{7.6.16}$$

where

$$\theta_j^i = \omega_j^i - \delta_j^i\omega \tag{7.6.17}$$

are the forms of the canonical affine connection γ_{123} induced by the 3-subweb [1, 2, 3].

Comparing equations (7.6.16) with the structure equations (7.1.21) of a general web $W(4,2,r)$, we see that the torsion tensor of the $GW(4,2,r)$ has the form

$$a_{jk}^i = a_{[j}\delta_{k]}^i. \tag{7.6.18}$$

Exterior differentiation of (7.6.3), (7.6.10), (7.6.11) and application of Cartan's lemma give

$$\begin{cases} \omega_i^{r+1} = f_{ij}\omega_0^j, \quad \omega_i^0 = -g_{ij}\omega_{r+1}^j, \\ \nabla a_i = (f_{ij} - h_{ij})\omega_0^j + (g_{ij} - h_{ij})\omega_{r+1}^j, \\ \nabla b_i = [b_i(b_j - a_j) + \lambda(f_{ij} - k_{ij})]\omega_0^j + (g_{ij} - k_{ij})\omega_{r+1}^j, \end{cases} \tag{7.6.19}$$

where f_{ij}, g_{ij}, h_{ij}, and k_{ij} are symmetric in i and j and ∇ is the sign of the covariant differential in the canonical connection γ_{123}:

$$\nabla a_i = da_i - a_j\theta_i^j, \quad \nabla b_i = db_i - b_j\theta_i^j.$$

Using (7.6.16), (7.6.19), and (7.6.15), we find the second group of the structure equations of $GW(4,2,r)$:

$$d\theta_j^i - \theta_j^k \wedge \theta_k^i = b_{jkl}^i\omega_0^k \wedge \omega_{r+1}^l, \tag{7.6.20}$$

where the curvature tensor b_{jkl}^i of the $GW(4,2,r)$ has the form:

$$b_{jkl}^i = f_{ik}\delta_l^i + g_{lj}\delta_k^i + h_{kl}\delta_j^i. \tag{7.6.21}$$

In the sequel we will need two consequtive extensions of equations (7.6.19). Exterior differentiation of (7.6.19) and application of Cartan's lemma give

$$\begin{cases} \nabla f_{ij} = (f_{ijk} - f_{ij}a_k)\omega_0^k + f_{ij}a_k\omega_{r+1}^k, \\ \nabla g_{ij} = -g_{ij}a_k\omega_0^k + (g_{ijk} + g_{ij}a_k)\omega_{r+1}^k, \\ \nabla h_{ij} = (h_{ijk} + 3h_{(ij}a_{k)} - h_{ij}a_k)\omega_0^k + (h_{ijk} + h_{ij}a_k)\omega_{r+1}^k, \\ \nabla k_{ij} = (\lambda k_{ijk} + 3k_{(ij}b_{k)} - k_{ij}a_k)\omega_0^k + (k_{ijk} + k_{ij}a_k)\omega_{r+1}^k, \end{cases} \tag{7.6.22}$$

where f_{ijk}, g_{ijk}, h_{ijk}, and k_{ijk} are symmetric in i, j and k and ∇ is the sign of the covariant differential in the connection γ_{123} : for example, ∇f_{ij} has the form $\nabla f_{ij} = df_{ij} - f_{kj}\theta_i^k - f_{ik}\theta_j^k$.

Equations (7.6.22) show that the quantities f_{ij}, g_{ij}, h_{ij}, and k_{ij} are components of $(0,2)$-tensors (see [L 53]).

Finally, exterior differentiation of (7.6.22) and application of Cartan's lemma give

$$\begin{cases} \nabla f_{ijk} = (f_{ijkm} - 2f_{ijk}a_m)\omega_0^m + [f_{ijk}a_m + 3f_{(ij}(f_{k)m} + g_{k)m})]\omega_{r+1}^m, \\ \nabla g_{ijk} = -[g_{ijk}a_m + 3g_{(ij}(f_{k)m} + g_{k)m})]\omega_0^m + (g_{ijkm} + 2g_{ijk}a_m)\omega_{r+1}^m, \\ \nabla h_{ijk} = (h_{ijkm} + h_{ijm}a_k + 3h_{(ij}f_{m)k})\omega_0^m + [h_{ijkm} + 3h_{ijk}a_m \\ \qquad -3h_{k(ij}a_{m)} + 3h_{(ij}(f_{k)m} + f_{m)k} + h_{k)m})]\omega_{r+1}^m, \\ \nabla k_{ijk} = (\lambda k_{ijkm} - k_{ijk}a_m + 3k_{(ij}f_{m)k})\omega_0^m + [k_{ijkm} + k_{ijk}a_m - (4/\lambda)k_{(ijk}b_{m)} \\ \qquad +(3/\lambda)k_{(ij}(f_{k)m} + f_{m)k} + k_{k)m})]\omega_{r+1}^m, \end{cases} \tag{7.6.23}$$

where f_{ijkm}, g_{ijkm}, h_{ijkm}, and k_{ijkm} are symmetric in the indices i, j, k, and m and, for example, ∇f_{ijk} has the form

$$\nabla f_{ijk} = df_{ijk} - f_{mjk}\theta_i^m - f_{imk}\theta_j^m - f_{ijm}\theta_k^m.$$

Equations (7.6.23) show that the quantities f_{ijk}, g_{ijk}, h_{ijk}, and k_{ijk} are components of $(0,3)$-tensors.

Since

$$\begin{cases} d^2 A_0 \equiv \omega_0^i \omega_i^{r+1} A_{r+1} \mod A_0,\ A_i, \\ d^2 A_{r+1} \equiv \omega_{r+1}^i \omega_i^0 A_0 \mod A_{r+1},\ A_i, \\ d^2 M \equiv (\omega_0^i + \omega_{r+1}^i)(\nabla a_i + \omega_i^0 - \omega_i^{r+1}) A_0 \mod M,\ M_i, \\ d^2 N \equiv (\lambda\omega_0^i + \omega_{r+1}^i)[\nabla b_i + b_i(a_k - b_k)\omega_0^k + \omega_i^0 - \lambda\omega_i^{r+1}] A_0 \mod N,\ N_i, \end{cases}$$

the second fundamental forms of the hypersurfaces U_1, U_2, U_3, and U_4 have the form

$$\begin{cases} \phi_1 = f_{ij}\omega_0^i\omega_0^j, & \phi_3 = -h_{ij}(\omega_0^i + \omega_{r+1}^i)(\omega_0^j + \omega_{r+1}^j), \\ \phi_2 = -g_{ij}\omega_{r+1}^i\omega_{r+1}^j, & \phi_4 = -k_{ij}(\lambda\omega_0^i + \omega_{r+1}^i)(\lambda\omega_0^j + \omega_{r+1}^j). \end{cases} \tag{7.6.24}$$

Comparing (7.6.14) with the similar equations (7.1.11) for a general web $W(4,2,r)$, we conclude that for the $GW(4,2,r)$ the basis affinor λ_j^i has the form:

$$\lambda_j^i = \delta_j^i \lambda \tag{7.6.25}$$

and the inequalities (7.1.8) give that $\lambda \neq 0, 1$.

For the general $W(4,2,r)$ the basis affinor λ_j^i satisfies the differential equations (7.1.27), (7.1.31) and the relations (7.1.28) and (7.1.32).

From (7.6.25), (7.1.27), (7.1.31), (7.6.11), and (7.6.19) it follows that

$$\lambda_{jk}^i = \delta_j^i \lambda(b_k - a_k), \qquad \mu_{jk}^i = \delta_j^i(b_k - \lambda a_k); \tag{7.6.26}$$

$$\begin{cases} \lambda_{jkl}^i = \delta_j^i\lambda(2b_k - a_k)(b_l - a_l) + \delta_j^i\lambda^2(f_{kl} - k_{kl}) - \delta_j^i\lambda(f_{kl} - h_{kl}), \\ \beta_{jkl}^i = \delta_j^i(b_k - a_k)(b_l - \lambda a_l) + \delta_j^i\lambda(h_{kl} - k_{kl}), \\ \alpha_{jkl}^i = \delta_j^i(b_l - a_l)(b_k - \lambda a_k) + \delta_j^i\lambda(h_{kl} - k_{kl}), \\ \mu_{jkl}^i = -\delta_j^i a_k(b_l - \lambda a_l) - \delta_j^i\lambda(g_{kl} - h_{kl}) + \delta_j^i(g_{kl} - k_{kl}). \end{cases} \tag{7.6.27}$$

The relations (7.1.28), (7.1.32), and (7.1.24) are satisfied identically for the $GW(4,2,r)$.

7.6.4 The Connection Forms and the Fundamental Tensors of 3-Subwebs of a Grassmann Web $GW(4,2,r)$

The 3-subwebs $[\xi, \eta, \zeta]$ of a Grassmann web $GW(4,2,r)$ are formed by the bundles of straight lines of the space P^{r+1} with the vertices located on U_ξ, U_η, U_ζ.

For a general web $W(4,2,r)$ the forms $\underset{\xi\eta\zeta}{\theta}{}^i_j$ of the canonical connections induced by the 3-subwebs $[\xi,\eta,\zeta]$ have the form (7.2.13), (7.2.21), and (7.2.28). The torsion and curvature tensors $\underset{\xi\eta\zeta}{a}{}^i_{jk}$ and $\underset{\xi\eta\zeta}{b}{}^i_{jkl}$ of these connections are expressed by the formulae (7.2.14), (7.2.22), (7.2.29) and (7.2.16), (7.2.23), (7.2.30).

For the Grassmann web $GW(4,2,r)$ the forms $\underset{123}{\theta}{}^i_j$ of the canonical connection γ_{123} are determined by (7.6.17), and the torsion and curvature tensors $\underset{123}{a}{}^i_{jk}$ and $\underset{123}{b}{}^i_{jkl}$ are given by (7.6.18) and (7.6.21).

For the $GW(4,2,r)$ we obtain from (7.6.25) that

$$\tilde{\lambda}^i_j = \delta^i_j/\lambda, \quad \tilde{\tilde{\lambda}}^i_j = \delta^i_j/(1-\lambda). \tag{7.6.28}$$

Applying the formulae mentioned above for the canonical connections $\gamma_{\xi\eta\zeta}$ and using the relations (7.6.18), (7.6.20), (7.6.21), (7.6.25), (7.6.26), (7.6.27), and (7.6.28), we obtain for 3-subwebs of $GW(4,2,r)$ the following expressions of the connection forms and the torsion and curvature tensors:

$$\left\{\begin{array}{l} \underset{124}{\theta}{}^i_j = \theta^i_j - \delta^i_j(b_k - \lambda a_k)\omega^k_{r+1}/\lambda, \\ \underset{314}{\theta}{}^i_j = \theta^i_j + a_j\omega^i_0 + \delta^i_j\{(b_k - a_k)\omega^k_0 + (b_k - \lambda a_k)\omega^k_{r+1}\}/(1-\lambda), \\ \underset{234}{\theta}{}^i_j = \theta^i_j - a_j\omega^i_{r+1} + \delta^i_j\{\lambda(b_k - a_k)\omega^k_0 + (2\lambda - 1)(b_k - \lambda a_k)\omega^k_{r+1}/\lambda\}/(1-\lambda); \end{array}\right. \tag{7.6.29}$$

$$\left\{\begin{array}{l} \underset{124}{a}{}^i_{jk} = \delta^i_{[k} b_{j]}/\lambda, \\ \underset{314}{a}{}^i_{jk} = \delta^i_{[k}(b_{j]} - a_{j]})/(1-\lambda), \\ \underset{234}{a}{}^i_{jk} = \delta^i_{[k}(\lambda a_{j]} - b_{j]})/[\lambda(1-\lambda)]; \end{array}\right. \tag{7.6.30}$$

$$\left\{\begin{array}{l} \underset{124}{b}{}^i_{jkl} = (\delta^i_l f_{jk} + \delta^i_k g_{lj} + \delta^i_j k_{kl})/\lambda, \\ \underset{314}{b}{}^i_{jkl} = (\delta^i_l h_{jk} + \delta^i_k f_{lj} + \delta^i_j k_{kl})/(1-\lambda), \\ \underset{234}{b}{}^i_{jkl} = (\delta^i_l g_{jk} + \delta^i_k h_{lj} + \delta^i_j k_{kl})[\lambda(1-\lambda)]. \end{array}\right. \tag{7.6.31}$$

To obtain similar quantities for other canonical connections we must use the relations (7.3.21).

7.7 Grassmann Webs $GW(4,2,r)$ with Algebraic 3-Subwebs

In this section we will study Grassmann webs $GW(4,2,r)$ for which one 3-subweb is an algebraic three-web of general or some special type (Bol, group or parallelisable).

Theorem 7.7.1 *In a domain of regularity of a Grassmann web $GW(4,2,r)$ the following statements are equivalent:*

(i) *The 3-subweb* (a) $[1,2,3]$, (b) $[1,2,4]$, (c) $[3,1,4]$, (d) $[2,3,4]$ *is hexagonal.*

(ii) *The second fundamental tensors of the hypersurfaces U_ξ satisfy the condition*

$$\begin{aligned} &\text{(a)}\ \ f_{ij}+g_{ij}+h_{ij}=0, &&\text{(b)}\ \ f_{ij}+g_{ij}+k_{ij}=0,\\ &\text{(c)}\ \ f_{ij}+h_{ij}+k_{ij}=0, &&\text{(d)}\ \ g_{ij}+h_{ij}+k_{ij}=0. \end{aligned}$$

(iii) *The form*

$$\begin{aligned} &\text{(a)}\quad 2\omega_0^0+\omega_{r+1}^{r+1}-a_i\omega_0^i\ \ (or\ \ \omega_i^i-\omega_0^0+a_i\omega_0^i=\theta_i^i+(r-1)\omega),\\ &\text{(b)}\quad 2\omega_0^0+\omega_{r+1}^{r+1}-b_i\omega_0^i,\ \ \text{(c)}\ 2\omega_0^0+\omega_{r+1}^{r+1}-(a_i+b_i)\omega_0^i,\ \ \text{(d)}\ 3\omega_0^0-(a_i+b_i)\omega_0^i \end{aligned}$$

is a total differential.

(iv) *Either the hypersurfaces* (a) $U_1,\ U_2,\ U_3$; (b) $U_1,\ U_2,\ U_3$; (c) U_1, U_3, U_4 ; (d) $U_2,\ U_3,\ U_4$ *belong to a cubic hypersurface, or the 3-subweb* (a) $[1,2,3]$, (b) $[1,2,4]$, (c) $[3,1,4]$, (d) $[2,3,4]$ *is algebraic.*

Proof. Regarding (i), (ii), (iv), their equivalence for the three-web $[1,2,3]$ was established in Section 2.6, and [Ak 73]. For the other 3-subwebs $[\xi,\eta,\zeta]$ the proof is similar.

To complete the proof, we note, for example, that in the case (a) we have

$$d(2\omega_0^0+\omega_{r-1}^{r-1}-a_i\omega_0^i)=-d[\theta_i^i+(r-1)\omega]=-(f_{ij}+g_{ij}+h_{ij})\omega_0^i\wedge\omega_{r+1}^j.$$

Hence the form $2\omega_0^0+\omega_{r-1}^{r-1}-a_i\omega_0^i$ (or $\theta_i^i+(r-1)\omega$) is a total differential if and only if we have $f_{ij}+g_{ij}+h_{ij}=0$. Therefore, the conditions (ii) and (iii) are equivalent. ■

Theorem 7.7.2 *Grassmann webs $GW(4,2,r)$ one 3-subweb of which is algebraic of general type exist and the set of such 4-webs depends on one function of r variables.*

Proof. Let us take, for example, the case (a). In this case we have

$$f_{ij}+g_{ij}+h_{ij}=0. \tag{7.7.1}$$

Equations (7.7.1) and (7.6.22) imply

$$g_{ijk}=-h_{ijk},\qquad f_{ijk}=-h_{ijk}-3h_{(ij}a_{k)}. \tag{7.7.2}$$

Replacing in the first two equations of (7.6.22) g_{ijk} and f_{ijk} by their values from (7.7.2), taking exterior derivatives of the resulting equations and applying Cartan's lemma to the two exterior quadratic equations obtained, we have

$$\begin{aligned} \nabla h_{ijk}=&(-h_{ijk}a_m+3g_{(ij}g_{m)k}+3g_{(ij}f_{k)m}-g_{ij}g_{mk}-2g_{ij}f_{mk})\omega_0^m\\ &+[h_{ijk}a_m-3a_{(k}h_{ij)m}+2f_{ij}h_{mk}-3(f_{(ij}f_{m)k}-g_{(ij}g_{m)k}-h_{(ij}h_{m)k})]\omega_{r+1}^m. \end{aligned} \tag{7.7.3}$$

The exterior differentiation of (7.7.3) leads to the identity.

Therefore, the closed system determining the web $GW(4,2,r)$ under consideration consists of the Pfaffian equations (7.6.3), (7.6.10), (7.6.11), (7.6.19), (7.6.22), and (7.7.3), and the exterior quadratic equation

$$\{\nabla k_{ijk} + (k_{ijk}a_m - 3k_{(ij}f_{m)k})\omega_0^m + [-k_{ijk}a_m - 4k_{(ijk}b_{m)}/\lambda \\ -3k_{(ij}(f_{k)m} + f_{m)k} + k_{k)m})/\lambda]\omega_{r+1}^m\} \wedge (\lambda\omega_0^k + \omega_{r+1}^k) = 0. \qquad (7.7.4)$$

The number of unknown functions $(\nabla k_{ijk} + \ldots)$ is equal to $q = r(r+1)(r+2)/6$. The consecutive characters ([Ca 45]) are $s_1 = r(r+1)/2,\ \ s_2 = r(r-1)/2, \ldots, s_r = 1$ and Cartan's number Q is given by

$$Q = s_1 + 2s_2 + \ldots + rs_r = \frac{1}{24}r(r+1)(r+2)(r+3).$$

Application of Cartan's lemma to (7.7.4) gives

$$\nabla k_{ijk} + \ldots = k_{ijkm}(\lambda\omega_0^m + \omega_{r+1}^m)$$

where the k_{ijkm} are symmetric in all indices. Therefore, the general r-dimensional integral element depends on $N = r(r+1)(r+2)(r+3)/24$ arbitrary parameters k_{ijkm}.

Because $Q = N$, the system is in involution (see [Ca 45] or [GJ 87]), and its general solution depends on one function of r variables. Note that a function defining the hypersurface U_4 can be considered as this arbitrary function. ■

We now will consider special cases when an algebraic 3-subweb indicated in Theorem 7.7.1 is not of general type.

Theorem 7.7.3 *In a domain of regularity of a Grassmann web $GW(4,2,r)$ the following statements are equivalent:*

(i) *The 3-subweb* (a) [1, 2, 3], (b) [1, 2, 4], (c) [3, 1, 4], (d) [2, 3, 4] *is a Bol three-web of the type* (B_l), (B_m) *or* (B_r).

(ii) *The second fundamental tensors of the hypersurfaces* U_ξ *satisfy the conditions*

(a_1) $f_{ij} = 0,\ g_{ij} + h_{ij} = 0$, (a_2) $h_{ij} = 0,\ f_{ij} + g_{ij} = 0$, (a_3) $g_{ij} = 0,\ f_{ij} + h_{ij} = 0$,

(b_1) $f_{ij} = 0,\ g_{ij} + k_{ij} = 0$, (b_2) $k_{ij} = 0,\ f_{ij} + g_{ij} = 0$, (b_3) $g_{ij} = 0,\ f_{ij} + k_{ij} = 0$,

(c_1) $h_{ij} = 0,\ f_{ij} + k_{ij} = 0$, (c_2) $k_{ij} = 0,\ f_{ij} + h_{ij} = 0$, (c_3) $f_{ij} = 0,\ h_{ij} + k_{ij} = 0$,

(d_1) $g_{ij} = 0,\ h_{ij} + k_{ij} = 0$, (d_2) $k_{ij} = 0,\ g_{ij} + h_{ij} = 0$, (d_3) $h_{ij} = 0,\ g_{ij} + k_{ij} = 0$.

(iii) *The forms* (a_1) ω_{r+1}^{r+1} *and* $2\omega_0^0 - a_i\omega_0^i$, (a_2) $\omega_0^0 - a_i\omega_0^i$ *and* $\omega_0^0 + \omega_{r+1}^{r+1}$, (a_3) ω_0^0 *and* $\omega_{r+1}^{r+1} - a_i\omega_0^i$ *etc. are total differentials.*

(iv) *The cubic hypersurface of Theorem* 7.7.1 *decomposes into a hyperplane and a hyperquadric.*

Proof. The equivalence of (i), (ii), and (iv) follows from [Ak 73]. As for the equivalence of (iii) and (ii) in the case (a_1), for example, it follows from the equations

$$d\omega_{r+1}^{r+1} = -f_{ij}\omega_0^i \wedge \omega_{r+1}^j, \quad d(2\omega_0^0 - a_i\omega_0^i) = -(g_{ij} + h_{ij})\omega_0^i \wedge \omega_{r+1}^j. \quad \blacksquare$$

Theorem 7.7.4 *Grassmann webs $GW(4,2,r)$ one of whose 3-subwebs is a Bol three-web, exist and the set of such 4-webs depends on one function of r variables.*

Proof. When (a) is the case, we have

$$f_{ij} = 0, \quad g_{ij} + h_{ij} = 0. \tag{7.7.5}$$

From (7.7.5) and (7.6.22) we obtain

$$f_{ijk} = 0, \quad -h_{ijk} = g_{ijk} = -3g_{(ij}a_{k)}. \tag{7.7.6}$$

It follows from (7.7.6) and (7.6.22) that

$$\nabla g_{ij} = g_{ij}a_k(\omega_{r+1}^k - \omega_0^k) - 3g_{(ij}a_{k)}\omega_{r+1}^k. \tag{7.7.7}$$

The exterior differentiation of (7.7.7) reduces to the identity. Therefore, the closed system determining the Grassmann webs $GW(4,2,r)$ under consideration consists of the Pfaffian equations (7.6.3), (7.6.10), (7.6.11), (7.6.19), (7.7.7) and the exterior quadratic equation

$$[\nabla k_{ij} + (k_{ij}a_k - 3k_{(ij}b_{k)})\omega_0^k - k_{ij}a_k\omega_{r+1}^k] \wedge (\lambda\omega_0^j + \omega_{r+1}^j) = 0. \tag{7.7.8}$$

In this case we have

$$q = \frac{1}{2}r(r+1), \quad s_1 = r, \quad s_2 = r-1, \ldots, s_r = 1, \quad Q = N = \frac{1}{6}r(r+1)(r+2).$$

The system is in involution, and its general solution depends on one function of r variables. $\blacksquare$

Theorem 7.7.5 *In a domain of regularity of a Grassmann web $GW(4,2,r)$ the following statements are equivalent:*

(i) *The 3-subweb* (a) [1,2,3], (b) [1,2,4], (c) [3,1,4], (d) [2,3,4] *is a group non-parallelisable three-web.*

(ii) *Three second fundamental tensors of the hypersurfaces U_ξ vanish:*

$$\text{(a) } f_{ij} = g_{ij} = h_{ij} = 0, \quad \text{(b) } f_{ij} = g_{ij} = k_{ij} = 0,$$
$$\text{(c) } f_{ij} = h_{ij} = k_{ij} = 0, \quad \text{(d) } g_{ij} = h_{ij} = k_{ij} = 0.$$

(iii) *The forms*

$$\begin{array}{ll}
\text{(a)}\ \omega^{r+1}_{r+1},\ \omega^0_0,\ \omega^0_0 - a_i\omega^i_0, & \text{(b)}\ \omega^{r+1}_{r+1},\ \omega^0_0,\ \omega^0_0 - b_i\omega^i_0, \\
\text{(c)}\ \omega^{r+1}_{r+1},\ \omega^0_0 - a_i\omega^i_0,\ \omega^0_0 - b_i\omega^i_0, & \text{(d)}\ \omega^0_0,\ \omega^0_0 - a_i\omega^i_0,\ \omega^0_0 - b_i\omega^i_0,
\end{array}$$

are total differentials.

(iv) *The hypersurfaces* (a) U_1, U_2, U_3, (b) U_1, U_2, U_4, (c) U_1, U_3, U_4, (d) U_2, U_3, U_4 *are hyperplanes of general position.*

Proof. The proof follows from [Ak 73] and the form of the exterior differentials of the forms indicated in (iii). ■

As for the existence of such the $GW(4,2,r)$, we have in the case (a), for example, equations (7.6.3), (7.6.10), (7.6.11) in all of which $f_{ij} = g_{ij} = h_{ij} = 0$ and (7.7.8). These again show that the general solution depends on one function of r variables.

Theorem 7.7.6 *In a domain of regularity of a Grassmann web* $GW(4,2,r)$, $r > 1$, *the following statements are equivalent:*

(i) *The 3-subweb* (a) $[1,2,3]$, (b) $[1,2,4]$, (c) $[3,1,4]$, (d) $[2,3,4]$ *is parallelisable.*

(ii) *The tensor* (a) a_i, (b) b_i, (c) $b_i - a_i$, (d) $b_i - \lambda a_i$ *vanishes.*

(iii) *The hypersurfaces* (a) U_1, U_2, U_3, (b) U_1, U_2, U_4, (c) U_1, U_3, U_4, (d) U_2, U_3, U_4 *are hyperplanes of a pencil.*

(iv) *The basis affinor* λ^i_j *is covariantly constant on the foliation* (a) λ_4, (b) λ_3 , (c) λ_2, (d) λ_1 *in the canonical connection* γ_{123}

(v) *The basis affinor* λ^i_j *is covariantly constant on the foliation* (a) λ_4, (b) λ_3 , (c) λ_2 , (d) λ_1 *in the canonical connection* $\gamma_{\xi\eta\zeta}$ *where* (a) $\xi,\eta,\zeta \neq 4$, (b) $\xi,\eta,\zeta \neq 3$, (c) $\xi,\eta,\zeta \neq= 2$, (d) $\xi,\eta,\zeta \neq 1$.

(vi) *The leaves of the foliation* (a) λ_4 , (b) λ_3, (c) λ_2, (d) λ_1 *are totally geodesic in the connection* (a) γ_{123}, (b) γ_{124}, (c) γ_{314}, (d) γ_{234} .

(vii) *The transformation* $[\eta,\zeta]$ *(see Section 7.1) transferring a curve* $\Gamma_\xi \subset F_\xi$ *to a curve* $\Gamma^{\eta\zeta}_{\xi\kappa\xi} \subset F_\xi$ *where* $\xi = 4,3,2,1$; $\eta,\zeta,\kappa \neq \xi$; $\zeta \neq \kappa$; $\eta \neq \kappa$, *is a geodesic preserving transformation.*

Proof. For each of equivalences we consider only the case (a). (i) ⇔ (ii). If the 3-subweb $[1,2,3]$ is parallelisable, its torsion and curvature tensors vanish. Using (7.6.18), (7.6.21), (7.6.22), and the inequality $r > 1$, we obtain that in this case $a_i = 0$ and $f_{ij} = g_{ij} = h_{ij} = 0$. Conversely, if $a_i = 0$, equations (7.6.19) give $f_{ij} = g_{ij} = h_{ij}$ and consequently, by (7.6.22), $f_{ijk} = g_{ijk} = h_{ijk} = 0$ and $\nabla f_{ij} = 0$. The exterior differentiation of the last equation leads to the relation $f_{(ij}f_{k)m} = 0$. Setting $i = j = k = m$, we obtain that all $f_{ii} = 0$. Taking $k = i \neq j = m$, we obtain $f_{ij} = 0$, $i \neq j$.

(ii) ⇔ (iii). This follows from the first part and the fact that the equations $a_i = 0$, $f_{ij} = g_{ij} = h_{ij} = 0$ are necessary and sufficient for the hypersurfaces U_1, U_2, U_3 to be hyperplanes having the common $(r-1)$-plane $\{A_1, \ldots, A_r\}$ (see (7.6.1), (7.6.12), and (7.6.24)).

(ii) $\Leftrightarrow$ (iv). This follows from (7.6.11).

(ii) $\Leftrightarrow$ (v). This follows from (7.6.11) and the expressions (7.6.29) and (7.6.30) of the connection forms $\underset{\xi\eta\zeta}{\theta}{}^i_j$ and the torsion tensor $\underset{\xi\eta\zeta}{a}{}^i_{jk}$ of the 3-subwebs $[\xi, \eta, \zeta]$ of the Grassmann web $GW(4,2,r)$.

(ii) $\Leftrightarrow$ (vi). For a general $W(4,2,r)$ the necessary and sufficient conditions for the leaves of the foliation λ_4 to be totally geodesic in the connection γ_{123} are the conditions (7.1.40):

$$\lambda^i_{(jk)} = \lambda^p_{(j}\mu^i_{k)p}. \tag{7.7.9}$$

For a $GW(4,2,r)$, using (7.6.25) and (7.6.26), we obtain that (7.7.9) reduces to $a_i = 0$.

(ii) $\Leftrightarrow$ (vii). We consider the transformation $[3,4]$ on a leaf $F_1 \subset \lambda_1$ through a leaf $F_2 \subset \lambda_2$. So, we take a curve $\Gamma_1 \subset F_1$. The leaves of λ_3 through the points of Γ_1 intersect F_2 in points of a curve $\Gamma^3_{12} \subset F_2$. The leaves of λ_4 through the points of Γ^3_{12} intersect F_1 in points of a curve Γ^{34}_{121}. If Γ_1 is a geodesic line in γ_{123}, then the necessary and sufficient conditions for Γ^{34}_{121} to be geodesic in γ_{123} have the form (7.7.9). Using (7.6.25), (7.6.26), and (7.7.9), we again obtain $a_i = 0$. ■

As for the existence theorem for the webs $GW(4,2,r)$ discussed in Theorem 7.7.6, we again obtain that such webs $GW(4,2,r)$ exist and depend on one function of r variables.

In conclusion of this section we establish conditions for the canonical connections $\gamma_{\xi\eta\zeta}$ induced by 3-subwebs $[\xi, \eta, \zeta]$ to be equiaffine.

Theorem 7.7.7 *Necessary and sufficient conditions for the canonical connection* (a) γ_{123}, (b) γ_{234}, (c) γ_{341}, (d) γ_{412} *to be equiaffine are*

$$\text{(a)}\ f_{ij} + g_{ij} + rh_{ij} = 0, \qquad \text{(c)}\ rf_{ij} + h_{ij} + k_{ij} = 0,$$
$$\text{(b)}\ g_{ij} + h_{ij} + rk_{ij} = 0, \qquad \text{(d)}\ f_{ij} + rg_{ij} + k_{ij} = 0.$$

Proof. The connection $\gamma_{\xi\eta\zeta}$ is equiaffine if and only if the form $\underset{\xi\eta\zeta}{\theta}{}^i_j$ is closed (see [G 66]). In the case (a) we obtain from (7.6.20) that it is equivalent to the equations

$$\underset{123}{b}{}^i_{ikl} = 0. \tag{7.7.10}$$

Equations (7.7.10) and (7.6.21) give

$$f_{ij} + g_{ij} + rh_{ij} = 0. \tag{7.7.11}$$

To prove (b), (c), and (d), we must use the formulae (7.6.31) and the equations $\underset{\xi\eta\zeta}{b}{}^i_{mjk} = \underset{\zeta\xi\eta}{b}{}^i_{jkm}$ (cf. (7.3.21)). ■

Corollary 7.7.8 *If the 3-subweb* $[1,2,3]$ *is a group three-web or a Bol three-web of the type* B_l, *then the connection* γ_{123} *is equiaffine.*

Proof. By Theorems 7.7.5 and 7.7.3, the 3-subweb $[1,2,3]$ is a group or Bol three-web if and only if we have correspondingly $f_{ij} = g_{ij} = h_{ij} = 0$ or $f_{ij} + g_{ij} = h_{ij} = 0$. Each of these conditions implies (7.7.11). ■

7.8 Algebraic Webs $AW(4,2,r)$

In this section we shall study algebraic webs $AW(4,2,r)$ and their special classes. According to Definition 2.6.2 considered for $d = 4$, a Grassmann web $GW(4,2,r)$ is called algebraic if the hypersurfaces U_ξ generating it belong to an r-dimensional quartic $Q = V_4^r$. Such algebraic webs were denoted by $AW(4,2,r)$.

First we will establish necessary and sufficient conditions for a Grassmann web $GW(4,2,r)$ to be a general algebraic web $AW(4,2,r)$. The first of them is a particular case of the condition (2.6.22) which we found in Corollary 2.6.14. We will prove it here using the method which is different from the method we have used in Section 2.6. As was mentioned in Section 2.6, each of these conditions is necessary and sufficient for the hypersurfaces U_ξ to belong to a hyperquartic V_4^r of the space P^{r+1}.

Theorem 7.8.1 *In a domain of regularity of a Grassmann web* $GW(4,2,r)$ *the following statements are equivalent:*

(i) *The second fundamental tensors* f_{ij}, g_{ij}, h_{ij}, *and* k_{ij} *satisfy the equation*

$$f_{ij} + g_{ij} + h_{ij} + k_{ij} = 0. \tag{7.8.1}$$

(ii) *The form*

$$\phi = 3\omega_0^0 + \omega_{r+1}^{r+1} - (a_i + b_i)\omega_0^i \tag{7.8.2}$$

is a total differential.

(iii) *The curvature tensors* $\underset{\xi\eta\zeta}{b}{}^i_{jkl}$ *of the 3-subwebs* $[\xi,\eta,\zeta]$ *are connected by the equation*

$$\underset{123}{b}{}^i_{(jkl)} + (\lambda^2 - \lambda)\underset{234}{b}{}^i_{(jkl)} + (1-\lambda)\underset{314}{b}{}^i_{(jkl)} + \lambda\underset{124}{b}{}^i_{(jkl)} = 0. \tag{7.8.3}$$

(iv) *A web* $GW(4,2,r)$ *is algebraic. Moreover the* r*-dimensional quartic generating such an* $AW(4,2,r)$ *in some special selected moving frame has the equation*

$$\begin{aligned}4A_{000\,r+1}(x^0)^3x^{r+1} + 4A_{0\,r+1\,r+1\,r+1}x^0(x^{r+1})^3 \\ + 6A_{00\,r+1\,r+1}(x^0)^2(x^{r+1})^2 + 6A_{00ij}(x^0)^2x^ix^j \\ + 6A_{r+1\,r+1\,ij}(x^{r+1})^2x^ix^j + 12A_{00i\,r+1}(x^0)^2x^ix^{r+1} \\ + 12A_{0i\,r+1\,r+1}x^0x^i(x^{r+1})^2 + 4A_{00ijk}x^0x^ix^jx^k + 12A_{0ij\,r+1}x^0x^ix^jx^{r+1} \\ + 4A_{ijk\,r+1}x^ix^jx^kx^{r+1} + A_{ijkm}x^ix^jx^kx^m = 0,\end{aligned} \tag{7.8.4}$$

where the coefficients in (7.8.4) *are the only non-vanishing coefficients which are functions of* λ, a_i, b_i, f_{ij}, g_{ij}, h_{ij}, k_{ij}, f_{ijk}, g_{ijk}, h_{ijk}, k_{ijk} *and the covariant derivatives* k_{ijkm} *of* k_{ijk} *(we will determine all these coefficients in the process of our proof).*

Proof. (i) $\Leftrightarrow$ (ii). From (7.8.2), (7.6.2), and (7.6.19) it follows that

$$d\phi = -(f_{ij} + g_{ij} + h_{ij} + k_{ij})\omega_0^i \wedge \omega_{r+1}^j.$$

Hence ϕ is a total differential if and only if (7.8.1) holds.

(i) $\Leftrightarrow$ (iii). In fact, it follows from (7.6.21) and (7.6.31) that relation (7.8.3) is equivalent to

$$\delta^i_{(j}(f_{kl)} + g_{kl)} + h_{kl)} + k_{kl)}) = 0. \tag{7.8.5}$$

Thus, (i) implies (iii). Conversely, contracting (7.8.5) with respect to the indices i and j, we obtain $(r+2)(f_{ij} + g_{ij} + h_{ij} + k_{ij}) = 0$, i.e., we have (7.8.1).

(i) $\Leftrightarrow$ (iv). Let Q be an r-dimensional quartic and

$$A_{uvwz}x^u x^v x^w x^z = 0, \qquad u, v, w, z = 0, 1, \ldots, r+1, \tag{7.8.6}$$

be the equation of Q in the moving frame $\{A_0,\ A_1, \ldots, A_r,\ A_{r+1}\}$

The conditions for the immobility of Q have the form

$$\nabla A_{uvwz} = \theta A_{uvwz}, \tag{7.8.7}$$

where

$$\nabla A_{uvwz} = dA_{uvwz} - A_{xvwz}\omega_u^x - A_{uxwz}\omega_v^x - A_{uvxz}\omega_w^x - A_{uvwx}\omega_z^x.$$

Let $A_0 \in Q$ and $A_{r+1} \in Q$. Since these points have the coordinates $(1, 0, \ldots, 0)$ and $(0, \ldots, 0, 1)$, it follows from (7.8.6) that in this case

$$A_{0000} = 0, \quad A_{r+1\,r+1\,r+1\,r+1} = 0. \tag{7.8.8}$$

Let $M = A_0 + A_{r+1} \in Q$ and $N = \lambda A_0 + A_{r+1} \in Q$. Then we obtain from (7.8.6) that

$$\begin{cases} 2A_{0\,r+1\,r+1\,r+1} + 2A_{000\,r+1} + 3A_{00\,r+1\,r+1} = 0, \\ 2A_{0\,r+1\,r+1\,r+1} + 2\lambda^2 A_{000\,r+1} + 3\lambda A_{00\,r+1\,r+1} = 0. \end{cases} \tag{7.8.9}$$

Setting

$$A_{000\,r+1} = -3, \tag{7.8.10}$$

we obtain from (7.8.9) that

$$A_{0\,r+1\,r+1\,r+1} = -3\lambda, \quad A_{00\,r+1\,r+1} = 2(1+\lambda). \tag{7.8.11}$$

Equations (7.8.7) and (7.8.8) give

$$A_{000i} = 0, \quad A_{i\,r+1\,r+1\,r+1} = 0. \tag{7.8.12}$$

Equations (7.8.12), (7.6.19), and (7.8.7) imply that

$$A_{00ij} = f_{ij}, \quad A_{r+1\,r+1\,ij} = -\lambda g_{ij}. \tag{7.8.13}$$

From equations (7.8.7), (7.8.10), (7.8.11), and (7.6.15) we obtain

$$\begin{cases} -A_{00i\,r+1}\omega_0^i + \theta + 4\omega = a_i(\omega_{r+1}^i - 3\omega_0^i), \\ -A_{0i\,r+1\,r+1}\omega_{r+1}^i + \lambda(\theta + 4\omega) = \lambda(b_i - 2a_i)\omega_0^i + (b_i + 2\lambda a_i)\omega_{r+1}^i, \\ -A_{00i\,r+1}\omega_{r+1}^i - A_{0i\,r+1\,r+1}\omega_0^i - (1+\lambda)(\theta + 4\omega) \\ = [(2+3\lambda)a_i - \lambda b_i]\omega_0^i - [(\lambda+2)a_i + b_i]\omega_{r+1}^i. \end{cases} \tag{7.8.14}$$

The solution of the system (7.8.14) is

$$A_{00i\,r+1} = b_i + a_i, \quad A_{0i\,r+1\,r+1} = -(b_i + \lambda a_i), \tag{7.8.15}$$

$$\theta + 4\omega = (b_i - 2a_i)\omega_0^i + a_i\omega_{r+1}^i. \tag{7.8.16}$$

Equations (7.8.7), (7.8.13), (7.8.15), (7.6.22), and (7.6.15) allow one to obtain

$$\begin{cases} 2A_{0ijk} = f_{ijk} - 3f_{(ij}(a_{k)} + b_{k)}), \\ 2A_{ijk\,r+1} = -\lambda g_{ijk} - 3g_{ijk} - 3g_{(ij}(\lambda a_{k)} + b_{k)}). \end{cases} \tag{7.8.17}$$

Equations (7.8.15), (7.8.7), (7.8.16), (7.6.15), (7.8.13), (7.8.11), and (7.6.22) show first of all that for an existence of the desired r-quartic Q we must have the equations (7.8.1) and secondly these equations give two expressions for A_{0ijr+1} :

$$\begin{aligned} 2A_{0ijr+1} &= -2a_{(i}b_{j)} - \lambda(f_{ij} + k_{ij}) - (f_{ij} + h_{ij}), \\ 2A_{0ijr+1} &= -2a_{(i}b_{j)} + \lambda(g_{ij} + h_{ij}) + (g_{ij} + k_{ij}), \end{aligned} \tag{7.8.18}$$

which coincide by virtue of (7.8.1). Equations (7.8.18) and (7.8.7) again imply (7.8.17).

In order to obtain the last coefficients A_{ijkm} , we need differential consequences of (7.8.1). They can be obtained by differentiation of (7.8.1) and use of (7.6.22) and have the form:

$$\begin{cases} f_{ijk} + h_{ijk} + \lambda k_{ijk} = -3h_{(ij}a_{k)} - 3k_{(ij}b_{k)}, \\ g_{ijk} + h_{ijk} + k_{ijk} = 0. \end{cases} \tag{7.8.19}$$

Differentiating (7.8.19) and using (7.6.23), we get

$$\begin{cases} h_{ijkm} = -\lambda k_{ijkm} - 2h_{ij(k}a_{m)} + 3g_{(ij}(f_{k)m} + g_{k)m}) + 3(f_{(ij} + g_{(ij})f_{m)k}, \\ f_{ijkm} = (\lambda - \lambda^2)k_{ijkm} - 4h_{(ijk}a_{m)} - 4\lambda k_{(ijk}b_{m)} - 12h_{(ij}a_k a_{m)} - 12k_{(ij}b_k b_{m)} \\ \qquad +3(1-\lambda)k_{(ij}(f_{k)m} + f_{m)k}) + 3(f_{(ij}f_{k)m} - g_{(ij}g_{k)m} - h_{(ij}h_{k)m} - \lambda k_{(ij}k_{k)m}), \\ g_{ijkm} = (\lambda - 1)k_{ijkm} - 4h_{(ijk}a_{m)} + (4/\lambda)k_{(ijk}b_{m)} + 3(1 - 1/\lambda)k_{(ij}(f_{k)m} + f_{m)k}) \\ \qquad +3(f_{(ij}f_{k)m} - g_{(ij}g_{k)m} - h_{(ij}h_{k)m} - (1/\lambda)k_{(ij}k_{k)m}). \end{cases} \tag{7.8.20}$$

Substituting (7.8.20) into (7.6.23), we obtain only two linearly independent equations among (7.6.23). Their differentiation gives two exterior quadratic equations from which it follows that

$$\nabla k_{ijkm} = l_{ijkml}\omega_0^l + m_{ijkml}\omega_{r+1}^l, \tag{7.8.21}$$

where the l_{ijkml} and m_{ijkml} are given by

$$\left\{\begin{aligned} l_{ijkml} &= \{1/(1-\lambda)\}\{(\lambda b_l - a_l)k_{ijkm} + 4(b_{(m} - \lambda a_{(m})k_{ijk)l} \\ &+10(g_{(ijk}g_{lm)} - h_{(ijk}h_{lm)} - k_{(ijk}k_{lm)} + f_{(ijk}f_{lm)}) \\ &+4(\lambda-1)k_{(ijk}f_{m)l} + 3(\lambda-1)k_{l(ij}(f_{k)m} + f_{m)k}) \\ &+3(1-1/\lambda)k_{(ij}(f_{k)ml} + f_{m)kl}) + 15a_m(-f_{(ij}f_{k)l} + g_{(ij}g_{k)l} - h_{(ij}h_{k)l}) \\ &-(15/\lambda)b_{(m}k_{ij}k_{k)l} - 16a_{(m}k_{ij}f_{k)l} + (12+4/\lambda)b_{(m}k_{ij}f_{k)l} \\ &+(8/\lambda)b_{(m}f_{ij}k_{k)l} + 6(1-1/\lambda)k_{l(i}b_j(f_{k)m} + f_{m)k}) \\ &+3(b_l - a_l)k_{(ij}(f_{k)m} + f_{m)k})\}, \\ m_{ijkl} &= \{1/(\lambda^2-\lambda)\}\{4\lambda(a_{(m} - b_{(m})k_{ijk)l} + k_{ijkm}[(1-2\lambda)b_l + (3\lambda^2-2\lambda)a_l] \\ &+10(-f_{(ijk}f_{lm)} + h_{(ijk}h_{lm)} + \lambda k_{(ijk}k_{lm)} - 3g_{l(ij}g_{k)m} - 3g_{lm(i}g_{jk)} \\ &-4f_{(ijk}g_{m)l} + 3(1-\lambda)k_{l(ij}(f_{k)m} + f_{m)k}) - 12g_{l(m}(h_{ij}a_{k)} + k_{ij}b_{k)}) \\ &+8(a_{(m} - b_{(m})(2k_{ij}f_{k)l} + f_{ij}k_{k)l}) - 3(b_l - a_l)k_{(ij}(f_{k)m} + f_{m)k}) \\ &+15b_{(m}k_{ij}k_{kl)} + 15a_{(m}(f_{ij}f_{kl)} - g_{ij}g_{kl)} + h_{ij}h_{kl)})\}. \end{aligned}\right. \tag{7.8.22}$$

Equations (7.8.17), (7.8.7), (7.6.23), (7.8.20), (7.8.16), (7.8.17), (7.8.18), and (7.6.15) yield expressions for A_{ijkm} :

$$\begin{aligned} 2A_{ijkm} &= (\lambda-\lambda^2)k_{ijkm} + 4\lambda k_{(ijk}a_{m)} + 4h_{(ijk}b_{m)} - 12g_{(ij}a_k b_{m)} \\ &+3(1-\lambda)k_{(ij}f_{m)k} + 3f_{(ij}(h_{k)m} + \lambda k_{k)m}) + 3f_{m(i}(h_{jk)} + k_{jk)}) \\ &+3(f_{(ij}f_{k)m} - g_{(ij}g_{k)m} + h_{(ij}h_{k)m} + \lambda k_{(ij}k_{k)m}). \end{aligned} \tag{7.8.23}$$

Finally equations (7.8.23) and (7.8.7) imply (7.8.21) and (7.8.22). The exterior differentiation of equations (7.8.21) leads to the identity.

Thus, equations (7.8.1) are necessary and sufficient conditions for the hypersurfaces U_ξ to belong to an r-dimensional quartic Q.

The equation of Q has the form (7.8.3) where coefficients are determined by (7.8.10), (7.8.11), (7.8.13), (7.8.15), (7.8.17), (7.8.18), and (7.8.23). ■

Corollary 7.8.2 *An algebraic web $AW(4,2,r)$ exists and depends on $(r+1)(r^3+13r^2+58r+96)/24$ arbitrary constants.*

Proof. In fact, we saw that such a 4-web is determined by the completely integrable system of Pfaffian equations (7.6.3), (7.6.10), (7.6.11), $(7.6.22)_{1,2,3}$, $(7.6.23)_{1,2}$, where f_{ijkm} and g_{ijkm} have the form (7.8.20) and (7.8.21). The number of linearly independent equations of this system is equal to

$$4 + 4r + 3 \cdot \frac{1}{2} r(r+1) + 2 \cdot \frac{1}{6} r(r+1)(r+2) + \frac{1}{24} r(r+1)(r+2)(r+3)$$
$$= \frac{1}{24}(r^3 + 13r^2 + 58r + 96)(r+1). \quad \blacksquare$$

Remark 7.8.3 It follows from equations (7.8.22) that the third covariant derivatives of the curvature tensor and the fourth covariant derivatives of the torsion tensor of an $AW(4,2,r)$ are expressed in terms of these tensors and their covariant derivatives. According to the terminology introduced in [Ak 75a], a G-structure associated with such a web (cf. Section 1.2.1) is a *closed G-structure of the first kind and the fourth class.*

We consider now algebraic webs $AW(4,2,r)$ for which an r-quartic Q decomposes into a cubic hypersurface and a hyperplane or two hyperquadrics.

Since a hyperplane must coincide with one of the hypersurfaces U_ξ, it follows from (7.6.24) and (7.8.1) that the first kind of degeneracy of Q is possible only in the following four cases:

$$\begin{aligned} f_{ij} &= 0, & g_{ij} + h_{ij} + k_{ij} &= 0, \\ g_{ij} &= 0, & f_{ij} + h_{ij} + k_{ij} &= 0, \\ h_{ij} &= 0, & f_{ij} + g_{ij} + k_{ij} &= 0, \\ k_{ij} &= 0, & f_{ij} + g_{ij} + h_{ij} &= 0, \end{aligned} \tag{7.8.24}$$

In the first case we obtain from (7.8.24), (7.8.13), (7.8.16), (7.8.23), (7.8.20), and (7.6.22) that

$$f_{ijk} = 0, \quad f_{ijkm} = 0, \quad A_{00ij} = A_{0ijk} = A_{ijkm} = 0. \tag{7.8.25}$$

Substituting (7.8.25) into (7.8.4), we find that the r-quartic Q decomposes into the hyperplane $x^{r+1} = 0$ and the cubic hypersurface

$$\begin{aligned} &-6(x^0)^3 - 6\lambda x^0(x^{r+1})^2 + 6(1+\lambda)(x^0)^2 x^{r+1} - 3\lambda g_{ij} x^i x^j x^{r+1} \\ &-6(2a_{(i}b_{j)} + \lambda k_{ij} + h_{ij}) x^0 x^i x^j + 6(a_i + b_i)(x^0)^2 x^i - 6(b_i + \lambda a_i) x^0 x^i x^{r+1} \\ &-[\lambda g_{ijk} + g_{(ij}(a_{k)} + b_{k)})] x^i x^j x^k = 0. \end{aligned} \tag{7.8.26}$$

The second and the third cases of (7.8.24) are similar to the first one. In the fourth case, using (7.8.24), (7.6.22), and (7.6.23), we obtain that $k_{ijk} = k_{ijkm} = 0$. Then it is easy to see that the r-quartic (7.8.4) decomposes into the hyperplane $x^0 - b_i x^i - \lambda x^{r+1} = 0$ and the cubic hypersurface

$$-6(x^0)^2 x^{r+1} + 6x^0(x^{r+1})^2 + 3f_{ij}x^0x^ix^j + 3g_{ij}x^ix^jx^{r+1}$$
$$+6a_jx^0x^jx^{r+1} + (f_{ijk} - 3f_{(ij}a_{k)})x^ix^jx^k = 0.$$

In each of these four cases the second covariant derivatives of the curvature tensor and the third covariant derivatives of the torsion tensor of the $AW(4,2,r)$ under consideration are expressed in terms of these tensors and their covariant derivatives of lower orders. The system determining such the $AW(4,2,r)$ is completely integrable and its general solution depends on $(r+1)(r^2+8r+2)$ constants. A G-structure associated with such the $AW(4,2,r)$ is a *closed G-structure of the first kind and the third class.*

There are three possibilities for degeneracy of Q of the second kind:

$$\begin{array}{lll} \text{(a)} & f_{ij} + g_{ij} = 0, & h_{ij} + k_{ij} = 0, \\ \text{(b)} & f_{ij} + h_{ij} = 0, & g_{ij} + k_{ij} = 0, \\ \text{(c)} & f_{ij} + k_{ij} = 0, & g_{ij} + h_{ij} = 0. \end{array} \tag{7.8.27}$$

It is easy to see from (7.8.27) and (7.8.4) that in these cases Q decomposes respectively into the following two hyperquadrics:

$$\begin{array}{ll} \text{(a)} & f_{ij}x^ix^j + 2x^0x^{r+1} = 0 \ \text{ and} \\ & 2(x^0)^2 + 2\lambda(x^{r+1})^2 - 2(1+\lambda)x^0x^{r+1} + 2(b_i - \lambda a_i)x^{r+1}x^i - 2(h_i + a_i)x^0x^i \\ & \quad + (h_{ij} + \lambda k_{ij} + 2a_{(i}b_{j)})x^ix^j = 0, \\ \text{(b)} & f_{ij}x^ix^j - 2x^0x^{r+1} + 2(x^{r+1})^2 + 2a_ix^ix^{r+1} = 0 \ \text{ and} \\ & -2(x^0)^2 + 2\lambda x^0x^{r+1} + 2b_ix^0x^i + \lambda g_{ij}x^ix^j = 0, \\ \text{(c)} & -2x^0x^{r+1} + 2\lambda(x^{r+1})^2 + 2b_ix^ix^{r+1} + f_{ij}x^ix^j = 0 \ \text{ and} \\ & -2x^0x^{r+1} + 2(x^0)^2 - 2a_ix^0x^i - g_{ij}x^ix^j = 0. \end{array}$$

In each of these three cases a G-structure associated with such the $AW(4,2,r)$ is a *closed G-structure of the first kind and the second class.* Each of the webs $AW(4,2,r)$ in these cases depends on $(r+1)(r+4)$ constants.

In fact, to give an example, in the case (a) it follows from (7.8.27) and (7.6.22) that

$$f_{ijk} = g_{ijk} = 0, \quad k_{ijk} = 3(h_{(ij}a_{k)} + k_{(ij}b_{k)})/(1-\lambda). \tag{7.8.28}$$

Hence the first covariant derivatives of the curvature tensor and the second covariant derivatives of the torsion tensor of the $AW(4,2,r)$ are expressed in terms of these tensors and their covariant derivatives of lower orders. In this case we have the completely integrable system of Pfaffian equations (7.6.3), (7.6.10), (7.6.11), (7.6.19), and $(7.6.22)_{1,4}$ in which we must substitute f_{ijk} and k_{ijk} by their values from (7.8.28). The number of linearly independent Pfaffian equations of this system is equal to $4 + 4r + r(r+1) = (r+1)(r+4)$.

In the next three theorems we give necessary and sufficient conditions for a decomposition of Q into a hyperplane and a cubic hypersurface, into two hyperquadrics and into two hyperplanes and a hyperquadric.

Theorem 7.8.4 *Let an $AW(4,2,r)$ be an algebraic 4-web and Q be an r-quartic generating it. Then in a domain of regularity of the $AW(4,2,r)$ the following statements are equivalent:*

(i) *Q decomposes into a hyperplane and a cubic hypersurface.*

(ii) *A 3-subweb $[\xi, \eta, \zeta]$ is hexagonal.*

(iii) *One of the forms*

$$\begin{array}{ll} \text{(a)} & \omega_{r+1}^{r+1} \ \textit{or} \ 3\omega_0^0 - (a_i + b_i)\omega_0^i, \\ \text{(b)} & \omega_0^0 \ \textit{or} \ 2\omega_0^0 + \omega_{r+1}^{r+1} - (a_i + b_i)\omega_0^i, \\ \text{(c)} & \omega = \omega_0^0 - a_i\omega_0^i \ \textit{or} \ 2\omega_0^0 + \omega_{r+1}^{r+1} - b_i\omega_0^i, \\ \text{(d)} & \omega_0^0 - b_i\omega_0^i \ \textit{or} \ 2\omega_0^0 + \omega_{r+1}^{r+1} - a_i\omega_0^0 \\ & (\textit{or} \ \omega_i^i - \omega_0^0 + a_i\omega_0^i = \theta_i^i + (r-1)\omega) \end{array}$$

is a total differential. Both forms in (a), (b), (c), *or* (d) *are total differential if and only if one of the hypersurfaces U_ξ is a hyperplane and three others belong to a cubic hypersurface.*

Proof. (i) $\leftrightarrow$ (ii). We consider, for example, the 3-subweb $[1,2,3]$. It is hexagonal if and only if $\underset{123}{b}{}^i_{(jkl)} = 0$ (see Table 1.1 in Section 1.2), or, by (7.6.21), if

$$f_{ij} + g_{ij} + h_{ij} = 0. \tag{7.8.29}$$

Equations (7.8.29) and (7.8.1) give $k_{ij} = 0$.

(i) $\leftrightarrow$ (iii). In the case (a), using (7.6.2), and (7.6.19), we obtain

$$\begin{aligned} d\omega_{r+1}^{r+1} &= -f_{ij}\omega_0^i \wedge \omega_{r+1}^j, \\ d[3\omega_0^0 - (b_i + a_i)\omega_0^i] &= -(g_{ij} + h_{ij} + k_{ij})\omega_0^i \wedge \omega_{r+1}^j. \end{aligned} \tag{7.8.30}$$

Equations (7.8.30) show that the first conditions in (7.8.24) are necessary and sufficient for the forms ω_{r+1}^{r+1} and $3\omega_0^0 - (b_i + a_i)\omega_0^i$ to be total differentials.

The last statement of Theorem 7.8.4 follows from the fact that any two of three conditions

$$f_{ij} = 0, \quad g_{ij} + h_{ij} + k_{ij} = 0 \quad \text{and} \quad f_{ij} + g_{ij} + h_{ij} + k_{ij} = 0$$

imply the third one. The proofs of (b), (c), and (d) are similar. ■

Theorem 7.8.5 *Let an $AW(4,2,r)$ be an algebraic 4-web and Q be an r-quartic generating it. Then in a domain of regularity of the $AW(4,2,r)$ the following statements are equivalent:*

(i) *Q decomposes into two hyperquadrics.*

(ii) *The distribution of hyperplanes determined by the equations*

$$\begin{array}{ll}\text{(a)} & a_i\omega^i_{r+1} - b_i\omega^i_0 = 0 \quad (or\ \ b_i\omega^i_{r+1} - \lambda a_i\omega^i_0 = 0),\\ \text{(b)} & b_i\omega^i_{r+1} + \lambda a_i\omega^i_0 = 0,\\ \text{(c)} & a_i\omega^i_{r+1} + b_i\omega^i_0 = 0\end{array}$$

is integrable.

(iii) *One of the forms*

$$\begin{array}{ll}\text{(a)} & \omega^0_0 + \omega^{r+1}_{r+1},\ \ \omega^i_i,\ \ and\ \ 2\omega^0_0 - (a_i + b_i)\omega^i_0,\\ \text{(b)} & \omega^0_0 + \omega^{r+1}_{r+1} - a_i\omega^i_0,\ \ \omega^i_i + a_i\omega^i_0,\ \ and\ \ 2\omega^0_0 - b_i\omega^i_0,\\ \text{(c)} & \omega^0_0 + \omega^{r+1}_{r+1} - b_i\omega^i_0,\ \ \omega^i_i + b_i\omega^i_0,\ \ and\ \ 2\omega^0_0 - a_i\omega^i_0,\end{array}$$

is a total differential. The third form and one of the first two forms in (a), (b), *and* (c) *are total differentials if and only if two of the hypersurfaces* U_ξ *belong to a hyperquadric and two others belong to another hyperquadric.*

Proof. (i) $\leftrightarrow$ (ii). In the case (a), using (7.6.16) and (7.6.19), we obtain

$$d(a_i\omega^i_{r+1} - b_i\omega^i_0) = (f_{ij} + g_{ij} - h_{ij} - k_{ij})\omega^i_0 \wedge \omega^i_{r+1}. \tag{7.8.31}$$

It follows from (7.8.31) that the distribution $a_i\omega^i_{r+1} - b_i\omega^i_0 = 0$ is integrable if and only if

$$f_{ij} + g_{ij} = h_{ij} + k_{ij}. \tag{7.8.32}$$

Equations (7.8.32) and (7.8.1) give the conditions (a) in (7.8.27).

(i) $\leftrightarrow$ (iii). Using (7.6.2) and (7.6.19), we obtain

$$\left\{\begin{array}{lcl} d(\omega^0_0 + \omega^{r+1}_{r+1}) & = & -d\omega^i_i = -(f_{ij} + g_{ij})\omega^i_0 \wedge \omega^j_{r+1},\\ d[2\omega^0_0 - (a_i + b_i)\omega^i_0] & = & -(h_{ij} + k_{ij})\omega^i_0 \wedge \omega^j_{r+1}.\end{array}\right. \tag{7.8.33}$$

Equations (7.8.33) show that the conditions (a) in (7.8.27) are necessary and sufficient for the forms $\omega^0_0 + \omega^{r+1}_{r+1}$ and $2\omega^0_0 - (a_i + b_i)\omega^i_0$ to be total differentials.

The last statement of Theorem 7.8.5 follows from the fact that any two of three conditions

$$f_{ij} + g_{ij} = 0, \quad h_{ij} + k_{ij} = 0, \quad f_{ij} + g_{ij} + h_{ij} + k_{ij} = 0$$

imply the third one. The proof of (b) and (c) is similar. ■

Theorem 7.8.6 *Let an* $AW(4,2,r)$ *be an algebraic* 4*-web and* Q *be an* r*-quartic generating it. Then in a domain of regularity of the* $AW(4,2,r)$ *the following statements are equivalent:*

(i) Q *decomposes into two hyperplanes and one of six hyperquadrics indicated in Theorem* 7.8.5.

(ii)

$$
\begin{array}{lll}
\text{(a)} & f_{ij}+g_{ij}=0, & h_{ij}+k_{ij}=0,\\
\text{(b)} & h_{ij}+k_{ij}=0, & f_{ij}+g_{ij}=0,\\
\text{(c)} & f_{ij}+h_{ij}=0, & g_{ij}+k_{ij}=0,\\
\text{(d)} & g_{ij}+k_{ij}=0, & f_{ij}+h_{ij}=0,\\
\text{(e)} & f_{ij}+k_{ij}=0, & g_{ij}+h_{ij}=0,\\
\text{(f)} & g_{ij}+h_{ij}=0, & f_{ij}+k_{ij}=0.
\end{array}
$$

(iii) *Two of the four 3-subwebs* $[\xi,\eta,\zeta]$ *are hexagonal.*
(iv) *One of the four 3-subwebs* $[\xi,\eta,\zeta]$ *is a Bol three-web.*
(v) *Two of the three forms*

$$
\begin{array}{llll}
\text{(a)} & \omega_0^0-a_i\omega_0^i, & \omega_0^0-b_i\omega_0^i, & \omega_0^0+\omega_{r+1}^{r+1},\\
\text{(b)} & \omega_{r+1}^{r+1}, & \omega_0^0, & 2\omega_0^0-(a_i+b_i)\omega_0^i,\\
\text{(c)} & \omega_0^0, & \omega_0^0-b_i\omega_0^i, & \omega_0^0+\omega_{r+1}^{r+1}-a_i\omega_0^i,\\
\text{(d)} & \omega_{r+1}^{r+1}, & \omega_0^0-a_i\omega_0^i, & 2\omega_0^0-b_i\omega_0^i,\\
\text{(e)} & \omega_0^0, & \omega_0^0-a_i\omega_0^i, & \omega_0^0+\omega_{r+1}^{r+1}-b_i\omega_0^i,\\
\text{(f)} & \omega_{r+1}^{r+1}, & \omega_0^0-b_i\omega_0^i, & 2\omega_0^0-a_i\omega_0^i
\end{array}
$$

are total differentials. All three forms in (a)–(e) *are total differentials if and only if two of the hypersurfaces* U_ξ *belong to a hyperquadric and two others are hyperplanes.*
Such webs $AW(4,2,r)$ *depend on* $(r+1)(r+8)/2$ *constants.*

Proof. It is easy to see that (i), (iii), (iv), and (v) are equivalent to (ii). Such $AW(4,2,r)$ are determned in the case (a), for example, by the completely integrable system of Pfaffian equations (7.6.3), (7.6.10), (7.6.11), (7.6.19), (7.6.22) , where $g_{ij}=-f_{ij}$, $h_{ij}=k_{ij}=0$, $f_{ijk}=0$. The number of linearly independent equations of this system is equal to $(r+1)(r+8)/2$. The G-structure associated with the $AW(4,2,r)$ again is a *closed G-structure of the first kind and the second class.* ■

In the last three theorems we shall study algebraic webs $AW(4,2,r)$ for which the hypersurfaces U_ξ are hyperplanes of general or some special position.

Theorem 7.8.7 *In a domain of regularity of a Grassmann web* $GW(4,2,r)$ *the following statements are equivalent:*
(i) *The hypersurfaces* $U_\xi, \xi=1,2,3,4$, *are hyperplanes of general position.*
(ii) *The second fundamental tensors of* U_ξ *vanish.*
(iii) *All 3-subwebs* $[\xi,\eta,\zeta]$ *are group non-parallelisable three-webs.*
(iv) *The forms* ω_{r+1}^{r+1}, $\omega_0^0-a_i\omega_0^i$, *and* $\omega_0^0-b_i\omega_0^i$ *are total differentials.*
(v) *Three of the four 3-subwebs* $[\xi,\eta,\zeta]$ *are hexagonal.*
(vi) *Two of the four 3-subwebs* $[\xi,\eta,\zeta]$ *are Bol three-webs.*

(vii) *The 3-subweb* $[1,2,4]$ *(or* $[3,1,4]$*) is hexagonal and the 3-subweb* $[1,2,3]$ *is a Bol web of the type* B_l.

(viii) *The connections* $\gamma_{\xi\eta\zeta}$, $\gamma_{\eta\zeta\kappa}$, $\gamma_{\xi\zeta\kappa}$, *and* $\gamma_{\xi\eta\kappa}$ *are equiaffine.*

Proof. It is easy to check that all listed conditions are equivalent to (ii). We note only that in the case (v) if, for example, the 3-subwebs $[1,\xi,\eta]$ are hexagonal, we have $h_{ij} = g_{ij} = k_{ij} = -f_{ij}/2$ and also, by means of (7.6.22), $f_{ijk} = g_{ijk} = h_{ijk} = k_{ijk} = 0$ and $f_{(ij}a_{k)} = f_{(ij}b_{k)} = 0$. From this follows that $f_{(ij}f_{k)m} = 0$ and therefore (see the proof of Theorem 7.7.5) $f_{ij} = g_{ij} = h_{ij} = k_{ij} = 0$. ■

The webs $AW(4,2,r)$ under consideration are determined by the completely integrable system of Pfaffian equations (7.6.3), (7.6.10), (7.6.11), and (7.6.19) where $f_{ij} = g_{ij} = h_{ij} = k_{ij} = 0$. Its general solution depends on $4(1+r)$ constants.

Theorem 7.8.8 *In a domain of regularity of a Grassmann web* $GW(4,2,r)$ *the following statements are equivalent:*

(i) *The hypersurfaces* U_ξ, $\xi = 1,2,3,4$, *are hyperplanes and* (a) U_1, U_2, U_3, (b) U_1, U_2, U_4, (c) U_1, U_3, U_4, (d) U_2, U_3, U_4 *belong to a pencil with an* $(r-1)$-*dimensional axes.*

(ii) *The second fundamental tensors of* U_ξ *vanish and*

$$\text{(a) } a_i = 0, \quad \text{(b) } b_i = 0, \quad \text{(c) } b_i = a_i, \quad \text{(d) } b_i = \lambda a_i.$$

(iii) *One of the four 3-subwebs* (a) $[1,2,3]$, (b) $[1,2,4]$, (c) $[3,1,4]$, (d) $[2,3,4]$ *is parallelisable and the others are group three-webs.*

(iv) *The basis affinor* $\lambda^i_j = \delta^i_j\lambda$ *is covariantly constant on the foliation* λ_α *in the connection* γ_{123} *and one of the 3-subwebs* $[\xi,\eta,\zeta]$ *is hexagonal (or one of the connections* $\gamma_{\xi\eta\zeta}$ *is equiaffine).*

(v) *The Desargues figures* (D_1) *or* (D_{12}) *is closed on a* $GW(4,2,r)$.

Proof. It is easy to show that all listed conditions are equivalent to (ii). We will prove this only for (v) in the cases of (D_1) and (D_{12}). As we saw in Theorem 7.4.3, for webs $W(4,2,r)$ the condition (D_1) is equivalent to

$$\mu^i_{jk} = 0, \quad \underset{123}{b}{}^i_{jkl} = 0. \tag{7.8.34}$$

By (7.6.26) and (7.6.31), for Grassmann webs $GW(4,2,r)$ we obtain that conditions (7.8.34) are equivalent to

$$b_i = \lambda a_i, \quad f_{ij} = g_{ij} = h_{ij} = k_{ij} = 0. \tag{7.8.35}$$

In the proof of Theorem 7.4.10 we saw that for webs $W(4,2,r)$ the condition (D_{12}) is equivalent to

$$\lambda^p_{(j}\mu^i_{k)p} = 0, \quad \underset{124}{b}{}^i_{(jkl)} = \underset{123}{b}{}^i_{(jkl)}. \tag{7.8.36}$$

By (7.6.26) and (7.6.31), for Grassmann webs we again obtain that conditions (7.8.36) are equivalent to (7.8.35). ■

The completely integrable system determining this class of webs $AW(4,2,r)$ consists of equations (7.6.3), (7.6.10), (7.6.11), and (7.6.19) where $f_{ij} = g_{ij} = h_{ij} = k_{ij} = 0$ and, for example $b_i = \lambda a_i$. Its general solution depends on $3r+4$ constants.

Theorem 7.8.9 *In a domain of regularity of a Grassmann web $GW(4,2,r)$ the following statements are equivalent:*

(i) *The hypersurfaces U_ξ, $\xi = 1,2,3,4$, are hyperplanes of a pencil with an $(r-1)$-dimensional axes.*

(ii) *The second fundamental tensors of U_ξ vanish and $a_i = b_i = 0$.*

(iii) *The web $GW(4,2,r)$ is parallelisable.*

(iv) *The basis affinor $\lambda^i_j = \delta^i_j \lambda$ is covariantly constant on the web $GW(4,2,r)$ in the connection γ_{123}.*

Proof. It is easy to check that all listed conditions are equivalent to (ii). ■

The solution of the completely integrable system determining this class of webs $AW(4,2,r)$ depends on $2r+4$ constants.

Note that the G-structures associated with the webs $AW(4,2,r)$ discussed in Theorems 7.8.7, 7.8.8, and 7.8.9 are *closed G-structures of the first kind and the first class.*

Corollary 7.8.10 *A group web $GW(4,2,r)$ is parallelisable.*

Proof. In fact, for a group web $W(4,2,r)$ all 3-subwebs are group three-webs and its basis affinor λ^i_j is covariantly constant on X^{2r} , i.e., for such a web we have $\underset{\xi\eta\zeta}{b}{}^i_{jkl} = 0$, $\lambda^i_{jk} = \mu^i_{jk} = 0$. By (7.6.26) and (7.6.31), for Grassmann webs $GW(4,2,r)$ we obtain that these conditions are equivalent to

$$f_{ij} = g_{ij} = h_{ij} = k_{ij} = a_i = b_i = 0. \tag{7.8.37}$$

By Theorem 7.8.9, part (ii), equalitions (7.8.37) mean that the web $GW(4,2,r)$ is parallelisable. ■

Corollary 7.8.10 means that a group non-parallelisable web $GW(4,2,r)$, and therefore $AW(4,2,r)$, does not exist.

In conclusion of this section we consider algebraic webs $AW(4,2,r)$ studied in Theorem 7.8.7, 7.8.8, and 7.8.9. For such webs all their 3-subwebs $[\xi,\eta,\zeta]$ are group three-webs but the whole webs $AW(4,2,r)$ are not group 4-webs since their basis affinors $\lambda^i_j = \delta^i_j\lambda$ are not covariantly constant on X^{2r}.

There are three possibilities:

(1) None of the 3-subwebs $[\xi, \eta, \zeta]$ is parallelisable. We have here the case of Theorem 7.8.7. In this case each Lie group determined by a group 3-subweb $[\xi, \eta, \zeta]$ has an $(r-1)$-parametric abelian subgroup [Ak 73].

(2) One of four 3-subwebs $[\xi, \eta, \zeta]$ is parallelisable. We have here the case of Theorem 7.8.8. The Lie groups determined by the parallelisable 3-subweb is an abelian group, and each of three other Lie groups determined by the non-parallelisable 3-subwebs has an $(r-1)$-dimensional abelian subgroup.

(3) All 3-subwebs $[\xi, \eta, \zeta]$ are parallelisable. In this case all four Lie groups determined by the 3-subwebs of the $AW(4, 2, r)$ are abelian groups.

NOTES

7.1–7.2. These sections are from [G 77, 80] except Section 7.2.5 which is from [To 85].

7.3–7.4. The results are due to the author [G 82c].

7.5. This section is from [G 82a].

7.6–7.8. The results are due to the author [G 82d]. Theorem 7.8.1 which has been presented at the University of California at Berkeley in 1981 inspired J.A. Wood to write his thesis [Wo 82] where he, using the same method of proof, generalised this theorem for webs $W(d, 2, r)$, $d > 4$. As we mentioned early, J.A. Wood in [Wo 84] gave another much simplier proof of this result (see Section 2.6 and Notes to this section).

Chapter 8

Rank Problems For Webs $W(d, 2, r)$

8.1 Almost Grassmannisable and Almost Algebraisable Webs $W(d, 2, r)$

8.1.1 Basic Notions and Equations for a Web $W(d, 2, r)$, $d \geq 3$

In Chapter 8 we shall study almost Grassmannisable and almost algebraisable webs $W(d, 2, r)$, find their relationship to each other and to Grassmann and the algebraic webs $W(d, 2, r)$, and demonstrate how these kinds of webs arise naturally when one studies 1- and r-rank problems for the webs $W(d, 2, r)$.

So, we consider now a general web $W(d, 2, r)$, $d \geq 3$ (see Definition 1.1.1 where $n = 2$). The foliations λ_ξ, $\xi = 1, \ldots, d$, of the web $W(d, 2, r)$ are given by d completely integrable systems of Pfaffian equations

$$\underset{\xi}{\omega}^i = 0, \qquad \xi = 1, \ldots, d; \qquad i = 1, \ldots, r, \tag{8.1.1}$$

where the forms $\underset{1}{\omega}^i$ and $\underset{2}{\omega}^i$ are the basis forms of the manifold X^{2r} carrying the web $W(d, 2, r)$. Using the same method which we used in Section 7.1 for a web $W(4, 2, r)$, we can reduce the dependences among the forms $\underset{\xi}{\omega}^i$ to the following relations:

$$-\underset{3}{\omega}^i = \underset{1}{\omega}^i + \underset{2}{\omega}^i, \quad \underset{\alpha}{\omega}^i = \underset{\alpha}{\lambda}{}^i_j \underset{1}{\omega}^j + \underset{2}{\omega}^i, \quad \alpha = 4, \ldots, d, \tag{8.1.2}$$

$$\det(\underset{\alpha}{\lambda}{}^i_j) \neq 0, \quad \det(\delta^i_j - \underset{\alpha}{\lambda}{}^i_j) \neq 0, \quad \det(\underset{\alpha}{\lambda}{}^i_j - \underset{\beta}{\lambda}{}^i_j) \neq 0, \quad \alpha \neq \beta, \quad \alpha, \beta = 4, \ldots, d \tag{8.1.3}$$

where the quantities $\underset{\alpha}{\lambda}{}^i_j$, $i, j = 1, \ldots, r$, form an $(1,1)$-tensor for any $\alpha = 4, \ldots, d$, and these $d - 3$ tensors $\underset{\alpha}{\lambda}{}^i_j$ are distinct.

Definition 8.1.1 The tensors $\underset{\alpha}{\lambda}{}^i_j$ are called the *basis affinors* of a web $W(d, 2, r)$.

As in Section 7.1, we have the following equations for the web $W(d,2,r)$:

$$\begin{cases} d\underset{1}{\omega}^i = \underset{1}{\omega}^j \wedge \omega_j^i + a_{jk}^i \underset{1}{\omega}^j \wedge \underset{1}{\omega}^k, \\ d\underset{2}{\omega}^i = \underset{2}{\omega}^j \wedge \omega_j^i - a_{jk}^i \underset{2}{\omega}^j \wedge \underset{2}{\omega}^k, \end{cases} \tag{8.1.4}$$

$$d\omega_j^i - \omega_j^k \wedge \omega_k^i = b_{jkl}^i \underset{1}{\omega}^k \wedge \underset{2}{\omega}^l, \tag{8.1.5}$$

$$\nabla a_{jk}^i = b_{[j|l|k]}^i \underset{1}{\omega}^k + b_{[jk]l}^i \underset{2}{\omega}^l, \tag{8.1.6}$$

$$a_{jk}^i = -a_{kj}^i, \tag{8.1.7}$$

$$b_{[jkl]}^i = 2a_{[jk}^m a_{|m|l]}^i, \tag{8.1.8}$$

$$\nabla \underset{\alpha}{\lambda}_j^i = \underset{\alpha}{\lambda}_{jk}^i \underset{1}{\omega}^k + \underset{\alpha}{\mu}_{jk}^i \underset{2}{\omega}^k, \quad \alpha = 4, \ldots, d, \tag{8.1.9}$$

$$\underset{\alpha}{\lambda}_{[jk]}^i - \underset{\alpha}{\mu}_{[j|p|}^i \underset{\alpha}{\lambda}_{k]}^p = \underset{\alpha}{\lambda}_p^i a_{jk}^p + \underset{\alpha}{\lambda}_{[k}^p \underset{\alpha}{\lambda}_{j]}^q a_{pq}^i, \quad \alpha = 4, \ldots, d, \tag{8.1.10}$$

where the quantities a_{jk}^i and b_{jkl}^i form the torsion and curvature tensor of the 3-subweb $[1,2,3]$ and ∇ is the symbol of the covariant differentiation in the canonical connection γ_{123} with connection forms ω_j^i, so that

$$\nabla a_{jk}^i = da_{jk}^i - a_{mk}^i \omega_j^m - a_{jm}^i \omega_k^m + a_{jk}^m \omega_m^i.$$

Note that equations (8.1.9) and (8.1.10) are similar to equations (7.1.27) and (7.1.28) for a web $W(4,2,r)$ when α takes only one value 4.

Definition 8.1.2 The tensors $\{a_{jk}^i\}$ and $\{b_{jkl}^i\}$ are said to be the *torsion and curvature tensors* of the web $W(d,2,r)$.

Later on we will need the expressions of the connection forms of the canonical connections $\gamma_{12\alpha}$, $\gamma_{31\alpha}$, and $\gamma_{23\alpha}$, $\alpha = 4, \ldots, d$. They have expressions similar to those in (7.2.13), (7.2.28), (7.2.21) and can be obtained by the same procedure that we used in Section 7.2:

$$\begin{cases} \underset{12\alpha}{\omega}{}_j^i = \omega_j^i - \underset{\alpha}{\mu}_{mk}^i \underset{\alpha}{\tilde{\lambda}}_j^m \underset{2}{\omega}^k, \\ \underset{31\alpha}{\omega}{}_j^i = \omega_j^i + 2a_{jk}^i \underset{1}{\omega}^k - \underset{\alpha}{\mu}_{mk}^i \underset{\alpha}{\tilde{\tilde{\lambda}}}_j^m \underset{3}{\omega}^k, \\ \underset{23\alpha}{\omega}{}_j^i = \omega_j^i - \underset{\alpha}{\tilde{\lambda}}_j^l \underset{\alpha}{\lambda}_{lk}^i \underset{3}{\omega}^k + \underset{\alpha}{\tilde{\lambda}}_j^l (\underset{\alpha}{\lambda}_{lk}^i - \underset{\alpha}{\mu}_{lk}^i - 2\underset{\alpha}{\lambda}_p^i a_{lk}^p) \underset{2}{\omega}^k. \end{cases} \tag{8.1.11}$$

In (8.1.11) the matrices $(\underset{\alpha}{\tilde{\lambda}}_j^i)$ and $(\underset{\alpha}{\tilde{\tilde{\lambda}}}_j^i)$ are the inverse matrices of $(\underset{\alpha}{\lambda}_j^i)$ and $(\delta_j^i - \underset{\alpha}{\lambda}_j^i)$.

8.1.2 Almost Grassmannisable Webs $AGW(d,2,r)$, $d > 3$

Definition 8.1.3 A web $W(d,2,r)$, $d > 3$, all of whose basis affinors $\underset{\alpha}{\lambda}{}^i_j$ are scalar:

$$\underset{\alpha}{\lambda}{}^i_j = \delta^i_j \underset{\alpha}{\lambda} \tag{8.1.12}$$

is said to be an *almost Grassmannisable* web.

We will denote such webs by $AGW(d,2,r)$.

Let us consider an isoclinic r-surface V^r of the manifold X^{2r} which is determined by the following system of Pfaffian equations (cf. Section 1.11 for $n = 2$):

$$\lambda \underset{1}{\omega}{}^i + \underset{2}{\omega}{}^i = 0 \tag{8.1.13}$$

where λ is a function of a point $x \in X^{2r}$. For a tangent vector to X^{2r} we have (7.1.9):

$$dx = \underset{1}{\omega}{}^i e^1_i + \underset{2}{\omega}{}^i e^2_i. \tag{8.1.14}$$

As we saw in Section 7.1, it follows from (8.1.14) that the vectors e^2_i, e^1_i, $e^3_i = e^1_i - e^2_i$, and $e^4_i = e^1_i - \underset{\alpha}{\lambda}{}^j_i e^2_j$, $\alpha = 4, \ldots, d$, are tangent to the leaves F_1, F_2, F_3, and F_α passing through the point p. On the surface V^r we have

$$dx = \underset{1}{\omega}{}^i (e^1_i - \lambda e^2_i). \tag{8.1.15}$$

The transversal vectors $\eta^\xi = \eta^i e^\xi_i$, $\xi = 1, \ldots, d$, are tangent to the leaves F_ξ at the point p. It is easy to see that for an $AGW(d,2,r)$ the vectors η^ξ lie in a 2-plane.

Definition 8.1.4 The transversal bivector $\eta^1 \wedge \eta^2$ determined by η^i is said to be a *transversal bivector* of $AGW(d,2,r)$.

It follows from (8.1.15) that the tangent r-plane of V^r intersects the transversal bivector $\eta^1 \wedge \eta^2$ in the direction of the vector $\eta^i(e^1_i - \lambda e^2_i)$. The cross ratio of this vector and the three vectors η^1, η^2, η^3 (η^α) is equal to λ (respectively λ/λ_α). This cross ratio does not depend on quantities η^i giving the direction of $\eta^1 \wedge \eta^2$. This is the reason for the following definition:

Definition 8.1.5 The surfaces V^r defined by (8.1.13) are called *isoclinic* for a web $AGW(d,2,r)$. A web $AGW(d,2,r)$ is said to be *isoclinic* if there exists a one-parameter family of isoclinic surfaces through any point $p \in D$.

Note that the definition of an isoclinic web $AGW(d,2,r)$ is exactly the same as that for its 3-subweb $[1,2,3]$ (see [Ak 74]). Because of this, an analytical criterion for an $AGW(d,2,r)$ be isoclinic is the same as that for its 3-subweb $[1,2,3]$.

In Section 1.11 we found the analytical condition for a web $W(n+1,n,r)$, $n>2$, to be isoclinic. This has the form (1.11.7). For a web $W(3,2,r)$, $r>2$, the necessary and sufficient condition of isoclinity has a similar form (see [Ak 74]):

$$a^i_{jk} = a_{[j}\delta^i_{k]}. \tag{8.1.16}$$

As we noted above, the same condition (8.1.16) is necessary and sufficient for a web $AGW(d,2,r)$, $d>3$, $r>2$, to be isoclinic.

For later use we will need necessary and sufficient conditions for a web $W(d,2,2)$ to be isoclinic. Thus we set $r=2$. In this case the torsion tensor a^i_{jk} always has the form (8.1.16). Because of this, the structure equations (8.1.4) can be written in the form

$$\begin{cases} d\underset{1}{\omega}^i = \underset{1}{\omega}^j \wedge \omega^i_j + a_j\underset{1}{\omega}^j \wedge \underset{1}{\omega}^i, \\ d\underset{2}{\omega}^i = \underset{2}{\omega}^j \wedge \omega^i_j - a_j\underset{2}{\omega}^j \wedge \underset{2}{\omega}^i. \end{cases} \tag{8.1.17}$$

Exterior differentiation of (8.1.17) gives

$$\begin{cases} \Omega^i_j \wedge \underset{1}{\omega}^j - \nabla a_j \wedge \underset{1}{\omega}^j \wedge \underset{1}{\omega}^i = 0, \\ \Omega^i_j \wedge \underset{2}{\omega}^j + \nabla a_j \wedge \underset{2}{\omega}^j \wedge \underset{2}{\omega}^i = 0, \end{cases} \tag{8.1.18}$$

where

$$\Omega^i_j = d\omega^i_j - \omega^k_j \wedge \omega^i_k, \quad \nabla a_j = da_j - a_k\omega^k_j.$$

In the case $r=2$ the general form of Ω^i_j and ∇a_j satisfying (8.1.18) is (8.1.5) and

$$\nabla a_i = p_{ij}\underset{1}{\omega}^j + q_{ij}\underset{2}{\omega}^j \tag{8.1.19}$$

where by means of (8.1.6) and (8.1.16) we have

$$b^i_{[j|l|k]} = \delta^i_{[k}p_{j]l}, \quad b^i_{[jk]l} = \delta^i_{[k}q_{j]l}. \tag{8.1.20}$$

Note that in the case $r=2$ the equations (8.1.18) do not imply a symmetry of p_{ij} and q_{ij}.

Theorem 8.1.6 *A web $AGW(d,2,2)$, $d>3$, is isoclinic if and only if the quantities p_{ij} and q_{ij} in (8.1.19) are symmetric.*

Proof. The equations (8.1.13) and their differential consequences are satisfied on a surface V^r. Exterior differentiation of (8.1.13) by means of (8.1.13) and (8.1.17) leads to

$$[d\lambda + (\lambda - \lambda^2)a_j\underset{1}{\omega}^j] \wedge \underset{1}{\omega}^i = 0. \tag{8.1.21}$$

On V^r the differential $d\lambda$ is expressed in terms of the forms $\underset{1}{\omega}^i$ only. This and (8.1.21) give

$$d\lambda = (\lambda^2 - \lambda)a_j\underset{1}{\omega}^j. \tag{8.1.22}$$

Exterior differentiation of (8.1.22) by means of (8.1.13), (8.1.22), and (8.1.19) gives

$$p_{[jk]} - \lambda q_{[jk]} = 0. \tag{8.1.23}$$

Since λ is arbitrary, it follows from (8.1.23) that

$$p_{jk} = p_{kj}, \quad q_{jk} = q_{kj}. \tag{8.1.24}$$

Conversely, the exterior differentiation of (8.1.22) leads to the identitity using (8.1.19) and (8.1.24). ■

Since, by (8.1.24), p_{jk} and q_{jk} are symmetric for an isoclinic web $AGW(d,2,2)$, for the webs $AGW(d,2,r)$, $r \geq 2$, we can use the same procedure which was used in [Ak 74] in the case $r > 2$ for the isoclinic webs $W(d,2,r)$. Let

$$p_{jk} = f_{jk} - h_{jk}, \quad q_{jk} = g_{jk} - h_{jk}. \tag{8.1.25}$$

where f_{jk}, g_{jk}, and h_{jk} are symmetric $(0,2)$-tensors. Using (8.1.25), we can rewrite equations (8.1.19) and (8.1.20) in the form

$$\nabla a_i = (f_{ij} - h_{ij})\underset{1}{\omega}^j + (g_{ij} - h_{ij})\underset{2}{\omega}^j, \tag{8.1.26}$$

$$b^i_{[j|l|k]} = (f_{l[j} - h_{l[j})\delta^i_{k]}, \quad b^i_{[jk]l} = (g_{l[j} - h_{l[j})\delta^i_{k]}. \tag{8.1.27}$$

For any tensor b^i_{jkl} the following identity

$$b^i_{jkl} = b^i_{(jkl)} + \frac{1}{3}b^i_{[jk]l} + \frac{1}{3}b^i_{[jl]k} + b^i_{[lk]j} + \frac{4}{3}b^i_{[j|k|l]} + \frac{2}{3}b^i_{[k|l|j]} \tag{8.1.28}$$

can be checked by inspection. Using (8.1.27), we can write the identity (8.1.28) in the form

$$b^i_{jkl} = a^i_{jkl} + f_{jk}\delta^i_l + g_{lj}\delta^i_k + h_{kl}\delta^i_j \tag{8.1.29}$$

where

$$a^i_{jkl} = b^i_{jkl} - (f_{(jk} + g_{(jk} + h_{(jk})\delta^i_{l)}. \tag{8.1.30}$$

If we impose the following restriction for the tensor a^i_{jkl} :

$$a^i_{ikl} = 0, \tag{8.1.31}$$

it allows us to determine h_{kl}. In fact, it follows from (8.1.30) and (8.1.31) that

$$h_{kl} = \frac{3}{4}b^i_{(ikl)} - (f_{kl} + g_{kl}). \tag{8.1.32}$$

Note that (8.1.32) is equivalent to (8.1.31).

Exterior differentiation of (8.1.5), where b^i_{jkl} is substituted from (8.1.29), and the application of Cartan's lemma give the following equations

$$\begin{cases} \dot{\nabla} f_{ij} = \underset{1}{f}_{ijk}\underset{1}{\omega}^k + \underset{2}{f}_{ijk}\underset{2}{\omega}^k, \\ \dot{\nabla} g_{ij} = \underset{1}{g}_{ijk}\underset{1}{\omega}^k + \underset{2}{g}_{ijk}\underset{2}{\omega}^k, \\ \dot{\nabla} h_{ij} = \underset{1}{h}_{ijk}\underset{1}{\omega}^k + \underset{2}{h}_{ijk}\underset{2}{\omega}^k, \\ \dot{\nabla} a_{jkl} = \underset{1}{a}^i_{jklm}\underset{1}{\omega}^m + \underset{2}{a}_{jklm}\underset{2}{\omega}^m, \end{cases} \tag{8.1.33}$$

where we introduced the new differential operator (see [Ak 74])

$$\dot{\nabla} = \nabla + a_m(\underset{1}{\omega}^m - \underset{2}{\omega}^m), \tag{8.1.34}$$

$\underset{s}{f}_{ijk}$, $\underset{s}{g}_{ijk}$, $\underset{s}{h}_{ijk}$, $s = 1, 2$, are symmetric with respect to i, j, k, and

$$\underset{1}{a}^i_{jl[km]} = \underset{1}{g}_{lj[k}\delta^i_{m]}, \quad \underset{2}{a}^i_{jl[km]} = \underset{2}{f}_{jl[k}\delta^i_{m]}, \tag{8.1.35}$$

$$a_m a^m_{ijk} = \underset{2}{f}_{ijk} - \underset{1}{g}^i_{ijk} + \underset{2}{h}_{ijk} - 3a_{(i}h_{jk)}. \tag{8.1.36}$$

Definition 8.1.7 A web $AGW(d,2,r)$ is *transversally geodesic* if for any bivector $\eta^1 \wedge \eta^2$ there exists a two-dimensional surface V^2 tangent to $\eta^1 \wedge \eta^2$ at p and each bivector $\eta^1 \wedge \eta^2$ is tangent to one and only one V^2.

Note again that this definition of a transversal geodesic web $AGW(d,2,r)$ is the same as that for its 3-subweb $[1,2,3]$.

Proposition 8.1.8 *An isoclinic web $AGW(d,2,r)$, $d \geq 3$, $r \geq 2$, is transversally geodesic if and only if*

$$a^i_{jkl} = 0. \tag{8.1.37}$$

Proof. In [Ak 69a] it was proved that the 3-subweb $[1,2,3]$ of an $AGW(d,2,r)$, and therefore, as we noted above, the whole web $AGW(d,2,r)$ is transversally geodesic if and only if

$$b^i_{(jkl)} = b_{(jk}\delta^i_{l)}. \tag{8.1.38}$$

In the case of an isoclinic $AGW(d,2,r)$ we have (8.1.29) which implies

$$b^i_{(jkl)} = a^i_{jkl} + (f_{(jk} + g_{(jk} + h_{(jk})\delta^i_{l)}. \tag{8.1.39}$$

The *sufficiency* of our proposition readily follows from comparison of (8.1.38) and (8.1.39).

To prove its *necessity* we will use (8.1.38) to write (8.1.39) in the form

$$a^i_{jkl} = a_{(jk}\delta^i_{l)}, \tag{8.1.40}$$

where $a_{jk} = b_{jk} - f_{jk} - g_{jk} - h_{jk}$. Contracting (8.1.40) in i and j and using the condition (8.1.31), we obtain

$$(r+2)a_{kl} = 0. \tag{8.1.41}$$

Equation (8.1.41) implies $a_{kl} = 0$ and, by (8.1.40), this leads to (8.1.37). ■

Corollary 8.1.9 *A web $AGW(d,2,2)$ is Grassmannisable if and only if the Pfaffian derivatives p_{jk} and q_{jk} of its torsion tensor are symmetric and the quantities a^i_{jkl} vanish.*

Proof. In fact, by Theorem 8.1.6 and Propositions 8.1.8, the conditions described in Corollary 8.1.9 are equivalent to the fact that the web $AGW(d,2,2)$ is isoclinic and transversally geodesic. If $d=3$, by Theorem 2.6.4, this will be the case if and only if the web $AGW(3,2,2)$ is Grassmannisable. If $d>3$, it is easy to see that under the imposed conditions all the almost Grassmann structures $AG(1,3)$ on X^{2r} induced by 3-subwebs of $AGW(d,2,2)$ coincide and the 3-subweb $[1,2,3]$ is both transversally geodesic and isoclinic, and conversely. By Theorem 2.6.5, this again will be the case if and only if the web $AGW(d,2,2)$ is Grassmannisable. ■

It follows from (8.1.37), (8.1.33), and (8.1.35) that for a Grassmannisable web $GW(d,2,2)$ we have

$$\underset{1}{g}_{lj[k}\delta^i_{m]} = 0, \quad \underset{2}{f}_{jl[k}\delta^i_{m]} = 0$$

i.e., for $r \geq 2$,

$$\underset{1}{g}_{ijk} = 0, \quad \underset{2}{f}_{ijk} = 0. \tag{8.1.42}$$

In addition, in this case it follows from (8.1.36), (8.1.37), and (8.1.42) that

$$\underset{1}{h}_{ijk} = \underset{2}{h}_{ijk} + 3a_{(i}h_{jk)}. \tag{8.1.43}$$

In the next theorem we will prove that, with one exception, the webs $AGW(d,2,r)$ are always isoclinic. This theorem and especially its part concerning the exceptional webs $AGW(4,2,2)$ will play very important role in Sections 8.4 – 8.7 when we will describe the webs of maximum r-rank.

Theorem 8.1.10 *An almost Grassmannisable web $AGW(d,2,r)$ is isoclinic if $r>2$, $d \geq 4$ or $r=2$, $d>4$. A web $AGW(4,2,2)$ can be isoclinic or non-isoclinic.*

Proof. Suppose that a web $W(d,2,r)$ is almost Grassmannisable. Then its basis affinors have the structure (8.1.12) where, by (8.1.3),

$$\underset{\alpha}{\lambda} \neq 0, 1. \tag{8.1.44}$$

It follows from (8.1.12) and (8.1.9) that

$$d\underset{\alpha}{\lambda} = \underset{\alpha 1}{\lambda_i}\underset{1}{\omega^i} + \underset{\alpha 2}{\lambda_i}\underset{2}{\omega^i} \tag{8.1.45}$$

Equations (8.1.12), (8.1.45), and (8.1.9) imply

$$\underset{\alpha}{\lambda}{}^i_{jk} = \delta^i_j \underset{\alpha 1}{\lambda}{}_k, \quad \underset{\alpha}{\mu}{}^i_{jk} = \delta^i_j \underset{\alpha 2}{\lambda}{}_k. \tag{8.1.46}$$

Substituting (8.1.46) and (8.1.12) into (8.1.10), we obtain (8.1.16) where

$$a_i = \left(-\underset{\alpha 1}{\lambda}{}_i + \underset{\alpha}{\lambda}\,\underset{\alpha 2}{\lambda}{}_i\right)\Big/\left(\underset{\alpha}{\lambda} - \underset{\alpha}{\lambda}{}^2\right). \tag{8.1.47}$$

This proves the theorem for $r > 2$. To prove it for $r = 2$, we note that equations (8.1.47) imply

$$\underset{\alpha 1}{\lambda}{}_i = \underset{\alpha}{\lambda}\,\underset{\alpha}{\lambda}{}_i + \left(\underset{\alpha}{\lambda}{}^2 - \underset{\alpha}{\lambda}\right)a_i. \tag{8.1.48}$$

where we denoted $\underset{\alpha 2}{\lambda}{}_i$ by $\underset{\alpha}{\lambda}{}_i$. Using equations (8.1.48), we can rewrite equations (8.1.45) in the form

$$d\underset{\alpha}{\lambda} = \underset{\alpha}{\lambda}\left(\underset{\alpha}{b}{}_i - a_i\right)\underset{1}{\omega}{}^i + \left(\underset{\alpha}{b}{}_i - \underset{\alpha}{\lambda} a_i\right)\underset{2}{\omega}{}^i \tag{8.1.49}$$

where $\underset{\alpha}{b}{}_i = \underset{\alpha}{\lambda}{}_i + \underset{\alpha}{\lambda} a_i$.

Exterior differentiation of equation (8.1.49) gives

$$\begin{aligned}[\nabla \underset{\alpha}{b}{}_i &+ (\underset{\alpha}{b}{}_i a_j - p_{[ij]} - \underset{\alpha}{\lambda} p_{(ij)} + \underset{\alpha}{\lambda} q_{(ij)} - \underset{\alpha}{b}{}_i \underset{\alpha}{b}{}_j)\underset{1}{\omega}{}^j \\ &- \underset{\alpha}{\lambda} q_{[ij]}\underset{2}{\omega}{}^j] \wedge (\underset{\alpha}{\lambda}\underset{1}{\omega}{}^i + \underset{2}{\omega}{}^i) + (1 - \underset{\alpha}{\lambda})(p_{[ij]} + \underset{\alpha}{\lambda} q_{[ji]}\underset{1}{\omega}{}^j \wedge \underset{1}{\omega}{}^i = 0\end{aligned} \tag{8.1.50}$$

where $\nabla \underset{\alpha}{b}{}_i = d\underset{\alpha}{b}{}_i - \underset{\alpha}{b}{}_j \omega^j_i$. Equation (8.1.50) shows that $\nabla \underset{\alpha}{b}{}_i$ is expressed in terms of $\underset{1}{\omega}{}^k$ and $\underset{2}{\omega}{}^k$:

$$\nabla \underset{\alpha}{b}{}_i = \underset{\alpha}{s}{}_{ik}\underset{1}{\omega}{}^k + \underset{\alpha}{b}{}_{ik}\underset{2}{\omega}{}^k. \tag{8.1.51}$$

Substitution of (8.1.51) into (8.1.50) gives by means of linear independence of $\underset{1}{\omega}{}^i \wedge \underset{1}{\omega}{}^j$, $\underset{2}{\omega}{}^i \wedge \underset{2}{\omega}{}^j$ and $\underset{1}{\omega}{}^i \wedge \underset{2}{\omega}{}^j$ the following relations:

$$\underset{\alpha}{s}{}_{[ij]} = p_{[ij]} - \underset{\alpha}{b}{}_{[i} a_{j]}, \tag{8.1.52}$$

$$\underset{\alpha}{b}{}_{[ij]} = \underset{\alpha}{\lambda} q_{[ij]}, \tag{8.1.53}$$

$$\underset{\alpha}{s}{}_{ij} = \underset{\alpha\alpha}{\lambda b}{}_{ji} + \underset{\alpha}{b}{}_i(\underset{\alpha}{b}{}_j - a_j) + \underset{\alpha}{\lambda}(p_{ji} - q_{ji}). \tag{8.1.54}$$

Equations (8.1.52)–(8.1.54) imply by means of the inequalities (8.1.45) the relations

$$p_{[ij]} - \underset{\alpha}{\lambda} q_{[ij]} = 0. \tag{8.1.55}$$

The equations (8.1.55) are satisfied identically for $r > 2$ because in this case the quantities p_{ij} and q_{ij} are symmetric.

Suppose that $r = 2$. Then, if $d > 4$, the relations (8.1.55) imply (8.1.24), and, by Theorem 8.1.6, almost Grassmannisable webs $AGW(d,2,2)$, $d > 4$, are isoclinic.

For the almost Grassmannisable webs $AGW(4,2,2)$ there are two possibilities:

(a) $p_{[ij]} = q_{[ij]} = 0$. In this case the webs $AGW(4,2,2)$ are isoclinic.

(b) $p_{[ij]} \neq 0$, $q_{[ij]} \neq 0$. In this case we have a non-isoclinic almost Grassmannisable web. For such a web we have (8.1.12), (8.1.19), and the equations

$$\left\{\begin{array}{lcl} \nabla \underset{4}{b}_i & = & [\underset{4}{\lambda} \underset{4}{b}_{ji} + \underset{4}{b}_i(\underset{4}{b}_j - a_j) + \underset{4}{\lambda}(p_{ij} - q_{ji})]\underset{1}{\omega}^j + \underset{4}{b}_{ij}\underset{2}{\omega}^j, \\ \underset{4}{b}_{[ij]} & = & \underset{4}{\lambda} q_{[ij]}, \quad p_{[ij]} = \underset{4}{\lambda} q_{[ij]}. \end{array}\right. \tag{8.1.56}$$

which follow from (8.1.51), (8.1.53)–(8.1.55).

Remark 8.1.11 Theorem 8.1.10 extends for $n = 2$, $r \geq 2$ Corollary 2.4.3 which has been proved for $n \geq 3$ and $r \geq 3$ (see [Ak 81]).

8.1.3 Isoclinic Almost Grassmannisable Webs $AGW(d,2,r)$

For the isoclinic almost Grassmannisable webs $AGW(d,2,r)$, $d > 4$, the equations (8.1.24), (8.1.25), and (8.1.54) allow us to write the equations (8.1.51) in the form

$$\nabla \underset{\alpha}{b}_i = \underset{\alpha}{b}_{ij}(\underset{\alpha}{\lambda}\underset{1}{\omega}^j + \underset{2}{\omega}^j) + [\underset{\alpha}{b}_i(\underset{\alpha}{b}_j - a_j) + \underset{\alpha}{\lambda}(f_{ij} - g_{ij})]\underset{1}{\omega}^j \tag{8.1.57}$$

where $\underset{\alpha}{b}_{ij} = \underset{\alpha}{b}_{ji}$. We introduce new quantities $\underset{\alpha}{k}_{ij}$, $\underset{\alpha}{k}_{ij} = \underset{\alpha}{k}_{ji}$, such that

$$\underset{\alpha}{b}_{ij} = g_{ij} - \underset{\alpha}{k}_{ij}. \tag{8.1.58}$$

Using (8.1.58), we can write the equations (8.1.57) in the form

$$\nabla \underset{\alpha}{b}_i = [\underset{\alpha}{b}_i(\underset{\alpha}{b}_j - a_j) + \underset{\alpha}{\lambda}(f_{ij} - \underset{\alpha}{k}_{ij})]\underset{1}{\omega}^j + (g_{ij} - \underset{\alpha}{k}_{ij})\underset{1}{\omega}^k. \tag{8.1.59}$$

Exterior differentiation of (8.1.59) and the application of Cartan's lemma give

$$\dot{\nabla}\, \underset{\alpha}{k}_{ij} = [\underset{\alpha}{k}_{ijk}(\underset{\alpha}{\lambda}\underset{1}{\omega}^k + \underset{2}{\omega}^k) + (\underset{\alpha}{b}_t a^t_{ijk} + \underset{1}{g}_{ijk} - \underset{\alpha}{\lambda}\underset{2}{f}_{ijk} + 3\underset{\alpha}{b}_{(k}\underset{\alpha}{k}_{ij)})\underset{1}{\omega}^k. \tag{8.1.60}$$

Thus, in the case of an isoclinic $AGW(d,2,r)$ we have (8.1.12), (8.1.44), (8.1.16), (8.1.17), (8.1.26), (8.1.27), (8.1.29), (8.1.31), (8.1.33), (8.1.35), (8.1.36), and (8.1.49). Equations (8.1.12) and (8.1.49) imply

$$\left\{\begin{array}{l} \underset{\alpha}{\lambda}{}^i_{jk} = \delta^i_j \underset{\alpha}{\lambda}/(\underset{\alpha}{b}_k - a_k), \quad \underset{\alpha}{\mu}{}^i_{jk} = \delta^i_j(\underset{\alpha}{b}_k - \underset{\alpha}{\lambda} a_k), \\ \underset{\alpha}{\tilde{\lambda}}{}^i_j = \delta^i_j/\underset{\alpha}{\lambda}, \quad \underset{\alpha}{\tilde{\tilde{\lambda}}}{}^i_j = \delta^i_j/(1 - \underset{\alpha}{\lambda}), \quad \alpha = 4, \ldots, d. \end{array}\right. \tag{8.1.61}$$

By means of (8.1.12) and (8.1.61), the equations (8.1.11) can be written in the form

$$\left\{\begin{array}{l} \underset{12\alpha}{\omega}{}^i_j = \omega^i_j + \delta^i_j(a_k - \underset{\alpha}{b}_k/\underset{\alpha}{\lambda})\underset{2}{\omega}^k, \\ \underset{31\alpha}{\omega}{}^i_j = \omega^i_j + a_j\underset{1}{\omega}^i - \delta^i_j a_k\underset{1}{\omega}^k + (\underset{\alpha}{b}_k - \underset{\alpha}{\lambda} a_k)(\underset{1}{\omega}^k + \underset{2}{\omega}^k)/(1 - \underset{\alpha}{\lambda}), \\ \underset{23\alpha}{\omega}{}^i_j = \omega^i_j + \delta^i_j\underset{\alpha}{\lambda}(\underset{\alpha}{b}_k - a_k)(\underset{1}{\omega}^k + \underset{2}{\omega}^k)/(1 - \underset{\alpha}{\lambda}) + [\delta^i_j\underset{\alpha}{b}_k(1 - 1/\underset{\alpha}{\lambda}) + 2\delta^i_{[j}a_{k]}]\underset{2}{\omega}^k. \end{array}\right. \tag{8.1.62}$$

Lemma 8.1.12 *If* $\underset{\xi\eta\zeta}{\omega}{}^i_j$ *are the connection forms of the canonical affine connection* $\gamma_{\xi\eta\zeta}$ *induced by the 3-subweb* $[\xi,\eta,\zeta]$ *then*

$$d\underset{\xi\eta\zeta}{\omega}{}^i_j - \underset{\xi\eta\zeta}{\omega}{}^k_j \wedge \underset{\xi\eta\zeta}{\omega}{}^i_k = (a^i_{jkl} + \delta^i_l \underset{\xi}{k}_{jk} + \delta^i_k \underset{\eta}{k}_{jl} + \delta^i_j \underset{\zeta}{k}_{kl})\underset{1}{\omega}{}^k \wedge \underset{2}{\omega}{}^l \qquad (8.1.63)$$

where

$$\underset{1}{k}_{ij} = f_{ij}, \quad \underset{2}{k}_{ij} = g_{ij}, \quad \underset{3}{k}_{ij} = h_{ij}. \qquad (8.1.64)$$

Proof. We shall distinguish two cases:

i) *At least two out of the three indices* ξ, η, ζ *are equal to* 1, 2, *or* 3. In this case we have the connection forms $\underset{123}{\omega}{}^i_j = \omega^i_j$, $\underset{12\alpha}{\omega}{}^i_j$, $\underset{31\alpha}{\omega}{}^i_j$, $\underset{23\alpha}{\omega}{}^i_j$, $\alpha = 4,\dots,d$, and 20 forms obtained from these forms by permutations of lower indices. For the first four forms the equation (8.1.63) can be proved by the straightforward exterior differentiation of ω^i_j and (8.1.62).

To prove (8.1.63) for the 20 other forms, first we must find the expressions of these forms. Using (7.3.21), (7.2.14), and (7.2.28), we find that

$$\begin{cases} \underset{\xi\eta\zeta}{\omega}{}^i_j = \underset{\eta\xi\zeta}{\omega}{}^i_j, \\ \underset{1\alpha 2}{\omega}{}^i_j = \underset{21\alpha}{\omega}{}^i_j - 2\underset{21\alpha}{a}{}^i_{jk}\underset{1}{\omega}{}^k, \\ \underset{3\alpha 1}{\omega}{}^i_j = \underset{31\alpha}{\omega}{}^i_j + 2\underset{31\alpha}{a}{}^i_{jk}(\underset{1}{\omega}{}^k + \underset{2}{\omega}{}^k). \end{cases} \qquad (8.1.65)$$

The equations (7.3.21) and (8.1.65) allow us to find the expressions of the connection forms $\underset{1\alpha 2}{\omega}{}^i_j$, $\underset{3\alpha 1}{\omega}{}^i_j$:

$$\begin{cases} \underset{\xi\eta\zeta}{\omega}{}^i_j = \underset{\eta\xi\zeta}{\omega}{}^i_j, \\ \underset{1\alpha 2}{\omega}{}^i_j = \underset{12\alpha}{\omega}{}^i_j + 2\underset{\alpha}{b}_{j[}\delta^i_{k]}\underset{1}{\omega}{}^k, \\ \underset{3\alpha 1}{\omega}{}^i_j = \underset{31\alpha}{\omega}{}^i_j - 2\delta^i_{[k}(\underset{\alpha}{b}_{j]} - a_{j]})(\underset{1}{\omega}{}^k + \underset{2}{\omega}{}^k)/(1 - \underset{\alpha}{\lambda}). \end{cases} \qquad (8.1.66)$$

Taking exterior derivatives of (8.1.66) and using (7.3.21), we can get the other 20 relations (8.1.63).

ii) *All the indices* ξ, η, ζ *are greater than* 3. The foliations λ_α, λ_β, and λ_γ of a 3-subweb $[\alpha,\beta,\gamma]$ are determined by the equations $\underset{a}{\omega}{}^i = 0$ where

$$\underset{a}{\omega}{}^i = -(\underset{a}{\lambda}\underset{1}{\omega}{}^i + \underset{2}{\omega}{}^i), \quad a = \alpha,\beta,\gamma. \qquad (8.1.67)$$

Eliminating $\underset{1}{\omega}{}^i$ and $\underset{2}{\omega}{}^i$ from the three equations (8.1.67), we obtain

$$(\underset{\beta}{\lambda} - \underset{\gamma}{\lambda})\underset{\alpha}{\omega}{}^i + (\underset{\gamma}{\lambda} - \underset{\alpha}{\lambda})\underset{\beta}{\omega}{}^i + (\underset{\alpha}{\lambda} - \underset{\beta}{\lambda})\underset{\gamma}{\omega}{}^i = 0. \qquad (8.1.68)$$

The equation (8.1.68) can be written in the form

$$\underset{\alpha}{\bar{\omega}}{}^i + \underset{\beta}{\bar{\omega}}{}^i + \underset{\gamma}{\bar{\omega}}{}^i = 0 \qquad (8.1.69)$$

if we use the substitution

$$\underset{\alpha}{\bar\omega}^i = \underset{\beta\gamma}{A}\underset{\alpha}{\omega}^i, \quad \underset{\beta}{\bar\omega}^i = \underset{\gamma\alpha}{A}\underset{\beta}{\omega}^i, \quad \underset{\gamma}{\bar\omega}^i = \underset{\alpha\beta}{A}\underset{\gamma}{\omega}^i \tag{8.1.70}$$

where $\underset{\beta\gamma}{A} = \underset{\beta}{\lambda} - \underset{\gamma}{\lambda}$ etc. The equations (8.1.69) for the three-web $[\alpha,\beta,\gamma]$ are similar to the equations $\underset{1}{\omega}^i + \underset{2}{\omega}^i + \underset{3}{\omega}^i = 0$ for the three-web $[1,2,3]$.

Exterior differentiation of the first two equations of (8.1.70) gives

$$\begin{cases} d\underset{\alpha}{\bar\omega}^i = \underset{\alpha}{\bar\omega}^k \wedge \underset{\alpha\beta\gamma}{\omega}{}^i_k + \underset{\alpha\beta\gamma}{a}{}_j \underset{\alpha}{\bar\omega}^j \wedge \underset{\alpha}{\bar\omega}^i, \\ d\underset{\beta}{\bar\omega}^i = \underset{\beta}{\bar\omega}^k \wedge \underset{\alpha\beta\gamma}{\omega}{}^i_k - \underset{\alpha\beta\gamma}{a}{}_j \underset{\beta}{\bar\omega}^j \wedge \underset{\beta}{\bar\omega}^i, \end{cases} \tag{8.1.71}$$

where

$$\begin{aligned} \underset{\alpha\beta\gamma}{\omega}{}^i_k = {} & \omega^i_k + \delta^i_k a_j(2\underset{2}{\omega}^j + \underset{1}{\omega}^j) - \underset{\alpha\beta\gamma}{A}(\underset{\alpha\beta}{A}\delta^i_k \underset{\alpha}{b}_j + \underset{\gamma\alpha}{A}\delta^i_j \underset{\beta}{b}_k + \underset{\beta\gamma}{A}\delta^i_k \underset{\gamma}{b}_j)\underset{\alpha}{\bar\omega}^j \\ & + \underset{\alpha\beta\gamma}{A}(\underset{\beta\gamma}{A}\delta^i_j \underset{\alpha}{b}_k + \underset{\alpha\beta}{A}\delta^i_k \underset{\gamma}{b}_j + \underset{\gamma\alpha}{A}\delta^i_k \underset{\gamma}{b}_j)\underset{\beta}{\bar\omega}^j, \end{aligned} \tag{8.1.72}$$

$$\underset{\alpha\beta\gamma}{a}{}_j = \underset{\alpha\beta\gamma}{A}(\underset{\beta\gamma}{A}\underset{\alpha}{b}_j + \underset{\gamma\alpha}{A}\underset{\beta}{b}_j + \underset{\alpha\beta}{A}\underset{\gamma}{b}_j), \tag{8.1.73}$$

$$\underset{\alpha\beta\gamma}{A} = [(\underset{\alpha}{\lambda} - \underset{\beta}{\lambda})(\underset{\beta}{\lambda} - \underset{\gamma}{\lambda})(\underset{\gamma}{\lambda} - \underset{\alpha}{\lambda})]^{-1}. \tag{8.1.74}$$

Comparison of (8.1.71) and (8.1.17) shows that the forms $\underset{\alpha\beta\gamma}{\omega}{}^i_j$ are the connection forms of the canonical affine connection induced by the 3-subweb $[\alpha,\beta,\gamma]$ and

$$\underset{\alpha\beta\gamma}{a}{}^i_{jk} = \underset{\alpha\beta\gamma}{a}{}_{[j}\delta^i_{k]}, \tag{8.1.75}$$

where the $\underset{\alpha\beta\gamma}{a}{}_j$ are determined by (8.1.73), is the torsion tensor of $[\alpha,\beta,\gamma]$. Exterior differentiation of (8.1.72) leads now to (8.1.63). ■

Corollary 8.1.13 *The curvature tensor of the three-subweb* $[\alpha,\beta,\gamma]$ *is*

$$\underset{\alpha\beta\gamma}{b}{}^i_{jkl} = \underset{\alpha\beta\gamma}{A}(a^i_{jkl} + \delta^i_l \underset{\alpha}{k}_{jk} + \delta^i_k \underset{\beta}{k}_{jl} + \delta^i_j \underset{\gamma}{k}_{kl}). \tag{8.1.76}$$

Proof. In fact, it follows from (8.1.70) that

$$\begin{cases} \underset{1}{\omega}^i = (-\underset{\alpha}{\bar\omega}^i/\underset{\beta\gamma}{A} + \underset{\beta}{\bar\omega}^i/\underset{\gamma\alpha}{A})/\underset{\alpha\beta}{A}, \\ \underset{2}{\omega}^i = (\underset{\beta}{\lambda}\underset{\alpha}{\bar\omega}^i/\underset{\beta\gamma}{A} - \underset{\alpha}{\lambda}\underset{\beta}{\bar\omega}^i/\underset{\gamma\alpha}{A})/\underset{\alpha\beta}{A}. \end{cases} \tag{8.1.77}$$

Substituting (8.1.77) into (8.1.63), we obtain

$$d\underset{\alpha\beta\gamma}{\omega}{}^i_j - \underset{\alpha\beta\gamma}{\omega}{}^k_j \wedge \underset{\alpha\beta\gamma}{\omega}{}^i_k = \underset{\alpha\beta\gamma}{A}(a^i_{jkl} + \delta^i_l \underset{\alpha}{k}_{jk} + \delta^i_k \underset{\beta}{k}_{jl} + \delta^i_j \underset{\gamma}{k}_{kl})\underset{\alpha}{\omega}^k \wedge \underset{\beta}{\omega}^l. \tag{8.1.78}$$

Comparison of (8.1.78) and (8.1.5) gives (8.1.76). ■

Corollary 8.1.14 *The exterior differential of the contracted connection forms* $\underset{\xi\eta\zeta}{\omega}{}^i_i$ *is*

$$d\underset{\xi\eta\zeta}{\omega}{}^i_i = (\underset{\xi}{k}_{kl} + \underset{\eta}{k}_{kl} + r\underset{\zeta}{k}_{kl})\underset{1}{\omega}{}^k \wedge \underset{2}{\omega}{}^l. \tag{8.1.79}$$

Proof. This result immediately follows from (8.1.63). ■

Corollary 8.1.15 *The connection* $\gamma_{\xi\eta\zeta}$ *is equiaffine if and only if*

$$\underset{\xi}{k}_{ij} + \underset{\eta}{k}_{ij} + r\underset{\zeta}{k}_{ij} = 0. \tag{8.1.80}$$

Proof. In fact the connection forms $\underset{\xi\eta\zeta}{\omega}{}^I_J$ of $\gamma_{\xi\eta\zeta}$ are

$$(\underset{\xi\eta\zeta}{\omega}{}^I_J) = \begin{pmatrix} \underset{\xi\eta\zeta}{\omega}{}^i_j & 0 \\ 0 & \underset{\xi\eta\zeta}{\omega}{}^i_j \end{pmatrix}, \qquad I,J = 1,\ldots,2r; \quad i,j = 1,\ldots,r.$$

Because of this, we have $\underset{\xi\eta\zeta}{\omega}{}^I_I = 2\underset{\xi\eta\zeta}{\omega}{}^i_i$. As we already indicated in the proof of Theorem 7.7.7, the connection $\gamma_{\xi\eta\zeta}$ is equiaffine if and only if $d\underset{\xi\eta\zeta}{\omega}{}^I_I = 0$. In our case $d\underset{\xi\eta\zeta}{\omega}{}^I_I = 2\underset{\xi\eta\zeta}{\omega}{}^i_i$, and it follows from (8.1.79) that this expression vanishes if and only if we have (8.1.80). ■

8.1.4 Almost Algebraisable Webs $AAW(d,2,r)$

Definition 8.1.16 A web $W(d,2,r)$ is said to be *almost algebraisable* if it is almost Grassmannisable and its tensors satisfy the following relation

$$K_{ij} = \sum_{\xi=1}^{d} \underset{\xi}{k}_{ij} = 0. \tag{8.1.81}$$

We will denote such webs by $AAW(d,2,r)$. Note that for a three-web $W(3,2,r)$ the condition (8.1.81) has the form $f_{ij} + g_{ij} + h_{ij} = 0$ and this is the necessary and sufficient condition for hexagonality (and algebraisability) for an isoclinic $W(3,2,r)$ (see [AS 81]). Therefore, an almost algebraisable web $W(3,2,r)$ is algebraisable.

In the case $d > 3$ the conditions (8.1.81) do not imply (8.1.37), i.e., an almost algebraisable web is not necessarily transversally geodesic and consequently it is not necessarily algebraisable.

We will find now under what condition an $AAW(d,2,r)$ must be algebraisable.

Theorem 8.1.17 *An almost algebraisable web* $AAW(d,2,r)$, $d > 3$, *is algebraisable if and only if it is transversally geodesic.*

Proof. The *necessity* of the theorem follows from Theorem 2.6.15 according to which an algebraisable web $AW(d,2,r)$ is Grassmannisable and consequently, by Theorem 2.6.5, is transversally geodesic. To prove the *sufficiency*, we note that an almost algebraisable transversally geodesic web $AGW(d,2,r)$ is a Grassmannisable web for which the conditions (8.1.83) hold. By Theorem 2.6.15, such a Grassmannisable web is algebraisable. ■

The next theorem establishes some criteria for an isoclinic $AGW(d,2,r)$ to be almost algebraisable.

Theorem 8.1.18 *An isoclinic almost Grassmannisable web $AGW(d,2,r)$ is almost algebraisable if and only if one of the following affine connections is equiaffine:*

(i) *The middle connection of the canonical affine connections $\gamma_{123}, \gamma_{234}, \ldots, \gamma_{d12}$.*

(ii) *The middle connection of all the $3!\binom{d}{3}$ affine connections $\Gamma_{\xi\eta\zeta}$.*

(iii) *For $r = d-2$ the middle connection of the affine connections $\Gamma_{\xi_0\eta_0\zeta}$ where ξ_0 and η_0 are fixed.*

(iv) *For $r = d-2$ the middle connection of the affine connections $\gamma_{\xi_0\eta\zeta}$ where ξ_0 is fixed and the pairs η, ζ are all neighbouring pairs of the sequence $1, 2, \ldots,$ $\xi_0 - 1,\ \xi_0 + 1, \ldots, d$.*

Proof. All these statements are equivalent to the condition (8.1.81). In fact, the contracted connection forms in these cases respectively are

$$\text{(i)}\quad \underset{1}{\Omega}{}_I^I = 2(\underset{123}{\omega}{}_i^i + \underset{234}{\omega}{}_i^i + \ldots + \underset{d12}{\omega}{}_i^i)/d,$$

$$\text{(ii)}\quad \underset{2}{\Omega}{}_I^I = 2\sum_{\xi,\eta,\zeta} \underset{\xi\eta\zeta}{\Omega}{}_i^i/[d(d-1)(d-2)],$$

$$\text{(iii)}\quad \underset{3}{\Omega}{}_I^I = 2\sum_{\xi} \underset{\xi_0\eta_0\zeta}{\omega}{}_i^i/d,$$

$$\text{(iv)}\quad \underset{4}{\Omega}{}_I^I = 2(\underset{\xi_0 12}{\omega}{}_i^i + \underset{\xi_0 23}{\omega}{}_i^i + \ldots + \underset{\xi_0,\xi_0-1,\xi_0+1}{\omega}{}_i^i + \ldots + \underset{\xi_0 d1}{\omega}{}_i^i)/(d-1).$$

Using (8.1.79) and the condition $r = d-2$ for (iii) and (iv), we obtain

$$d\underset{s}{\Omega}{}_I^I = \underset{s}{a} K_{ij} \underset{1}{\omega}{}^i \wedge \underset{2}{\omega}{}^j, \quad s = 1,2,3,4, \tag{8.1.82}$$

where

$$K_{ij} = \sum_{\xi} \underset{\xi}{k}_{ij} \quad \text{and} \quad \underset{1}{a} = 2(r+2)/d, \quad \underset{2}{a} = r+2, \quad \underset{3}{a} = 2r/(r+2).$$

The relation (8.1.82) shows that each of the equations $d\underset{s}{\Omega}{}_I^I = 0$ is equivalent to (8.1.81). ■

Combining the results of Theorems 8.1.17 and 8.1.18, we obtain the following condition of algebraisability of an isoclinic $AGW(d,2,r)$:

Corollary 8.1.19 *An isoclinic almost Grassmannisable web* $AGW(d,2,r)$ *is algebraisable if and only if it is transversally geodesic and one of the affine connections listed in Theorem* 8.1.17 *is equiaffine.* ■

Using Definition 8.1.3 and Corollary 8.1.19, we obtain

Corollary 8.1.20 *A web* $W(d,2,r)$ *is algebraisable if and only if it is transversally geodesic, isoclinic, has scalar basis affinors and one of the affine connections listed in Theorem* 8.1.17 *is equiaffine.* ■

Note that if $d=3$, then

$$d\underset{1}{\Omega}{}_I^I = d\underset{2}{\Omega}{}_I^I = \frac{2}{3}(f_{ij}+g_{ij}+h_{ij})\underset{1}{\omega}{}^i \wedge \underset{2}{\omega}{}^j. \tag{8.1.83}$$

As we mentioned before, the condition $f_{ij}+g_{ij}+h_{ij}=0$ is a necessary and sufficient condition for hexagonality (and consequently algebraisability) of an isoclinic web $W(3,2,r)$. Using Theorem 8.1.17, equation (8.1.83) and the above remark, we obtain the following two propositions for a three-web $W(3,2,r)$:

Proposition 8.1.21 *An isoclinic web* $W(3,2,r)$ *is hexagonal if and only if the middle connection* $\gamma = \frac{1}{3}(\gamma_{123}+\gamma_{231}+\gamma_{312})$ *is equiaffine.* ■

Proposition 8.1.22 *A web* $W(3,2,r)$ *is algebraisable if and only if it is isoclinic and the middle connection* γ *is equiaffine.* ■

These two results are new for three-webs $W(3,2,r)$.

We will conclude this subsection by giving analytical consequences of the condition (8.1.81) for almost algebraisability and the conditions (8.1.81) and (8.1.37) for algebraisability of an isoclinic $AGW(d,2,r)$.

Differentiating (8.1.81) and using (8.1.33) and (8.1.60), we obtain by means of the linear independence of $\underset{1}{\omega}{}^i$ and $\underset{2}{\omega}{}^i$ that

$$\underset{1}{K}{}_{ijk} + \underset{\alpha}{b}{}_t a^t_{ijk} + \underset{1}{g}{}_{ijk} - \underset{\alpha}{\lambda}\underset{2}{f}{}_{ijk} = 0, \quad \underset{2}{K}{}_{ijk} = 0, \tag{8.1.84}$$

where

$$\begin{cases} \underset{1}{K}{}_{ijk} = \underset{1}{f}{}_{ijk} + \underset{1}{g}{}_{ijk} + \underset{1}{h}{}_{ijk} + \sum\limits_{\alpha} \underset{\alpha}{\lambda}\underset{\alpha}{k}{}_{ijk} + 3\sum\limits_{\alpha} \underset{\alpha}{b}{}_{(i}\underset{\alpha}{k}{}_{jk)}, \\ \underset{2}{K}{}_{ijk} = \underset{2}{f}{}_{ijk} + \underset{2}{g}{}_{ijk} + \underset{2}{h}{}_{ijk} + \sum\limits_{\alpha} \underset{\alpha}{k}{}_{ijk}. \end{cases} \tag{8.1.85}$$

If, in addition to (8.1.81), we have (8.1.37), this gives us (8.1.42), and the equations (8.1.84) and (8.1.85) become

$$\underset{1}{K}{}_{ijk} = 0, \quad \underset{2}{K}{}_{ijk} = 0, \tag{8.1.86}$$

$$\left\{\begin{array}{l} \underset{1}{K}_{ijk} = f_{ijk} + h_{ijk} + \sum\limits_{\alpha} \underset{\alpha}{\lambda}\underset{\alpha}{k}_{ijk} + 3h_{(i}a_{jk)} + 3\sum\limits_{\alpha} \underset{\alpha}{b}_{(i}\underset{\alpha}{k}_{jk)}, \\ \underset{2}{K}_{ijk} = g_{ijk} + h_{ijk} + \sum\limits_{\alpha} \underset{\alpha}{k}_{ijk} \end{array}\right. \tag{8.1.87}$$

where $\underset{1}{f}_{ijk} = f_{ijk}$, $\underset{2}{g}_{ijk} = g_{ijk}$, $\underset{2}{h}_{ijk} = h_{ijk}$.

In Section 8.4 we will need the following identity which is obtained from (8.1.86) and (8.1.87):

$$f_{ijk} - \underset{\alpha_0}{\lambda} g_{ijk} + (1 - \underset{\alpha_0}{\lambda}) h_{ijk} + \sum_{\alpha} (\underset{\alpha}{\lambda} - \underset{\alpha_0}{\lambda}) \underset{\alpha}{k}_{ijk} = -3 \sum_{\alpha} \underset{\alpha}{b}_{(i}\underset{\alpha}{k}_{jk)} - 3a_{(i}h_{jk)} \tag{8.1.88}$$

where α_0 is fixed.

8.1.5 Non-Isoclinic Almost Grassmannisable Webs $AGW(4,2,2)$

According to Theorem 8.1.10, among the webs $AGW(d,2,r)$ only the webs $AGW(4,2,2)$ can be of both kinds: isoclinic and non-isoclinic. We considered the non-isoclinic case for any $r > 2$ in the subsection 4. Now we will study non-isoclinic webs $AGW(4,2,2)$.

Let $AGW(4,2,2)$ be a non-isoclinic almost Grassmannisable web. For such a web we have the equations (8.1.17), (8.1.5), (8.1.19), (8.1.20), (8.1.49) and (8.1.56). Denote $p_{[12]}$ by p and $q_{[12]}$ by q:

$$p_{[12]} = p, \quad q_{[12]} = q. \tag{8.1.89}$$

Since $\lambda_4 \neq \lambda_s$, $s = 1,2,3$, we have

$$p \neq 0, \quad q \neq 0, \quad p \neq q. \tag{8.1.90}$$

The equation (8.1.56) shows that

$$\lambda = p/q. \tag{8.1.91}$$

It means that the foliation λ_4 of a non-isoclinic web $AGW(4,2,2)$ is uniquely determined by its 3-subweb $[1,2,3]$ provided that the inequalities (8.1.90) hold.

Let us find out under what conditions a non-isoclinic three-web $W(3,2,2)$ can be extended to a non-isoclinic almost Grassmannisable four-web $AGW(4,2,2)$.

For this we need the prolongations of equations (8.1.19). Exterior differentiation of (8.1.19) by means of (8.1.17), (8.1.5), and (8.1.19) and application of Cartan's lemma lead to the following equations:

$$\nabla p_{ij} = \underset{1}{p}_{ijk}\underset{1}{\omega}^k + \underset{2}{p}_{ijk}\underset{2}{\omega}^k, \quad \nabla q_{ij} = \underset{1}{q}_{ijk}\underset{1}{\omega}^k + \underset{2}{q}_{ijk}\underset{2}{\omega}^k \tag{8.1.92}$$

where

$$\begin{cases} \underset{1}{p}_{i[jk]} + p_{i[j}a_{k]} = 0, \\ \underset{1}{q}_{i[jk]} - q_{i[j}a_{k]} = 0, \\ \underset{2}{p}_{ijk} = \underset{1}{q}_{ikj} + a_l b^l_{ijk} = 0, \end{cases} \tag{8.1.93}$$

and $\nabla p_{ij} = dp_{ij} - p_{kj}\omega_i^k - p_{ik}\omega_j^k$, $\nabla q_{ij} = dq_{ij} - q_{kj}\omega_i^k - q_{ik}\omega_j^k$. It follows from (8.1.92) and (8.1.89) that

$$\begin{cases} dp = p\omega_i^i + \underset{1}{p}_i \underset{1}{\omega}^i + \underset{2}{p}_i \underset{2}{\omega}^i, \\ dq = q\omega_i^i + \underset{1}{q}_i \underset{1}{\omega}^i + \underset{2}{q}_i \underset{2}{\omega}^i \end{cases} \tag{8.1.94}$$

where

$$\underset{k}{p}_i = \underset{k}{p}_{[12]i}, \quad \underset{k}{q}_i = \underset{k}{q}_{[12]i}, \quad i,k = 1,2. \tag{8.1.95}$$

Differentiating (8.1.91) by means of (8.1.94) and (8.1.49), we get

$$\underset{1}{p}_i - \lambda \underset{1}{q}_i = p(b_i - a_i), \qquad \underset{2}{p}_i - \lambda \underset{2}{q}_i = qb_i - pa_i. \tag{8.1.96}$$

Theorem 8.1.23 *A non-isoclinic three-web $W(3,2,2)$ given in X^4 and satisfying the condition*

$$q(q\underset{1}{p}_i - p\underset{1}{q}_i) - p(q\underset{2}{p}_i - p\underset{2}{q}_i) - pq(p-q)a_i \tag{8.1.97}$$

and the inequalities (8.1.90) *can be uniquely extended to a non-isoclinic almost Grassmannisable four-web $AGW(4,2,2)$. Such a web $AGW(4,2,2)$ is determined by the equations* (8.1.17), (8.1.5), (8.1.19), (8.1.20), (8.1.49), (8.1.56), (8.1.92), (8.1.96), (8.1.97) *and* (8.1.90).

Proof. In fact, if we eliminate b_i from (8.1.96), we will get the condition (8.1.97). This condition must be satisfied for a non-isoclinic web $W(3,2,2)$ which can be extended to a non-isoclinic $AGW(4,2,2)$. In addition, such a web $W(3,2,2)$ has to satisfy the conditions $\lambda \neq 0, \infty, 1$, which, by virtue of (8.1.91), give (8.1.90). For the web $AGW(4,2,2)$ we can find the quantities b_i from (8.1.96) and b_{ij} from (8.1.56). ■

Note that if we depart from any 3-subweb $[\xi, \eta, \zeta]$ of a non-isoclinic $AGW(4,2,r)$, then its extension to the four-web is the given web $AGW(4,2,2)$.

To show this, we need to calculate for $[\xi, \eta, \zeta]$ the quantities similar to $\underset{1}{\omega}^i$, $\underset{2}{\omega}^i$, $\underset{4}{\omega}^i$, λ, ω_j^i, a_i, b_i, p_{ij}, q_{ij}, and b_{ij} and check that for them the equations (8.1.56) hold.

Let us consider, for example, the 3-subweb $[1,2,4]$. If for this subweb one denotes all the quantities mentioned above by the same letters with bar, then it is easy to see that

$$\begin{cases} \underset{1}{\bar{\omega}}^i = \lambda \underset{1}{\omega}^i, \quad \underset{2}{\bar{\omega}}^i = \underset{1}{\omega}^i, \quad \underset{4}{\bar{\omega}}^i = \underset{3}{\omega}^i, \quad \bar{\lambda} = 1/\lambda, \\ \bar{\omega}_j^i = \omega_j^i - \delta_j^i (b_k - \lambda a_k)\underset{2}{\omega}^k/\lambda, \quad \bar{a}_i = b_i/\lambda, \quad \bar{b}_i = a_i/\lambda, \\ \bar{p}_{ij} = (b_{ji} + p_{ij} - q_{ji})/\lambda, \quad \bar{q}_{ij} = b_{ij}/\lambda, \quad \bar{b}_{ij} = q_{ji}/\lambda. \end{cases} \tag{8.1.98}$$

Using (8.1.98), one can check that $\bar{b}_{[ij]} = \bar{p}_{[ij]} = \lambda \bar{q}_{[ij]}$ provided that (8.1.56) holds for the 3-subweb $[1,2,3]$. A similar calculation can be done for any $[\xi, \eta, \eta]$.

The following proposition give the analytical condition under which for a non-isoclinic web $AGW(4,2,2)$ all four affine connections mentioned in Theorem 8.1.18 are equiaffine.

Theorem 8.1.24 *For a non-isoclinic almost Grassmannisable web $AGW(4,2,2)$ the following statements are equivalent:*

(i) *Each of the four affine connections indicated in Theorem 8.1.18 is equiaffine.*

(ii) *The form $\omega_k^k + (a_k - \underset{4}{b_k}/\underset{4}{\lambda})\underset{2}{\omega^k}$ is a total differential.*

(iii) *The curvature tensor b^i_{jkl} of $AGW(4,2,2)$ satisfies the following equation:*

$$b^k_{kij} = \underset{4}{b_{ij}} - q_{ij}. \tag{8.1.99}$$

Proof. Using (8.1.62) and (8.1.66), we find the expressions of the forms $\underset{\xi}{\Omega^I_I}$, $\xi = 1,2,3,4$ (see the proof of Theorem 8.1.18):

$$\begin{cases} \underset{1}{\Omega^I_I} &= 2\omega^i_i + 2(a_i - \underset{4}{b}/\underset{4}{\lambda})\underset{2}{\omega^i} - d\ln(1-\underset{4}{\lambda})^{3/2}, \\ \underset{2}{\Omega^I_I} &= 2\omega^i_i + 2(a_i - \underset{4}{b_i}/\underset{4}{\lambda})\underset{2}{\omega^i} - d\ln[\underset{4}{\lambda}/(1-\underset{4}{\lambda})^5], \\ \underset{3}{\Omega^I_I} &= 2\omega^i_i + 2(a_i - \underset{4}{b_i}/\underset{4}{\lambda})\underset{2}{\omega^i}, \\ \underset{4}{\Omega^I_I} &= 2\omega^i_i + 2(a_i - \underset{4}{b_i}/\underset{4}{\lambda})\underset{2}{\omega^i} + 2d\ln[\underset{4}{\lambda}/(1-\underset{4}{\lambda})], \end{cases} \tag{8.1.100}$$

where $\underset{3}{\Omega^I_I}$ and $\underset{4}{\Omega^I_I}$ are calculated correspondingly for $\xi_0 = 1$, $\eta_0 = 2$, and $\xi_0 = 1$.

It is easy to check that $d\underset{\xi}{\Omega^I_I} = 0$ if and only if the form $\omega^i_i + (a_i - \underset{4}{b_i}/\underset{4}{\lambda})\underset{2}{\omega^i}$ is a total differential. Using (8.1.19) and (8.1.56), we obtain for this form:

$$d[\omega^i_i + (a_i - \underset{4}{b_i}/\underset{4}{\lambda})] = (b^i_{ijk} - \underset{4}{b_{jk}} + q_{jk})\underset{1}{\omega^j} \wedge \underset{2}{\omega^k}. \tag{8.1.101}$$

The equation (8.1.101) shows that the form $\omega^i_i + (a_i - \underset{4}{b_i}/\underset{4}{\lambda})\underset{2}{\omega^i}$ is a total differential if and only if the condition (8.1.99) holds. ■

8.1.6 Examples of Non-Extendable Non-Isoclinic Webs $W(3,2,2)$

We will consider now examples of three-webs $W(3,2,2)$ which can not be extended to a non-isoclinic $AGW(4,2,2)$ since they do not satisfy conditions of Theorem 8.1.23. In Section 8.4 we will also construct webs $W(3,2,2)$ which can be extended to a non-isoclinic $AGW(4,2,2)$.

In each example we define a $W(3,2,2)$ by its closed form equations $z^i = f^i(x^j,\ y^k)$, calculate its torsion tensor a^i_{jk} and the connection forms ω^i_j by means of the following formulae which are similar to formulas obtained in Section 3.3 for a web $W(n+1,n,r)$, $n>2$ (see [AS 71a], [AS 81]):

$$a^i_{jk} = \Gamma^i_{[jk]}, \quad \omega^i_j = \Gamma^i_{kj}\underset{1}{\omega^k} + \Gamma^i_{jk}\underset{2}{\omega^k} \tag{8.1.102}$$

where

$$\Gamma^i_{jk} = -\frac{\partial^2 f^i}{\partial x^l \partial y^m}\bar{g}^l_j \tilde{g}^m_k \tag{8.1.103}$$

and $(\bar{g}^l_j)$ and $(\tilde{g}^m_k)$ are the inverse matrices of the matrices $(\bar{f}^i_j) = (\partial f^i/\partial x^j)$ and $(\tilde{f}^i_j) = (\partial f^i/\partial y^j)$.

Then using the equations (8.1.16), (8.1.19), (8.1.89), (8.1.103) and (8.1.102), we find consecutively a_i, p_{ij}, q_{ij}, p and q.

Example 8.1.25 (see [B 35]). A web $W(3,2,2)$ is given by

$$f^1 = x^1 + y^1, \quad f^2 = (x^2+y^2)(y^1-x^1). \tag{8.1.104}$$

For this web we have

$$a_1 = 2/(y^1 - x^1), \quad a_2 = 0, \quad p_{12} = q_{12} = p_{21} = q_{21} = 0. \tag{8.1.105}$$

Because of (8.1.105), the web (8.1.104) is isoclinic.

Example 8.1.26 Consider the web $W(3,2,2)$ given by

$$f^1 = x^1y^1 - x^2y^2, \quad f^2 = x^1y^2 + x^2y^1. \tag{8.1.106}$$

Calculations give

$$\begin{aligned} a_1 &= 2(x^1y^1 + x^2y^2)/(\Delta_1\Delta_2), \\ a_2 &= 2(x^2y^1 - x^1y^2)/(\Delta_1\Delta_2), \quad p = 0, \quad q = -4y^1y^2/(\Delta_1^2\Delta_2), \end{aligned} \tag{8.1.107}$$

where $\Delta_1 = y_1^2 + y_2^2$, $\Delta_2 = x_1^2 + x_2^2$.

Example 8.1.27 A web $W(3,2,2)$ is given by

$$f^1 = x^2 e^{x^1y^1}, \quad f^2 = x^2 + y^2. \tag{8.1.108}$$

We have

$$\begin{cases} a_1 = 0, \quad a_2 = -(x^1+1)/(x^1x^2), \\ p = -[2(x^1x^2)^2 y^1 e^{x^1y^1}]^{-1}, \quad q = 0. \end{cases} \tag{8.1.109}$$

Example 8.1.28 Suppose that a web $W(3,2,2)$ is given by

$$f^1 = x^1 + y^1, \quad f^2 = x^1y^1 + x^2y^2. \tag{8.1.110}$$

In this case

$$a_1 = (x^1 - y^1)/(x^2y^2), \quad a_2 = 0, \quad p = q = (y^1 - x^1)/[2(x^2y^2)^2]. \tag{8.1.111}$$

It follows from (8.1.107), (8.1.109), and (8.1.111) that the webs (8.1.106), (8.1.108), and (8.1.110) are non-isoclinic (the quantities p and q do not vanish simultaneously) but they can not be extended to to a non-isoclinic $AGW(4,2,2)$ since the inequalities (8.1.90) do not hold for them.

Example 8.1.29 A web $W(3,2,r)$ is given by

$$f^1 = (x^1 + y^1)^3/6 + [(x^1)^2 + (y^1)^2 + 2x^2y^2]/2, \quad f^2 = x^2 + y^2. \tag{8.1.112}$$

The calculations give

$$a_1 = 0, \quad a_2 = -(x^1 + y^1)(x^2 - y^2)/(\Delta_1\Delta_2) \tag{8.1.113}$$

where

$$\Delta_1 = \frac{1}{2}(x^1 + y^1)^2 + x^1, \quad \Delta_2 = \frac{1}{2}(x^1 + y^1)^2 + y^1 \tag{8.1.114}$$

and

$$\left\{\begin{array}{rcl} 2p & = & \{1 - (x^1 + y^1)[(x^1 + y^1)^3 + 3(x^1 + y^1)^2/2 + y^1]/(\Delta_1\Delta_2)\}\cdot \\ & & \cdot(x^2 - y^2)/(\Delta_1^2\Delta_2), \\ 2q & = & \{1 - (x^1 + y^1)[(x^1 + y^1)^3 + 3(x^1 + y^1)^2/2 + x^1]/(\Delta_1\Delta_2)\}\cdot \\ & & \cdot(x^2 - y^2)/(\Delta_1\Delta_2^2). \end{array}\right. \tag{8.1.115}$$

We can see from (8.1.114) and (8.1.115) that the three-web (8.1.113) is non-isoclinic. However, this web can not be extended to a non-isoclinic $AGW(4,2,2)$ since it does not satisfy the conditions (8.1.97). To see this, we note that the equations (8.1.97) for the web (8.1.113) are equivalent to the equations

$$\frac{\partial\lambda}{\partial x^1}\Delta_2 = (x^1 + y^1)(\lambda - \lambda^2), \quad \frac{\partial\lambda}{\partial y^1}\Delta_1 = (x^1 + y^1)(1 - \lambda) \tag{8.1.116}$$

where $\lambda = p/q$ and p, q are determined by (8.1.115). A straightforward calculation shows that the conditions (8.1.97) are not satisfied for the web (8.1.113).

8.2 1-rank Problems for Almost Grassmannisable Webs $AGW(d,2,r)$

8.2.1 Basic Equations for a Web $W(d,2,r)$ of Non-Zero 1-rank

In this section we shall study 1-rank problems for the webs $W(d,2,r)$. Let us first to define the 1-rank of such webs.

Definition 8.2.1 The 1-*rank* of a web $W(d,2,r)$ is the maximal number of linearly independent d-tuples of functions $\varphi_1,\ldots,\varphi_d$ which are not simultaneously constant on foliations $\lambda_1,\ldots,\lambda_d$ and satisfy the equation

$$\sum_{\xi=1}^{d}\varphi_\xi = \text{const.} \tag{8.2.1}$$

It follows from (8.2.1) that

$$\sum_{\xi=1}^{d} d\varphi_\xi = 0. \tag{8.2.2}$$

Definition 8.2.2 An equation of the form (8.2.2) is said to be an *abelian 1-equation.*

It is obvious that the 1-rank of a web, as it was defined in Definition 8.2.1, is the same as the maximal number of linearly independent abelian equations (8.2.2) admitted by a web $W(d,2,r)$. The webs admitting abelian equations are not general webs. A first question about such webs is whether or not the 1-rank is finite for them, and if so, then it is natural to find an upper bound for this rank and describe the webs of maximum 1-rank.

Suppose that a web $W(d,2,r)$ is given by the systems (8.1.1). In this case we have the equations (8.1.2)–(8.1.10).

Since φ_ξ, ξ fixed, is constant on λ_ξ, we have

$$\begin{cases} d\varphi_1 = \alpha_i \underset{1}{\omega}^i, & d\varphi_3 = \gamma_i(\underset{1}{\omega}^i + \underset{2}{\omega}^i), \\ d\varphi_2 = \beta_i \underset{2}{\omega}^i, & d\varphi_\alpha = \underset{\alpha}{\sigma}_i(\underset{\alpha}{\lambda}{}^i_j \underset{1}{\omega}^j + \underset{2}{\omega}^i), \quad \alpha = 4,\ldots,d. \end{cases} \tag{8.2.3}$$

Substituting (8.2.3) into (8.2.2), we obtain by means of the linear independence of the forms $\underset{1}{\omega}^i$ and $\underset{2}{\omega}^i$ that

$$\alpha_i = -(\gamma_i + \sum_{\alpha=4}^{d} \underset{\alpha}{\sigma}_j \underset{\alpha}{\lambda}{}^j_i), \quad \beta_i = -(\gamma_i + \sum_{\alpha=4}^{d} \underset{\alpha}{\sigma}_i). \tag{8.2.4}$$

Using (8.2.4), we can rewrite the equations (8.2.3) in the form

$$\begin{cases} -d\varphi_1 = -(\gamma_i + \sum\limits_{\alpha=4}^{d} \underset{\alpha}{\sigma}_j \underset{\alpha}{\lambda}{}^j_i)\underset{1}{\omega}{}^i, & d\varphi_3 = \gamma_i(\underset{1}{\omega}{}^i + \underset{2}{\omega}{}^i), \\ -d\varphi_2 = -(\gamma_i + \sum\limits_{\alpha=4}^{d} \underset{\alpha}{\sigma}_i)\underset{2}{\omega}{}^i, & d\varphi_\alpha = \underset{\alpha}{\sigma}_i(\underset{\alpha}{\lambda}{}^i_j \underset{1}{\omega}{}^j + \underset{2}{\omega}{}^i), \quad \alpha = 4, \ldots, d. \end{cases} \tag{8.2.5}$$

Taking exterior derivatives of (8.2.5) and using (8.1.4) and (8.1.9), we obtain the following linearly independent quadratic exterior equations:

$$\begin{cases} [\nabla\gamma_m + \gamma_l a^l_{jm}(\underset{1}{\omega}{}^j - \underset{2}{\omega}{}^j)] \wedge (\underset{1}{\omega}{}^m + \underset{2}{\omega}{}^m) = 0, \\ [\nabla\underset{\alpha}{\sigma}_m - \underset{\alpha}{\sigma}_p a^p_{jm}\underset{2}{\omega}{}^j - \underset{\alpha}{\sigma}_p(\underset{\alpha}{\mu}{}^p_{jm} + a^p_{mq}\underset{\alpha}{\lambda}{}^q_j)\underset{1}{\omega}{}^j] \wedge (\underset{\alpha}{\lambda}{}^m_k \underset{1}{\omega}{}^k + \underset{2}{\omega}{}^m) = 0, \\ [\nabla\gamma_m + \sum\limits_{\alpha=4}^{d} \nabla\underset{\alpha}{\sigma}_m - (\gamma_p + \sum\limits_{\alpha=4}^{d} \underset{\alpha}{\sigma}_p) a^p_{jm}\underset{2}{\omega}{}^j] \wedge \underset{2}{\omega}{}^m = 0, \end{cases} \tag{8.2.6}$$

where

$$\nabla\gamma_m = d\gamma_m - \gamma_p\omega^p_m, \quad \nabla\underset{\alpha}{\sigma}_m = d\underset{\alpha}{\sigma}_m - \underset{\alpha}{\sigma}_p\omega^p_m.$$

The application of the Cartan lemma to the first two equations of (8.2.6) gives

$$\begin{cases} \nabla\gamma_i = (u_{ij} - \gamma_l a^l_{ji})\underset{1}{\omega}{}^j + (u_{ij} + \gamma_l a^l_{ji})\underset{2}{\omega}{}^j, \\ \nabla\underset{\alpha}{\sigma}_i = [\underset{\alpha}{v}_{il}\underset{\alpha}{\lambda}{}^l_j + \underset{\alpha}{\sigma}_p(\underset{\alpha}{\mu}{}^p_{ji} + a^p_{iq}\underset{\alpha}{\lambda}{}^q_j)]\underset{1}{\omega}{}^j + (\underset{\alpha}{v}_{ij} + \underset{\alpha}{\sigma}_p a^p_{ji})\underset{2}{\omega}{}^j, \end{cases} \tag{8.2.7}$$

where $u_{ij} = u_{ji}$, $\underset{\alpha}{v}_{ij} = \underset{\alpha}{v}_{ji}$.

Substituting (8.2.7) into the third equation of (8.2.6), we get

$$\begin{cases} u_{ij} + \gamma_l a^l_{ji} + \sum_\alpha \underset{\alpha}{v}_{jl}\underset{\alpha}{\lambda}{}^l_i + \sum\limits_{\alpha} \underset{\alpha}{\sigma}_p(\underset{\alpha}{\lambda}{}^l_j a^p_{il} + \underset{\alpha}{\mu}{}^p_{ij}) = 0, \\ u_{ij} - \gamma_l a^l_{ji} + \sum_\alpha \underset{\alpha}{v}_{il}\underset{\alpha}{\lambda}{}^l_j + \sum\limits_{\alpha} \underset{\alpha}{\sigma}_p(\underset{\alpha}{\lambda}{}^l_j a^p_{il} + \underset{\alpha}{\mu}{}^p_{ij}) = 0. \end{cases} \tag{8.2.8}$$

Symmetrising and alternating equations (8.2.8), we obtain

$$u_{ij} = -\sum_{\alpha=4}^{d} [\underset{\alpha}{\lambda}{}^p_{(i}\underset{\underline{\alpha}}{v}_{j)p} + (\underset{\alpha}{\lambda}{}^p_{(i} a^q_{j)p} + \underset{\alpha}{\mu}{}^q_{(ij)})\underset{\alpha}{\sigma}_q], \tag{8.2.9}$$

$$0 = \sum_{\alpha=4}^{d} \underset{\alpha}{\lambda}{}^p_{[i}\underset{\underline{\alpha}}{v}_{j]p} + \gamma_p a^p_{ji} + \sum_{\alpha=4}^{d} (\underset{\alpha}{\lambda}{}^p_{[i} a^q_{j]p} + \underset{\alpha}{\mu}{}^q_{[ij]})\underset{\alpha}{\sigma}_q. \tag{8.2.10}$$

We underline the index α to show that the symmetrisation and alternation operations are not concerned with it.

The equations (8.1.9) give expressions of u_{ij} in terms of $\underset{\alpha}{v}_{ij}$, $\underset{\alpha}{\lambda}{}^i_j$, a^i_{jk}, $\underset{\alpha}{\mu}{}^i_{jk}$ and $\underset{\alpha}{\sigma}_i$. Thus, a d-tuple of functions φ_ξ, $\xi = 1, \ldots, d$, satisfying (8.2.1) or (8.2.2) is determined by the functions γ_i, $\underset{\alpha}{\sigma}_i$, and $\underset{\alpha}{v}_{ij}$.

8.2.2 The Upper Bound for the 1-rank of an Almost Grassmannisable Web $GW(d,2,r)$, $r>1$

If $r=1$, we do not have condition (8.2.10). In the sequel we shall suppose that $r>1$. The case $r=1$ has been extensively studied in the 1930's by Blaschke and his school [BB 38]. In particular, Blaschke proved that the upper bound for webs $W(d,2,1)$ is $(d-1)(d-2)/2$ and that the webs $W(4,2,1)$ of maximum 1-rank 3 are algebraisable. Bol ([BB 38], [B 36]) proved that this is not true for $d>5$: he gave an example of the web $W(5,2,1)$ which has the maximum 1-rank 6 and is not algebraisable. Some of these results can be also obtained by the methods which we will develop in this section.

Before we will try to find the upper bound $\pi(d,2,r)$, let us outline the method that we are going to apply. The 1-rank R_1 was defined as the maximal number of linearly independent d-tuples $(\varphi_1,\ldots,\varphi_d)$ of functions φ_ξ satisfying (8.2.1). As we saw in Section 8.2.1, these d-tuples are determined by $s=r(d-2)$ independent functions γ_i, $\underset{\alpha}{\sigma}_i$, $\alpha=4,\ldots,d$, from equations (8.2.5). An equivalent definition of the rank R_1 is: the number of arbitrary constants on which the s-tuples γ_i, $\underset{\alpha}{\sigma}_i$ depend provided that for such s-tuples the forms (8.2.5) are total differentials. As we have already mentioned, in general the rank R_1 is not necessarily finite. However, as we will prove, for almost Grassmannisable webs it is finite.

In the case $r>1$ equations (8.2.10) in general permit us to express $r(r-1)/2$ quantities $\underset{\alpha}{v}_{ij}$, $i\neq j$, α fixed (out of the total number $r(r+1)(d-3)/2$) in terms of $\underset{\alpha}{v}_{ii}$, γ_i, $\underset{\alpha}{\sigma}_i$, and a^i_{jk}, $\underset{\alpha}{\lambda}^i_j$, $\underset{\alpha}{\mu}^i_{jk}$. In this general case we need to find consequtive extensions of (8.2.10) and investigate their compability with the other equations.

In this book we will restrict ourselves to a special case which must be considered separately from the general case. This is the case when the first term of (8.2.10) vanishes:

$$\underset{\alpha}{\lambda}^p_i\,\underset{\alpha}{v}_{jp}-\underset{\alpha}{\lambda}^p_j\,\underset{\alpha}{v}_{ip}=0. \tag{8.2.11}$$

In this case by means of the linear independence of the forms λ_i and $\underset{\alpha}{\sigma}_i$ we deduce from (8.2.10) that

$$a^k_{ij}=0, \tag{8.2.12}$$

$$\underset{\alpha}{\mu}^p_{[ij]}=0. \tag{8.2.13}$$

This case is special because the equations (8.2.10) are satisfied identically and we do not need to investigate their consequtive prolongations.

Note further that the equations (8.2.12) imply that a web $W(d,2,r)$ is torsion-free.

We will study only this special case and, in addition, we shall suppose that the quantities $\underset{\alpha}{v}_{ij}$, $i\leq j$, are linearly independent.

It is easy to see that a very wide class of webs $W(d,2,r)$, namely almost Grassmannisable webs $AGW(d,2,r)$, satisfy the conditions (8.2.11). To see this, one should check by inspection that the equations (8.2.11) are satisfied indentically if the basis affinors are of the form (8.1.12). The following lemma gives an additional condition which together with the equations (8.2.11) imply that a web $W(d,2,r)$, $r > 1$, of non-zero 1-rank is almost Grassmannisable.

Lemma 8.2.3 *If $r > 1$ and the quantities $\underset{\alpha}{v}_{ij}$, $i \leq j$, are linearly independent, then a web $W(d,2,r)$ of non-zero 1-rank is almost Grassmannisable.*

Proof. Using the symmetry and independence of $\underset{\alpha}{v}_{ij}$ and the fact that $r > 1$, we can get from (8.2.11) that

$$\underset{\alpha}{\lambda}{}^i_j = \delta^i_j \underset{\alpha}{\lambda}, \tag{8.2.14}$$

where $\underset{\alpha}{\lambda} \neq \underset{\beta}{\lambda}$ and $\underset{\alpha}{\lambda} \neq 0, 1$ (see (8.1.3)). As we know from Definition 8.1.3, the conditions (8.2.14) mean that a web $W(d,2,r)$ is an almost Grassmannisable web $AGW(d,2,r)$. ■

From now on we shall study almost Grassmannisable webs $AGW(d,2,r)$, $r > 1$, of non-zero 1-rank.

Theorem 8.2.4 *The 1-rank of almost Grassmannisable webs $AGW(d,2,r)$, $r > 1$, is bounded: $R_1 < \pi_1(d,2,r)$ where*

$$\pi_1(d,2,r) = \binom{d+r+1}{r+1} - (d-1). \tag{8.2.15}$$

The webs $AGW(d,2,r)$, $r > 1$, of maximum 1-rank are parallelisable.

Proof. As we know from Section 8.1, for the webs $AGW(d,2,r)$ the $d\underset{\alpha}{\lambda}$ is expressed in terms of the $\underset{1}{\omega}{}^i$ and $\underset{2}{\omega}{}^i$ by equations (8.1.45) and the $\underset{\alpha}{\lambda}{}^i_{jk}$, $\underset{\alpha}{\mu}{}^i_{jk}$, $\underset{\alpha 1}{\lambda}{}_i$, $\underset{\alpha 2}{\lambda}{}_i$ are connected by (8.1.46). The conditions (8.2.13) and (8.1.46) imply

$$\underset{\alpha 2}{\lambda}{}_i = 0. \tag{8.2.16}$$

The conditions (8.1.10), (8.2.15), and (8.2.16) show that

$$\underset{\alpha 1}{\lambda}{}_i = 0. \tag{8.2.17}$$

It follows from (8.2.16), (8.2.17), and Proposition 7.1.8 that the basis affinors are constant on the whole of X^{2r} and that the foliation λ_α, $\alpha = 4, \ldots, d$, is determined by the system $\underset{\alpha}{\lambda}\underset{1}{\omega}{}^i + \underset{2}{\omega}{}^i = 0$. In addition, the equations (8.2.9), (8.2.12), (8.2.14), (8.2.15), and (8.2.16) give

$$u_{ij} = \sum_{\alpha=4}^{d} \underset{\alpha}{\lambda}\underset{\alpha}{v}_{ij}. \tag{8.2.18}$$

Thus, if $r > 1$, for the webs $AGW(d,2,r)$ the conditions (8.2.12), (8.2.15), (8.2.16), and (8.2.18) allow us to rewrite the equations (8.2.7) in the form:

$$\left\{\begin{array}{lcl} \nabla\gamma_i & = & -\sum_{\alpha=4}^{d} \underset{\alpha}{\lambda} \underset{\alpha}{v}_{ij}(\underset{1}{\omega}^j + \underset{2}{\omega}^j), \\ \nabla\underset{\alpha}{\sigma}_i & = & \underset{\alpha}{v}_{ij}(\underset{\alpha}{\lambda}\underset{1}{\omega}^j + \underset{2}{\omega}^j), \quad \alpha = 4, \ldots, d. \end{array}\right. \tag{8.2.19}$$

Moreover, it follows from (8.2.12) and (8.1.6) that the curvature tensor b^i_{jkl} satisfies the conditions

$$b^i_{[jk]l} = b^i_{[j|k|l]} = 0, \tag{8.2.20}$$

i.e. it is symmetric with respect to all its lower indices. Using (8.1.12) and (8.2.20), we can show that the equations (8.1.8) are valid.

To continue the proof, we need the extensions of the equations (8.1.5). Taking their exterior derivatives, we obtain by means of (8.2.12) the following exterior cubic equations:

$$\nabla b^i_{jkl} \wedge \underset{1}{\omega}^k \wedge \underset{2}{\omega}^l = 0, \tag{8.2.21}$$

where $\nabla b^i_{jkl} = db^i_{jkl} - b^i_{mkl}\omega^m_j - b^i_{jml}\omega^m_k - b^i_{jkm}\omega^m_l + b^m_{jkl}\omega^i_m$. It follows from (8.2.21) that

$$\nabla b^i_{jkl} = \vec{c}^{\,i}_{jklm}\underset{1}{\omega}^m + \tilde{c}^{\,i}_{jklm}\underset{2}{\omega}^m, \tag{8.2.22}$$

$$\vec{c}^{\,i}_{jk[lm]} = 0, \quad \tilde{c}^{\,i}_{jk[lm]} = 0. \tag{8.2.23}$$

The conditions (8.2.20) and (8.2.23) show that $\vec{c}^{\,i}_{jklm}$ and $\tilde{c}^{\,i}_{jklm}$ are symmetric with respect to all their lower indices.

Exterior differentiation of (8.2.19) gives the following exterior quadratic equations:

$$\left\{\begin{array}{l} \sum_{\alpha=4}^{d} \underset{\alpha}{\lambda} \nabla \underset{\alpha}{v}_{ij} \wedge (\underset{1}{\omega}^j + \underset{2}{\omega}^j) - \gamma_m b^m_{ikl}\underset{1}{\omega}^k \wedge \underset{2}{\omega}^l = 0, \\ \nabla\underset{\alpha}{v}_{ij} \wedge (\underset{\alpha}{\lambda}\underset{1}{\omega}^j + \underset{2}{\omega}^j) + \underset{\alpha}{\sigma}_m b^m_{ikl}\underset{1}{\omega}^k \wedge \underset{2}{\omega}^l = 0 \end{array}\right. \tag{8.2.24}$$

where $\nabla \underset{\alpha}{v}_{ij} = d\underset{\alpha}{v}_{ij} - \underset{\alpha}{v}_{kj}\omega^k_i - \underset{\alpha}{v}_{ik}\omega^k_j$. It follows from (8.2.24) that

$$\nabla \underset{\alpha}{v}_{ij} = \underset{\alpha}{v}_{ijk}(\underset{\alpha}{\lambda}\underset{1}{\omega}^k + \underset{2}{\omega}^k) - \underset{\alpha}{\sigma}_p b^p_{ijk}\underset{1}{\omega}^k, \tag{8.2.25}$$

$$\sum_{\alpha=4}^{d} (\underset{\alpha}{\lambda}^2 - \underset{\alpha}{\lambda})\underset{\alpha}{v}_{ijk} = (\gamma_p + \sum_{\alpha=4}^{d} \underset{\alpha}{\lambda}\underset{\alpha}{\sigma}_p) b^p_{ijk}, \tag{8.2.26}$$

where the $\underset{\alpha}{v}_{ijk}$ are symmetric with respect to all lower indices.

Exterior differentiation of (8.2.25) and (8.2.26) and further application of Cartan's lemma lead to

$$\nabla \underset{\alpha}{v}_{ijk} = \underset{\alpha}{v}_{ijkm}\left(\underset{\alpha}{\lambda}\underset{1}{\omega}^m + \underset{2}{\omega}^m\right) - \left(3\underset{\alpha}{v}_{p(i}b^p_{jk)m} + \underset{\alpha}{\sigma}_p \vec{c}^{\,p}_{ijkm}\right)\underset{1}{\omega}^m, \tag{8.2.27}$$

$$
\left\{
\begin{array}{l}
\displaystyle\sum_{\alpha=4}^{d}(\underset{\alpha}{\lambda}^2 - \underset{\alpha}{\lambda})\underset{\alpha}{v}_{ijkm} = (\gamma_p + \sum_{\alpha=4}^{d} \underset{\alpha}{\sigma}_p)\tilde{c}^p_{ijkm}, \\
\displaystyle\sum_{\alpha=4}^{d}(\underset{\alpha}{\lambda}^3 - \underset{\alpha}{\lambda}^2)\underset{\alpha}{v}_{ijkm} = (\gamma_p + \sum_{\alpha=4}^{d} \underset{\alpha}{\lambda}\underset{\alpha}{\sigma}_p)\bar{c}^p_{ijkm} \\
\displaystyle+\sum_{\alpha=4}^{d}(\underset{\alpha}{\lambda}^2 - \underset{\alpha}{\lambda}^{)}\underset{\alpha}{\sigma}_p\tilde{c}^p_{ijkm} + 4\sum_{\alpha=4}^{d}(\underset{\alpha}{\lambda}^2 - \underset{\alpha}{\lambda})\underset{\alpha}{v}_{p(i}b^p_{jkm)},
\end{array}
\right.
\tag{8.2.28}
$$

where the $\underset{\alpha}{v}_{ijkm}$ are symmetric with respect to all lower indices.

We have $r(d-2)$ linearly independent quantities γ_i and $\underset{\alpha}{\sigma}_i$, $\alpha = 4, \ldots, d$; $r(r+1)(d-3)/2$ quantities $\underset{\alpha}{v}_{ij}$ which are linearly independent themselves and together with γ_i, $\underset{\alpha}{\sigma}_i$. The equations (8.2.26) allow us to determine $r(r+1)(r+2)/3!$ quantities $\underset{\alpha}{v}_{ijk}$ out of the total number of $r(r+1)(r+2)(d-3)/3!$. The equations (8.2.28) determine $2r(r+1)(r+2)(r+3)/4!$ quantities $\underset{\alpha}{v}_{ijkm}$ out of the total number of $r(r+1)(r+2)(r+3)(d-3)/4!$. Thus, we have respectively $r(r+1)(r+2)(d-4)/3!$ and $r(r+1)(r+2)(r+3)(d-5)/4!$ quantities $\underset{\alpha}{v}_{ijk}$ and $\underset{\alpha}{v}_{ijkm}$ which are linearly independent themselves and together with all previous v's. If we continue this process, at the next step we can get $3r(r+1)(r+2)(r+3)(r+4)/5!$ quantities $\underset{\alpha}{v}_{i_1i_2i_3i_4i_5}$ out of the total number of $r(r+1)(r+2)(r+3)(r+4)(d-3)/5!$ from the system similar to (8.2.26) and (8.2.28). If we consider the deduction of (8.2.25) and (8.2.26) as the first step, after p steps we get $pr(r+1)\ldots(r+p+1)/(p+2)!$ quantities $\underset{\alpha}{v}_{i_1i_2\ldots i_{p+2}}$ out of the total number $r(r+1)\ldots(r+p+1)(d-3)/(p+2)!$. When $p = d-3$, we get all the quantities $\underset{\alpha}{v}_{i_1i_2\ldots i_{d-1}}$. Note that the determinant of the last system as well as the determinants of all the previous systems which determine a part of the quantities $\underset{\alpha}{v}_{i_1\ldots i_p}$, $p < d-1$), is a multiple of the Vandermonde. It is equal to

$$
\prod_{\alpha=1}^{d}(\underset{\alpha}{\lambda}^2 - \underset{\alpha}{\lambda}) \times \prod_{\substack{(\alpha,\beta)\\(\alpha<\beta)}}((\underset{\alpha}{\lambda} - \underset{\beta}{\lambda})
$$

and is different from zero because $\underset{\alpha}{\lambda} \neq \underset{\beta}{\lambda}$, $\underset{\alpha}{\lambda} \neq 0, 1$ (see (8.2.14)). If we substitute all the $\underset{\alpha}{v}_{i_1\ldots i_{d-1}}$ obtained in the differential equations for $\underset{\alpha}{v}_{i_1\ldots i_{d-2}}$, we shall have in these equations only quantities whose differentials are known. Exterior differentiation of these equations gives an equation of the form

$$
A_{i_1\ldots i_{d-2}jk}\underset{1}{\omega}^j \wedge \underset{2}{\omega}^k = 0,
$$

where the $A_{i_1\ldots i_{d-2}jk}$ are linear functions of the γ_i, $\underset{\alpha}{\sigma}_i$, $\underset{\alpha}{v}_{ij}$, $\underset{\alpha}{v}_{ijk}, \ldots$ and $\underset{\alpha}{v}_{i_1\ldots i_{d-2}}$. Since all the products $\underset{1}{\omega}^j \wedge \underset{2}{\omega}^k$ are linearly independent, we have

$$
A_{i_1\ldots i_{d-2}jk} = 0. \tag{8.2.29}
$$

For example, if $d = 4$, then for the web $AGW(4,2,r)$ we have the following equation for $\underset{4}{v}_{ij} = v_{ij}$:

$$\nabla v_{ij} = (\gamma_m + \sigma_m) b^m_{ijk} \underset{1}{\omega}^k/(\lambda - 1) + (\gamma_m + \lambda\sigma_m) b^m_{ijk} \underset{2}{\omega}^k/(\lambda^2 - \lambda)$$

where $\lambda = \underset{4}{\lambda}$, $\sigma_m = \underset{4}{\sigma}_m$, $v_{ij} = \underset{4}{v}_{ij}$ and $\nabla v_{ij} = dv_{ij} - v_{kj}\omega^k_i - v_{ik}\omega^k_j$. The equations (8.2.29) in this case have the following form:

$$4b^p_{(ijk}v_{m)p} + (\bar{c}^p_{ijkm} - \lambda\tilde{c}^p_{ijkm})\gamma_p/(\lambda^2 - \lambda) + (\bar{c}^p_{ijkm} + \tilde{c}^p_{ijkm})\sigma_p/(\lambda - 1) = 0. \quad (8.2.30)$$

For $d > 4$ these equations have a similar form but involve Pfaffian derivatives of γ_i and $\underset{\alpha}{\sigma}_i$ of higher orders. The term with the highest order derivative is $b^m_{(ijk} \underset{\alpha}{v}_{i_1 \ldots i_{d-3})m}$. The terms with the lower order derivatives of γ_i and $\underset{\alpha}{\sigma}_i$ have coefficients which are linear combinations of b^m_{ijk} and their consecutive Pfaffian derivatives up to $(d-3)$th order.

The consecutive extension of (8.2.29) leads to new relations which are linear with respect to γ_i, $\underset{\alpha}{\sigma}_i$, $\underset{\alpha}{v}_{ij}, \ldots, \underset{\alpha}{v}_{i_1 \ldots i_{d-2}}$. The system (8.2.19), (8.2.25), (8.2.27), (8.2.26, (8.2.28), ... contains the same number of independent differential equations and unknown functions. This number is equal to $N - s$, where N is the number of independent differential equations in the system mentioned above and s is the number of independent equations in the linear system which we discussed above.

This number is maximal if $s = 0$. In this case we shall denote it by $\pi_1(d,2,r)$. It is easy to see that the necessary and sufficient condition for this maximality is

$$b^m_{ijk} = 0. \quad (8.2.31)$$

The equations (8.2.31) mean that the web $AGW(d,2,r)$ of maximum 1-rank is not only torsion-free but also curvature-free.

Thus the 1 rank of the webs $AGW(d,2,r)$, $r > 1$, cannot be greater than $N = \pi_1(d,2,r)$. It is equal to $\pi_1(d,2,r)$ if and only if the conditions (8.2.31) hold and the system of Pfaffian equations is completely integrable. In this case it follows from (8.2.31) and (8.2.22) that $\bar{c}^m_{ijkl} = \tilde{c}^m_{ijkl} = 0$, and we have the same situation for all the following Pfaffian derivatives of b^m_{ijk} of higher orders. Because of this, all consequences of (8.2.29) (not only equations similar to (8.2.40)) are satisfied identically: their left members are linear combinations of γ_i, $\underset{\alpha}{\sigma}_i$, $\underset{\alpha}{v}_{ij}, \ldots, \underset{\alpha}{v}_{i_1 \ldots i_{d-2}}$, whose coefficients are functions of b^m_{ijk} and their Pfaffian derivatives vanishing by means of (8.2.31).

For the webs $AGW(d,2,r)$, $r > 1$, of maximum 1-rank the Pfaffian equations (8.2.25), (8.2.27) and similar ones have the form:

$$\nabla \underset{\alpha}{v}_{i_1 \ldots i_s} = \underset{\alpha}{v}_{i_1 \ldots i_s k}(\underset{\alpha}{\lambda} \underset{1}{\omega}^k + \underset{2}{\omega}^k), \quad s = 2, \ldots, d-3, \quad d \geq 5, \quad (8.2.32)$$

$$\nabla \underset{\alpha}{v}_{i_1 \ldots i_{d-2}} = 0, \quad d \geq 4. \quad (8.2.33)$$

The functions γ_i, $\underset{\alpha}{\sigma}_i$, $\underset{\alpha}{v}_{ij}$, $\underset{\alpha}{v}_{i_1\ldots i_{d-2}}$ are connected by the relations

$$\begin{cases} \sum_{\alpha=4}^{d}(\underset{\alpha}{\lambda}^2 - \underset{\alpha}{\lambda})\underset{\alpha}{v}_{i_1\ldots i_s} = 0, \\ \sum_{\alpha=4}^{d}(\underset{\alpha}{\lambda}^3 - \underset{\alpha}{\lambda}^2)\underset{\alpha}{v}_{i_1\ldots i_s} = 0, \\ \cdots\cdots\cdots\cdots\cdots\cdots\cdots\cdots\cdots\cdots \\ \sum_{\alpha=4}^{d}(\underset{\alpha}{\lambda}^{s-1} - \underset{\alpha}{\lambda}^{s-2})\underset{\alpha}{v}_{i_1\ldots i_s} = 0, \quad s = 3,\ldots,d-2. \end{cases} \tag{8.2.34}$$

Note that $\underset{\alpha}{v}_{i_1\ldots i_{d-2}} = 0$. The system (8.2.19), (8.2.32), (8.2.33) is completely integrable. The maximum 1-rank $\pi_1(d,2,r)$ is equal to the number of independent constants in its solution. Therefore, the 1-rank $\pi_1(d,2,r)$ is equal to the number of linearly independent functions γ_i, $\underset{\alpha}{\sigma}_i$, $\underset{\alpha}{v}_{ij}$, $\underset{\alpha}{v}_{i_1\ldots i_{d-2}}$, i.e., it is equal to

$$\pi_1(d,2,r) = r(d-2) + \frac{1}{2!}r(r+1)(d-3) + \frac{1}{3!}r(r+1)(r+2)(d-4) + \ldots$$
$$+\frac{1}{(d-2)!}r(r+1)\ldots(r+d-3)\cdot 1$$

or briefly

$$\pi_1(d,2,r) = \sum_{n=1}^{d-2}\binom{n+r-1}{n}(d-n-1) = \binom{d+r+1}{r+1} - (d-1). \tag{8.2.35}$$

The last equality in (8.2.35) is obtained using the well-known identity

$$\sum_{k=0}^{p}\binom{k+a}{a} = \binom{a+p+1}{a+1}$$

in the following equations

$$\sum_{n=1}^{d-2}\binom{n+r-1}{r-1}(d-n-1) \;=\; \sum_{n=0}^{d-2}\binom{n+r-1}{r-1}(d-n-1) - (d-1)$$
$$= \sum_{m=0}^{d-2}\sum_{n=0}^{m}\binom{n+r-1}{r-1} - (d-1) \;=\; \sum_{m=0}^{d-2}\binom{m+r}{r} - (d-1) = \binom{d+r-1}{r+1} - (d-1).$$

We proved in Theorem 7.2.2 that the necessary and sufficient conditions for a web $W(4,2,r)$ to be parallelisable are the parallelisability of its 3-subweb [1,2,3] and the constancy of the basis affinor on the whole manifold X^{2r}. The same proof can be performed for webs $W(d,2,r)$, $d > 4$. The only difference is that all the basis affinors of $W(d,2,r)$ must be constant on X^{2r}.

Using this result and conditions (8.2.12), (8.2.16), (8.2.17), and (8.2.31), we obtain that a web $AGW(d,2,r)$, $r > 1$, of maximum 1-rank is parallelisable. However, it is not an arbitrary parallelisable web $W(d,2,r)$ because its basis affinors $\underset{\alpha}{\lambda}{}^i_j$ have a special structure (8.2.14). In the next subsection we will give a complete geometric characteristic of such webs.

Corollary 8.2.5 *A web $AGW(d,2,r)$, $r > 1$, of maximum 1-rank is determined by the completely integrable system of equations*

$$\begin{cases} d\underset{1}{\omega}^i = \underset{1}{\omega}^j \wedge \omega_j^i, & d\underset{2}{\omega}^i = \underset{2}{\omega}^j \wedge \omega_j^i, \\ d\omega_j^i - \omega_j^k \wedge \omega_k^i = 0, & d\underset{\alpha}{\lambda} = 0 \end{cases} \tag{8.2.36}$$

and the equations (8.2.19), (8.2.32), (8.2.33), *and* (8.2.34).

Remark 8.2.6 The following recursive formulae for calculation of the $\pi_1(d,2,r)$ determined by equation (8.2.35) can be verified by the direct calculation:

$$\pi_1(d,2,r) = \pi_1(d,2,r-1) + \pi_1(d-1,2,r) + d - 2, \tag{8.2.37}$$

$$\pi_1(d+1,2,r) = \pi_1(d,2,r) + \sum_{k=1}^{d-1} \binom{k+r-1}{k}. \tag{8.2.38}$$

Since

$$\pi_1(3,2,r) = r, \quad \pi_1(d,2,1) = \frac{1}{2}(d-1)(d-2), \tag{8.2.39}$$

formulae (8.2.37) and (8.2.38) allow us to calculate $\pi_1(d,2,r)$ for any $d = 3,4,\ldots$ and $r = 1,2,\ldots$. For instance, we can obtain

$$\begin{aligned} \pi_1(d,2,2) &= \frac{1}{3!}(d-1)(d-2)(d+3), \\ \pi_1(d,2,3) &= \frac{1}{4!}(d-1)(d-2)(d^2+5d+12), \\ \pi_1(d,2,4) &= \frac{1}{5!}(d-1)(d-2)(d^3+8d^2+27d+60). \end{aligned}$$

Remark 8.2.7 In the case $d = 3$ we do not have the quantities $\underset{\alpha}{\sigma}$, $\alpha = 4,\ldots,d$, and the equations (8.2.10), (8.2.19), (8.2.29) have the form

$$\gamma_m a_{ij}^m = 0, \quad \nabla\gamma_i = 0, \quad \gamma_m b_{ijk}^m = 0.$$

Thus we have (8.2.12) and (8.2.31). These are all the conditions for a general $W(3,2,r)$ to be of maximum 1-rank. Our three-web is parallelisable. In this case a triple of functions φ_1, φ_2, φ_3 is uniquely determined by γ_i, $i = 1,\ldots,r$, only. It corresponds to the formula (8.2.35) which gives for $d = 3 : \pi_1(3,2,r) = 3$.

Note also that in this case it is easy to see that *a general web $W(3,2,r)$, $r > 1$, is of maximum 1-rank $\pi_1(3,2,r) = r$ if and only if it is parallelisable.*

Remark 8.2.8 In the case $d = 4$ we do not have equations (8.2.32) and (8.2.34). In this case for v_{ij} we have $\nabla v_{ij} = 0$ (we are omitting the subscript 4 everywhere), and all $v_{ijk} = 0$. Thus, $\pi_1(4,2,r) = \frac{1}{5}r(5+r)$.

In the case $d = 5$ the equations (8.2.32), (8.2.33), and (8.2.34) have the form

$$\begin{cases} \nabla v_{ij} = \underset{\alpha}{v}_{ijk}(\underset{\alpha}{\lambda}\underset{1}{\omega}^k + \underset{2}{\omega}^k), \\ \nabla \underset{\alpha}{v}_{ijk} = 0, \quad \alpha = 4,5, \\ \sum_{\alpha=4}^{5} (\underset{\alpha}{\lambda}^2 - \underset{\alpha}{\lambda}) \underset{\alpha}{v}_{ijk} = 0. \end{cases}$$

In this case $\pi_1(5,2,r) = \frac{1}{2}r(r^2 + 9r + 26)$.

8.2.3 Description of the Webs $AGW(d,2,r)$, $d \geq 4$, $r > 1$, of Maximum 1-Rank

In this part we shall give a complete geometric characteristic of the webs $AGW(d,2,r)$, $d \geq 4$, $r > 1$, of maximum 1-rank.

Theorem 8.2.9 *An almost Grassmannisable web $AGW(d,2,r)$, $d \geq 4$, $r > 1$, of maximum 1-rank is equivalent to an algebraic web $AW(d,2,r)$ for which the algebraic hypersurface V_d^r of degree d generating the $AW(d,2,r)$ decomposes into d hyperplanes of a pencil with an $(r-1)$-dimensional vertex.*

Proof. First we note that algebraic webs $W(d,n,r)$ were defined in Definition 2.6.2 (see also Example 1.1.5). We will consider an algebraic web $AW(d,2,r)$ in a projective space P^{r+1}. We associate the space P^{r+1} with the moving frame $\{A_u\}$, $u = 0,1,\ldots,r+1$. The equations of infinitesimal displacements of this frame are

$$dA_u = \theta_u^v A_v, \quad u,v = 0,1,\ldots,r+1, \tag{8.2.40}$$

where the Pfaffian forms θ_u^v satisfy the structure equations of P^{r+1} :

$$d\theta_u^v = \theta_u^w \wedge \theta_w^v, \quad u,v,w = 0,1,\ldots,r+1. \tag{8.2.41}$$

(cf. (7.6.1) and (7.6.2)).

Let us take such moving frames $\{A_u\}$ connected with the corresponding points K, L, M, N_α of the generating hypersurfaces U_1, U_2, U_3, U_α, $\alpha = 4,\ldots,d$, for which

$$K = A_0, \quad L = A_{r+1}, \quad M = A_0 + A_{r+1}, \quad N_\alpha = \underset{\alpha}{\lambda} A_0 + A_{r+1}$$

and the forms θ_0^i, θ_{r+1}^i, $i = 1,\ldots,r$, are the basis forms of $AW(d,2,r)$ (see Section 7.6 where we studied the case $d = 4$).

Suppose now that the algebraic hypersurface V_d^r generating the $AW(d,2,r)$ to which the generating hypersurfaces U_ξ belong, decomposes into d hyperplanes (these hyperplanes are parts of U_ξ) belonging to a pencil with an $(r-1)$-dimensional vertex. If we place the points A_i, $i=1,\ldots,r$, of our moving frame at this vertex, then the web $AW(d,2,r)$ being considered is determined by the following completely integrable system (see Theorem 7.8.9 where the case $d=4$ was discussed, and equations (7.6.3), (7.6.10), (7.6.11), (7.6.19) etc. where $a_i=b_i=0$):

$$\theta_0^{r+1}=0,\quad \theta_{r+1}^0=0,\quad \theta_0^0-\theta_{r+1}^{r+1}=0,\quad \theta_i^{r+1}=0,\quad \theta_i^0=0,\quad d\underset{\alpha}{\tilde{\lambda}}=0. \tag{8.2.42}$$

As to the bundles of straight lines with the vertices on U_ξ which are web leaves, in the adapted frame they are defined by the systems:

$$\theta_0^i=0,\quad \theta_{r+1}^i=0,\quad \theta_i^0+\theta_{r+1}^i=0,\quad \underset{\alpha}{\tilde{\lambda}}\theta_0^i+\theta_{r+1}^i=0,\quad \alpha=4,\ldots,d, \tag{8.2.43}$$

(cf. (7.6.14) for $d=4$). As in Chapter 7, we will restrict ourselves to a domain $\tilde{D}\subset P^{r+1}$ where the straight lines of the bundles with the vertices on U_ξ cut each of the hypersurfaces $U_\xi, \xi=1,\ldots,d$, at a single point.

To prove the equivalence stated in the theorem, we consider a mapping φ of a domain D of X^{2r} where the web $AGW(d,2,r)$ of maximum 1-rank is defined into a domain $\tilde{D}$ of P^{r+1} where the algebraic web $AW(d,2,r)$ is defined. Such a mapping is given by

$$\underset{1}{\omega}^i=\theta_0^i,\quad \underset{2}{\omega}^i=\theta_{r+1}^i. \tag{8.2.44}$$

The mapping (8.2.44) transfers a point $p\in D\subset X^{2r}$ defined by the completely integrable system $\underset{1}{\omega}^i=0$, $\underset{2}{\omega}^i=0$ into a straight line $l\in\tilde{D}\subset P^{r+1}$ defined by the completely integrable system $\theta_0^i=0$, $\theta_{r+1}^i=0$. It follows from (8.2.40) and (8.2.43) that the forms θ_0^i, θ_{r+1}^i, $\theta_0^i+\theta_{r+1}^i$ and $\underset{\alpha}{\tilde{\lambda}}\theta_0^i+\theta_{r+1}^i$ are the basis forms of the cotangent spaces $PT^*_{A_0}(U_1)$, $PT^*_{A_{r+1}}(U_2)$, $PT^*_M(U_3)$, and $PT^*_{N_\alpha}(U_\alpha)$ where $M=A_0+A_{r+1}$, $\tilde{N}_\alpha=\underset{\alpha}{\tilde{\lambda}}A_0+A_{r+1}$. Along the leaves $F_1\subset\lambda_1$ and $F_2\subset\lambda_2$ we have correspondingly $\underset{1}{\omega}^i=0$ and $\underset{2}{\omega}^i=0$. Then from (8.2.44) it follows that $\theta_0^i=0$ and $\theta_{r+1}^i=0$. Therefore, a leaf $F_3\subset\lambda_s$, $s=1,2$, is transferred by φ into the bundle of straight lines with the vertex $A_s\in U_s$. In the same way we can prove that a leaf $F_3\subset\lambda_3$ is transferred by φ into the bundle of straight lines with the vertex $M\in U_3$. As for $F_\alpha\subset\lambda_\alpha$, it is transferred by φ into the bundle of straight lines with the vertex $N_\alpha\subset U_\alpha$ if and only if (see (8.1.1), (8.2.43), and (8.2.44)):

$$\underset{\alpha}{\lambda}=\underset{\alpha}{\tilde{\lambda}}. \tag{8.2.45}$$

Thus, the images of d leaves F_ξ, $\xi=1,\ldots,d$, through a point $p\in D$ are d bundles of straight lines whose vertices are the points of U_ξ, $\xi=1,\ldots,d$, belonging to a straight line l. Therefore, $\varphi(\lambda_\xi)=\tilde{\lambda}_\xi$, $\xi=1,\ldots,d$.

We continue to study the mapping φ. Taking the exterior derivatives of (8.2.44) by means (8.2.36), (8.2.41), and (8.2.42), we obtain

$$\theta_0^j\wedge(\theta_j^i-\delta_j^i\theta-\omega_j^i)=0,\quad \theta_{r+1}^j\wedge(\theta_j^i-\delta_j^i\theta-\omega_j^i)=0 \tag{8.2.46}$$

where $\theta = \theta_0^0 = \theta_{r+1}^{r+1}$. It follows from (8.2.46) that

$$\theta_j^i - \delta_j^i \theta = \omega_j^i. \tag{8.2.47}$$

Taking into account that, by (8.2.41) and (8.2.42), we have

$$d\theta = 0, \tag{8.2.48}$$

we can easily check, using (8.2.36), (8.2.41), and (8.2.42), that exterior differentiation of (8.2.47) leads to the identity.

Therefore, the mapping φ is determined by equations (8.2.44), (8.2.42), (8.2.45), (8.2.47), and (8.2.48). It is easy to see from the structure equations (8.2.41) and (8.2.36) that this system is completely integrable and the forms θ_u^v determined by it satisfy the structure equations of a projective space P^{r+1}. This shows that the mapping φ is a local diffeomorphism which maps a web $AGW(d,2,r)$ of maximum 1-rank onto an algebraic web $AW(d,2,r)$ for which the algebraic hypersurface V_d^r generating the $AW(d,2,r)$ decomposes into d hyperplanes of a pencil with an $(r-1)$-dimensional vertex. ∎

8.2.4 Explicit Expressions of the Functions φ_ξ and Description of Their Level Sets

We can explicitly determine the functions φ_ξ, $\xi = 1, \ldots, d$ appearing in the 1-rank problem (see (8.2.1)).

First of all, it follows from (8.2.31) that the forms $\omega_j^i = \theta_j^i - \delta_j^i \theta$ satisfy the equations

$$d\omega_j^i = \omega_j^k \wedge \omega_k^i. \tag{8.2.49}$$

Equations (8.2.49) show that a moving frame $\{A_u\}$ can be chosen in such a way that $\omega_j^i = 0$ (cf. Section 4.3). Then we obtain from (8.2.36) that

$$d\underset{1}{\omega}^i = 0, \quad d\underset{2}{\omega}^i = 0. \tag{8.2.50}$$

Integration of (8.2.50) leads to

$$\underset{1}{\omega}^i = dx_1^i, \quad \underset{2}{\omega}^i = dx_2^i. \tag{8.2.51}$$

By means of $\omega_j^i = 0$, equations (8.2.33) become $d\underset{\alpha}{v}_{i_1 \ldots i_{d-2}} = 0$, i.e. $\underset{\alpha}{v}_{i_1 \ldots i_{d-2}}$ are constants as well as $\underset{\alpha}{\lambda}$ (see (8.2.36)).

Integrating the systems (8.2.43) and using (8.2.45) and (8.2.51), we obtain the equations of the leaves of d foliations of a web $GW(d,2,r)$, $d > 3$, $r > 1$, of maximum 1-rank in the form:

$$x_1^i = C_1^i, \quad x_2^i = C_2^i, \quad x_1^i + x_2^i = C_3^i, \quad \underset{\alpha}{\lambda} x_1^i + x_2^i = C_\alpha^i, \quad \alpha = 4, \ldots, d, \tag{8.2.52}$$

where C_1^i, C_2^i, C_3^i and C_α^i are constants.

Because of $\omega_j^i = 0$ and (8.2.44), (8.2.51), equations (8.2.19), (8.2.32), and (8.2.33) can be written in the form:

$$\begin{cases} d\gamma_i = -\sum\limits_{\alpha=4}^{d} \underset{\alpha}{\lambda} \underset{\alpha}{v}_{ij}(dx_1^j + dx_2^j), \\ d\underset{\alpha}{\sigma}_i = \underset{\alpha}{v}_{ij}(\underset{\alpha}{\lambda} dx_1^j + dx_2^j), \\ d\underset{\alpha}{v}_{i_1\ldots i_s} = \underset{\alpha}{v}_{i_1\ldots i_s j}(\underset{\alpha}{\lambda} dx_1^j + dx_2^j), \quad s = 2, \ldots, d-3, \quad d \geq 5, \\ d\underset{\alpha}{v}_{i_1\ldots i_{d-2}} = 0, \quad d \geq 4. \end{cases} \tag{8.2.53}$$

Integration of (8.2.53) gives

$$\begin{cases} \underset{\alpha}{v}_{i_1\ldots i_{d-2}} = \underset{\alpha}{v}^0_{i_1\ldots i_{d-2}}, \quad d \geq 4, \\ \underset{\alpha}{v}_{i_1 i_2 \ldots i_s} = \sum\limits_{q=1}^{d-2-s} \underset{\alpha}{v}^0_{i_1 i_2\ldots i_s j_1\ldots j_q}(\underset{\alpha}{\lambda} x_1^{j_1} + x_2^{j_1}) \ldots (\underset{\alpha}{\lambda}_1^{j_q} + x_2^{j_q}) + \underset{\alpha}{v}^0_{i_1 i_2\ldots i_s}, \\ \quad s = 2, \ldots, d-3, \\ \underset{\alpha}{\sigma}_i = \sum\limits_{q=1}^{d-3} \underset{\alpha}{v}_{i j_1\ldots j_q}(\underset{\alpha}{\lambda} x_1^{j_1} + x_2^{j_1}) \ldots (\underset{\alpha}{\lambda} x_1^{j_q} + x_2^{j_q}) + \underset{\alpha}{\sigma}^0_i, \\ \gamma_i = -\sum\limits_{s=4}^{d} \lambda_s \Big\{ \sum\limits_{q=1}^{d-4} \underset{\alpha}{v}^0_{ikj_1\ldots j_q}(\underset{\alpha}{\lambda} x_1^{j_1} + x_2^{j_1}) \ldots (\underset{\alpha}{\lambda} x_1^{j_q} + x_2^{j_q}) + \underset{\alpha}{v}^0_{ik} \Big\} \cdot \\ \cdot (x_1^k + x_2^k) + \gamma_i^0, \end{cases} \tag{8.2.54}$$

where γ_i^0, $\underset{\alpha}{\sigma}^0_i$, $\underset{\alpha}{v}^0_{ij}, \ldots, \underset{\alpha}{v}_{i_1\ldots i_{d-2}}$ are constants and $\underset{\alpha}{v}_{i_1\ldots i_s}$, $s = 3, \ldots, d-2$, satisfy the equations (8.2.34). The last statement can be easily proved by the substitution of $\underset{\alpha}{v}_{i_1\ldots i_s}$, $s = 3, \ldots, d-2$, into (8.2.34) and taking into account the arbitrariness of x_1^i and x_2^i.

Conditions (8.2.51) and (8.2.14) allow us to write the expressions of $d\varphi_\xi$ from (8.2.5) in the form:

$$\begin{cases} d\varphi_1 = -(\gamma_i + \sum\limits_{\alpha=4}^{d} \underset{\alpha}{\lambda} \underset{\alpha}{\sigma}_i) dx_1^i, \quad d\varphi_3 = \gamma_i(dx_1^i + dx_2^i), \\ d\varphi_2 = -(\gamma_i + \sum\limits_{\alpha=4}^{d} \underset{\alpha}{\sigma}_i) dx_2^i, \quad d\varphi_\alpha = \underset{\alpha}{\sigma}_i(\underset{\alpha}{\lambda} dx_1^i + dx_2^i), \quad \alpha = 4, \ldots, d. \end{cases} \tag{8.2.55}$$

Integrating (8.2.55) and using (8.2.13) and (8.2.34), we obtain

$$\left\{\begin{array}{l} -\varphi_1 = \sum_{\alpha=4}^{d} \sum_{s=2}^{d-2} (\underset{\alpha}{\lambda}^s - \underset{\alpha}{\lambda}^{s-1}) \underset{\alpha}{v}{}^0_{i_1 \ldots i_s} x_1^{i_1} \ldots x_1^{i_s} + (\gamma_i^0 + \sum_{\alpha=4}^{d} \underset{\alpha}{\lambda} \underset{\alpha}{\sigma}{}_i^0) x_1^i - \varphi_1^0, \\ -\varphi_2 = \sum_{\alpha=4}^{d} \sum_{s=2}^{d-2} (\underset{\alpha}{\lambda}^{s-1} - \underset{\alpha}{\lambda}^{s-2}) \underset{\alpha}{v}{}^0_{i_1 \ldots i_s} x_2^{i_2} \ldots x_2^{i_s} + (\gamma_i^0 + \sum_{\alpha=4}^{d} \underset{\alpha}{\sigma}{}_i^0) x_2^i - \varphi_2^0, \\ \varphi_3 = -\sum_{s=2}^{d-2} \underset{\alpha}{v}{}^0_{i_1 \ldots i_s} (x_1^{i_1} + x_2^{i_1}) \ldots (x_1^{i_s} + x_2^{i_s}) + \gamma_i^0 (x_1^i + x_2^i) + \varphi_3^0, \\ \varphi_\alpha = \sum_{s=2}^{d-2} \underset{\alpha}{v}{}^0_{i_1 \ldots i_s} (\underset{\alpha}{\lambda} x_1^{i_1} + x_2^{i_1}) \ldots (\underset{\alpha}{\lambda} x_1^{i_s} + x_2^{i_s}) + \underset{\alpha}{\sigma}{}_i^0 (\underset{\alpha}{\lambda} x_1^i + x_2^i) + \varphi_\alpha^0, \end{array}\right. \tag{8.2.56}$$

where φ_ξ^0, $\xi = 1, \ldots, d$, are constants.

Considering x_1^i and x_2^i as coordinates on the Grassmannian $G(1, r+1)$, we see that the equations (8.2.52) give the equations of the leaves of the d foliations of the web $AGW(d,2,r)$ of maximum 1-rank. We have proved

Theorem 8.2.10 *The level sets of the functions φ_ξ, $\xi = 1, \ldots, d$, determined by equations (8.2.56) are algebraic hypercones of order $d-2$ with $(r-2)$-dimensional vertices on the Grassmannian $G(1, r+1)$. For each leaf there exists a corresponding hypercone passing through it.* ■

Remark 8.2.11 In the case of $W(3,2,r)$, the equations (8.2.54) give only $\gamma_i = \gamma_i^0$, the equations (8.2.55) become

$$\begin{aligned} -\varphi_1 &= \gamma_i^0 x_1^i - \varphi_1^0, \\ -\varphi_2 &= \gamma_i^0 x_2^i - \varphi_2^0, \\ \varphi_3 &= \gamma_i^0 (x_1^i + x_2^i) + \varphi_3^0, \end{aligned}$$

and the hypercones mentioned in Theorem 8.2.10 are hyperplanes without common points with the corresponding leaves of $W(3,2,r)$.

8.3 r-Rank Problems for Almost Grassmannisable Webs $AGW(d,2,r)$

8.3.1 The r-Rank of Webs $W(d,n,r)$

In this section we shall study the r-rank problems for webs $W(d,2,r)$. Let us first to define the r-rank of webs $W(d,n,r)$.

Suppose that the leaves of the ξth foliation of a web $W(d,n,r)$ are given as level sets of functions $u_\xi(x)$:

$$u_\xi^i(x) = \text{const.}, \qquad \xi = 1, \ldots, d; \tag{8.3.1}$$

The functions $u_\xi(x)$ are defined up to a local diffeomorphism in the space of $u_\xi(x)$.

Definition 8.3.1 An exterior r-equation of the form

$$\sum_{\xi=1}^{d} f_\xi(u_\xi^j) du_\xi^1 \wedge \ldots \wedge du_\xi^r = 0, \quad j, i, \ldots, i_r = 1, \ldots, r, \tag{8.3.2}$$

is said to be an *abelian r-equation*. The maximum number R_r of linearly independent abelian r-equations admitted by the $W(d,n,r)$ is called the *r-rank* of the web $W(d,n,r)$.

It follows from the definition that the coefficients f_ξ are constant on the leaves of the ξth foliation of $W(d,n,r)$.

If there exists an upper bound $\pi_r(d,n,r)$ of R_r, then $R_r \leq \pi_r(d,n,r)$.

P.A. Griffiths [Gr 77] gave a more general definition of the q-rank for the webs of codimension r, $1 \leq q \leq r$. We will give this definition for the r-rank of a web $W(d,n,r)$.

Definition 8.3.2 Suppose that the leaves of the ξth foliation of a web $W(d,n,r)$ are given by a completely integrable system $\underset{\xi}{\omega} = 0$, $i = 1, \ldots, r$; $\xi = 1, \ldots, d$. An *abelian r-equation* is given by a linear relation

$$\sum_\xi F_\xi \underset{\xi}{\omega}^1 \wedge \ldots \wedge \underset{\xi}{\omega}^r = 0 \tag{8.3.3}$$

where

$$dF_\xi \equiv 0 \ \ \{\omega_\xi^1, \ldots, \omega_\xi^r\}. \tag{8.3.4}$$

Note that condition (8.3.4) is equivalent to the fact that dF_ξ is expressed as a linear combination of $\underset{\xi}{\omega}^i$, i.e., it is equivalent to the condition

$$dF_\xi \wedge \underset{\xi}{\omega}^1 \wedge \ldots \wedge \underset{\xi}{\omega}^r = 0, \tag{8.3.5}$$

or to the fact that the exterior differential of each term of (8.3.3) vanishes:

$$d(F_\xi \underset{\xi}{\omega}^1 \wedge \ldots \wedge \underset{\xi}{\omega}^r) = 0. \tag{8.3.6}$$

The following lemma proves that Definitions 8.3.1 and 8.3.2 are equivalent.

Lemma 8.3.3 *Definitions* 8.3.1 *and* 8.3.2 *are equivalent.*

Proof. First of all we have

$$\underset{\xi}{\omega}^i = \underset{\xi}{A}{}_j^i du_\xi^j, \quad \det(\underset{\xi}{A}{}_j^i) \neq 0, \quad \xi = 1, \ldots, d. \tag{8.3.7}$$

If equation (8.3.3) is an abelian r-equation, then after substitution (8.3.7) ξth term of the sum (8.3.3) has the form $f_\xi du_\xi^1 \wedge \ldots \wedge du_\xi^r$. According to (8.3.6), its exterior differential vanishes: $df_\xi \wedge du^1_\xi \wedge \ldots \wedge du^r_\xi = 0$. It follows from this that df_ξ is a linear combination of du^i_ξ only, f_ξ depends only on u^j_ξ and (8.3.3) is an abelian r-equation in the sense of Definition 8.3.1.

Conversely, suppose we have an equation (8.3.2). It is obvious that the exterior differential of each of its terms vanishes. After we substitute du^i_ξ expressed in terms of $\underset{\xi}{\omega}^i$, each term of the equation (8.3.2) will turn into the corresponding term of (8.3.3), and its exterior differential vanishes since it is still the same term. ■

Note that there are two differences between (8.3.1) and (8.3.2): F_ξ does not only depend on the u^j_ξ and, in general, the $d\underset{\xi}{\omega}^i$ do not vanish.

Note also that we will use Definition 8.3.2, i.e., we will consider an abelian r-equation in the form (8.3.3) provided that condition (8.3.6) holds.

There are two fundamental problems in the study of r-rank problems:

1) *Finding an upper bound $\pi_r(d,n,r)$ for the r-rank if it exists.*

2) *The description of the maximum r-rank webs $W(d,n,r)$.*

As for the first problem for the webs $W(d,n,r)$, it was solved by Chern and Griffiths [CG 78b]. They proved the theorem which we shall only formulate here without giving its proof:

Theorem 8.3.4 *For webs $W(d,n,r)$, $2 \leq r \leq n$, there exists an upper bound $\pi_r(d,n,r)$ for the r-rank R_r and this upper bound is equal to*

$$\pi_r(d,n,r) = \sum_{\mu \geq 0} max\left(\binom{r+\mu-1}{\mu}[d-(r+\mu)n+r-1+\mu],\, 0\right). \tag{8.3.8}$$

Corollary 8.3.5 *A web $W(d,n,r)$ does not admit abelian r-equations, i.e., $\pi_r(d,n,r) = 0$ if and only if*

$$d \leq nr - r + 1 \tag{8.3.9}$$

Proof. It follows from (8.3.8). ■

In this book we will study the r-rank problems for webs $W(d,2,r)$, i.e., we will take $n=2$. Suppose again that the leaves of the foliations λ_ξ of a web $W(d,2,r)$ are given by the equations (8.1.1). Using (8.1.1), we can write an abelian r-equation for the web $W(d,2,r)$ in the form:

$$\alpha \underset{1}{\omega}^1 \wedge \ldots \wedge \underset{1}{\omega}^r + \beta \underset{2}{\omega}^1 \wedge \ldots \wedge \underset{2}{\omega}^r + \gamma(\underset{1}{\omega}^1 + \underset{2}{\omega}^1) \wedge \ldots \wedge (\underset{1}{\omega}^r + \underset{2}{\omega}^r)$$
$$+ \sum_{\alpha=4}^{d} \underset{\alpha}{\sigma}(\underset{\alpha}{\lambda}{}^1_j \underset{1}{\omega}^j + \underset{2}{\omega}^1) \wedge \ldots \wedge (\underset{\alpha}{\lambda}{}^r_j \underset{1}{\omega}^j + \underset{2}{\omega}^r) = 0, \tag{8.3.10}$$

where, according to Definition 8.3.2, each term is a closed r-form.

Equating to zero coefficients of $\underset{1}{\omega}^i \wedge \underset{2}{\omega}^2 \wedge \ldots \underset{2}{\omega}^r$, $i \neq 1$; $\underset{2}{\omega}^1 \wedge \underset{1}{\omega}^j \wedge \ldots \wedge \underset{2}{\omega}^r$, $j \neq 2; \ldots, \underset{1}{\omega}^1 \wedge \underset{2}{\omega}^2 \wedge \ldots \wedge \underset{2}{\omega}^r$, $\underset{2}{\omega}^1 \wedge \underset{1}{\omega}^2 \wedge \ldots \wedge \underset{2}{\omega}^r, \ldots$ in (8.3.10) we obtain

$$\sum_\alpha \underset{\alpha}{\sigma} \underset{\alpha}{\lambda}^i_j = 0, \quad i \neq j, \quad \gamma = -\sum_\alpha \underset{\alpha}{\sigma} \underset{\alpha}{\lambda}^i_i \quad \text{(no summation in } i\text{)}, \tag{8.3.11}$$

$$\sum_\alpha \underset{\alpha}{\sigma}(\underset{\alpha}{\lambda}^i_i - \underset{\alpha}{\lambda}^j_j) = 0, \quad i \neq j; \quad \text{(no summation in } i \text{ and } j\text{)}. \tag{8.3.12}$$

We will state now the theorem recently proved by J.B. Little [Lit 86].

Theorem 8.3.6 *Every maximum r-rank web $W(d,n,r)$, $r \geq 2$, $d > r(n-1)+2$, is almost Grassmannisable.* ■

Applying Theorem 8.3.6 to the webs $W(d,2,r)$, i.e., to the case $n=2$, we see that the webs $W(d,2,r)$, $r \geq 2$, $d > r+2$, are almost Grassmannisable. It follows from this result that in all these cases $d > 4$ since $r \geq 2$. So, to cover all possible cases of webs $W(d,2,r)$ of maximum r-rank, we must consider separately the cases $d=4$ and $d \leq r+2$.

For webs $W(4,2,r)$ we shall prove that such webs are almost Grassmannisable if they admit at least one abelian r-equation, i.e., they are of non-zero (not necessarily maximum) r-rank.

Proposition 8.3.7 *Webs $W(4,2,r)$ admitting at least one abelian r-equation are almost Grassmannisable.*

Proof. Suppose that $d=4$. If $\underset{4}{\sigma} = 0$, it follows from (8.3.10) that $\alpha = \beta = \gamma = 0$. In this case the $W(4,2,r)$ does not admit abelian r-equations. Suppose that $\underset{4}{\sigma} \neq 0$. The equations (8.3.11) and (8.3.12) imply (8.2.14) where $\underset{4}{\lambda} = \underset{4}{\lambda}^1_1 = \ldots = \underset{4}{\lambda}^r_r$. This means that webs $W(4,2,r)$ admitting at least one abelian r-equation are almost Grassmannisable. ■

If $d > 4$, the equations (8.3.11) and (8.3.12) are satisfied identically for almost Grassmannisable webs $AGW(d,2,r)$ of non-zero r-rank.

From now on we will restrict ourselves to almost Grassmannisable webs $AGW(d,2,r)$ of maximum r-rank. In them the equations (8.3.11) and (8.3.12) are satisfied identically. By Theorem 8.3.6 and Proposition 8.3.7, this assumption is natural for webs $W(d,2,r)$, $r \geq 2$, $d > r+2$, and webs $W(4,2,r)$ of maximum r-rank. It is still unknown whether the webs $W(d,2,r)$, $d = r+2$, of maximum r-rank are almost Grassmannisable or not.

Proposition 8.3.8 *The r-rank of a web $AGW(d,2,r)$ is equal to zero if $r > 2$ and $d \leq r+1$.*

Proof. For webs $AGW(d,2,r)$ we have the equations (8.2.14) and the equation

$$\gamma = -\sum_\alpha \underset{\alpha}{\sigma}\underset{\alpha}{\lambda} \qquad (8.3.13)$$

which follows from (8.3.11) and (8.2.14). In addition, comparison of coefficients of $\underset{1}{\omega}^1 \wedge \ldots \wedge \underset{1}{\omega}^r$ and $\underset{2}{\omega}^1 \wedge \ldots \wedge \underset{2}{\omega}^r$ gives by means of (8.3.13) and (8.2.14) that

$$\alpha = -\sum_\alpha \underset{\alpha}{\sigma}(\underset{\alpha}{\lambda} - \underset{\alpha}{\lambda}^r), \quad \beta = -\sum_\alpha \underset{\alpha}{\sigma}(\underset{\alpha}{\lambda} - 1). \qquad (8.3.14)$$

Let $r > 2$. Comparing coefficients of $\underset{1}{\omega}^1 \wedge \underset{1}{\omega}^2 \wedge \underset{2}{\omega}^3 \wedge \ldots \wedge \underset{2}{\omega}^r$, $\underset{1}{\omega}^1 \wedge \underset{1}{\omega}^2 \wedge \underset{1}{\omega}^3 \wedge \underset{2}{\omega}^4 \wedge \ldots \wedge \underset{2}{\omega}^r, \ldots$, we obtain the system

$$\sum_\alpha \underset{\alpha}{\sigma}(\underset{\alpha}{\lambda}^k - \underset{\alpha}{\lambda}) = 0, \quad k = 2, \ldots, r-1, \qquad (8.3.15)$$

of $r-2$ equations in $d-3$ quantities $\underset{\alpha}{\sigma}$. It is easy to see that if $d \leq r+1$, then rank $(\underset{\alpha}{\lambda}^k - \underset{\alpha}{\lambda}) = d-3$, and the system (8.3.15) has only the trivial solution. ■

By Proposition 8.3.8, almost Grassmannisable webs $AGW(d,2,r)$, $r > 2$, may have a non-zero r-rank if $r = 2$ or $r > 2$, $d > r+1$. We will study here only the case $r = 2$.

8.3.2 Almost Grassmannisable Webs $AGW(d,2,2)$ of Maximum 2-Rank

For webs $AGW(d,2,2)$ of non-zero r-rank we have equations (8.2.14), (8.3.13), and (8.3.14). In addition, we note that for $r = 2$ the torsion tensor a^i_{jk} has always the form (8.2.14) (see Section 8.1) and the structure equations (8.1.4) become (8.1.22). Further, as it was proved in Section 8.1, equations (8.2.14) imply (8.1.49) and (8.1.57).

In order to find the maximum 2-rank webs $AGW(d,2,2)$, we note that the left member of the abelian 2-equation (8.3.10) is the sum of d summands:

$$\sum_{\xi=1}^{d} \Omega_\xi = 0. \qquad (8.3.16)$$

In the case $r = 2$ we have (see (8.3.12), (8.3.13), and (8.3.14))

$$\begin{cases} \Omega_1 = \sum_\alpha \underset{\alpha}{\sigma}(\underset{\alpha}{\lambda} - \underset{\alpha}{\lambda}^2)\underset{1}{\omega}^1 \wedge \underset{1}{\omega}^2, \\ \Omega_2 = \sum_\alpha \underset{\alpha}{\sigma}(\underset{\alpha}{\lambda} - 1)\underset{2}{\omega}^1 \wedge \underset{2}{\omega}^2, \\ \Omega_3 = \sum_\alpha \underset{\alpha}{\sigma}\underset{\alpha}{\lambda}(\underset{1}{\omega}^1 + \underset{2}{\omega}^1) \wedge (\underset{1}{\omega}^2 + \underset{2}{\omega}^2), \\ \Omega_\alpha = \sum_\alpha \underset{\alpha}{\sigma}(\underset{\alpha}{\lambda}\underset{1}{\omega}^1 + \underset{2}{\omega}^1)(\underset{\alpha}{\lambda}\underset{1}{\omega}^2 + \underset{2}{\omega}^2), \quad \alpha = 4, \ldots, d \end{cases} \qquad (8.3.17)$$

where according to Definition 8.3.2, Ω_ξ, $\xi = 1, \dots, d$, are closed 2-forms.

Exterior differentiation of (8.3.17) gives the following independent cubic exterior equations:

$$\begin{cases} \sum_\alpha[(\underset{\alpha}{\lambda} - \underset{\alpha}{\lambda}^2)(d\underset{\alpha}{\sigma} - \underset{\alpha}{\sigma}\omega_i^i) - \underset{\alpha}{\sigma}(\underset{\alpha}{\lambda} - 1)(\underset{\alpha}{b_i} - \underset{\alpha}{\lambda} a_i)\underset{2}{\omega^i}] \wedge \underset{1}{\omega^1} \wedge \underset{1}{\omega^2} = 0, \\ \sum_\alpha[(\underset{\alpha}{\lambda} - 1)(d\underset{\alpha}{\sigma} - \underset{\alpha}{\sigma}\omega_i^i) + \underset{\alpha}{\sigma}\underset{\alpha}{\lambda}(\underset{a}{b_i} - a_i)\underset{1}{\omega^i}] \wedge \underset{2}{\omega^1} \wedge \underset{2}{\omega^2} = 0, \\ d\underset{\alpha}{\sigma} - \underset{\alpha}{\sigma}(\omega_k^k + 2a_i\underset{2}{\omega^i} + \underset{\alpha}{b_i}\underset{1}{\omega^i}) \wedge (\underset{\alpha}{\lambda}\underset{1}{\omega^1} + \underset{2}{\omega^1}) \wedge (\underset{\alpha}{\lambda}\underset{1}{\omega^2} + \underset{2}{\omega^2}) = 0. \end{cases} \tag{8.3.18}$$

It follows from (8.3.18) that

$$d\underset{\alpha}{\sigma} - \underset{\alpha}{\sigma}\omega_k^k - \underset{\alpha}{\sigma}(\underset{\alpha}{b_i}\underset{1}{\omega^i} + 2a_i\underset{2}{\omega^i}) = \underset{\alpha}{\sigma_i}(\underset{\alpha}{\lambda}\underset{1}{\omega^i} + \underset{2}{\omega^i}), \tag{8.3.19}$$

$$\sum_\alpha(\underset{\alpha}{\lambda} - \underset{\alpha}{\lambda}^2)\underset{\alpha}{\sigma_i} = \sum_\alpha \underset{\alpha}{\sigma}[(2\underset{\alpha}{\lambda} - 1)\underset{\alpha}{b_i} - \underset{\alpha}{\lambda} a_i]. \tag{8.3.20}$$

Note that two quantities $\underset{\alpha_0}{\sigma_i}$, α_0 fixed, out of the total number $2(d-3)$ quantities $\underset{\alpha}{\sigma_i}$, can be determined from (8.3.20) because $\underset{\alpha}{\sigma} \neq 0$ and $\lambda \neq 0, 1$.

Exterior differentiation of (8.3.19) and the application of Cartan's lemma give

$$\nabla\underset{\alpha}{\sigma_i} - \underset{\alpha}{\sigma_i}\omega_k^k + \underset{\alpha}{\sigma}(3f_{ik} + \underset{\alpha}{k_{ik}} - 2\underset{\alpha}{\sigma}_{(k}\underset{\alpha}{b_{i)}})\underset{1}{\omega^k} - 3\underset{\alpha}{\sigma_i}a_k\underset{2}{\omega^k} = \underset{\alpha}{\sigma_{ik}}(\underset{\alpha}{\lambda}\underset{1}{\omega^k} + \underset{2}{\omega^k}) \tag{8.3.21}$$

where $\nabla\underset{\alpha}{\sigma_i} = d\underset{\alpha}{\sigma_i} - \underset{\alpha}{\sigma_j}\omega_i^j$ and the $\underset{\alpha}{\sigma_{ij}}$ are symmetric with respect to i and j.

Differentiating (8.3.20) and using (8.3.21), we obtain, by means of the linear independence of the $\underset{1}{\omega^k}$ and $\underset{2}{\omega^k}$, the following equations:

$$\begin{cases} \sum_\alpha(\underset{\alpha}{\lambda}^2 - \underset{\alpha}{\lambda}^3)\underset{\alpha}{\sigma_{ik}} = \sum_\alpha\{(3\underset{\alpha}{\lambda}^2 - 2\underset{\alpha}{\lambda})(2\underset{\alpha}{\sigma}_{(i}\underset{\alpha}{b_{k)}} - \underset{\alpha}{\sigma}\underset{\alpha}{k_{ik}} + \underset{\alpha}{\sigma}(6\underset{\alpha}{\lambda} - 2)\underset{\alpha}{b_i}\underset{\alpha}{b_k} \\ \qquad -2\underset{\alpha}{\lambda}^2\underset{\alpha}{\sigma}_{(i}a_{k)} - 4\underset{\alpha}{\sigma}\underset{\alpha}{\lambda}^2\underset{\alpha}{b}_{(i}a_{k)} + \underset{\alpha}{\sigma}(\underset{\alpha}{\lambda} - \underset{\alpha}{\lambda}^2)f_{ik} + \underset{\alpha}{\sigma}\underset{\alpha}{\lambda}h_{ik}\}, \\ \sum_\alpha(\underset{\alpha}{\lambda} - \underset{\alpha}{\lambda}^2)\underset{\alpha}{\sigma_{ik}} = \sum_\alpha\{(2\underset{\alpha}{\lambda} - 1)[\underset{\alpha}{\sigma}(g_{ik} - \underset{\alpha}{k_{ik}}) + 2\underset{\alpha}{\sigma}_{(i}\underset{\alpha}{b_{k)}}] \\ \qquad +2\underset{\alpha}{\sigma}(\underset{\alpha}{b_i}\underset{\alpha}{b_k} - \underset{\alpha}{b}_{(i}a_{k)}) + \underset{\alpha}{\lambda}[\underset{\alpha}{\sigma}(h_{ik} - g_{ik}) - 2\underset{\alpha}{\sigma}_{(i}a_{k)}]\}. \end{cases} \tag{8.3.22}$$

Note that six quantities $\underset{\alpha_0}{\sigma_{ik}}$, $\underset{\beta_0}{\sigma_{ik}}$, $\alpha_0 \neq \beta_0$, α_0, β_0 fixed, out of the total number $3(d-3)$ quantities $\underset{\alpha}{\sigma_{ik}}$ can be determined from (8.3.22) since the corresponding determinant is equal to

$$\underset{\alpha_0}{\sigma}\underset{\beta_0}{\sigma}(\underset{\alpha_0}{\lambda} - \underset{\alpha_0}{\lambda}^2)(\underset{\beta_0}{\lambda} - \underset{\beta_0}{\lambda}^2)(\underset{\beta_0}{\lambda} - \underset{\alpha_0}{\lambda})$$

and different from zero because of $\underset{\alpha}{\sigma} \neq 0$, $\underset{\alpha_0}{\lambda} \neq \underset{\beta_0}{\lambda}$ and $\underset{\alpha}{\lambda} \neq 0, 1$.

Using the same procedure, we can get $3 \cdot 4$ out of the total number $4(d-3)$ quantities $\underset{\alpha}{\sigma_{ijk}}$, $4 \cdot 5$ out of the total number $5(d-3)$ quantities $\underset{\alpha}{\sigma_{ijkl}}$, etc. and all $(d-2)(d-3)$ quantities $\underset{\alpha}{\sigma_{i_1 i_2 \dots i_{d-3}}}$.

If we substitute all the $\underset{\alpha}{\sigma}_{i_1\ldots i_{d-3}}$ obtained from the differential equations for $\underset{\alpha}{\sigma}_{i_1\ldots i_{d-4}}$, we will have in these equations only functions whose differentials are known. Exterior differentiation of these equations gives equations of the form

$$A_{i_1\ldots i_{d-4}jk}\underset{1}{\omega}^j \wedge \underset{2}{\omega}^k = 0 \tag{8.3.23}$$

where $A_{i_1\ldots i_{d-4}jk}$ are linear functions of $\underset{\alpha}{\sigma}$, $\underset{\alpha}{\sigma}_i$, $\underset{\alpha}{\sigma}_{jk}$, $\ldots, \underset{\alpha}{\sigma}_{i_1\ldots i_{d-4}}$. Since all products $\underset{1}{\omega}^i \wedge \underset{2}{\omega}^k$ are linearly independent, we have

$$A_{i_1\ldots i_{d-4}jk} = 0. \tag{8.3.24}$$

Differentiation of (8.3.22) gives by means of the linear independence of the $\underset{1}{\omega}^j$ and $\underset{2}{\omega}^j$ two equations similar to (8.3.24), i.e., linear equations in $\underset{\alpha}{\sigma}$, $\underset{\alpha}{\sigma}_{i_1}, \ldots, \underset{\alpha}{\sigma}_{i_1\ldots i_{d-4}}$. Differentiation of these two equations leads to four new equations linear in $\underset{\alpha}{\sigma}$, $\underset{\alpha}{\sigma}_{i_1}, \ldots, \underset{\alpha}{\sigma}_{i_1\ldots i_{d-4}}$.

The 2-rank R_2 of a web $AGW(d,2,2)$ is equal to the number of independent Pfaffian equations (8.3.19), (8.3.21), ... which is

$$(d-3)\sum_{k=0}^{d-4}(k+1) - \sum_{k=1}^{d-4} k(k+1) - \varepsilon$$

where the first term is the number of equations (8.3.19), (8.3.21), ... , the second term is the number of relations (8.3.20), (8.3.22), ... and the last term is the number of independent equations in the system consisting of (8.3.24) and its differential consequences where all the quantities $\underset{\alpha}{\sigma}_{i_1}$, $\underset{\alpha}{\sigma}_{i_1 i_2}, \ldots, \underset{\alpha}{\sigma}_{i_1\ldots i_{d-4}}$ which can be found from (8.3.20), (8.3.22),.. are substituted.

This 2-rank has the maximum value $\pi_2(d,2,2)$ if and only if ε is minimal. It can be proved that the minimal value for ε is 0. This means that all coefficients of (8.3.24) vanish. In fact, we will see that the vanishing of the coefficient in $\underset{\alpha}{\sigma}_{i_1\ldots i_{d-4}}$ implies (8.1.81) if $d=4$ and (8.1.81), (8.1.37) if $d>4$, and the vanishing of all the other coefficients of (8.3.24) is a differential consequence of (8.1.81). Therefore, the maximum 2-rank

$$\pi_2(d,2,2) = (d-3)\sum_{k=0}^{d-4}(k+1) - \sum_{k=1}^{d-4} k(k+1) = (d-1)(d-2)(d-3)/6.$$

This matches Theorem 8.3.4 in the case $n=r=2$.

To clarify our considerations, we consider the cases $d=4$ and $d>4$ separately because, as we mentioned before, the conditions for webs $W(4,2,2)$ and $W(d,2,2)$, $d>4$ to be of maximum 2-rank differ for these two cases.

8.3.3 Webs $W(d,2,2)$, $d > 4$, of Maximum 2-Rank

Theorem 8.3.9 *A web $W(d,2,2)$, $d > 4$, is of maximum 2-rank $\pi_2(d,2,2)$ if and only if it is algebraisable.*

Proof. First, note that by Theorem 8.3.6, webs $W(d,2,2)$, $d > 4$, of maximum 2-rank are almost Grassmannisable. We will consider webs $W(5,2,2)$ – the general case $d > 4$ is similar. For $W(5,2,2)$ we have equations (8.3.20) and (8.3.22) where $\alpha = 4,5$. We can find $\underset{5}{\sigma}_i$ (or $\underset{4}{\sigma}_i$), $\underset{4}{\sigma}_{ik}$, and $\underset{5}{\sigma}_{ik}$ from these equations:

$$\underset{\alpha}{\sigma}_i = \Big\{\sum_{\gamma=4}^{5}\Big[(2\underset{\gamma}{\lambda} - 1)\underset{\gamma}{b} - \underset{\gamma}{\lambda} a_i\Big]\underset{\gamma}{\sigma} - (\underset{\beta}{\lambda} - \underset{\beta}{\lambda}^2)\underset{\beta}{\sigma}_i\Big\}/(\underset{\alpha}{\lambda} - \underset{\alpha}{\lambda}^2), \tag{8.3.25}$$

$$\begin{aligned}\underset{\alpha}{\sigma}_{ik} = \Big\{ & f_{ik}\sum_{\gamma}\underset{\gamma}{\sigma}(\underset{\gamma}{\lambda} - \underset{\gamma}{\lambda}^2) + g_{ik}\underset{\beta}{\lambda}\sum_{\gamma}\underset{\gamma}{\sigma}(1 - \underset{\gamma}{\lambda}) + h_{ik}(1 - \underset{\beta}{\lambda})\sum_{\gamma}\underset{\gamma}{\sigma}\underset{\gamma}{\lambda} \\ & +(\underset{\beta}{\sigma}\underset{\beta}{k}_{ik} - 2\underset{\beta}{\sigma}_i\underset{\beta}{b}_k)(\underset{\beta}{\lambda} - \underset{\beta}{\lambda}^2) + (3\underset{\alpha}{\lambda}^2 - 2\underset{\alpha}{\lambda} - 2\underset{\alpha}{\lambda}\underset{\beta}{\lambda} + \underset{\beta}{\lambda})(2\underset{\alpha}{\sigma}_{(i}\underset{\alpha}{b}_{k)} - \underset{\alpha}{\sigma}\underset{\alpha}{k}_{ik}) \\ & +\underset{\beta}{\sigma}\underset{\beta}{b}_i\underset{\beta}{b}_k(4\underset{\beta}{\lambda} - 2) + \underset{\alpha}{\sigma}\underset{\alpha}{b}_i\underset{\alpha}{b}_k(6\underset{\alpha}{\lambda} - 2\underset{\beta}{\lambda} - 2) + 2\underset{\alpha}{\sigma}_{(i}\underset{\alpha}{a}_{k)}(\underset{\alpha}{\lambda} - \underset{\beta}{\lambda}) \\ & -2\underset{\beta}{\sigma}\underset{\beta}{\lambda}\underset{\beta}{b}_{(i}a_{k)} + \underset{\alpha}{\sigma}\underset{\alpha}{b}_{(i}a_{k)}(2\underset{\beta}{\lambda} - 4\underset{\alpha}{\lambda})\Big\}/[(\underset{\alpha}{\lambda} - \underset{\alpha}{\lambda}^2)(\underset{\alpha}{\lambda} - \underset{\beta}{\lambda})]\end{aligned} \tag{8.3.26}$$

where $\beta \neq \alpha$. We have now only two independent equations (8.3.21) for $\underset{4}{\sigma}_i$ – two others are their consequences because of (8.3.25). Substituting $\underset{4}{\sigma}_{ik}$ from (8.3.26) into these equations and taking exterior derivatives of the equations obtained, we get equations of the form (8.3.23) and (8.3.24). The latter one is:

$$\begin{aligned}&\underset{4}{\sigma}_t(a^t_{ijk} + 3\delta^{(t}_i\delta^l_j\delta^{m)}_k K_{lm})(\underset{4}{\lambda} - \underset{4}{\lambda}^2)(\underset{4}{\lambda} - \underset{5}{\lambda}) \\ &\quad +\underset{4}{\sigma}[\underset{1}{\phi}_{ijk} - \underset{5}{\lambda}\underset{2}{\phi}_{ijk} + 3(\underset{4}{\lambda} - \underset{5}{\lambda})(1 - 2\underset{4}{\lambda})(\underset{4}{b}_{(i}K_{jk)} - \underset{4}{b}_{(i}\underset{4}{k}_{jk)}) \\ &\quad -3(\underset{4}{\lambda} - \underset{4}{\lambda}^2)(\underset{5}{b}_{(i}K_{jk)} - \underset{5}{b}_{(i}\underset{5}{k}_{jk)} + 3\underset{1}{\lambda}(\underset{1}{\lambda} - \underset{5}{\lambda})(a_{(i}K_{jk)} - a_{(i}h_{jk)})] \\ &\quad +\underset{5}{\sigma}(\underset{5}{\lambda} - \underset{5}{\lambda}^2)[\underset{1}{\theta}_{ijk} - \underset{4}{\lambda}\underset{2}{\theta}_{ijk} - 3\underset{4}{b}_{(i}K_{jk)} + 3\underset{4}{b}_{(i}\underset{4}{k}_{jk)}] = 0\end{aligned} \tag{8.3.27}$$

where

$$\begin{cases}\underset{1}{\phi}_{ijk} = (\underset{4}{\lambda} - \underset{4}{\lambda}^2)\underset{1}{f}_{ijk} + \underset{5}{\lambda}(1 - \underset{4}{\lambda})\underset{1}{g}_{ijk} + \underset{4}{\lambda}(1 - \underset{5}{\lambda})\underset{1}{h}_{ijk} \\ \qquad +(-3\underset{4}{\lambda}^2 + 2\underset{4}{\lambda} + 2\underset{4}{\lambda}\underset{5}{\lambda} - \underset{5}{\lambda})(\underset{4}{\lambda}\underset{4}{k}_{ijk} + 3\underset{4}{b}_{(i}\underset{4}{k}_{jk)} + \underset{4}{b}_t a^t_{ijk} + \underset{1}{g}_{ijk} - \underset{4}{\lambda}\underset{2}{f}_{ijk}), \\ \underset{2}{\phi}_{ijk} = \underset{5}{\lambda}^{-1}\Big[(3\underset{5}{\lambda} - 2\underset{4}{\lambda})(\underset{4}{\lambda} - \underset{4}{\lambda}^2)\underset{2}{f}_{ijk} + \underset{5}{\lambda}(\underset{4}{\lambda} - \underset{4}{\lambda}^2)\underset{2}{g}_{ijk} + \underset{4}{\lambda}^2(1 - \underset{5}{\lambda})\underset{2}{h}_{ijk} \\ \qquad -\underset{4}{\lambda}^2(2\underset{4}{\lambda} - \underset{5}{\lambda} - 1)\underset{4}{k}_{ijk}\Big], \\ \underset{1}{\theta}_{ijk} = \underset{1}{f}_{ijk} + \underset{1}{g}_{ijk} + \underset{1}{h}_{ijk} + \underset{5}{\lambda}\underset{5}{k}_{ijk} + 3\underset{5}{b}_{(i}\underset{5}{k}_{jk)} + \underset{5}{b}_t a^t_{ijk} + \underset{1}{g}_{ijk} - \underset{5}{\lambda}\underset{2}{f}_{ijk}, \\ \underset{2}{\theta}_{ijk} = \underset{2}{f}_{ijk} + \underset{2}{g}_{ijk} + \underset{2}{h}_{ijk} + \underset{5}{k}_{ijk}.\end{cases} \tag{8.3.28}$$

As we mentioned earlier, differentiation of (8.3.28) gives us two new equations linear in $\underset{4}{\sigma}_i$, $\underset{4}{\sigma}$, $\underset{5}{\sigma}$. Their differentiation gives four new equations, etc. For a web $AGW(5,2,2)$ of maximum 2-rank $\underset{4}{\sigma}$ and $\underset{5}{\sigma}$ are independent and our system is a system for $\underset{4}{\sigma}_i$ only. The 2-rank of $AGW(5,2,2)$ is equal to the number of independent equations from (8.3.19) and (8.3.21). By means of (8.3.25), the number of independent equations from (8.3.21) is $4-\varepsilon$ where ε is the rank of the matrix of the coefficients of the system in $\underset{4}{\sigma}_i$ which we discussed above. It is obvious that $0 \leq \varepsilon \leq 2$.

We obtain the conditions for a maximum 2-rank web $AGW(5,2,2)$ requiring $\varepsilon = 0$ if possible. This means that the equation (8.3.27) must vanish. It follows from (8.3.27) that in this case we have first of all

$$\underset{4}{\sigma}_t (a^t_{ijk} + 3\delta^{(t}_i \delta^k_j \delta^{m)}_k K_{lm}) = 0. \tag{8.3.29}$$

Consider the eight equations (8.3.29) for all possible values $i,j,k = 1,2$. We obtain the following conditions:

$$\begin{cases} a^1_{111} + 3K_{11} = 0, & a^2_{111} = 0, \\ a^2_{222} + 3K_{22} = 0, & a^1_{222} = 0, \\ a^1_{112} + 2K_{12} = 0, & a^1_{211} + K_{11} = 0, \\ a^2_{212} + 2K_{12} = 0, & a^1_{122} + K_{22} = 0. \end{cases} \tag{8.3.30}$$

The relations (8.1.31) show that conditions (8.3.30) imply

$$K_{ij} = 0, \qquad a^i_{jkl} = 0, \tag{8.3.31}$$

i.e., conditions (8.1.81) and (8.1.37).

Let us show that under conditions (8.3.31) two other terms of (8.3.27) vanish. In fact, (8.3.31) implies (8.1.42), (8.1.86), (8.1.87), (8.1.88), and therefore

$$\begin{cases} \underset{1}{\phi}_{ijk} - \underset{5}{\lambda}\underset{2}{\phi}_{ijk} = 3\underset{5}{\lambda}(\underset{4}{\lambda} - \underset{5}{\lambda}) a_{(i}h_{jk)} + 3(\underset{4}{\lambda} - \underset{5}{\lambda})(1 - 2\underset{5}{\lambda}) \underset{4}{b}_{(i}\underset{4}{k}_{jk)} - 3(\underset{4}{\lambda} - \underset{4}{\lambda}^2) \underset{5}{b}_{(i}\underset{5}{k}_{jk)} \\ \underset{1}{\theta}_{ijk} - \underset{4}{\lambda}\underset{2}{\theta}_{ijk} = -3\underset{4}{b}_{(i}\underset{4}{k}_{jk)}. \end{cases} \tag{8.3.32}$$

The equations (8.3.32) prove the vanishing two last terms of (8.3.27).

Thus, conditions (8.3.31) are necessary and sufficient for a web $W(5,2,2)$ to be of maximum 2-rank $\pi_2(5,2,2) = 4$. The conditions (8.3.31) together with Theorem 8.1.17 prove that a web $W(5,2,2)$ of maximum 2-rank is algebraisable. ■

8.3.4 Four-Webs $W(4,2,2)$ of Maximum 2-Rank

Theorem 8.3.10 *An isoclinic web $W(4,2,2)$ is of maximum 2-rank if and only if it is almost algebraisable. A non-isoclinic web $W(4,2,2)$ is of maximum 2-rank if and only if it is almost Grassmannisable web and any of the four affine connections mentioned in Theorem* 8.1.18 *is equiaffine.*

Proof. For webs $W(4,2,2)$ of maximum 2-rank we have $\underset{4}{\sigma} \neq 0$ and we can express all $\underset{4}{\sigma}_i$ from (8.3.20). Substituting them into (8.3.19), we obtain

$$d\underset{4}{\sigma} - \underset{4}{\sigma}\omega_k^k - \underset{4}{\sigma}(\underset{4}{b}_i\underset{1}{\omega}^i + 2a_i\underset{2}{\omega}^i) = \underset{4}{\sigma}[(2\underset{4}{\lambda} - 1)\underset{4}{b}_i - \underset{4}{\lambda}a_i](\underset{4}{\lambda}\underset{1}{\omega}^i + \underset{2}{\omega}^i)/(\underset{4}{\lambda} - \underset{4}{\lambda}^2)$$

or

$$d\ln\underset{4}{\sigma} = \omega_k^k - d\ln(\underset{4}{\lambda} - 1) + (a_i - \underset{4}{b}_i/\underset{4}{\lambda})\underset{2}{\omega}^i. \tag{8.3.33}$$

By Proposition 8.3.7, webs $W(4,2,2)$ of maximum 2-rank are almost Grassmannisable. According to Theorem 8.1.10, we must distinguish two cases:

a) *A web $AGW(4,2,2)$ is isoclinic.* In this case we have (8.1.26), (8.1.59), (8.1.33), and (8.1.60). Exterior differentiation of (8.3.33) gives

$$(b_{kij}^k - h_{ij} + \underset{4}{k}_{ij})\underset{1}{\omega}^i \wedge \underset{2}{\omega}^j = 0. \tag{8.3.34}$$

Since the $\underset{1}{\omega}^i \wedge \underset{2}{\omega}^j$ are linearly independent, it follows from (8.3.34) that

$$b_{kij}^k - h_{ij} + \underset{4}{k}_{ij} = 0. \tag{8.3.35}$$

But (8.1.29) and (8.1.31) show that

$$b_{kij}^k = f_{ij} + g_{ij} + 2h_{ij}. \tag{8.3.36}$$

It follows from (8.3.36) that condition (8.3.35) can be written in the form (8.1.81). Therefore, isoclinic webs $W(4,2,2)$ of maximum 2-rank are characterised by (8.2.14), (8.1.26), (8.1.57), and (8.1.81). Under these conditions the equation (8.3.33) is completely integrable and its solution depends on one constant. This means that the maximum 2-rank of isoclinic webs $W(4,2,2)$ is equal to one: $\pi_2(4,2,2) = 1$, and an isoclinic web $W(4,2,2)$ is of maximum 2-rank if and only if it is almost algebraisable.

b) *A web $W(4,2,2)$ is not isoclinic.* In this case we have (8.1.19) and (8.1.56). Exterior differentiation of (8.3.33) gives

$$(b_{kij}^k - \underset{4}{b}_{ij} + q_{ij})\underset{1}{\omega}^i \wedge \underset{2}{\omega}^j = 0. \tag{8.3.37}$$

Equation (8.3.37) implies (8.1.99). Therefore, non-isoclinic webs $W(4,2,2)$ of maximum 2-rank are characterised by equations (8.2.14), (8.1.19), (8.1.56), and (8.1.99). Under these conditions equation (8.3.33) is completely integrable and maximal 2-rank is again equal to one: $\pi_2(4,2,2) = 1$.

According to Theorem 8.1.24, a non-isoclinic web $W(4,2,2)$ is of maximum 2-rank if and only if any of four affine connections of Theorem 8.1.18 is equiaffine. ■

Note that in both cases which were considered in the proof of Theorem 8.3.10, we can write equation (8.3.19) in the following form:

$$d\ln[\underset{4}{\sigma}(\underset{4}{\lambda} - 1)] = \omega_k^k + (a_i - \underset{4}{b}_i/\underset{4}{\lambda})\underset{2}{\omega}^i. \tag{8.3.38}$$

Under conditions (8.1.49), (8.1.26), (8.1.59), and (8.1.81) or (8.1.19), (8.1.56), and (8.1.99) the exterior differential of the right member of (8.3.38) vanishes. This means that under any of these conditions equation (8.3.38) is completely integrable. The conditions mentioned above distinguish two classes of webs $W(4,2,2)$ of maximum 2-rank.

We are able to write now the only abelian 2-equation for both kinds of webs $W(4,2,2)$ of maximum 2-rank:

$$(\underset{4}{\lambda} - \underset{4}{\lambda}^2)\underset{4}{\sigma}\underset{1}{\omega}^1 \wedge \underset{1}{\omega}^2 + (\underset{4}{\lambda} - 1)\underset{4}{\sigma}\underset{2}{\omega}^1 \wedge \underset{2}{\omega}^2 - \underset{4}{\lambda}\underset{4}{\sigma}(\underset{1}{\omega}^1 + \underset{2}{\omega}^1) \wedge (\underset{1}{\omega}^2 + \underset{2}{\omega}^2)$$
$$+\underset{4}{\sigma}(\underset{4}{\lambda}\underset{1}{\omega}^1 + \underset{2}{\omega}^1) \wedge (\underset{4}{\lambda}\underset{1}{\omega}^2 + \underset{2}{\omega}^2) = 0. \qquad (8.3.39)$$

This equation holds for any $W(4,2,2)$ – it is an identity – but it is an abelian 2-equation if and only if the function $\underset{4}{\sigma}$ is a solution of (8.3.33) and a web $W(4,2,2)$ is almost algebraisable (satisfies (8.2.14), (8.1.49), (8.1.26), (8.1.59), (8.1.81)) or a web $W(4,2,2)$ satisfies conditions (8.2.14), (8.1.19), (8.1.56), (8.1.99).

This situation is similar to one for $W(3,2,r)$: the equation $\underset{1}{\omega}^i + \underset{2}{\omega}^i + \underset{3}{\omega}^i = 0$ holds for any $W(3,2,r)$. However, it will be an abelian 1-equation only for parallelisable webs – for them this equation can be reduced to the form $du_1^i + du_2^i + du_3^i = 0$ (if $r = 1$, the equation $\underset{1}{\omega} + \underset{2}{\omega} + \underset{3}{\omega} = 0$ can be reduced to the form $du_1 + du_2 + du_3 = 0$ if and only if a web $W(3,2,1)$ is hexagonal).

We conclude this section by the following remark:

Remark 8.3.11 The case $d = 4$ for isoclinic webs differs from the case $d > 4$ since the final Pfaffian equations which do not contain new functions (equations (8.3.33) for $d = 4$ and (8.3.21), (8.3.27) for $d = 5$) contain only the contracted connection form ω_k^k for $d = 4$ and contain the forms ω_j^i and ω_k^k for $d > 4$. After exterior differentiation in the first case only the contracted curvature tensor b_{kij}^k appears whose expression does not contain the tensor a_{jkl}^i (because of (8.1.37)) while in the second case the general curvature tensor b_{jkl}^i appears whose expression contains the tensor a_{jkl}^i.

The webs $W(4,2,2)$ of maximum 2-rank are exceptional in the sense that they are not necessarily algebraisable while the webs $W(d,2,2)$, $d > 4$, of maximum 2-rank are algebraisable.

In Section 8.4 we shall prove the existence of webs $W(4,2,2)$ of maximum 2-rank by giving examples of both kinds (isoclinic and non-isoclinic) of webs $W(4,2,2)$ of maximum 2-rank. Their existence disproves P.A. Griffiths' conjecture that webs $W(d,n,r)$ of maximum r-rank are algebraisable.

8.4 Examples of Webs $W(4,2,2)$ of Maximum 2-Rank

8.4.1 The Isoclinic Case

We will construct here examples of isoclinic webs $W(4,2,2)$ of maximum 2-rank. By Theorem 8.3.4, such webs are almost Grassmannisable. We found in Section 8.1 that the leaves of the foliation λ_ξ of an almost Grassmannisable web $AGW(4,2,2)$ are determined by the equations $\underset{\xi}{\omega}^i = 0$ where

$$-\underset{3}{\omega}^i = \underset{1}{\omega}^i + \underset{2}{\omega}^i, \quad -\underset{4}{\omega}^i = \lambda \underset{1}{\omega}^i + \underset{2}{\omega}^i, \quad \lambda \neq 0, 1. \tag{8.4.1}$$

In addition, if a web $AGW(4,2,2)$ is isoclinic, we found that we have for it equations (8.1.17), (8.1.5), (8.1.26), (8.1.59), (8.1.31), and (8.1.32).

Using the notations (8.1.64):

$$\underset{1}{k}_{ij} = f_{ij}, \ \underset{2}{k}_{ij} = g_{ij}, \ \underset{3}{k}_{ij} = h_{ij},$$

we will write here all these equations:

$$\begin{cases} d\underset{1}{\omega}^i = \underset{1}{\omega}^j \wedge \omega_j^i + a_j \underset{1}{\omega}^j \wedge \underset{1}{\omega}^i, \\ d\underset{2}{\omega}^i = \underset{2}{\omega}^j \wedge \omega_j^i - a_j \underset{2}{\omega}^j \wedge \underset{2}{\omega}^i, \end{cases} \tag{8.4.2}$$

$$d\omega_j^i - \omega_j^k \wedge \omega_k^i = b^i_{jkl} \underset{1}{\omega}^k \wedge \underset{1}{\omega}^l, \tag{8.4.3}$$

$$d\lambda = \lambda(b_i - a_i)\underset{1}{\omega}^i + (b_i - \lambda a_i)\underset{2}{\omega}^i, \tag{8.4.4}$$

$$da_i - a_j \omega_i^j = (\underset{1}{k}_{ij} - \underset{3}{k}_{ij})\underset{1}{\omega}^j + (\underset{2}{k}_{ij} - \underset{3}{k}_{ij})\underset{2}{\omega}^j, \tag{8.4.5}$$

$$db_i - b_j \omega_i^j = [b_i(b_j - a_j) + \lambda(\underset{1}{k}_{ij} - \underset{4}{k}_{ij})]\underset{1}{\omega}^j + (\underset{2}{k}_{ij} - \underset{4}{k}_{ij})\underset{2}{\omega}^j, \tag{8.4.6}$$

$$a^i_{ikl} = 0, \tag{8.4.7}$$

$$\underset{3}{k}_{ij} = \frac{3}{4} b^k_{kij} - \underset{1}{k}_{ij} - \underset{2}{k}_{ij}. \tag{8.4.8}$$

Recall that the $\underset{\xi}{k}_{ij}$, $i,j = 1,2,;\ \xi = 1,2,3,4$, are symmetric in i and j and the quantities

$$a^i_{jk} = a_{[j}\delta^i_{k]}, \tag{8.4.9}$$

$$b^i_{jkl} = a^i_{jkl} + \underset{1}{k}_{jk}\delta^i_l + \underset{2}{k}_{lj}\delta^i_k + \underset{3}{k}_{kl}\delta^i_j, \tag{8.4.10}$$

are the torsion and curvature tensors of such $W(4,2,2)$ as well as of its 3-subweb $[1,2,3]$.

Exterior differentiation of (8.4.5) and (8.4.3) where b^i_{jkl} are substituted from (8.4.10) gives

$$(\nabla \underset{1}{k}_{ij} - \nabla \underset{3}{k}_{ij}) \wedge \underset{1}{\omega}^j + (\nabla \underset{2}{k}_{ij} - \nabla \underset{3}{k}_{ij}) \wedge \underset{2}{\omega}^j + a_k b^k_{ijm} \underset{1}{\omega}^j \wedge \underset{2}{\omega}^m$$
$$+(\underset{1}{k}_{ij} - \underset{3}{k}_{ij}) a_m \underset{1}{\omega}^m \wedge \underset{1}{\omega}^j - (\underset{2}{k}_{ij} - \underset{3}{k}_{ij}) a_m \underset{2}{\omega}^m \wedge \underset{2}{\omega}^j = 0, \qquad (8.4.11)$$

$$[\nabla a^i_{sjm} + \nabla \underset{1}{k}_{sj} \delta^i_m + \nabla \underset{2}{k}_{ms} \delta^i_j + \underset{3}{k}_{jm} \delta^i_s + a_l b^i_{sjm} (\underset{1}{\omega}^l - \underset{2}{\omega}^l)] \underset{1}{\omega}^j \wedge \underset{2}{\omega}^m = 0, \qquad (8.4.12)$$

where

$$\nabla \underset{\alpha}{k}_{ij} = d\underset{\alpha}{k}_{ij} - \underset{\alpha}{k}_{mj} \omega^m_i - \underset{\alpha}{k}_{im} \omega^m_j, \quad \alpha = 1,2,3,$$
$$\nabla a^i_{jkm} = da^i_{jkm} - a^i_{lkm} \omega^l_j - a^i_{jlm} \omega^l_k - a^i_{jkl} \omega^l_m + a^l_{jkm} \omega^i_l.$$

Contracting (8.4.12) with respect to i and j and using (8.4.7), we obtain

$$[\nabla \underset{1}{k}_{jm} + \nabla \underset{2}{k}_{jm} + 2\nabla \underset{3}{k}_{jm} + a_l (\underset{1}{k}_{jm} + \underset{2}{k}_{jm} + 2\underset{3}{k}_{jm})(\underset{1}{\omega}^l - \underset{2}{\omega}^l)] \wedge \underset{1}{\omega}^j \wedge \underset{2}{\omega}^m = 0. \quad (8.4.13)$$

It follows from (8.4.11) and (8.4.13) that the $\underset{\alpha}{k}_{ij}$ have the form:

$$\nabla \underset{\alpha}{k}_{ij} = \underset{\alpha 1}{k}_{ijm} \omega^m + \underset{\alpha 2}{k}_{ijm} \omega^m, \quad \alpha = 1,2,3, \qquad (8.4.14)$$

where

$$\underset{11}{k}_{i[jm]} - \underset{31}{k}_{i[jm]} + (\underset{1}{k}_{i[j} - \underset{3}{k}_{i[j}) a_{m]} = 0, \qquad (8.4.15)$$

$$\underset{22}{k}_{i[jm]} - \underset{32}{k}_{i[jm]} - (\underset{2}{k}_{i[j} - \underset{3}{k}_{i[j}) a_{m]} = 0, \qquad (8.4.16)$$

$$a_s b^s_{ijm} = \underset{12}{k}_{ijm} - \underset{32}{k}_{ijm} - \underset{21}{k}_{imj} + \underset{31}{k}_{imj}, \qquad (8.4.17)$$

$$\underset{11}{k}_{i[jm]} + \underset{21}{k}_{i[jm]} + 2\underset{31}{k}_{i[jm]} + (\underset{1}{k}_{i[j} + \underset{2}{k}_{i[j} + 2\underset{3}{k}_{i[j}) a_{m]} = 0, \qquad (8.4.18)$$

$$\underset{12}{k}_{i[jm]} + \underset{22}{k}_{i[jm]} + 2\underset{32}{k}_{i[jm]} - (\underset{1}{k}_{i[j} + \underset{2}{k}_{i[j} + 2\underset{3}{k}_{i[j}) a_{m]} = 0. \qquad (8.4.19)$$

Alternating (8.4.17) first with respect to i and k and next with respect to i and m and using (8.4.10), we obtain respectively

$$\underset{21}{k}_{m[ij]} - \underset{31}{k}_{m[ij]} = a_{[i}(\underset{2}{k}_{j]m} - \underset{3}{k}_{j]m}), \qquad (8.4.20)$$
$$\underset{12}{k}_{j[im]} - \underset{32}{k}_{j[im]} = a_{[i}(-\underset{1}{k}_{m]j} + \underset{3}{k}_{m]j}). \qquad (8.4.21)$$

Substituting $\underset{1s}{k}_{i[jm]}$ and $\underset{2s}{k}_{i[jm]}$ from (8.4.15), (8.4.16), (8.4.20), and (8.4.21) into (8.4.18) and (8.4.19), we get

$$\underset{31}{k}_{i[jm]} = -\underset{3}{k}_{i[j} a_{m]}, \quad \underset{32}{k}_{i[jm]} = \underset{3}{k}_{i[j} a_{m]}. \qquad (8.4.22)$$

Equations (8.4.22), (8.4.15), (8.4.16), (8.4.20), and (8.4.21) give

$$\underset{\alpha 1}{k}{}_{i[jm]} = -\underset{\alpha}{k}{}_{i[j}a_{m]}, \quad \underset{\alpha 2}{k}{}_{i[jm]} = \underset{\alpha}{k}{}_{i[j}a_{m]}, \alpha = 1,2,3. \tag{8.4.23}$$

Now, because of (8.4.23), all the equations (8.4.15), (8.4.16), (8.4.18), (8.4.19), (8.4.20), and (8.4.21) become identities.

Equations (8.4.12), (8.4.14), and (8.4.23) imply

$$\nabla a^i_{sjm} = \underset{1}{a}{}^i_{sjml}\underset{1}{\omega}{}^l + \underset{2}{a}{}^i_{sjml}\underset{2}{\omega}{}^l, \tag{8.4.24}$$

where

$$\underset{1}{a}{}^i_{sm[jl]} + a^i_{sm[j}a_{l]} = (\underset{21}{k}{}_{ms[j} + \underset{2}{k}{}_{sm}a_{[j})\delta^i_{l]}, \tag{8.4.25}$$

$$\underset{2}{a}{}^i_{sj[ml]} - a^i_{sj[m}a_{l]} = (\underset{12}{k}{}_{sj[m} - \underset{1}{k}{}_{sj}a_{[m})\delta^i_{l]}, \tag{8.4.26}$$

By Proposition 8.1.8, an isoclinic web $AGW(4,2,2)$ is transversally geodesic (and therefore Grassmannisable) if and only if condition (8.1.37) holds:

$$a^i_{jkl} = 0. \tag{8.4.27}$$

For a Grassmannisable web $W(4,2,2)$ the equations (8.4.24) and (8.4.27) imply

$$\underset{1}{a}{}^i_{jklm} = 0, \quad \underset{2}{a}{}^i_{jklm} = 0. \tag{8.4.28}$$

The equations (8.4.25), (8.4.26), (8.4.27), and (8.4.28) show that for a Grassmannisable web $W(4,2,2)$ we have

$$(\underset{1}{k}{}_{ij[m} - \underset{1}{k}{}_{ij}a_{[m})\delta^i_{l]} = 0, \quad (\underset{2}{k}{}_{ij[m} + \underset{2}{k}{}_{ij}a_{[m})\delta^i_{l]} = 0. \tag{8.4.29}$$

It follows from (8.4.29) that for a Grassmannisable $W(4,2,2)$ the following identities hold:

$$\underset{1}{k}{}_{ijm} = \underset{1}{k}{}_{ij}a_m, \quad \underset{2}{k}{}_{ijm} = -\underset{2}{k}{}_{ij}a_m. \tag{8.4.30}$$

Note that starting from (8.4.11), we were dealing with the isoclinic 3-subweb [1,2,3] of an isoclinic $AGW(4,2,2)$ formed by the foliations λ_1, λ_2, and λ_3. Note also that in Section 8.1 we discussed similar conditions for an isoclinic web $AGW(d,2,r)$. However, in Section 8.1 we used slightly different notation.

In Section 8.3 we proved that an isoclinic web $W(4,2,2)$ is of maximum 2-rank if and only if it is almost algebraisable, its maximum 2-rank is equal to one (Theorem 8.3.10), and that the only abelian 2-equation for a web $W(4,2,2)$ of maximum 2-rank has the form (8.3.39):

$$(\lambda - \lambda^2)\sigma\underset{1}{\omega}{}^1 \wedge \underset{1}{\omega}{}^2 + (\lambda - 1)\sigma\underset{2}{\omega}{}^1 \wedge \underset{2}{\omega}{}^2 - \lambda\sigma(\underset{1}{\omega}{}^1 + \underset{2}{\omega}{}^1) \wedge (\underset{1}{\omega}{}^2 + \underset{2}{\omega}{}^2)$$
$$+\sigma(\lambda\underset{1}{\omega}{}^1 + \underset{2}{\omega}{}^1) \wedge (\lambda\underset{1}{\omega}{}^2 + \underset{2}{\omega}{}^2) = 0, \tag{8.4.31}$$

where σ is a solution of the completely integrable equation (8.3.33):

$$d\ln[\sigma(\lambda-1)] = \omega_i^i + (a_i - b_i/\lambda)\underset{2}{\omega}^i. \tag{8.4.32}$$

Note that the equation (8.4.32) is an identity, and it is an abelian 2-equation for an isoclinic web $W(4,2,2)$ only under the condition (8.4.32) and the condition (8.1.81) of almost algebraisability which for $d=4$ has the form:

$$\sum_{\xi=1}^{4} \underset{\xi}{k_{ij}} = 0. \tag{8.4.33}$$

In our construction of examples of webs $W(4,2,2)$ of maximum 2-rank we will depart from a given isoclinic three-web $W(3,2,2)$ and extend it to an almost algebraisable web $AAW(4,2,2)$ which, as we indicated above, will be a web $W(4,2,2)$ of maximum 2-rank.

If an isoclinic three-web $W(3,2,2)$ is given, this means that the forms $\underset{1}{\omega}^i$, $\underset{2}{\omega}^i$, $\underset{3}{\omega}^i$, ω_j^i and functions a_i, $\underset{\alpha}{k_{ij}}$, $\alpha = 1,2,3$, a_{jk}^i, b_{jkl}^i, a_{jkl}^i satisfying equations (8.4.1), (8.4.2), (8.4.3), (8.4.5), (8.4.9), (8.4.10), (8.4.7), (8.4.8), (8.4.14), (8.4.17), (8.4.23), (8.4.24), (8.4.25), and (8.4.26) are given.

To construct an $AAW(4,2,2)$, we should find functions λ, b_i, and $\underset{4}{k_{ij}}$ satisfying (8.4.4), (8.4.6), and (8.4.33) and eventually find equations of the fourth foliation λ_4 integrating the system

$$\lambda\underset{1}{\omega}^i + \underset{2}{\omega}^i = 0. \tag{8.4.34}$$

Suppose that three foliations λ_1, λ_2, and λ_3 of the isoclinic three-web are given as the level sets $u_\alpha^i = \text{const.}$, $\alpha = 1,2,3$, of the following functions:

$$\lambda_1 : u_1^i = x^i; \quad \lambda_2 : u_2^i = y^i; \quad \lambda_3 : u_3^i = f^i(x^j, y^k), \quad i,j,k = 1,2. \tag{8.4.35}$$

Let us now indicate four steps which we will perform to extend the isoclinic three-web (8.4.35) to an $AAW(4,2,2)$.

Step 1. Find the forms $\underset{\alpha}{\omega}^i$, $\alpha = 1,2,3$, ω_j^i, and the functions a_{jk}^i, b_{jkl}^i, a_i, a_{jkl}^i, $\underset{\alpha}{k_{ij}}$.

The forms $\underset{\alpha}{\omega}^i$, ω_j^i and the functions a_{jk}^i and b_{jkl}^i can be found my means of the following formulae (see [AS 71a] or [AS 81], cf. also formulae (8.1.102) and (8.1.103)):

$$\underset{1}{\omega}^i = \bar{f}_j^i dx^j, \quad \underset{2}{\omega}^i = \tilde{f}_j^i dy^j, \quad \underset{3}{\omega}^i = -du_3^i, \tag{8.4.36}$$

where

$$\bar{f}_j^i = \partial f^i/\partial x^j, \quad \tilde{f}_j^i = \partial f^i/\partial y^j, \quad \det(\bar{f}_j^i) \neq 0, \quad \det(\tilde{f}_j^i) \neq 0,$$

and

$$d\underset{1}{\omega}^i = -d\underset{2}{\omega}^i = \Gamma_{jk}^i \underset{1}{\omega}^j \wedge \underset{2}{\omega}^k, \tag{8.4.37}$$

$$\Gamma_{jk}^i = (-\partial^2 f^i/\partial x^l \partial y^m)\bar{g}_j^l \tilde{g}_k^m, \tag{8.4.38}$$

$$\omega_j^i = \Gamma_{kj}^i \underset{1}{\omega^k} + \Gamma_{jk}^i \underset{2}{\omega^k}, \tag{8.4.39}$$

$$a_{jk}^i = \Gamma_{[jk]}^i, \tag{8.4.40}$$

$$\begin{aligned} b_{jkl}^i &= \frac{1}{2}\Big(\frac{\partial \Gamma_{kl}^i}{\partial x^m}\bar{g}_j^m + \frac{\partial \Gamma_{jl}^i}{\partial x^m}\bar{g}_k^m - \frac{\partial \Gamma_{kj}^i}{\partial y^m}\tilde{g}_l^m - \frac{\partial \Gamma_{kl}^i}{\partial y^m}\tilde{g}_j^m \\ &\quad + \Gamma_{jl}^m \Gamma_{km}^i - \Gamma_{kj}^m \Gamma_{ml}^i + 2\Gamma_{kl}^m a_{mj}^i\Big). \end{aligned} \tag{8.4.41}$$

As for the functions a_i, $\underset{\alpha}{k_{ij}}$, $\alpha = 1,2,3$, and a_{jkl}^i, they can be easily calculated using (8.4.11), (8.4.5), (8.4.8), and (8.4.10).

Step 2. Find $\underset{4}{k_{ij}}$, λ, and b_i.

The functions $\underset{4}{k_{ij}}$ can be found from (8.4.33). In order to find λ and b_i, we will take exterior derivatives of (8.4.6) using (8.4.2)–(8.4.6) and (8.4.14). Equating to zero the coefficients in $\underset{1}{\omega^j} \wedge \underset{1}{\omega^k}$, $\underset{2}{\omega^j} \wedge \underset{2}{\omega^k}$ and $\underset{1}{\omega^j} \wedge \underset{2}{\omega^k}$, by means of (8.4.23) we get two identities and

$$\begin{aligned} \lambda[(\underset{1}{k_{ij}} - \underset{4}{k_{ij}})a_l - (\underset{12}{k_{ijl}} - \underset{42}{k_{ijl}})] &= -b_m a_{ijl}^m - 3b_{(j}\underset{4}{k_{il)}} \\ &\quad - a_j(\underset{2}{k_{il}} - \underset{4}{k_{il}}) - (\underset{21}{k_{ilj}} - \underset{41}{k_{ilj}}). \end{aligned} \tag{8.4.42}$$

In general, the equation (8.4.42) gives a dependence among λ, b_1, and b_2. Differentiating it by means of (8.4.4)–(8.4.6), (8.4.24), (8.4.14) and their prolongations and equating to zero coefficients in linearly independent forms $\underset{1}{\omega^i}$ and $\underset{2}{\omega^i}$, we get new relations among λ, b_1, and b_2. Some of them may be satisfied identically. Others should be checked for their compatibility among each other and with (8.4.42). If no contradiction exists, the same procedure which has been applied to (8.4.42) should be applied to these new relations until all three quantities λ, b_1, and b_2 will be found or no new relations among them will appear.

The following cases are possible:

i) The relations obtained among λ, b_1, and b_2 are not compatible. This means that the given isoclinic three-web (8.4.35) cannot be extended to an $AAW(4,2,2)$.

ii) The relations among λ, b_1, and b_2 allow us to find s of these functions, $s = 0,1,2,3$. In this case other $3-s$ of these functions should be found by integrating the completely integrable system (8.4.4), (8.4.6). Its solution will depend on $3-s$ constants. In particular, if $s=0$, all the functions, λ, b_1, and b_2, will be uniquely determined and the given isoclinic three-web (8.4.35) can be uniquely extended to an $AAW(4,2,2)$.

Step 3. Find the equations of the fourth foliation λ_4 of an $AAW(4,2,2)$ by integrating the completely integrable system (8.4.34) where the $\underset{1}{\omega^i}$ and $\underset{2}{\omega^i}$ are determined by (8.4.36) and λ is determined in Step 2.

Step 4. Find the only abelian 2-equation of the web $AAW(4,2,2)$. For this:

(a) find σ by integrating (8.4.32);

(b) write the abelian 2-equation in the form (8.4.31) substituting λ from Step 2 and σ from (a) into (8.4.31); and

(c) write the abelian 2-equation in the form of (8.3.2) expressing $\underset{\xi}{\omega^1} \wedge \underset{\xi}{\omega^2}$, $\xi = 1,2,3,4$, in terms of $du^1_\xi \wedge du^2_\xi$.

Example 8.4.1 Suppose that a given isoclinic three-web is algebraisable. In this case we have (8.4.27), (8.4.28), (8.4.30), and (8.1.81) where $d = 3$:

$$\underset{1}{k_{ij}} + \underset{2}{k_{ij}} + \underset{3}{k_{ij}} = 0. \tag{8.4.43}$$

Equations (8.4.43) and (8.4.33) imply that

$$\underset{4}{k_{ij}} = 0. \tag{8.4.44}$$

It is clear that the web $AAW(4,2,2)$ – the extension of the given $W(3,2,2)$ – is algebraisable since the given three-web and consequently its extension, the four-web $AAW(4,2,2)$, are transversally geodesic. Moreover, by Theorem 7.7.1, it follows from the equations (8.4.43) and (8.4.44) that, up to equivalence, our algebraisable web $AW(4,2,r)$ is generated by an algebraic surface V_4^2 of degree four which is decomposed into a cubic surface V_3^2 and a plane V_1^2.

Conditions (8.4.44), (8.4.27), (8.4.28), (8.4.30), and (8.4.43) show that equation (8.4.42) has the form $0 \cdot \lambda = 0$. Thus, the extension $AW(4,2,2)$ of the given algebraisable three-web $AW(3,2,2)$ is determined by the completely integrable system (8.4.4), (8.4.6). Therefore, it depends on three constants. One can consider the coefficients of the equation of V_1^2 as these constants.

Thus, *an algebraisable web $AW(3,2,2)$ can be extended to an algebraisable web $AW(4,2,2)$ of a special kind.*

Example 8.4.2 Consider the three-web $W(3,2,2)$ defined by (see [B 35]):

$$\begin{cases} \lambda_1 : x^1 = \text{const.}, \quad x^2 = \text{const.}; \\ \lambda_2 : y^1 = \text{const.}, \quad y^2 = \text{const.}; \\ \lambda_3 : u_3^1 = x^1 + y^1 = \text{const.}, \quad u_3^2 = (x^2+y^2)(y^1 - x^1) = \text{const.} \end{cases} \tag{8.4.45}$$

Step 1. Using (8.4.36)–(8.4.41), (8.4.9), (8.4.10), and (8.4.8), we obtain for a web

(8.4.45):

$$\begin{cases} \Gamma^1_{ij} = \Gamma^2_{22} = 0, \quad \Gamma^2_{11} = 2(x^2+y^2)/(x^1-y^1), \quad \Gamma^2_{21} = -\Gamma^2_{12} = 1/(x^1-y^1); \\ \omega^1_i = 0, \quad \omega^2_1 = (dx^1+dy^1)(x^2+y^2)/(x^1-y^1) - (dx^2-dy^2), \quad \omega^2_2 = -d\ln(x^1-y^1); \\ a_1 = 2/(y^1-x^1), \quad a_2 = 0, \quad p_{2i} = q_{2i} = 0, \quad p_{11} = -q_{11} = 2/(x^1-y^1)^2; \\ b^1_{ijk} = b^2_{222} = b^2_{211} = b^2_{122} = b^2_{212} = b^2_{211} = 0, \\ b^2_{112} = -b^2_{121} = 2/(x^1-y^1)^2, \\ b^2_{111} = -8(x^2+y^2)/(x^1-y^1)^2, \\ \underset{3}{k}_{ij} = 0, \quad \underset{1}{k}_{11} = -\underset{2}{k}_{11} = 2/(x^1-y^1)^2, \\ \underset{1}{k}_{ij} = \underset{2}{k}_{ij} = 0, \quad (i,j) \neq (1,1); \\ a^1_{ijk} = 0, \quad a^2_{111} = b^2_{111}, \quad a^2_{ijk} = 0, \quad (i,j,k) \neq (1,1,1). \end{cases} \tag{8.4.46}$$

Equations (8.4.46) show that the three-web (8.4.45) is isoclinic and not transversally geodesic since the tensor a_i does not vanish. Since for this web we have

$$\underset{1}{k}_{ij} + \underset{2}{k}_{ij} = 0, \quad \underset{3}{k}_{ij} = 0, \tag{8.4.47}$$

it is natural to call such a web an *almost Bol three-web* (cf. Theorem 7.7.3 where algebraic Bol 3-subwebs were discussed).

Step 2. It follows from (8.4.33) and (8.4.47) that

$$\underset{4}{k}_{ij} = 0. \tag{8.4.48}$$

The equations (8.4.46) and (8.4.47) imply

$$\begin{cases} \underset{3s}{k}_{ijm} = \underset{4s}{k}_{ijm} = 0, \quad s = 1,2; \quad \underset{1s}{k}_{ijk} = \underset{2s}{k}_{ijk} = 0, \quad (i,j,k) \neq (1,1,1), \\ \underset{11}{k}_{111} = -\underset{12}{k}_{111} = -\underset{21}{k}_{111} = \underset{22}{k}_{111} = 4/(y^1-x^1)^3. \end{cases} \tag{8.4.49}$$

By means of (8.4.46), (8.4.48), and (8.4.49), the equations (8.4.42) can be written as

$$\lambda = 1 + b_2 u_3^2, \tag{8.4.50}$$

where u_3^2 is defined by (8.4.45). Differentiation of (8.4.50) implies three identities and

$$b_1 = (2 + b_2 u_3^2)/(y^1 - x^1). \tag{8.4.51}$$

Differentiation of (8.4.51) gives no new relations among λ, b_1, and b_2.

Equation (8.4.6) for $i = 2$ can be written by means of (8.4.46), (8.4.48), (8.4.50), and (8.4.51) in the form

$$d\ln[b_2(x^1-y^1)] = b_2(y^1-x^1)dx^2. \tag{8.4.52}$$

Integration of (8.4.52) gives

$$b_2 = 1/[(x^1-y^1)(x^2+C)], \tag{8.4.53}$$

where C is a constant. The equations (8.4.50), (8.4.51), and (8.4.53) imply

$$\lambda = (C - y^2)/(x^2 + C), \tag{8.4.54}$$

$$b_1 = (2C + x^2 - y^2)/[(x^2 + C)(y^1 - x^1)]. \tag{8.4.55}$$

We can see that *an extension of the isoclinic three-web* (8.4.45) *to an* $AAW(4,2,2)$ *depends on one constant.*

Step 3. Substituting λ from (8.4.54) and $\underset{1}{\omega}^i$, $\underset{2}{\omega}^i$ from (8.4.36) into (8.4.34), we have

$$-\alpha dx^1 + dy^1 = 0, \quad 2(1+\alpha)dy^1 + (y^1 - x^1)d\alpha = 0, \tag{8.4.56}$$

where $\alpha = (y^2 - C)/(x^2 + C)$. It follows from (8.4.56) that

$$\frac{dx^1}{y^1 - x^1} = \frac{dy^1}{\alpha(y^1 - x^1)} = \frac{d\alpha}{-2\alpha(1+\alpha)}. \tag{8.4.57}$$

The two independent first integrals of (8.4.57) give a system defining the foliation λ_4 of the $AAW(4,2,2)$:

$$\begin{cases} u_4^1 = (x^1 - y^1)^2(x^2 + y^2)^2/[(x^2 + C)(y^2 - C)] = \text{const.}, \\ u_4^2 = x^1 + y^1 + [(y^1 - x^1)(x^2 + y^2)/\sqrt{(x^2 + C)(y^2 - C)}] \cdot \\ \cdot \arctan \sqrt{(y^2 - C)(x^2 + C)} = \text{const.} \end{cases} \tag{8.4.58}$$

Step 4. Using (8.4.46), (8.4.54), (8.4.55), and (8.4.53), we can write equation (8.4.32) in the form:

$$d \ln[\sigma(x^2 + y^2)/(x^2 + C)] = -d \ln[(y^1 - x^1)(C - y^2)]. \tag{8.4.59}$$

It follows from (8.4.59) that

$$\sigma = A(x^2 + C)/[(x^2 + y^2)(y^1 - x^1)(C - y^2)], \tag{8.4.60}$$

where A is a constant. Taking $A = 1$, we get from (8.4.60) that

$$\sigma = (x^2 + C)/[(x^2 + y^2)(y^1 - x^1)(C - y^2)]. \tag{8.4.61}$$

By means of (8.4.61) and (8.4.54), the only abelian 2-equation (8.4.31) for our web $AAW(4,2,2)$ can be written in the form

$$\Omega_1 + \Omega_2 + \Omega_2 + \Omega_4 = 0 \tag{8.4.62}$$

where

$$\begin{cases} \Omega_1 = [1/((y^1 - x^1)(x^2 + C))]\underset{1}{\omega}^1 \wedge \underset{1}{\omega}^2, \\ \Omega_2 = [1/((y^1 - x^1)(y^2 - C))]\underset{2}{\omega}^1 \wedge \underset{2}{\omega}^2, \\ \Omega_3 = [((x^1 - y^1)(x^2 + y^2))]\underset{3}{\omega}^1 \wedge \underset{3}{\omega}^2, \\ \Omega_4 = [(x^2 + C)/((y^1 - x^1)(C - y^2)(x^2 + y^2))]\underset{4}{\omega}^1 \wedge \underset{4}{\omega}^2, \end{cases}$$

and each of the Ω_ξ, $\xi = 1, 2, 3, 4$, is a closed 2-form (see Definition 8.3.2).

Using (8.4.45), (8.4.46), (8.4.1), and (8.4.54), we find

$$\begin{cases} \underset{1}{\omega}^1 \wedge \underset{1}{\omega}^2 = (y^1 - x^1) dx^1 \wedge dx^2, \\ \underset{2}{\omega}^1 \wedge \underset{2}{\omega}^2 = (y^1 - x^1) dy^1 \wedge dy^2, \\ \underset{3}{\omega}^1 \wedge \underset{3}{\omega}^2 = du_3^1 \wedge du_3^2, \\ \underset{4}{\omega}^1 \wedge \underset{4}{\omega}^2 = [(y^2 - C)^2/(2(y^1 - x^1)(x^2 + y^2))] du_4^1 \wedge du_4^2. \end{cases} \tag{8.4.63}$$

The equations (8.4.63) allow us to write the abelian 2-equation (8.4.62) in the form (8.3.2):

$$(1/x^2)dx^1 \wedge dx^2 + (1/y^2)dy^1 \wedge dy^2 - (1/u_3^2)du_3^1 \wedge du_3^2 - (1/(2u^1))du_4^1 \wedge du_4^2 = 0. \tag{8.4.64}$$

Example 8.4.3 Suppose that a three-web $W(3,2,2)$ is given by

$$\begin{cases} \lambda_1 : \ x^1 = \text{const.}, \quad x^2 = \text{const.}; \\ \lambda_2 : \ y^1 = \text{const.}, \quad y^2 = \text{const.}; \\ \lambda_3 : \ u_3^1 = x^1 + y^1 = \text{const.}, \quad u_3^2 = -x^1 y^2 + x^2 y^1 = \text{const.} \end{cases} \tag{8.4.65}$$

Step 1. By means of (8.4.36)–(8.4.41), (8.4.9), (8.4.10), and (8.4.8), we obtain for a web (8.4.65):

$$\begin{cases} \Gamma_{ij}^1 = \Gamma_{22}^2 = 0, \quad \Gamma_{11}^2 = x^2/x^1 - y^2/y^1, \quad \Gamma_{12}^2 = -1/x^1, \quad \Gamma_{21}^2 = -1/y^1; \\ a_1 = 1/y^1 - 1/x^1, \quad a_2 = 0, \quad p_{12} = q_{12} = p_{21} = q_{21} = p_{22} = q_{22} = 0, \\ p_{11} = 1/(x^1)^2, \ \ q_{11} = -1/(y^1)^2; \quad b_{111}^2 = (1/y^1 - 1/x^1)(x^2/x^1 - y^2/y^1), \\ b_{121}^2 = -1/(y^1)^2, \ b_{112}^2 = -1/(x^1)^2, \ \ b_{ijk}^1 = b_{211}^2 = b_{122}^2 = b_{212}^2 = b_{221}^2 = 0, \\ \underset{3}{k}_{11} = (1/y^1)^2 - 1/(x^1)^2)/4, \ \ \underset{1}{k}_{11} = (3/(x^1)^2 + 1/(y^1)^2)4, \\ \underset{2}{k}_{11} = -(1/(x^1)^2 + 3/(y^1)^2)/4, \ \ \underset{\alpha}{k}_{ij} = 0, \ \ (i,j) \neq (1,1); b_{222}^2 = 0, \\ a_{111}^1 = -a_{211}^2 = -a_{111}^2 = -a_{112}^2 = (1/(y^1)^2 - 1/(x^1)^2)/4, \\ a_{111}^2 = b_{111}^2, \ \ a_{ijk}^1 = a_{122}^2 = a_{212}^2 = a_{221}^2 = a_{222}^2 = 0, (i,j,k) \neq (1,1,1). \end{cases} \tag{8.4.66}$$

It follows from (8.4.66) that the web (8.4.65) is isoclinic and not transversally geodesic. Moreover, since the $b_{(jkl)}^i \neq 0$, it is not a hexagonal web (see Table 1.1, Section 1.2).

Step 2. The equations (8.4.33) and (8.4.66) imply that

$$\underset{4}{k}_{11} = \underset{3}{k}_{11}, \quad \underset{4}{k}_{ij} = 0, \quad (i,j) \neq (1,1) \tag{8.4.67}$$

Equations (8.4.66) and (8.4.67) give

$$\begin{cases} \underset{21}{k}_{111} = \underset{31}{k}_{111} = \underset{41}{k}_{111} = -\dfrac{1}{3} \underset{11}{k}_{111} = 1/(2(x^1)^3), \\ \underset{22}{k}_{111} = \underset{32}{k}_{111} = \underset{42}{k}_{111} = -\dfrac{1}{3} \underset{12}{k}_{111} = 1/(2(y^1)^3), \\ \underset{as}{k}_{ijk} = 0, \quad a = 1, 2, 3, 4; \quad s = 1, 2; \quad (i,j,k) \neq (1,1,1). \end{cases} \tag{8.4.68}$$

By virtue of (8.4.66), (8.4.67), and (8.4.68), the equations (8.4.42) can be written in the form

$$\lambda = (x^1)^2[1/(y^1)^2 - b_1(1/x^1 + 1/y^1) + b_2(x^2/x^1 - y^2/y^1)]. \tag{8.4.69}$$

Differentiation of (8.4.69) leads to an identity.

Therefore, the extension $AAW(4,2,2)$ of the given isoclinic three-web (8.4.65) is defined by the completely integrable system (8.4.4), (8.4.6). Equation (8.4.69) shows that system contains only two independent equations (8.4.6).

Thus, *the extension* $AAW(4,2,2)$ *depends on two constants.*

We will integrate the system (8.4.4), (8.4.6), (8.4.69) and find explicit expressions of λ, b_1, and b_2. For this, using (8.4.66) and (8.4.69), we write (8.4.6) in the form:

$$\begin{aligned} db_1 &= [1/(y^1)^2 + b_1(b_1 - 2/y^1) - (b_1 - 1/y^1)b_2y^2]dx^1 \\ &\quad +(b_1 - 1/y^1)b_2y^1dx^2 - (b_2y^2 + 1/y^1)dy^1/y^1 + b_2dy_2, \end{aligned} \tag{8.4.70}$$

$$db_2 = b_2(b_1 - b_2y^2 - 1/y^1)dx^1 + b_2^2y^1dx^2 - b_2dy^1/y^1. \tag{8.4.71}$$

It follows from (8.4.70) and (8.4.71) that

$$d[(b_1 - 1/y^1)/(b_2y^1)] = d(y^2/y^1). \tag{8.4.72}$$

Equation (8.4.72) gives

$$b_1 = b_2(y^2 + C_1y^1) + 1/y^1 \tag{8.4.73}$$

where C_1 is a constant.

Substituting b_1 from (8.4.73) into (8.4.71), we easily obtain

$$d(b_2y^1)/(b_2y^1)^2 = C_1dx^1 + dx^2. \tag{8.4.74}$$

It follows from (8.4.74) that

$$b_2 = -(x_2 + C_1x^1 + C_2)^{-1}/y^1 \tag{8.4.75}$$

where C_2 is a constant.

Equation (8.4.75) allows us to express b_1 determined by (8.4.73) in the form

$$b_1 = [C_2 + C_1(x^1 - y^1) + x^2 - y^2](x^2 + C_1x^1 + C_2)^{-1}/y^1. \tag{8.4.76}$$

The equations (8.4.69), (8.4.75), (8.4.76) give the following expression for λ :

$$\lambda = x^1(y^2 + C_1y^1 - C_2)(x^2 + C_1x^1 + C_2)^{-1}/y^1. \tag{8.4.77}$$

If we take $C_1 = C_2 = 0$, the equations (8.4.75), (8.4.76), and (8.4.77) become

$$b_2 = -1/(x^2y^1), \quad b_1 = (1 - y^2/x^2)/y^1 \quad \lambda = x^1y^2/(y^1x^2). \tag{8.4.78}$$

Step 3. Define α by

$$\alpha = (y^2 + C_1 y^1 - C_2)/(x^2 + C_1 x^1 + C_2). \tag{8.4.79}$$

Then $\lambda = \alpha x^1/y^1$, and using (8.4.66), we can write (8.4.34) in the form

$$\begin{cases} (\alpha x^1/y^1)dx^1 + dy^1 = 0, \\ (1+\alpha)dy^1 - x^1 d\alpha = 0. \end{cases} \tag{8.4.80}$$

The equations (8.4.80) can be also written in the form

$$\frac{dx^1}{-y^1} = \frac{dy^1}{x^1\alpha} = \frac{d\alpha}{\alpha(1+\alpha)}. \tag{8.4.81}$$

If we denote the common value of the expressions in (8.4.81) by dt/t, then we find

$$t = \alpha/(1+\alpha) \tag{8.4.82}$$

and

$$\frac{dx^1}{dt} = -\frac{y^1}{t}, \quad \frac{dy^1}{dt} = \frac{x^1}{1-t}. \tag{8.4.83}$$

Eliminating y^1, we get from (8.4.83)

$$t(t-1)\frac{d^2x^1}{dt^2} + (t-1)\frac{dx^1}{dt} - x^1 = 0, \tag{8.4.84}$$

or

$$\frac{d}{dt}[t(t-1)\frac{dx^1}{dt} - tx^1] = 0. \tag{8.4.85}$$

The equations (8.4.83), (8.4.85), and (8.4.82) give

$$\begin{cases} x^1 = -A(1 + \frac{1}{1+\alpha}\ln|\alpha|) - B\frac{1}{1+\alpha}, \\ y^1 = A(1 - \frac{\alpha}{1+\alpha}\ln|\alpha|) - B\frac{1}{1+\alpha}, \end{cases} \tag{8.4.86}$$

where A and B are arbitrary constants.

Solving (8.4.86) for A and B, we obtain two independent first integrals of (8.4.80) defining the foliation λ_4 of the web $AAW(4,2,2)$:

$$\begin{cases} u_4^1 = (u_3^2 + C_1 u_3^1)/(x^2 + y^2 + C_1 u_3^2) = \text{const.}, \\ u_4^2 = -u_4^1 \ln|(y^2 + C_1 y^1 - C_2)/(x^2 + C_1 x^1 + C_2)| - u_3^1 = \text{const.} \end{cases} \tag{8.4.87}$$

Step 4. The equations (8.4.66), (8.4.75), (8.4.76), and (8.4.77) allow us to write equation (8.4.32) in the form

$$d\ln[x^1 y^1(\lambda - 1)\sigma] = d\ln[y^1/(y^2 + C_1 y^1 - C_2)]. \tag{8.4.88}$$

Integrating (8.4.88) and taking an appropriate constant of integration, we get

$$\sigma = 1/[x^1(\lambda - 1)(y^2 + C_1 y^1 - C_2)]. \tag{8.4.89}$$

By means of (8.4.89) and (8.4.77), the only abelian 2-equation (8.4.31) for our web $AAW(4,2,2)$ has the form

$$\Omega_1 + \Omega_2 + \Omega_3 + \Omega_4 = 0 \tag{8.4.90}$$

where

$$\Omega_1 = -[1/(y^1(x^2 + C_1 x^1 + C_2))]\underset{1}{\omega^1} \wedge \underset{1}{\omega^2}, \quad \Omega_3 = (u_3^2 + C_2 u_3^1)^{-1}\underset{3}{\omega^1} \wedge \underset{3}{\omega^2},$$
$$\Omega_2 = [1/(x^1(y^2 + C_1 y^1 - C_2))]\underset{2}{\omega^1} \wedge \underset{2}{\omega^2}, \quad \Omega_4 = -(u_3^2 + C_2 u_3^1)^{-1}\lambda^{-1}\underset{4}{\omega^1} \wedge \underset{4}{\omega^2},$$

and each of the Ω_ξ, $\xi = 1,2,3,4$, is a closed 2-form (see Definition 8.3.2).

Using (8.4.65), (8.4.66), (8.4.1), (8.4.77), and (8.4.87), we find

$$\begin{cases} \underset{1}{\omega^1} \wedge \underset{1}{\omega^2} = y^1 dx^1 \wedge dx^2, & \underset{3}{\omega^1} \wedge \underset{3}{\omega^2} = du_3^1 \wedge du_3^2, \\ \underset{2}{\omega^1} \wedge \underset{2}{\omega^2} = -x^1 dy^1 \wedge dy^2, & \underset{4}{\omega^1} \wedge \underset{4}{\omega^2} = \lambda(x^2 + y^2 + C_1 u_3^1) du_4^1 \wedge du_4^2. \end{cases} \tag{8.4.91}$$

The equations (8.4.91) allow us to write the abelian 2-equation (8.4.90) in the form of (8.3.2):

$$-\frac{1}{x^2 + C_1 x^1 + C_2} dx^1 \wedge dx^2 - \frac{1}{y^2 + C_1 y^1 - C_2} dy^1 \wedge dy^2$$
$$+ \frac{1}{u_3^2 + C_2 u_3^1} du_3^1 \wedge du_3^2 - \frac{1}{u_4^1} du_4^1 \wedge du_4^2 = 0. \tag{8.4.92}$$

8.4.2 The Non-Isoclinic Case

Now we shall construct an example of a non-isoclinic web $W(4,2,2)$ of maximum 2-rank. By Theorem 8.3.10, a non-isoclinic web $W(4,2,2)$ is of maximum 2-rank if and only if it is almost Grassmannisable and any of the four affine connections indicated in Theorem 8.3.10 is equiaffine. So, for such a web $W(4,2,2)$ we again have the equations (8.4.1)–(8.4.4) where $b^i_{[jkl]} = 0$ and the equations and inequalities (8.1.19), (8.1.20), (8.1.56), (8.1.90), (8.1.92)–(8.1.94),(8.1.96), (8.1.97) and (8.1.99). Let us write the latter equations:

$$da_i - a_j \omega_i^j = p_{ij}\underset{1}{\omega^j} + q_{ij}\underset{2}{\omega^j}, \tag{8.4.93}$$

$$b^i_{[j|l|k]} = \delta^i_{[k} p_{j]l}, \quad b^i_{[jk]l} = \delta^i_{[k} q_{j]l}, \tag{8.4.94}$$

$$db_i - b_j \omega_i^j = [b_i(b_j - a_j) + \lambda(b_{ji} + p_{ij} - q_{ji})]\underset{1}{\omega^j} + b_{ij}\underset{2}{\omega^j}, \tag{8.4.95}$$

$$b_{[ij]} = p_{[ij]} = \lambda q_{[ij]}, \tag{8.4.96}$$

$$\nabla p_{ij} = \underset{1}{p}_{ijk}\underset{1}{\omega}^k + \underset{2}{p}_{ijk}\underset{2}{\omega}^k, \quad \nabla q_{ij} = \underset{1}{q}_{ijk}\underset{1}{\omega}^k + \underset{2}{q}_{ijk}\underset{2}{\omega}^k, \tag{8.4.97}$$

$$\begin{cases} \underset{1}{p}_{i[jk]} + p_{i[j}a_{k]} = 0, \quad \underset{2}{q}_{i[jk]} - q_{i[j}a_{k]} = 0, \\ \underset{2}{p}_{ijk} - \underset{2}{q}_{ijk} + a_m b^m_{ijk} = 0, \end{cases} \tag{8.4.98}$$

$$dp = p\omega^i_i + \underset{1}{p}_i\underset{1}{\omega}^i + \underset{2}{p}_i\underset{2}{\omega}^i, \quad dq = q\omega^i_i + \underset{1}{q}_i\underset{1}{\omega}^i + \underset{2}{q}_i\underset{2}{\omega}^i, \tag{8.4.99}$$

$$p \neq 0, \ \ q \neq 0, \ \ p \neq q, \tag{8.4.100}$$

$$q(q\underset{1}{p}_i - p\underset{1}{q}_i) - p(q\underset{2}{p}_i - p\underset{2}{q}_i) = pq(p-q)a_i, \tag{8.4.101}$$

$$b^k_{kij} = b_{ij} - q_{ij}, \tag{8.4.102}$$

where

$$\nabla p_{ij} = dp_{ij} - p_{kj}\omega^k_i - p_{ik}\omega^k_j, \quad \nabla q_{ij} = dq_{ij} - q_{kj}\omega^k_i - q_{ik}\omega^k_j,$$

$$p_{[12]} = p \qquad q_{[12]} = q,$$

$$\underset{k}{p}_i = \underset{k}{p}_{[12]i}, \quad \underset{k}{q}_i = \underset{k}{q}_{[12]i}.$$

Recall that the conditions (8.4.100) and (8.4.101) guarantee that a non-isoclinic three-web $W(3,2,2)$ can be uniquely extended to a non-isoclinic $AGW(4,2,2)$ and that the condition (8.4.102) is necessary and sufficient for the $AGW(4,2,2)$ to be of maximum 2-rank.

As was so for isoclinic webs $W(4,2,2)$ of maximum 2-rank, the only abelian 2-equation for a non-isoclinic web $AW(4,2,2)$ of maximum 2-rank has the form (8.4.31) where σ is a solution of the completely integrable equation (8.4.32). Note that the equation (8.4.32) is an identity, and it is an abelian 2-equation for a non-isoclinic web $W(4,2,2)$ only under the conditions (8.4.32) and (8.4.102).

Let us describe the consecutive steps which should be performed for finding a non-isoclinic web $W(4,2,2)$ of maximum 2-rank.

Step 1. *Find a non-isoclinic three-web* $W(3,2,2)$ *satisfying* (8.4.100) *and* (8.4.101).

As in the isoclinic case, we will suppose that three foliations λ_1, λ_2 and λ_3 of the non-isoclinic web $W(3,2,2)$ are given as level sets $u^i_\alpha = \text{const}$, $\alpha = 1,2,3$, of the functions indicated in (8.4.35). The forms $\underset{\alpha}{\omega}^i$, ω^i_j and the functions a^i_{jk} and b^i_{jkl} can be found my means of the formulae (8.4.36)–(8.4.41). The functions a_i, p_{ij}, q_{ij}, p, q, $\underset{k}{p}_i$, and $\underset{k}{q}_i$ can be easily calculated from (8.4.9), (8.4.93), (8.1.89), and (8.4.99).

Step 2. *Extend the web from Step* 1 *to a non-isoclinic web* $W(4,2,2)$ by finding λ, b_1, and b_2 from (8.4.95), (8.4.5), and (8.4.94), and eventually by finding the equations of the foliation λ_4 from the system $\lambda\underset{1}{\omega}^i + \underset{2}{\omega}^i = 0$ (cf. (8.4.34)) where the forms $\underset{1}{\omega}^i$ and $\underset{2}{\omega}^i$ are determined in Step 1 and the function λ in Step 2.

Step 3. *Check whether the condition* (8.4.102) *is satisfied or not.* If the answer is yes, find the only abelian 2-equation of the web $W(4,2,2)$.

For this:

(a) find σ by integrating (8.4.32);

(b) write the abelian 2-equation in the form (8.4.31) substituting λ from Step 2 and σ from (a) into (8.4.31); and

(c) write the abelian 2-equation in the form of (8.3.2) expressing $\underset{\xi}{\omega^1} \wedge \underset{\xi}{\omega^2}$, $\xi = 1,2,3,4$, in terms of $du^1_\xi \wedge du^2_\xi$.

We have already found examples of non-isoclinic three-webs $W(3,2,2)$ which cannot be extended to a non-isoclinic four-webs $W(4,2,2)$ since for them one of the conditions (8.4.100) or (8.4.101) fails (see Examples 8.1.26–8.1.29).

We will give one more example for which the condition $p \neq q$ fails. This example includes a wide class of three-webs.

Example 8.4.4 *Polynomial three-webs.* We will call a web $W(3,2,2)$ *polynomial* if it is defined by

$$X_3 : u^i_3 = x^i + y^i + c^i_{jk} x^j y^k = \text{const.}, \quad c^i_{jk} = \text{const.} \tag{8.4.103}$$

Note that the Taylor expansions of the functions $u^i_3 = f^i(x^j, y^k)$ can be always reduced to a canonical form (see [Ak 69b]; cf. Section 3.5). The equation (8.4.103) is a particular case of this canonical form when it contains only terms of the first and the second degree and the coefficients are constants.

For a polynomial web, using (8.4.36)–(8.4.40) and (8.4.9), we obtain

$$\left\{\begin{aligned} a_1 = &\ [c^2_{21} - c^2_{12} + (c^2_{11}c^2_{22} - c^2_{12}c^2_{21})(y^1 - x^1) + (c^1_{12}c^2_{11} - c^1_{11}c^2_{12})x^1 \\ &+ (c^1_{22}c^2_{11} - c^1_{21}c^2_{12})x^2 + (c^1_{11}c^2_{21} - c^1_{21}c^2_{11})y^1 \\ &+ (c^1_{12}c^2_{21} - c^1_{22}c^2_{11})y^2]/(\Delta_1\Delta_2), \\ a_2 = &\ [c^1_{12} - c^1_{21} + (c^1_{11}c^1_{22} - c^1_{12}c^1_{21})(y^2 - x^2) + (c^1_{22}c^2_{11} - c^1_{21}c^2_{12})x^1 \\ &+ (c^1_{22}c^2_{21} - c^1_{21}c^2_{22})x^2 + (c^1_{12}c^2_{21} - c^1_{22}c^2_{11})y^1 \\ &+ (c^1_{12}c^2_{22} - c^1_{22}c^2_{12})y^2]/(\Delta_1\Delta_2), \end{aligned}\right. \tag{8.4.104}$$

where

$$\begin{aligned} \Delta_1 &= (1 + c^1_{1i}y^i)(1 + c^2_{2j}y^j) - c^2_{1i}c^1_{2j}y^i y^j, \\ \Delta_2 &= (1 + c^1_{i1}x^i)(1 + c^2_{j2}x^j) - c^2_{i1}c^1_{j2}x^i x^j. \end{aligned}$$

Using (8.1.89) and (8.4.104), we calculate p and q and conclude that the condition

$$(c^2_{22} + c^1_{12})(c^2_{12} - c^2_{21}) + (c^1_{11} + c^2_{21})(c^1_{12} - c^1_{21}) \neq 0 \tag{8.4.105}$$

is sufficient to satisfy $p \neq 0$ and $q \neq 0$. Therefore, a polynomial web (8.4.103) satisfying (8.4.105) is non-isoclinic.

However, for any c^i_{jk} we have $p = q$, the condition (8.4.100) fails, and *the polynomial three-webs* (8.4.103) *cannot be extended to a non-isoclinic* $W(4,2,2)$.

In conclusion we give an example of a non-isoclinic $W(3,2,2)$ that can be extended to a non-isoclinic $W(4,2,2)$ and the $W(4,2,2)$ is of maximum 2-rank.

Example 8.4.5 Suppose that the foliation λ_3 of $W(3,2,2)$ is defined by

$$\lambda_3 : u_3^1 = x^1 + y^1 + (x^1)^2 y^2/2 = \text{const.}, \quad u_3^2 = x^2 + y^2 - x^1(y^2)^2/2 = \text{const.} \tag{8.4.106}$$

Again using (8.4.36)-(8.4.40) and (8.4.9), we obtain (see Step 1)

$$a_1 = y^2/[\Delta(2-\Delta)], \quad a_2 = x^1/[\Delta(2-\Delta)], \tag{8.4.107}$$

$$\begin{cases} p_{11} = 2(y^2)^2(\Delta - 1)/[\Delta^3(2-\Delta)^2], \\ p_{21} = (\Delta^2 - 2\Delta + 2)/[\Delta^3(2-\Delta)^2], \\ q_{22} = 2(x^1)^2(\Delta - 1)/[\Delta^2(2-\Delta)^3], \\ q_{12} = (\Delta^2 - 2\Delta + 2)/[\Delta^2(2-\Delta)^3], \\ p_{12} = q_{11} = p_{22} = q_{21} = 0, \end{cases} \tag{8.4.108}$$

$$p = -(\Delta^2 - 2\Delta + 2)/[2\Delta^3(2-\Delta)^2], \quad q = (\Delta^2 - 2\Delta + 2)/[2\Delta^2(2-\Delta)^3], \tag{8.4.109}$$

$$\begin{cases} \underset{1}{p_2} = \underset{1}{q_2} = \underset{2}{p_1} = \underset{2}{q_1} = 0, \\ \underset{1}{p_1} = y^2(-\Delta^3 + 4\Delta^2 - 8\Delta + 6)/[\Delta^5(2-\Delta)^3], \\ \underset{2}{p_2} = x^1(-\Delta^3 + 3\Delta^2 - 6\Delta + 4)/[\Delta^4(2-\Delta)^4], \\ \underset{1}{q_1} = y^2(\Delta^3 - 4\Delta^2 + 8\Delta - 6)/[\Delta^4(2-\Delta)^4], \\ \underset{2}{q_2} = x^1(\Delta^3 - 2\Delta^2 + 4\Delta - 2)/[\Delta^3(2-\Delta)^5], \end{cases} \tag{8.4.110}$$

where $\Delta = 1 + x^1 y^2$. It follows from (8.4.109) and (8.4.110) that the conditions (8.4.100) and (8.4.101) are satisfied. Hence the equations (8.4.106) define a *non-isoclinic three-web $W(3,2,2)$ that can be extended to a non-isoclinic four-web $W(4,2,2)$*.

To find the extension (Step 2), we determine from (8.4.95) and (8.4.109) that

$$\lambda = 1 - 2/\Delta \tag{8.4.111}$$

and from (8.4.5) and (8.4.94) that

$$b_1 = -y^2/\Delta^2, \quad b_2 = x^1/(2\Delta - \Delta^2), \tag{8.4.112}$$

$$b_{11} = b_{21} = 0, \quad b_{12} = -p_{21}, \quad b_{22} = q_{22}, \tag{8.4.113}$$

where the p_{ij} and q_{ij} are given by (8.4.108).

Using (8.4.111), (8.4.36), and (8.4.106), we can write the equations (8.4.34) of the foliation λ_4 in the form

$$d[x^1(x^1y^2/2 - 1) + y^1] = 0, \quad d[-y^2(x^1y^2/2 + 1) + x^2] = 0. \tag{8.4.114}$$

It follows from (8.4.114) that the foliation λ_4 is defined by

$$\lambda_4 : u_4^1 = -x^1 + y^1 + (x^1)^2 y^2/2 = \text{const.}, \quad u_4^2 = x^2 - y^2 - x^1(y^2)^2/2 = \text{const.} \tag{8.4.115}$$

To check whether the web defined by (8.4.106) and (8.4.115) is of maximum 2-rank or not (Step 3), we find $d\omega_k^k$ by means of (8.4.39) and compare it with (8.4.3). This gives

$$b_{k11}^k = b_{k21}^k = b_{k22}^k = 0, \quad b_{k12}^k = 2(-\Delta^2 + 2\Delta - 2)/[\Delta^3(2-\Delta)^3]. \qquad (8.4.116)$$

Equations (8.4.116), (8.4.113), and (8.4.108) show that condition (8.4.102) is satisfied. Thus, *the four-web defined by* (8.4.106) *and* (8.4.115) *is of maximum* 2*-rank.*

To find the only abelian 2-equation admitted by this web, we integrate (8.4.32) where λ, a_i, and b_i are defined by (8.4.111), (8.4.39), (8.4.107), and (8.4.112). Up to a constant factor, the solution is

$$\sigma = \Delta/[2(\Delta - 2)]. \qquad (8.4.117)$$

Substituting λ from (8.4.111) and σ from (8.4.117) into (8.4.31), we obtain the only abelian 2-equation in the form

$$(2/\Delta)\underset{1}{\omega}^1 \wedge \underset{1}{\omega}^2 - (2/(\Delta-2))\underset{2}{\omega}^1 \wedge \underset{2}{\omega}^1 - \underset{3}{\omega}^1 \wedge \underset{3}{\omega}^2 + (\Delta/(\Delta-2))\underset{4}{\omega}^1 \wedge \underset{4}{\omega}^2 = 0, \quad (8.4.118)$$

where each term is a closed 2-form (see Definition 8.3.2). Using (8.4.36), we can write the equation (8.4.118) in the form of (8.3.2):

$$2dx^1 \wedge dx^2 + 2dy^1 \wedge dy^2 - du_3^1 \wedge du_3^2 + du_4^1 \wedge du_4^2 = 0. \qquad (8.4.119)$$

Note that *the four-web constructed in this example represents the first and only known example of a non-isoclinic web* $W(4,2,2)$ *of maximum* 2*-rank.*

8.5 The Geometry of the Exceptional Webs $W(4,2,2)$ of Maximum 2-Rank

8.5.1 Double Fibrations and Webs

Definition 8.5.1 A *double fibration* (abbreviation DF) is a diagram

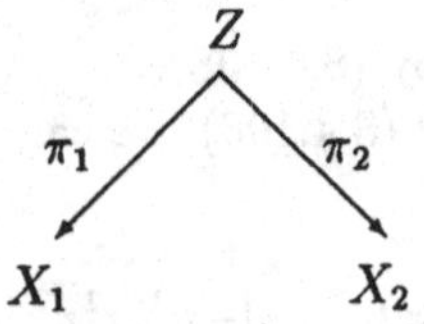

where Z, X_1, and X_2 are smooth manifolds and

a) Z is a smooth fibration with respect to π_1 and π_2 ;

b) $\pi_1 \times \pi_2 : Z \to X_1 \times X_2$ is a non-degenerate injective diffeomorphism;

c) for any $x_1,\ x_2 \in X_1,\ x_1 \neq x_2$, and $\xi_1,\ \xi_2 \in X_2,\ \xi_1 \neq \xi_2$, we have respectively $\pi_2 \cdot \pi_1^{-1} x_1 \neq \pi_2 \cdot \pi_1^{-1} x_2,\ \pi_1 \cdot \pi_2^{-1} \xi_1 \neq \pi_1 \cdot \pi_2^{-1} \xi_2$.

The idea of DF can be found in the paper [C 42] of Chern. It was extensively used by Gelfand and his collaborators in their work in integral geometry (see for example [GS 83]).

In [GS 83] the authors write that "π_1 and π_2 enable us to carry the various analytical objects (functions, forms etc.) from X_1 to X_2 by first lifting them from X_1 to Z and subsequently descending them to X_2".

Every three-web $W(3,2,r)$ defines an r-parameter family of DF. To get a DF, one has to take the first two foliations λ_1 and λ_2 of $W(3,2,r)$ as X_1 and X_2, fix a leaf F_3 of the foliation λ_3 and consider Z as the manifold of pairs (F_1, F_2), $F_1 \subset X_1,\ F_2 \subset X_2,\ F_1 \cap F_2 = p \in F_3$. It is easy to see that $Z \to X_1 \times X_2$ is a smooth embedding. Note also that one gets more DF taking any pair of foliations of $W(3,2,r)$ as X_1 and X_2.

For a d-web $W(d,2,r)$, $d > 3$, one can construct $d-2$ r-parameter families of DF. In general, for a web $W(d,n,r)$, $d \geq n+1$, one can get $d-n$ r-parameter families of n-fold foliations taking the first foliations as $X_1, \ldots, X_n$.

In this last section of the book we will study the geometry of exceptional webs $W(4,2,2)$ of maximum 2-rank. In particular, we will introduce some analytical objects – functions, vector fields, forms – given on the different foliations. The program outlined in [GS 83] can be applied to these objects.

8.5.2 Interior Products Associated with an Exceptional Four-Web

For a web $W(4,2,2)$ a two-dimensional flat generator $\zeta_p(2)$ of the Segre cone $C_p(2,2)$ is determined by a transversally geodesic bivector $E^1 \wedge E^2$ where $E^1 = \xi^i e_i^1$, $E^2 = \xi^i e_i^2$ and ξ^i satisfies the equation (1.9.3):

$$d\xi^i + \xi^j \omega_j^i = \varphi \xi^i. \tag{8.5.1}$$

This bivector intersects the tangent planes $T_p(V_\xi)$ to leaves V_ξ along the directions parallel to the vectors

$$W_1 = -\xi^i e_i^2,\ \ W_2 = \xi^i e_i^1,\ \ W_3 = \xi^i (e_i^2 - e_i^1),\ \ W_4 = \xi^i (e_i^1 - \lambda e_i^2). \tag{8.5.2}$$

Recall that the only abelian 2-equation for exceptional webs $W(4,2,2)$ of maximum 2-rank has the form (see (8.4.31)):

$$\sum_{a=1}^{4} \Omega_a = 0 \tag{8.5.3}$$

where

$$\begin{cases} \Omega_1 = (\lambda - \lambda^2)\sigma \underset{1}{\omega}^1 \wedge \underset{1}{\omega}^2, & \Omega_3 = -\lambda\sigma(\underset{1}{\omega}^1 + \underset{2}{\omega}^1) \wedge (\underset{1}{\omega}^2 + \underset{2}{\omega}^2), \\ \Omega_2 = (\lambda - 1)\sigma \underset{2}{\omega}^1 \wedge \underset{2}{\omega}^2, & \Omega_4 = \sigma(\lambda\underset{1}{\omega}^1 + \underset{2}{\omega}^1) \wedge (\lambda\underset{1}{\omega}^2 + \underset{2}{\omega}^2) \end{cases} \tag{8.5.4}$$

are closed 2-forms, σ satisfies (8.4.32), and conditions (8.4.33) or (8.4.102) are respectively satisfied for isoclinic and non-isoclinic four-webs.

Proposition 8.5.2 *For an exceptional web $W(4,2,2)$ of maximum 2-rank there are the following relations among the interior products of the forms Ω_ξ with respect to the vector fields W_η :*

$$i_{W_\xi}\Omega_\xi = 0, \tag{8.5.5}$$

$$\sum_\eta i_{W_\xi}\Omega_\eta = 0. \tag{8.5.6}$$

Proof. First of all, for each vector field W the interior product i_W is defined as a skew-derivation by the following formulae: $i_W f = 0$ for every function f, $i_W\omega = \omega(W)$ for every 1-form ω, and

$$(i_W\omega)(Y_1, \ldots, Y_{r-1}) = r \cdot \omega(W, Y_1, \ldots, Y_{r-1}) \tag{8.5.7}$$

for every r-form ω where $Y_1, \ldots, Y_{r-1}$ are arbitrary vector fields (see [KN 63]). For an r-form ω and an s-form ω' the following formula holds:

$$i_W(\omega \wedge \omega') = i_W\omega \wedge \omega' + (-1)^r \omega \wedge i_W\omega' \tag{8.5.8}$$

(see [KN 63]).

The proof of (8.5.5) and (8.5.6) is straightforward. Applying (8.5.7) and (8.5.8) to the vector field W_4 and the 2-forms defined respectively by (8.5.2) and (8.5.4), we easily obtain that

$$\begin{cases} \alpha_1 = i_{W_4}\Omega_1 = (\lambda - \lambda^2)\sigma(\xi^1\underset{1}{\omega}^2 - \xi^2\underset{1}{\omega}^1), \\ \alpha_2 = i_{W_4}\Omega_2 = (\lambda^2 - \lambda)\sigma(-\xi^1\underset{2}{\omega}^2 + \xi^2\underset{2}{\omega}^1), \\ \alpha_3 = i_{W_4}\Omega_3 = (\lambda^2 - \lambda)\sigma[\xi^1(\underset{1}{\omega}^2 + \underset{2}{\omega}^2) - \xi^2(\underset{1}{\omega}^1 + \underset{2}{\omega}^1)], \\ \alpha_4 = i_{W_4}\Omega_4 = 0. \end{cases} \tag{8.5.9}$$

The relations (8.5.5) and (8.5.6) for W_4 follow from (8.5.9). The proof for W_1, W_2, and W_3 is similar. ■

Remark 8.5.3 If one changes W_ξ for $\tilde{W}_\xi = kW_\xi$, then $i_{\tilde{W}_\xi}\Omega_\eta$ will be factored by k and (8.5.5) and (8.5.6) will still hold for $\tilde{W}_\xi$.

8.5.3 Exterior 3-Forms Associated with an Exceptional Four-Web

Let us consider now two pairings (α_1, Ω_2) and (α_2, Ω_1) where α_1, α_2 and Ω_1, Ω_2 are defined by (8.5.9) and (8.5.4). The following exterior cubic forms are associated with these pairings:

$$\begin{cases} \psi_1 = \alpha_1 \wedge \Omega_2 = \sigma^2\lambda(1-\lambda)^2(\xi^2 \underset{1}{\omega^1} - \xi^1 \underset{1}{\omega^2}) \wedge \underset{2}{\omega^1} \wedge \underset{2}{\omega^2}, \\ \psi_2 = \alpha_2 \wedge \Omega_1 = \sigma^2(\lambda-\lambda^2)^2(\xi^2 \underset{2}{\omega^1} - \xi^1 \underset{2}{\omega^2}) \wedge \underset{1}{\omega^1} \wedge \underset{1}{\omega^2}. \end{cases} \tag{8.5.10}$$

It is known that a vector field ξ is an *infinitesimal conformal transformation* (abbreviation i.c.t) of an exterior form ω if the Lie derivative $L_\xi \omega$ is proportional to ω:

$$L_\xi \omega = k\omega, \tag{8.5.11}$$

and ξ is an *infinitesimal automorpism* (abbreviation i.a.) of ω if

$$L_\xi \omega = 0 \tag{8.5.12}$$

(see [YK 84]).

The next definition is a generalisation of these two notions.

Definition 8.5.4 We will say that two exterior s-forms ω_1 and ω_2 define a *quasi-recurrent Lie derivative pairing* generated by a vector field ξ if

$$L_\xi \omega_i = a_i^j \omega_j, \quad i, j = 1, 2. \tag{8.5.13}$$

Theorem 8.5.5 *The exterior cubic forms ψ_1 and ψ_2 determined by* (8.5.10) *are exterior recurrent and define a quasi-recurrent Lie derivative pairing generated by the vector field W_ξ. In addition, the vector fields W_1, W_2, and W_3 are i.c.t of both forms ψ_1 and ψ_2, and the vector field W_4 is an i.c.t. of the form ψ_2.*

Proof. First of all, it follows from (8.5.10), (8.4.32), and (8.5.1) that

$$d\psi_1 = \delta_1 \wedge \psi_1, \quad d\psi_2 = \delta_2 \wedge \psi_2 \tag{8.5.14}$$

where

$$\delta_1 = b_i \underset{1}{\omega^i}, \quad \delta_2 = (b_i - a_i)\underset{1}{\omega^i} - a_i \underset{2}{\omega^i}. \tag{8.5.15}$$

The equations (8.5.14) show that the forms ψ_1 and ψ_2 are exterior recurrent [Da 82] with δ_1 and δ_2 respectively as recurrence forms.

To calculate the Lie derivatives $L_{W_4}\psi_1$ and $L_{W_4}\psi_2$, we need the following formulae:

$$\begin{cases} L_{W_4}\psi_i = (i_{W_4}\sigma_i)\psi_i - \sigma_i \wedge i_{W_4}\psi_i + d(i_{W_4}\psi_i), \quad i = 1, 2, \\ i_{W_4}\sigma_1 = \xi^i b_i, \quad i_{W_4}\sigma_2 = \xi^i[b_i + (\lambda - 1)a_i], \\ i_{W_4}\psi_1 = -i_{W_4}\psi_2 = -\delta_1 \wedge \delta_2, \\ d(i_{W_4}\psi_1) = -d(i_{W_4}\psi_2) = [a_i - 2b_i)\underset{1}{\omega^i} + a_i \underset{2}{\omega^i}] \wedge \delta_1 \wedge \delta_2. \end{cases} \tag{8.5.16}$$

Using (8.5.16), we find that

$$\begin{cases} L_{W_4}\psi_1 = (b_i + \lambda a_i)\xi^i\psi_1 + (a_i - b_i)\xi^i\psi_2, \\ L_{W_4}\psi_2 = [2b_i + (\lambda - 1)a_i]\xi^i\psi_2. \end{cases} \tag{8.5.17}$$

Similarly one can prove that

$$\begin{cases} L_{W_1}\psi_1 = (b_i\xi^i/\lambda)\psi_1, \quad L_{W_1}\psi_2 = a_i\xi^i\psi_2, \\ L_{W_2}\psi_1 = b_i\xi^i\psi_1, \quad L_{W_2}\psi_2 = (2b_i - a_i)\xi^i\psi_2, \\ L_{W_3}\psi_1 = -(1 + 1/\lambda)b_i\xi^i\psi_1, \quad L_{W_2}\psi_2 = -2b_i\xi^i\psi_2. \end{cases} \tag{8.5.18}$$

Equations (8.5.17) and (8.5.18) prove Theorem 8.5.5. ■

8.5.4 Infinitesimal Automorphisms of Exterior Cubic Forms Associated with an Exceptional Four-Web

Let us find under what conditions the vector fields W_ξ defined by (8.5.2) are i.a. of a cubic form

$$\psi = f\psi_1 + g\psi_2, \quad f, g \in C^\infty(M^4). \tag{8.5.19}$$

The differentials of f and g have the form

$$df = f_i\underset{1}{\omega}^i + \tilde{f}_i\underset{2}{\omega}^i, \quad dg = g_i\underset{2}{\omega}^i + \tilde{g}_i\underset{2}{\omega}^i. \tag{8.5.20}$$

For the vector field W_1 we have

$$i_{W_1}f = -\xi^i\tilde{f}_i, \quad i_{W_1}g = -\xi^i\tilde{g}_i. \tag{8.5.21}$$

Using (8.5.21), we calculate the Lie derivative $L_{W_1}\xi$:

$$\begin{aligned} L_{W_1}\psi &= (i_{W_1}f)\psi_1 + fL_{W_1}\psi_1 + (i_{W_1}g)\psi_2 + gL_{W_1}\psi_2 \\ &= (-\tilde{f}_i + b_if/\lambda)\xi^i\psi_1 + (-\tilde{g}_i + a_ig)\xi^i\psi_2. \end{aligned} \tag{8.5.22}$$

The vector field W_1 is an i.a. of ψ for any ξ^i if and only if

$$\tilde{f}_i = b_if/\lambda, \quad \tilde{g}_i = a_ig. \tag{8.5.23}$$

Because of (8.5.23), the equations (8.5.19) can be written in the form

$$df = f_i\underset{2}{\omega}^i + fb_i\underset{2}{\omega}^i/\lambda, \quad dg = g_i\underset{1}{\omega}^i + ga_i\underset{2}{\omega}^i. \tag{8.5.24}$$

Similarly one can find that the vector fields W_2, W_3, and W_4 are i.a. of ψ for any ξ^i if and only if the functions f and g satisfy respectively the following equations:

$$df = -fb_i\underset{1}{\omega}^i + \tilde{f}_i\underset{2}{\omega}^i, \quad dg = g(a_i - 2b_i)\underset{2}{\omega}^i + \tilde{g}_i\underset{2}{\omega}^i, \tag{8.5.25}$$

$$df = f_i(\underset{1}{\omega^i} + \underset{2}{\omega^i}) + (1 + 1/\lambda) f b_i \underset{2}{\omega^i}, \quad dg = g_i(\underset{1}{\omega^i} + \underset{2}{\omega^i}) + 2g b_i \underset{2}{\omega^i}, \tag{8.5.26}$$

$$\begin{cases} df = \tilde{f}_i(\lambda \underset{1}{\omega^i} + \underset{2}{\omega^i}) - (\lambda a_i + b_i) f \underset{2}{\omega^i}, \\ dg = g_i(\lambda \underset{1}{\omega^i} + \underset{2}{\omega^i}) + [f(b_i - a_i) - g(2b_i + (1-\lambda)a_i)] \underset{1}{\omega^i}. \end{cases} \tag{8.5.27}$$

We have proved the following proposition:

Proposition 8.5.6 *The vector field W_ξ, $\xi = 1, 2, 3,$ or 4, defined by* (8.5.2) *are i.a. of the cubic form ψ defined by* (8.5.19) *for any ξ^i if and only if the functions f and g in* (8.5.19) *satisfy respectively the equations* (8.5.24), (8.5.25), (8.5.26), *and* (8.5.27). ■

We have to study now the compability of equations (8.5.24)–(8.5.27) with the equations (8.4.2)–(8.4.6), (8.4.33) for isoclinic exceptional webs $W(4,2,2)$ of maximum 2-rank and with the equations (8.4.2)–(8.4.4), (8.4.93)–(8.4.95), (8.4.102) for nonisoclinic exceptional webs $W(4,2,2)$ of maximum 2-rank.

Theorem 8.5.7 *The vector fields W_1 and W_2 (for any ξ^i) cannot be i.a. of the cubic form ψ defined by* (8.5.19) *if an exceptional web is nonisoclinic. In all other cases the set of forms ψ for which the vector field W_ξ, $\xi = 1, 2, 3,$ or 4, is i.a. of for any ξ^i depends on two arbitrary functions of two independent variables.*

Proof. In the case of W_1 we have (8.5.24). If an exceptional web is nonisoclinic, the exterior differential of (8.5.24) by means of (8.4.2)–(8.4.4) and (8.4.93)–(8.4.95) leads to two exterior quadratic equations. The first of them is

$$\{df_i - f_j \omega_i^j + f_{[i} a_{j]} \underset{1}{\omega^j} - [b_j f_i/\lambda + f(b_{ij} + p_{ji} - q_{ij})] \underset{2}{\omega^j}\} \wedge \underset{1}{\omega^i} + f q_{ij} \underset{2}{\omega^j} \wedge \underset{2}{\omega^i} = 0. \tag{8.5.28}$$

It follows from (8.5.28) that $q_{[ij]} = 0$ and this contradicts the non-isoclinity of an exceptional web (see (8.4.95)).

If an exceptional web is isoclinic, the exterior differentiation of (8.5.24) by means of (8.4.2)–(8.4.6) gives the following exterior quadratic equations:

$$\Delta f_i \wedge \underset{1}{\omega^i} = 0, \quad \Delta g_i \wedge \underset{1}{\omega^i} = 0 \tag{8.5.29}$$

where

$$\begin{aligned} \Delta f_i &= df_i - f_j \omega_i^j + f_{[i} a_{j]} \underset{1}{\omega^j} - [b_j f_i/\lambda + f(\underset{1}{k_{ij}} - \underset{4}{k_{ij}})] \underset{2}{\omega^j}, \\ \Delta g_i &= dg_i - g_j \omega_i^j + g_{[i} a_{j]} \underset{1}{\omega^j} + [a_j g_i - g(\underset{1}{k_{ij}} - \underset{3}{k_{ij}})] \underset{2}{\omega^j}. \end{aligned}$$

The number of unknown functions (Δf_i and Δg_i) is equal to $q = 4$. The consecutive Cartan characters ([Ca 45]) are $s_1 = 2$, $s_2 = 2$, $s_3 = 0$, and the Cartan number $Q = s_1 + 2s_2 = 6$. It follows from (8.5.29) that the general two-dimensional integral element depends on $N = 6$ parameters. Because $Q = N$, the system (8.5.24), (8.5.29) is in involution and its solution depends on two functions of two variables.

The proof for the vector fields W_2, W_3, and W_4 is similar. ■

8.5.5 Infinitesimal Conformal Transformations of Exterior Cubic Forms Associated with an Exceptional Four-Web

According to Theorem 8.5.5, the vector fields W_1, W_2, and W_4 are i.c.t. of ψ_1 and ψ_2, and W_4 is that of ψ_2. We will find under what conditions the vector field W_ξ, $\xi = 1,2,3$, or 4, is an i.c.t. of the cubic form ψ defined by (8.5.19).

We will suppose that $f \neq 0$ and $g \neq 0$ and consider for example the vector field W_1.

One can find from (8.5.11) and (8.5.24) that the W_1 for any ξ^i is an i.c.t. of ψ if and only if

$$\frac{\tilde{f}_i}{f} - \frac{\tilde{g}_i}{g} = \frac{1}{\lambda}(b_i - \lambda a_i). \tag{8.5.30}$$

For the system (8.5.20), (8.5.30) for both isoclinic and non-isoclinic exceptional webs one gets: $q = 6$, $s_1 = s_2 = s_3 = 2$, $s_4 = 0$, $Q = s_1 + 2s_2 + 3s_3 = 12$, $N = 12$. We have proved the theorem:

Theorem 8.5.8 *The vector field W_ξ, $\xi = 1,2,3$, or 4, for any ξ^i is an i.c.t. of the exterior cubic form ψ defined by* (8.5.19) *if and only if the functions f and g in* (8.5.19) *satisfy* (8.5.20) *and respectively* (8.5.30) *and equations similar to* (8.5.30). *In each of these four cases the set of forms ψ depends on two functions of three independent variables.* ■

NOTES

8.1. This section is from [G 83, 85b]. It is worth to add to the result of Theorem 8.1.23 that in [G 85b] Goldberg gave a geometrical description of the location of the foliation λ_4 of the extended web $AGW(4,2,2)$ with respect to the foliations λ_1, λ_2, and λ_3 in terms of null α-planes (see [P 68] or [AHS 78]) of a pseudo-conformal structure $CO(2,2)$ which is associated with any three-web W(3,2,2) (see [Ak 83a]).

8.2. This section is from [G 84].

Chern in his recent paper [C 85] emphasised the importance of the rank problems considerations. He wrote: "The high-dimensional abelian equations constitute a subject which has contacts with many branches of mathematics (such as functional equations, partial differential equations, combinatorial characteristic classes, algebraic K-theory etc.) and which should have a very promising future."

As we mentioned in Section 8.3, Griffiths [Gr 77] defined the q-rank for a web $W(d,n,r)$. An equation of the form

$$\sum_{\xi=1}^{d} \underset{\xi}{A}_{i_1 \ldots i_q} \underset{\xi}{\omega}^{i_1} \wedge \ldots \wedge \underset{\xi}{\omega}^{i_q} = 0$$

is said to be an *abelian q-equation.* The maximum number of abelian q-equations admitted by a web $W(d,n,r)$ is called the *q-rank* of the web $W(d,n,r)$, $1 \leq q \leq r$.

In the case $r = 1$ there is only one possible value for $q : q = 1$. Both problems indicated in Section 8.3.1 (finding the upper bound and description of webs of maximum rank) were solved for webs $W(d,2,1)$ and $W(d,3,1)$ as well for d-webs of curves in three-dimensional space during the intensive development of web geometry in the 1930's (see [BB 38] and [Bl 55]). In particular, as we mentioned in Section 8.2.2, it was proved that while webs W(4,2,1) of maximum 1-rank are algebraisable, this is not true for webs $W(d,2,1)$, $d > 5$: Bol [B 36] constructed a famous example of a 5-web consisting of four pencils of straight lines whose centres are in general position and an one-parameter family of conics passing through these four centres. This 5-web is of maximum rank 6 and is not algebraisable.

Chern [C 36a] found an upper bound for the 1-rank of a web $W(d,n,1)$. Chern and Griffiths described so-called normal webs $W(d,n,1)$ of maximum 1-rank (see [CG 77, 78a, 81] and [C 82]) and found an upper bound for the r-rank of the webs $W(d,n,r)$. However, they did not describe the webs $W(d,n,r)$ of maximum r-rank. The author's results presented in Section 8.3 give the description of almost Grassmannisable webs $AGW(d,2,r)$ and, in the cases $d > r+2$ (see Theorem 8.3.6 and [Lit 86]) and $d = 4$, $r = 2$, general webs $W(d,2,r)$ of maximum r-rank.

The author's results presented in Section 8.2 represent the first study of q-rank problems for webs $W(d,n,r)$ with $q < r$, namely 1-rank problems are considered for webs $W(d,2,r)$, i.e. the extreme case $q = 1$ was taken there.

Note in conclusion of this brief description of the history of development of rank problems for webs that Damiano [D 83] solved both problems mentioned above for d-webs of curves in an n-dimensional space taking into consideration abelian $(n-1)$-equations. In particular, he proved that a so-called exceptional $(n+3)$-web of curves in $\mathbf{R}^n$ generalising the Bol 5-web mentioned above is of maximum $(n-1)$-rank, and it is the unique non-algebraisable (non-linearisable) web among the quadrilateral d-webs, $d > n > 3$. A *quadrilateral* d-web is defined as a d-web of curves for which every n-subweb generates a coordinate system in the ambient n-space.

8.3. The results are due to the author [G 83, 85b].

8.4. Chern in [C 85] wrote: "In general, the determination of all webs of maximum rank will remain a fundamental problem in web geometry and non-algebraic ones if there are any, will be most interesting."

The examples 8.4.2, 8.4.3, and 8.4.5 of isoclinic and non-isoclinic webs $W(4,2,2)$ of maximum 2-rank presented in this section are the first known examples of non-algebraic webs $W(4,2,2)$ of maximum 2-rank. These examples were constructed by the author in [G 85a,86,87].

8.5. This section is from [GR 87]. Note that in Definition 8.5.4 we used the term "quasi-recurrent" that has been used in [Ros 76] in a similar situation for a pair of vector fields.

Bibliography

[Ac 65] Aczel, J.: *Quasigroups, nets and nomograms.* Adv. in Math. 1 (1965), no. 3, 383–450. (MR #1395.)

[Ak 69a] Akivis, M.A.: *Three-webs of multidimensional surfaces.* (Russian) Trudy Geometr. Sem. 2 (1969), 7–31. (MR 40 #7967.)

[Ak 69b] Akivis, M.A.: *The canonical expansions of the equations of a local analytic quasigroup.* (Russian) Dokl. Akad. Nauk SSSR 188 (1969), no. 5, 967–970. English translation: Soviet Math. Dokl. 10 (1969), no. 5, 1200-1203. (MR 41 #7021.)

[Ak 73] Akivis, M.A.: *The local differentiable quasigroups and three-webs that are determined by a triple of hypersurfaces.* (Russian) Sibirsk. Mat. Zh. 14 (1973), no. 3, 467–474. English translation: Siberian Math. J. 14 (1973), no. 3, 319–324. (MR 48 #2911.)

[Ak 74] Akivis, M.A.: *On isocline three-webs and their interpretation in a ruled space of projective connection.* (Russian) Sibirsk. Mat. Zh. 15 (1974), no. 1, 3–15. English translation: Siberian Math. J. 15 (1974), no. 1, 1–9. (MR 50 #3129.)

[Ak 75a] Akivis, M.A.: *Closed G-structures on a differentiable manifold.* (Russian) Problems in geometry, Vol. 7, 69–79. Akad. Nauk SSSR Vsesoyuz. Inst. Nauchn. i Tekhn. Inform., Moscow, 1975. (MR 57 #17549.)

[Ak 75b] Akivis, M.A.: *The almost complex structure associated to a three-web of multidimensional surfaces.* (Russian) Trudy Geom. Sem. Kazan. Gos. Univ. Vyp. 8 (1975), 11–15. (MR 54 #13772.)

[Ak 76] Akivis, M.A.: *The local algebras of a multidimensional three-web.* (Russian) Sibirsk. Mat. Zh. 17 (1976), no. 1, 5–11. English translation: Siberian Math. J. 17 (1976), no. 1, 3–8. (MR 53 #9055.)

[Ak 78] Akivis, M.A.: *Geodesic loops and local triple systems in a space with an affine connection.* (Russian) Sibirsk. Mat. Zh. 19 (1978), no. 2, 243–253. English translation: Siberian Math. J. 19 (1978), no. 2, 171–178. (MR 58 #7438.)

[Ak 80] Akivis, M.A.: *Webs and almost Grassmann structures.* (Russian) Dokl. Akad. Nauk SSSR 252 (1980), no. 2, 267–270. English translation: Soviet Math. Dokl. 21 (1980), no. 3, 707-709. (MR 82a:53016.)

[Ak 81] Akivis, M.A.: *A geometric condition of isoclinity of a multidimensional web.* (Russian) Webs and Quasigroups, Kalinin. Gos. Univ., Kalinin, 1981, 3–7. (MR 83e:53010.)

[Ak 82] Akivis, M.A.: *Webs and almost Grassmann structures.* (Russian) Sibirsk. Mat. Zh. 23 (1982), no. 6, 6–15. English translation: Siberian Math. J. 23 (1982), no. 6, 763–770. (MR 84b:53018.)

[Ak 83a] Akivis, M.A.: *Completely isotropic submanifolds of a four-dimensional pseudo-conformal structure.* (Russian) Izv. Vyssh. Uchebn. Zaved. Mat. 1983, no. 1(248), 3–11. English translation: Soviet Math. (Iz. VUZ) 27 (1983), no. 1, 1–11. (MR 84i:53016.)

[Ak 83b] Akivis, M.A.: *A local condition of algebraizability of a system of submanifolds of a real projective space.* (Russian) Dokl. Akad. Nauk SSSR 272 (1983), no. 6, 1289–1291. English translation: Soviet Math. Dokl. 28 (1983), no. 2, 507–509. (MR 85c:53018.)

[Ak 83c] Akivis, M.A.: *Differential geometry of webs.* (Russian) Problems in Geometry, Vol. 15, 187–213, Itogi Nauki i Tekhniki, Akad. Nauk SSSR, Vsesoyuz. Inst. Nauchn. i Tekhn. Informatsii, Moscow, 1983. English translation: J. Soviet Math. 29(1985), no. 5, 1631–1647. (MR 85i:53019.)

[AGe 86] Akivis, M.A.; Gerasimenko, S.A.: *Multidimensional Bol webs.* (Russian) Problems in Geometry, Vol. 18, 73–103, Itogi Nauki i Tekhniki, Akad. Nauk SSSR, Vsesoyuz. Inst. Nauchn. i Tekhn. Informatsii, Moscow, 1986.

[AG 74] Akivis, M.A.; Goldberg, V.V.: *The four-web and the local differentiable ternary quasigroup that are determined by a quadruple of surfaces of codimension two.* (Russian) Izv. Vyssh. Uchebn. Zaved. Mat. 1974, no. 5(144), 12–24. English translation: Soviet Math. (Iz. VUZ) 18 (1974), no. 5, 9–19. (MR 50 #8321.)

[AS 71a] Akivis, M.A.; Shelekhov, A.M.: *The computation of the curvature and torsion tensors of a multidimensional three-web and of the associator of the local quasigroup that is connected with it.* (Russian) Sibirsk. Mat. Zh. 12 (1971), no. 5, 953–960. English translation: Siberian Math. J. 12 (1971), no. 5, 685–689. (MR 44 #5876.)

[AS 71b] Akivis, M.A.; Shelekhov, A.M.: *Local differentiable quasigroups and connections that are associated with a three-web of multidimensional surfaces.* (Russian) Sibirsk. Mat. Zh. 12 (1971), no. 6, 1181–1191. English translation: Siberian Math. J. 12 (1971), no. 6, 845–892. (MR 44 #5877.)

[AS 81] Akivis, M.A.; Shelekhov, A.M.: "Foundations of the theory of webs." (Russian) Kalinin. Gos. Univ., Kalinin, 1981, 88 pp. (MR 83h:53001.)

[Al 43] Albert, A.A.: *Quasigroups* I. Trans. Amer. Math. Soc. 54 (1943), 507–519. (MR 5, 229.)

[AHS 78] Atiyah, M.F.; Hitchin, N.I.; Singer, I.M.: *Self-duality in four-dimensional Riemannian geometry.* Proc. Roy. Soc. London Ser. A 362 (1978), 425–461. (MR 80d:53023.)

[Au 38] Aue, H.: *Hyperflächenscharen im n-dimensionalen Raum R_n* . Mitt. Math. Ges. Hamburg 7 (1938), 367–399.

[Bar 59] Bartoshevich, M.A.: *Plane-parallel webs of hypersurfaces.* (Russian) Dokl. Akad. Nauk SSSR 124 (1959), no. 5, 970–972. (MR 21 #3870.)

[Ba 51a] Bartsch, H.: *Übertragung der Achtflachengewebeeigenschaften auf Hyperflächengeweben des n-dimensionalen Raumes.* Abh. Math. Sem. Univ. Hamburg 17 (1951), 1–21. (MR 13, 277.)

[Ba 51b] Bartsch, H.: *Hyperflächengewebe des n-dimensionalen Raumes.* Ann. Mat. Pura Appl (4) 32 (1951), 249–269. (MR 13, 775.)

[Ba 53] Bartsch, H.: *Über eine Klasse von Hyperflächengeweben.* Ann. Mat. Pura Appl. (4) 34 (1953), 349–364. (MR 14, 1119.)

[Bau 82] Baumann, J.: "Starrheit und Nichtäquivalenz von analytischen Polyedergebieten." Schriftenreihe Math. Inst. Univ. Münster, 2. Serie, Heft 24, 1982, v+182 pp.

[BT 70] Behnke, H; Thullen, P.: "Theorie der Funktionen mehrerer komplexer Veränderlicher." 2. Auflage, Springer-Verlag, Berlin-Heidelberg-New York, 1970, xvi+225 pp. (MR 42 #6274.)

[Be 67] Belousov, V.D.: "Foundations of the theory of quasigroups and loops." (Russian) Izdat. "Nauka", Moscow, 1967, 223 pp. (MR 36 #1569.)

[Be 71] Belousov, V.D.: "Algebraic nets and quasigroups." (Russian) Izdat. "Shtiintsa", Kishinev, 1971, 165 pp. (MR 49 #5214.)

[Be 72] Belousov, V.D.: "n-ary quasigroups."(Russian) Izdat. "Shtiintsa", Kishinev, 1972, 227 pp. (MR 50 #7369.)

[BS 66] Belousov, V.D.; Sandik, M.D.: *n-ary quasigroups and loops.* (Russian) Sibirsk. Mat. Zh. 7 (1966), no. 1, 31–54. English translation: Siberian Math. J. 7 (1966), no. 1, 24–42. (MR 34 #4403.)

[Ber 24] Bertini, E.: "Einführung in die projective Geometrie mehrdimensionalen Räume." L.W. Seidel, Wien, 1924, xxii+480 pp.

[Bl 55] Blaschke, W.: "Einführung in die Geometrie der Waben." Birkhäuser-Verlag, Basel-Stuttgart, 1955, 108 pp. (MR 17, 780.)

[BB 38] Blaschke, W.; Bol, G.: "Geometrie der Gewebe." Springer-Verlag, Berlin, 1938, viii+339 pp.

[BD 28] Blaschke, W.; Dubourdieu, J.: *Invarianten von Kurvengeweben.* Abh. Math. Sem. Univ. Hamburg 6 (1928), 198-215.

[B 35] Bol, G.: *Über 3-Gewebe in vierdimensionalen Raum.* Math. Ann. 110 (1935), 431–463.

[B 36] Bol, G.: *Über ein bemerkenswertes 5-Gewebe in der Ebene.* Abh. Math. Sem. Univ. Hamburg 11(1936), 387–393.

[B 37] Bol, G.: *Gewebe und Gruppen.* Math. Ann. 114 (1937), 414–431.

[B 67] Bol, G.: "Projektive Differentialgeometrie," 3. Teil, Vandenhoeck & Ruprecht, Göttingen-Zurich, 1967, viii+527 pp. (MR 37 #840.)

[Bo 82] Bolodurin, V.S.: *On the invariant theory of point correspondences of three projective spaces.* (Russian) Izv. Vyssh. Uchebn. Zaved. Mat. 1982, no. 5 (240), 9–15. English translation: Soviet Math. (Iz. VUZ) 26 (1982), no. 5, 8–16. (MR 84b:53010.)

[Bo 83] Bolodurin, V.S.: *Some projective-differential properties of point correspondences of four straight lines.* (Russian) Izv. Vyssh. Uchebn. Zaved. Mat. 1983, no. 1(248), 27–35. English translation: Soviet Math. (Iz. VUZ) 27 (1983), no. 1, 28–38. (MR 85c:53031.)

[Bo 84] Bolodurin, V.S.: *Point correspondences between projective straight lines.* (Russian) Izv. Vyssh. Uchebn. Zaved. Mat. 1984, no. 12(271), 17–23. English translation: Soviet Math. (Iz. VUZ) 28 (1984), no. 12, 20–26. (MR 86h:53012.)

[Bo 85] Bolodurin, V.S.: *On the theory of point correspondences of $n + 1$ projective spaces.* (Russian) Izv. Vyh 13-21. (MR 87g:53021)

[Car 83] Carneiro, M.J.D.: *Singularities of envelope of families of submanifolds in* $\mathbf{R}^N$. Ann. Sci. École Norm. Sup. (4) 16 (1983), no. 2, 173–192. (MR 85h:58023.)

[Ca 08] Cartan, É.: *Les sous groupes des groupes continus de transformations.* Ann. École Norm. (3) 25 (1908), 57–194.

[Ca 23] Cartan, É.: *Sur les variétés à connexion affine et la théorie de la relativité généralisée.* Ann. École Norm. 40 (1923), 325–412.

[Ca 27] Cartan, É.: *La géométrie des groupes de transformations.* J. Math. Pures Appl. 6 (1927), 1–119.

[Ca 71] Cartan, É.: "Les systèmes différentiels extérieurs et leurs applications géométriques," 2nd ed. Hermann, Paris, 1971, 214 pp.

[CaH 31] Cartan, H.: *Les fonctions de deux variables complexes et le probléme de la représentation analytique.* J. Math. Pures Appl. (9) 10 (1931), 1–114.

[CPS 88] Chein, O.; Pflugfelder, H.; Smith, J.D.H. (eds): "Quasigroups and loops: theory and applications." Heldermann-Verlag, Berlin, 1988.

[Ch 73] Chen, B.Y.: "Geometry of submanifolds." Marcel Dekker, Inc., New York, 1973, vii+298 pp. (MR 50 #5697.)

[C 36a] Chern, S.S.: *Abzählungen für Gewebe.* Abh. Math. Sem. Univ. Hamburg 11 (1936), no. 1–2, 163–170.

[C 36b] Chern, S.S.: *Eine Invariantentheorie der Dreigewebe aus r-dimensionalen Mannigfältigkeiten in* $\mathbf{R}_{2r}$. Abh. Math. Sem. Univ. Hamburg 11 (1936), no. 1–2, 333–358.

[C 42] Chern, S.S.: *On integral geometry in Klein spaces.* Ann. Math. 43 (1942), 178–189.

[C 82] Chern, S.S.: *Web geometry.* Bull. Amer. Math. Soc. (N.S.) 6 (1982), no. 1, 1–8. (MR 84g:53024.)

[C 85] Chern, S.S.: *Wilhelm Blaschke and web geometry.* In Wilhelm Blaschke Gesammelte Werke, Thales-Verlag, Essen, Vol. 5 (1985), 25–28.

[CG 77] Chern, S.S.; Griffiths, P.A.: *Linearization of webs of codimension one and maximum rank.* Proc. Intern. Symp. Algebraic Geometry 1977, Kyoto, Japan, 85–91. (MR 81k:53010.)

[CG 78a] Chern, S.S.; Griffiths, P.A.: *Abel's theorem and webs.* Jahresber. Deutsch. Math.-Verein. 80 (1978), no. 1–2, 13–110. (MR 80b:53008.)

[CG 78b] Chern, S.S.; Griffiths, P.A.: *An inequality for the rank of a web and webs of maximum rank.* Ann. Scuola Norm. Sup. Pisa Cl. Sci. (4) 5 (1978), no. 3, 539–557. (MR 80b:53009.)

[CG 81] Chern, S.S.; Griffiths, P.A.: *Corrections and addenda to our paper "Abel's theorem and webs".* Jahresber. Deutsch. Math.-Verein. 83 (1981), 78–83. (MR 82k:53030.)

[CM 74] Chern, S.S.; Moser, J.: *Real hypersurfaces in complex manifolds.* Acta Math. 133 (1974), 219–271. (MR 54 #13112.)

[CDD 82] Choquet-Bruhat, Y.; DeWitt-Morette, C.; Dillard -Bleick, M.: "Analysis, manifolds and physics." North-Holland, Amsterdam-New York-Oxford, 1982, xx+630 pp. (MR 84a:58002.)

[D 83] Damiano, D.B.: *Webs and characteristic forms on Grassmann manifolds.* Amer. J. Math. 105 (1983), 1325–1345. (MR 85g:53014.)

[Da 82] Datta, D.K.: *Exterior recurrent forms on a manifold.* Tensor (N.S.) 36 (1982), no. 1, 115–120. (MR 86m:53083.)

[DFN 85] Dubrovin, B.A.; Fomenko, A.T., Novikov, S.P.: "Modern geometry – methods and applications." Part II. The geometry and topology of manifolds. Springer-Verlag, New York–Berlin, 1985, xv+430 pp. (MR 86m:53001.)

[F 78] Fedorova, V.I.: *A condition defining multidimensional Bol's three-webs.* (Russian) Sibirsk. Mat. Zh. 19 (1978), no. 4, 922–928. English translation: Siberian Math. J. 19 (1978), no. 4, 657–661. (MR 58 #24036.)

[Gei 67] Geidel'man, R.M.: *Differential geometry of families of subspaces in multidimensional homogeneous spaces.* (Russian) Itogi Nauki; Algebra, Topology, Geometry 1965, 323–374. Akad. Nauk SSSR, Vsesoyuz. Inst. Nauchn. i Tekhn. Inform., Moscow, 1967. (MR 35 #7224.)

[GS 83] Gel'fand, I.M.; Shmelev, G.S.: *Geometric structures of double fibrations and their connection with certain problems of integral geometry.* (Russian) Funktsional Anal. i Prilozhen. 17 (1983), no. 2, 7–22. English translation: Functional Anal. Appl. 17 (1983), no. 2, 84–96. (MR 85f:53061.)

[Ge 84a] Gerasimenko, S.A.: *On certain relations for the curvature tensor of a multidimensional* $(n+1)$*-web.* (Russian) Webs and Quasigroups, 52–56, Kalinin. Gos. Univ., Kalinin, 1984.

[Ge 84b] Gerasimenko, S.A.: *The computation of the components of the curvature tensor of an* $(n+1)$*-web.* (Russian) VI Pribalt. Geom. Conf. Tezisy Dokl., Tallin, 1984, 34.

[Ge 85a] Gerasimenko, S.A.: *Transversally geodesic* $(n+1)$*-webs.* (Russian) Problems of the theory of webs and quasigroups, 148–154, Kalinin. Gos. Univ., Kalinin, 1985.

[Ge 85b] Gerasimenko, S.A.: *Multidimensional Bol* $(n+1)$*-webs.* (Russian) Preprint, 76 pp., bibl. 18 titles. Mosk. Gos. Pedag. Inst., Moscow, 1985. Dep. in VINITI 10/4/85 under no. 7082-B85.

[Gl 61] Glagolev, N.A.: "Course of nomography." (Russian) Vyssh. Shkola, Moscow, 1961, 268 pp.

[Glu 64] Gluskin, L.M.: *On positional operatives.* (Russian) Dokl. Akad. Nauk SSSR 157 (1964), no. 4, 757–760. English translation: Soviet Math. Dokl. 5 (1964), no. 4, 1001–1004. (MR 29 #2206.)

[G 66] Goldberg, V.V.: *On a normalisation of p-conjugate systems of an n-dimensional projective space.* (Russian) Trudy Geom. Sem. 1 (1966), 89–109. (MR 35 #900.)

[G 73] Goldberg, V.V.: $(n+1)$*-webs of multidimensional surfaces.* (Russian) Dokl. Akad. Nauk SSSR 210 (1973), no. 4, 756–759. English translation: Soviet Math. Dokl. 14 (1973), no. 3, 795–799. (MR 48 #2919.)

[G 74a] Goldberg, V.V.: $(n+1)$*-webs of multidimensional surfaces.* (Russian) Bulgar. Akad. Nauk Izv. Mat. Inst. 15 (1974), 405–424. (MR 51 #13889.)

[G 74b] Goldberg, V.V.: *Isocline $(n+1)$-webs of multidimensional surfaces.* (Russian) Dokl. Akad. Nauk SSSR 218(1974), no. 5, 1005–1008. English translation: Soviet Math. Dokl. 15 (1974), no. 5, 1437–1441. (MR 52 #11763.)

[G 75a] Goldberg, V.V.: *An invariant characterization of certain closure conditions in ternary quasigroups.* (Russian) Sibirsk. Mat. Zh. 16 (1975), no. 1, 29–43. English translation: Siberian Math. J. 16 (1975), no. 1, 23–34. (MR 51 #6619.)

[G 75b] Goldberg, V.V.: *Local ternary quasigroups that are connected with a four-web of multidimensional surfaces.* (Russian) Sibirsk. Mat. Zh. 16 (1975), no. 2, 247–263. English translation: Siberian Math. J. 16 (1975), no. 2, 190–202. (MR 51 #8318.)

[G 75c] Goldberg, V.V.: *The almost Grassmann manifold that is connected with an $(n+1)$-web of multidimensional surfaces.* (Russian) Izv. Vyssh. Uchebn. Zaved. Mat. 1975, no. 8(159), 29–38. English translation: Soviet Math. (Iz. VUZ) 19 (1975), no. 8, 23–31. (MR 54 #11226.)

[G 75d] Goldberg, V.V.: *A certain property of webs with zero curvature.* (Russian) Izv. Vyssh. Uchebn. Zaved. Mat. 1975, no. 9(160), 10–13. English translation: Soviet Math. (Iz. VUZ) 19 (1975), no. 9, 7–10. (MR 54 #6007.)

[G 75e] Goldberg, V.V.: *The $(n+1)$-webs defined by $n+1$ surfaces of codimension $n-1$.* (Russian) Problems in Geometry, Vol. 7, 173–195. Akad. Nauk SSSR Vsesoyuz. Inst. Nauchn. i Tekhn. Informatsii, Moscow, 1975. (MR 57 #17537.)

[G 75f] Goldberg, V.V.: *The diagonal four-web formed by four pencils of multidimensional planes in a projective space.* (Russian) Problems in Geometry, Vol. 7, 197–213. Akad. Nauk SSSR Vsesoyuz. Inst. Nauchn. i Tekhn. Inform., Moscow, 1975. (MR 58 #7422.)

[G 76] Goldberg, V.V.: *Reducible, group, and $(2n+2)$-hedral $(n+1)$-webs of multidimensional surfaces.* (Russian) Sibirsk. Mat. Zh. 17 (1976), no. 1, 44–57. English translation: Siberian Math. J. 16 (1975), no. 2, 190–202. (MR 54 #5986.)

[G 77] Goldberg, V.V.: *On the theory of four-webs of multidimensional surfaces on a differentiable manifold X_{2r}.* (Russian) Izv. Vyssh. Uchebn. Zaved. Mat. 1977, no. 11(186), 15–22. English translation: Soviet Math. (Iz. VUZ) 21 (1977), no. 11, 97–100. (MR 58 #30859.)

[G 80] Goldberg, V.V.: *On the theory of four-webs of multidimensional surfaces on a differentiable manifold X_{2r}.* (Russian) Serdica 6 (1980), no. 2, 105–119. (MR 82f:53023.)

[G 82a] Goldberg, V.V.: *A classification of six-dimensional group four-webs of multidimensional surfaces.* Tensor (N.S.) 36 (1982), no. 1, 1–8. (MR 87b:53024.)

[G 82b] Goldberg, V.V.: *The solutions of the Grassmannization and algebraization problems for $(n+1)$-webs of codimension r on a differentiable manifold of dimension nr.* Tensor (N.S.) 36 (1982), no. 1, 9–21. (MR 87a:53027.)

[G 82c] Goldberg, V.V.: *Multidimensional four-webs on which the Desargues and triangle figures are closed.* Geom. Dedicata 12 (1982), no. 3, 267–285. (MR 83i:53031.)

[G 82d] Goldberg, V.V.: *Grassmann and algebraic four-webs in a projective space.* Tensor (N.S.) 38 (1982), 179–197. (MR 87e:53024.)

[G 83a] Goldberg, V.V.: *r-Rank problems for webs* $W(d,2,r)$. Math. Sci. Research Inst., Berkeley, California, 83–046, 1983, 8 pp.

[G 83b] Goldberg, V.V.: *Tissus de codimension r et de r-rang maximum.* C. R. Acad. Sci. Paris Sér. I Math 297 (1983), no. 6, 339–342. (MR 85f:53020.)

[G 84] Goldberg, V.V.: *An inequality for the 1-rank of a scalar web SW(d,2,r) and scalar webs of maximum 1-rank.* Geom. Dedicata 17 (1984), no. 2, 109–129. (MR 86f:53014.)

[G 85a] Goldberg, V.V.: *4-tissus isoclines exceptionnels de codimension deux et de 2-rang maximal.* C.R. Acad. Sci. Paris Ser. I Math 301 (1985), no. 11, 593–596. (MR 87b:53025.)

[G 85b] Goldberg, V.V.: *r-Rank problems for a web* $W(d,2,r)$. Preprint, 50 pp., 1985.

[G 86] Goldberg, V.V.: *Isoclinic webs* $W(4,2,2)$ *of maximum 2-rank.* Differential Geometry, Peniscola 1985, 168–183. Lecture Notes in Math., 1209, Springer-Verlag, Berlin-New York, 1986.

[G 87] Goldberg, V.V.: *Nonisoclinic 2-codimensional 4-webs of maximum 2-rank.* Proc. Amer. Math. Soc. 100 (1987), no. 4, 701–708.

[GR 87] Goldberg, V.V.; Rosca, R.: *Geometry of exceptional webs* $EW(4,2,2)$ *of maximum 2-rank.* Preprint, 18 pp., 1987.

[Go 99] Goursat, E.: *Sur les équations du second ordre à n variables, analogues à l'équation de Monge–Ampère.* Bull. Soc. Math. France 27 (1899), 1–34.

[GF 76] Grauert, H.; Fritzsche, K.: "Several complex variables." Springer-Verlag, New York-Heidelberg-Berlin, 1976, viii+207 pp. (MR 54 #3004.)

[GRe 84] Grauert, H.; Remmert, R.: "Coherent analytic sheaves." Springer-Verlag, Berlin-Heidelberg-New York-Tokyo, 1984, xviii+249 pp. (MR 86a:32001.)

[Gr 76] Griffiths, P.A.: *Variations on a theorem of Abel.* Invent. Math. 35 (1976), 321–390. (MR 55 #8036.)

[Gr 77] Griffiths, P.A.: *On Abel's differential equations.* Algebraic Geometry, J.J. Sylvester Sympos., Johns Hopkins Univ., Baltimore, Md., 1976, 26–51. Johns Hopkins Univ. Press, Baltimore, Md, 1977. (MR 58 #655.)

[GJ 87] Griffiths, P.A.; Jensen, G.R.: "Differential systems and isometric embeddings." Princeton Univ. Press, Princeton, N.J., 1987, xii+225 pp.

[GuR 65] Gunning, R.C.; Rossi, H.: "Analytic functions of several complex variables." Prentice Hall, Inc., Englewood Cliffs, N.J., 1965, xii+317 pp. (MR 31 #4927.)

[H 75] Hall, M.: "Combinatorial theory," 2nd ed. Wiley & Sons, New York, 1986, xv+440 pp.

[Ha 66] Hangan, Th.: *Géométrie différentielle grassmannienne.* Rev. Roumaine Math. Pures Appl. 11 (1966), no. 5, 519–531. (MR 34 #744.)

[Ha 80] Hangan, Th.: *Sur l'intégrabilité des structures tangentes produits tensoriels réels.* Ann. Mat. Pura Appl. (4) 126 (1980), 149–185. (MR 82e:53051.)

[He 78] Helgason, S.: "Differential geometry, Lie groups, and symmetric spaces." Academic Press, New York-London, 1978, xv+628 pp. (MR 80k:53081.)

[Her 75] Herstein, I.N.: "Topics in algebra," 2nd ed. Xerox College Publishing, Lexington, Mass.-Toronto, Ont., 1975, xi+388 pp. (MR 50 #9456.)

[HP 52] Hodge, W.V.D.; Pedoe, D.: "Methods of algebraic geometry," Vol. 2. Cambridge Univ. Press, Cambridge, 1952. (MR 13, 972.)

[HS 86a] Hofmann, K.H.; Strambach, K.: *Lie's fundamental theorems for local analytical loops.* Pacific J. Math. 123 (1986), no. 2, 301–327.

[HS 86b] Hofmann, K.H.; Strambach, K.: *The Akivis algebra of a homogeneous loop.* Mathematika 33 (1986), no. 1, 87–95.

[Ho 63] Hosszu, M.: *On the explicit form n-group operations.* Publ. Math. Debrecen 10 (1963), no. 1–4, 436–440. (MR 29 #4816.)

[Hu 66] Husemoller, D.: "Fibre bundles," 2nd ed. Springer-Verlag, New York–Heidelberg, 1975, xv+327 pp. (MR 51 #6805.)

[Ja 62] Jacobson, N.: "Lie algebras," Wiley-Interscience, New York, 1962, ix+331 pp. (MR 26 #1345.)

[J 50] Jeger, M.: *Projektive Methoden in der Gewebegeometrie.* Comment. Math. Helv. 24 (1950), 260-290. (MR 12, 857.)

[K 34] Kähler, E.: "Einführung in die Theorie der Systeme von Differentialgleichungen." Teubner-Verlag, Leipzig-Berlin, 1934, iv+78 pp. Reprint Chelsea Publishing Co., New York, 1949.

[Ki 84] Kilp, H.: *Geometry of quasilinear systems of differential equations and m-webs.* (Russian) Tartu Riikl. Ul. Toimetised no. 665 (1984), 14–22. (MR 85g:53015.)

[Kn 32] Kneser, H.: *Gewebe und Gruppen.* Abh. Math. Sem. Univ. Hamburg 9 (1932), 147–151.

[KN 63] Kobayashi, S.; Nomizu, K.: "Foundations of differential geometry," Vol. 1. Wiley–Interscience, New York-London, 1963, xi+329 pp. (MR 27 #2945.)

[Kr 73] Kramareva, R.F.: "Investigations in the theory of n-quasigroups." Ph. D. Dissertation, V.I. Lenin Moscow Pedagogical Institute, 1973.

[Ku 70] Kuz'min, E.N.: *Mal'cev algebras of dimension five over a field of zero characteristic.* (Russian) Algebra i Logika 9 (1970), no. 5, 691–700. English translation: Algebra and Logic 9 (1970), no. 5, 416–421. (MR 44 #266.)

[L 53] Laptev, G.F.: *Differential geometry of imbedded manifolds.* (Russian) Trudy Moskov. Mat. Obshch. 2 (1953), 275–382. (MR 15, 254.)

[La 74] Lawson, H. Blaire, Jr.: *Foliations.* Bull. Amer. Math. Soc. (N.S.) 80 (1974), no. 3, 369–418. (MR 49 #8031.)

[Li 1893] Lie, S.: "Vorlesungen über kontinuierliche Gruppen mit geometrishen und anderen Anwendungen." Teubner, Leipzig, 1893, xii + 810 pp. Reprint Chelsea Publishing Co., Bronx, N.Y., 1971. (MR 52 #13275.)

[Lit 83] Little, J.B.: *Translation manifolds and the converse of Abel's theorem.* Compositio Math. 49 (1983), 147–171. (MR 85d:14041.)

[Lit 86] Little, J.B.: *On webs of maximum rank.* Preprint, 30 pp., 1986.

[Ma 55] Mal'cev, A.I.: *Analytical loops.* (Russian). Mat. Sb. 36 (1955), 569–576. (MR 16, 997.)

[Mal 76] Malgrange, B.: *Frobenius avec singularites-*I. *Codimension un.* Inst. Hautes Etudes Sci. Publ. Math. 46 (1976), 163–173. (MR 58 #22685a.)

[Mi 78] Mikhailov, Ju.I.: *The structure of almost Grassmann manifolds.* (Russian) Izv. Vyssh. Uchebn. Zaved. Mat. 1978, no. 2(160), 62–72. English translation: Soviet Math. (Iz. VUZ) 22 (1978), no. 2, 54–63. (MR 81e:53031.)

[Mo 35] Moufang, R.: *Zur Struktur von Alternativ Körpern.* Math. Ann. 110 (1935), 416–430.

[Na 73] Narasimhan, R.: "Analysis on real and complex manifolds." 2nd ed. Amer. Elsevier Publishing Co., New York, 1973, x+246 pp. (MR 49 #11576.)

[N 80] Nishimori, T.: *Octahedral webs on closed manifolds.* Tôhoku Math. J. 32 (1980), no. 3, 399–410. (MR 82h:53025.)

[N 81] Nishimori, T.: *Some remarks on octahedral webs.* Japan J. Math. (N.S.) 7 (1981), no. 1, 169–179. (MR 85h:53018.)

[P 68] Penrose, R.: *Structure of space-time.* Battelle rencontres, 1967 Lectures in Mathematics and Physics, Chapter VII, 121–135, Benjamin, New York-Amsterdam, 1968. (MR 38 #955.)

[R 60] Rado, F.: *Generalisation of space webs for certain algebraic structures.* (Roumanian) Studia Univ. Babeş–Bolyai Ser. Math. Phys. 1960, no. 1, 41–55. (MR 32 #395.)

[Ra 59] Raschewski, P.K.: "Riemannsche Geometrie and Tensoranalysis." VEB Deutscher Verlag der Wissenschaften, Berlin, 1959, 606 pp. (MR 21 #2258.)

[Rei 28] Reidemeister, K.: *Gewebe und Gruppen.* Math. Z. 29 (1928), 427–435.

[RS 60] Remmert, R.; Stein, K.: *Eigentliche holomorphe Abbildungen.* Math. Z. 73 (1960), 159–189. (MR 23 #A1840.)

[Ri 64a] Rischel, H.: *Ein Satz über eigentliche holomorphe Abbildungen von analytischen Polyedergebieten.* Math. Scand. 14 (1964), 220–224. (MR 30 #3238.)

[Ri 64b] Rischel, H.: *Holomorphe Überlagerungskorrespondenzen.* Math. Scand. 15 (1964), 49–63. (MR 31 #2418.)

[Ros 76] Rosca, R.: *Couple de champs vectorials quasirécurrent réciproques.* C.R. Acad. Sci. Paris Sér. A–B 282 (1976), no. 13, A699-A701. (MR 54 #6027.)

[Ro 35] Rothstein, W.: "Zur Theorie der analytischen Abbildungen im Raume zweier komplexer Veränderlicher: Das Verhalten der Abbildung auf glatten analytischen Randhyperflächen." Dissertation, Münster, 1935.

[S 65a] Sandik, M.D.: *Completely reducible quasigroups.* (Russian) Izv. Akad. Nauk Moldav. SSR Ser. Fiz.-Tekhn. Mat. Nauk 7 (1965), 55–67. (MR 33 #2752.)

[S 65b] Sandik, M.D.: *The uniqueness of representation of an n-quasigroup.* (Russian) Studies in General Algebra, Akad. Nauk Moldav. SSR, Kishinev, 1965, 123–135. (MR 34 #5731.)

[Sa 25] Sauer, R.: *Die Raumteilungen, welche durch Ebenen erzeugt werden.* Sitzungsber. Bayer. Akad., Math.-Naturwiss. 1925, 41–56.

[Sc 84] Scheiderer, C.: "Gewebegeometrie 10.6 bis 16.6.1984." Tagungsbericht 27/1984, Mathematisches Forschungsinstitut Oberwolfach 1984.

[SS 38] Schouten, J.A.; Struik, D.J.: "Einführung in die neueren Methoden der Differentialgeometrie, 2 Band "Geometrie"," 2. Auflage, P. Noordhoff, Groningen-Batavia, 1938.

[Sh 66] Shulikovsky, V.I.: *On intrinsic connections of N-webs.* (Russian) Travaux Sci. École Norm. Sup. Plovdiv, Math. IV (1966), Fasc. 1, 13–22.

[Sm 88] Smith, J.D.H.: *Multilinear algebras and Lie's theorem for formal n-loops.* Arch. Math. (Basel) 1988 (to appear).

[Ste 58] Stein, K.: *Die Existenz komplexer Basen zu holomorphen Abbildungen.* Math. Ann. 136 (1958), 1–8. (MR 20 #4657.)

[St 83] Sternberg, S.: "Lectures on differential geometry," 2nd ed. Chelsea Publishing Co., New York, 1983, xviii+442 pp.

[Ti 75] Timoshenko, V.V.: *On three-webs over commutative associative algebras.* (Russian) Izv. Vyssh. Uchebn. Zaved. Mat. 1975, no. 11 (133), 109–112. English translation: Soviet Math. (Iz. VUZ) 16 (1975), no. 11, 93–96. (MR 54 #8516.)

[Ti 77] Timoshenko, V.V.: *Three-webs over an algebra the curvature of which is a divisor of zero.* (Russian) Izv. Vyssh. Uchebn. Zaved. Mat. 1977, no. 3 (149), 116–118. English translation: Soviet Math. (Iz. VUZ) 21 (1977), no. 3, 92–94. (MR 57 #17527.)

[To 85] Tolstikhina, G.A.: *On a property of a 4-web carrying a group three-subweb.* (Russian) Problems of the theory of webs and quasigroups, 121–128, Kalinin. Gos. Univ., Kalinin, 1985.

[V 83] Vasiljev, A.M.: *Linear differential systems and invariant realisation of differential geometric structures.* (Russian) Izv. Vyssh. Uchebn. Zaved. Mat. 1984, no 7(266), 22–34. English translation: Soviet Math. (Iz. VUZ) 1984, no. 7, 26–40. (MR 86a:58117.)

[Va 69] Vasiljeva, M.V.: "Lie groups of transformations." Mosk. Gos. Pedag. Inst., Moscow, 1969, 175 pp.

[We 64] Weyl, H.: "The concept of a Riemann surface." Translation from German, 3d ed., Addison Wesley Publishing Co., Reading, Mass., 1964, x+191 pp. (MR 29 #3628.)

[Wh 72] Whitney, H.: "Complex analytic varieties." Addison-Wesley Publishing Co., Reading, Mass.-London-Don Mills, Ont. 1972, xii+399 pp. (MR 52 #8473.)

[Wo 82] Wood, J.A.: "An algebraization theorem for local hypersurfaces in projective space." Ph. D. Dissertation, Univ. of California, Berkeley, 1982, 87 pp.

[Wo 84] Wood, J.A.: *A simple criterion for local hypersurfaces to be algebraic.* Duke Math. J. 51 (1984), no. 1, 235–237. (MR 85d:14069.)

[YK 84] Yano, K.; Kon, M.: "Structures on manifolds." World Scientific, Singapore, 1984, ix+508 pp. (MR 86g:53001)

SYMBOLS FREQUENTLY USED

The list below contains many of the symbols whose meaning is usually fixed throughout the book.

$\underset{\alpha\beta}{a}{}^{i}_{jk}$: torsion tensor of a web $W(n+1,n,r)$, 14
$AAW(d,2,r)$: almost algebraisable d-web of codimension r given on X^{2r}, 385
$AG(n-1,r+n-1)$: almost Grassmann structure, 73
$AGW(d,2,r)$: almost Grassmannisable d-web of codimension r given on X^{2r}, 375
A^{nr}: affine space of dimension nr, 5
$AW(d,n,r)$: algebraic d-web of codimension r given on X^{nr}, 90
(B_l): left Bol closure condition, 17
(B_m): middle Bol closure condition, 17
(B_r): right Bol closure condition, 17
$B_\xi(n+1,n,r)$: Bol $(n+1)$-web of codimension r defined by the foliation λ_ξ on X^{nr}, 162
$\underset{\alpha\beta}{b}{}^{i}_{jk}$: curvature tensor of a web $W(n+1,n,r)$, 14
$\mathbf{C}$: field of complex numbers, 265
$\mathbf{C}^n$: complex n-space, 266
$\mathbf{CP}^2$: complex projective plane, 5
$\mathbf{CP}^{2*}$: space dual to the complex projective plane $\mathbf{CP}^2$, 5
$C(r,n)$: Segre cone, 70
e: unit of a loop, 102
F: leaf of a foliation (fibre of a fibration), 1
$G(k,N)$: Grassmannian of k-planes in a projective N-space P^N, 5, 99, 100
$\mathbf{GL}(r)$: general linear group (complex or real), 10, 71, 74
γ_ξ: affine connection induced by the foliation λ_ξ on X^{nr}, 21
$\gamma_{\alpha\beta}$: affine connection of the 3-subweb $[n+1,\alpha,\beta]$ induced by λ_{n+1}, 24
$\tilde{\gamma}_{\alpha\beta}$: canonical affine connection of the 3-subweb $[n+1,\alpha,\beta]$, 25
$GW(d,n,r)$: Grassmann web of codimension r given on X^{nr}, 90
$L(a_1,\dots,a_n)$: local n-loop, 106
λ^i_j: basis affinor of $W(4,2,r)$, 302
λ_ξ: foliations of $W(d,n,r)$, 4
(M): Moufang 3-web, 18
$M(n+1,n,r)$: Moufang $(n+1)$-web of codimension r given on X^{nr}, 159
$[n+1,\alpha,\beta]$: 3-subweb of $W(n+1,n,r)$ generated by foliations $\lambda_{n+1},\lambda_\alpha,\lambda_\beta$, 24
∇: covariant differential, 4, 354, 375
$\pi_q(d,n,r)$: upper bound for q-rank of a web $W(d,n,r)$, 396, 407
P^N: projective N-space, 5
P^{N*}: space dual to a projective N-space P^N, 69
PT: projectivisation of a space T, 70
$PW(d,n,r)$: parallelisable d-web of codimension r given on X^{nr}, 5
Q_r: r-parameter local differentiable n-quasigroup, 102
$\mathbf{R}$: field of real numbers, 265
$\mathbf{R}^N$: Euclidean N-space, 6
$(\mathbf{R},[\ ,\ ],(\ ,\ ,\ ))$: Akivis algebra, 133
$(\mathbf{R},[\ ,\ ],<\ ,\ ,\ >)$: comtrans algebra, 137
$\mathbf{RP}^2$: real projective plane, 5
$\mathbf{SL}(n)$: special linear group (complex or projective), 70, 74
$S(r-1,n-1)$: Segre variety, 69

$T_p(S)$: tangent space to a manifold S at a point p, 3
Tg-surface: transversally geodesic surface, 48
V_d^r: algebraic variety of dimension r and degree d, 5
$W(d, n, r)$: d-web of codimension r given on X^{nr}, 4
X^{nr}: differentiable manifold of dimension nr, 4
$\wedge$: exterior multiplication, 2
$\{ , \ldots , \}$: alternator, 108
$(, ,)$: associator, 132
$[,]$: commutator, 132
$[, ,]$: commutator, 133, 134
$< , , >$: translator, 133, 134
$/ , , /$: bogus product, 135

Index

D

T

U

V

W